Principles of Sustainable Energy Systems

Principles of Sustainable Energy Systems provides students with a fundamental and practical understanding of the energy transition. It discusses the design, production, and economics of energy conversion and storage technologies, as well as requirements and technologies for the end-use sectors of transportation, buildings, and industry.

This book begins by introducing students to the important field of sustainability and then presents comprehensive coverage of solar, wind, hydropower, biomass and biofuels, geothermal, nuclear, and ocean-based energy technologies. This new edition features recent advances in batteries and other storage technologies, electricity transmission, electric vehicles, and beneficial electrification and demand response in buildings, as well as approaches for reducing emissions from shipping and aviation. It introduces new material on low-carbon building materials, heat pumps, and the practical design aspects of solar photovoltaic systems. This book also covers economics and energy systems analysis methods such as life cycle assessment and greenhouse gas accounting, including detailed examples of design and financial analysis using the System Advisor Model (SAM).

This book is intended for upper-level undergraduate and graduate engineering students taking courses in Renewable Energy, Energy Systems, and Energy Conversion.

Instructors will have access to a Solutions Manual and Figure Slides for their course.

Mechanical and Aerospace Engineering

A Series of Textbooks and Reference Books

Founding Editor

Frank Kreith

Power Plant Synthesis
Dimitris Al. Katsaprakakis

Air Distribution in Buildings
Essam E. Khalil

Nanotechnology: Understanding Small Systems, Third Edition
Ben Rogers, Jesse Adams, and Sumita Pennathur

Introduction to Biofuels
David M. Mousdale

Energy Conversion
D. Yogi Goswami and Frank Kreith

CRC Handbook of Thermal Engineering
Raj P. Chhabra

Energy Efficiency and Renewable Energy Handbook, Second Edition
D. Yogi Goswami and Frank Kreith

Energy, the Environment, and Sustainability
Efstathios E. Michaelides

Principles of Sustainable Energy Systems, Third Edition
Charles F. Kutscher, Jana B. Milford, and Frank Kreith

Fluid Power Circuits and Controls: Fundamentals and Applications, Second Edition
John S. Cundiff and Michael F. Kocher

Principles of Sustainable Energy Systems, Fourth Edition
Charles F. Kutscher, Jana B. Milford, and Frank Kreith

For more information about this series, please visit: https://www.crcpress.com/Mechanical-and-Aerospace-Engineering-Series/book-series/CRCMECAERENG

Principles of Sustainable Energy Systems

Fourth Edition

Charles F. Kutscher, Jana B. Milford, and
Frank Kreith

CRC Press
Taylor & Francis Group
Boca Raton London New York

CRC Press is an imprint of the
Taylor & Francis Group, an **informa** business

Designed cover image: Dennis Schroeder, NREL

Fourth edition published 2026
by CRC Press
2385 NW Executive Center Drive, Suite 320, Boca Raton FL 33431

and by CRC Press
4 Park Square, Milton Park, Abingdon, Oxon, OX14 4RN

CRC Press is an imprint of Taylor & Francis Group, LLC

First edition published by CRC Press 2010
Second edition published by CRC Press 2014
Third edition published by CRC Press 2019

Library of Congress Cataloging-in-Publication Data
Names: Kutscher, Charles F. author | Milford, Jana B. author | Kreith, Frank author
Title: Principles of sustainable energy systems / Charles F. Kutscher, Jana B. Milford, Frank Kreith.
Description: Fourth edition. | Boca Raton, FL : CRC Press, 2026. | Series: Mechanical and aerospace engineering series | Revised edition of: Principles of sustainable energy systems / Frank Kreith, editor, University of Colorado (retired), Susan Krumdieck, co-editor, University of Canterbury; and Principles of sustainable energy / Frank Kreith, Jan F. Kreider. | Includes bibliographical references and index.
Identifiers: LCCN 2025009305 (print) | LCCN 2025009306 (ebook) | ISBN 9781032957951 hbk | ISBN 9781032957968 pbk | ISBN 9781003586616 ebk
Subjects: LCSH: Renewable energy sources | Electric power production
Classification: LCC TJ808 .K79 2026 (print) | LCC TJ808 (ebook) | DDC 621.042—dc23/eng/20250401
LC record available at https://lccn.loc.gov/2025009305
LC ebook record available at https://lccn.loc.gov/2025009306

ISBN: 978-1-032-95795-1 (hbk)
ISBN: 978-1-032-95796-8 (pbk)
ISBN: 978-1-003-58661-6 (ebk)

DOI: 10.1201/9781003586616

Typeset in Palatino
by codeMantra

Access the Support Material: Routledge.com/9781032957951

This book is dedicated to the memory of our colleague, Frank Kreith,

with the hope that it will advance his inspiring vision for

protecting our planet through sustainable energy.

Contents

We have the technologies we need to slash emissions. Renewables are booming as costs plummet and governments realize the benefits of cleaner air, good jobs, energy security, and increased access to power.

United Nations Secretary-General António Guterres
Address on World Environment Day, June 5, 2024

Foreword

As with the previous editions, this fourth edition of *Principles of Sustainable Energy Systems* provides additional clarity and insights to the foundational understanding of our energy challenges and opportunities. As we witness the devastating effects of climate change, it is clearer today than ever that our global community must accelerate the transition to a low-carbon energy economy. Previous editions acknowledged this need by emphasizing that a better understanding of all energy generation and end-use sources and options is required. To support this, those editions added relevant chapters on energy efficiency, as well as fossil and nuclear energy, and expanded the text to include strategies for integrating variable renewable energy into the existing electricity grid. This fourth edition introduces important new sections on life cycle assessment and the social cost of carbon—both essential for understanding the broader environmental and economic impacts of energy systems. It also includes expanded coverage of energy systems models, beneficial building electrification, transmission needs, the growth of offshore wind, new battery chemistries, and advances in transportation, including potential solutions for the hard-to-abate sectors of long-distance shipping and aviation. These additions ensure that the text remains comprehensive and reflects the latest advancements and challenges in the energy field.

The energy landscape has changed dramatically in recent years. Solar and wind energy have dropped significantly in cost and are now widely recognized as the lowest-cost electric generators. Renewable energy dominates new generation additions, and it is a testament to the progress in the field that while it took decades for solar PV to reach 1 TW of global deployment, it took only two years to double to 2 TW. End-use sectors are electrifying to take advantage of this clean, renewable electricity. Battery electric vehicles (BEVs) have achieved cost parity with gasoline vehicles on a lifetime basis and are nearing parity on a purchase-cost basis. High-performance cold-climate heat pumps are increasingly replacing fossil fuel home heating.

Despite these advances, engineers face many remaining challenges in achieving the energy transition. These include addressing the rapid grid expansion needed to meet growing data center demands, speeding the interconnection of renewables to the grid, finding the best solutions for long-duration storage, tapping the potential of virtual power plants, and electrifying end-use sectors in an affordable way, among others. Tackling these challenges will require engineers to have a fundamental knowledge of the principles of sustainable energy systems, which makes this text more vital than ever.

Additionally, this book examines the practical aspects surrounding the widespread adoption of energy efficiency and renewable energy systems in the context of new technological developments, policies, and evolving global markets. While support for clean energy alternatives has never been higher, the obstacles to achieving a true energy transformation that appropriately values low-carbon alternatives remain formidable.

Once again, I can think of no better team to preserve and update Professor Kreith's original book than the authors of this updated edition. The team for this edition was led by Dr. Chuck Kutscher of the Renewable and Sustainable Energy Institute, with whom I had the pleasure of working for many years. This edition also includes the perspective and significant contributions from accomplished Professor Jana Milford, an expert on the environmental effects of energy systems. And as someone who has followed and actively engaged in renewable energy research, development, and deployment for decades,

Professor Kreith's contributions in this arena are legendary—starting for me when I was a student and studied from his books. All three authors bring deep and complementary expertise in renewable energy technology and sustainability assessment.

To those who choose renewable energy and sustainability as an academic field or career path, this text—and the insights you gain from it—will likely become fundamental to the achievements that await you. But this text is not merely academic; it serves a valuable purpose in its clear-headed look at the energy picture from the ground up, and the environmental, economic, and sustainability benefits that renewable energy systems can provide.

Principles of Sustainable Energy Systems builds on previous editions and combines the expertise earned from decades of practical and scholarly research with the most up-to-date analysis of the energy scene. It provides an important guide for practitioners worldwide. Today, sustainable energy is no longer just a topic discussed at the margins of the energy sector—it is central to it. Its study also provides incredible opportunities for practitioners and decision-makers to create a future that will serve humanity for generations to come.

Dan E. Arvizu, Ph.D., NAE
Director Emeritus, National Renewable Energy Laboratory
Former Chair, National Science Board (NSB)
Former Member of President's Council of Advisors on Science and Technology

Preface to the Fourth Edition

Driven by changing energy costs and the need to reduce greenhouse gas and air pollution emissions and their environmental impacts, the world is rapidly transitioning from fossil fuels to renewable energy. Solar and wind now provide the cheapest electricity and are the largest sources of new electricity generation globally. To take advantage of cleaner sources of electricity, battery electric vehicles are replacing gasoline cars, and heat pumps are replacing fossil fuel-fired furnaces for buildings and industrial process heat.

Despite these advances, fossil fuels still supply approximately 80% of the energy used in the global economy. Mainly due to the combustion of fossil fuels over the past century, humans have increased the amount of heat-trapping carbon dioxide in the atmosphere by over 50%. 2024 was the first year that the global average temperature exceeded 1.5°C above the pre-industrial level, highlighting the need to sharply cut greenhouse gas emissions to meet an international goal of keeping the global temperature increase below 2°C. The temperature rise to date has resulted in more frequent and higher intensity weather-related disasters including heat waves, droughts, wildfires, floods, and extreme precipitation events.

The transition to sustainable energy is well underway but must be accelerated. The purpose of this book is to provide engineering and science students with a fundamental understanding of sustainable energy systems that they will need to participate in and ultimately help lead this critical endeavor.

Chapter 1 introduces the concept of sustainability and explores how energy intersects with other sustainability issues, including food and water supplies and climate change. This chapter provides an overview of the main energy resources and technologies that are used now or on the near horizon, including solar, wind, hydropower, ocean and nuclear power, biofuels, hydrogen, and energy efficiency. This fourth edition provides updated energy data and addresses the new challenges that have emerged as renewable energy has become the fastest-growing contributor to the world's energy supply.

Chapters 2 and 3 present analysis methods and metrics for assessing energy supply and conservation measures. Chapter 2 covers the basics of financial analysis for energy projects and introduces the concept of external costs of energy, exemplified by the social costs of carbon dioxide and other greenhouse gases. Chapter 3 introduces methods for using a life cycle perspective to assess and compare the materials, energy, and environmental costs of energy systems. The fourth edition includes new life cycle analysis examples and new estimates of embodied energy and emissions for a wide range of materials and equipment.

Chapter 4 covers energy use and efficiency in the buildings and industry sectors. It contains a new section on beneficial building electrification, including the growing use of air-source, cold-climate heat pumps and geothermal (ground-coupled) heat pump districts. There are also new sections on low-carbon building materials and industrial electrification. Chapter 5 covers the electric grid, including descriptions of the different electricity financial markets and the various ways to accommodate the variability of solar and wind electricity supplies. The fourth edition contains a new section on technical options for improving transmission.

Chapter 6 provides an overview of the production and use of fossil fuels as a baseline for understanding the transition to renewable energy. This chapter describes resources, extraction, and energy conversion methods for coal, natural gas, and petroleum, along with information on the health and climate impacts of their emissions. Nuclear energy, which is covered in Chapter 7, plays an important role in reducing global carbon dioxide emissions, although high capital costs, long deployment times, and waste management are recognized challenges with current technology. Chapter 7 includes updated nuclear power costs and has an expanded section on next-generation nuclear power technologies with special emphasis on small modular reactors (SMRs).

The fundamentals of wind power are explained in Chapter 8, including wind characteristics and resource assessment, turbine technology, and environmental benefits and impacts of wind energy. This chapter also contains new information on the growing deployment of offshore wind. Chapter 9 covers energy from biomass, including feedstock characterization and availability and technologies for converting biomass to gaseous or liquid fuels and electricity. This chapter also discusses land use, health, and environmental impacts of biomass energy.

Chapters 10, 11, and 12 cover, respectively, the fundamentals of solar radiation, solar photovoltaic technology, and solar thermal applications. Chapter 11 has been expanded to cover not only the fundamental principles but also provide more detail on practical PV system design, including a new example of how to commission a PV system. It also contains new information on recent PV advances including bifacial panels, agrivoltaics, and perovskite cells. Chapter 12 includes sections on the design, characterization, and testing of solar thermal collectors, as well as applications for building heating, hot water, industrial process heat, and power generation. It also contains the latest solar thermal deployment data.

Chapter 13 covers ocean, hydropower, and geothermal energy technologies. For ocean energy, this chapter includes ocean thermal energy conversion (OTEC), for which some island communities have exhibited renewed interest, along with tidal and wave energy. In addition to conventional hydropower and geothermal technologies, there is new coverage of some of the latest geothermal power technologies under development.

Chapter 14 covers storage technologies, which are critical for electric vehicles and have become increasingly important for firming variable renewables on the grid. This chapter surveys pumped hydropower and compressed air energy storage (CAES), battery and fuel cell electrical storage, and thermal storage technologies for buildings and concentrating solar power. The fourth edition describes new battery technologies, including sodium-ion and iron-air chemistries. There is also important new material on lithium-ion battery advances, life cycle impacts, and recycling.

Chapter 15 covers the transportation sector, starting with the need to address transportation equity and the health and environmental impacts of transportation systems. This chapter addresses the growing deployment of battery electric vehicles and improvements in charging infrastructure and discusses tradeoffs of battery and fuel cell electric options for large trucks. It also reviews alternative fuels and other options for reducing emissions from aviation and transoceanic shipping.

As with the third edition, this fourth edition includes detailed examples of financial and performance analysis for renewable energy technologies using the System Advisor Model (SAM) developed by the National Renewable Energy Laboratory and Sandia National Laboratories. SAM is a free software tool that simplifies the analysis of renewable energy projects. SAM is designed to easily access information on weather, solar radiation, and utility rate data for many locations, as well as information on various types of energy conversion and storage equipment. SAM examples appear for solar hot water, wind power, solar photovoltaics, and concentrating solar power.

This is a critical and exciting time for engineering and science students to enter the energy field. As the impacts of climate change continue to be felt around the world, we believe every effort must be made to speed the transition to carbon-free energy technologies. Beyond addressing climate change, the rewards of that effort will include millions of new jobs created, a more secure electric grid, cleaner air and water, and greater energy access and independence around the world.

Charles F. Kutscher
Jana B. Milford
Boulder, Colorado
January 2025

Acknowledgments

The authors would like to thank our contributing authors: Susan Krumdieck, Chapter 1 (Introduction to Sustainable Energy); Gary Pawlas, Chapter 8 (Wind Energy); Robert C. Brown and Mark M. Wright, Chapter 9 (Capturing Solar Energy through Biomass); Jan Kreider, Chapter 10 (Fundamentals of Solar Radiation); and Jan Kreider and Jeffrey H. Morehouse, Chapter 12 (Solar Thermal Collectors and Systems).

The authors sincerely thank the following individuals for their important contributions to the fourth edition. For Chapter 3 (Energy Systems Analysis Methodologies), Brittany Smith and Wesley Ingwersen shared insights on life cycle assessment and embodied energy, respectively. For Chapter 5 (Electricity Supply Systems), Randy Huston provided excellent suggestions, and Julie Kemp offered valuable insights on a new Berkeley Lab transmission study. Pradyumna Rao of CU Boulder provided a valuable review of Chapter 7 (Nuclear Power). We are especially grateful to Solar Energy International, one of the oldest solar energy training facilities in the country, for generously providing photovoltaic technology schematics for Chapter 11 (Photovoltaics). Mark Mehos provided expert assistance with a SAM concentrating solar power example in Chapter 12.

We would like to thank our many colleagues at NREL and the University of Colorado Boulder for all the collaborations and insightful energy discussions over the years. We also thank our mechanical and environmental engineering students who have taken Sustainable Energy at the University of Colorado Boulder for their feedback and insights, as well as our teaching assistants who provided much useful input on lectures and homework problems: Colin Harkins, David Pfotenhauer, Nick Sanchez-Jans, Joseph Wilson, and Pray Shah. Charles Kutscher and Jana Milford would especially like to express our enduring appreciation for Frank Kreith, the original author of this textbook, who invited us to participate in this project and made countless contributions to renewable energy and student education over the course of his lifetime.

Last but not least, we want to express our deep thanks to our respective spouses, Karen Thompson and Rick Adcock, for their constant encouragement and support throughout the many months of manuscript preparation.

Authors

Charles F. Kutscher is a fellow of the Renewable and Sustainable Energy Institute (RASEI), a joint institute between the University of Colorado Boulder (CU-Boulder) and the National Renewable Energy Laboratory (NREL). He served as the Director of the Buildings and Thermal Sciences Center at NREL from 2013 until his retirement in 2018. He has worked in the field of renewable energy for over four decades, during which time he has led research investigations on solar heating and cooling, building energy efficiency, solar industrial process heat, geothermal power, and concentrating solar power. He is a fellow of the American Solar Energy Society (ASES) and served as the Society's chair in 2000 and 2001. He was the chair of two major conferences: the SOLAR 2006 National Solar Energy Conference and the 2012 World Renewable Energy Forum. He led the ASES study, *Tackling Climate Change in the U.S.*, which detailed how energy efficiency and six renewable energy technologies could greatly reduce U.S. carbon emissions by 2030. He also served as the lead author of the RASEI study, *Accelerating the US Clean Energy Transformation: Challenges and Solutions by Sector.* He has served as an Adjunct Professor at the University of Colorado Boulder and the Colorado School of Mines. He earned a BS in Physics from the State University of New York at Albany, an MS in Nuclear Engineering from the University of Illinois at Urbana-Champaign, and a PhD in Mechanical Engineering from the University of Colorado Boulder. He is a registered professional engineer in California and Colorado.

Jana B. Milford is an Emeritus Professor in the Department of Mechanical Engineering and the Environmental Engineering Program at the University of Colorado at Boulder, where she worked from 1994 to 2022. Her research and teaching interests have focused on air quality modeling and data analysis, environmental impacts of energy systems, and environmental law and policy. She has also worked as a congressional fellow and analyst at the Congressional Office of Technology Assessment, Assistant Professor in the Department of Civil Engineering at the University of Connecticut, and a senior scientist and staff attorney at the Environmental Defense Fund. She has served as a member of the Colorado Air Quality Control Commission, the Board of Environmental Studies and Toxicology of the National Academy of Sciences, the Science Advisory Board of the U.S. Environmental Protection Agency, and the Review Committee of the Health Effects Institute. She holds a BS in Engineering Science from Iowa State University, a MS in Civil Engineering from Carnegie Mellon University, a PhD in Engineering and Public Policy from Carnegie Mellon University, and a JD from the University of Colorado School of Law.

Frank Kreith taught at the University of California, Lehigh University, and the University of Colorado, as a Professor Emeritus of Engineering. From 1988 to 2001, he was the American Society of Mechanical Engineers International (ASME) legislative fellow for Energy and Environment at the National Conference of State Legislatures (NCSL) where he provided assistance on energy, transportation, and environmental protection to legislators in all 50 state governments. Prior to joining NCSL in 1988, Kreith was the chief of thermal research at the Solar Energy Research Institute (SERI), currently the National Renewable Energy Laboratory. During his tenure at SERI, he participated in the Presidential Domestic Energy Review, served as an energy advisor to the Governor of Colorado, and was the editor of the ASME *Journal of Solar Energy Engineering*. He was the author of over a hundred peer-reviewed articles and textbooks on heat transfer, solar energy, and transportation. He was the recipient of the Charles Greeley Abbot Award from the American Solar Energy Society and the Max Jakob Award from ASME-AIChE, and in 1997, he received the Washington Award for "unselfish and preeminent service in advancing human progress." In 1998, Kreith was awarded the ASME Medal for research, publications, and public service, and in 2004, he was named ASME Honorary Member. In 2005, the ASME established the Frank Kreith Energy Award in recognition of Kreith's contribution to heat transfer and renewable energy.

Contributors

Robert C. Brown
Department of Mechanical Engineering
Iowa State University
Ames, IA, USA

Jan F. Kreider
Boulder, CO, USA
(deceased)

Susan Krumdieck
School of Energy, Geoscience, Infrastructure and Society
Heriot-Watt University
Scotland, United Kingdom

Jeffrey H. Morehouse
Department of Mechanical Engineering
Molinaroli College of Engineering and Computing
University of South Carolina
Columbia, SC, USA

Gary Pawlas
Varia Fitness
Louisville, CO, USA

Mark Mba-Wright
Department of Mechanical Engineering
Iowa State University
Ames, IA, USA

1

Introduction to Sustainable Energy

It is evident that the fortunes of the world's human population, for better or for worse, are inextricably interrelated with the use that is made of energy resources.

M. King Hubbert (1969)

The development of human society and our quality of life are inextricably linked to our use of energy. The emergence of coal in the eighteenth century made possible the Industrial Revolution, and widespread electrification in the twentieth century brought with it all manner of labor-saving devices—from washing machines to clothes driers to refrigerators. Similarly, the advent of internal combustion engine vehicles in the early 1900s, followed by the airplane, turned us into a highly mobile society. All of this was made possible by fossil fuels. But after more than a century of growth and technological advances powered by these fuels, we have come to realize that we humans have made a large-scale Faustian bargain.

In exchange for the wonders of these fossil fuels, we are now coming to grips with their serious side effects. Air pollution from the burning of coal and oil has led to a wide range of health problems—from childhood asthma to life-threatening lung diseases. Acid rain has contaminated lakes, and mercury emissions have poisoned fish and so compromised our food supply. Most critically, the carbon dioxide (CO_2) emissions from power plant smokestacks and vehicle tailpipes are increasing the infrared radiation-absorbing blanket in our atmosphere, which is heating the Earth and causing it to rapidly depart from the 10,000-year period of stable temperature that allowed human civilization to develop. We are already experiencing the effects of the climate crisis, which include melting glaciers and ice sheets (see Figure 1.1), rising sea levels, and an increase in extreme weather events.

Although measures have been taken to reduce air pollution by cleaning up stack emissions from power plants and capturing tailpipe emissions from cars, these have only been temporary solutions to an energy system with a fundamental flaw. The burning of fossil fuels is not sustainable. Fossil fuels are the result of millions of years of nature sequestering atmospheric CO_2 in the ground, and we are releasing all that carbon in a timescale of decades. Although it is technically feasible to capture the carbon in fossil fuels and sequester it in the ground again, the fundamental solution to these issues is to convert our unsustainable, carbon-emitting energy sources to sustainable, carbon-free ones. To accomplish that transition, we need to simultaneously reduce our energy needs by using energy much more efficiently and producing energy from noncarbon sources. This book introduces the knowledge and tools energy professionals and informed citizens will need to lead this energy transition.

The purpose of this chapter is to give an overview of the principles of sustainability in the context of energy engineering. Section 1.1 gives a historical review of sustainability principles. Section 1.2 presents the critical issues for sustainability—population, water, food, and growth—and the relationships between these systems and energy use. Section 1.3 presents some context for the complex nature of sustainable energy by examining the

DOI: 10.1201/9781003586616-1

FIGURE 1.1
People in front of a melting glacier, Kangerlussuaq, Greenland.

world energy system of the past, present, and future and the urgent challenge of addressing climate change. Section 1.4 explains some of the most important considerations for analyzing sustainability, including energy return on energy invested (EROI) and the costs of energy. Section 1.5 explains the most cost-effective of all sustainable energy strategies—reducing energy demand by improving efficiency. Section 1.6 gives an overview of conventional fossil and nuclear energy. Section 1.7 reviews the technology, history, current status, and future prospects for renewable resources including geothermal, hydro, wind, solar, biomass, and ocean energy. Section 1.8 discusses the issues involved in the concept of a hydrogen economy.

A unique feature of this book is that it makes use of energy system design and analysis tools developed by the National Renewable Energy Laboratory (NREL). It includes detailed examples of the analysis and design of renewable energy systems using the freely available System Advisor Model (SAM). An introduction to SAM is presented in Section 1.9, with SAM examples for specific technologies included in subsequent chapters. NREL building energy modeling tools are demonstrated in Chapter 4.

1.1 Sustainability Principles

This book focuses on the engineering challenges confronting our efforts to achieve sustainability, but it is important to keep in mind that the development of energy sustainability is taking place in an era when other relevant forces intersect with one another. The energy challenge also involves societal issues such as demographic change, resource depletion, and environmental degradation. As shown in Figure 1.2, these issues impinge upon the engineering design and economics of a sustainable energy future. Although this book does not directly cover these other issues, it is important to keep them in mind because, unless political and social systems support the roadmap toward a sustainable

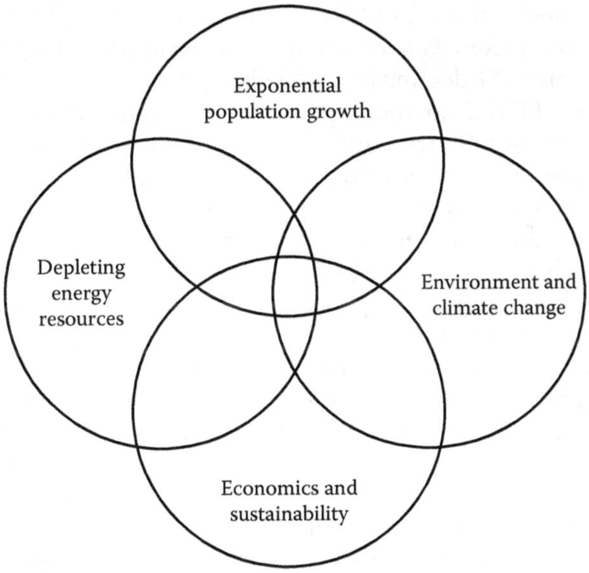

FIGURE 1.2
The interrelated nature of engineering, social, and environmental issues. (From Alliance for Water Efficiency and American Council for an Energy-Efficient Economy [1].)

energy future, social obstacles can impede the implementation of long-term solutions that are technically reasonable as well as economically acceptable.

The energy choices made today are among the most important of any choices in human history. Although energy development questions may appear to be mostly technical and economic, there are broader considerations. Sustainability considerations reflect priorities in our society as well as our attitude toward future generations. Development of more renewable energy is not a panacea for all of our energy problems. A sustainable energy future appears feasible only if we develop an overall social and political strategy that includes renewable energy development combined with energy conservation and efficiency. In some parts of the world, access to energy is still a pressing issue, while in other places efforts are needed to reduce per capita energy consumption.

1.1.1 Energy Crisis: Security Issues

The world oil production capacity increased apace with the demand for liquid fuels for most of the last 100 years. Consider the perspective of people in 1950. At that time, the potential future oil supply would have been much larger than any conceivable demand. Oil was abundant and cheap, and few were thinking about global, irreversible environmental impacts. The production of oil in the United States has been controlled at affordable prices by various organizations since 1910. In this context, the decisions made about building roads, dismantling urban trams, underinvestment in rail, and promotion of low-density land use reflect the perception that the oil supply would last forever.

The average price of oil remained below \$3/barrel until 1970. On October 5, 1973, the Yom Kippur War started when Syria and Egypt attacked Israel. The United States and most Western countries showed support for Israel. Several Arab nations, members of the Organization of Petroleum Exporting Countries (OPEC), plus Iran imposed an oil embargo on Western nations and curtailed their production by 5 million barrels/day. There had been no spare production in the United States, the largest oil-producing nation, since 1971, and other non-OPEC producers were able to increase production by only 1 million barrels/day, resulting in an overall 7% decline in world oil supply. By the end of 1974, the price of oil was over $12/barrel. OPEC did not increase production again after the crisis, but there was a rapid increase in world supply from other areas like the North Sea, Mexico, and Alaska. Despite the increased supply, the price remained in the new range of $12.50–14.50/barrel. The Iranian revolution followed shortly after the invasion by Iraq in September 1980 and led to a decrease in production of 6.5 million barrels/day from the two warring nations, representing a further 10% decline in world supply compared to 1979. This led to a second oil supply shortage with long lines at gas stations and scarcity around the world. By 1981, crude oil prices had more than doubled to $35/barrel, and the global economic recession of the early 1980s was the most severe economic crisis since the Great Depression.

In the United States, this series of events caused a serious energy crisis in the 1980s. Electricity demand had grown rapidly since World War II with a building boom and an explosion of new energy uses from conveniences like air conditioning, refrigeration, washing machines, and other appliances. With no spare oil production capacity in the United States, the OPEC Oil Embargo also caused electricity price increases and shortages in some places. The energy crisis generated great interest in alternative energy sources, spurred research and development, and resulted in new standards and regulations on energy efficiency for buildings, appliances, and vehicles. The energy crisis also started a political obsession with energy independence as a national security objective that persists to this day. The realization that all critical activities relied on finite fossil fuels like oil and natural gas influenced thinking about sustainability to extend beyond environmental impacts and to look at the future implications of depletion. Energy engineering emerged as a new field, primarily within mechanical engineering with courses on energy conversion, air pollution, and energy management taught at many major universities.

However, as new oil fields in the North Sea and other non-OPEC nations were developed, world oil production capacity increased again, the oil price declined to a new price range below $30/barrel, and it became politically unpopular to raise concerns about energy shortages. In fact, the discussion around fossil fuel use and the development of the resources in remote and higher-risk environments became politically polarized and framed in terms of the environment versus the economy, including debates over appropriate locations for oil and gas development. Beginning in the mid-2000s, U.S. producers began to exploit directional drilling and hydraulic fracturing technology to access shale deposits for oil and natural gas. Increasing production from shale resources led to the United States becoming a net exporter of natural gas in 2017 and a net exporter of crude oil starting in 2021. More recently, the COVID-19 pandemic temporarily reduced global demand for oil, and Russia's February 2022 invasion of Ukraine shifted trade patterns and spurred an increase in renewable energy development in Western Europe. Both are reminders of the hard-to-predict events that can disrupt the global energy system.

The imperative for a rapid transition to a more sustainable energy system is now being driven by the need to address the health issues and climate change impacts linked to the burning of fossil fuels. At the same time, dramatic improvements in the cost-effectiveness of renewable energy technologies, especially solar photovoltaic energy and wind power, are accelerating a transition to these technologies. This transition began in the electricity sector, with renewables displacing coal and natural gas, and is now moving to the transportation sector with rapidly growing adoption of hybrid and battery electric vehicles (see Figure 1.3), as well as the electrification of buildings with heat pumps. These new uses of electricity, together with the projected growth of data centers for artificial intelligence

FIGURE 1.3
The 2017 Chevy Bolt was the first electric vehicle with an electric range of over 200 miles (320 km) (the Environmental Protection Agency range was 238 miles or 381 km) and a retail price (after U.S. federal tax credit) of under $30,000. (Photo by C. Kutscher.)

and cryptocurrency, are expected to spur a rapid expansion of electricity generation. This marks a significant change in the United States, following a 15-year period of nearly constant electricity supply.

1.1.2 Sustainable Development

In 1983, the United Nations set up an independent body headed by Gro Harlem Brundtland, the former prime minister of Norway, called the World Commission on Environment and Development. The commission was asked to examine the critical environmental and economic development problems facing the planet and to formulate realistic proposals to solve them. The proposed solution was to ensure that human progress could be sustained without bankrupting the resources of future generations. The outcome of this study was an important book entitled *Our Common Future* [2]. The Brundtland Commission concluded that worldwide efforts to guard and maintain human progress to meet human needs and to realize human ambition are unsustainable, both in developed and developing nations, because they rely on the ever-increasing use of already overdrawn environmental resources. The commission provided a widely quoted definition of sustainability.

> Sustainable development should meet the needs of the present without compromising the ability of future generations to meet their own needs [2].

While this definition of sustainability is widely used, it is worth noting that it reflects an anthropocentric viewpoint. It does not address the broader responsibility of human beings to be stewards of the planet, including all species, even those that do not appear to benefit human existence. This is the concept of ecological sustainability, for which one possible definition is "conducting human activities in ways that do not diminish biodiversity, allowing ecosystems to maintain their functionality and preserving natural processes over the long term" [3]. A U.S. law that reflects this principle is the Endangered Species Act of 1973, which protects species that may have little or no interaction with human populations.

The Brundtland Commission stated that sustainable global development requires that those who are more affluent adopt lifestyles that can be accommodated by the planet's ecological means, particularly their use of energy. Furthermore, a rapidly growing population increases the pressure on resources. Sustainable development can, therefore, only be achieved if population size and the population growth rate are limited to a point where they no longer exceed the productive potential and maintenance level of the Earth's ecosystems. The commission realized that future economic growth must be less energy-intensive than it has been in the past. Energy efficiency policies must be the primary vehicle of national energy strategies for sustainable development, but the commission also realized that energy efficiency could only buy time for the world to develop energy paths based on renewable sources, and that this goal should be the foundation of any global energy strategy for the twenty-first century.

In 2015, all member states of the United Nations (UN) adopted the 2030 Agenda for Sustainable Development, which lists 17 urgent goals for global collaboration [4]. These goals include ending poverty and hunger; achieving gender equality; universal access to clean water and sanitation, affordable and clean energy, and quality education; combating climate change; and protecting biodiversity in the oceans and on land. The UN publishes annual reports on progress toward meeting these goals. Progress reported to date remains slow, although the UN reports success in improving access to clean water, sanitation, and energy as well as in reducing infant mortality. The Sustainable Development Goals provide a valuable framework for placing sustainable energy in the broader context of efforts needed to support human flourishing and protect ecosystems and biodiversity on the planet.

1.1.3 Sustainability Principles in Practice

Managing the use of common resources is an ancient challenge for sustainability. In a famous essay, Garrett Hardin described the "Tragedy of the Commons," whereby individuals with access to a shared resource, such as a common grazing area, are prone to overuse it [5]. Individuals want to maximize their personal benefit from the resource, leading to overuse and degradation when too many people have the same individual incentive. Grazing lands, ocean fisheries, and public parks are examples of common-pool resources cited in Hardin's essay, as is an unpolluted environment.

Historically in Western countries, the primary approach to managing land, water, or minerals that might have been viewed as common resources has been to privatize the resource, whether by conquest, grant, or sale. When governments hold or protect resources in common for public benefit, as with national parks, fisheries, and minerals on public land, overuse may be limited by lotteries, entrance fees, licenses, or leases. Laws and regulations limiting discharges and emissions are the principal mechanisms used to protect the environmental commons from excess pollution.

As industries grew up around the United States in the twentieth century, states, counties, and cities worked with health boards to monitor and regulate pollution discharged into lakes, rivers, and the atmosphere. The 1970s saw a raft of federal environmental legislation in the United States, including the Clean Air Act Amendments of 1970 and the Safe Drinking Water Act (1974), which resulted in great improvements in air and water quality. In 1987, the Montreal Protocol on Substances that Deplete the Ozone Layer was ratified by most of the nations of the world and required corporations to stop producing certain widely used refrigerants and chemicals that were very useful and profitable, but were shown by scientists to be depleting the Earth's protective ozone layer. The United Nations Framework Convention on Climate Change was agreed upon in 1992, initiating decades of international efforts to manage the essential common resource of a stable global climate. These efforts to combat climate change have yielded some progress but remain insufficient.

Corporations usually focus on "the bottom line," meaning the financial balance sheet. However, major corporations around the world are voluntarily addressing not just the financial benefits of their operations but the social and environmental impacts as well. This is the idea behind "triple bottom line" (TBL) accounting, which accounts for the financial, social, and environmental impacts of corporate operations on workers, customers, and people who live near factories or mines. Social and environmental bottom lines are not as easy to quantify as the financial bottom line, so corporations have set out principles for their practices so that they do not exploit resources or people in a way that negatively impacts their well-being. For example, fair trade certification has been developed to ensure that suppliers in less developed countries receive a fair price for their goods and labor. Ecological footprint, life cycle assessment, cradle-to-grave, and cradle-to-cradle (zero waste) analyses of goods are techniques that have been developed to quantify the energy and material use and emissions associated with products.

1.1.4 Challenges for Sustainability Engineering

A key challenge for engineering professionals is to institute sustainability considerations in all projects and operations in much the same way that safety issues are included. Safety engineering requires honesty with employees and the public about the previous safety failures, the potential hazards, and effective risk management measures. Energy engineers will need to similarly be honest with business leaders, policymakers, and the public about the issues of raw material supply security, the potential for alternative technologies, and environmental risks. One challenge is the perceived conflict between economic benefit and sustainability. Again, energy engineers can follow the lead of safety engineers, who are educated to value health and well-being over profit.

1.2 Carrying Capacity and Exponential Growth

1.2.1 Population Issue

World energy use and the associated carbon emissions are clearly linked to the world population. The world population started to increase rapidly at the end of the eighteenth century. It took approximately 120 years for the population to double from about 1 billion in 1800 to 2 billion in 1920 and then only 40 years to double from 3 to 6 billion [6]. The annual population growth rate peaked above 2.2% in the early 1960s [7]. The world population reached 8 billion in 2022. The UN predicts that the world population may reach 9 billion by 2040, as shown in Figure 1.4 [7].

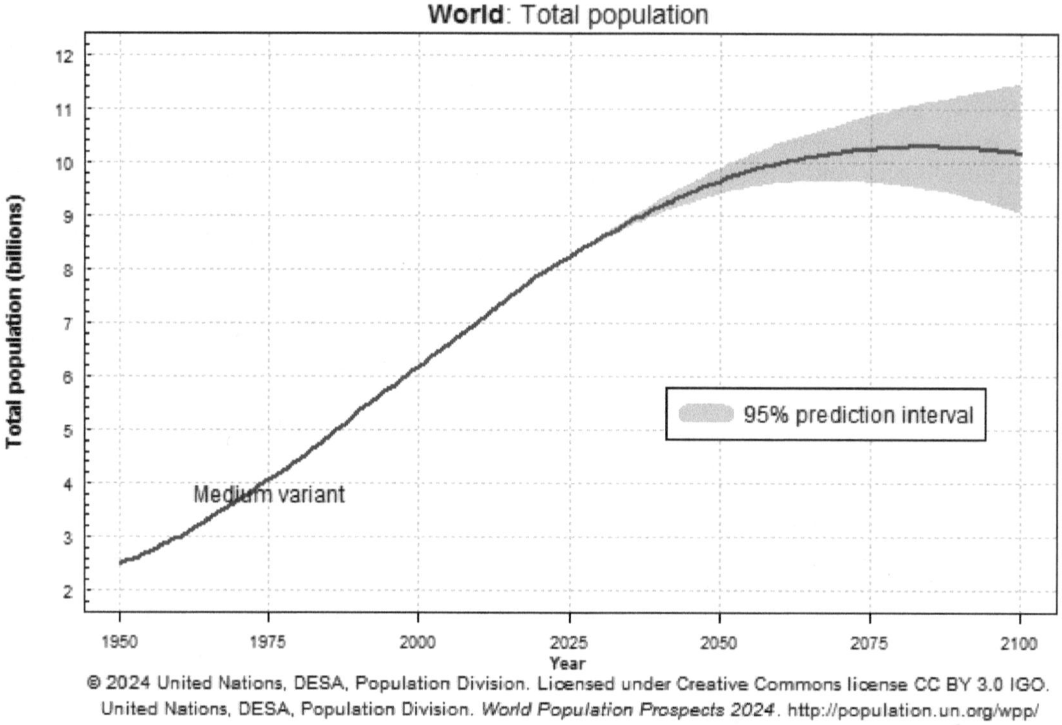

FIGURE 1.4
Growth of world population with estimates through 2023 and projections to 2100. (From Population Division of the Department of Economic and Social Affairs of the United Nations Secretariat [7].)

Current population is a function of the number of people alive last year, adding the number who were born and subtracting the number who died this year. The number who die in a particular year is related to life expectancy, which accounts for the deaths at all ages, including people who die in childhood. A higher life expectancy means more elderly people, but it also means more females are living to childbearing age. Life expectancy at birth is currently around 79 years in Europe and Northern America, 78 in China, 72 in India, and 60 years in Sub-Saharan Africa [7]. Consider that in 1900, the life expectancy in the United States was 46–48 years.

The average fertility rate is expressed as the expected number of children a woman would have during her lifetime. The dramatic increase in world population in the last century was due to reduced infant and childhood mortality, an increasing number of fertile women, and increased longevity. The rate of population growth has slowed appreciably in many countries around the world but remains higher in some parts of Africa, the Middle East, and parts of Southeast Asia. Globally, the total fertility rate dropped from just over 5 live births per woman in 1950 to 2.25 in 2023 [7]. According to the UN, in more than 60 countries, the size of the population peaked before 2024 and is now stable or declining [7]. Up-to-date information on population, life expectancy, and fertility rates is available from the United Nations Department of Economic and Social Affairs, Population Division, at https://esa.un.org/unpd/wpp.

A large body of research (see, for example, Bongaarts [8]) has concluded that if women have socially acceptable alternatives to childbearing, they will voluntarily limit the number of children they have. Education has been shown to be the most effective way to reduce population growth. In locations where girls are given the chance to go to school, the average age of first pregnancy shifts upward. And in places where women have the opportunity for employment, as occurs with increased urbanization, the number of children per woman declines. Adding the availability of contraception and social acceptability of small families can bring the population growth rate down to replacement levels. On the other hand, providing support for women and families, including through parental leave, affordable childcare, and equitable sharing of household responsibilities, can encourage childbearing in countries where a declining population is a concern.

1.2.2 Water Issue

Less than 1% of the water on the planet is suitable for drinking and agriculture; the rest is saltwater, brackish, or frozen. Freshwater refers to rivers and lakes fed by seasonal precipitation. Aquifers are underground freshwater reservoirs in permeable gravel or sand. Unconfined aquifers are replenished by surface precipitation, but confined aquifers like the massive Ogallala aquifer under the Great Plains in the United States hold fossil water, deposited over 1 million years ago. The sustainable water consumption in any given location depends on the precipitation rate, storage systems, and the requirements to sustain local ecosystems. Confined aquifers have a finite lifetime for use rather than a sustainable usage rate. In Texas, artesian wells were plentiful in the early twentieth century, but wells must now be pumped and in some places have run dry. A study published in 2012 found that 30% of the Southern Great Plains will have exhausted the groundwater resource within 30 years [9].

The World Health Organization stipulates that the basic requirement for water is that 20 liters per day (L/day) per person should be accessible within 1 km of the home. In an industrial society such as the United States, personal water consumption is 10 times as large, somewhere between 200 and 300 L/day for household uses. But if industrial and energy production usage are added, freshwater usage can exceed 5,000 L/day on a per capita basis. The World Bank [10] suggests that an indicator of water scarcity is an annual availability of less than 1000 m^3 (or 1 million L) per person. By comparing a country's total freshwater resources to this number, a country's water-carrying capacity can be determined, that is, how many millions of people the water resources can support. Many countries in the world, especially in the Middle East, already have populations that exceed the country's water-carrying capacity.

There are more than 16,000 desalination plants in operation worldwide, with 60% located in the Middle East. The levelized cost of water produced by various desalination plants is between \$0.60/m^3 for large plants and \$3.00/m^3 for small remote plants. The energy requirement varies greatly with the type of process used and is much lower for brackish water than for seawater. Thermal processes like multistage flash desalination use 10–15 kilowatt hours of electricity per cubic meter of water (kWh/m^3) while reverse osmosis uses 4–13 kWh/m^3 of electricity for seawater and 0.5–2.5 kWh/m^3 for brackish water [11]. Desalination is currently an option only in those parts of the world where energy supplies are plentiful and the value of water is extremely high. For example, in Saudi Arabia, the world's largest desalination plant produces 128 million gallons/day (0.5 million m^3/day) of potable water.

It is estimated that around 3% of the electricity demand in the United States (56 billion kWh/year) is for treating and pumping freshwater. California has a large agricultural sector and a sizable population living far from freshwater sources, resulting in 19% of electricity demand associated with the provision of water. The biggest use of freshwater from lakes and rivers is for cooling thermoelectric power plants. Figure 1.5 shows a breakdown of the water withdrawal and consumption in the United States according to [1]. (Water withdrawal refers to the sum of water that is consumed without being returned to the source and the water that is returned to the source.) Power generation and agriculture are the largest sources of water withdrawal, and agriculture is by far the largest consumer of water.

As discussed in Chapter 6, thermal power plants operating on a Rankine cycle reject heat in a condenser. The amount of water consumed or withdrawn depends on the kind of cooling system used for the condenser. In coastal locations or where a large river or lake is available and where the water temperature is cool year-round, the withdrawn water is returned to the freshwater body. In hot weather, the freshwater is sprayed over evaporation pads, the air is blown through the cooling tower, and the evaporation cools the water stream. In this process, some of the freshwater is consumed and exhausted into the atmosphere. Table 1.1 gives the amount of water consumed by various types of thermal electric power plants. The U.S. average water consumption for thermoelectric power generation was about 23 gal (87 L) per kWh in 2005 for all types of cooling [12]. Water use in electricity generation is an issue of long-term sustainability. In an extensive report prepared for the U.S. Congress in 2006, the U.S. Department of Energy [13] stated that:

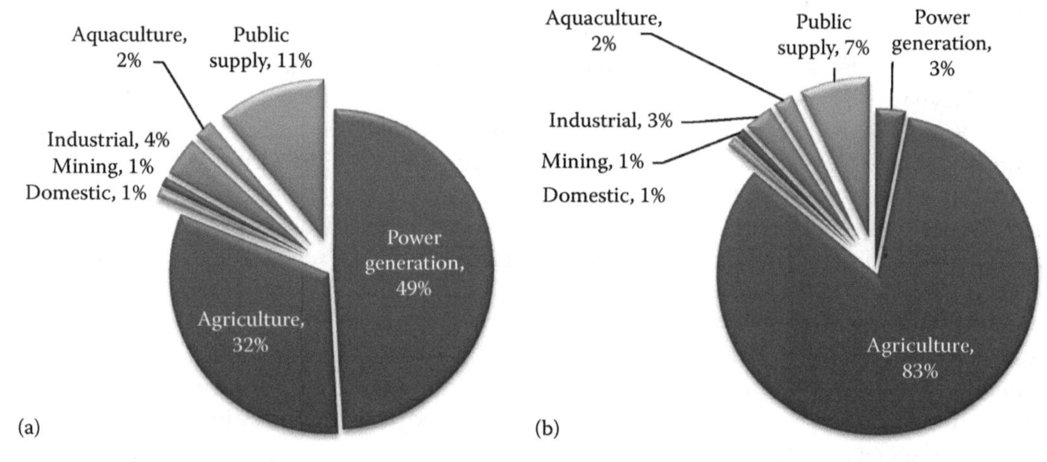

FIGURE 1.5
(a) U.S. freshwater withdrawal and (b) U.S. freshwater consumption. (From Alliance for Water Efficiency and American Council for an Energy-Efficient Economy [1].)

TABLE 1.1
Water Consumption for Different Types of Power Plants

Plant Type	Water Consumption per Unit of Electricity (L/MWh$_e$)
Solar thermal	2,600–3,800
Nuclear	2,600–3,000
Subcritical pulverized coal	2,000
Supercritical pulverized coal	1,700
Integrated gasification combined cycle	1,200
Natural gas combined cycle	720

> Available surface water supplies have not increased in 20 years, and groundwater tables and supplies are dropping at an alarming rate. New ecological water demands and changing climate could reduce available freshwater supplies even more [13].

As a result, more and more power plants are being air-cooled. Air-cooled condensers require a large heat exchange surface area and considerable fan power to reject heat to the ambient air. Air-cooled plants can suffer significantly reduced performance on hot summer afternoons—the time when electricity is most highly valued. To address this problem, some plants use a combination of air and evaporative cooling during these periods.

Production of biofuels also places pressure on water resources. Ethanol processing from corn requires 3 L of water per L of fuel and anywhere from 10 to 330 L of water per L of fuel for agriculture depending on irrigation requirements [14]. American petroleum production requires 2–6 L of water per L of crude oil depending on the amount of recycling of produced water as injection fluid. Water consumption and pollution are major issues with the extraction of oil from Canadian tar sands. Webber [15] provides an in-depth discussion of the linkages between energy and water.

1.2.3 Food Supply Issues

To have sustainable development, global food production must be adequate to feed the world's population. Food comes from the sea and the land. It has been estimated that the Earth has 8.2 billion acres (ac) (33 million km^2) of potentially arable land. According to a study by the Millennium Institute [16], this acreage is divided into the following levels of productivity:

- World Agricultural Land Endowment:
  - Highly productive—1.1 billion ac
  - Total somewhat productive—2.2 billion ac
  - Total slightly productive—4.9 billion ac
  - Currently productive—3.6 billion ac
  - Potentially productive—8.2 billion ac

In the United States, about 1.5 ac (0.006 km^2) of land is used to produce food for one person per year, but a modest per capita global footprint based on diets lower in meat and higher in grains is estimated to require only 1 ac/person. At this modest per capita global footprint, the global carrying capacity would be somewhere between 3.6 and 8.2 billion people. In 2015, the United Nations Food and Agriculture Organization estimated that 795 million people were undernourished globally, most of whom live in Asia and Africa [17]. Global constraints on food and water availability suggest that more than 9 or 10 billion people would be an unsustainable global population.

The food supply system in the United States and other industrial countries currently runs on fossil fuel. The food supply system for the typical American diet has been estimated to require approximately 2,000 L of oil equivalent per year per person and account for 19% of the total national energy use, with 14% for food production, processing, packaging and storage, and 5% for transportation [18].

Competition between agriculture for food and for biofuels is an international issue. Corn, soybeans, wheat, and sorghum are the basis of the American food production system. In 2014, the United States produced about 17% of the total world grain supply.

Furthermore, only about 12% of the domestic corn market is directly used for food. Most of the corn grown in the United States is used for animal feed and ethanol. Biofuel production grew rapidly after 2005 in response to U.S. government mandates and incentives. The consequences of the bioethanol boom were that the percentage of corn crops used for food in the United States dropped from 13% in 2000 to 11% in 2011, exports dropped from 20% to 13%, and feed for U.S. cattle and chickens plummeted from 60% to 36% of the annual crop. In 2016, ethanol processing consumed 40% of the total U.S. corn crop. According to the U.S. Energy Information Administration, biodiesel expanded from 0.5 million gallons in 1999 to 1.7 billion gallons in 2023 [19]. Despite this growth, biodiesel accounted for a little less than 1% of U.S. highway fuel in 2023.

1.2.4 Per Capita Energy Use

Unlike food and water, the basic level of energy to sustain a given human population is more difficult to specify. The daily energy requirement for a family living a traditional lifestyle on a remote island in Fiji is two coconut husks dried for 2 days in the sun that are used to cook the day's taro in a pit fire and fry the day's fish catch on a simple grill. On the same island, there are homes connected to village electric grids and the island water supply. Their daily energy demand also includes 72 Wh for two small lights for 6 h/day when the village generator functions and a small share of the 20 L of diesel fuel used to run a water pump for 3 hours each day. Also, on the same island, a home built by an Australian has a much different lifestyle with continuous electricity, a refrigerator, TV, lights in every room, and air conditioning. This home generates its own 11 kWh of electricity per day by solar panels, a small wind generator, a large battery bank, and several diesel generators. Compressed natural gas is used for cooking. Although the needs of this house are only about half the average in Australia, they are still vastly different from their neighbors on the island [20]. The standard of living for all of the island residents is roughly similar to the extent that they all have adequate nutrition, access to the same quality of air and water, access to medical treatment and education, and the same life expectancy. Clearly, the energy use of the traditional lifestyle contrasts sharply with that of the Australian lifestyle, which depends heavily on finite resources. Even though about half of the energy for the Australian home is provided by renewable sources, some fossil fuels were consumed in the manufacture of the renewable technologies.

The UN proposed a measure of standard of living, termed the Human Development Index, based on various criteria [21]. Figure 1.6 [22] is a plot of energy consumption per capita for various countries as a function of a composite indicator, developed by the United Nations Development Programme, to show countries' relative well-being using the Human Development Index. It is apparent that the United States and Iceland have the highest energy consumption per capita (although Iceland uses 100% renewable energy composed of geothermal and hydropower), but their human satisfaction level is not much different from other developed countries such as France or Germany, which provide their populations with a good standard of living at much lower energy consumption per capita.

There are large variations in the amount and type of food and the amount of water used according to lifestyle, just as there are wide variations in energy use. We can set physiological requirements for food and water; however, it is hard to set such a requirement for energy. In considering the data in Figure 1.6, 1,000 kg of oil equivalent (41.9 GJ) can be considered a modest energy requirement, or the fossil energy budget plus an allowance for shelter in more extreme climates. This level of energy use may seem draconian in relation to the current U.S. average per capita consumption of over 9,000 (377 GJ).

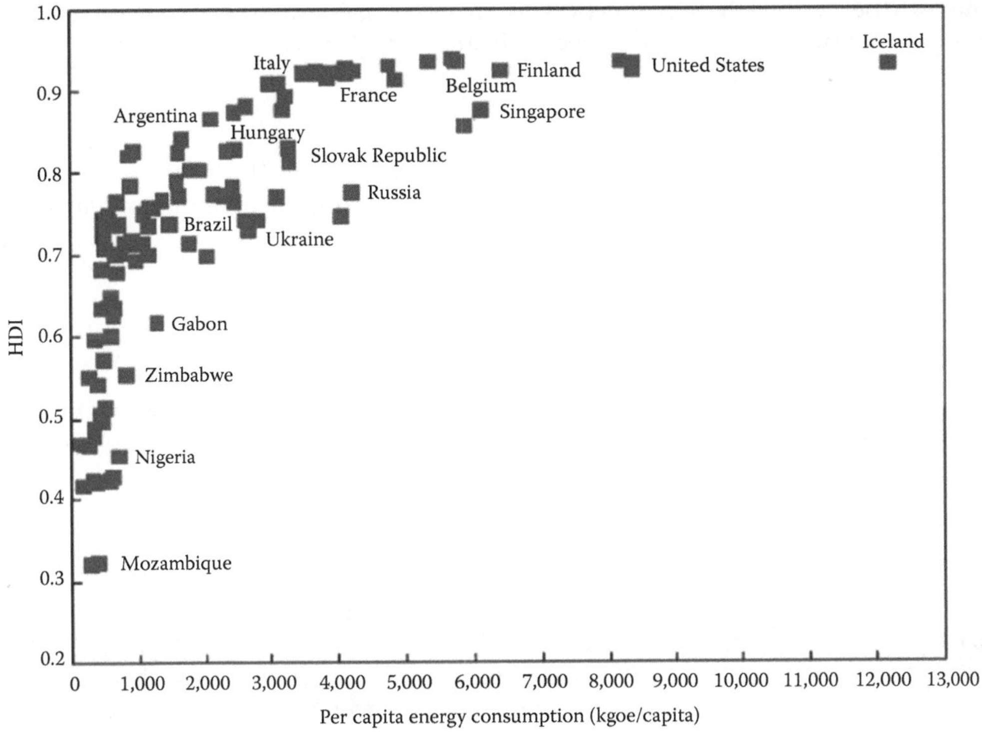

FIGURE 1.6
Relationship between Human Development Index and per capita energy use, 1999–2000. (From Goldemberg and Johansson [22].)

However, in quantifying carrying capacity, we are not considering lifestyle; rather, we are considering necessity.

1.2.5 Mathematics of Exponential Growth

Economic growth is a complex phenomenon, which is conventionally measured in terms of year-over-year increases in the value of goods and services produced in a particular country (i.e., the gross national product). The factors that influence economic growth include increases in capital stock, advances in technology and education, and growth in workforce population. One important limitation of standard economic measures of growth such as change in gross national product is that they fail to capture changes in the distribution of wealth or income across the society. Another important limitation, especially relevant to this textbook, is that they ignore the long-term economic and environmental consequences of depleting natural resources.

For much of the past 150 years, energy use has increased relentlessly along with economic activity, population, and water, food, and land use. An important goal of sustainable energy engineering is to relax the historical linkage between improved quality of life and increased consumption of nonrenewable energy and natural resources. This goal can be accomplished by switching to renewable resources, using technological advances to improve efficiency, and reducing consumption in the developed world. The importance of taking steps to curtail overconsumption of finite resources is clearly illustrated using the simple arithmetic of compound growth.

Suppose the quantity, N, periodically increases from an initial value, N_0, at a fixed growth rate, r, per unit time, t. The future value of the quantity, $N(t)$, is given by

$$N(t) = N_0 (1+r)^t \tag{1.1}$$

After one period, the quantity will be equal to $N(1)=N_0(1+r)$, after two periods $N(2)=N_0(1+r)^2$, and so on. The time it would take for the original quantity to double, or the doubling time, D, is of interest in the studies of growth. Setting $N(t)=2N_0$ in Equation 1.1 and solving for t gives

$$D = t = \frac{\ln(2)}{\ln(1+r)} \tag{1.2}$$

For most practical situations, r is small, say less than 10%, and the quantity $\ln(1+r) \cong r$. Noting that ln(2) is equal to 0.693, Equation 1.2 simplifies to the rule-of-thumb approximation for doubling time:

$$D \cong \frac{0.7}{r} \tag{1.3}$$

Thus, for an annual growth rate of 7% ($r=0.07$), the doubling time would be about 10 years.

Now, suppose that a finite resource would have a total useable quantity, N_T. The initial consumption rate per year is N_0. If the consumption rate does not change, then the lifetime of the resource, t_L, would be found by $t_L = N_T/N_0$. However, when the consumption rate of the resource is increasing every year with an annual fractional growth rate, r, the lifetime of the resource is given by

$$t_L = \frac{\ln\left\{\frac{N_T \ln(1+r)}{N_0} + 1\right\}}{\ln(1+r)} \tag{1.4}$$

Economic value or money is a human construct and not a natural resource, so the supply of money can indeed grow as long as all the participants in the economy continue to agree on its value as a proxy for their goods and labor. However, when we are considering the carrying capacity of a region or the lifetime of a finite resource, we need to be able to perform quantitative analysis to assess the depletion risks and the time frames associated with these issues.

Example 1.1: Continuous Growth, Finite Resources, and Perception

In 1950, the city of Fort Collins, Colorado, had a population of 15,000 and occupied 3,750 ac. The city was surrounded by 50,000 ac of prime farmland within a 20-minute drive, which grew most of the food sold in the city markets plus grains, fruits, vegetables, and dairy that were transported to the capital city of Denver. Like the rest of the country, Fort Collins was about to experience a prolonged period of population growth at an annual rate of 4.2%. New housing for the population was typical suburban single-family homes, and 158 ac was converted from agriculture to urban use in 1950.

In 1950, people started to wonder: if this current rate of land conversion were to continue, how long until the farmland is consumed? The answer (50,000/158) is 317 years. No worries! But the next year, 164 ac was developed, and the next year, 171 ac. A professor at Colorado State University used Equation 1.1 and determined that the city

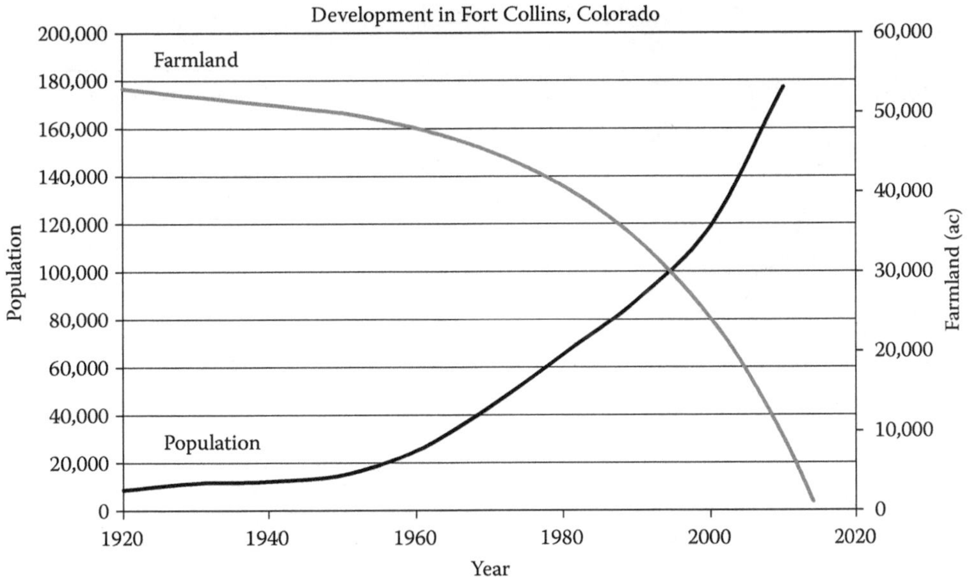

FIGURE 1.7
Population growth and farmland reduction as a function of time for Fort Collins assuming an annual population growth rate of 4.2%.

seemed to be increasing its land use by 4.2% each year, the same rate as the population growth. If the 3,750 ac of urban land in 1950 were to increase apace with population, then Equation 1.4 gives the year that the city would have grown to 53,750 ac and consumed all of the farmland as 64 years later, in 2014, as shown in Figure 1.7. The professor in 1953 would surely have looked at the map and the farmland around his city and not really been able to picture 118,000 people living there and all of the farmland gone. The mathematical model is clear, but the professor's perception of the issues associated with future growth would likely be limited by his lack of experience with this kind of situation. He might think that something would change in the future to slow the growth or manage the preservation of some of the agricultural productivity.

The professor would have been called an alarmist if he suggested that population and land use were an issue. The mayor would not appreciate the suggestion that population growth could be a problem when it was needed to pay for the new schools, sewer system, and library to meet the needs of previous growth. In 1997, there was still half of the farmland remaining around the city. But in fact, the steady, "healthy" growth rate of the city continued, and all of the original farmland is now within urban city limits.

1.3 Context for Sustainable Energy

1.3.1 Historical Energy Development in the United States

The United States has made two major energy transitions since the mid-1850s as shown in Figure 1.8 [23] and is now embarking on its third transition. Before the Civil War, U.S. industry depended on wood, wind, water, and animals for its heat, energy, and power. Charcoal from hardwoods has been used for processing metals and ceramics since ancient times. The

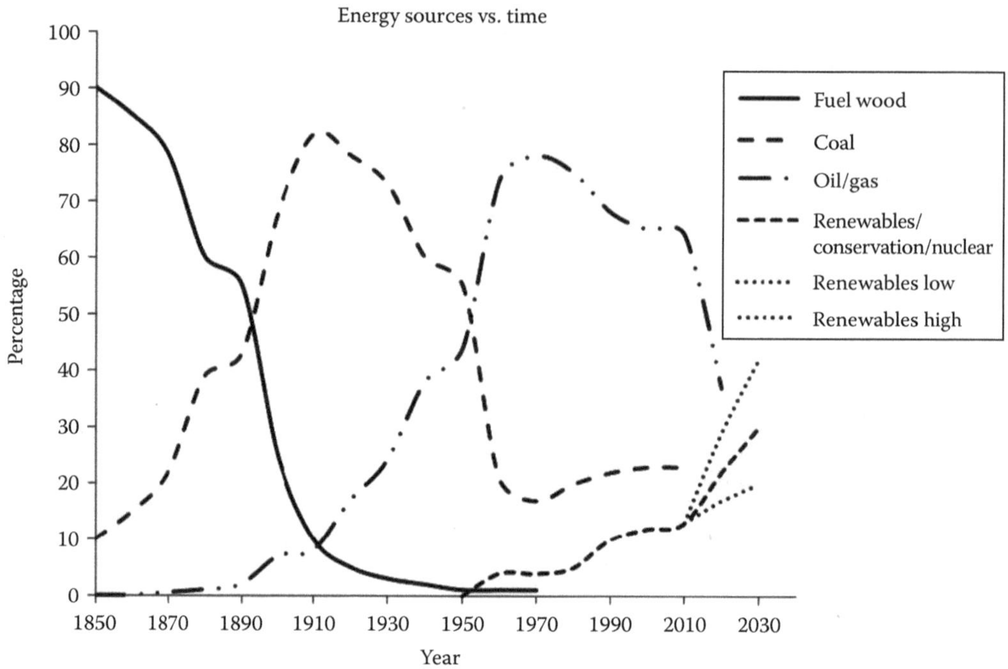

FIGURE 1.8
Diagram illustrating energy transitions in share of total energy supplied by wood, coal natural gas, and renewables. (From U.S. Energy Information Administration [23].)

Industrial Revolution outstripped the wood supply in Europe and contributed to the clearing of forests in North America. In the first energy transition, coal became the major supply for U.S. energy needs between 1885 and the start of World War II in 1941. New steam engines that could pump water and ventilate coal mines stimulated a transition to coal followed by a dramatic increase in total energy consumption. Coal was found to be a vastly superior fuel to wood in both running the new steam engines and high materials processing. However, the air pollution from coal burning in factories and homes, and gasification of coal for lighting, was a serious public health issue in most cities around the turn of the century.

The second transition from coal was not due to a shortage of coal; rather, there were great advantages to using oil and gas, especially with the rising share of energy consumed by transportation. Although the share of consumer energy provided by coal declined after the post-war surge in energy use in the 1950s, the total amount of coal production increased until 2008 when comparatively cheap natural gas from shale increased dramatically and displaced coal. Beginning in the 1970s, the United States transitioned from being a net exporter of oil to being an importer. Total U.S. energy production had leveled off and conversion efficiency had improved, but consumption continued to grow. The United States became a net exporter again in the last decade, based on the development of oil and gas resources found in shale reservoirs. In the 1960s, it was thought that nuclear energy would be the next major energy source of the future. However, the costs of nuclear energy turned out to be much higher than expected, and the anticipated safe long-term waste storage solution did not emerge. In the 1980s, concerns about nuclear plant safety caused public opposition, and excessive cost overruns derailed new nuclear power developments.

The third energy transition in the United States and around the world is today's transition to renewable energy, driven by environmental concerns and enabled by large cost reductions in solar and wind electricity. The needed transition includes the use of renewable or other low-carbon resources for electricity generation combined with the use of low-carbon electricity for transportation and heat in buildings and industrial processes.

1.3.2 Current Energy Use

Although tapping into unconventional sources of fossil fuels has spurred periods of increased production subject to international economic conditions, ultimately, the imperative of combating climate change will require a transition to other energy sources, namely renewables and nuclear (fission in the near term and likely fusion in the long term). Even though we are not yet running out of oil, the transition to other forms of energy is already occurring. Former Saudi Oil Minister, Sheik Ahmed Zaki Yamani, was famously quoted as saying "The Stone Age didn't end because we ran out of stones." The rapid energy transition that is now occurring is driven not by declining fossil fuel reserves but by the growing concern for climate change, as well as by the pollution and health concerns in China (see Figure 1.9) and India associated with burning coal.

To get an idea of the scale of the transition from current energy systems to renewable systems, the detailed pattern of energy supply and consumption needs to be examined. It is useful to look at end-use consumption by dividing it into four major sectors: residential, commercial, industrial, and transportation. These sectors consume, respectively, 19.7%, 17.2%, 33.2%, and 29.9% of total U.S. energy, as shown in Figure 1.10 [24]. Total U.S.

FIGURE 1.9
Surgical masks are a common sight on the streets of Beijing, China, where air pollution has been a major health concern.

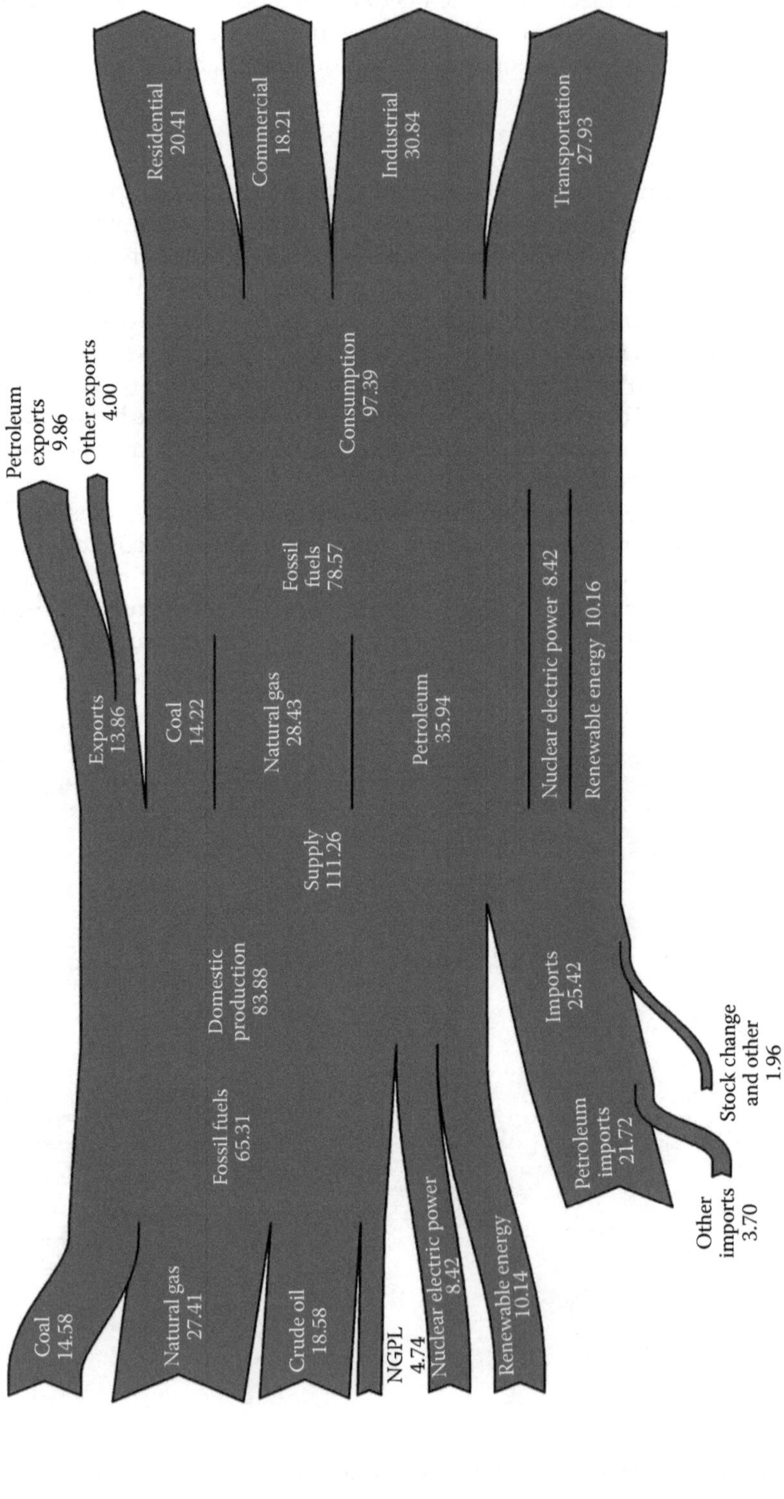

FIGURE 1.10
U.S. primary energy flow in 2023 in quadrillion Btus. (From U.S. Energy Information Administration [241].)

consumption in 2023 (excluding exports) was 94 quads (99 exajoules), which is about one-fifth of the world's energy use, with fossil fuels accounting for about 82% of the total. The total contribution of renewable energy in 2023 was 8.8%. Renewable energy includes biomass (wood, waste, and biofuels, 61%), hydropower (10%), wind (17%), solar (11%), and geothermal (1%) [25].

Because of the significant legacy reliance on fossil fuels, the transition to renewable energy systems represents a very large-scale shift in primary energy supply. In the United States, renewable energy technologies have been incentivized with tax credit policies and state renewable portfolio standards (which require that a certain percentage of electricity be supplied by renewable sources). Although these policy measures have helped give a boost to an emerging industry, recent large increases in the deployment of wind and solar have stemmed from large cost reductions. Wind energy costs have dropped because of the trend toward taller turbine towers that yield higher wind speeds and steadier conditions. Solar photovoltaic modules have dropped in price as a result of Chinese investments in low-cost manufacturing. As both technologies have seen a jump in deployment, learning curve effects have provided positive feedback to drive down costs even further. As a result of these low costs, more than half of the new electric-generating capacity added in the United States over the past decade has been comprised of wind and solar. In 2024, solar, wind, and battery storage were expected to make up more than 90% of electric-generating capacity additions [26].

The manufacture of renewable energy technologies initially requires some fossil fuel investment with its associated carbon emissions. This impact is accounted for in life cycle assessments. Also, although utilities are experienced at handling variations in electricity demand, the transition to more wind and solar on the grid means that utilities must also address variations in energy supply. These and other issues are covered in later chapters.

1.3.3 Future Energy Scenarios for the United States

Future scenarios are typically developed using historical information, interpreting trends, and modeling future behavior under different assumptions. Nearly every nation has a governmental energy modeling unit dedicated to developing the national energy outlook. In its Annual Energy Outlook 2023 reference scenario, the U.S. Energy Information Administration (EIA) projects that electricity generation from natural gas and coal will decline through 2050 while generation from non-hydro renewables increases [27]. Notably, while EIA is an excellent source for current and historical energy data, its past reference case projections have underpredicted future renewable energy deployment. A transition away from fossil fuels is needed to combat climate change. The Intergovernmental Panel on Climate Change (IPCC) calls for net-zero carbon emissions by 2050, and this is reflected in an International Energy Agency net-zero energy roadmap that is discussed in the next section.

1.4 Key Sustainability Considerations

1.4.1 The Challenge of Climate Change

The average surface temperature of the Earth is determined by a heat balance with outer space. Absorbed incoming solar energy is balanced by outgoing infrared radiation (IR). The Earth's surface reflectivity, or albedo, plays a role in how much sunlight is absorbed. Deforestation, carbon soot from diesel fuel burning, and melting ice all affect the albedo. The atmosphere also plays a very large role by absorbing or reflecting some of the solar radiation (aerosols from man-made air pollution play a significant role in reducing the sunlight that makes it through the atmosphere) and by absorbing and reemitting much of the outgoing IR. Gases in the atmosphere that absorb and emit IR are called "greenhouse gases"—a reference to the window glass used in many greenhouses, which absorbs infrared radiation. Thanks to the greenhouse gases that are naturally present in the atmosphere—primarily water vapor, CO_2, and methane—that trap outgoing IR, the Earth is about 33°C warmer than it otherwise would be, thus making the widespread presence of liquid water possible.

However, since we began burning fossil fuels on a large scale in the late nineteenth century, we have increased the CO_2 content of the atmosphere by over 50%—or about 37% just since 1958 when the American scientist Charles David Keeling began measuring it with high accuracy atop Mauna Loa in Hawaii. This CO_2 can be correlated with the known amount of fossil fuel burned, and the carbon has the same isotopic composition as that contained in fossil fuels. Basic calculations and ever-improving detailed computer models indicate that this has upset the Earth's heat balance. It is estimated that there is now about 1 W/m^2 of the Earth's surface area more solar energy being absorbed than there is IR radiation going out. (This is warming the Earth with an amount of energy equivalent to about eight 15 kiloton atom bombs exploding every single second.) This warming is manifest in many scientific observations: increasing surface land and sea temperatures, decreasing sea ice volume and land-based ice mass, rising sea levels, and shifting flora and fauna. Because warmer air holds more moisture, we are seeing more extreme precipitation events. And because the heating of the planet's surface has been nonuniform, weather patterns have changed. All other possible explanations for the warming have been exhaustively investigated by scientists and eliminated. (For example, accurately measured solar output does not correlate to heating.) On the other hand, specific observed changes are unique signatures of greenhouse gas warming. These include the temperature distributions in the ocean and the atmosphere and the preferential warming in the Arctic.

The main human CO_2 additions to the atmosphere are from burning fossil fuels and from the manufacture of cement, plus forest clearance and soil disturbance. In 2024, the CO_2 emissions from fossil fuels and cement manufacturing were 37.4 Gt/year (where Gt refers to billion metric tons). The oceans are estimated to be absorbing about 25% of the total CO_2 emitted, which is causing acidification that, in addition to ocean warming and other man-made impacts, is harming coral reefs and other marine species. About 30% of human emissions are absorbed by land ecosystems (plants and soils), although this amount varies and is subject to debate. Thus, roughly half of the CO_2 emitted from fossil fuels and cement manufacturing winds up in the atmosphere. This has led to the current atmospheric CO_2 concentration, which is higher than it has been in 14 million years. This has resulted in a long-term average surface temperature rise since preindustrial times. The temperature can vary from one year to the next, but 2024 was the first calendar year for which the global mean temperature exceeded a 1.5°C (2.7°F) temperature rise.

It is important to put this seemingly small temperature rise into context. Our species, Homo sapiens, has existed for approximately 300,000 years. However, it is only in about the last 10,000 years that permanent cities developed. This development of civilization coincides with a period of very stable temperature and sea level following the last glacial period. During this time, the global average temperature stayed within a ±0.5°C band. The global average temperature rise of 1.5°C already puts us outside that stable temperature band. If high global emissions continue, the temperature rise could be as high as 6°C by 2100. If current climate policies are actually followed, the rise could be limited to about 3°C—but that is still too high to avoid very damaging consequences. In other words, much more action is needed.

Thus, man-made climate change—caused mostly by the burning of fossil fuels—is rapidly subjecting us to climates that human civilization has never before experienced. The 1.5°C change to date has already resulted in increased probability of extreme weather events such as floods, droughts, heat waves, and severe storms. Numerous studies have concluded that the future economic costs of climate change are much greater than the cost of making an energy transition to carbon-free fuels. That is, it is much more cost-effective to prevent further climate change than to clean up the damage if we don't address it.

Global CO_2 emissions have continued to rise every year except for years of economic downturns, such as in 2009 due to the financial crisis and in 2020 due to COVID-19. Atmospheric levels of methane (which, like CO_2, is a powerful greenhouse gas) have also been rising, and it is estimated that two-thirds of the methane emissions are due to human activities. Of the human methane emissions, the major contributors are agriculture (40%), production of fossil fuels (34%), and waste (19%).

Each unit of energy produced by the combustion of fossil fuels results in the emission of CO_2. According to EIA, the energy productions associated with each kilogram of CO_2 emitted for the three main fossil fuel products are:

Natural gas	19.9 MJ/kg CO_2
Gasoline[a]	13.9 MJ/kg CO_2
Coal[b]	11.0 MJ/kg CO_2

[a] U.S. motor gasoline excluding ethanol
[b] Average U.S. coal

The issue of climate disturbance from fossil fuel combustion has taken on highly political and social aspects, even though the effects of man-made emissions and land-use changes have long been well understood. The number one thing that must be done to address climate change is to transition from fossil fuel energy sources to sustainable, low-carbon sources.

In June 2016, following the December 2015 COP21 agreement in Paris, the International Energy Agency (IEA) published Energy Technology Perspectives: Towards Sustainable Urban Energy Systems [28], in which they provided projections of what must be done to meet carbon reduction goals. The worst-case scenario for climate change this century is a 6°C global average temperature rise above the preindustrial level; whereas the IPCC target is 2°C to prevent the worst impacts of climate change. (A 2°C increase is by no means "safe," however, as we are already experiencing a wide range of consequences at today's 1.5°C rise.) The dramatic drop in the costs of solar and wind electricity in recent years has led to a key strategy for addressing climate change: convert electric power generation from coal and gas to solar and wind (with help from batteries and more transmission) and replace fossil fuels with electricity in the end-use sectors of transportation, buildings, and industry.

IEA developed a roadmap for achieving net-zero global emissions by 2050 [29]. Net zero means that any emissions that still exist from hard-to-abate sectors will be compensated by methods that remove carbon dioxide from the atmosphere. The IEA roadmap is shown in Figure 1.11. Key strategies are to replace fossil fuel-powered electricity generation primarily with solar and wind along with some nuclear, convert ground transportation to electric power, convert building heating and cooling and a large fraction of industrial process heat to heat pumps, transition aviation and shipping to low-carbon or carbon-neutral fuels, apply energy efficiency measures across the board, and deploy carbon capture and removal methods coupled with carbon storage to compensate for remaining emissions.

Figure 1.12 is from IEA's updated (2023) net zero by 2050 roadmap [30]. It shows the changes in global energy production (in exajoules) for the various energy sources for each

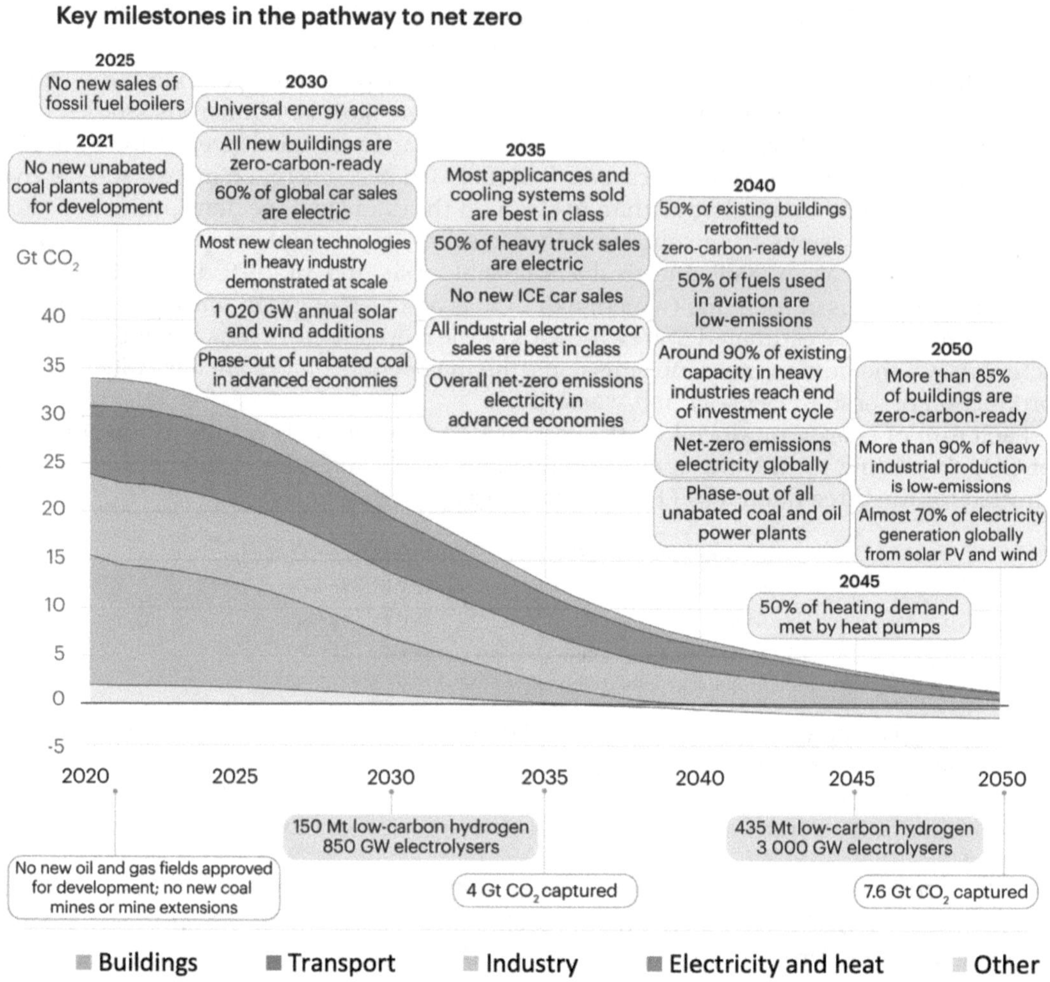

FIGURE 1.11
IEA key milestones for the different energy use sectors on a path to net-zero global carbon emissions by 2050. The layers of the graph and the descriptions in bubbles are ordered from top to bottom in the sequence of Buildings-Transport-Industry-Electricity and Heat-Other. (From International Energy Agency [29]. Licensed under Creative Commons license CC BY 4.0.)

year compared to 2022 for years out to 2050. Energy sources below the zero line decrease with time, and those above the zero line increase. The amounts of coal, oil, and natural gas (all without carbon capture), and traditional biomass combustion all decrease with time. Moving upward from the zero line are coal and natural gas with CCUS (carbon capture, utilization, and storage), nuclear power, hydropower, biofuels, wind power, solar power, and other renewables (e.g., geothermal and ocean power). Thus, there is an overall transition from fossil fuels and conventional biomass burning to a suite of renewables and nuclear power together with a small amount of coal and natural gas with CCUS. The round data points indicate the total energy production, which initially decreases with time due to energy efficiency measures and then increases with population and economic growth. Of course, the IEA roadmap illustrates just one path to net-zero emissions by

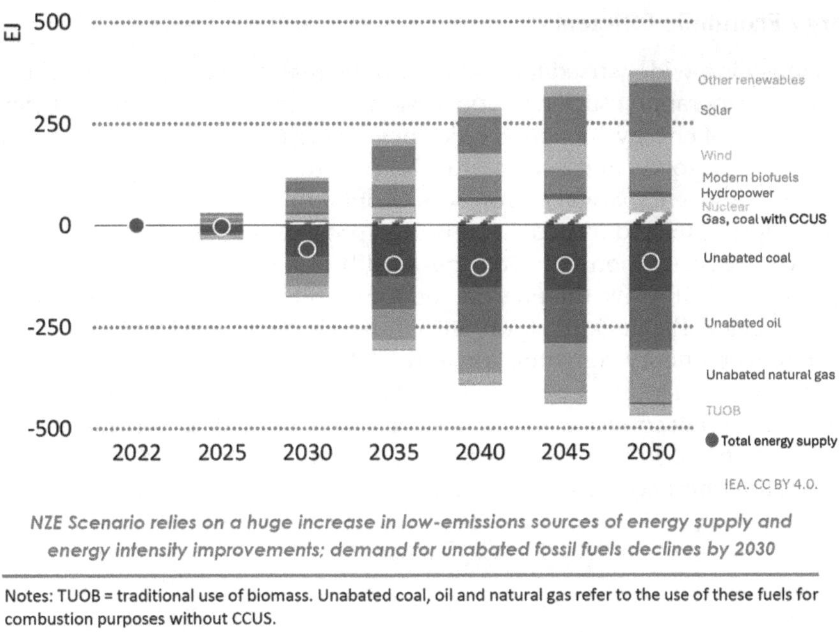

FIGURE 1.12
Changes in total energy supply in EJ by source in the IEA Net-Zero Energy Scenario for each year compared to 2022 for years out to 2050. (From International Energy Agency [30]. Licensed under Creative Commons license CC BY 4.0.)

mid-century. Debate is ongoing regarding the optimum mix of efficiency, renewables, nuclear, and CCUS.

In December 2015, 195 countries signed the landmark COP21 Paris agreement. (COP21 stands for the 21st Annual United Nations Conference of Parties to the United Nations Framework Convention on Climate Change.) (Only Syria and Nicaragua did not join the Paris Agreement. Syria was engulfed in civil war, and Nicaraguan representatives stated that the agreement was not strong enough.) Each country made individual voluntary pledges to reduce carbon emissions to support the goal of keeping the global average temperature rise below 2°C. One of the hallmarks of the Paris Agreement is that it addresses a long-held criticism of previous climate agreements, namely that they excluded developing nations. At the Paris meeting, all of the 195 nations pledged to make reductions, although follow-through on pledges has varied. U.S. support has varied depending on the political party in power, but many states and cities have carried out programs aimed at pursuing the Paris goals. In particular, the State of California, which is now by itself the world's fourth-largest economy, is committed to going beyond the Paris Agreement. Thus, it appears that the United States will continue to make important contributions to this international agreement, albeit not always at the federal level.

The transition to an energy economy dominated by renewables will require addressing new issues. For example, high-value wind resources are often located in sparsely inhabited regions requiring investment in upgraded transmission systems to transport the power from the source to where it is needed in industrial parts of the country. The same applies, although to a lesser extent, to good solar radiation sources, many of which are located some distance from industrial centers and large cities.

1.4.2 Energy Economic Efficiency

Energy efficiency is a widely used term in the public realm. In engineering, there are technical meanings: generation efficiency, conversion efficiency, or appliance efficiency. There is also the concept of energy economic efficiency. Energy economic efficiency refers to the ratio of the value of goods and services to the resources and energy needed to produce them. Economic efficiency always improves profitability and is a necessary part of the strategy for adapting to resource constraints in a prosperous manner.

There is a direct correlation between the wealth of a country and its energy use. The data in Figure 1.13 [31] show the energy consumption per capita as a function of gross domestic product (GDP) per capita for many countries. Although there is a strong positive correlation between energy consumption and GDP, several countries have a similar GDP but lower energy use. The United States is at the high end of energy use per capita and is exceeded only by a few countries, most of which have large domestic fossil fuel reserves. The United States has 2.5 times the energy use per capita as China and 3.5 times the GDP per capita. See the referenced website for a table comparing countries.

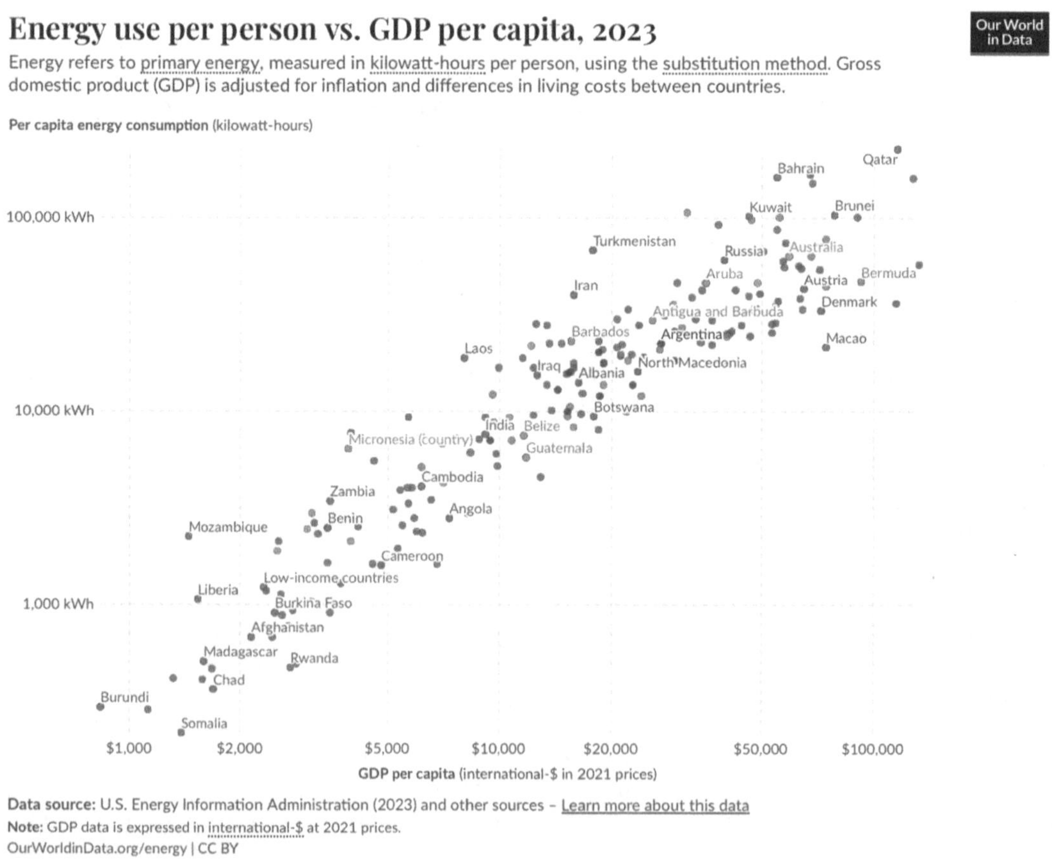

FIGURE 1.13
Per capita energy consumption versus per capita gross domestic product. (From Our World in Data [31], Licensed under Creative Commons license CCBY 4.0.)

1.4.3 Energy Return on Energy Invested

Energy transformation systems take natural resources from the environment and produce fuels and electricity for sale in the economy. They initially require an investment in money and materials as well as in energy for manufacture and installation. Once installed, the energy production system will use or consume natural resources and fuel to generate electricity, or it will extract and process resources into fuel. During the lifetime of the energy production plant, there will also be maintenance and replacement requirements. If we put all of these inputs and outputs in terms of energy, we can compare the consumer energy produced for the market to the energy required either from the economy or parasitic internal use. The main point of examining the EROI is to compare different energy transformation systems in terms of their effectiveness. Also, in calculating an EROI, one must explore the appropriate system boundaries and, in doing so, it is possible to identify potential improvements.

Figure 1.14 gives EROI values of different electric generation systems [32]. One difference between fossil fuel electric generation and renewables is that the fossil EROI is often calculated on an annual basis using annualized values of embedded energy in equipment, whereas the EROI for wind and solar is determined over the generating lifetime. The EROI for coal power generation is on the order of 10. For nuclear power plants, the EROI would depend greatly on the lifetime of the plant and whether the energy needed for the storage of spent fuel and decommissioning is counted. Nuclear power plants are very energy-intensive to build, but the EROI not counting the uranium processing and waste storage is in the range of 10–15. However, even a small energy requirement for long-term safe storage of spent fuel, given the very long storage period, can considerably reduce the lifetime EROI of nuclear power.

EROI is a quantitative means to understand the past, present, and future net flows of energy to the economy for different energy technologies. But it depends on a variety of items such as the sources and amounts of energy used in manufacturing a facility and transporting a fuel, the lifetime of a facility, and in the case of renewable energy, the quality of the local resource. Even though EROI is valuable for comparing the efficiency of different means of providing marketable energy, it does not consider the impact of carbon emissions. Although coal-fired electric power generation has a higher EROI than electricity generation by solar photovoltaics (PV) or geothermal power, these renewable technologies have dramatically lower life cycle carbon emissions and thus are among the technologies of choice for limiting global temperature rise. EROI can be a useful metric for comparing fossil fuel technologies to each other or comparing renewable technologies to each other.

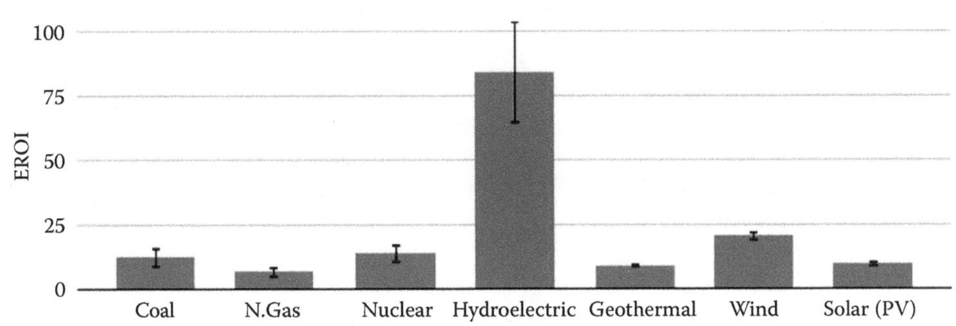

FIGURE 1.14
EROIs for different electricity generation technologies. (From Hall et al. [32].)

However, life cycle greenhouse gas emissions or the carbon payback period (discussed in Section 1.7) and the levelized cost of the energy produced are more useful for comparing conventional and renewable sources and determining the best path forward to address climate change. EROI and other metrics of the sustainability of energy systems are presented in detail in Chapter 3.

1.4.4 Cost of Energy Production

The levelized cost of energy (LCOE) is a way to compare the cost of building and operating an energy production system to the market value of the energy produced. Energy production costs include the capital cost, which is the cost of engineering and building the plant, plus costs of maintenance and fuel. For an electricity-generating plant, the energy revenue depends on the plant capacity, its capacity factor (the amount of energy it produces over a year divided by the amount of energy it would produce if it always operated at capacity), and its capacity credit (which is a measure of the ability of the plant to generate on demand during peak times). A broad perspective of the capital and levelized energy costs of renewables, as well as more traditional energy technologies, has been prepared by Lazard, a well-known financial advisory and asset management firm. The firm provides advice on mergers and acquisitions, as well as asset management services for large institutions and governments. The technology assessments shown here were prepared to advise its customers about investments in energy technologies. The details of the approach taken by Lazard are available in [33]. Chapter 2 explains how to make such estimates and describes critical assumptions such as the cost of money or the assumed life of the system.

The cost figures provided by Lazard will be useful for preliminary estimates, and particularly for making what are called "back-of-the-envelope" calculations. Lazard has estimated ranges of capital costs (i.e., the cost of installing a given technology in $/kW) and LCOE (in $/MWh) for the different types of electricity-generating systems. A summary of these costs is given in Table 1.2 [33]. They do not include the benefit of subsidies for alternative sources (e.g., investment and production tax credits). Therefore, actual costs after subsidies for renewable technologies are often lower. These estimates also do not include social and environmental costs, which are much higher for conventional technologies.

TABLE 1.2

Capital Costs and Levelized Costs of Energy (LCOE) for Conventional and Renewable Electric Power Plants

	Technology	Capital Cost ($/kW)	LCOE ($/MWh)
Conventional	Coal	3,200–6,775	68–166
	Gas combined cycle	650–1,300	39–101
	Gas turbine	700–1,150	115–221
	Nuclear	8,475–13,925	141–221
Renewable	Rooftop PV	2,230–4,150	117–282
	Community PV	1,200–2,850	49–185
	Utility PV	700–1,400	24–96
	Utility PV+battery storage	1,075–1,600	46–102
	Onshore wind	1,025–1,700	24–75
	Onshore wind+battery storage	1,375–2,250	42–114
	Offshore wind	3,000–5,000	72–140
	Geothermal	4,700–6,075	61–102

Source: Assembled from data contained in Lazard [33].

Solar photovoltaic technology is now the world's fastest-growing electricity source. Table 1.2 includes costs for different types of solar PV installations. The cost per megawatt hour (MWh) of utility-scale solar (without battery storage) is about 50% lower than community scale solar, and the cost of residential rooftop PV is about 70% higher than community scale. It is important to note that these estimates are based on current generation technologies for the conventional systems, which are considered mature. It should also be noted that the conventional systems, such as coal, nuclear, and natural gas, can operate continuously, whereas solar and wind are variable. Moreover, the performance of renewable technologies is very dependent on geographic location.

1.4.5 Other Costs of Energy Development

Land has many values, so the amount of land required by different electric-generating technologies is of interest. Figure 1.15 shows the total land requirements in units of m²/MWh

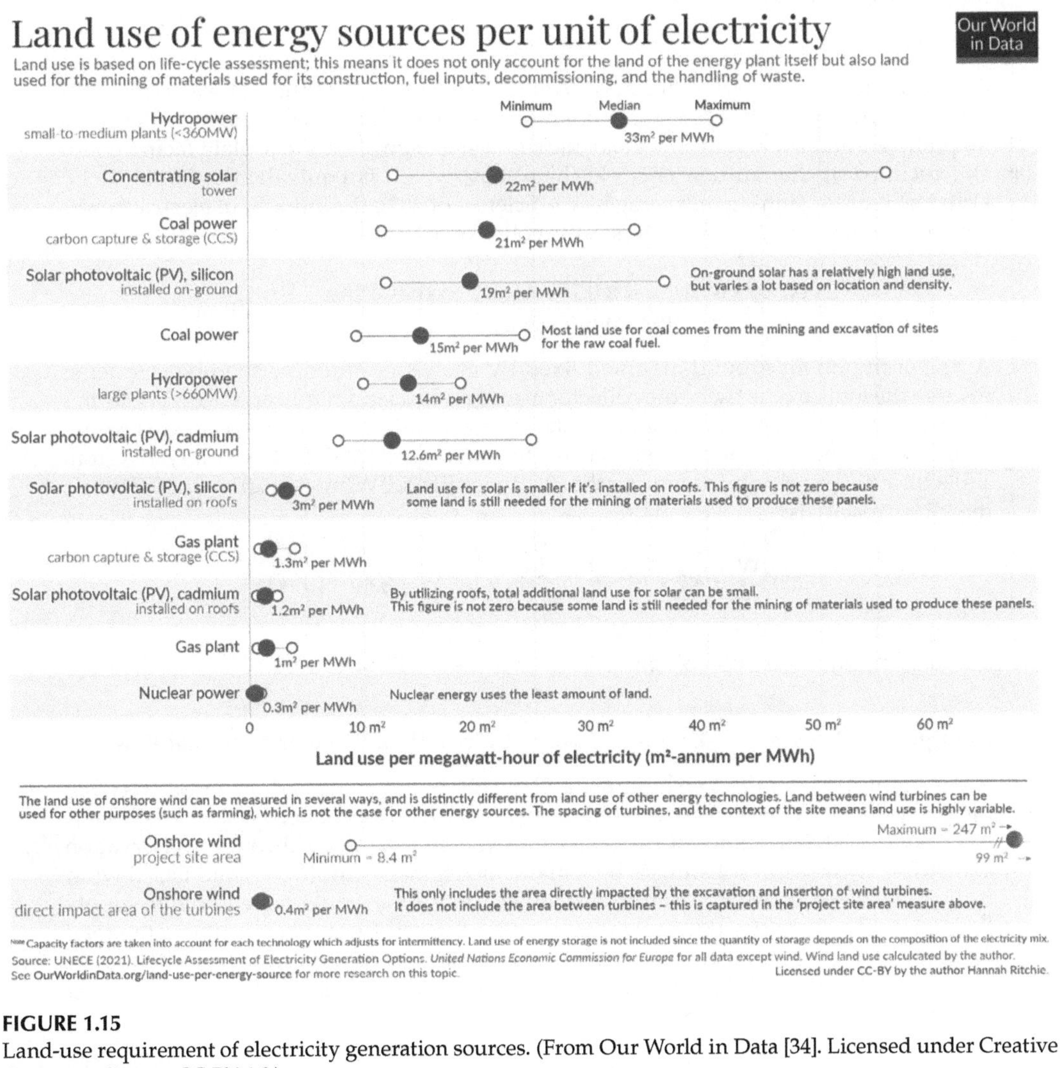

FIGURE 1.15
Land-use requirement of electricity generation sources. (From Our World in Data [34]. Licensed under Creative Commons license CC BY 4.0.)

for various electric-generating technologies [34]. Fossil fuel plants themselves have a very large energy density, but fossil fuels require mining and pipelines. When these upstream requirements are accounted for, the total land areas needed for coal plants are similar to those needed for ground-mounted solar power plants but much greater than the land areas needed to mine the materials for roof-mounted solar photovoltaic arrays. In the case of onshore wind power, the land associated with the turbines themselves and service roads is very low because the land between turbines can be used for agriculture, grazing, or other purposes. Nuclear power requires the least amount of land because even though the fuel must be mined, nuclear fuel has an extremely large energy density.

Example 1.2: Back-of-the-Envelope Energy Calculation: Solar Power

The Arizona legislature has passed a law requiring that an additional 15% of the state's electric energy generation must come from renewable sources. Calculate the collector area required, the cost, and land area for this target to be met by stationary arrays of utility-scale solar PV.

Solution:

According to the Energy Information Administration in 2023 the net electricity generation in Arizona was 111,839 GWh. From the NREL U.S. Solar Radiation Resource Map, the annual average of the solar radiation impinging on a flat plate facing south and inclined at the latitude is 5.5 kWh/m^2-day, which is equivalent to about 2×10^3 kWh/m^2-year. Assuming an average PV efficiency of 20%, the amount of electric power delivered is 400 kWh/m^2-year. Hence, the collector area is

$$A = \frac{111{,}839\ \text{GWh/year} \times 0.15 \times 1 \times 10^6\ \text{kWh/GWh}}{400\ \text{kWh/year-m}^2} = 4.2 \times 10^7\ \text{m}^2$$

A rule of thumb for ground-mounted fixed PV arrays to minimize shading is spacing rows so the total area is twice the collector area, giving a land requirement of 8.4×10^7 m^2. The land area of Arizona is 29.5×10^{10} m^2; hence, the required PV power plant area would be 0.029% of the state of Arizona. Using the average capital cost of utility PV from Table 1.2 of $1,050/kW and assuming the PV capacity is 200 W/m^2 (i.e., 20% of 1,000 W/m^2), the total capacity is:

$$200\frac{\text{W}}{\text{m}^2} \times 4.2 \times 10^7\ \text{m}^2 = 8.40 \times 10^9\ \text{W} = 8.40 \times 10^6\ \text{kW}$$

Multiplying this by $1,050/kW gives a cost about $8.8 billion.

Example 1.3: Back-of-the-Envelope Energy Calculation: Wind and Nuclear Power

A new administration in Washington, D.C. wants to eliminate carbon emissions from the portion of U.S. electricity generation that is currently produced using fossil fuels. Assume that 20% of the fossil fuel emissions can be addressed through conservation and efficiency measures. Estimate how many new 8-megawatt (MW) land-based wind turbines (assume an average yearly capacity factor of 40%) would be required to supply the remaining electricity currently produced from fossil fuels. Then estimate the total costs and land area of these wind turbines. Next, estimate how many 1,000-MW nuclear power plants with a 92% capacity factor would be required, their land requirement, and what the capital cost would be.

Solution:

According to the U.S. Energy Information Administration,

Total U.S. electricity generation in 2023: 4.25×10^9 MWh/year

Fossil-based electricity generation in 2023 (60% of total): 2.55×10^9 MWh/year

Assuming conservation and efficiency can reduce the electricity needs by 20%, we are left with $0.8 \times 2.55 \times 10^9$ MWh/year or 2.04×10^9 MWh/year of electric energy that must be produced by renewable sources.

Wind power: Given a 40% annual capacity factor, the 8-MW wind turbine would be expected to generate

$$(0.40) \times (8\ \text{MW}) \times \left(8{,}760\ \frac{\text{h}}{\text{year}}\right) = 2.80 \times 10^4\ \frac{\text{MWh}}{\text{year}}$$

Dividing this into the total energy needed of 2.04×10^9 MWh/year, we find that 72,900 wind turbines would be needed. Using the total wind farm area from Figure 1.15, we obtain:

$$99\ \frac{\text{m}^2}{\frac{\text{MWh}}{\text{year}}} \times 2.80 \times 10^4\ \frac{\text{MWh}}{\text{year}} \times 72{,}900 = 2.02 \times 10^{11}\ \text{m}^2$$

or 50 million ac. This is 2.2% of the U.S. land area. However, land between the turbines in wind farms is used for other purposes. According to Figure 1.15, if we only consider the land area directly impacted by the turbine installations, that is only $0.4\,\text{m}^2$/MWh/year. Using that value instead of $99\,\text{m}^2$/MWh/year, the land area utilized is only 200,000 ac or 0.009% of the total U.S. land area.

From Table 1.2, the average capital cost of wind turbines is $1,362.50/kW. Thus, the capital needed for the installation of these farms is:

$$\frac{\$1{,}362.5}{\text{kW}} \times 1{,}000\ \frac{\text{kW}}{\text{MW}} \times 8\ \frac{\text{MW}}{\text{turbine}} \times 72{,}900\ \text{turbines} = \$795\ \text{billion}$$

If we include the cost of battery storage, then from Table 1.2 the average capital cost is $1,812.50/kW. Using that value for capital cost in the equation above, the total cost becomes $1.06 trillion.

Nuclear power: The yearly electric energy produced by one 1,000-MW nuclear power plant operating at 92% capacity factor is:

$$0.92 \times 1{,}000\ \text{MW} \times 8{,}760\ \frac{\text{h}}{\text{year}} = 8.06 \times 10^6\ \text{MWh/year}$$

Dividing this into the total energy needed, we find that this would require 253 power plants. At an average capital cost of $11,200/kW for nuclear plants (from Table 1.2), we determine the capital investment for the construction of these plants:

$$\frac{\$11{,}200}{\text{kW}} \times 1{,}000\ \frac{\text{kW}}{\text{MW}} \times 1{,}000\ \frac{\text{MW}}{\text{plant}} \times 253\ \text{plants} = \$2.8\ \text{trillion}$$

Thus, the nuclear power plant cost is about 2.6 times the cost of the wind farms with battery storage. We can see from Figure 1.15 that the nuclear plants would occupy only about 75% of the directly impacted area of the wind farms.

Both the nuclear power plants and wind farms will provide carbon-free electricity (ignoring emissions associated with manufacturing them). The nuclear plants will provide base-load electricity 24 h/day, which, even with battery storage, the wind farms cannot do. However, the total power output of so many wind farms spread out over a large geographic area will be reasonably steady, and the batteries provide some flexibility in terms of meeting electricity demand. Wind farms can also be built roughly ten times faster than nuclear power plants, thus displacing fossil fuel emissions much faster, which results in lower total carbon emissions.

1.5 Energy Efficiency and Conservation

Reducing energy consumption through efficiency and conservation provides important and economically strategic prospects for transition to a sustainable energy system. In 2023 the per capita primary energy consumption in the United States was 294 GJ, which is equivalent to a constant energy consumption of about 9 kW/person. The average consumption in Europe is 4.2 kW, and the average for the whole world is only 2.5 kW. The IEA's Net Zero by 2050 roadmap had the annual global energy intensity (energy use per GDP) improving at a rate of about 2% per year in the decade from 2010 to 2020 and 4% per year in the decade from 2020 to 2030. However, this level of improvement has not been seen to date, with the one bright spot being reported as the transition to electric vehicles.

1.5.1 Energy End-Use Demand Reduction in Buildings

When electrical system energy losses are included, buildings (not including those in the industrial sector) account for about 37% of all end-use energy demand in the United States. Buildings use 74% of the electric power and are responsible for approximately 33% of the nation's greenhouse gas emissions. Residential buildings account for about half of the building energy use. Over half of all energy used in residences is for heating and cooling, and a further 20% is for heating water. Residential air conditioning is a main contributor to the peak demand on the power grid. Peak power is the most expensive to provide, as the grid must be able to meet the peak load. However, the peaking power plants only run, and thus only earn revenues, during the peak periods, resulting in very low utilization factors. Economic residential energy efficiency and conservation programs focus on reducing space conditioning and water usage. Figure 1.16 shows the average relative effectiveness of a range of residential energy retrofit options (although effectiveness varies with location and building vintage). Windows are a high source of conductive heat loss as shown in Figure 1.17.

Commercial buildings are demolished and new ones are built on a shorter cycle than residential buildings, with around 1% of buildings turned over each year. Heating, ventilating, air conditioning, and lighting are major loads in commercial buildings. Professional energy auditing and energy management are required for commercial buildings. Commercial sites such as shopping centers, offices, medical centers, and schools also have a transportation impact because they become attractors for travel demand. Thus, the location and the context of commercial buildings are part of the overall U.S. Green Building Council's Leadership in Energy and Environmental Design (LEED) framework. The framework is a building certification that identifies and implements green design,

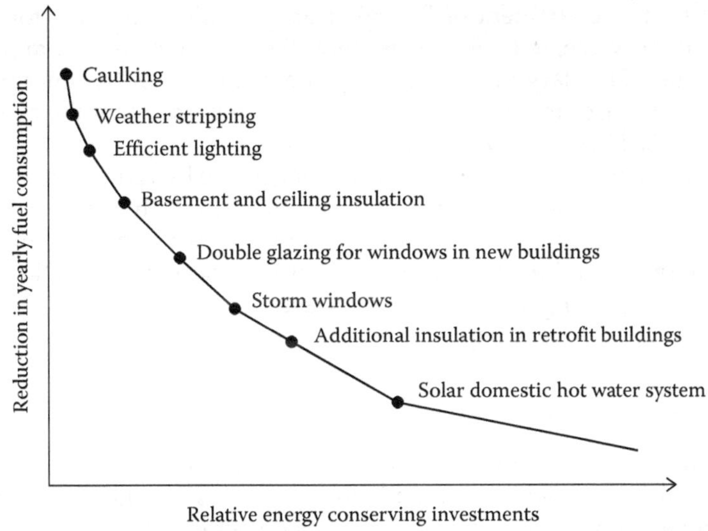

FIGURE 1.16
Relative energy savings of various energy efficiency measures.

FIGURE 1.17
An infrared image of a house. The lightest colored windows indicate the highest heat loss. (Courtesy of John Avenson.)

construction, operation, and maintenance measures yet is flexible for each existing or new project. The approach is to consider the building life cycle as well as the surrounding community and environment.

1.5.2 Energy End-Use Demand Reduction in Transportation

Of every 6 dollars spent in the United States, $1 is estimated to be related to transport: fuel, vehicles, road, and parking infrastructure or related services. The U.S. personal transportation system has many features that make it the ideal opportunity for energy reduction.

According to the U.S. Department of Energy, transportation accounts for approximately 30% of total U.S. energy use, with approximately 90% of that energy from petroleum. Cars and light trucks use 54%, large trucks 25%, and aircraft 9% of the energy consumed in transport. The U.S. Environmental Protection Agency reported that the average fuel economy of model-year 2022 vehicles was 26.0 miles/gallon.

Various programs can reduce transportation energy use by reducing vehicle miles traveled (VMT). Car-pooling and car-sharing programs address this, as does the acceptance of telecommuting. The trend toward urbanization and walkable communities directly impacts VMT. Increased public transportation, construction of bike paths, and bike-sharing programs are also having an increasing impact.

Transportation is covered in Chapter 15.

1.5.3 Energy Management in Industry

Industry is a prime candidate for energy management projects. A good way for energy engineers to tap into the energy savings in this sector is to study the best examples of energy management projects that have already been achieved and develop new projects for those companies that do not yet have progressive energy management practices. For example, at 3M engineering teams are rewarded based on meeting energy efficiency targets. From 2016 through 2021, 3M reduced CO_2-equivalent emissions by 301,000 metric tons and saved 1,758,000 GJ of energy, which saved $24 million [35]. The U.S. Department of Energy has a Better Buildings program [36] and a Better Plants program [37] aimed at reducing energy use in buildings and industry.

Energy efficiency in buildings and industry is covered in Chapter 4.

1.6 Conventional Energy

1.6.1 Fossil Fuels

The production rate of fossil fuels is, from a historical perspective, truly enormous. The amount of fossil fuels in the Earth is finite, and in the first decade of the twenty-first century, there was much discussion of the prospect of "peak oil," whereby the world production of oil would finally peak and then decline, with attendant economic consequences. However, the application of hydrofracking and directional drilling in shale reserves led to a temporary resurgence in U.S. oil and gas production. Thus, the near-term supply issues with fossil fuels are not so much depletion, but rather the increased costs associated with pursuing less conventional reserves. The long-term sustainability issues with fossil fuels arise from environmental degradation and particularly the need to drastically reduce carbon emissions.

Burning fossil fuels causes widespread air pollution problems. Soot from coal and diesel and nitrogen oxide emissions from combustion of coal, petroleum, and natural gas directly cause negative health effects. Combustion science and air pollution control engineering have provided great improvements in air quality in many countries since the 1970s, but severe air quality problems persist in other regions. McDuffie et al. [38] estimated that combustion of coal, oil, and natural gas contributes to roughly 1 million deaths a year around the world, with the majority of those occurring in Asia. Air pollution from fossil fuel combustion also damages ecosystems and biodiversity through acidification of surface water, over-fertilization by nitrogen compounds, and deposition of toxic mercury. While improved control technology applied to power plants and motor vehicles can address many of these issues, the biggest sustainability challenge for the use of fossil fuels is their disruptive impact on climate. In principle, emissions of carbon dioxide from fossil fuel combustion in power plants can be abated using carbon capture and sequestration (CCS), but this technology is expensive and is not available for motor vehicles. Production of fossil fuels (oil, natural gas, and coal) is also a major source of methane, a potent greenhouse gas that accounts for about 30% of the global warming experienced since the Industrial Revolution. Technology exists to reduce methane emissions from fossil fuel production activities, but it will need to be deployed at millions of well sites and mines globally to achieve the needed cuts.

Fossil fuels are covered in detail in Chapter 6.

1.6.2 Nuclear Power

Nuclear power plants utilize the energy release of nuclear fission, the splitting of U-235 atoms, to generate steam in thermal power cycles. In 2024, there were 440 reactor units in operation in 33 countries contributing about 9% of the world's electric energy production. The majority of reactors are located in the United States (94 units), China (56), France (56 units), Russia (36), and Japan (33 units) [39]. Nuclear reactors are typically large, base-load power producers—on the order of 1,000 MW. Unlike the burning of fossil fuels, energy from nuclear fission does not result in the emission of greenhouse gases. Thus, nuclear power plants offer a means to produce carbon-free electricity on a large scale, and so nuclear power is a favored technology by many climate scientists.

As one of the potential solutions to climate change, however, nuclear power presents a number of challenges. Nuclear power plant waste presents problems with long-term storage and nuclear proliferation. Although nuclear power generally has a good track record for reliability and safety, accidents can have serious consequences. Most notably, the meltdown of the graphite-moderated reactor in Chernobyl, Ukraine, resulted in the evacuation of the city and caused both short-term and long-term deaths. And the meltdown of three reactor units in Fukushima, Japan, in March 2011, which resulted from an earthquake and subsequent tsunami, is expected to take decades to clean up.

The biggest challenge facing further nuclear power development, however, is its high capital cost. Large cost overruns have occurred for new plants being built in the United States. Future U.S. plants are estimated to cost $14,000/kW or more. In addition to the high capital cost, nuclear reactors have experienced high operating costs, causing a number of plants to be shut down. Some countries heavily depend on nuclear power for their future energy needs, but low electricity growth for many decades and rapidly falling costs of wind and solar power have put further pressure on the nuclear industry in the United States. The recent growth of data centers, which operate 24 h/day, has resulted in new interest in nuclear power.

Compared to the carbon-free renewable energy options, nuclear power does maintain the advantage of being able to provide base-load power 24 h/day with high availability. And the operating nuclear fleet in the United States supplies about 20% of U.S. electricity, thus providing enormous savings in carbon emissions. If one properly accounts for the cost of climate change damage, an argument can be made to keep the existing nuclear plants in the United States operating for as long as is safely possible. Countries that have limited renewable sources, such as Japan and South Korea, may depend on nuclear power for much of their future energy.

Nuclear power is covered in Chapter 7.

1.7 Renewable Energy

By definition, the term "reserves" does not apply to renewable resources. So, we need to look at the annual potential of each resource. Figure 1.18 from the IPCC Special Report on Renewable Energy [40], summarizes the range of technical resource potentials for each of the renewable energy technologies. The maximum and minimum values of potential, which are derived from different studies, are shown, respectively, as the top and bottom of the bars and are also given numerically in the table at the bottom of the figure. A key observation is that direct solar energy has a technical potential that is one to two orders of magnitude greater than for the other renewable technologies and vastly exceeds the current global primary energy supply. The IPCC defines renewable energy technical potential as "the amount of RE output obtainable by full implementation of demonstrated and likely to develop technologies or practices" [40]. Although it does not explicitly refer to costs, economic considerations may come into play in determining the technical potential.

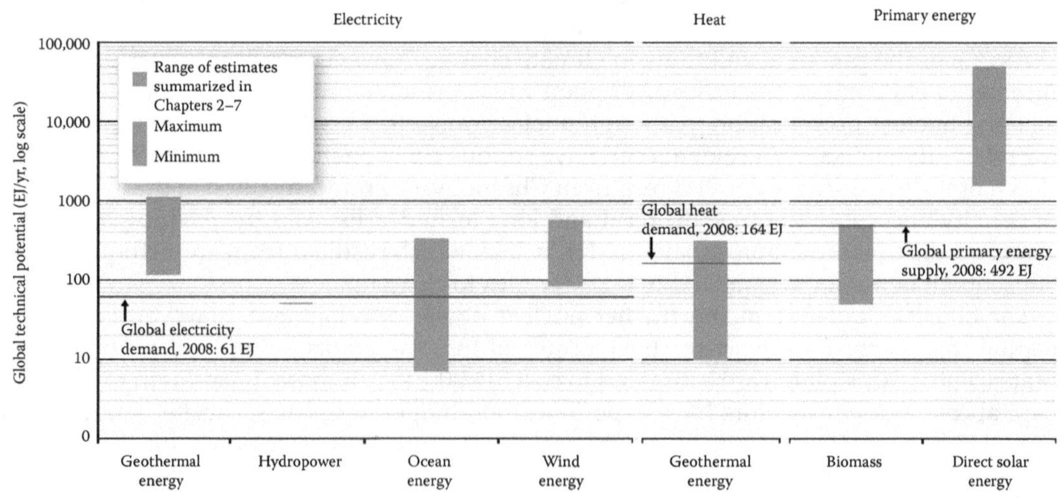

Range of estimates of global technical potentials

	Geothermal energy	Hydropower	Ocean energy	Wind energy	Geothermal energy	Biomass	Direct solar energy
Max (in EJ/yr)	1109	52	331	580	312	500	49837
Min (in EJ/yr)	118	50	7	85	10	50	1575

FIGURE 1.18
Ranges of global technical potentials of RE sources derived from studies presented in Chapters 2–7 of Renewable Energy Sources and Climate Change Mitigation, Special Report of the Intergovernmental Panel on Climate Change (IPCC). Biomass and solar are shown as primary energy due to their multiple uses. Note that the figure is presented in logarithmic scale due to the wide range of assessed data. (From Intergovernmental Panel on Climate Change, Working Group III [40].)

Renewable energy sources are not completely free of carbon emissions, with some CO_2 emissions occurring in their manufacture and construction. (Nuclear power also has associated emissions resulting from mining, concrete manufacture, and plant construction.) In determining the overall impact of an energy transition on carbon emissions, it is thus valuable to know how the various renewable and nuclear technologies compare in terms of life cycle greenhouse gas emissions, typically expressed as grams of CO_2 equivalent emissions per kWh of energy produced. Renewable energy and nuclear power have very low life cycle emissions compared to the fossil fuel technologies, although there is a range for each technology, depending on what manufacturing techniques are employed and, in the case of renewables, the quality of the resource in which the technology is deployed. For example, the manufacture of thin-film photovoltaic cells generally results in lower greenhouse gas emissions than the manufacture of crystalline silicon PV. Also, a solar PV system will have lower life cycle emissions per kWh produced if deployed in Arizona than in New Jersey. Technologies are also sometimes compared in terms of how long it would take (compared to fossil fuel technologies) to save an amount of carbon emissions equivalent to those created during the manufacture. For most renewable technologies, the carbon payback period is typically 1–3 years or less, but it depends on the fuel mix in the country where the technology was manufactured as well as both the fuel mix that it is replacing and the resource quality in the location where it is deployed.

Of all the renewable technologies, photovoltaics and wind have exhibited the most rapid cost reduction and market penetration. In the following sections, a brief introduction of each of the renewable energy technologies is provided, with greater detail to follow in the individual chapters.

1.7.1 Wind Energy

Wind energy technology has progressed significantly over the last two decades. Wind turbine towers have gotten taller and taller, thereby reaching stronger and steadier winds. As a result of this and other technology improvements, the levelized costs have come down greatly. In 2016, the levelized cost of wind power was one-quarter what it was in 1990, and in 2023, levelized costs were as low as 2.4 cents/kWh. Wind power is thus very economical at locations with good wind resources. By 2024, there were over 150,000 MW of wind installed in the United States—more than a doubling over 10 years, and there were over 1 million MW installed in the world. Although offshore wind farms are considerably more expensive than land-based systems, they offer the advantage of higher wind speeds and steadier winds (see Figure 1.19).

The biggest issue with wind generation is the variability and inability to dispatch electricity on demand. When the wind slows suddenly, a large drop in wind generation can cause instability on the grid. When system operators harness wind farms over a wider area, the total wind output is much steadier than with a single wind farm. A high level of cooperation among power generators and network operators on the grid can help manage variable generation. Firming capacity can be provided by natural gas turbines, which provide rapid start-up and only operate for short periods of time, as well as by hydroelectricity. The addition of energy storage to the grid can help enable larger wind penetrations. Another means for accommodating variable electricity supply is to control the electric demand. So-called demand response measures in buildings can be very effective. Wind energy is covered in Chapter 8, while approaches to managing supply variability are discussed in Chapter 5.

FIGURE 1.19
The 160-MW Horns Rev offshore wind farm in the North Sea. Water vapor streams show how turbine wakes affect downwind turbines. Optimizing wind farm geometries and individual turbine controls to address wake effects and maximize total wind farm output is a major research area for wind energy. (Aeolus Energy Group, used with permission.)

1.7.2 Solar Photovoltaics

Solar photovoltaic modules convert solar energy directly to electricity. They are installed on rooftops or ground-mounted in community solar gardens or in large utility-owned installations. They have typically been installed as fixed arrays, but ground-mounted arrays can provide 15%–20% more energy if oriented in north–south rows that actively track the sun from east to west, and so these are becoming more common. Although there are many different PV cell types, crystalline silicon modules dominate the market with over 90% of installations. PV modules are reliable and, unlike thermal power generation, do not require cooling systems.

The solar-to-electric conversion efficiency of modules has increased significantly over the years. Today's commercial PV modules that utilize crystalline silicon have efficiencies in the range of 16%–23%. Solar PV panels have come down greatly in cost from about $30/W to under 30 cents/W in the last three decades. As a result of the rapid drop in costs, solar PV growth has soared. By the end of 2023, there was a cumulative total of over 137,500 MW_{AC} of installed PV in the United States—an 11-fold increase over 10 years! It took many decades to reach 1 TW of installed global PV capacity in 2022 and then just 2 years to double that capacity to 2 TW in 2024 [41]. While residential rooftop PV has grown rapidly, utility-scale PV is even bigger, owing to its lower cost. PV is also highly versatile. For example, small PV systems are being deployed in developing nations around the world to provide indoor lighting and to charge cell phone batteries (see Figure 1.20).

1.7.3 Solar Thermal

Concentrating (or concentrated) solar power (CSP) uses reflectors to concentrate sunlight onto a fluid-carrying pipe (line-focus technology) or a tower-mounted receiver (power tower technology) to provide the high temperatures needed to produce steam in an efficient Rankine power cycle. CSP is the first solar technology that demonstrated its grid power potential. The Solar Electric-Generating Systems (SEGS) parabolic trough power plants in the Mojave Desert reliably provided over 350 MW of electricity to southern California beginning in the 1980s. Progress in solar thermal power stalled after that time because of a decrease in government incentives and low natural gas prices. However, a resurgence of interest in this technology began with the completion of the 64-MW Nevada Solar One parabolic trough plant near Las Vegas in 2007. In addition, a feed-in tariff in Spain stimulated the development of a significant number of CSP plants. At the end of 2023, there were about 6,800 MW of CSP installed in the world, about 1,500 MW of which were in the United States [42].

FIGURE 1.20
In a small Brazilian village, each 50-W PV system can power two fluorescent bulbs, allowing students to study and do homework in the evening. (Roger Taylor, NREL.)

Compared to PV, CSP provided the capability to include thermal energy storage, which offered a cheaper and more efficient storage option than batteries (see Figure 1.21). Also, fuels such as natural gas or biogas may be used as backup to ensure continuous operation. Thus, CSP offers the ability to dispatch electricity when it is needed. To improve the economic competitiveness of CSP, advanced power towers have been designed to achieve higher temperatures and thus higher power cycle efficiencies. Also, CSP can utilize solar thermal storage in such a way that electricity is sold primarily when rates are at their highest. In recent years, however, the costs of both PV modules and batteries have fallen so much that it is more difficult for CSP to compete.

Besides electricity, solar energy has long provided heat as the final product. Low-temperature solar thermal collectors to heat domestic hot water and provide industrial process heat are well developed. By the end of 2023, an installed capacity of approximately 560 GW_{th} of low-temperature solar thermal energy was in operation around the world [43], a vast majority of this being used in China for domestic hot water (see Figure 1.22). The improvements made in parabolic trough technology for CSP could be leveraged to better utilize solar thermal energy for industrial process heat.

1.7.4 Ocean and Geothermal Energy

Ocean energy is of two types: thermal and hydrokinetic. An ocean thermal energy conversion (OTEC) plant uses the temperature difference between ocean water at the surface and at depth to drive a Rankine power cycle. This can be done via a closed cycle using ammonia as the working fluid or via an open cycle in which surface water is flashed to

FIGURE 1.21
The 250-MW Solana parabolic trough power plant with 6 hours of thermal storage in Gila Bend, Arizona, provides electricity for Phoenix. (Courtesy of Atlantica Yield.)

FIGURE 1.22
China has the most solar hot water systems in the world. Simple rooftop thermosyphon systems like those shown here are the most widely used type.

vapor in a vacuum chamber. The latter cycle has the benefit of also producing desalinated water. Because of the very low temperature difference involved, OTEC plants operate at very low thermodynamic efficiencies, which makes economic power production difficult. As a result, only small experimental OTEC plants have been built.

Ocean hydrokinetic technologies harness the kinetic energy of moving ocean water associated with tidal flows, wave action, and ocean currents. A wide variety of different technologies have been developed to tap into these various energy sources. Although hydrokinetic projects have been built, this does not yet represent a significant commercial source of world electricity and is under development. Greater effort has been spent on the development of offshore wind.

Geothermal power has been successfully deployed around the world. Geothermal heat originates from two sources: residual heat generated in the formation of the planet by gravitational collapse and heat generated by the decay of radioactive isotopes. Commercially viable geothermal resources occur along spreading or subduction zones between tectonic plates and at other locations where magma from the mantle has been pushed up toward the surface. Commercial geothermal resources have depended on the presence of water. In these hydrothermal resources, the water is heated by convection as it flows through cracks in the hot rock zone. Geothermal resources are tapped by drilling into the high-pressure zone, and electricity is produced from brine at temperatures of 120°C–400°C. Steam is produced either directly from the resource or through flashing liquid brine, and this then drives a steam turbine. For resource temperatures below about 175°C, binary-cycle plants are employed (see Figure 1.23). In this type of plant, an organic fluid such as pentane or isobutane is boiled and used to power an organic Rankine cycle (ORC). An organic Rankine

FIGURE 1.23
The 30-MW, binary-cycle, geothermal power plant in Mammoth Lakes, California, with large arrays of air-cooled condensers. The plant owner has experimented with systems that evaporatively precool the air on hot summer days. (Courtesy of Ted J. Clutter.)

cycle can also be used as a bottoming cycle in a flash steam system where it can extract additional energy from the liquid brine remaining from the flash process. Geothermal power plants are considered base load and are (on average) available more than 90% of the time.

Because of the limited number of hydrothermal resources suitable for electric power production, there is increased interest in creating artificial reservoirs in hot dry rock. This approach is called enhanced geothermal systems (or EGS, also sometimes referred to as engineered geothermal systems) and can tap into a large amount of the Earth's stored thermal energy at great depth. The EGS technology attempts to drill into the Earth at depths of up to 10,000 m. Injected water is used in a hydrofracking operation to create porosity and establish a water reservoir in the rock. Production wells are then tapped to bring the hot water to the surface and power either a flash system or a binary-cycle plant. EGS has so far been limited to experimental and small early commercial projects, and efforts are underway to lower the costs.

1.7.5 Biomass and Biofuel

According to the IEA, bioenergy accounts for about 8%, or 50 EJ, of the world's primary energy supply [44]. About 40% of this is used for cooking and heating, and in developing countries, it has serious consequences for human health (as a result of open flames used indoors) and deforestation. Even a small improvement in conversion efficiency for cookstoves could reduce wood fuel demand and indoor air pollution for millions of people.

Various efforts are being made to develop better cookstoves. The Clean Cooking Alliance [45] is an organization hosted by the UN Foundation with a mission of increasing accessibility to clean cookstoves.

Many governments of developed countries produce liquid biofuels from biomass. Most ethanol in the United States is made from corn via fermentation. Between 40% and 70% of the U.S. corn crop, depending on weather and market conditions, can be processed into ethanol. From a climate change standpoint, this is only marginally better than gasoline. Efforts to transition from corn to cellulosic feedstocks such as crop residues, which have a much lower carbon footprint, have not achieved commercial success. Ethanol in the United States today is used mainly in a 10% blend with gasoline (E10) to comply with the federal renewable fuel standard.

In 2022, the world's annual ethanol production had reached about 28.2 billion gallons, of which 54% was in the United States. Brazil, whose transportation systems rely heavily on ethanol from sugar cane, was second with 26% of the production [46].

Biofuel production depends primarily on agricultural production and therefore on land use. The major feedstocks for ethanol and their average production rates in liters per hectare are sugar cane (6,000), beets (5,000), and corn (4,300). The largest global feedstock for biodiesel is oil palms, although soybean oil is the largest feedstock in the United States. The high economic value of palm oil for a variety of uses has resulted in enormous amounts of forestland in Indonesia being burned to make way for oil palm plantations. (Indonesia is the world's largest producer of biodiesel followed by the United States and Brazil.) This has caused a significant increase in greenhouse gas emissions and negatively impacted the wildlife through massive habitat destruction.

The proliferation of zero-emission vehicle mandates (i.e., new vehicles are required to have no harmful tailpipe emissions by a certain date, such as 2035) in California and countries around the world has spurred the production of battery electric vehicles and lowered the future prospects for biofuels in light-duty vehicles. However, bio-derived fuels may play an important future role for ships and planes.

Besides producing biofuels, biomass feedstocks can also be burned to generate electricity. In 2024, there were 4,800 biomass power plants in the world producing a total of 81.5 GW_e [47]. Although biomass power is generally considered carbon-neutral (CO_2 is absorbed by growing feedstock and then emitted when it is burned), it is possible to gasify the biomass and capture and sequester the carbon. Such a system, called BECCS (Bio Energy with Carbon Capture and Storage), is carbon-negative and could play a role in drawing down atmospheric CO_2.

Biomass resources and conversion technologies are described in Chapter 9.

1.7.6 Hydroelectric Generation

According to EIA, in 2023, 21.4% of U.S. utility-scale electricity generation was renewable and 26.6% of that (or 5.7% of the total electricity) was hydroelectric [48]. On a worldwide basis, in 2022, hydropower provided 14.9% of total electricity and 51% of the world's renewable electricity [49]. There are two types of hydropower: reservoir systems, in which reservoir water is released through turbines in a dam, and run-of-the-river systems, which rely on the kinetic energy of river flow. In addition, load shifting can be provided by pumped hydropower storage systems, in which water is pumped to a high-elevation reservoir and released during times of peak electricity demand. Hydropower is dominated by the first type of system. However, small run-of-the-river projects are being pursued in many rural areas, and there is increased interest in pumped storage as a means to firm variable wind and PV on the grid.

Large-scale hydropower can produce electricity at a very low cost and high EROI. The enormous hydropower public works projects undertaken in the Columbia River Basin during the Great Depression have been credited with allowing the United States to produce large quantities of aluminum at a low cost, which was needed for the manufacture of aircraft during World War II. Hydropower dams have the ability to provide both base-load power and also follow the load when needed. And once built, hydropower is a carbon-free form of energy generation. Figure 1.24 shows a Pacific Northwest dam.

There are a number of drawbacks to large-scale hydropower, however. Large reservoirs can displace human and animal populations and damage the ecosystem, including fish and other aquatic species. The buried vegetation will decompose and release methane, especially in tropical regions, thus contributing to climate change. This consequence must be considered when assessing the overall impact of large hydropower on climate change. The creation of a reservoir will increase the evaporation of the river water and water salinity, an issue in dry regions. When a river that carries a large amount of silt is dammed, the silt will build up at the bottom of the reservoir over time, thus decreasing the reservoir volume until it is no longer viable.

FIGURE 1.24
Bonneville Dam spillway on the Columbia River.

1.8 Hydrogen

When considering a hydrogen economy, it is important to recognize that hydrogen is an energy carrier, much like electricity, and not an energy source (although, if nuclear fusion technology develops and is implemented, hydrogen is a potential feedstock). However, hydrogen differs from grid electricity because its production does not instantaneously balance demand, that is, hydrogen can be stored more readily than electricity. This ability makes hydrogen attractive as a fuel. It also allows hydrogen-powered vehicles to be refilled in a time frame similar to that of gasoline vehicles, whereas electric battery vehicles require longer charge times.

Figure 1.25 shows the major pathways to produce molecular hydrogen as an energy carrier [50]. The top row of Figure 1.25 shows the primary energy sources: fossil fuels, nuclear, and renewable sources. The next three rows show the major processing steps for the conversion of the primary energy into hydrogen. Following the hydrogen row are the two methods of using it in energy applications: one is to combust hydrogen to produce heat for various applications and the other is to generate electricity from hydrogen by means of a fuel cell.

IEA reports that global hydrogen production in 2023 was 97 million metric tons, and less than 1% was low emissions [51]. Over 10 million metric tons of molecular hydrogen are used annually in the United States for hydrotreating and hydrocracking in petroleum refineries and to produce ammonia. It is important to note that approximately 95% of the hydrogen produced in the United States is via the steam reforming of natural gas, which contributes to climate change. For a hydrogen economy to address climate

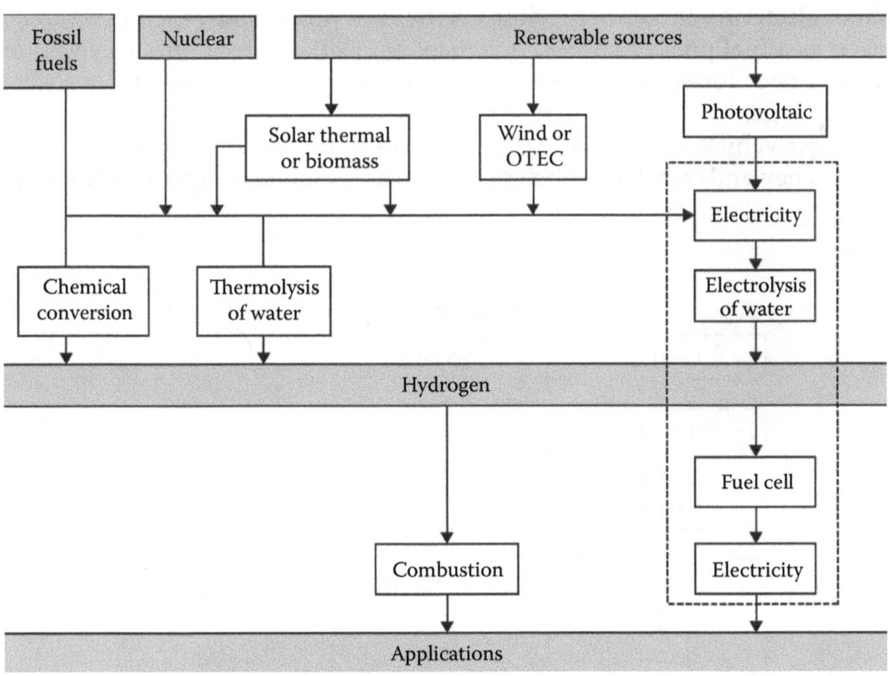

FIGURE 1.25
Pathways for hydrogen production and utilization. (From Kreith and West [50].)

change, the hydrogen must be produced by nuclear or renewable energy. Potential pathways are electrolysis from nuclear or renewable electricity or thermolysis of water (breaking down water into hydrogen and oxygen at high temperatures) via nuclear or solar thermal energy.

Water provides a large resource for hydrogen; however, water splitting requires a substantial energy input. The following reaction describes water splitting into hydrogen and oxygen:

$$H_2O \rightarrow H_2 + \frac{1}{2}O_2 \tag{1.5}$$

To split water via this reaction, 120 MJ/kg-hydrogen is needed (all gases at 25°C). It is useful to examine the round-trip efficiency of hydrogen production and utilization quantitatively. Figure 1.26 shows approximations for the round-trip efficiency of electricity to hydrogen via water electrolysis and conversion of that hydrogen back to electricity via a fuel cell. The key steps are shown with current and target efficiencies [52].

With current technology, it takes approximately 3–3.5 kWh of electricity input to produce 1 kWh of electricity output, and with advanced efficiency values, it might require approximately 2–2.5 kWh to yield 1 kWh of output. Thus, the cost of electrolysis is dependent upon the power system having a large number of hours in which the price of electricity is very low or zero (as could happen at high penetrations of wind and solar PV generation). But using hydrogen for transportation means that much more grid electricity would be needed.

In summary, a round-trip energy analysis shows that electrolysis is energy inefficient compared to using the electricity directly. (In the case of building heating, using electricity directly to power heat pumps can provide between 2 and 4 kWh of thermal energy for each kWh of electricity input, depending on the heat pump source temperature.) The use of hydrogen as a fuel presents other challenges as well. No distribution system exists for hydrogen as it does for electricity. Also, hydrogen is very leak-prone and is a strong indirect greenhouse gas.

For passenger vehicles, battery electric technology is superior to fuel cell technology in terms of efficiency and cost. Nevertheless, several automobile companies are still pursuing

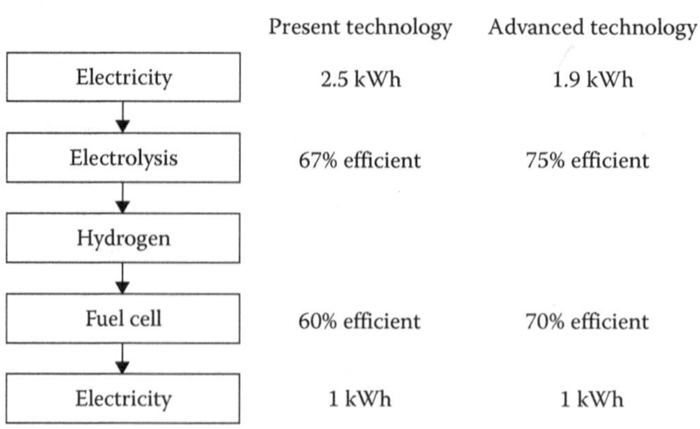

FIGURE 1.26
Inefficiencies associated with the generation of electricity via hydrogen that is produced from electrolysis [52].

hydrogen fuel cell passenger cars. Decreases in battery costs, the ability of battery electric vehicles to connect directly to the grid (allowing charging at home), and their efficient use of electricity have led to their dominance in the marketplace. Hydrogen as a transportation fuel may prove to be competitive in sectors such as ships and aircraft for which CO_2 emissions are hard to abate. Hydrogen might also serve a role for large-scale, long-duration grid storage. As our energy system is decarbonized, clean hydrogen will clearly be needed to replace hydrogen that is currently being produced from fossil fuels and used to produce ammonia and other useful products.

1.9 NREL System Advisor Model

An important feature of this book is the utilization of efficiency and renewable energy tools developed by NREL. Foremost among these tools is the System Advisor Model (SAM) [53], which provides the ability to analyze a variety of renewable energy systems. Estimating the performance and cost of most renewable energy systems is difficult because the availability of the resource, be it wind or solar, varies with time of day, time of year, the local weather, and location. To make realistic estimates of performance and cost, it is necessary to understand and be able to simulate the basic theory of the energy generation system using local environmental data. SAM calculates the performance and cost of generating electricity or heat using data for the project's location, installation, operating costs, financing, and physical system specifications. SAM's performance model makes hour-by-hour calculations for various types of renewable systems' thermal or electric output, generating a set of 8,760 hourly values that present the system's heat or energy production during a typical year. SAM then uses the hourly performance data to estimate the system's total annual output and capacity factors for practical performance evaluations. Throughout this text, you will find calculations for major technologies based on engineering analysis, as well as calculations using SAM for similar systems. The examples demonstrate the use of the model based on typical performance calculations at a given location using default parameters that are built on the experience with and data from previous installations. To use the model properly, however, it is necessary to first understand the engineering basis for the given technology. Then default parameters can be replaced with parameters that are more specific to the system being analyzed.

As of 2024, SAM provides performance models for the following technologies:

- Photovoltaic systems (flat and concentrating systems; residential to utility-scale)
- Parabolic trough, power tower, and linear Fresnel CSP systems for electricity generation
- Parabolic trough and linear Fresnel concentrating solar systems for industrial process heat
- Wind power electric turbines
- Marine wave and tidal energy
- Geothermal power
- Biomass power

Fuel cells

Battery storage (lithium-ion, lead acid, and flow batteries)

Flat-plate collectors for domestic or commercial hot water heating

A simple model for other technologies, including fossil fuel thermal power plants.

SAM's financial models allow users to analyze financial performance for distributed generation as in residential solar, or for utility-scale projects whereby developers deliver electricity to the grid for sale at pre-negotiated or market prices.

SAM requires information about the location, type of equipment used in the system, various operating parameters related to the resource, geometry of the energy generation system, cost of installation, and operating and fuel costs during its life, as well as economic parameters such as interest and inflation rates. Running the model for wind or solar technologies requires a weather file for the project's location. These files are generally used at hourly resolution for a typical meteorological year (TMY). SAM links to several libraries for weather and solar resource files, including NREL's National Solar Radiation Database, Wind Integration National Dataset (WIND) Toolkit, and Biofuels Atlas. Typical meteorological year files are provided for some other countries besides the United States. Users can also supply their own meteorological data as text files that follow SAM's weather file format requirements. SAM provides, for each technology, a set of sample files that contain complete sets of sample cost and performance input data. However, for more accurate results users will need to modify the input data as appropriate for the design of the system they are working on.

Finally, SAM displays modeling results in tables and graphs, displaying first-year annual energy production, LCOE, net present value, and many other performance and financial metrics. A built-in graphic tool displays a set of default graphs and allows for the creation of custom graphs that fit each individual system and its requirements. SAM also includes built-in simulation capabilities for sensitivity analysis, parametric analysis, and Monte Carlo simulations. The availability of SAM makes it possible for the designer or operator to quickly assess the effect of changing parameters and avoid the tedious calculations previously associated with inputting the data and carrying out hour-by-hour calculations to obtain the daily, monthly, or yearly performance of the system. It should be emphasized, however, that using SAM without a thorough knowledge of the basic engineering information associated with the system can lead to serious errors.

SAM is available for free at https://sam.nrel.gov. Versions are available for installation on Windows, OS X, and Linux operating systems. NREL and its partners continue to develop SAM, including adding new technology performance models and updating cost and performance estimates, so it is important to keep abreast of the website. In addition to new model releases, the website provides access to legacy versions, which are useful for being able to replicate earlier work. Because of updates, use of later versions to solve problems presented in this book will likely result in modest departures from the solutions shown.

1.9.1 Energy Units and Conversion Factors

In reading the literature pertaining to energy, an overwhelming number of units that are not usually used in conventional engineering calculations are encountered. To assist the reader in handling data from various sources that include unusual units such as barrels of oil equivalent, conversion factors are presented in Table 1.3.

TABLE 1.3

Energy Conversion Factors

	Btus	Quads	Calories	kWh	MWy
Btus	1	10^{-15}	252	2.93×10^{-4}	3.35×10^{-11}
Quads	10^{15}	1	2.52×10^{17}	2.93×10^{11}	3.35×10^{4}
Calories	3.97×10^{-3}	3.97×10^{-18}	1	1.16×10^{-6}	1.33×10^{-13}
kWh	3413	3.41×10^{-12}	8.60×10^{5}	1	1.14×10^{-7}
MWy	2.99×10^{10}	2.99×10^{-5}	7.53×10^{12}	8.76×10^{6}	1
bbls oil	5.50×10^{6}	5.50×10^{-9}	1.38×10^{9}	1,612	1.84×10^{-4}
Tonnes oil	4.04×10^{7}	4.04×10^{-8}	1.02×10^{10}	1.18×10^{4}	1.35×10^{-3}
kg coal	2.78×10^{4}	2.78×10^{-11}	7×10^{6}	8.14	9.29×10^{-7}
Tonnes coal	2.78×10^{7}	2.78×10^{-8}	7×10^{9}	8139	9.29×10^{-4}
MCF gas	10^{6}	10^{-9}	2.52×10^{8}	293	3.35×10^{-5}
Joules	9.48×10^{-4}	9.48×10^{-19}	0.239	2.78×10^{-7}	3.17×10^{-14}
EJ	9.48×10^{14}	0.948	2.39×10^{17}	2.78×10^{11}	3.17×10^{4}

	bbls Oil Equivalent	Tonnes Oil Equivalent	kg Coal Equivalent	Tonnes Coal Equivalent	MCF Gas Equivalent	Joules	EJ
Btus	1.82×10^{-7}	2.48×10^{-8}	3.6×10^{-5}	3.6×10^{-8}	10^{-6}	1055	1.06×10^{-15}
Quads	1.82×10^{8}	2.48×10^{7}	3.6×10^{10}	3.6×10^{10}	10^{9}	1.06×10^{18}	1.06
Calories	7.21×10^{-10}	9.82×10^{-11}	1.43×10^{-7}	1.43×10^{-10}	3.97×10^{-9}	4.19	4.19×10^{-18}
kWh	6.20×10^{-4}	8.45×10^{-5}	0.123	1.23×10^{-4}	3.41×10^{-3}	3.6×10^{6}	3.6×10^{-12}
MWy	5435	740	1.08×10^{6}	1076	2.99×10^{4}	3.15×10^{13}	3.15×10^{-5}
bbls oil	1	0.136	198	0.198	5.50	5.80×10^{9}	5.80×10^{9}
Tonnes oil	7.35	1	1,455	1.45	40.4	4.26×10^{10}	4.26×10^{-8}
kg coal	5.05×10^{-3}	6.88×10^{-4}	1	0.001	0.0278	2.93×10^{7}	2.93×10^{-11}
Tonnes coal	5.05	0.688	1,000	1	27.8	2.93×10^{10}	2.93×10^{-8}
MCF gas	0.182	0.0248	36	0.036	1	1.06×10^{9}	1.06×10^{-9}
Joules	1.72×10^{-10}	2.35×10^{-11}	3.41×10^{-8}	3.41×10^{-11}	9.48×10^{-10}	1	10^{-18}
EJ	1.72×10^{8}	2.35×10^{7}	3.41×10^{10}	3.41×10^{7}	9.48×10^{8}	10^{18}	1

Notes: To convert from the first column units to other units, multiply by the factors shown in the appropriate row (e.g., 1 Btu=252 cal). MWy=megawatt-year; bbls=barrels; tonnes=metric tons=1,000 kg=2,204.6 lb; MCF=thousand cubic feet; EJ=exajoule=10^{18} J. Nominal calorific values assumed for coal, oil, and gas.

Problems

1.1 The hallways of a three-story building in Denver, Colorado, use 600 lights per floor, and each light is an incandescent bulb rated at 60 W.

a. Estimate the amount of energy that could be saved in August by turning the lights on when the sun sets and turning them off when it becomes light. Assume the number of daylight hours in Denver in August is 425 hours.

b. Assuming the power for the lights comes from a 30% efficient coal-fired power plant, estimate the amount of CO_2 that would be saved with this conservation measure. Assume that the chemical composition of the coal is $CH_{0.8}$ and that coal has a heating value of 29 MJ/kg.

c. Prepare a brief report with recommendations for additional conservation measures.

1.2 Two different fuels are being considered for a 1-MW (net output) heat engine, which can operate between the highest temperature produced during the burning of the fuels and an atmospheric temperature of 300 K. Fuel A burns at 2,500 K, delivering 50 MJ/kg (heating value), and costs $2.00/kg. Fuel B burns at 1,500 K, delivering 40 MJ/kg, and costs $1.50/kg. Compare the fuel costs per hour of fuel A and fuel B, assuming the heat engine operates:

a. At Carnot efficiency.

b. At 50% of Carnot efficiency.

1.3 A solar energy collector produces a maximum temperature of 100°C. The collected energy is used in a cyclic heat engine that operates in a 5°C environment. Estimate the thermal efficiency if the engine operates at 40% of Carnot efficiency. How would the answer change if the collector were redesigned to focus the incoming solar radiation to enhance the maximum temperatures to 400°C?

1.4 Because meat is higher on the food chain than grain, the more calories you get from meat, the more agricultural land you use. Assume that 5.6 GJ metabolized energy is produced annually per hectare for meat and that 41.4 GJ of metabolized energy is produced annually per hectare for grain. Use the arable land area in your state or province and estimate:

a. How many vegetarians could you feed with land available in your state or province? Determine this as a number and percentage of the current population.

b. Repeat this estimate for someone who gets two-thirds of their calories from meat and one-third of their calories from grain.

1.5 Estimate the power produced in kWh/day per capita on average if 10% of the state of Colorado's 270,000 km^2 land area was covered with PV collectors. Assume the average insolation is 6 $kWh/m^2/day$ and the efficiency of the PV system is 16%. If 10% of the 5.9 million people living in Colorado had to drive 100 km/day to commute to work, how many kWh/day would they consume? Assume they commute in gasoline vehicles that average 28 miles/gallon. Discuss the implications of these two estimates. How many 3 m×10 m PV collectors would be required to produce the energy for commuting with an all-electric car with 60% efficient batteries?

1.6 Assuming that a heat pump has a coefficient of performance equal to 4, how much heat could this device deliver to the interior of a building if the heat pump used 1 kWh of electricity?

1.7 Estimate the increase in the price of a barrel of oil if a carbon tax of $250/short ton (1 U.S. short ton=907.2 kg) of CO_2 were imposed on fossil fuels.

1.8 Back in 2009, the administration of U.S. President Barack Obama instituted a Cash for Clunkers program. You are to evaluate the effectiveness of the program, which was intended to stimulate the economy and reduce the import of oil. According to the program announcement, an average of $4,000 per car in rebate was offered if a consumer brought in a "clunker" and exchanged it for a new car that would have a fuel efficiency equal to at least 10 miles/gallon better than the car they traded. The amount of money allocated to the program was $3 billion/year.

a. How many barrels of oil per year were saved by the program? Assume the "clunkers" get 20 miles/gallon and the cars (old and new) are driven 13,500 miles a year.

b. Assuming that imported oil at the time cost $55/barrel, how many years would the new cars need to replace the old ones for fuel cost savings to pay for the program?

c. The energy required to build a new car, called the embedded energy in car manufacturing, has been estimated to be 30,500 kWh/car. The embedded energy is mostly coal and nuclear used to produce electricity, whereas the energy saved by the improved efficiency is mostly oil imported over the life of the car. Estimate how many years it would take to recoup the energy invested in the manufacture of the car by the savings in energy from the improved mileage of the new vehicle.

d. What reduction in CO_2 emissions did the program achieve? Discuss the effectiveness of the program from an energy perspective and what changes in the program you would recommend to make it more effective.

1.9 A large nuclear power plant has a rated capacity of 1 GW. Its actual output is estimated to average about 80% of capacity because of the need for maintenance and imperfect match between supply and demand. Assuming that year-round average residential electricity needs are about 1 kW electric per housing unit and there are three residents per unit, estimate how many power plants would be needed to satisfy the demand of a metropolitan area with a population of 30 million.

1.10 The yearly average solar flux incident on a horizontal surface in the United States is about 180 W/m^2. The average total energy demand in the United States is approximately 100 quads (~100×10^{18} J) per year. Assuming an average efficiency of PV cells of 14% and a biomass conversion efficiency around 1%, estimate the percentage of land required to provide the energy of the country by these two renewable technologies. (The total land area of the United States is about 9.2×10^6 km^2.)

1.11 Calculate the doubling time for annual percentage increases of 2% and 10%, assuming an exponential growth function.

1.12 If current global coal consumption is 7.8×10^9 metric tons/year and estimated reserves are 1.14×10^{12} metric tons, estimate how long the coal would last at the current rate of use and how long it would last if the usage rate increased by 5% per year and by 10% per year.

1.13 Derive Equation 1.4 for the lifetime, t_L, of a finite resource, that has a total usable quantity, N_T, and a consumption rate, $N(t)$, that increases every year with an annual fractional growth rate, r. (Hint: Integrate $N(t)$ given in Equation 1.1.) What would be a reasonable approximation for Equation 1.4 if the annual fractional growth rate, r, is small?

1.14 Globally, electricity generation from wind was about 2,304 TWh in 2023, up from 104 TWh in 2005. Electricity generation from solar PV was 1,630 TWh in 2023, up from 3.95 TWh in 2005. Electricity generation from coal in 2023 was 10,500 TWh.

a. Assuming electricity production from wind grew exponentially with a constant annual growth rate from 2005 to 2023, what would the annual growth rate have been?
b. With the same assumption as in part a, what would the annual growth rate have been for solar PV?
c. Based on the annual growth rate you estimated for wind in part a, and assuming it stays constant going forward, in what year would electricity generation from wind match the generation from coal that was observed in 2023?

Discussion Questions

Discussion of Sustainability Principles

Do some research on a resource management regulation in your area. There are fishing regulations in most areas. Many national parks have limits on the number of visitors and how they can access sensitive areas. Regulations may exist for grazing or logging on public lands. Do all of these existing regulations arise because of previously unsustainable exploitation? Do they all involve limits on what people can take so that the resource can continue into the future?

Discussion of Land Uses

Consider an acre of land. At today's solar PV efficiencies and assuming a ground cover ratio of 50%, what percent of the solar radiation could be converted to electricity? Research practical biomass conversion efficiencies and compare this to PV. If you wanted to use the land for energy production (electricity or biofuels), what things would you consider besides solar conversion efficiency? Consider the many potential uses of land. What issues do you think are important in making decisions about how to use land and how do these differ from one country to another? Finally consider agrivoltaics, which combines rows of PV with land use for either growing crops or grazing sheep. Under what circumstances do you think these dual uses make sense?

Discussion of Energy Content of Corn and Transportation

If the American sport utility vehicle driver rode his bicycle the 225 miles that the tank of E10 gasoline would have provided, pedaling at an average power output of 150 W, how many kg of corn would he need to eat to fuel his transportation? Discuss the difference in corn consumption by the human body vs. a car running on E10 gasoline.

Discussion of CO_2 Emissions

Make a spreadsheet of CO_2 emissions out to the year 2100. Start with the current figure (the emissions were 37.4 Gt/year at the end of 2024). Think about different scenarios and try plotting them. Make sure to keep track of the sum total of CO_2 emitted. What year would a 1,000 Gt remaining CO_2 budget (to stay under a 2°C temperature rise relative to preindustrial) be breached if growth in fossil fuel production continues at 1.7% per year? What if growth were 1% per year? What if an aggressive demand management program were driven by the engineering profession worldwide to stop the growth of fossil fuels and maintain consumption at the current level?

Discussion of Instability Caused by Growth

Can you think of some examples of civilizations that have faced risks from hitting the limits of population growth? What were the implications? Did any of these civilizations know they were on an unsustainable path? What did they do about it? Most engineers are employed in industry or public service. Can you think of any examples of how engineering work has been influenced by resource limits that conflict with population growth? Can you think of any ideas of how engineering work could help to manage the risks of growth beyond carrying capacity limits?

References

1. Alliance for Water Efficiency and American Council for an Energy-Efficient Economy, May 2011. Addressing the energy-water nexus: A blueprint for action and policy agenda; ASME, ETP, Energy–water nexus—Cross-cutting impacts.
2. Brundtland, G.H., 1987. *Our Common Future, World Commission on Environment and Development*, Oxford University Press, New York.
3. Tagg, N., and Jones, M., 2023. Ecological sustainability for the future of all life on Earth, October 31, Open Access News: Environment, https://www.openaccessgovernment.org/ecological-sustainability-for-the-future-of-all-life-on-earth/169341/.
4. United Nations, 2024. Sustainable development: The 17 goals, Department of Economic and Social Affairs, https://sdgs.un.org/goals.
5. Hardin, G., 1968. The tragedy of the commons, *Science* 168, 1243.
6. Population Division of the Department of Economic and Social Affairs of the United Nations Secretariat, March 2009. World population prospects: The 2008 revision.
7. Population Division of the Department of Economic and Social Affairs of the United Nations Secretariat, July 2024. World population prospects: The 2024 revision.
8. Bongaarts, J., 2003. Completing the fertility transition in the developing world: The role of educational differences and fertility preferences, *Population Studies* 57(3), 321–335, doi:10.1080/0032472032000137835.
9. Scanlon, B., Faunt, C., Longuevergne, L., Reedy, R., Alley, W., McGuire, V., and McMahon, P., June 2012. Groundwater depletion and sustainability of irrigation in the US High Plains and Central Valley, *Proceedings of the National Academy of Sciences* 109(24), 9320–9325. doi:10.1073/pnas.1200311109.
10. World Bank, 2003. Development data and statistics, www.worldbank.org/data.
11. Loupasis, S., May 2002. Technical analysis of existing RES desalination schemes—RE driven desalination systems REDDES, Report, Contract # 4.1030/Z/01-081/2001.
12. Kenny, J., Barber, N., Huston, S., Linsey, K., Lovelace, J., and Maupin, A., 2005. Estimated use of water in the United States in 2005, Circular 1344, U.S. Geological Survey, Washington, DC, pp. 38–41.

13. U.S. DOE, 2006. *Energy Demands on Water Resources, Report to Congress*, U.S. Department of Energy, Washington, DC.
14. Wu, M., Mintz, M., Wang, M., and Arora, S., January 2009. *Consumptive Water Use in the Production of Ethanol and Petroleum Gasoline*, Argonne National Laboratory, Argonne, IL.
15. Webber, M.E., 2016. *Thirst for Power: Energy, Water, and Human Survival*, Yale University Press, New Haven, CT and London.
16. Barney, G.O., Blewett, J., and Barney, K.R., 1993. *Global 2000 Revisited: What Shall We Do?* The Millennium Institute, Alexandria, VA, p. 11.
17. United Nations Food and Agriculture Organization, 2015. The state of food insecurity in the world 2015, https://www.fao.org/3/a-i4646e.pdf.
18. Pimentel, D., Williamson, S., Alexander, C., Gonzalez-Pagan, O., Kontak, C., and Mulkey, S., 2008. Reducing energy inputs in the US food system, *Human Ecology* 36, 459–471.
19. Alternative Fuels Data Center, 2024. U.S. Biodiesel production, exports, and consumption, https://afdc.energy.gov/data/10325.
20. Krumdieck, S., and Hamm, A., 2009. Strategic analysis methodology for energy systems with remote island case study, *Energy Policy* 37(9), 3301–3313.
21. United Nations Development Programme Human Development Reports, About human development, https://hdr.undp.org/en/humandev/ (last accessed April 26, 2025).
22. Goldemberg, J., and Johansson, T.B., eds., 2004. *World Energy Assessment: Energy and the Challenge of Sustainability*, United Nations Development Programme, New York, https://www.undp.org/content/undp/en/home/librarypage/environment-energy/sustainable_energy/world_energy_assessmentoverview2004update.html.
23. U.S. Energy Information Administration, 2013. History of energy consumption in the United States, 1776–2012, https://www.eia.gov/todayinenergy/detail.php?id=11951.
24. U.S. Energy Information Administration, 2024. U.S. energy flow 2023, https://www.eia.gov/totalenergy/data/flow-graphs/total-energy.php.
25. U.S. Energy Information Administration, 2024. Renewable energy consumption, *Monthly Energy Review*, November, https://www.eia.gov/totalenergy/data/monthly/index.php#renewable.
26. U.S. Energy Information Administration, 2024. Solar and battery storage to make up 81% of new U.S. electric generating capacity in 2024, Today in Energy, February 15, https://www.eia.gov/todayinenergy/detail.php?id=61424.
27. U.S. Energy Information Administration, 2023. Annual energy outlook 2023, https://www.eia.gov/outlooks/aeo/.
28. International Energy Agency, 2016. Energy technology perspectives 2016: Towards sustainable urban energy systems, https://www.iea.org/reports/energy-technology-perspectives-2016.
29. International Energy Agency, 2021. Net zero by 2050: A roadmap for the global energy sector, Pg. 20, https://iea.blob.core.windows.net/assets/deebef5d-0c34-4539-9d0c-10b13d840027/NetZeroby2050-ARoadmapfortheGlobalEnergySector_CORR.pdf.
30. International Energy Agency, 2023. Net zero roadmap: A global pathway to keep the 1.5C goal in reach, 2023 Update, Pg. 73, https://iea.blob.core.windows.net/assets/8ad619b9-17aa-473d-8a2f-4b90846f5c19/NetZeroRoadmap_AGlobalPathwaytoKeepthe1.5CGoalinReach-2023Update.pdf.
31. Our World in Data, 2023. Energy use per person vs. GDP per capita, 2023, https://ourworldindata.org/grapher/energy-use-per-person-vs-gdp-per-capita.
32. Hall, C.A.S., Lambert, J.G., and Balogh, S.B., 2014. EROI of different fuels and the implications for society, *Energy Policy* 64, 141–152.
33. Lazard, 2023. LCOE+ levelized cost of energy +, April, https://www.lazard.com/research-insights/2023-levelized-cost-of-energyplus/.
34. Our World in Data, 2022. How does the land use of different electricity sources compare? https://ourworldindata.org/land-use-per-energy-source.

35. U.S. Department of Energy, 2022. DOE recognizes 3M company for leadership in energy management, September 19, https://www.energy.gov/eere/amo/articles/doe-recognizes-3m-company-leadership-energy-management#:~:text=Thanks%20to%20its%20dedicated%20energy,strengthened%20the%20company's%20brand%20image.
36. U.S. Department of Energy, 2024. Better buildings, https://betterbuildingssolutioncenter.energy.gov.
37. U.S. Department of Energy, 2024. Better plants, https://energy.gov/eere/amo/better-plants.
38. McDuffie, E., Martin, R., Yin, H., and Brauer, M., 2021. Global burden of disease from major air pollution sources (GBD MAPS): A global approach, *Research Report, Health Effects Institute* 2021(210), 1–45, Boston, MA, https://pmc.ncbi.nlm.nih.gov/articles/PMC9501767/.
39. Statista, 2024. Number of operable nuclear power reactors worldwide as of May 2024, by country, https://www.statista.com/statistics/267158/number-of-nuclear-reactors-in-operation-by-country/.
40. Intergovernmental Panel on Climate Change, Working Group III, 2012. Renewable energy sources and climate change mitigation, Special Report of the Intergovernmental Panel on Climate Change.
41. Global Solar Council, 2024. Global Solar Council announces 2 terawatt milestone achieve forsolar,November8,https://www.globalsolarcouncil.org/news/global-solar-council-announces-2-terawatt-milestone-achieved-for-solar/.
42. Statista, 2025. Leading countries in installed concentrated solar power (CSP) in 2023, https://www.statista.com/statistics/494169/global-installed-concentrated-solar-power-csp-capacity-by-key-country/#:~:text=Global%20installed%20concentrated%20solar%20power%20capacity%20by,with%201.5%20gigawatts%20and%20596%20megawatts%2C%20respectively.
43. International Energy Agency, 2024. Solar heat worldwide, https://www.iea-shc.org/solar-heat-worldwide.
44. International Energy Agency, 2024. Bioenergy, https://www.iea.org/energy-system/renewables/bioenergy.
45. Clean Cooking Alliance, 2024. https://cleancooking.org.
46. Alternative Fuels Data Center, 2024. Global ethanol production by country and region, sourced from Renewable Fuels Association, https://afdc.energy.gov/data/10331.
47. Gill, T., and Birch, T., 2024. A guide to biomass power plants, December 4, https://www.theecoexperts.co.uk/news/biomass-power-plant#:~:text=Are%20biomass%20power%20plants%20popular,meaning%20as%20gigawatts%20per%20hour.
48. U.S. Energy Information Administration, 2024. What is U.S. electricity generation by energy source? https://www.eia.gov/tools/faqs/faq.php?id=427&t=3.
49. International Renewable Energy Agency (IRENA), 2024. Electricity generation by energy source, https://www.irena.org/-/media/Files/IRENA/Agency/Publication/2024/Jul/Renewable_energy_highlights_FINAL_July_2024.pdf.
50. Kreith, F., and West, R.E., 2004. Fallacies of a hydrogen economy: A critical analysis of hydrogen production and utilization, *Journal of Energy Resources Technology* 126(4), 249–257, doi:10.1115/1.1834851.
51. International Energy Agency, 2024. Hydrogen production, Global Hydrogen Review 2024, https://www.iea.org/reports/global-hydrogen-review-2024/hydrogen-production.
52. U.S. Department of Energy, 2024. Hydrogen and fuel cell technologies office, technical targets for high temperature electrolysis, https://www.energy.gov/eere/fuelcells/technical-targets-high-temperature-electrolysis.
53. NREL, 2024. System Advisor Model (SAM), https://sam.nrel.gov.

2

Economics of Energy Generation and Conservation Systems

This chapter and the next one present useful analysis tools that engineers should understand. These include a basic process for economic assessment, energy flow analysis, and resource accounting methods for complex energy systems. The objective of economic assessment, covered in this chapter, is to provide useful financial insight and enable a comparison between various energy options. The standard methods presented here are equally applicable to renewable and nonrenewable energy systems as well as to energy conservation projects. The objectives of energy flow analysis and resource accounting, presented in Chapter 3, are to conceptualize and compare different energy system investments, to account for life cycle resource and external costs, and to identify sustainability issues.

As the world undergoes a transition from heavy reliance on fossil fuels to more sustainable alternatives, the approach of sustainable energy engineering is to use quantitative analysis to understand the end use, environmental and social requirements, and resource availability and then help make development decisions based on analysis of the dynamics of energy systems into the future. This analysis should lead to adaptive design innovations that meet the environmental, social, and economic requirements, manage risks, and provide benefits. This chapter and the next one explore several different ways to understand the energy/environmental/economic system and the ways that this understanding can inform investment and development decisions.

2.1 Unit Cost of Energy

The source of energy in renewable energy (RE) systems, such as wind turbines and solar collectors is essentially free, but so are oil, coal, and gas in the current economic system that only requires purchase of mineral rights and a small royalty on mining as the price of access to fossil resources. The costs for energy prospecting, extraction, refining, and generation are largely associated with the equipment purchased and the fuels or electricity used [1]. Neglecting interest charges on capital and operating costs, the unit cost of energy produced by a system, C_s dollars, with initial construction and installation cost, C_o dollars, over the life of the system, t years, and annual energy production, Q units of energy, is given by

$$C_s = \frac{C_o}{Q \cdot t} \tag{2.1}$$

For example, consider a residential solar photovoltaics (PV) system with an installed cost of \$600 per 1.6 m^2 panel (\$375/m^2). The panels have a 10-year warranty for 90% of nominal power output and a 25-year warranty for 80% of power output. The performance

DOI: 10.1201/9781003586616-2

specifications are measured at standard test conditions (25°C, 1,000 W/m²) giving conversion efficiency of 15%, and the panel is installed in a location where the mean horizontal surface irradiance is 200 W/m². The electricity production can be estimated for a year as $Q = 0.2\ (\text{kW/m}^2) \times 15\% \times 8{,}760\ (\text{h/year}) = 263\ (\text{kWh/year/m}^2)$. The cost of solar electricity can be estimated by using Equation 2.1:

$$C_s = \frac{375\left(\$/\text{m}^2\right)}{263\left(\text{kWh/year}\cdot\text{m}^2\right)\times\left(0.9\times 10\ \text{year} + 0.8\times 15\ \text{year}\right)} = \$0.068/\text{kWh}$$

Note that we generally use three significant figures when evaluating energy costs, as the methods do not have accuracy to justify using more. The sensitivity of energy production cost to the technical factors and resource factors can be explored using Equation 2.1. If the solar PV panel were installed in a location with only 100 W/m² mean solar radiation, although the system cost would be the same, the electricity production cost would increase to $0.136/kWh. If advances in materials, manufacturing and installation experience for the solar PV system could increase efficiency to 20%, reduce the cost to $300/m², and improve lifetime and performance degradation to 90% power output over 30 years, then the electricity production cost with 200 W/m² average irradiance would be reduced to just $0.032/kWh.

2.2 Payback Period

A common method of evaluating the value of an energy generation or conservation system is to determine the time required for the system to pay back the initial costs. The payback period (PP) is determined by comparing the total cost of equipment and installation, C_o, to the annual benefit in either value of net energy produced or net energy saved per year. The value of the energy saved equals the amount of energy saved per year, B_i (kWh/year) times the unit cost of energy, C_s ($/kWh). Once the yearly value of the net saved energy and the cost of installing a conservation measure are known, the simple PP in years is

$$\text{PP} = \frac{C_o\,(\$)}{B_i C_s\,(\$/\text{year})} \tag{2.2}$$

This approach is acceptable for preliminary estimates if the PP is short, say less than 4 years. For a more precise estimate, or for longer horizons, the time value of money, the inflation rate, and the escalation in fuel costs or maintenance costs must be considered.

2.3 Time Value of Money

If money is deposited in a bank, it earns interest. Similarly, when money (capital) is invested in a project or a business that is financed through a sale of bonds or stock, a certain rate of return on the investment is expected to compensate the owners of the invested capital for both the opportunity cost of having their money temporarily unavailable for other uses, and also for the risk of loss, since few investments are "sure things." Conversely, if you borrow money from a bank, you must pay interest as the cost of the money lent to you. These issues are known as the *time value of money*.

Since it is convenient to conduct economic analyses in a unit of fixed monetary value, a concept known as *present worth* is used. The present worth of a future cash flow or a future payment is the future value of that flow, accounting for the time value of the money. For example, the future value F, of a sum of money P, invested today at an annual interest rate i is

$$F = P(1+i)^t \tag{2.3}$$

where t is the time interval in years or other time unit corresponding to the time basis of the interest rate i expressed as a fraction. This equation essentially states that an initial sum of money P appreciates by a multiplier factor, $(1+i)$ each year, disregarding monetary inflation.

Stated alternatively, a future sum of money F has a *present worth* P given by

$$P = \frac{F}{(1+i)^t} \tag{2.4}$$

This indicates that the present worth of a given amount of money in the future is discounted, in constant dollars, by a factor of $(1+i)^{-t}$ for each year in the future. In this context, i is generally called the discount rate, but is also referred to as the cost of money or rate of return. Note that this approach assumes i remains constant over the period of concern. In real life, however, it may well not be constant and constitutes a source of uncertainty.

The present worth factor [PWF(i, t)] is given by

$$\text{PWF}(i,t) = \frac{1}{(1+i)^t} \tag{2.5}$$

To obtain the present value of a cash flow F that would come t years into the future, it is multiplied by PWF(i, t).

As an example, consider a discount rate of 5% and a cash inflow of \$100, 1 year from now. Using Equation 2.4, the present worth is only \$95. Conversely, if \$95 were invested at 5% interest today, it would be worth approximately \$100 a year from now. The same approach can be used to evaluate cash flows at any future period.

The present worth concept is the key for a discounted cash flow analysis. Its use permits all calculations to be made in terms of *present discounted money*. Every business or organization has a discount rate it uses that reflects the rate of return it expects on investments. Remember, the company could invest capital in making more products and expect a certain rate of return, determined by their experience with past investments. If they instead invest that capital in an energy conservation project, they want to know how that investment compares to other options. The mathematics of the discounted value of money demonstrates why revenues realized in the future are worth less than money earned at the present. For example, \$100 of revenue realized 40 years in the future at an annual discount rate i of 10% has a present worth of only \$2.21 today. This explains why the far future has little influence on conventional business planning and why many businesses shy away from investments in RE systems or conservation measures that have high upfront costs and long-term benefits.

When future cash flows are fixed in size, *A*, and regularly occur over a specific number of periods, the situation is known as an annuity. For an annuity, the present worth is given by Equation 2.6.

$$P = A\left[\frac{(1+i)^t - 1}{i(1+i)^t}\right] \tag{2.6}$$

For example, assume a cash flow of \$100/year at the end of each year for the next 5 years at a discount rate of 10%. According to Equation 2.6, the present value of the \$500 of inflow is only \$379.

$$P = \$100 \times \frac{(1+0.10)^5 - 1}{0.10(1+0.10)^5} = \$379$$

The present worth is most commonly used to compare different options for investment in different energy management or energy conservation opportunities as illustrated in the following example.

Example 2.1

A company has a discount rate of 10% for its investments. The company has two different energy conservation opportunities. Assume that opportunity A generates benefits equal to \$100, \$150, and \$200 at the end of years 1, 2, and 3, respectively. Assume that opportunity B yields benefits of \$225 in year 2 and \$225 in year 3. Over 3 years, both opportunities A and B yield benefits of \$450. However, the timing of the benefits received is different in each case. By using the present worth technique, the two benefit flows can be viewed in terms of today's dollar value.

Solution:

STEP 1

Compute the PWF using Equation 2.5 for the discount rate of $i = 10\%$

$$\text{PWF}(10,1) = \frac{1}{(1+0.1)^1} = 0.909$$

$$\text{PWF}(10,2) = \frac{1}{(1+0.1)^2} = 0.826$$

$$\text{PWF}(10,3) = \frac{1}{(1+0.1)^3} = 0.751$$

STEP 2

Compute the present worth of each opportunity benefit flow by multiplying the PWF by the annual benefit amount. Note that we use the same significant figures to report results as were given in the cost and benefit estimates.

		Annual Benefit ($)		Present Worth ($)	
Year	**PWF**	**A**	**B**	**A**	**B**
1	0.909	100	0	91	0
2	0.826	150	225	124	186
3	0.751	200	225	150	169
Total		450	450	365	355

Although the total benefits are the same, the present worth for opportunity A is higher because the annual benefits occur sooner. Since opportunity A provides cash sooner, that money can be invested for additional financial return, which is more beneficial for the investor.

2.4 Inflation

The cost and revenues of a project can be expressed either in "current dollars" or "constant dollars." Actual cash flows observed in the marketplace at a given time are called current dollar cash flows. They are the actual number of dollars required or acquired in the year the cost or benefit is incurred. The base year is defined as the year around which an analysis is structured. Most commonly, it is the initial period, which may be denoted by 0. In the base year, constant and current dollar cash flows are the same. The future current dollar equivalent, F_t, of a current amount, F_0 dollars, after t years at inflation rate of j percent per year is

$$F_t = F_0\left(1+j\right)^t \tag{2.7}$$

For example, energy savings due to efficiency improvement valued at $100 in current 2025 dollars, assuming 4% inflation over the next 10 years, will be valued at $148 in 2035 dollars. The effect of inflation on the future money is to reduce its real value or purchasing power by a factor $(1+j)^{-1}$/year. The future inflation rate will not be constant and will not be known, which constitutes a source of uncertainty in financial analysis. A simple way to deal with inflation risk is to add an extra percentage to the future annual inflation rate.

In practice, it is important to distinguish between real and nominal discount rates. A real discount rate excludes inflation, whereas a nominal value includes inflation:

$$i' \sim i + j \tag{2.8}$$

where

i' is the nominal discount rate (including the effect of inflation)
i is the real discount rate (with inflation factored out)
j is the inflation rate.

Equation 2.8 is a close approximation when the rate of inflation is low. In general, the discount rate relationship is more precisely calculated as

$$i' = (1+i)(1+j) - 1 \quad \text{or} \quad i = \frac{1+i'}{1+j} - 1 \tag{2.9}$$

For example, given an inflation rate of 6% and a nominal discount rate of 10%, Equation 2.8 approximates the real discount rate as 4%. Equation 2.9 gives a value of 3.8%. For evaluating power generation projects, a typical real discount rate is 3%–4%.

When calculating a present value, it is important to remember that the real discount rate must be used if cash flows are to be in constant dollars, whereas a nominal discount rate is to be used if the cash flow is in current dollars.

If it is necessary to convert a stream of cash flows from current dollars to constant dollars or vice versa, it is useful to refer to an inflation rate index. There are numerous inflation rate indices, for example, the consumer price index (CPI). The CPI measures the average change in prices of goods and services at the retail end for urban consumers. The gross domestic purchases (GDP) index accounts for purchases by government and businesses, in addition to retail consumers. Table 2.1 presents the U.S. GDP and CPI index values from 1990 to 2023. Additional information, including up-to-date data and inflation indexes for specific categories of goods and services, can be obtained from the U.S. Department of Commerce, Bureau of Economic Analysis (https://www.bea.gov) and U.S. Bureau of Labor Statistics (https://www.bls.gov).

The effect of inflation can be integrated into the present value analysis by using the nominal discount rate:

$$\mathrm{PV}[F_0] = \frac{F_t}{(1+i')^t} \tag{2.10}$$

It should be noted that if all cash flows are inflated at the same rate and are discounted with the appropriate discount rate, then working with either all constant or all current dollars gives the same result for PV, as demonstrated below in Examples 2.2 and 2.3.

2.5 Total Life Cycle Costs

A total life cycle cost (TLCC) analysis should consider all significant dollar costs over the life of a project. These costs are then discounted to a base year value using present value analysis. Any revenue generated from the resale of an investment is also discounted to the base year and subtracted from present value costs. TLCC analysis includes many different costs and must use the appropriate form of the cash flows (current or constant dollars) so that the correct discount rate is applied. If the cash flows are not discounted with the appropriate rate, the analysis will be flawed. The TLCC analysis is normally done to evaluate differences in the costs and the timing of these costs between alternative projects. The formula for calculating the present value of TLCC is

$$\mathrm{TLCC} = \sum_{t=0}^{N} \frac{C_t}{(1+i')^t} \tag{2.11}$$

where

C_t is the cost in current dollars in period t, including finance charges, operating and maintenance (O&M) costs, and energy costs

N is the analysis period

i' is the nominal discount rate.

TABLE 2.1

Selected Inflation Indices

Year	Gross Domestic Purchases Price Index (2017=100)[a]	GDP Price Index Annual Inflation Rate (%)	Consumer Price Index (1982–1984=100)[b]	CPI Annual Inflation Rate (%)
1990	60.1	3.96	130.7	5.40
1991	62.0	3.14	136.2	4.21
1992	63.5	2.33	140.3	3.01
1993	64.9	2.26	144.5	2.99
1994	66.3	2.10	148.2	2.56
1995	67.7	2.16	152.4	2.83
1996	68.9	1.74	156.9	2.95
1997	69.9	1.48	160.5	2.29
1998	70.3	0.67	163.0	1.56
1999	71.4	1.52	166.6	2.21
2000	73.3	2.60	172.2	3.36
2001	74.7	1.94	177.1	2.85
2002	75.7	1.37	179.9	1.58
2003	77.4	2.17	184.0	2.28
2004	79.6	2.87	188.9	2.66
2005	82.3	3.49	195.3	3.39
2006	85.0	3.22	201.6	3.23
2007	87.3	2.72	207.3	2.85
2008	89.8	2.84	215.3	3.84
2009	89.4	–0.43	214.5	–0.36
2010	90.7	1.50	218.1	1.64
2011	92.9	2.41	224.9	3.16
2012	94.5	1.75	229.6	2.07
2013	95.9	1.44	233.0	1.46
2014	97.4	1.56	236.7	1.62
2015	97.6	0.19	237.0	0.12
2016	98.2	0.66	240.0	1.26
2017	100.0	1.79	245.1	2.13
2018	102.2	2.22	251.1	2.44
2019	103.7	1.45	255.7	1.81
2020	105.0	1.29	258.8	1.23
2021	109.5	4.24	271.0	4.70
2022	116.9	6.78	292.7	8.00
2023	120.9	3.41	304.7	4.12

[a] Data from http://www.bea.gov, Table 1.6.4 Price Indexes for Gross Domestic Purchases (accessed September 3, 2024).

[b] Data from http://www.bls.gov/cpi, Historical CPI-U, July 2024 (accessed September 3, 2024).

Since the major costs associated with most conservation and RE technologies are the initial investment costs and O&M costs (including fuel), the formula for TLCC for residential, government, and nonprofit organizations often simplifies to

$$\mathrm{TLCC} = C_0 + \mathrm{PVOM} \tag{2.12}$$

where

C_0 is the initial investment

PVOM is the present value of all O&M costs

$$\text{PVOM} = \sum_{t=0}^{N} \frac{(\text{O\&M})_t}{(1+i')^t}$$

If the investment is by a for-profit organization, taxes must be included in the calculations as follows:

$$\text{TLCC} = C_0 + \text{PVOM} - T(\text{PVDEP} + \text{PVOM}) \quad (2.13)$$

where

T is the income tax rate

PVDEP is the present value of depreciation.

The simple tax treatment represented in Equation 2.13 assumes that businesses deduct costs from income on which taxes are owed. Upfront equipment purchases are not treated as costs for tax accounting purposes, as the exchange of payment for the new equipment is assumed to be an even exchange of value. Rather, costs are incurred as the value of the equipment depreciates over time. Various depreciation formulas are used in tax accounting, including "straight line" depreciation in equal increments over the life of the equipment and the "double-declining balance" formula, which is given by Equation 2.14 and illustrated in Example 2.3 below:

$$\text{DEP}_t = \frac{2}{N}\text{BV}_{t-1} \quad (2.14)$$

where

DEP_t is the depreciation taken in year t

N is the equipment life

BV_{t-1} is the book value of the equipment in year $t-1$, equal to the original value less the depreciation taken up to that date.

Example 2.2

A nonprofit organization invests $10,000 in new double-paned windows and estimates that O&M costs will be $1,300 in the first year and each year thereafter, and the windows will have no salvage value. Assuming an annual inflation rate of 3% and a nominal discount rate of 12%, Table 2.2 demonstrates the TLCC evaluation, both in current dollars and in constant dollars. Note that the TLCC is the same, but the discount rate and cash flows are different.

Example 2.3

Suppose the investor of Example 2.2 is a private company in the 34% federal income bracket but pays no state income tax. The initial investment is $10,000, O&M is $1,300/year in current dollars in year zero with an inflation rate of 3%, and the nominal discount rate is 12%. The investor uses a "5-year double-declining depreciation schedule" as shown in column 4. There is no salvage value. The results are shown for each of the steps for evaluating the after-tax TLCC (Table 2.3).

TABLE 2.2

Current and Constant Dollar TLCC Evaluation

Year	Investment	Discounted Investment	Current Dollar O&M Costs	Discounted O&M Costs (*w*/12% Nominal)	Constant Dollar O&M Costs	Discounted O&M Costs (*w*/8.74% Real)
0	$10,000	$10,000	$0	$0	$0	$0
1	0	0	$1,339	$1,196	$1,300	$1,196
2	0	0	$1,379	$1,100	$1,300	$1,100
3	0	0	$1,421	$1,011	$1,300	$1,011
4	0	0	$1,463	$930	$1,300	$930
5	0	0	$1,507	$855	$1,300	$855
NPV		$10,000		$5,091		$5,091

TLCC=$15,091.

TABLE 2.3

After-Tax TLCC Evaluation

Year	Investment ($)	Discounted Investment ($)	Depreciation ($)	Discounted Depreciation ($)	O&M Costs ($)	Discounted O&M Costs ($)
0	10,000	10,000			0	0
1	0	0	4,000	3,571	1,339	1,196
2	0	0	2,400	1,913	1,379	1,100
3	0	0	1,440	1,025	1,421	1,011
4	0	0	864	550	1,463	930
5	0	0	518	294	1,507	855
NPV		10,000		7,353		5,091

TLCC=$10,000−(0.34×$7,353)+($5,091×0.66)=$10,860.

2.6 Internal Rate of Return

The internal rate of return (IRR) analysis is a convenient procedure to compare a variety of investment activities. It is commonly used for accept/reject decisions by comparing the IRR with a minimal acceptable rate, often referred to as the "hurdle rate." The IRR of an investment with a series of future cash flows (F_0, F_1, …, F_t) is the rate of return on investment that causes the net present value (NPV) of cash flows to equal zero. IRR implicitly assumes reinvestments of any return at the IRR rate and is therefore not recommended for ranking projects. However, it has the advantage of giving an investor a quick comparison of the after-tax return on various financial instruments such as bonds and providing the investor with a quick assessment of energy management project options.

The NPV can be expressed in equation form as

$$\mathrm{NPV} = \sum_{t=0}^{N} \frac{F_t}{(1+i')^t} \tag{2.15}$$

where

NPV is the net present value of the capital investment
F_t is the net cash flow received at time t in current dollars
i' is the nominal discount rate.

As mentioned before, IRR is the rate of return that will make the NPV equal to zero. In equation form,

$$\text{NPV} = 0 = \sum_{t=0}^{N} \frac{F_t}{(1+\text{IRR})^t} \tag{2.16}$$

Thus, IRR is the rate of return that equates the present value of positive and negative cash flows when used as the discount rate.

Example 2.4

As in Example 2.3, an investment of $10,000 is made initially. The O&M cost in the first year is $1,300, and it inflates at a rate of 3%/year thereafter. A 5-year double-declining depreciation schedule is used with a marginal federal income tax rate of 34%. There is no state income tax and no salvage value. The project saves 1,000 units of energy in its first year of operation, 950 in the second, 925 in the third, and 900 in the fourth and fifth years. The energy is sold at $10/unit at year zero and inflates at 3%/year thereafter. The results of the IRR analysis calculations are shown in Table 2.4. A trial-and-error method may be used to obtain the appropriate IRR. The last column in the table demonstrates that if a return of 61% were earned from the after-tax cash flow each year, the total cash flow would equal $10,000, which is the initial investment. Thus, for this example, if the hurdle rate were higher than 61%, then an investment in another project would be appropriate, but if the hurdle rate is lower than 61%, the investment in this project represents a good opportunity.

Potential problems with IRR analyses are described in [2]. One of them is that IRR might overstate profitability because it assumes all interim proceeds are reinvested at a rate equal to the IRR. Hence, for projects of different scales and lives, it is possible for different ranking criteria such as NPV and IRR to produce conflicting results as a consequence of the differing reinvestment assumptions. The IRR method can, however, be modified, and this modified IRR analysis method can be found in [2].

TABLE 2.4
After-Tax IRR

Year	Investment (*a*) ($)	O&M Costs (*b*) ($)	Revenue (*c*) ($)	Depreciation (*d*) ($)	Net Taxable Income $(e)=(c-b-d)$ ($)	Federal Income Tax $(f)=e\times\text{Tax}$ ($)	After-Tax Cash Flow $F_n(g)=(c-a-b-f)$ ($)	$F_n/(1.61)^n$ ($)
0	10,000	0	0	0	N/A	N/A	−10,000	
1	0	1,339	10,300	4,000	4,961	1,687	7,274	4,516
2	0	1,379	10,079	2,400	6,300	2,142	6,558	2,530
3	0	1,421	10,108	1,440	7,247	2,464	6,223	1,490
4	0	1,463	10,130	864	7,802	2,653	6,014	895
5	0	1,507	10,434	518	8,408	2,859	6,068	561

Example 2.5

Estimate the payback time and the IRR for a PV system with rated production capacity of 3 kW (@25°C, 1,000 W/m^2) installed in Boulder, Colorado (40°N latitude, 105°W longitude), in 2008 (see Figure 2.1). The cost of the system was $25,294. The electric utility serving the customer offered a rebate of $4.50/W in order to meet a RE mandate in the Colorado constitution which required that 20% of all electric power be from renewable sources by 2020. Moreover, the federal government offered a tax credit capped at $2,000 for the 2008 tax year.

Use the following assumptions:
Price of electricity in year 0: $0.152/kWh
Average annual rate of power cost increase: 5%/year (nominal terms)
Tilt: 20°
Azimuth: 180°
Productive life: 20 years
Discount rate (nominal): 6%
Inflation rate: 3%

The system's annual energy output was estimated by inputting the system parameters at the National Renewable Energy Laboratory's PVWatts website: http://pvwatts.nrel.gov.

Solution:

To find the value of the annual savings, we first multiply the annual energy output by the price of electricity. We assume annual savings start 1 year after the initial investment is made at year 0. Since the cost of electricity will increase 5%/year, we must project the cost of energy for 20 years. The future cost of energy, F_t, is found from Equation 2.7, using

FIGURE 2.1
Kreith residence, 3 kW Sanyo high-efficiency photovoltaic system. (See also Figure 11.1.)

a 5% annual increase from the original value. In the first year, $F_1=0.152(1+0.05)^1=\$0.16$. Then in the second year, as for each successive year, the value of F_t for the previous year becomes the value of F_{t-1}. So in the 20th year, electricity will cost \$0.40/kWh. Then, the electricity costs for each year are used to calculate annual savings by multiplying by 4,207 kWh/year.

The future costs of energy for each of the next 20 years are shown in Table 2.5.

The annual values of electricity produced by the PV system are found by multiplying each year's respective cost of energy by the annual energy output. The results are shown in Table 2.6.

The savings this electricity production represents will be cumulative and must be adjusted for the time value of money. To find the cumulative savings, simply add all of the annual savings values. After 20 years of production, the cumulative savings is \$22,813. With future savings in current dollars, the nominal discount rate of 6% is used in Equation 2.4 to adjust the yearly savings back to the NPV of money. After each year's savings are adjusted for the time value of money, they are added together to find the cumulative NPV savings. The cumulative NPV savings of each year is shown in Table 2.7. After 20 years, the total cumulative NPV savings is \$12,219.

After the deduction of the utility's rebate and the federal tax credit, the net capital cost to the customer is \$25,294 − (\$4.50 × 3,000) − \$2,000 = \$9,794. After 20 years, assuming a 6% discount rate the customer has made a profit of \$2,425. The PP is calculated by comparing the cost of the system to the customer and the value of the energy saved per year. From Table 2.7, we can see that the adjusted cumulative savings (the cost of the energy saved) equals the net cost of the system to the customer between the 15th and 16th years of operation. Therefore, the payback time of the system for the customer with the

TABLE 2.5

Future Cost of Energy

Year	**1**	**2**	**3**	**4**	**5**	**6**	**7**	**8**	**9**	**10**
Price/kWh (\$)	0.15	0.16	0.17	0.18	0.18	0.19	0.20	0.21	0.22	0.24
Year	**11**	**12**	**13**	**14**	**15**	**16**	**17**	**18**	**19**	**20**
Price/kWh (\$)	0.25	0.26	0.27	0.29	0.30	0.32	0.33	0.35	0.37	0.38

TABLE 2.6

Value of Electricity Produced

Year	**1**	**2**	**3**	**4**	**5**	**6**	**7**
kWh	4,207	4,207	4,207	4,207	4,207	4,207	4,207
Price/kWh (\$)	0.16	0.17	0.18	0.18	0.19	0.20	0.21
Annual savings (\$)	670.55	704.08	739.29	776.25	815.06	855.82	898.61
Year	**8**	**9**	**10**	**11**	**12**	**13**	**14**
kWh	4,207	4,207	4,207	4,207	4,207	4,207	4,207
Price/kWh (\$)	0.22	0.24	0.25	0.26	0.27	0.29	0.30
Annual savings (\$)	943.54	990.71	1,040.25	1,092.26	1,146.87	1,204.22	1,264.43
Year	**15**	**16**	**17**	**18**	**19**	**20**	
kWh	4,207	4,207	4,207	4,207	4,207	4,207	
Price/kWh (\$)	0.32	0.33	0.35	0.37	0.38	0.40	
Annual savings (\$)	1,327.65	1,394.03	1,463.73	1,536.92	1,613.77	1,696.69	

TABLE 2.7
Annual Adjusted Net Savings

Year	1	2	3	4	5
Annual savings ($)	670.55	704.08	739.25	776.25	815.06
Cumulative savings ($)	1,309.17	2,013.25	2,752.50	3,528.75	4,343.81
Adjusted annual savings ($)	632.59	626.63	620.69	614.86	609.06
Adjusted cumulative savings ($)	1,271.21	1,897.84	2,518.53	3,133.39	3,742.45
Year	**6**	**7**	**8**	**9**	**10**
Annual savings ($)	855.82	898.61	943.54	990.71	1,040.25
Cumulative savings ($)	5,199.63	6,098.24	7,041.78	8,032.49	9,072.74
Adjusted annual savings ($)	603.32	597.63	591.99	586.40	580.87
Adjusted cumulative savings ($)	4,345.77	4,943.40	5,535.39	6,121.79	6,702.66
Year	**11**	**12**	**13**	**14**	**15**
Annual savings ($)	1,092.26	1,146.87	1,204.22	1,264.43	1,327.65
Cumulative savings ($)	10,165.00	11,311.87	12,516.09	13,780.52	15,108.17
Adjusted annual savings ($)	575.39	569.96	564.59	559.26	553.98
Adjusted cumulative savings ($)	7,278.05	7,848.01	8,412.59	8,971.85	9,525.83
Year	**16**	**17**	**18**	**19**	**20**
Annual savings ($)	1,394.03	1,463.73	1,536.92	1,613.77	1,696.69
Cumulative savings ($)	16,502.20	17,965.93	19,502.85	21,116.62	22,813.31
Adjusted annual savings ($)	548.75	543.58	538.45	533.37	529.04
Adjusted cumulative savings ($)	10,074.59	10,618.17	11,156.62	11,689.99	12,219.02

applied incentive and tax rebate is about 15.5 years. Note that the cost of the system is not paid back to the owner through energy savings without the incentives.

The IRR, found by using Equation 2.16, represents the summation of the discounted cash flows that will equal the initial investment. It is found by iterating on the discount rate until the summation of the yearly discounted cash flows equals zero. The results are shown in Table 2.8, which shows the calculation for a nominal discount rate of 7.81%, representing the IRR.

2.7 Capital Recovery Factor

In many financial transactions, a *series* of annual or monthly sums is invested or used to pay off a loan. The series of equal periodic payments necessary to pay off such a loan is determined as follows. If a uniform payment P_{ann} is invested each year at interest rate i, the present worth of the *sum* of these payments S is

$$S = \frac{P_{ann}}{1+i} + \frac{P_{ann}}{(1+i)^2} + \frac{P_{ann}}{(1+i)^3} + \cdots, \tag{2.17}$$

or

$$S = P_{ann}\left[(1+i)^{-1} + (1+i)^{-2} + (1+i)^{-3} + \cdots + (1+i)^{-t}\right]$$

TABLE 2.8

Yearly Cash Flows for PV System IRR

Year	Net Investment ($)	O&M Cost ($)	Annual Savings ($)	Cash Flow Discounted at 7.81% ($)
0	9,794	0	–9,794	–9,794.00
1	0	0	670.55	621.97
2	0	0	704.08	605.76
3	0	0	739.25	589.95
4	0	0	776.25	574.60
5	0	0	815.06	559.62
6	0	0	855.82	545.04
7	0	0	898.61	530.83
8	0	0	943.54	517.00
9	0	0	990.71	503.52
10	0	0	1,040.25	490.40
11	0	0	1,092.26	477.61
12	0	0	1,146.87	465.16
13	0	0	1,204.22	453.04
14	0	0	1,264.43	441.23
15	0	0	1,327.65	429.73
16	0	0	1,394.03	418.53
17	0	0	1,463.73	407.62
18	0	0	1,536.92	397.00
19	0	0	1,613.77	386.65
20	0	0	1,696.69	377.07
Summation of cash flows discounted at 7.81%				~0

The expression in brackets is a geometric series with the first term $P_{ann}/(1+i)$ and ratio $(1+i)^{-n}$. It converts equal payments made at the end of each of t consecutive periods to the equivalent amount in the present using an interest or discount rate i. It follows from the expression for the sum of such a series that

$$P_{ann} = S\left[\frac{i}{1-(1+i)^{-t}}\right] \quad (2.18)$$

where P_{ann} is the annual payment required to achieve a sum S, after t years. The term in brackets in Equation 2.18 is called the *capital recovery factor* (CRF(i, t)), which can be rearranged as

$$\mathrm{CRF}(i,t) = \left[\frac{i(1+i)^t}{(1+i)^t-1}\right] \quad (2.19)$$

Table 2.9 is a summary tabulation of CRFs. The term P_{ann} is the annual or periodic payment on a self-amortizing loan of value S. Each payment is a mix of interest and principal repayment. Early payments are mostly interest because of the large outstanding balance; later payments are primarily principal repayment. Most solar systems owned by private firms or individuals are purchased with self-amortizing loans. This repayment process is also widely used in paying for the mortgage loan on a house.

TABLE 2.9

Capital Recovery Factors (CRF[*i*, *t*])

	Interest Rate, *i*									
t	0%	2%	4%	6%	8%	10%	12%	15%	20%	25%
1	1.00000	1.02000	1.04000	1.06000	1.08000	1.10000	1.12000	1.15000	1.20000	1.25000
2	0.50000	0.51505	0.53020	0.54544	0.56077	0.57619	0.59170	0.16152	0.65455	0.69444
3	0.33333	0.34675	0.36035	0.37411	0.38803	0.40211	0.41635	0.43798	0.47473	0.51230
4	0.25000	0.26262	0.27549	0.28859	0.30192	0.31547	0.32923	0.35027	0.38629	0.42344
5	0.20000	0.21216	0.22463	0.23740	0.25046	0.26380	0.27741	0.29832	0.33438	0.37184
6	0.16667	0.17853	0.19076	0.20336	0.21632	0.22961	0.24323	0.26424	0.30071	0.33882
7	0.14286	0.15451	0.16661	0.17914	0.19207	0.20541	0.21912	0.24036	0.27742	0.31634
8	0.12500	0.13651	0.14853	0.16101	0.17401	0.18744	0.20130	0.22285	0.26061	0.30040
9	0.11111	0.12252	0.13449	0.14702	0.16008	0.17364	0.18768	0.20957	0.24808	0.28876
10	0.10000	0.11133	0.12329	0.13587	0.14903	0.16275	0.17698	0.19925	0.23852	0.28007
11	0.09091	0.10218	0.11415	0.12679	0.14008	0.15396	0.16842	0.19107	0.23110	0.27349
12	0.08333	0.09156	0.10655	0.11928	0.13270	0.14676	0.16144	0.18148	0.22526	0.26845
13	0.07692	0.08812	0.10014	0.11296	0.12652	0.14078	0.15568	0.17911	0.22062	0.26454
14	0.07143	0.08260	0.09467	0.10758	0.12130	0.13575	0.15087	0.17469	0.21689	0.26150
15	0.06667	0.07783	0.08994	0.10296	0.11683	0.13147	0.14682	0.17102	0.21388	0.25912
16	0.06250	0.07365	0.08582	0.09895	0.11298	0.12782	0.14339	0.16795	0.21144	0.25724
17	0.05882	0.06997	0.08220	0.09544	0.10963	0.12466	0.14046	0.16537	0.20944	0.25576
18	0.05556	0.06670	0.07899	0.09236	0.10670	0.12193	0.13794	0.16319	0.20781	0.25459
19	0.05263	0.06378	0.07614	0.08962	0.10413	0.11955	0.13576	0.16134	0.20646	0.25366
20	0.05000	0.06116	0.07358	0.08718	0.10185	0.11746	0.13388	0.15976	0.20536	0.25292
25	0.04000	0.05122	0.06401	0.07823	0.09368	0.11017	0.12750	0.15470	0.20212	0.25095
30	0.03333	0.04465	0.05783	0.07265	0.08883	0.10608	0.12414	0.15230	0.20085	0.25031
40	0.02500	0.03656	0.05052	0.06646	0.08386	0.10226	0.12130	0.15056	0.20014	0.25003
50	0.02000	0.03182	0.04655	0.06344	0.08174	0.10086	0.12042	0.15014	0.20002	0.25000
100	0.01000	0.02320	0.04081	0.06018	0.08004	0.10001	0.12000	0.15000	0.20000	0.25000
∞		0.02000	0.04000	0.06000	0.08000	0.10000	0.12000	0.15000	0.20000	0.25000

Note: For interest rates *i* from 0% to 25% and for periods of analysis *n* from 1 to 100 years.

2.8 Levelized Cost of Energy

The levelized cost of energy (LCOE) is widely used for comparing different energy generation technologies. For example, LCOE could be used to compare the cost of electric energy from a renewable source with that from a nuclear generating unit. The LCOE is defined as the cost that, if assigned to every unit of energy produced (or saved) by the system over the analysis period, will equal the TLCC when discounted back to the base year. The LCOE is the present value of all the investment costs plus operation and maintenance costs (including fuel) in each future year per unit power generation, which is also discounted to the present time:

$$\text{LCOE} = \frac{\sum_{t=1}^{N} \frac{C_t}{(1+i)^t}}{\sum_{t=1}^{N} \frac{Q_t}{(1+i)^t}} \tag{2.20}$$

where

Q_t is the energy produced by the system (e.g., electricity in kWh) in year t (or the energy savings from conservation measures)

N is the analysis period in years

C_t is the project cost in year t, including finance charges, operating and maintenance costs and fuel costs, less tax savings and incentives as applicable.

LCOE can use the real discount rate or the nominal discount rate. The nominal discount rate is needed to compare the LCOE to future energy prices in current dollars, while the real discount rate applies for comparisons in constant dollars. Substituting Equation 2.11, we can calculate the LCOE in terms of the TLCC:

$$\text{LCOE} = \frac{\text{TLCC}}{\sum_{t=1}^{N}\left(Q_t/(1+i)^t\right)} \tag{2.21}$$

Note that, if the system energy output, Q_t (or the energy savings), is constant over the time of analysis, the relation for LCOE simplifies to

$$\text{LCOE} = \left(\frac{\text{TLCC}}{Q}\right) \times \text{CRF} \tag{2.22}$$

where Q is the annual energy output or energy saving in the case of a conservation system. This form of the equation illuminates the term "levelized cost" as the CRF takes the TLCC of the project and distributes it as uniform annual cost over the project lifetime.

As one illustration, using the data from Example 2.2 and assuming the windows save 120,000 kWh/year, the uniform capital recovery factor for $i'=0.12$ and $t=5$ is 0.277 (Table 2.9) and, therefore, the nominal LCOE for energy saved by the windows is

$$\text{LCOE} = (\$15{,}091/120{,}000) \times 0.277 = \$0.0348/\text{kWh}$$

The next example illustrates the use of LCOE for comparing a conservation measure to a less efficient alternative.

Example 2.6

Calculate the TLCC and LCOE for replacing a 75 W incandescent light bulb with a 40 W fluorescent bulb. Assume the incandescent bulb operates 6 hours per night and needs to be replaced every year at a cost of $1.00, while the fluorescent bulb lasts for 5 years with an initial cost of $15. The cost of electricity over this period is $0.08/kWh. Assume a discount rate of 12%.

Solution:

The yearly energy requirement for the incandescent lamp is 164 kWh, while the 5-year discounted O&M cost, that is, the total discounted cost of annual replacement, is

$$\$1\left[1+\left(\frac{1}{1.12}\right)+\left(\frac{1}{1.12}\right)^2+\left(\frac{1}{1.12}\right)^3+\left(\frac{1}{1.12}\right)^4\right] = \$4.03$$

The discounted cost of electricity at the end of the first year is

$$\text{Cost of Electricity} = \frac{164\text{kWh}}{\text{year}} \times \frac{\$0.08}{\text{kWh}} \times \left(\frac{1}{1.12}\right)^1 = \$11.71$$

The following table shows the discounted cost of electricity after every year and the total discounted cost of electricity assuming that the electricity is paid at the end of each year.

75 W Incandescent Light Bulb						
Year	**1**	**2**	**3**	**4**	**5**	**Total**
Cost of electricity	\$11.71	\$10.46	\$9.34	\$8.34	\$7.44	\$47.29

The TLCC for the incandescent light bulb can now be calculated to be

$$\text{TLCC} = \$4.03 + \$47.29 = \$51.32$$

The 40 W fluorescent bulb will require 87.6 kWh/year. The discounted electric power cost is \$25.26 if the electricity is also paid at the end of each year. The following table shows the same information as the earlier table, but for the fluorescent bulb.

40 W Fluorescent Light Bulb						
Year	**1**	**2**	**3**	**4**	**5**	**Total**
Cost of electricity	\$6.26	\$5.59	\$4.99	\$4.45	\$3.98	\$25.26

Therefore, the TLCC for the fluorescent light bulb is

$$\text{TLCC} = \$15 + \$25.26 = \$40.26$$

Thus, the net 5-year savings is \$11.06 and the CRF from Table 2.9 is 0.277. Hence, the levelized cost of the saved energy is

$$\text{LCOE} = \frac{\$11.06}{76.4\text{kWh}} \times 0.277 = \$0.04/\text{kWh}$$

Note that LCOE for an energy conservation measure should be based on the incremental costs and savings for the system over the analysis period.

More extensive economic analyses, including taxes, incentives, depreciation, and salvage value, are presented in [2].

2.9 Societal and Environmental Costs

The use of conventional technologies, such as coal-fired power plants, to generate electricity creates environmental degradation and societal costs that are usually considerably higher than those from most renewable technologies. These hidden costs are called *externalities* and are not included in conventional energy economics. Externalities are defined as "an activity of one agent (for example, an individual or an organization, such as a company) that affects the well-being of another agent and occurs outside the market mechanism" [3]. Many different studies have estimated external costs of providing energy. The results vary widely because they depend on many factors that are difficult to pinpoint. For example, what is the appropriate discount rate for long-term effects? How does one assign a monetary value to items such as human health or life? How does one treat the effects of global warming?

A representative set of external cost estimates for electricity systems is shown in Figure 2.2 [4]. The estimates are from the European Commission's Externalities of Energy (ExternE) project [5]. The project applied a systematic method for estimating external costs for a range of electricity, heating, and transportation technologies. ExternE focused on the effects of air pollutant emissions, including human health effects and effects on crops, materials, and climate. Results in Figure 2.2 are representative of European electricity systems in the late 1990s. For coal, the external costs were estimated at about 6 cents/kWh. Forty to fifty percent of the costs were from climate change, with the balance mainly from premature mortality associated with particulate air pollution. A methodology for evaluating the economic cost of the environmental impacts of energy generation is presented in [6].

For the United States, a 2010 study by the National Research Council [3] found average external costs of coal-fired electricity generation of 3.2 cents/kWh, accounting only for conventional air pollutant effects, not climate change. Across the population of coal-fired power plants in the United States, these costs varied from 0.19 to 12 cents/kWh, depending on control technology and location relative to surrounding population. For comparison, residential electricity prices in the United States in 2010 averaged about 12 cents/kWh [7]. For conventional gasoline-fueled passenger vehicles, the same study [3] estimated that external health costs of air pollution (not considering climate change) averaged 1.3 cents/mile traveled, with a range from 0.3 to 5 cents/mile. Again, for comparison, at $3/gallon of gasoline, fuel costs for a car getting 27 miles/gallon would be about 10 cents/mile.

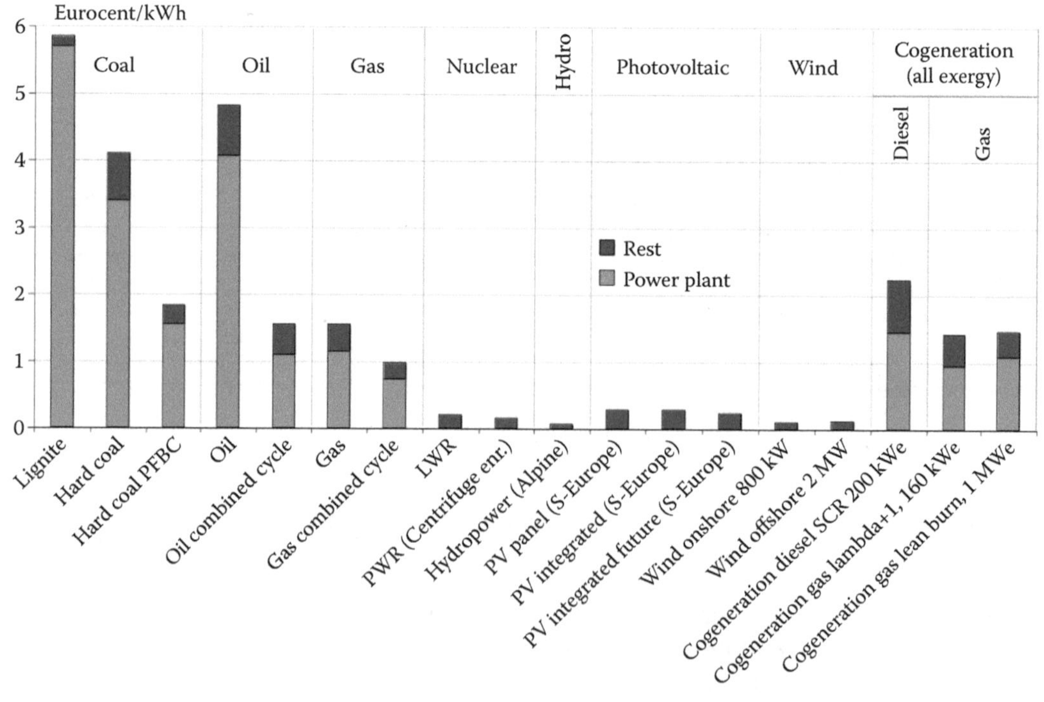

FIGURE 2.2
External costs (€/MWh) of current and more advanced electricity systems associated with emissions from the operation of the power plant and the rest of the fuel-supply chain [5]. "Rest" is the external cost related to the fuel cycle (1 € = 1.3 US$ approximately). (Reprinted from IPCC [4].)

As with direct costs used for project evaluation, external cost estimates are sensitive to the choice of discount rates used in the analysis. Discount rates come into play in placing a dollar value on externalities because the effects of pollution, such as climate change or cancer, may lag the activity that produces emissions by many years. Though this approach is not without controversy, economists use discounting to compare the benefits of an activity such as electricity generation versus the external costs it imposes across time, assuming near-term consequences are more costly to society than impacts that will not be felt until sometime in the future. Societal discount rates for U.S. government applications have been estimated based on the inflation-adjusted interest rate for 10-year U.S. Treasury Securities, resulting in real discount rates of 1.5%–3%. The ExternE study cited above used 3% for its central estimates.

The Social Costs of Greenhouse Gases (SC-GHG) are widely used external cost metrics for analysis of energy systems and environmental regulations. SC-GHG are estimates of the economic damages from emitting one additional metric ton of a greenhouse gas in a specific year. In the United States, SC-GHG estimates are used by local, state, and federal governments to inform policy and investment decisions. Other countries use similar metrics, although the values they adopt may differ. The SC of a greenhouse gas such as carbon dioxide or methane is estimated by predicting future greenhouse gas emissions based on population, economic growth, and other factors; modeling future climate responses, such as temperature and precipitation changes and sea level rise; assessing the economic impact of these changes on agriculture, health, energy use, and other aspects of the economy; then converting the future damages to a present value via discounting and summing them up. Each of these steps is highly complex and subject to significant uncertainty, so SC-GHG estimates are best viewed as a range of values rather than as a single number. SC-GHG values are also likely to change over time as modeling methodologies improve. Table 2.10 presents SC estimates for CO_2, CH_4, and N_2O (nitrous oxide) developed by the U.S. Environmental Protection Agency (EPA) for use in evaluating the economic benefits of reducing greenhouse gas emissions [8]. EPA notes these estimates likely underestimate the costs of GHG emissions (or benefits of reducing emissions) because current damage assessment models for climate change omit or undercount consequences of altered precipitation, changes in frequency or severity of extreme weather, ocean acidification, and biodiversity loss. Methods to account for these effects are being studied and are expected to lead to revised SC-GHG estimates in the future.

TABLE 2.10

Recent Estimates of the Social Cost of Greenhouse Gases

	SC-CO_2 (2020 US$ per Metric Ton CO_2)			SC-CH_4 (2020 US$ per Metric Ton CH_4)			SC-N_2O (2020 US$ per Metric Ton N_2O)		
	Near-Term Discount Rate			Near-Term Discount Rate			Near-Term Discount Rate		
Emission Year	2.5	2	1.5	2.5	2	1.5	2.5	2	1.5
2020	120	190	340	1,300	1,600	2,300	35,000	54,000	87,000
2030	140	230	380	1,900	2,400	3,200	45,000	66,000	100,000
2040	170	270	430	2,700	3,300	4,200	55,000	79,000	120,000

Source: U.S. Environmental Protection Agency [8].

Problems

2.1 What is the present value of a \$10,000 payment 5 years from now if the payment is in current dollars and the nominal discount rate is 7%?

2.2 Repeat the previous problem if the \$10,000 payment is in constant dollars, the nominal discount rate is 7% and the inflation rate is 4%/year.

2.3 Calculate the annual payments on a \$10,000 loan for a residential solar system. Assume that the interest rate on the loan is 8% and the mortgage term is 15 years.

2.4 A home buyer obtains a mortgage of \$100,000 at an interest rate of 8% over 20 years. Estimate the annual payments to repay this loan.

2.5 You take out a \$300,000 loan to buy a new home with an 8% interest rate and a 30-year term. Estimate the annual payments to repay the loan assuming fixed annual payments (i.e., a self-amortizing loan). (a) Show the solution by calculating the capital recovery factor from Equation 2.19. (b) Find the CRF from Table 2.9, and check your answer from part (a).

2.6 A rooftop solar hot water heater with collector area of 8 m^2 and storage of 640 L is to be installed in Phoenix, Arizona, on a house that uses 160 L/day of hot water at 60°C. Of the total energy required, 70% will be supplied by solar energy (the solar fraction is 70%) and 30% is supplemented with an existing electric heater. The solar system costs \$3,000 for installation. Assuming the electricity cost is \$0.10/kWh and does not change over time, estimate the simple payback time for the solar hot water heater. The temperature of the water in the main from which it is to be heated is 10°C.

2.7 Repeat the previous problem but assume the cost of installation is to be repaid at 7% interest over 15 years and that the electricity costs escalate at 10%/year. Calculate annual cash flows in current dollars accounting for the loan payments and electricity savings. Then identify the year when the NPV turns positive. Is the solar system a good investment relative to the payback time?

2.8 Compare the capital costs and the levelized costs of electricity production for a 1,000 MW nuclear power plant and a 1,000 MW PV system constructed in the Mojave Desert in California. The PV system is to use the best monocrystalline silicon technology currently available. Neglect the cost of nuclear fuel. Comment on the environmental impacts and availability of water.

2.9 Calculate and tabulate the annual cash flows associated with a solar system for a 10-year period of analysis if the initial solar cost is \$6,000 and if
Interest rate = 9%
Power cost for operating the system = \$30/year escalating at 10%/year
Property tax = 0
Income tax bracket = 32%
Maintenance = 0.5% of the capital per year (no escalation)
Scrap value = 50% of initial cost

Assume only interest payments on the loan (not principal payments, depreciation, or other expenses) can be deducted from taxes. If the solar system saves \$550/year in conventional fuel (escalating at 10%/year), is it cost-effective? Work the problem in current dollars.

2.10 Compute the annual payment of a self-amortizing loan for the solar system with the characteristics tabulated as follows:

Factor	Specification
Expected system lifetime t (year)	20
Effective interest rate (%)	8
Collector area A_c (m^2)	20
Collector cost (\$/m^2)	100
Storage cost (\$/m^2)	6.25
Cost of control system (\$)	100
Miscellaneous costs (e.g., pipes, pumps, motors, heat exchangers) (\$)	$200+(5A_c)$

2.11 A factory uses 700 40 W standard fluorescent lamps for its lighting. The lamps are to be replaced by 34 W high-efficiency lamps that can operate on the existing standard ballasts. The standard and high-efficiency fluorescent lamps cost \$1.80 and \$2.50 each, respectively. The factory operates 3,000 h/year, and all of the lamps operate during that period. Assuming the unit cost of electricity is 6 cents/kWh and the ballasts consume 10% of the rated power of the lamps, calculate how much energy and money will be saved as a result of switching to the high-efficiency fluorescent lamps. Then, estimate the PP assuming a nominal discount rate of 6% and an inflation rate of 3%. Finally, calculate the levelized cost of the saved energy in real terms assuming the lamps last for 2 years.

2.12 The hallways of a university are lit by 40 fluorescent lights, each containing two lamps rated at 60 W each. The lights are on all day and night during the year, but the building is only used from 7 a.m. to 7 p.m. 5 days/week during the year. If the price of electricity is 7 cents/kWh, estimate the amount of energy and money that could be saved by installing two motion sensors that turn off the lights when the building is not being used. Then, estimate the simple payback and the payback with a 6% effective interest rate if the price of a sensor is \$50 and the cost of installation is \$40.

2.13 Estimate the levelized cost of electricity in \$/kWh for a new natural gas fired power plant that has an initial capital cost of \$700 million, an expected lifetime of 35 years, and will produce 1.2 billion kWh/year of electricity. Assume there will be no salvage value or decommissioning costs. Total operating costs are \$20 million/year. The power plant developer needs a 12% rate of return to cover financing costs and the expected profit on the investment.

2.14 A farmers' cooperative in southern Colorado wants to install a solar PV farm to cover the electricity needs of their residences and those of the grain elevator the co-op runs. The project would have a 1.5 MW capacity with a capacity factor of 25% and an expected lifetime of 20 years. Assume the rate of inflation over the duration of the project will be 2% and that tax considerations can be neglected. The co-op uses a nominal discount rate of 7% for project decisions. The upfront cost is \$4.5 million for equipment and installation. The co-op has \$500,000 in reserves to use as a down payment for the project. The co-op will take out a 20-year loan at 5% interest for the remaining \$4 million of upfront costs. The loan is self-amortizing and will be paid back with interest in uniform annual payments. Annual O&M costs are expected to be \$2,000/year, escalating at the rate of inflation. After 10 years the co-op anticipates having to replace some of the inverters used in the project at an expected cost of $100,000 in constant (year 0) dollars. The expected decommissioning cost after 20 years is $200,000. (a) What is the amount of the annual "levelized" payment the co-op will need to make each year to pay back the loan? (b) If the nominal discount rate used by the co-op is 7%, what is the real discount rate? (c) Find the present value of the loan payments, annual O&M expenses, equipment replacement, and decommissioning costs. (d) Calculate the TLCC. (e) Calculate the nominal LCOE and the real LCOE. (f) If the co-op can avoid purchasing electricity at $0.11/kWh, escalating at 2%/year, what is the *NPV* for the project?

References

1. Ruegg, R.T., 1997. Economic methods, in: Kreith, F. and West, R.E. eds., *CRC Handbook of Energy Efficiency*, Chapter 3, pp. 101 – 124, CRC Press, Boca Raton, FL.
2. Tester, J.W., Drake, E.M., Golay, M.W., Driscoll, M.J., and Peters, W.A., 2005. *Sustainable Energy: Choosing among Options*, MIT Press, Cambridge, MA.
3. National Research Council Committee on Health, Environmental, and Other External Costs and Benefits of Energy Production and Consumption, 2010. *Hidden Costs of Energy: Unpriced Consequences of Energy Production and Use*, The National Academies Press, Washington, DC.
4. Sims, R.E.H., Schock, R.N., Adegbulugbe, A., Fenhann, J., Konstantinaviciute, I., Moomaw, W., Nimir, H.B., Schlamadinger, B., Torres-Martinez, J., Turner, C., Uchiyama, Y., Vuori, S.J.V., Wamukonya, N., and Zhang, X., 2007. Energy supply. in: Metz, B., Davidson, O.R., Bosch, P.R., Dave, R. and Meyer L.A. eds., *Climate Change 2007: Mitigation. Contribution of Working Group III to the Fourth Assessment Report of the Inter-Governmental Panel on Climate Change*, Chapter 4, pp. 253 – 322, Cambridge University Press, Cambridge, UK.
5. European Union, 2005. Externalities of energy: Extension of accounting framework and policy applications (Extern E-Pol), Final report of project, https://www.externe.info/externe_2006/expoltec.pdf.
6. Rabl, A., and Spadaro, J.V., 2007. Environmental impacts and cost of energy, in: Kreith, F. and Goswami, D.Y. eds., *CRC Handbook of Energy Efficiency and Renewable Energy*, Chapter 4, 4-1 – 4-20, CRC Press, Boca Raton, FL.
7. Energy Information Administration, 2016. Electric power monthly: Average price of electricity to ultimate consumers, https://www.eia.gov/electricity/monthly/epm_table_grapher.cfm?t=epmt_5_3.
8. U.S. Environmental Protection Agency, 2023. *Report on the Social Cost of Greenhouse Gases: Estimates Incorporating Recent Scientific Advances*, National Center for Environmental Economics, Washington, DC.

3

Energy Systems Analysis Methodologies

The objective of energy systems analysis is to understand the energy production and supply system and its relationships to consumer energy, the economy, and the environment. Energy systems analysis is necessary to identify and evaluate sustainability issues and potentially beneficial developments. Figure 3.1 demonstrates the flow of energy and the flow of money through the economy. Money, created and circulated in the economy, is not subject to any physical constraints. In contrast, the energy and resources supplied to the economy by energy transformation systems and the associated environmental impact are subject to both technological and resource constraints. For the development of sustainable energy systems, it is important to evaluate the potential of a proposed energy production system to substitute for fossil fuels. It is also important to evaluate the adaptive potential of the end use and the upstream effects of demand. The energy provided to the end user has to be "economical," meaning that it should be offered at an acceptable market price but must also cover the total production cost including any waste disposal or emissions costs.

Figure 3.1 can be used to conceptualize the entire energy system or one specific subsystem. The definitions listed below the figure can be calculated on an annual, unit, or cumulative basis. Some common characterizations of the energy and resource flows are given in Table 3.1. A flow of energy consumed by an energy transformation plant (S_1)

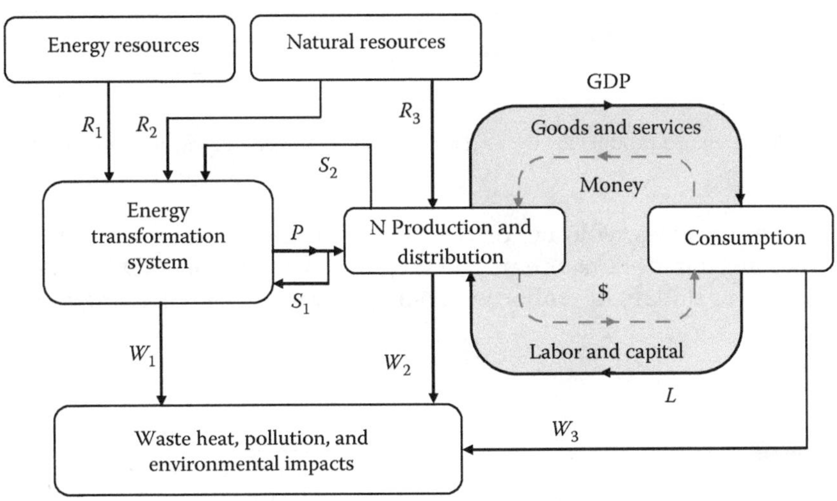

FIGURE 3.1
Schematic of the flows of energy, resources, goods, and money through the economy. R_i are the flows of primary energy and natural resources, P is the consumer energy produced for sale into the economy, S_1 is the energy (electricity or fuel) needed to run the energy transformation system, and S_2 is the embodied energy in equipment and materials needed to extract and process energy. N is the net energy to the economy, GDP refers to dollar or unit values of products and services, L includes the units of labor and investment or taxes, and W_i are the waste flows.

DOI: 10.1201/9781003586616-3

TABLE 3.1

Indicators of Energy and Resource Characteristics of the Economy, a Particular Sector, or a Particular Subsystem Using the Structure in Figure 3.1

Indicator	Parameters	Example
Energy conversion efficiency	$\eta_C = P/R_1$	MWh_e/ton coal
Energy resource intensity	P/R_2	Gal ethanol/gal water
Energy intensity of the economy	P/GDP	kWh/$
Energy return on energy invested	EROI $= P/(S_1 + S_2)$	MJ/MJ
Embodied energy	R_1/GDP	MJ/kg steel
Net energy to the economy	$N = P - (S_1 + S_2)$	GWh electricity or GJ fuels
Externality loading of conversion	W_1/P	Ton-CO_2/MWh
Life cycle assessment	ΣR/GDP	Resources/unit good or service
External costs	ΣW/GDP	Impacts/unit good or service
Impact footprint	W/GDP	Carbon footprint
Energy prosperity	N/P = 1-1/EROI	Rate of return on energy

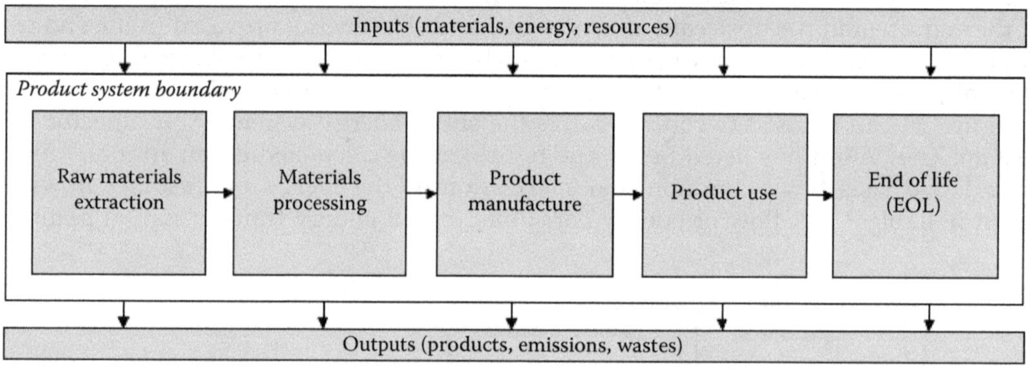

FIGURE 3.2
Cradle-to-grave depiction of a product life cycle. (From U.S. Environmental Protection Agency [1].)

can be measured in mass, volume, or the heating value. A flow of capital equipment, (S_2) can be measured in dollars or embodied energy. Production (GDP) can be measured in units of goods or dollars or embodied energy or resources according to the objective of the analysis.

3.1 Life Cycle Approach

Figure 3.1 implies an approach to energy systems analysis that considers the full life cycle of a product or service, starting from raw material extraction and processing through to disposal, reuse, or recycling at the end of the useful life. Figure 3.2 depicts this in a conventional "cradle-to-grave" format [1]. Though not shown, the goal of sustainability may be further analyzed in a circular "cradle-to-cradle" framework that eliminates end-of-life disposal by recycling or repurposing material. In either case, a life cycle perspective is useful in assessing and comparing economic costs of competing energy technologies or projects and in characterizing their physical attributes, including the use of fossil fuels and other resources and environmental impacts on the biosphere, land, air, and water. Environmental life cycle assessments (LCA) systematically inventory the energy and material requirements that go into producing a product (such as electricity) along with the environmental discharges and wastes requiring disposal, reuse, or recycling. Depending on the scope, LCAs may assess impacts on resource use, air and water quality, climate, or human and ecosystem health. Using a life cycle perspective helps avoid unintended consequences of alternatives that might appear favorable in their direct use phase but could cause harmful impacts upstream or at the point of disposal. Life cycle approaches are also valuable for helping identify opportunities for reducing energy or resource use or minimizing environmental or health impacts.

Some life cycle assessments reach beyond inputs and discharges that directly contribute to making, using, or disposing of a product, to include secondary inputs and outputs that contribute to making or doing something in the supply chain that supports the product of interest. For example, an expanded LCA for electricity generation from wind could go beyond accounting for energy and materials required to manufacture and install a wind turbine to consider the upstream energy and materials required to construct the equipment needed to manufacture the turbine. Extending in a different direction, "consequential" LCA studies examine how a new product or technology might alter energy and resource flows beyond those associated with its direct use, by displacing or enabling other products or technologies. Because LCA studies could theoretically extend without limit in many directions, analysts must carefully choose the study scope and boundaries and must clearly articulate these choices.

The International Organization for Standardization (ISO) has developed standards for conducting environmental LCA, which are published as ISO 14040:2006 [2] and ISO 14044:2006 [3]. The standards detail four key steps to LCA: (1) goal definition and scoping, (2) conducting a life cycle inventory of energy and material inputs and outputs, (3) assessment of the impacts associated with the energy and material flows, and (4) interpretation. The standards emphasize careful selection of a functional unit for the assessment, which should be chosen so that two alternative ways of delivering the same function can be compared on equal footing. For electricity generation technologies, MWh of electricity is a common functional unit; for freight shipping, an appropriate unit might be ton-km moved. For studies that will be used to make public statements comparing alternatives, the standards also emphasize the importance of full documentation of the data collection process, setting and adhering to data quality specifications, and peer review of the assessment.

Figure 3.3 shows estimated life cycle greenhouse gas (GHG) emissions for a range of electricity-generating technologies, which were compiled from hundreds of individual LCA studies published in the literature [4]. The central or median estimate for each technology is indicated by the heavy black bar and shows the significant advantage that renewable electricity technologies have over fossil fuel-based technologies on a life cycle basis. For coal, oil, and natural gas, most emissions come from combustion of the fuel in the power plant. For renewable technologies, life cycle greenhouse gas emissions are generally much lower than for fossil fuels but are not zero, due to emissions associated with other stages in the life cycle, including equipment manufacturing, installation, maintenance, and decommissioning.

Material and energy inventories required for LCA can be done using "bottom-up" process chain analysis (PCA) (also known as process flow analysis), or "top-down" input/output

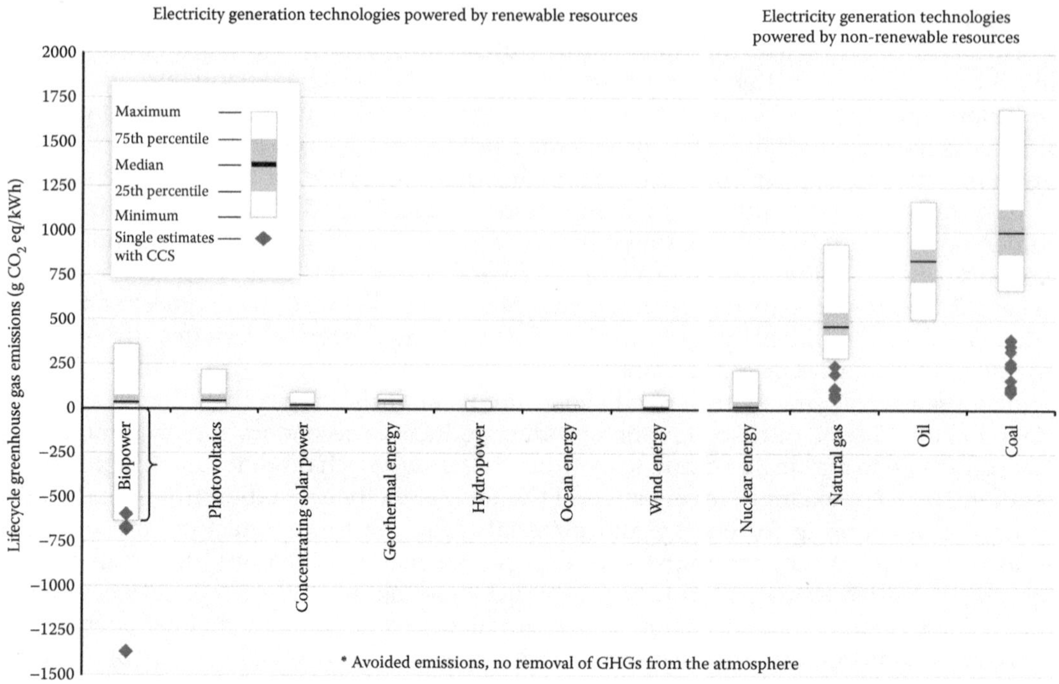

Count of estimates	222(+4)	124	42	8	28	10	126	125	83(+7)	24	169(+12)
Count of References	52(+0)	26	13	6	11	5	49	32	36(+4)	10	50(+10)

FIGURE 3.3
Estimates of life cycle GHG emissions (g CO_2eq/kWh) for broad categories of electricity generation technologies, plus some technologies integrated with carbon capture and sequestration (CCS). Land-use-related net changes in carbon stocks (mainly applicable to biopower and hydropower from reservoirs) and land management impacts are excluded; negative estimates for biopower are based on assumptions about avoided emissions from residues and wastes in landfill disposals and co-products. References and methods for the review are reported in Annex II of reference [4]. The number of estimates is greater than the number of references because many studies considered multiple scenarios. Numbers reported in parentheses pertain to additional references and estimates that evaluated technologies with CCS. Distributional information relates to estimates currently available in LCA literature, not necessarily to underlying theoretical or practical extrema or the true central tendency when considering all deployment conditions. Note 1. "Negative estimates" within the terminology of life cycle assessments presented in this report refer to avoided emissions. Unlike the case of bioenergy combined with CCS, avoided emissions do not remove GHGs from the atmosphere. (From Sathaye et al. [4].)

modeling (I/O), or a hybrid of the two. These approaches are briefly described in Sections 3.2 and 3.3, respectively. Matthews et al. [5] provide an excellent introduction to life cycle assessment that covers the bottom-up, top-down, and hybrid approaches in more detail. They also provide information on both free and commercial databases and software tools for conducting LCA. As a starting point for LCA studies, the National Renewable Energy Laboratory (NREL) provides a free U.S. Life Cycle Inventory (LCI) database that covers commonly used materials, products, and processes (http://www.nrel.gov/lci). NREL LCI data and other LCA data compiled by U.S. agencies can be accessed at the Federal LCA Commons (https://www.lcacommons.gov).

3.1.1 LCA for a Utility-Scale Photovoltaic System

A life cycle assessment for a utility-scale photovoltaic (UPV) system illustrates some of the analysis steps and key assumptions that go into LCA studies. Smith et al. examined life cycle energy demand and greenhouse gas emissions for utility-scale PV installations in the United States, considering three different installation locations representing a range of solar irradiance levels; low, medium, and high-carbon emissions manufacturing supply chains either from China and southeast Asia or from the United States; and end-of-life management scenarios ranging from landfilling to high-materials recovery recycling [6]. In each scenario, they considered a 100 MW-scale system with monocrystalline silicon PV panels having a module efficiency of 20.3%, single-axis tracking, and an assumed lifetime of 30 years. They specified kWh of electricity produced as the functional unit for the analysis. Smith et al. summarized their results in terms of four LCA metrics:

1. **Cumulative energy demand (CED):** the total primary energy consumed over the lifetime of the project, covering the full life cycle shown in Figure 3.2. (Note that the renewable energy in sunlight or wind is not counted in the CED for a renewable energy technology.) CED values are reported in this study as megajoules of oil equivalent primary energy ($MJ_{oil\text{-}eq}$).
2. **Cumulative greenhouse gas emissions:** the sum of greenhouse gases emitted over the full life cycle and lifetime of the project, reported as CO_2 equivalent (CO_2-e) emissions (see Section 3.6).
3. **Energy payback time (EPBT):** the time required for the PV installation to generate an amount of electricity equal to the CED.
4. **Carbon payback time (CPBT):** the time required for the PV system to offset the cumulative GHG emissions associated with the project, assuming the electricity it generates would displace more carbon-intensive electricity from the regional grid.

Figure 3.4 (Figure 2 from Smith et al.) shows the processes covered in the assessment. Smith et al. obtained energy consumption and emissions data for these processes from recent technical literature and original data gathering, in addition to using data from a commercial LCI database. They modeled the system's energy consumption and emissions using openLCA software. They used NREL's System Advisor Model (SAM) to estimate electricity generated by the UPV project. Calculation of displaced GHG emissions for the CPBT metric considered projected changes to the electricity grid mixes in each modeled location over the 30-year project lifetime [6].

Results from Smith et al.'s study are shown in Table 3.2. CED values (shown relative to UPV system electricity generation) and EPBT show that the electricity generation from the modeled UPV systems is highly efficient in terms of primary energy consumption, with a low ratio of primary energy consumed to electric energy generated and an EPBT that is much shorter than the system's lifetime. Lifecycle GHG emissions per kWh are also low in comparison to electricity-generating technologies using fossil fuels, as seen by comparing the results in Table 3.2 with those shown in Figure 3.3. The carbon payback reflects the GHG intensity of electricity from the UPV system relative to that produced from other generating resources on the grid. The CPBT for Seattle is relatively long (though still less than the system lifetime) due in part to the location having relatively low solar irradiance. However, the main reason is that electricity generation in Washington state already has a large contribution from hydroelectric power plants that makes it relatively clean in terms of GHG emissions compared to the mix of power plants supplying electricity in Arizona and Kansas.

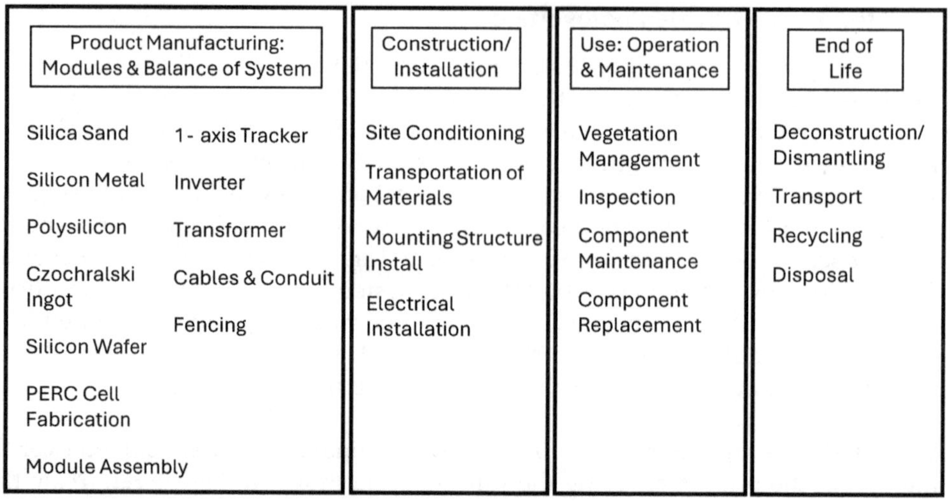

FIGURE 3.4
Processes Included in Life Cycle Assessment for Utility-Scale Photovoltaic Systems [6].

TABLE 3.2
Life Cycle Assessment Results for Utility-Scale Photovoltaic Systems [6]

	Installation Location	CED ($MJ_{oil\text{-}eq}$/ MJ_{UPV})	GHG Emissions (g CO_2-e/kWh)	EPBT (Years)	CPBT (Years)
Low-carbon import, high-EOL materials recovery	Phoenix, AZ	0.05	11	0.5	0.9
Average import, partial-EOL materials recycling	Fredonia, KS	0.07	19	0.6	2.1
High-carbon import, landfill at EOL	Seattle, WA	0.12	36	1.2	20
Low-carbon domestic, high-EOL materials recovery	Phoenix, AZ	0.05	10	0.5	0.8
Average domestic, partial-EOL materials recycling	Fredonia, KS	0.07	14	0.6	1.5
High-carbon domestic, landfill at EOL	Seattle, WA	0.1	30	1.1	16

Source: Adapted from Smith et al. [6].

Beyond the topline metrics shown in Table 3.2, an important result from the LCA is the breakdown of energy consumption and emissions by process within the life cycle. Across the scenarios, polysilicon production, module manufacturing, and tracker manufacturing make the largest contributions to CED and lifecycle GHG emissions, highlighting these processes as targets for future efficiency improvements. Landfilling as the end-of-life disposition of the UPV equipment is the fourth largest contributor unless recycling and materials recovery are assumed. Detailed results showing these contributions for each of the scenarios in Table 3.2 are presented in [6].

3.2 Process Chain Analysis

PCA is a basic tool that can be used for auditing any system or subsystem of a production process or product life cycle. The purpose of PCA is to take into account the energy and resources needed and the waste produced over the different production steps. The PCA approach is used in environmental LCA, embodied energy calculations, supply chain analysis, EROI calculations, and many other types of analysis. PCA can be used to provide conceptual understanding or to calculate specific energy and material flows and costs. In PCA, the production steps are separated, and for each of them, the relevant inputs and outputs are identified and evaluated according to the specific goals of the analysis.

Figure 3.5 shows a schematic of the PCA method applied to a coal fuel production example. The individual, sequential steps or unit operations in a process are separated out, and all inputs and outputs are identified for each process in the sequence. Inputs and outputs include the product as it is transformed, the utilities, U, services, S, materials, M,

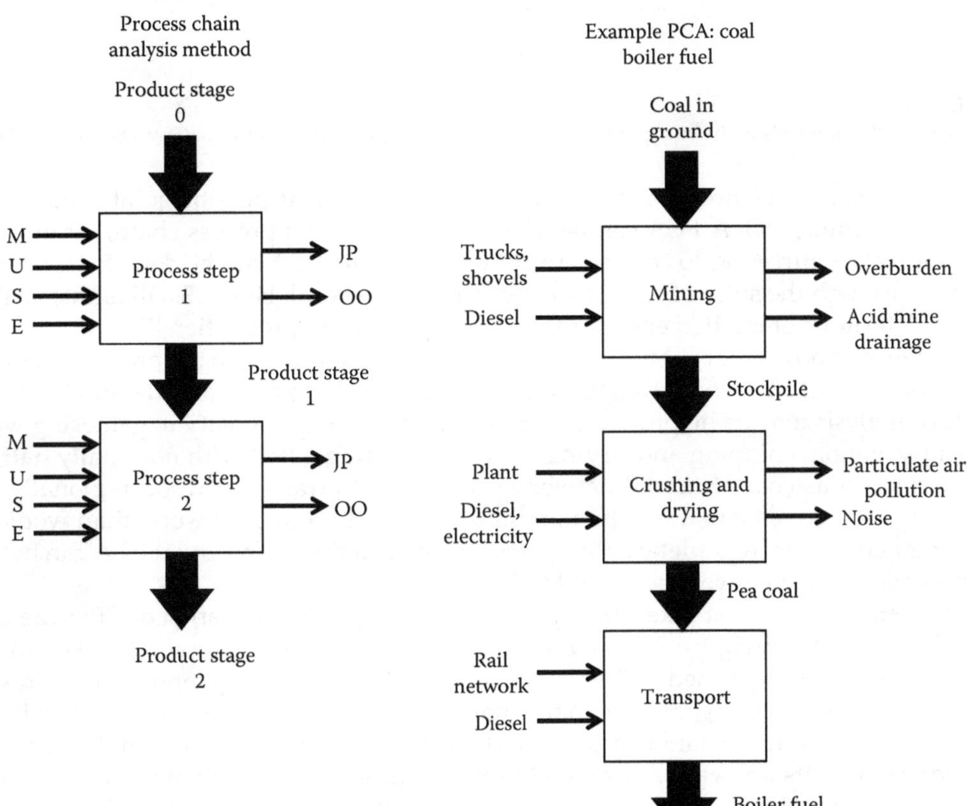

FIGURE 3.5
Illustration of the concept of process chain analysis (PCA) with an example of qualitative assessment of the transformation of coal deposits to delivered pea coal boiler fuel.

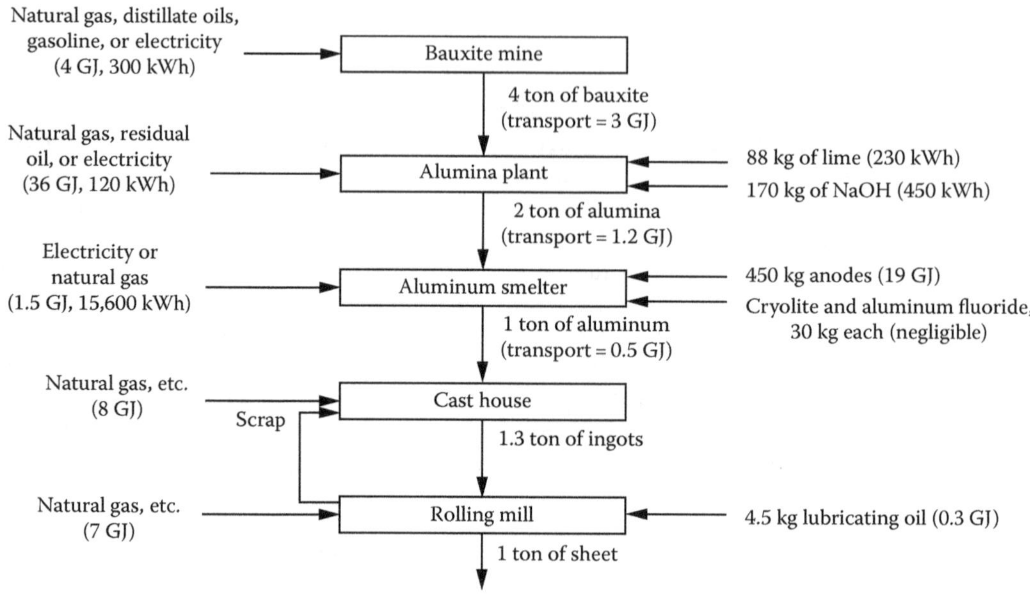

FIGURE 3.6
Quantitative PCA analysis of the energy use in the production of primary aluminum. (From Boercker [7].)

and equipment, E, that are required for production, as well as possibly joint products, JP, and other outputs, OO. A PCA can be done for any part of a process chain, for example from natural resources, R, to consumer products, P, or the PCA can be done for the whole life cycle through the steps of consumer end use and disposal. Figure 3.6 illustrates a PCA used to determine embodied energy for primary aluminum production [7].

The example shown for coal fuel production could be compared to the product chain for an alternative fuel like wood. Depending on the regional locations and the grade of coal, the PCA analysis may be helpful in understanding the energy penalty in gathering wood from a plantation, chipping and drying, and transporting a fuel with nominally half the energy density as coal. If no rail network is available to transport wood regionally, the way there is for coal, more diesel fuel might be needed to transport wood than would be needed for coal. This hypothetical illustrates how even a basic conceptual PCA can help to inform energy system investments and policy.

In PCA applications that take a life cycle approach, such as calculation of CED, the PCA approach depicted in Figure 3.5 is conventionally extended to the point that key inputs and outputs can be expressed as "elementary" flows [5]. Elementary inputs are defined as material or energy entering the system that has been drawn from the environment without previous human transformation (e.g., groundwater from an aquifer or coal in the ground). Elementary outputs are wastes released to the environment that will not undergo subsequent human transformation (e.g., CO_2 emissions). These flows would then be expressed in elementary units of energy or mass of material.

Whether in terms of elementary flows or intermediate products like electricity, evaluating the relevant inputs and outputs for each step of the process can be tedious and complicated if exact figures for a sector or specific product are desired. When PCA is used to evaluate hypothetical systems, assumptions must be made about possible technologies where none actually exists for reference. The results can vary widely depending on the assumptions made. For discussion of the assumptions and possible pitfalls of PCA, the reader is referred to [5]. The NREL LCI database and other databases described in [5] provide PCA data on material and energy requirements for thousands of finished and intermediate products and unit processes. These databases are excellent starting points for assembling PCA data for screening purposes and assessing whether a particular intermediate or unit operation warrants further study.

3.3 Input–Output (I/O) Analysis

Economists use input–output (I/O) analysis of money flowing through the economy to characterize the effect of production in one sector on the production in other sectors. Industries use the products of other industries to produce their own products that are in turn used by the other industries. In keeping with a life cycle approach to sustainability, it can be critical to account for these supply chain dependencies in considering the resource, energy, and material consequences of alternative products or technologies.

The results of a simple hypothetical input–output analysis illustrate the possible "blow-up" result of production chain loops between sectors. Consider a hypothetical isolated economy that wants to produce synthetic transportation fuel from its domestic coal supply. Let's say it takes 3 tons of coal to make a ton of synthetic fuel and that the direct synthetic fuel requirement for the mining, crushing, and transportation of coal is 0.2 tons of fuel per ton of coal. Assume the final demand for coal in the economy is 1,000 tons and the final demand for synthetic fuel is 10 tons. In order to make the synthetic fuel to meet the final demand, along with the fuel needed for coal mining, we actually need to produce 10 + (1,000)(0.2) = 210 tons of fuel. But making the extra 200 tons of fuel takes 600 tons of coal. Now that extra coal would require another (600)(0.2) = 120 tons of fuel, which would in turn require an additional (120)(3) = 360 tons of coal. This causal loop of internal production demand is shown in Figure 3.7, together with seven rounds of incremental accounting of the internal demand.

A more convenient way to assess the chain of indirect production requirements is to solve the causal loops mathematically. To do this in our two-input case, consider the following:

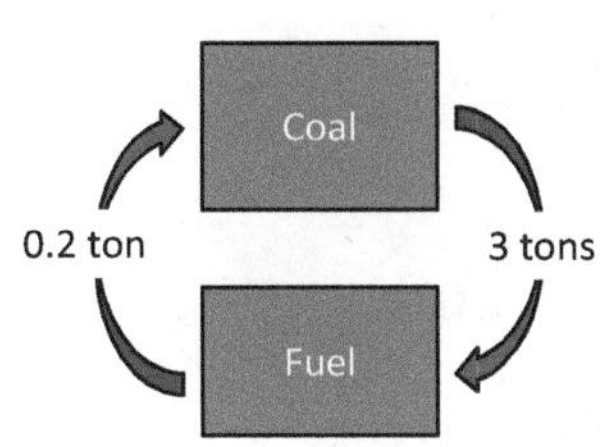

Coal	Fuel
1000	10
30	200
600	6
18	120
360	3.6
10.8	72
216	2.16
6.48	43.2
2241.3	456.9

FIGURE 3.7
Causal loop diagram for coal and synthetic fuel illustrating the possible "blow-up" effect for the baseline production rates of different sectors to produce a unit of output in another sector.

- The coal production process produces 1 ton of coal for every 0.2 tons of synthetic fuel as input
- The synthetic fuel production process produces 1 ton of fuel for every 3 tons of coal as input

We want to find the total amount of coal and synthetic fuel needed to produce a net final product of 1,000 tons of coal and 10 tons of fuel. We define two variables to solve for:

x_1 = total amount of fuel in tons that must be produced to achieve the net final product

x_2 = total amount of coal in tons that must be produced to achieve the net final product.

The two processes provide us with two linear equations in the two unknowns.

Coal production process:

$$\text{Amount of coal produced} - \text{amount of coal consumed} = \text{net coal product}$$

$$x_2 \text{ tons coal} - 3 \text{ tons coal/ton fuel} \times x_1 \text{ tons fuel} = 1{,}000 \text{ tons coal}$$

or

$$x_2 - 3x_1 = 1{,}000 \tag{3.1}$$

Synthetic fuel production process:

$$\text{Amount of fuel produced} - \text{amount of fuel consumed} = \text{net fuel product}$$

$$x_1 \text{ tons fuel} - 0.2 \text{ tons fuel/ton coal} \times x_2 \text{ tons coal} = 10 \text{ tons fuel}$$

or

$$x_1 - 0.2x_2 = 10 \tag{3.2}$$

Rearranging, the two equations are:

$$-3x_1 + x_2 = 1{,}000 \tag{3.3}$$

$$x_1 - 0.2x_2 = 10 \tag{3.4}$$

To solve these two linear simultaneous equations for the two unknowns, we can use the elimination method. Multiplying the second equation by 3 and adding to the first equation leaves us with a single equation in x_2 only:

$$0.4x_2 = 1{,}030 \tag{3.5}$$

$$x_2 = 2{,}575 \text{ tons coal} \tag{3.6}$$

Inserting the solution for x_2 into either equation and solving for x_1 yields:

$$x_1 = 525 \text{ tons fuel} \tag{3.7}$$

For more complex problems, a matrix solution is typically used. Solving this same problem in matrix form [5]:

$$\begin{bmatrix} -3 & 1 \\ 1 & -0.2 \end{bmatrix}\begin{bmatrix} x_1 \\ x_2 \end{bmatrix}=\begin{bmatrix} 1{,}000 \\ 10 \end{bmatrix} \tag{3.8}$$

In generalized matrix notation, this equation has the form $AX = Y$. The square matrix, A, is called the technology matrix. Requirements for synthetic fuel and coal to produce a specified net output for the economy can be found by replacing the entries in the total demand vector, Y, with the specified amounts, and solving for $X = A^{-1}Y$. In our case, the inverse matrix (which is determined by a method similar to the elimination method used above) is:

$$A^{-1}=\begin{bmatrix} 0.5 & 2.5 \\ 2.5 & 7.5 \end{bmatrix} \tag{3.9}$$

Production of 1,000 tons of coal for the economy and 10 tons of fuel requires:

$$\begin{bmatrix} x_1 \\ x_2 \end{bmatrix}=\begin{bmatrix} 0.5 & 2.5 \\ 2.5 & 7.5 \end{bmatrix}\begin{bmatrix} 1{,}000 \\ 10 \end{bmatrix}=\begin{bmatrix} 525 \\ 2{,}575 \end{bmatrix} \tag{3.10}$$

Thus, for this illustrative example to produce a net 1,000 tons of coal and 10 tons of synthetic fuel requires a total of 2,575 tons of coal from the coal production process and 525 tons of fuel from the synthetic fuel production process. Of the total coal produced, 2,575 – 1,000 = 1,575 tons goes to making fuel. Of the total fuel produced, 525 – 10 = 515 goes to making coal. The cumulative causal loop sequence in Figure 3.7 would eventually converge to the values shown in Table 3.10.

The implications of this analysis are interesting. If we just looked at the unit production requirement for synthetic fuel, we might reason that the production of 10 tons of fuel requires 30 tons of coal. But when the production chain analysis is taken to the limit with a final demand vector of $Y=\begin{bmatrix} 0 \\ 10 \end{bmatrix}$ we find that the total amount of fuel needed is $x_1 =$ 25 tons and the amount of coal needed is $x_2 = 75$ tons. The supply chain contributions of this simple example are exaggerated because of the numbers chosen. Nevertheless, supply chain effects in the economy are important to consider in assessing sustainability.

Following the approach developed by Wassily Leontief in the 1930s, government agencies around the world use economic input–output analysis to characterize the structure of their national economies. The transaction tables used in these analyses are expressed in monetary terms and have tens to hundreds of sectors, with table elements showing the value of inputs to each sector from every other sector that supplies it. In many countries, the tables are developed directly from industry surveys, with periodic updates. The U.S. Bureau of Economic Analysis publishes I/O tables for U.S. industry (https://www.bea.gov/data/industries/input-output-accounts-data). To understand these tables, it is useful to know the standard nomenclature, which can again be illustrated with a hypothetical two-sector example, in which corn is an input to make biofuel and biofuel is used in production of corn. The following example and explanation are adapted from a similar example provided for teaching purposes by instructors in the Green Design Institute at Carnegie Mellon University [5].

TABLE 3.3

Complete Transactions Table ($ Billions) for Two-Sector Example Economic Input–Output Analysis

	Corn	Biofuel	Final Demand	Total Output
Corn	55	70	30	155
Biofuel	20	15	170	205
Value added	80	120		
Total input	155	205		

TABLE 3.4

Direct Requirements Matrix for Two-Sector Input–Output Analysis

	Corn	Biofuel
Corn	0.355	0.341
Biofuel	0.129	0.073

Assume the two-sector economy produces $155 billion worth of corn, including $55 billion used in corn production, $70 billion used in biofuel production, and $30 billion sold to consumers. The economy produces $205 billion worth of biofuel, including $20 billion used in corn production, $15 billion used in biofuel production, and $170 billion sold to consumers. In producing corn, farmers add $80 billion worth of value through their knowledge, labor, and other inputs; similarly, biofuel manufacturers add $120 billion worth of value through process engineering, labor, and other inputs. The complete transactions table (Table 3.3) shows all of these exchanges. For a two-sector economy, the direct requirements matrix (Table 3.4) is a 2 × 2 matrix that shows the requirements from each sector needed to produce a unit (i.e., $1) of output for each sector and is obtained by dividing the entries in Table 3.3 by the total output for each commodity. In matrix notation, for an economy with n sectors, the requirements matrix A is an $n \times n$ matrix with elements a_{ij} = the amount of output from sector i required to produce a unit of output from sector j.

The value of input–output analysis is that the requirements matrix can be used to calculate not only the direct purchases required to produce a given product, but also the indirect purchases—that is, the corn used to produce the biofuel used to produce the corn, and back through the production chain. For a given level of total demand, Y, the vector of required production throughout the full supply chain, X, can be calculated as:

$$X = [I + A + AA + AAA + \cdots]Y \tag{3.11}$$

where I is the identity matrix. It turns out that the geometric series in the brackets has a simple approximation:

$$[I + A + AA + AAA + \cdots] = [I - A]^{-1} \tag{3.12}$$

In input–output analysis nomenclature, the matrix $[I - A]^{-1}$ is called the total requirements matrix. It is shown for our case in Table 3.5.

With the total requirements matrix from Equation 3.12 in hand, we can use Equation 3.11 to determine how production in each sector, X would need to change to meet additional

TABLE 3.5
Total Requirements Matrix for Two-Sector Input–Output Analysis

	Corn	Biofuel
Corn	1.67	0.62
Biofuel	0.23	1.16

demand, *Y*. For example, assume \$30 billion more biofuel is required due to a new policy promoting biofuels over petroleum-based fuels:

$$\begin{bmatrix} x_1 \\ x_2 \end{bmatrix} = \begin{bmatrix} 1.67 & 0.62 \\ 0.23 & 1.16 \end{bmatrix} \begin{bmatrix} 0 \\ \$\ 30 \text{ Billion} \end{bmatrix} = \begin{bmatrix} \$\ 18.5 \text{ Billion} \\ \$\ 34.9 \text{ Billion} \end{bmatrix} \tag{3.13}$$

where x_1 = total corn production and x_2 = total biofuel production, each in billions of dollars.

The new demand would require total outputs, accounting for the full supply chain, of \$18.5 billion more corn and \$34.9 billion more biofuels. In contrast, given the direct requirements matrix in Table 3.4, direct purchases by the biofuels suppliers to produce \$30 billion more biofuel for the market would be \$10.2 billion in corn and \$2.2 billion in biofuels.

As seen above, I/O analysis can be conducted in terms of material flows, as done in process matrix analysis, or in terms of money, as in economic I/O analysis. I/O analysis can also be conducted in terms of energy requirements or environmental burdens such as greenhouse gas emissions. The embodied energy or life cycle environmental impacts of production can be determined from an extended I/O analysis as:

$$E = RX = R[I - A]^{-1}[Y] \tag{3.14}$$

where *E* is the vector of energy or environmental effects, *X* is the vector of required production determined through LCA (in dollars), and *R* is a square matrix with diagonal elements that represent the direct energy use or environmental burden per dollar of production in the sector. The U.S. Environmental Protection Agency developed the U.S. Environmentally-Extended Input–Output (USEEIO) model based on I/O tables from the U.S. Bureau of Economic Analysis, together with sector-level data for energy and natural resource consumption and pollution releases that are routinely compiled by other U.S. agencies [8,9]. Note that because *X* represents the total requirements, the energy or environmental effects vector *E* reflects not only the direct energy use or environmental burdens of production activities within the specified sector but also accounts for those carried through the full supply chain.

3.4 Embodied Energy and Environmental Impacts

Embodied (or embedded) energy, as calculated through PCA or extended I/O analysis, is a quantitative measure of the energy cost associated with the processing or supply chain for a material or product, accounting for all the materials and energy used in the mining and manufacture of production inputs and equipment in addition to direct production steps.

Energy intensity factors (MJ/$ or kWh/$) give the energy required to produce a dollar's worth of a given product or service.

The embodied energy of a given quantity of a commodity such as steel or a piece of equipment like a new pickup truck is estimated by multiplying the cost of the purchase by the relevant intensity factor. Intensity factors are commonly estimated through extended I/O analysis, considering the energy use and activity of relatively broad sectors, in which case they reflect average values. With detailed analysis, PCA-based embodied energy or impact values can offer more precise results for specific products or production methods than I/O-based estimates, but the latter approach is more efficient in generating screening estimates for a wide range of products at once. With information on environmental outputs such as GHG emissions rather than energy inputs, the same approaches can be used to calculate embodied environmental impacts, such as embodied GHG emissions.

Table 3.6 presents energy, greenhouse gas, and land-use intensity estimates calculated with the EPA USEEIO model v. 2.0.1-411 for a selection of industries and services [9,10]. These intensity factors were calculated using I/O tables for 2012 from the U.S. Bureau of Economic Analysis, with adjustments to convert from producer prices to purchased prices, and sector-specific price adjustments to convert from 2012 to 2018 dollars [9]. Table 3.6

TABLE 3.6

Embodied Energy and Environmental Impact Intensities of Selected Production and Service Sectors, Expressed in Terms of Purchase Prices in 2018 USD

Commodity, Product, or Service	GHG Emissions per USD	Energy Use per USD	Land Use per USD
	kg CO_2-e/$	MJ/$	m^2/year/$
Fresh wheat, corn, rice, and other grains	3.95	14.51	17.23
Unrefined oil and gas (extraction)	1.06	159.34	0.63
Coal (mining)	1.61	320.05	0.37
Copper, nickel, lead, and zinc (mining)	1.52	14.44	0.47
Sand, gravel, clay, phosphate, other nonmetallic minerals (mining/quarrying)	0.55	14.49	0.19
Electricity (generation, transmission and distribution)	3.94	61.05	0.14
Lumber and treated lumber	0.27	8.11	22.47
Clay and ceramic products	0.35	6.40	0.25
Glass and glass products	0.48	8.67	0.16
Cement	4.56	18.91	0.10
Ready-mix concrete	0.80	7.10	0.13
Primary iron, steel, and ferroalloy products	1.02	24.58	0.27
Secondary steel products	0.62	13.12	0.21
Primary aluminum	1.49	18.82	0.27
Secondary aluminum	0.74	10.77	0.19

(Continued)

TABLE 3.6 (*Continued*)

Embodied Energy and Environmental Impact Intensities of Selected Production and Service Sectors, Expressed in Terms of Purchase Prices in 2018 USD

Commodity, Product, or Service	GHG Emissions per USD	Energy Use per USD	Land Use per USD
	kg CO_2-e/$	MJ/$	m^2/year/$
Construction machinery	0.22	4.04	0.21
Computers	0.04	0.62	0.02
Semiconductors	0.22	1.98	0.05
Storage batteries	0.28	3.96	0.09
Primary batteries	0.16	2.94	0.08
Automobiles	0.22	3.68	0.28
Pickup trucks, vans, and SUVs	0.29	5.06	0.44
Heavy duty trucks	0.32	5.02	0.53
Motorcycle, bicycle, and parts	0.17	3.25	0.29
Flours and malts	1.57	9.10	5.68
Vegetable oils and by-products	1.15	9.38	7.31
Packaged meat (except poultry)	1.79	5.15	18.68
Breweries and beer	0.31	4.04	0.71
Fabric	0.57	9.17	0.84
Clothing	0.07	1.12	0.24
Paper	0.74	30.56	4.37
Gasoline, fuels, and by-products of petroleum refining	1.00	108.31	0.47
Plastics	1.54	22.65	0.74
Fertilizers	2.17	10.45	0.23
Paints and coatings	0.56	8.44	0.58
Air transport	0.96	20.22	0.15
Rail transport	0.70	11.16	0.70
Truck transport	1.31	16.71	0.34

Source: Adapted from U.S. Environmental Protection Agency [10].

presents a small subset of the results from the USEEIO model, which covers 411 sectors and more than 20 impact indicators, with the 50 U.S. states modeled as a single region. Note that the results in Table 3.6 correspond to the production or provision of the good or service. Energy and environmental impacts of household or other final uses of the products are not included. For example, for motor vehicles, the estimates do not count the embodied energy or greenhouse gas emissions associated with the fuel used for operating the vehicle.

Scanning through the entries in Table 3.6 shows a wide range of energy and environmental impact intensities. Low intensity values can result from small impacts and/or high value added. This is demonstrated by comparing impact intensities for basic commodities versus finished products—e.g., comparing intensity values for fabric with those for clothing. Clothing has lower energy and impact intensities than fabric due in part to the value that creative design and marketing activities add to the price of the finished product.

Table 3.6 presents a small sample of intensity values for illustrative purposes. Researchers and practitioners will want to refer to more comprehensive results and updated estimates. Full results, documentation, and updates for the USEEIO model are available at https://www.epa.gov/land-research/us-environmentally-extended-input-output-useeio-technical-content. Links to EEIO models and data for other countries and regions are available from the openLCA Nexus website (nexus.openlca.org).

3.5 Energy Return on Energy Invested

Energy return on energy invested (EROI) is a valuable concept for sustainable energy engineers to understand and to communicate to the public and policy makers. In terms of the schematic in Figure 3.1, the EROI is given by the equation [11]:

$$\text{EROI} = P/(S_1 + S_2) \tag{3.15}$$

where P is the rate of energy production (e.g., kWh/analysis period) from the transformation system, S_1 is the conversion energy input (kWh/analysis period), and S_2 is the embodied energy in the various items used by the production system (kWh/analysis period). The EROI can be evaluated for a region's whole energy system in a given year, or it can be assessed for a particular energy production technology over the useful life of the plant.

The EROI is a measure of the energy profitability of an energy transformation system. Since the energy invested in the transformation system cannot be used by other consumers in the economy, the EROI is also an indication of the availability of energy to meet consumer demand and to provide the surplus needed for maintenance, replacement, new manufacturing, and construction. For the economy as a whole, EROI can be viewed as a determining factor for prosperity. In areas of primary production such as food, materials, and energy, prosperity is more fundamentally determined by the physical ability to return a positive net product to the economy than by the prices the economy attaches to commodities. Companies are not likely to invest in an energy platform with low EROI because it would be bad business. However, misguided government policy and mandates sometimes provide incentives and subsidies for platforms with poor EROI, reducing the overall EROI of the economy, and increasing external costs and the risk of exposure to supply security issues.

The main point of evaluating the EROI of a given energy supply system is the same as for financial analysis of different investments. EROI lets us compare different energy transformation platforms and make informed investment decisions. The objective of investing is to realize a profit. In the energy system, the "profit" for society is the net energy. The net energy yield, N, is given by Equation 3.16:

$$N = P - (S_1 + S_2) = P(1 - 1/\text{EROI}) \tag{3.16}$$

The net energy yield can be assessed for a particular energy conversion project or for the whole sector. The net energy yield can be evaluated on a year-by-year basis or over the lifetime of the energy conversion plant. On a year-by-year basis, the net energy yield may be negative at the early stage of a power plant project, and the point at which the cumulative net energy return reaches zero is also known as the EPBT.

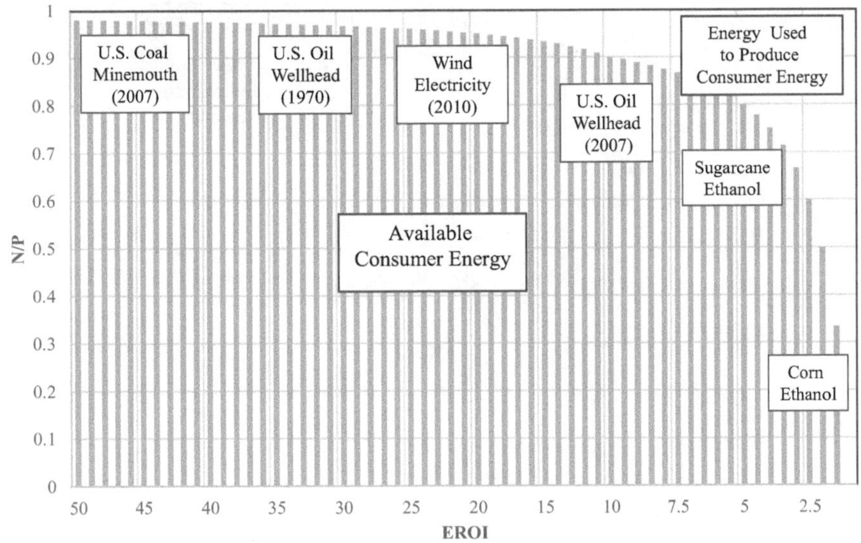

FIGURE 3.8
Height of the gray bars is the ratio of net energy available for the consumer economy to the energy used in the energy transformation system vs. EROI. EROI values on the chart are from Hall et al. [12] and represent central estimates from the literature at the time of their study, for standard EROI at the point of production based on thermal energy content.

The implication of low EROI energy investments is demonstrated in Figure 3.8, where the net energy return to the economy as a percentage of the total energy produced, $N/P = 1\text{-}1/\text{EROI}$, is plotted versus EROI. The ratio N/P is essentially a measure of the energy prosperity for a society. An N/P of 95% implies that most of the energy and resources used by the energy sector were returned to the economy as useful energy production. An N/P of 50% means that the energy transformation sector consumed half of the resources internally and only returned half to the economy. Inspection of Figure 3.8 shows that once EROI drops below 5, the available energy and energy prosperity decrease precipitously.

Energy budgeting is an important concept for planning a sustainable energy future. In an industrialized society, a certain amount of energy is used to maintain the energy infrastructure, which includes the pipelines and roads for distributing liquid fuel or natural gas, as well as the hydroelectric dams and power plants that generate electricity. An additional amount of energy must be set aside to find more energy to replace the energy being used up. The need to maintain a flow of energy requires drilling for oil or digging for coal and uranium in a fossil-based energy system or building wind turbines and solar PV in a renewable energy system.

The remaining available energy is divided into two parts. The first part is used for basic needs that support the standard of living such as health, shelter, water, and food. The second part is for discretionary uses associated with lifestyle such as entertainment, vacations, and the arts. This latter part can also be used for the growth of the economy. Figure 3.9 shows an energy flow diagram that accounts for the energy budget of the system depicted in Figure 3.1.

The energy budget diagram is useful in illustrating the impact on the economy of a trend to low EROI energy resources. Energy budget diagrams for an energy system with EROI = 25 and a system with EROI = 2.5 are shown in Figure 3.10a and b. In a simplified fashion, we can divide our energy budget into energy necessary to maintain the energy

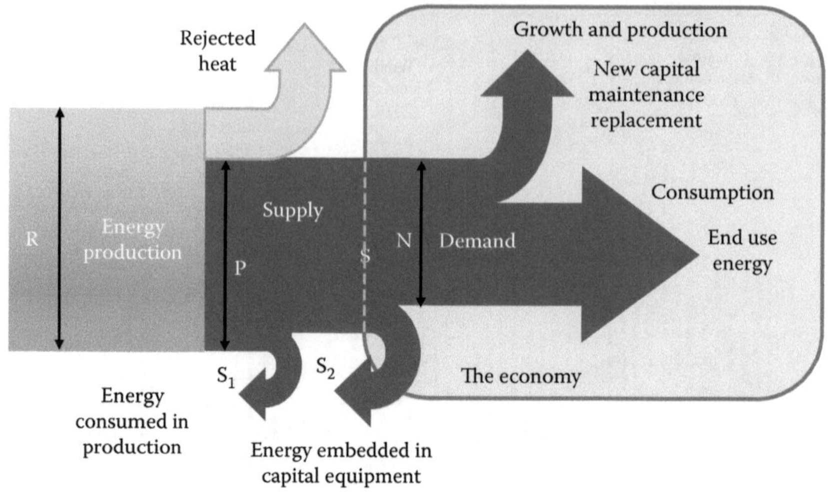

FIGURE 3.9
Energy flow diagram depicting the energy budget from extraction and transformation to end use.

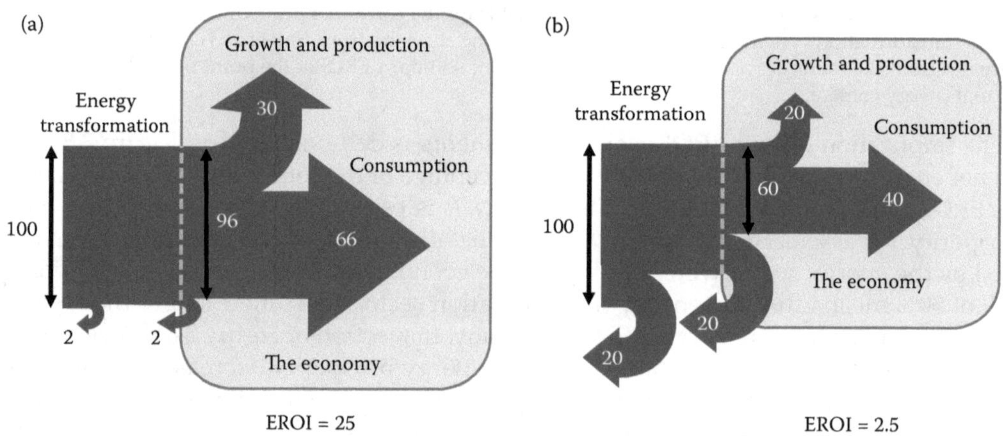

FIGURE 3.10
Energy budget diagrams for two energy transformation systems (a) have EROI = 100/(2 + 2) = 25 and supports a larger economy than (b) which has EROI = 100/(20 + 20) = 2.5 even though both have the same energy production.

flows, buildings, roads, and other basic needs, and energy that can be used for discretionary activities. If there is a surplus of energy, we can apply that toward growth and/or prosperity, but the amount is finite and cannot be fully allocated to both simultaneously.

3.5.1 Calculation of EROI

Energy analyses are useful supplements to standard economic procedures based on monetary assessments of costs and benefits for several reasons. All economic analyses require assumptions about the future, such as interest rates, future price of conventional fuels, and inflation rates. Also, assumptions about future government regulation, taxation rates, and political decisions can greatly influence the economic evaluation process. EROI is calculated by putting all inputs and outputs into energetic terms so the evaluation involves fewer economic assumptions, and the result can be interpreted in terms of the general potential for prosperity of the economy.

Although EROI is calculated in physical rather than monetary units, different time-scales and different assumptions about physical systems made in EROI analysis need to be considered in interpreting the results. The boundaries for the analysis are one important choice. As defined by Murphy et al. [13] "standard" (also known as "primary") EROI assesses energy production and inputs (including embodied energy) at the point of production (e.g., crude oil at the well head), typically excluding energy associated with labor and supporting economic services. "Point of use" (also known as "final") EROI would extend the analysis to include energy costs for refining and transporting fuel to the final user, for example to provide diesel fuel at a filling station. A third approach would extend the boundary to include end-use services (e.g., heat, light, or propulsion), accounting for the energy efficiency of end-use devices [14]. A related distinction made in EROI analysis is whether to adjust for energy quality, either using a thermodynamic measure such as exergy, or a broader price-based measure of quality that accounts for characteristics such as ease of fuel transport or lower combustion emissions that may be reflected in market prices. The sharp difference in quality between the energy content of coal versus that of the electricity produced by wind and solar is important to bear in mind in comparing EROI values across these energy resources.

Directly determining the embodied energy in power plant and fuel exploration and extraction equipment can be difficult, so back-of-the-envelope calculations may use embodied energy values for materials such as those in Table 3.6 in combination with estimates of the costs of fuel and equipment. It may also be necessary to construct a PCA of the energy transformation platform to define the energy and materials used for analysis.

Figure 3.11 schematically illustrates the EROI calculation method for a hypothetical energy transformation project. During the construction phase, two units of energy are used to build the plant; during its operational life, two units of energy are used; finally, one unit of energy is used to decommission the facility. The energy produced is 25 units and the net energy output for this project is 20 units. Thus, the EROI for this example is 5. In Figure 3.11, the shaded area of energy input during the operational lifetime would include the energy of the fuel input for a fossil or nuclear plant. If this were a renewable plant that did not require fuel to operate and decommission, the EROI would be 12.5.

A key difference between the EROI for fossil or nuclear-generating systems and renewable technologies, such as solar or wind, is the fact that the renewable systems depend on a primary energy source that not only varies with time of day and time of year but also with the location at which the system is deployed. Therefore, the evaluation of the energy produced must be carried out in considerably more detail for a renewable system than for a fossil or nuclear system. In the following chapters, we will present detailed methods for estimating the energy output of renewable systems.

The EROI for conservation measures and waste heat recovery is calculated using the same approach as for energy transformation systems. The energy being wasted or rejected to the environment is treated as a resource like wind or solar, and the energy recovered or saved is treated as the energy production, P.

3.5.2 EROI for a Wind Energy System

Determination of a project-level EROI is illustrated by the analysis performed by White and Kulcinski [15]. In the time since their paper was published, unit costs of wind equipment

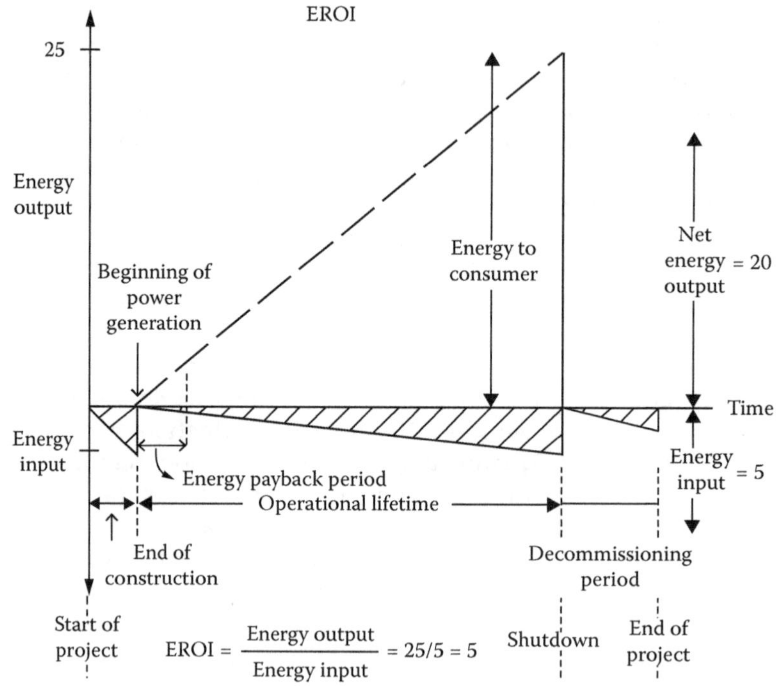

FIGURE 3.11
Energy input and output for a renewable energy generation system during its lifetime.

have come down and capacity factors have increased, but the paper still provides a useful illustration of the EROI calculation method. The analyzed project was built in 1997 and is located near the city of DePere, Wisconsin and will be referred to as the DePere Wind Project (DPWP). It consists of two Taike 600e turbines that have a combined electric power capacity of 1.2 MW_e with a capacity factor of 31% (predicted), and a predicted life for each of the nacelles of 20 years.

Table 3.7 shows the energy requirements to construct the two DePere wind turbines [15]. These requirements are obtained from the cost data shown in column 1, for example, the craning cost was \$75,000. Then the energy per dollar in 1977 dollars was obtained from I/O sector data analogous to that in Table 3.6 (but from an earlier I/O study), for example, 30,233 Btu per 1977 dollar for hoists and cranes. But the time at which the craning cost was \$75,000 was 1997, while the I/O tables are for 1977 dollars. Hence, we must adjust the cost figure for inflation. Conversion from 1997 dollars (the year of construction) to 1977 dollars (the year the I/O tables were developed) was done using the Consumer Price Index (CPI) ratio 1977:1997 = 60.6:160.5, obtained from the CPI of the Bureau of Labor Statistics. Multiplying the cost times the price index ratio and the Btu/\$ gives the embodied energy, that is, 8.56×10^8 Btu. A similar analysis for the other items in column 1 yields the total energy input required to make the turbines. Analysis of the energy embodied in on-site construction is laid out in Table 3.8 [15].

The lifetime and annual energy investments in the DPWP are shown in Table 3.9. Using these data, the *total energy input* per installed 1.2 MW_e is 13,933 GJ_{th} or 3.89×10^6 kWh_{th}. The power output is $1.2 \times 10^3\ kW_e \times 0.31 = 0.372 \times 10^3\ kW_e$, and for the assumed lifetime of the system of 20 years, the *total power output* is $0.372 \times 10^3\ kW_e \times 20$ year $\times$ 365 day/year $\times$ 24 h/day = $65.2 \times 10^6\ kWh_e$. Hence, EROI = 65.2/3.89 = 16.75.

TABLE 3.7

Energy Requirements to Construct Two Turbine Wind Farms at DePere, Wisconsin

	1997 Dollars	I/O Sector	Btu/1977 Dollars[a]	Btu	GJ_{th}/2 Turbines
Craning	75,000	Hoists, cranes	30,233	8.56E+08	903
Labor	25,000	Misc. business services	10,000	9.44E+07	100
Local equipment rental	3,000	Construction machinery	34,534	3.91E+07	41
Lodging and food for employees	8,000	AVI[b]	28,780	8.69E+07	92
Electrical grounding	12,000	NC, electrical utilities	30,648	1.39E+08	147
				Total: 1,283 GJ_{th}/2 turbines	

Source: White and Kulcinski [15].

[a] 1997 dollars were calculated from the CPI by the scale 1977/1997 = 60.6/160.5 = 0.3776.

[b] Sector AVI is an average of the I/O sectors eat and drink places and hotels.

TABLE 3.8

Costs and Energy Requirements for On-site Construction of the DePere Wind Power Plant

	1997 Dollars	I/O Sector	Btu	GJ
Foundation preparation	73,000	Ready-mix concrete	2.23E+09	2,353
Transformer (2 at $10,000 each)	20,000	Transformers	3.22E+08	339
Craning	75,000	Hoists, cranes	8.56E+08	903
Labor	25,000	Misc. business services	9.44E+07	100
Local equipment rental	3,000	Construction machinery	3.91E+07	41
Lodging and food for employees	8,000	AVI	8.69E+07	92
Electrical grounding	12,000	NC, electric utilities	1.39E+08	147
				1,282

Source: White and Kulcinski [15].

If the wind farm were located at a place where new transmission lines would have to be built to deliver the power to a user, additional energy input for grid extension could be included. If the wind-generated electricity were to be stored, as is often done for remote power installations, then the energy embedded in batteries would need to be included.

3.5.3 Representative EROI Values and Implications

Hall et al. [12] conducted an extensive review of published estimates and concluded that the EROI for globally important fuels and electricity sources is declining. Historically, hydroelectricity has had a relatively high EROI among electricity sources, with reported values above 80:1, but returns are highly variable depending on location, and many of the most productive sites have already been developed. Among fossil fuels, coal has had the highest EROIs on a thermal energy basis, with values above 50 for mine-mouth coal in the United States [12]. The EROI for coal mined in the United States apparently declined through the 1960s, 1970s, and 1980s, but then rose again as more production shifted to surface mining. Note that the high EROI values for coal and hydroelectricity bring into sharp

TABLE 3.9

Lifetime and Annual Energy Investments for the DePere Wind Power Plant

		Total Energy per Installed 1.2 MW_e	Annual Energy per GW_ey
Process	**Source**	GJ_{th}/Power Plant	GJ_{th}/GW_ey
Wind turbine (embodied)			
Blades	PCA	154	
Nacelles	PCA	2,273	
Inverter	I/O	339	
Wiring	PCA	209	
Tower	PCA	4,890	
Foundations	PCA	1,925	
Materials total		9,790	326
Transportation			
Blades		21	1
Nacelles		257	9
Towers		178	6
Concrete		23	
Control cabinets		0.29	0.01
Transportation totals		480	15
Construction	I/O	1,282	43
Maintenance	I/O	1,636	55
Decommissioning		641	21
Total required energy		13,933	

focus a critical limitation of the EROI metric for assessing sustainability, as it ignores the high external costs of damming rivers for large-scale hydropower, the ecological damage of surface coal mining, and air pollution and greenhouse gas emissions associated with coal combustion.

Hall et al.'s review shows that the EROI for oil and gas production declined in the United States from roughly 30:1 circa 1960 to 10:1 in 2010 [12]. Time series studies found an even sharper decline in the EROI for conventional oil and gas production in Canada, from over 60:1 in 1970 to 10–15:1 in 2010. Oil from the Canadian tar sands has an even lower EROI of approximately 4:1. This means that a very large amount of energy needs to be invested for a relatively small amount in return. Globally, the EROI trend for primary oil and gas is driven by the increasing challenge of accessing new resources, although this may sometimes be tempered by more efficient production techniques. Going beyond the primary stage, Aramendia et al. [14] assessed globally representative "final" and "useful" stage EROIs for fossil fuels, accounting for energy required to convert primary commodities (e.g., oil) to final products (roadway fuel) and then convert the products to useful energy services (roadway propulsion). Defined at the final product stage, the EROI for roadway fuels derived from oil was about 7.0. The useful stage EROI for roadway propulsion driven by oil products was about 2.0. There was little change in the useful stage EROI from 1971 to 2020, as improvements in refining and vehicle efficiency offset increased energy needs at the production stage.

Manufacture of a wind and solar plant requires investment of fossil fuels and energy-intensive materials, so EROI assessments are relevant for these technologies, too. Estimates of the EROI for wind and solar vary widely, depending on the resources, conversion efficiency, service life, and capacity factor assumed in the analysis. EROIs reported in the literature from 2017 to 2020 for wind-generated electricity delivered to the grid range from ~15 to 60 [16]. EROI values published over the same period for electricity from solar PV range from ~10 to 20 [16]. The levelized costs of energy for renewable energy systems (Table 1.4), as well as the energy return on investment, depend heavily on the assumed lifetime. In fact, since no fuel costs are associated with the operation of renewable systems, the EROI is directly proportional to the lifetime of the system. Therefore, for future energy systems using renewable energy sources, a key goal of the engineering design should be the longest possible lifetime.

Finally, another way to address the economy's energy budget is to invest in energy efficiency—changing existing energy end-use systems to reduce demand. The EROI for energy efficiency systems has the same definition as for energy production. The "low-hanging fruit" conservation investments, such as improving insulation in buildings, have very high lifetime EROI and are also profitable.

3.6 Greenhouse Gas Accounting

As international efforts to combat climate change began to accelerate in the late 1990s, the World Resources Institute and World Business Council for Sustainable Development identified the need for a standard protocol for businesses to perform greenhouse gas accounting. The Corporate Accounting and Reporting Standard they developed (ghgprotocol.org; hereinafter GHG Protocol) was first published in 2001, was revised in 2004, and is now used by most large companies around the world to track and report their GHG emissions and emissions reduction efforts. Companies use GHG accounting to identify opportunities to reduce emissions, fulfill requirements of either voluntary or mandatory reporting programs, support participation in carbon trading markets, and report to shareholders and the public. The corporate reporting protocol was adopted by the International Organization for Standardization in 2006 as the basis for ISO 14064-1: Specification with Guidance at the Organizational Level for Quantification and Reporting of Greenhouse Gas Emissions and Removals [17].

The GHG Protocol requires reporting of emissions of the seven greenhouse gases or groups of gases covered by the United Nations Framework Convention on Climate Change (UNFCCC): carbon dioxide (CO_2), methane (CH_4), nitrous oxide (N_2O), hydrofluorocarbons (HFCs), perfluorocarbons (PFCs), sulfur hexafluoride (SF_6), and nitrogen trifluoride (NF_3). Businesses are required to identify the basis for their reporting, as either emissions for operations in which they have an equity share (ownership) or operations under their control. The scope of reporting is split into three tiers:

Scope 1: Direct greenhouse gas emissions coming from sources that are owned or controlled by the company. Such emissions may come from combustion in the company's own furnaces or vehicles, from physical or chemical processing operations such as cement or fertilizer manufacture, or from fugitive emissions such as natural gas or refrigerant leaks.

Scope 2: Indirect emissions from generation of purchased electricity that is used in operations owned or controlled by the company.

Scope 3: Other indirect emissions that occur as a consequence of the company's activities, but not from sources that are owned or controlled by the company. Examples include emissions from extraction or processing of fuel used by the company, emissions associated with the use of the company's products, or from waste disposal.

Companies adhering to the GHG Protocol are required to report Scope 1 and 2 emissions, but tracking and quantifying Scope 3 emissions is optional. Following a revision to the Scope 2 requirements in 2015, companies must report emissions from purchased electricity in two ways: first, using grid-average emissions for their operational locations, and second, using a contract-based approach that accounts for electricity purchases from specific facilities or purchases of electricity generated with specific characteristics, such as from low carbon or renewable energy.

Emissions can be determined through direct measurements of concentrations and flow rates, or more commonly through engineering estimates. In the case of combustion emissions, these are often determined through mass balances and stoichiometry based on purchased quantities of fuel. Emissions can also be estimated using corporate records of operational activities (e.g., vehicle mileage) combined with published emissions factors. The GHG Protocol website includes spreadsheet tools for estimating emissions from stationary sources, mobile sources, air conditioning and refrigeration operations, and from a variety of manufacturing activities.

The GHG Protocol requires companies to report emissions of the seven UNFCCC greenhouse gases separately and to report combined emissions on a CO_2-equivalent basis using global warming potentials. The global warming potential (GWP) for a greenhouse gas is essentially the amount of energy that a unit mass of the gas will absorb over a given amount of time, relative to the amount of energy absorbed over the same period by a unit mass of CO_2. GWP values reflect the fact that greenhouse gases differ in their infrared radiation absorption efficiencies and in their atmospheric lifetimes. GWP values also reflect differences in the gases' indirect effects. For example, methane oxidation contributes to the formation of ozone, which is also a greenhouse gas. Because GWP depends on changing atmospheric conditions and because GWP calculations depend on evolving scientific understanding of atmospheric chemistry and dynamics, recommended GWP values have changed over time. For selected greenhouse gases, Table 3.10 gives GWP estimates integrated over 20-year and 100-year time horizons from the fifth Intergovernmental Panel on Climate Change (IPCC) Assessment Report (AR5), published in 2014 [18] and the sixth IPCC Assessment Report (AR6) published in 2021 [19]. The AR5 report presents GWP values estimated with and without accounting for the feedback effect of changes in the climate on the carbon cycle but recommends using values with climate-carbon feedbacks for consistency in the treatment of CO_2 versus the other greenhouse gases. AR6 values include climate feedbacks. Values for many other greenhouse gases are given in Table 8.A.1 of the AR5 report [18] and Table 7.SM.7 of the AR6 report [19].

The GHG Protocol requires the use of the most recent GWP values estimated by the IPCC over a 100-year time horizon to estimate CO_2-equivalent emissions. Reporting CO_2-equivalent emissions based on 20-year GWP values is optional under the protocol but is important for assessing near-term effects and mitigation opportunities. Following UNFCCC reporting guidelines, the U.S. EPA's 2024 Inventory of U.S. Greenhouse Gas Emissions and Sinks [20] used 100-year AR5 GWP values excluding climate-carbon feedbacks.

TABLE 3.10

Global Warming Potential (GWP) Values from IPCC AR5 [18] and AR6 [19]

	AR5 Lifetime (Years)	AR5 GWP_{20}	AR5 GWP_{100}	AR6 Lifetime (Years)	AR6 GWP_{20}	AR6 GWP_{100}
CH_4	12.4	86/84	34/28	11.8	81.2	27.9
HFC-134a	13.4	3,790/3,710	1,550/1,300	14	4,140	1,530
CFC-11	45	7,020/6,900	5,350/4,660	52	8,320	6,230
N_2O	121	268/264	298/265	109	273	273
SF_6	3,200	17,783/ 17,500	26,087/ 23,500	3,200	18,200	24,300
NF_3	500	13,000/ 12,800	17,885/ 16,100	569	13,400	17,400

Source: IPCC [18] and Forster et al. [19].

3.7 Energy Systems Modeling and Scenario Analysis

Energy systems modeling is a multi-disciplinary field that draws on economics, engineering, operations research and computational sciences. It encompasses a wide variety of approaches and application areas that have a common aim of evaluating the effects of economics, technical advances, resource constraints, and policy choices across the energy sector. Modeling approaches range from consumer-oriented agent-based simulation models that examine the emergent effects of scaling up choices by individual consumers, to top-down macroeconomic global equilibrium models that examine the impacts of energy costs on the rest of the economy and the effects of economic conditions on the energy sector.

Bottom-up macroscale models represent a widely used branch of energy systems models. These models are formulated to solve a constrained optimization problem to minimize costs or maximize benefits of supplying energy to meet demand for energy services at regional, national, or global scale. Though their assumptions and level of detail vary, they all contain representations of the current and future cost and performance of energy supply options that can be used to meet energy service demands over a particular time period and geographic region, subject to technical, economic, and regulatory constraints. Public and private sector decision makers and researchers use them to explore future scenarios and optimal pathways for energy system development and to assess the costs and benefits of various energy and environmental policies. For example, bottom-up macroscale energy system models might be used to help plan optimal deployment of new transmission infrastructure to support expanded solar energy development in southern Europe, or to analyze the effects of new U.S. tax incentives on the expansion of renewable energy capacity and associated GHG emissions reductions.

Some macroscale energy systems models cover all sectors of energy use, including building and industrial heating, electricity, and transportation fuels. Others are focused on a particular energy sector, most often the electric power sector. Within the power sector, "capacity expansion models" are run over time periods of years to decades using time steps of one to a few years, to investigate options for investing in generation, storage, and transmission capacity. In contrast, "production cost" models operate at hourly or sub-hourly timescales to assess which power plant units should be dispatched or taken offline to match forecast load.

Some energy or power system models specify the demand for electricity or energy services exogenously and solve for the constrained, least-cost means of meeting that demand. Other "partial equilibrium" energy models iterate between supply and demand modules, adjusting demand as electricity or fuel prices change based on energy supply results. The time horizon of optimization is another distinguishing characteristic. In "perfect foresight" or intertemporal models, the system is optimized over all segments of the modeled time horizon simultaneously, with cost and performance trends for all technologies specified in advance. In contrast, sequential models solve a sequence of optimization problems for each segment of time, incorporating the results from the prior time segment into the baseline for the next one. This sequential approach can produce sub-optimal results overall, as decisions for early time segments limit later choices, but it may match real-world decision-making more closely than the perfect foresight approach.

Due to their complexity and extensive input requirements, the application of energy systems models requires significant training and expert judgment. They are not off-the-shelf tools. The array of models in use today is too broad to cover here. However, we provide brief descriptions and links to further information for three models that are especially relevant to sustainable energy engineers: the Regional Energy Deployment System (ReEDS) model developed at the U.S. NREL; the National Energy Modeling System (NEMS) developed and used by the U.S. Energy Information Administration, and the International Energy Agency's (IEA's) Global Energy and Climate Model (GEC).

The ReEDS Model Version 2020 (ReEDS 2.0) [21] is a power sector capacity expansion model that has been used extensively to explore scenarios and policies to support deployment of renewable electricity technologies in the U.S. ReEDS contains separate optimization modules for electricity supply and demand, along with a variable renewable energy (VRE) module to simulate the contributions of wind, solar, and storage technologies to electricity supply capacity, based on hourly load, wind, and solar resource profiles. The model subdivides the United States into 134 balancing areas in which demand for electricity must be met. The supply module adds or retires generating and storage technologies and transmission infrastructure to minimize capital and operating costs for the electricity sector, subject to load balancing and to constraints on transmission capacity, renewable resource availability, grid operating requirements, emissions limits, and other policies. Natural gas, coal, and uranium prices are supplied exogenously. A wide array of renewable energy technologies, including wind, solar, geothermal and hydropower are represented in detail. The demand module maximizes consumer welfare (i.e., acquiring desired energy services at low cost, including capital and operating costs) with a choice of end-use energy service devices available based on customer category. ReEDS can be run with fixed demand forecasts input to the supply module, or iteratively with elastic demand adjusted in response to retail electricity prices determined based on output from the supply module. It can also be run sequentially, with perfect foresight, or with a sliding optimization window.

NEMS is a comprehensive model of the U.S. energy system encompassing coal, oil and gas, and renewable energy supplies, electricity and liquid fuel conversion technologies, and demand in the residential, commercial, industrial, and transportation sectors [22]. Energy system modules are linked to a general equilibrium model of the rest of the U.S. economy through a macroeconomic activity module. An international energy module links U.S. liquid fuel production and prices to the global trade in these commodities. NEMS cycles through supply, conversion, demand, and macroeconomic modules in sequence until the model converges to reach an equilibrium between supply and demand. Geographic

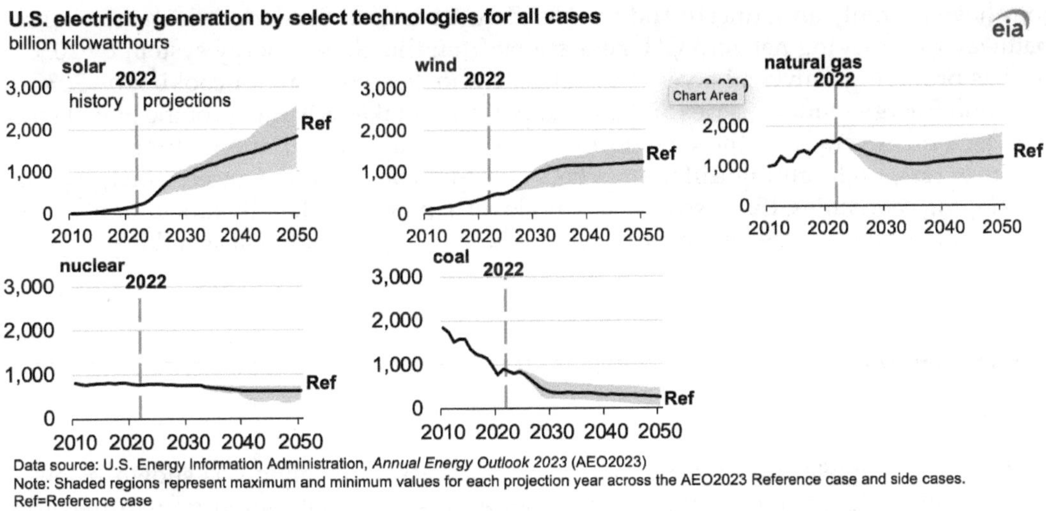

FIGURE 3.12
NEMS projections of U.S. electricity generation from the Annual Energy Outlook 2023 [23].

resolution depends on the module. For example, end-use demands are aggregated to the nine U.S. census divisions and the electricity market module to 25 supply regions. The U.S. Energy Information Administration uses NEMS to make multi-decadal projections that are published in its Annual Energy Outlook (AEO) series (https://www.eia.gov/outlooks/aeo/). Figure 3.12 is an example of NEMS results from the 2023 AEO, showing projected electricity generation by technology across time. The shaded area reflects the range of projections across EIA's "side cases," which include high and low oil prices, oil and gas supplies, zero-carbon technology costs, and economic growth. The results shown account for the Inflation Reduction Act passed in August 2022. However, EIA projections are conservative in that they only incorporate fully enacted policies. Consequently, they have historically underestimated the deployment of renewables and energy efficiency measures as later-adopted policies helped drive actual outcomes.

The GEC model is the International Energy Agency's primary tool for analysis of future energy scenarios [24]. The GEC model is a bottom-up, partial equilibrium model of the global energy system, geographically resolved into 29 regions. The model includes fossil fuel and renewable energy supplies, energy transformations to produce electricity, heat, liquid fuels and hydrogen, energy storage, and end-use demand in industry, transportation, buildings, and agriculture. The partial equilibrium model iterates between supply modules that output energy prices and price-responsive demand modules. Economic growth, demographic trends, and technological advances are exogenous inputs. Model outputs include GHG and local air pollution emissions from fuel combustion, oil and gas exploration and production and industrial processes. In its World Energy Outlook series (e.g., https://www.iea.org/reports/world-energy-outlook-2023), IEA presents GEC model results as trends through 2050 based on periodically updated scenarios that are defined based on GHG emissions reduction and sustainable development goals. These are the Stated Policies Scenario, which accounts for current policies to reduce GHG emissions from the energy system, including policies under development; the Announced Pledges Scenario, which assumes countries will meet the GHG emissions reduction commitments they have formally announced; and the Net Zero Emissions Scenario, which represents a pathway to achieving net zero GHG emissions from the global energy system by 2050 as well as providing universal access to electricity and clean energy for cooking by 2030. In its World Energy Outlook series and other reports, IEA takes advantage of the global scope and sectoral coverage of the GEC to examine international variations in energy needs and progress, highlighting differences between countries with advanced, emerging, and developing economies. IEA also uses the model to explore the globally linked implications of technological trends, economic disruptions, and changing political agendas.

Problems

3.1 The complete transactions matrix for a two-sector economy is shown below, with units in $. Find the output from each sector that would be required to meet $150 of additional consumer demand from sector 1 and $200 of additional consumer demand from sector 2, considering supply chain requirements along with direct requirements. Compare the total output requirements across the supply chain with the direct requirements.
To complete this problem, use the following steps:

a. Compute the technical requirements matrix

b. Compute the Leontief inverse matrix

c. Compute the total (including supply chain) output required for each sector to meet the new demand

d. Compute the direct (excluding supply chain) output required for each sector to meet the new demand.

	Sector 1	Sector 2	Final (Consumer) Demand	Total Output
Sector 1	300	350	500	1,150
Sector 2	200	300	1,000	1,500
Value added	650	850	1,500	
Total input	1,150	1,500		

3.2 Use information for ready-mix concrete and primary iron, steel, and ferrous alloy products from Table 3.6 to make a rough approximation of the energy embedded in the Golden Gate Bridge. The Bridge contains about 390,000 cubic yards of concrete and about 83,000 tons of structural steel. Assume concrete costs about $150 per cubic yard in 2018 dollars and steel costs about $680 per ton in 2018 dollars. Express your answer for GHG emissions in metric tons. Express your answer for energy in MJ and in barrels of oil equivalent (BOE), where 1 BOE = 5,800 MJ.

3.3 Research the typical price of a new gasoline-fueled motorcycle, sedan, or pickup sold in the United States, and then use information from Table 3.6 to estimate the embodied energy and embodied GHG emissions in that vehicle. (Use the CPI to adjust your prices to 2018 dollars to correspond to the values in the table.) Then research the typical fuel economy for the type of vehicle you chose and estimate in-use GHG emissions associated with fuel combustion over a 150,000-mile vehicle life. Compare the in-use emissions with embodied emissions for your vehicle.

3.4 A 200 MW coal-fired power plant operates with an 87% capacity factor and has a 45-year lifetime. The power plant's gross efficiency is 36%. The plant is located at the mine, thus minimizing the costs and energy requirements of transporting coal. Five percent of the total electricity produced by the plant is used internally. Find the EROI for the plant, assuming that the predominant upstream energy requirement is for the coal mining operation. You can neglect the energy embodied in equipment at the plant itself. Use an embodied energy estimate for coal mining operations of 56,700 Btu/$ to estimate the energy used to mine the coal, assuming coal costs of $0.65 per million Btu in equivalent dollars. In computing the denominator for the EROI, you should account for both the electrical energy used internally at the plant and the energy used for coal mining, but neglect the heat content of the coal itself, as is the convention for EROI calculations for electricity from fossil fuels. Use the following steps and units as you calculate the EROI:

a. Compute the net electrical energy to the consumer in kWh per year and in Btu per year (ignore transmission losses).

b. Compute the coal energy required by the plant in Btu per year and the cost of this coal.

c. Calculate the embodied energy for producing the coal in Btu per year.

d. Calculate the amount of energy consumed by the plant in its internal operations in Btu per year.

e. Calculate the EROI.

f. How would the EROI change if 25% of the plant's electricity has to be used internally, as might be the case with added carbon capture and sequestration? (Assume the gross capacity remains at 200 MW.)

3.5 You are an energy consultant who has been asked to assess the sustainability of a concentrating solar thermal power (CSP) plant with thermal energy storage. The specified plant has a capacity of 250 MW, a capacity factor of 0.50, and an expected lifetime of 30 years. Major inputs to plant construction include $190 million in machinery (turbine, pumps, etc.); $400 million in glass products; and $30 million in cement, all in 2018 dollars. The plant uses 15,000 MWh/year of external energy for its operations. Use embodied energy values from Table 3.6 for your calculations.

a. Based on only the specified inputs, and assuming continuous operation and no deterioration in performance over time, calculate the plant's EROI.

b. From a sustainable energy perspective, would you support building the CSP plant based on the EROI you calculated? Explain why or why not, by discussing the significance of the EROI metric and including a brief comparison of your EROI with those from other energy technologies or fuels.

3.6 A farmer in eastern Colorado is interested in quantifying the Scope 1 greenhouse gas emissions from her operations, as a starting point for considering how to reduce them. The farm has 250 dairy cows and 1,500 acres of irrigated wheat. The cows are estimated to emit 100 kg CH_4 per year per cow. Irrigated wheat is estimated to emit 1 kg N_2O per acre per year. The milking barn uses 8,000 kg/year of natural gas for heating and hot water. Calculate the farm's total GHG emissions on a CO_2 equivalent basis and find the fractions contributed by the cows, wheat, and natural gas heating. You may assume the natural gas is pure methane (CH_4, MW = 16 g/mol) and that it is combusted completely to CO_2 (MW = 44 g/mol). Use 100-year AR6 GWPs from Table 3.10.

3.7 Complete a preliminary GHG emissions report for a small natural food cooperative, accounting for Scope 1 and Scope 2 emissions. The food co-op is housed in a 5,000 ft^2 space. It uses 65,000 kWh of electricity per year for space cooling, refrigeration, lighting, and electronic equipment. Grid electricity in the area has average emissions of 1.8 lb CO_2 per kWh. Space heating needs are met by a natural gas furnace that uses 250 million Btu of NG per year. The co-op owns a fleet of five gasoline-fueled delivery vans with average fuel economy of 7 miles/gallon. Each vehicle is driven 25,000 miles/year. The co-op has an old refrigerator/freezer unit that uses HFC 134a. The co-op manager estimates the freezer is charged with 1 kg of refrigerant every year to make up for leaks. Assume NG has the composition of methane (CH_4) and a heat content (LHV) of 50 MJ/kg. Assume gasoline has a density of 0.74 kg/L and an effective molecular formula of C_7H_{16} (MW = 100 g/mol). Present results separately for each scope and each relevant GHG. Also, present results for each scope in terms of CO_2-equivalent emissions. What are some opportunities for the co-op to reduce its GHG emissions?

3.8 Access the EIA's Annual Energy Outlook for the most recent year of publication. Find the charts for projected electricity generation from solar, wind, natural gas, and coal and compare the projected levels in 2050 in the new AEO to those projected for 2050 in AEO2023, which are shown in Figure 3.12. If there are noticeable differences, what economic, technological, or policy changes incorporated into the NEMS model might have caused them?

References

1. U.S. Environmental Protection Agency, April 2003. Application of life cycle assessment to nanoscale technology: Lithium-ion batteries for electric vehicles, EPA 744-R-12-001.
2. ISO 14040:2006. Environmental management—Life cycle assessment—Principles and framework, https://www.iso.org/obp/ui/#iso:std:iso:14040:ed-2:v1:en.
3. ISO 14044:2006. Environmental management—Life cycle assessment—Requirements and guidelines, https://www.iso.org/obp/ui/#iso:std:iso:14044:ed-1:v1:en.
4. Sathaye, J., Lucon, O., Rahman, A., Christensen, J., Denton, F., Fujino, J., Heath, G., Kadner, S., Mirza, M., Rudnick, H., Schlaepfer, A., and Shmakin, A., 2011. Renewable energy in the context of sustainable development, in: Edenhofer, O., Pichs-Madruga, R., Sokona, Y., Seyboth, K., Matschoss, P., Kadner, S., Zwickel, T., Eickemeier, P., Hansen, G., Schlömer, S. and von Stechow, C. eds., *IPCC Special Report on Renewable Energy Sources and Climate Change Mitigation*, Chapter 9, pp. 707–790, Cambridge University Press, Cambridge, UK.
5. Matthews, H.S., Hendrickson, C.T., and Matthews, D.H., 2014. *Life Cycle Assessment: Quantitative Approaches for* Decisions That Matter, Open access textbook, https://www.lcatextbook.com.

6. Smith, B.L., Sekar, A., Mirletz, H., Heath, G., and Margolis, R., 2024. An updated life cycle assessment of utility-scale solar photovoltaic systems installed in the United States, National Renewable Energy Laboratory, NREL/TP-7A40-87372, https://www.nrel.gov/docs/fy24osti/87372.pdf.
7. Boercker, S.W., 1978. Energy use in the production of primary aluminum, ORAU/IEA-78-14, Oak Ridge Associated Universities, Oak Ridge, TN.
8. Yang, Y., Ingwersen, W.W., Hawkins, T.R., Srocka, M., and Meyer, D.E., 2017. USEEIO: A new and transparent United States environmentally-extended input-output model, *Journal of Cleaner Production* 158, 308–318, doi:10.1016/j.jclepro.2017.04.150.
9. Ingwersen, W.W., Li, M., Young, B., Vendries, J., and Birney, C., 2022. USEEIOv2.0, The US environmentally-extended input-output model v 2.0, *Scientific Data* 9, 194, doi:10.1038/s41597-022-01293-7.
10. U.S. Environmental Protection Agency, April 2024. US Environmentally-Extended Input-Output (USEEIO) Technical Content, https://www.epa.gov/land-research/us-environmentally-extended-input-output-useeio-technical-content.
11. Spreng, D.T., 1988. *Net Energy Analysis and the Energy Requirements of Energy Systems*, Praeger Publishing Company, New York.
12. Hall, C.A.S., Lambert, J.G., and Balogh, S.B., 2014. EROI of different fuels and the implications for society, *Energy Policy* 64, 141–152, doi:10.1016/j.enpol.2013.05.049.
13. Murphy, D.J., Hall, C.A.S., Dale, M., and Cleveland, C., 2011. Order from Chaos: A preliminary protocol for determining the EROI of fuels, *Sustainability* 3, 1888–1907, doi:10.3390/su3101888.
14. Aramendia, E., Brockway, P.E., Taylor, P.G., Norman, J.B., Heun, M.K., and Marshall, Z., 2024. Estimation of useful-stage energy returns on investment for fossil fuels and implications for renewable energy systems, *Nature Energy* 9, 803–816, doi:10.1038/s41560-024-01518-6.
15. White, S.W., and Kulcinski, G.L., December 1998. *Net Energy Payback and CO_2 Emissions from Wind-Generated Electricity in the Midwest*, Fusion Tech Inst., University of Wisconsin, Madison.
16. Murphy, D.J., Raugei, M., Carbajeles-Dale, M., and Rubio Estrada, B., 2022. Energy return on investment of major energy carriers: Review and harmonization, *Sustainability* 14, 7098, doi:10.3390/ su14127098.
17. ISO 14064-1:2006. Greenhouse gases—Part 1: Specification with guidance at the organization level for quantification and reporting of greenhouse gas emissions and removals, https://www.iso.org/standard/38381.html.
18. Myhre, G., Shindell, D., Bréon, F.-M., Collins, W., Fuglestvedt, J., Huang, J., Koch, D., Lamarque, J.-F., Lee, D., Mendoza, B., Nakajima, T., Robock, A., Stephens, G., Takemura, T., Zhang, H., 2013. Anthropogenic and natural radiative forcing, in: Stocker, T.F., Qin, D., Plattner, G.-K., Tignor, M., Allen, S.K., Boschung, J., Nauels, A., Xia, Y., Bex, V., and P.M. Midgley (eds.), *Climate Change 2013: The Physical Science Basis. Working Group I Contribution to the Fifth Assessment Report of the Intergovernmental Panel on Climate Change*, Cambridge University Press, Cambridge, UK, https://www.ipcc.ch/pdf/assessment-report/ar5/wg1/WG1AR5_Chapter08_FINAL.pdf.
19. Forster, P., Storelvmo, T., Armour, K., Collins, W., Dufresne, J.-L., Frame, D., Lunt, D.J., Mauritsen, T., Palmer, M.D., Watanabe, M., Wild, M., and Zhang, H., 2021. The earth's energy budget, climate feedbacks, and climate sensitivity, in: Masson-Delmotte, V., Zhai, P., Pirani, A., Connors, S.L., Pean, C., Berger, S., Caud, N., Chen, Y., Goldfarb, L., Gomis, M.I., Huang, M., Leitzell, K., Lonnoy, E., Matthews, J.B.R., Maycock, T.K., Waterfield, T., Yelekci, O., Yu, R. and Zhou, B. eds., *Climate Change 2021: The Physical Science Basis. Contribution of Working Group I to the Sixth Assessment Report of the Intergovernmental Panel on Climate Change*, Cambridge University Press, Cambridge, UK and New York, Table 7.SM.7, doi:10.1017/9781009157896.009.
20. U.S. Environmental Protection Agency, 2024. Inventory of U.S. greenhouse gas emissions and sinks: 1990-2022, EPA 430-R-24-004, https://www.epa.gov/ghgemissions/inventory-us-greenhouse-gas-emissions-and-sinks-1990-2022.
21. Ho, J., Becker, J., Brown, M., Brown, P., Chernyakhovskiy, I., Cohen, S., Cole, W., Corcoran, S., Eurek, K., Frazier, W., Gagnon, P., Gates, N., Greer, D., Jadun, P., Khanal, S., Machen, S., Macmillan, M., Mai, T., Mowers, M., Murphy, C., Rose, A., Schleifer, A., Sergi, B., Steinberg, D., Sun, Y., and

Zhou, E., 2020. Regional Energy Deployment System (ReEDS) model documentation: Version 2020, NREL/TP-6A20-78195, National Renewable Energy Laboratory, Golden, CO, https://www.nrel.gov/docs/fy21osti/78195.pdf.
22. U.S. Energy Information Administration, May 2023. *The National Energy Modeling System: An Overview*, U.S. Department of Energy, Washington, DC.
23. U.S. Energy Information Administration, March 2023. *Annual Energy Outlook AEO 2023*, U.S. Department of Energy, Washington, DC.
24. International Energy Agency, 2023. Global energy and climate model documentation 2023, https://www.iea.org/reports/global-energy-and-climate-model.

4

Energy Use and Efficiency in Buildings and Industry

Energy efficiency is one weapon in the arsenal against over-dependence on GHG-emitting fossil fuels. I believe efficiency is the best weapon: cheapest, safest, and most immediately achievable.

Art Rosenfeld [1]

4.1 Background

Energy conservation and efficiency have long been considered the low-hanging fruit in our efforts to improve our energy system. Saving energy is generally more cost-effective than generating new energy. In the words of energy futurist Amory Lovins, "negawatts" (i.e., watts of saved power) are better than megawatts. Energy *conservation* means using less energy or getting by with less. This can be done through active choices or interventions, such as actively turning down the lights or thermostat. A problem with conventional "energy conservation" measures is that they depend on human behavior and motivation and may be considered inconvenient. Increasingly, however, automated systems can take care of these measures for us, setting thermostats back or shutting off lights when a building or room is unoccupied. Energy *efficiency* entails using energy in a more efficient way that has little or no impact on the function or service provided. Thus, the replacement of incandescent bulbs with light-emitting diode (LED) bulbs can reduce energy consumption while providing the same illumination level and quality of light. Or a homeowner might purchase a higher efficiency furnace that provides the same level of heating with less energy required.

In the United States, total energy use in 2023 (on-site primary fuel use plus the primary fuel used to generate electricity) was divided among the end-use sectors as follows: 37% buildings (split 53:47 between residential and commercial), 33% industrial, and 30% transportation [2]. World energy use is more heavily skewed to industrial and is divided as follows: 20% buildings (split 66:34 between residential and commercial), 55% industrial, and 25% transportation [3].

Energy efficiency measures are not as conspicuous as big new power plants, and so people often do not appreciate the great strides that have been made. Since the oil embargoes of the 1970s, energy efficiency measures have greatly limited energy growth, even as standards of living have risen.

How much difference have efficiency measures made? In a 2013 International Energy Agency report covering 13 member countries, it was concluded that energy use would have been 65% higher in 2010 than it otherwise would have been if energy efficiency improvements since 1974 had not been made [4].

DOI: 10.1201/9781003586616-4

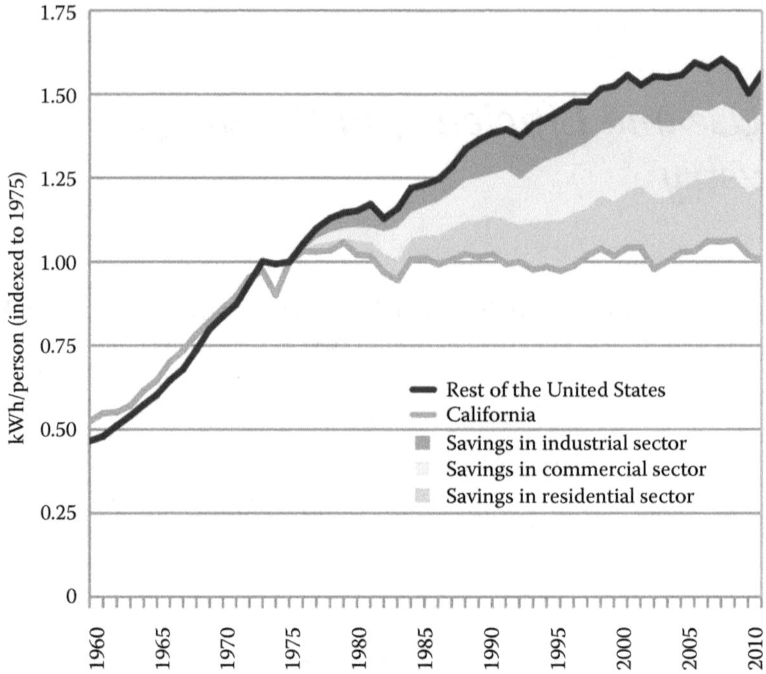

FIGURE 4.1
California electricity growth vs. rest of United States. (From NRDC Fact Sheet [5].)

A classic case in the United States is the state of California, where aggressive measures began in the 1970s to control energy use. As shown in Figure 4.1, California managed to keep per-capita electricity use essentially constant despite increasing the use of air conditioning, whereas per-capita electricity use increased in the United States as a whole [5].

Note that 80% of the electricity savings in California compared to the rest of the country are due to efficiency improvements in residential and commercial buildings. Much of this success is due to utility incentives and state policies that promote aggressive building energy efficiency and federal appliance efficiency standards. Despite these energy efficiency measures (or perhaps, in part, because of them) if California were a nation, it would rank as the world's fourth largest economy.

When reviewing historical improvements in energy consumption, a logical question is how much of the change is due to energy efficiency versus structural changes in industry. (Structural change involves shifts in economic production and employment, for example, from agriculture and manufacturing to software and services.) The American Council for an Energy-Efficient Economy (ACEEE) has estimated that approximately 40% of the improvement in total energy use in the United States has been due to structural changes as shown in Figure 4.2 [6]. However, the total energy use in the United States today is 60% less than it would otherwise be if energy efficiency improvements had not been made.

A great success story has been the impact of appliance and equipment efficiency standards in the United States. Figure 4.3 shows the relative improvement in the energy consumption of some new appliances and equipment in the United States from 1980 to 2018 [6].

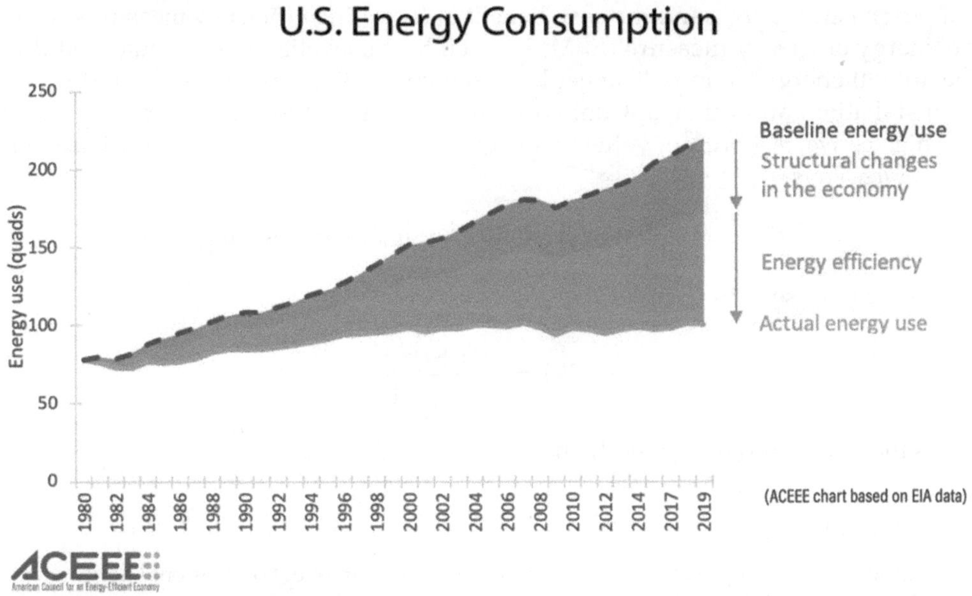

FIGURE 4.2
Total U.S. primary energy use changes due to structural change and energy efficiency. (Nadel, S. ACEEE [6].)

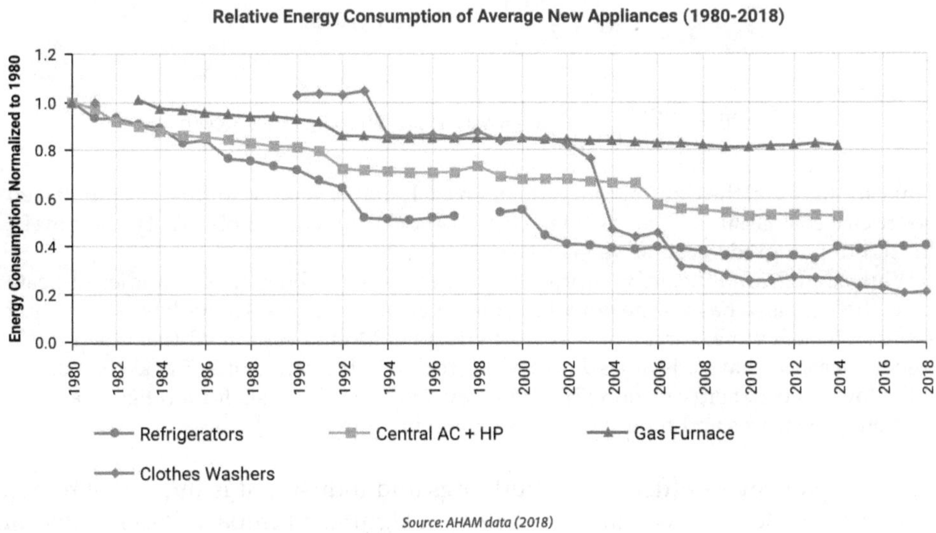

FIGURE 4.3
Relative average energy consumption of new appliances sold in the United States from 1980 to 2018. (From Nadel, S. ACEEE [6]; see also https://energyefficiencyimpact.org/buildings/.)

These improvements came with no downside. For example, today's refrigerators are lower in cost (in real dollars) and somewhat larger than the ones sold in 1980, despite using about one-quarter of the energy.

To calculate the cost of energy efficiency measures, a common metric used is the cost of conserved energy or CCE. CCE is useful because it allows a direct comparison between the cost of purchasing energy and the cost/benefit of installing efficiency measures. The CCE for an energy efficiency measure (EEM) is equal to the levelized net annual cost divided by the annual energy savings. The net levelized cost is the capital cost (including equipment, installation, and other upfront costs) multiplied by the capital recovery factor, or CRF, which converts a present value to an equivalent constant annual value based on the discount rate. Thus,

$$\text{CCE} = \text{Capital Cost} \times \text{CRF}/\text{Annual Energy Savings} \tag{4.1}$$

where, as we have seen in Chapter 2,

$$\text{CRF} = \frac{i(1+i)^t}{(1+i)^t - 1}$$

where i=the discount rate (e.g., 0.04) and t=number of years.

Example 4.1

An office lighting upgrade has a net capital cost premium over conventional lighting of \$2,000 but saves 2 kW of electric power during its 5,000 h/year of operation.

The annual savings=2 kW×5,000 hours=10,000 kWh

Assuming a 15-year lifetime (t=15) and a discount rate of 4% (i=0.04),

$$\text{CRF} = \frac{0.04(1+0.04)^{15}}{(1+0.04)^{15} - 1} = 0.08994$$

$$\text{CCE} = \frac{\$2,000 \times 0.08994}{10,000\ \text{kWh}} = \$0.018/\text{kWh} \quad \text{or} \quad 1.8\,¢/\text{kWh}$$

Thus, in this case, the lighting upgrade has a CCE that is about one-fifth the purchased electricity cost (assuming an average cost of 10 ¢/kWh). Higher electricity costs make the lighting upgrade even more favorable.

While significant efficiency improvements have already been made, studies indicate that efficiency also has a large potential going forward. Various studies have shown that energy efficiency measures are very cost-effective. Molina estimated that energy efficiency measures have a levelized cost of electricity in the range of 2–5 ¢/kWh, making it the lowest cost energy option [7]. McKinsey analyzed the costs for a range of energy efficiency measures and their results are shown in Figure 4.4 [8].

While this chapter covers efficiency in buildings and industry, it is important to note that standards for vehicle efficiency are also having a significant impact. Recent international historical and target miles per gallon and kilometers per liter for passenger vehicles are given at the Center for Climate and Energy Solutions [9]. Improved aerodynamics, lighter weight construction materials, low rolling resistance tires, internal combustion engine refinements, and hybrid and electric vehicles are all playing a role in achieving improved vehicle efficiency. The batteries in electric vehicles (EVs) will be increasingly important for allowing deeper penetration of variable renewables and zero-energy buildings on the grid. Transportation is covered in Chapter 15.

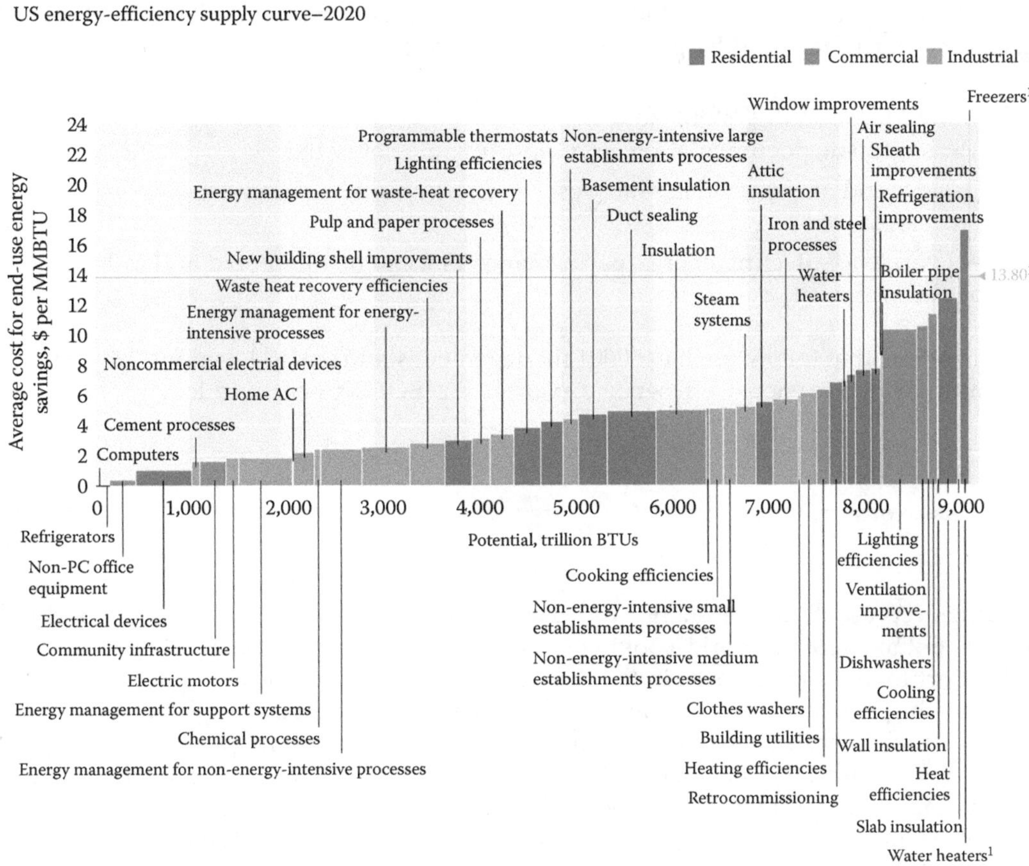

[1]Average price of avoided energy consumption at the industrial price; $35.60/MMBTU represents the highest regional electricity price used; new build cost based on AEO 2008 future construction costs.

FIGURE 4.4
CCE comparison of different energy efficiency measures. (From McKinsey & Company [8]. Copyright 2010 McKinsey & Company. All rights reserved. Reprinted by permission.)

4.2 Energy Audits and Energy Management

Achieving high levels of energy efficiency requires a focused, systematic approach. This has spawned the field of Strategic Energy Management, or SEM. SEM programs have become popular in states and utilities across the United States and Canada [10]. A typical SEM plan benchmarks current energy use including an energy audit, develops energy performance goals, evaluates the impact and cost of potential measures, establishes a measurement and verification plan, and schedules multi-year implementation.

An international Energy Management System standard, International Organization of Standardization (ISO) 50001, has been developed to provide an energy management framework for organizations to follow in improving energy efficiency. ISO 50001 is a voluntary standard that provides guidance on the following [11]:

- Energy use and consumption
- Measurement, documentation, and reporting of energy use
- Design and procurement practices for energy-using equipment, systems, and processes
- All variables affecting energy performance that can be monitored and influenced by the organization.

Organizations that follow the ISO 50001 management standard and demonstrate continuous improvement in energy efficiency can apply for ISO 50001 certification.

4.3 Buildings

Buildings represent 40% of energy use in the United States and 30% in the world [12]. Throughout much of the world, cooking is a major energy user in homes. There is a worldwide effort to make cookstoves in the developing world more efficient and safer. Clearly, space heating and water heating are major end uses of building energy throughout the world. Note that while space cooling is a major energy consumer in the United States, it is still not widely used in much of the world. Cooling tends to increase as countries develop economically and so is increasing around the world. Appliances, which include computers and office equipment, are an especially major load in nonresidential buildings.

4.3.1 Calculations of Heating and Hot Water Loads in Buildings

Energy requirements for space heating or service water heating can be calculated from basic conservation of energy principles. For example, the heat required to maintain the interior of a building at a specific temperature is the total of all heat transmission losses from the structure and heat required to warm and humidify the air exchange with the environment by infiltration and ventilation.

Comfort in buildings has long been a subject of investigation by the American Society of Heating, Refrigerating, and Air-Conditioning Engineers (ASHRAE). ASHRAE has developed extensive heat load calculation procedures embodied in the *ASHRAE Handbook of Fundamentals* [13]. The most frequently used load calculation procedures will be summarized in this section; the reader is referred to the ASHRAE handbook for details.

Figure 4.5 shows the combinations of temperature and humidity that are required for human comfort. The shaded area is the standard U.S. comfort level for sedentary persons. Many European countries have human comfort levels from 3°C to 7°C below U.S. levels. If activity of a continuous nature is anticipated, the comfort zone lies to the left of the shaded area; if extra clothing is worn, the comfort zone is displaced similarly.

4.3.1.1 Calculation of Heat Loss

It is outside the scope of this book to describe the details of the heat load calculations for buildings. However, the method is described briefly in this section. For details, one should

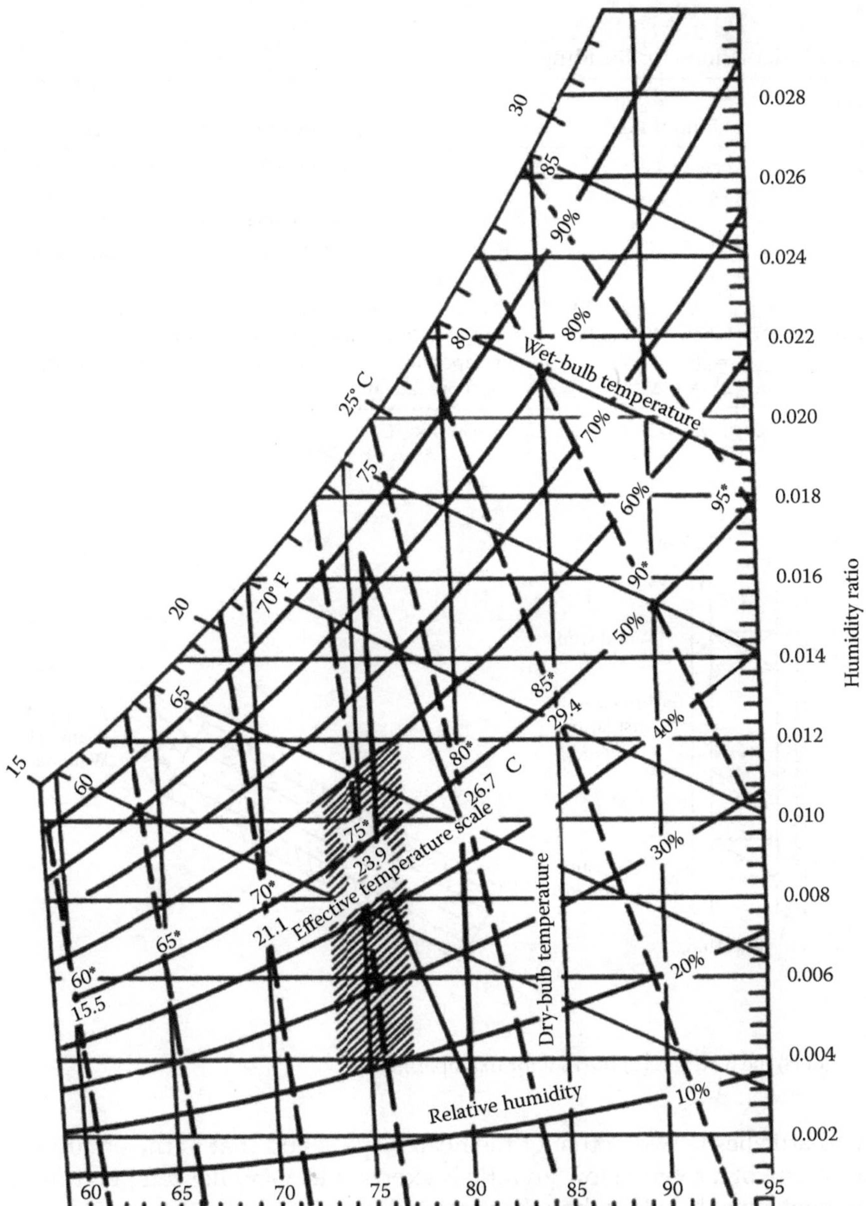

FIGURE 4.5
Psychrometric chart of moist air properties with the range of dry-bulb temperature and relative humidity conditions for human comfort.

refer to the *ASHRAE Handbook of Fundamentals* [13] or some other textbook on heating and air conditioning. See also a general heat transfer textbook, such as *Principles of Heat Transfer*, 8th edition, by Frank Kreith and Raj M. Manglik [14]. Table 4.1 lists the components of heat loss calculations of a building.

Transmission heat losses through attics, unheated basements, and the like, are buffered by the thermal resistance of the unheated space. For example, the temperature of

TABLE 4.1

Heating Load Calculations for Buildings

Heating Load Component	Equations	Descriptions/References
Walls, roof, ceilings, glass	$q=U{\cdot}A(T_i-T_o)$	T_i, T_o are inside and outside air temperatures, respectively. U values of composite section are calculated from the thermal properties of components given in standard property tables
Concrete floors on ground	$q_{fe}=F_eP_e(T_i-T_o)$	P_e is the perimeter of the slab. F_e is the perimeter heat loss value [13]
Infiltration and ventilation air	$q_{sensible}=Q\rho_aC_{pa}(T_i-T_o)$ or $=1{,}200\times Q(T_i-T_o)$ $q_{latent}=Q\rho_ah_{fg}\Delta W$ or $=2{,}808^{a}Q\Delta W$	Q is the volume of airflow in m³/s. ρ_a and C_{pa} are density and specific heat of air h_{fg} is the latent heat of water at room temperature. ΔW is humidity ratio difference between inside and outside air

[a] Assuming $\rho_a=1.2\,\text{kg/m}^3$; $h_{fg}=2{,}340$ kJ/kg, $C_{pa}=1{,}000$ J/kg°C.

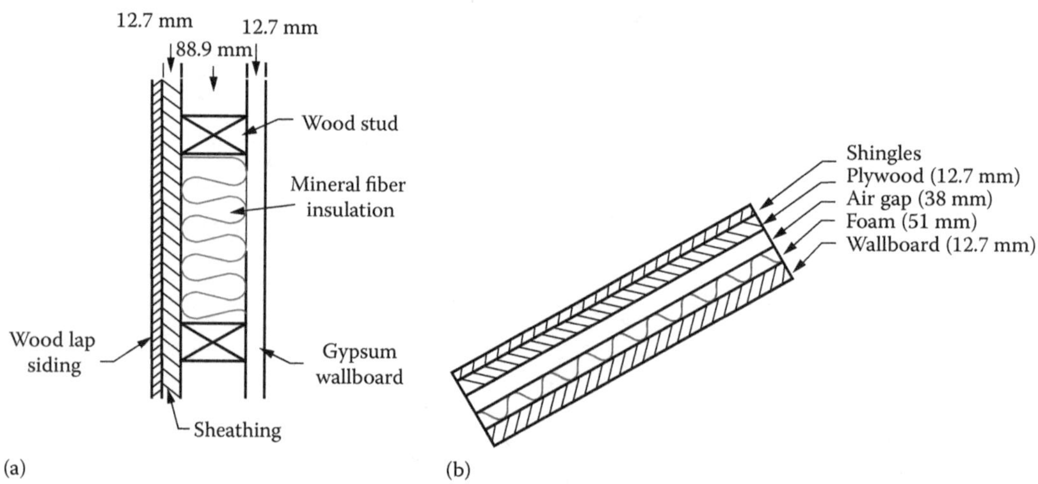

FIGURE 4.6
Cross sections of (a) the wall and (b) the roof for Example 4.2.

an unheated attic lies between that of the heated space and that of the environment. As a result, the ceiling of a room below an attic is exposed to a smaller temperature difference and consequent lower heat loss than the same ceiling without the attic would be. The effective conductance of thermal buffer spaces can easily be calculated by forming an energy balance on such spaces.

The following example is an illustration of the heat loss calculation method described in this section.

Example 4.2

Calculate the heat load on a house for which the wall area is 200 m², the floor area is 600 m², the roof area is 690 m², and the window area totals 100 m². Inside wall height is 3 m. The construction of the wall and the roof is shown in Figure 4.6.

Solution:

The thermal resistance of the wall shown in Figure 4.6 can be found by the electrical resistance analogy as

$$R_{wa} = R_{outside air} + R_{woodsiding} + R_{sheathing} + R_{comb} + R_{wallboard} + R_{inside air}$$

The wood stud and the insulation are in parallel, so as in the case of electrical resistors in parallel, to get the combined resistance we add them in parallel. (The *R*-value equals 1/U. The U values indicate the heat flow and are additive just like the currents in parallel resistors. Thus, the reciprocals of parallel resistors add.)

The wood stud is a 2 by 4 with an actual depth of 3.5″. Its *R*-value is 4.37 ft^2°F/(Btu/h). To convert from English to SI units, we multiply by 0.176. Thus, the *R*-value of the wood stud is 4.37×0.176=0.77 m^2°C/W.

The 3.5″ of mineral fiber insulation has an *R*-value of 11.02 ft^2°F/(Btu/h)=1.94 m^2°C/W.

Assume that the studs occupy 15% of the wall area and the insulation occupies the other 85%. Then

$$\frac{1}{R_{comb}} = \frac{0.15}{0.77} + \frac{0.85}{1.94}$$

or

$$R_{comb} = 1.58\ \text{m}^{2\circ}\text{C/W}$$

Therefore, the wall thermal resistance, R_{wa}, can be found as

Element	Thermal Resistance (m$^{2\circ}$C/W)
Outside air (6.7 m/s wind)	0.030
Wood bevel lap siding	0.14
12.7 mm sheathing	0.23
88.9 mm combined wood stud and mineral fiber insulation	1.58
1.7 mm gypsum wallboard	0.079
Inside air (still)	0.12
	R_{wa}=2.179

Therefore,

$$U_{wa} = \frac{1}{R_{wa}} = \frac{1}{2.179} = 0.46\ \text{W/m}^{2\circ}\text{C}$$

The heat loss through the windows depends on whether they are single- or double-glazed. In this example, single-glazed windows are installed, and a *U* factor equal to 4.7 W/m^2°C is used. (If double-glazed windows were installed, the *U* factor would be 2.4 W/m^2°C.)

The roof is constructed of 12.7 mm gypsum wallboard, 51 mm foam insulation board, 38 mm still air, 12.7 mm plywood, and asphalt shingles (wooden beams and roofing paper are neglected for the simplified calculations here). Therefore,

$$U_{rf} = \frac{1}{0.030_{\text{outside air}} + 0.077_{\text{Shingles}} + 0.11_{\text{Plywood}} + 0.17_{\text{Airgap}} + 2.53_{\text{Foam}} + 0.079_{\text{Wallboard}} + 0.1_{\text{inside air}}}$$

$$= 0.32\ \text{W/m}^{2\circ}\text{C}$$

If the respective areas and U factors are known, the rate of heat loss per hour for the walls, windows, and roof can be calculated, assuming that floor heat loss is negligible:

$$\text{Walls}: q_{wa} = \left(200 \text{ m}^2\right) \times \frac{0.46 \text{ W}}{\text{m}^{2\circ}\text{C}} = 92 \text{ W/°C}$$

$$\text{Windows}: q_{wi} = \left(100 \text{ m}^2\right) \times \frac{4.7 \text{ W}}{\text{m}^{2\circ}\text{C}} = 470 \text{ W/°C}$$

$$\text{Roof}: q_{rf} = \left(690 \text{ m}^2\right) \times \frac{0.32 \text{ W}}{\text{m}^{2\circ}\text{C}} = 220 \text{ W/°C}$$

$$\text{Total} = 782 \text{ W/°C}$$

If double-glazed windows were used, the total heat loss would be reduced to 552 W/°C.

The infiltration and ventilation rate Q for this building is assumed to be 0.5 air changes per hour (ACH). The sensible and latent heat loads of the infiltration air may be calculated using the equations given in Table 4.1. Therefore,

$$Q = 0.5 \times \left(600 \text{ m}^2 \times 3 \text{ m}\right) = 900 \text{ m}^3/\text{h} = 0.25 \text{ m}^3/\text{s}$$

$$q_{\text{sensible}} = 0.25 \text{ m}^3/\text{s} \times \left(1.2 \text{ kg/m}^3\right)\left(1{,}000 \text{ J / kg°C}\right) = 300 \text{ W/°C}$$

In residential buildings, humidification of the infiltration air is rarely done. Neglecting the latent heat, the total rate of heat loss q_{tot} is the sum of q_{sensible} and q_{tr}:

$$q_{\text{tot}} = (782 + 300) = 1{,}082 \text{ W/°C}$$

This calculation is simplified for purposes of illustration. Heat losses through the slab surface and edges have been neglected, for example.

More refined methods of calculating energy requirements on buildings do not use the steady-state assumption used earlier [15]. The thermal inertia of buildings may be expressly used as a load-leveling device. If so, the steady-state assumption is not met, and the energy capacitance of the structure must be considered for accurate results. Many adobe structures in the U.S. Southwest are built intentionally to use daytime sun absorbed by 1 ft thick walls for nighttime heating, for example.

4.3.1.2 Internal Heat Sources in Buildings

Heat supplied to a building to offset energy losses is derived from both the heating system and internal heat sources. Table 4.2 lists the common sources of internal heat generation for residences. Commercial buildings such as hospitals, computer facilities, or supermarkets will have large internal gains specific to their function. Internal heat gains tend to offset heat losses from a building but will add to the cooling load of an air-conditioning system. The magnitude of the reduction in heating system operation will be described in the next section.

4.3.1.3 Degree-Day Method

The preceding analysis of heat loss from buildings expresses the loss on a per-unit temperature difference basis (except for unexposed floor slabs). In order to calculate the

TABLE 4.2

Some Common Internal Sensible Heat Gains That Tend to Offset the Heating Requirements of Buildings

Type	Magnitude (W or J/s)
Incandescent lights	Total W
Fluorescent lights	Total W
Electric motors	746×(hp/efficiency)
Natural gas stove	$8.28 \times m^3/h$
Appliances	Total W
A dog	50–90
People	
Sitting	70
Walking	75
Dancing	90
Working hard	170

Source: From ASHRAE [13].

peak load and total annual load for a building, appropriate design temperatures must be defined for each. The outdoor design temperature is usually defined statistically, such that the actual outdoor temperature will exceed the design temperature 97.5% or 99% of the time over a long period. The design temperature difference (ΔT) is then the interior building temperature minus the outdoor design temperature. The design ΔT is used for rating non-solar heating systems but is not useful for the selection of solar thermal energy systems, since they rarely provide 100% of the energy demand of a building at peak conditions.

A more useful index of heating energy demand is the total annual energy requirement for a building. This quantity is somewhat more difficult to calculate than the peak load. It requires knowledge of day-to-day variations in ambient temperature during the heating season and the corresponding building heat load for each day. Building heat loads vary with ambient temperatures, as shown in Figure 4.7. The outdoor temperature, T_{nl}, above which no heat needs to be supplied to the building, is a few degrees below the required interior temperature, T_i, because of internal heat-generation effects.

The no-load temperature at which internal source generation, q_i, just balances transmission and infiltration losses can be determined from the energy balance:

$$q_i = \overline{UA_B}(T_i - T_{\text{nl}}) \tag{4.2}$$

Where $\overline{UA_B}$ is the overall heat loss coefficient for the building (W/°C). Then,

$$T_{\text{nl}} = T_i - \frac{q_i}{\overline{UA_B}} \tag{4.3}$$

The total annual heat load on the building, Q_T, can be expressed as follows:

$$Q_T = \int_{365\text{ days}} \overline{UA_B}(T_{\text{nl}} - T_a)^+ \, dt \tag{4.4}$$

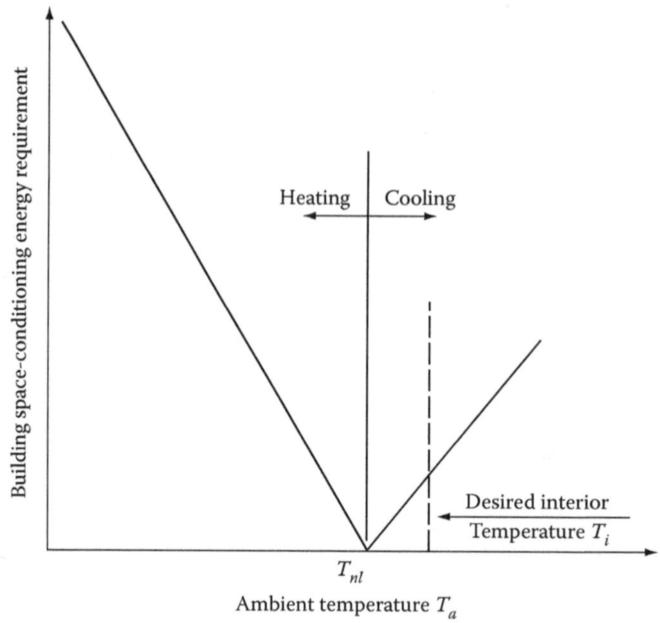

FIGURE 4.7
Building load profile versus ambient temperature showing no-load temperature, T_{nl}, and desired interior temperature, T_i.

in which all arguments of the integral are functions of time. The superscript + indicates that only positive values are considered. In practice, it is difficult to evaluate this integral; therefore, three simplifying assumptions are made:

1. $\overline{UA_B}$ is independent of time.
2. T_{nl} is independent of time.
3. The integral can be expressed by the sum.

$$\overline{UA_B}\sum_{n=1}^{365}\left(T_{nl}-\overline{T}_a\right)_n^+ \tag{4.5}$$

where n is the day number, and the daily average temperature $\overline{T_a}$ can be approximated by $1/2(T_{a,\max}+T_{a,\min})$, in which $T_{a,\max}$ and $T_{a,\min}$ are the daily maximum and minimum temperatures, respectively.

The quantity $(T_{nl}-T_a)^+$ is called the *degree-day unit*. For example, if the average ambient temperature for a day is 5°C and the no-load temperature is 20°C, 15°C-days are said to exist for that day. However, if the ambient temperature is 20°C or higher, 0°C-days exist, indicating 0 demand for heating that day. Degree-day totals for monthly $\Sigma_{\text{month}}\left(T_{nl}-T_a\right)^+$, and annual periods can be used directly in Equation 4.5 to calculate the monthly and annual heating energy requirements.

ASHRAE provides data on climatic design conditions around the world that include both heating and cooling degree-days [16]. Typically, a no-load outdoor temperature of 65.0°F, or 18.3°C, is used as a base for both, but the ASHRAE data also includes degree-days for an additional base value of 50.0°F, or 10°C, for buildings that have higher levels of internal heat generation.

TABLE 4.3

Monthly and Annual Energy Demands for Example 4.3

Month	°C-Days	Energy Demand[a] (GJ)
January	518	12.4
February	423	10.2
March	396	9.5
April	214	5.2
May	68	1.6
June	14	0.3
July	0	0
August	0	0
September	26	0.6
October	148	3.6
November	343	8.2
December	472	11.3
	2,622	62.9

[a] Energy demand equals $\overline{UA_B} \times$ °C-days $\times$ 24 h/day.

Example 4.3

A building located in Denver, CO, has a heat loss coefficient $\overline{UA_B}$ of 1,000 kJ/h°C and internal heat sources of 4,440 kJ/h. If the interior temperature is 20°C (68°F), what are the monthly and annual heating energy requirements? A gas furnace with 65% efficiency is used to heat the building.

Solution:

In order to determine the monthly degree-day totals, the no-load temperature (degree-day basis) must be evaluated from Equation 4.3:

$$T_{\text{nl}} = 20 - \frac{4{,}440}{1{,}000} = 15.6^{\circ}\text{C}\left(60^{\circ}\text{F}\right)$$

The monthly °C-days for Denver are taken from the U.S. NWS and given in Table 4.3.

The energy demand is calculated as

$$\text{Energy demand} = \overline{UA_B} \times 24\,\frac{\text{h}}{\text{day}} \times {^{\circ}\text{C}} - \text{days}$$

The monthly energy demand is given in Table 4.3.

If the annual energy demand of 62.9 GJ is delivered by a 65% efficient gas furnace, then

$$\text{Average annual purchased energy} = \frac{62.9}{0.65}\,\text{GJ} = 96.8\,\text{GJ}$$

4.3.1.4 Service Hot Water Load Calculation

Service hot water loads can be calculated precisely with the knowledge of only a few variables. The data required for the calculation of hot water demand are:

Water source temperature (T_s)
Water delivery temperature (T_d)
Volumetric demand rate (Q)

The energy requirement for service water heating q_{hw} is given by

$$q_{hw}(t) = \rho_w Q(t) c_{pw} \left[T_d - T_s(t) \right] \tag{4.6}$$

where

ρ_w is the water density

c_{pw} is the specific heat of water.

The demand rate, *Q*(*t*), varies in general with time of day and time of year; likewise, the source temperature varies seasonally. Source temperature data are not compiled in a single reference; local water authorities are the source of such temperature data.

Few generalized data exist with which to predict the demand rate *Q*. Table 4.4 indicates some typical usage rates for several common building types. Process water-heating rates are peculiar to each process and can be ascertained by reference to process specifications.

Example 4.4

Calculate the monthly energy required to heat water for a family of four in Nashville, TN. Monthly source temperatures for Nashville are shown in Table 4.5, and the water delivery temperature is 60°C (140°F).

TABLE 4.4

Approximate Service Hot Water Demand Rates

Usage Type	Demand per Person L/Day	Demand per Person Gal/Day
Retail store	2.8	0.75
Elementary school	5.7	1.5
Multifamily residence	76.0	20.0
Single-family residence	76.0	20.0
Office building	11.0	3.0

TABLE 4.5

Water-Heating Energy Demands for Example 4.4

Month	Days/Month	Demand (m^3/Month)	Source Temperature (°C)	Energy Requirement (GJ/Month)
January	31	9.3	8	2.0
February	28	8.4	8	1.8
March	31	9.3	12	1.9
April	30	9.0	19	1.5
May	31	9.3	17	1.7
June	30	9.0	21	1.5
July	31	9.3	22	1.5
August	31	9.3	24	1.4
September	30	9.0	24	1.4
October	31	9.3	22	1.5
November	30	9.0	14	1.7
December	31	9.3	12	1.9

Solution:

For a family of four, the demand rate Q may be found using a demand recommended from Table 4.4:

$$Q = 4 \times 76 \text{ L/day} \times \frac{1 \text{ m}^3}{1{,}000 \text{ L}} = 0.30 \text{ m}^3\text{/day}$$

The density of water can be taken as 1,000 kg/m³ and the specific heat as 4.18 kJ/kg°C. Monthly demands are given by

$$q_m = (Q \times \text{days/month})(\rho_w c_{pw})\left[T_d - T_s(t)\right] = (0.30 \times \text{days/month})(1{,}000 \times 4.18)\left[60 - T_s(t)\right]$$

The monthly energy demands are tabulated in Table 4.5.

4.3.2 Cooling Requirements for Buildings

The cooling load of a building is the rate at which heat must be removed to maintain the air in a building at a given temperature and humidity. It is usually calculated on the basis of the peak load expected during the cooling season. For a given building, the cooling load depends primarily on:

1. Inside and outside dry-bulb temperatures and relative humidities
2. Solar radiation heat load and wind speed
3. Infiltration and ventilation
4. Internal heat sources.

A method of calculating the cooling load is presented in detail in [13].

The steps in calculating the cooling load of a building are as follows:

1. **Specify the building characteristics:** wall area, type of construction, and surface characteristics; roof area, type of construction, and surface characteristics; window area, setback, and glass type; and building location and orientation.
2. Specify the outside and inside wet- and dry-bulb temperatures.
3. Specify the solar heat load and wind speed.
4. **Calculate building cooling load resulting from the following:** heat transfer through windows; heat transfer through walls; heat transfer through roof; sensible and latent heat gains resulting from infiltration and ventilation; and sensible and latent heat gains (water vapor) from internal sources, such as people, lights, cooking, etc.

Equations 4.7 through 4.13 may be used to calculate the various cooling loads for a building. Cooling loads resulting from lights, building occupants, and so on, may be estimated from [13]. For unshaded or partially shaded windows, the load is

$$\dot{Q}_{\text{wi}} = A_{\text{wi}}\left[F_{\text{sh}}\overline{\tau}_{b,\text{wi}}I_{h,b}\frac{\cos i}{\sin\alpha} + \overline{\tau}_{d,\text{wi}}I_{h,d} + \overline{\tau}_{r,\text{wi}}I_r + U_{\text{wi}}\left(T_{\text{out}} - T_{\text{in}}\right)\right] \tag{4.7}$$

For shaded windows, the load (neglecting sky diffuse and reflected radiations) is

$$\dot{Q}_{\text{wi,sh}} = A_{\text{wi,sh}}U_{\text{wi}}\left(T_{\text{out}} - T_{\text{in}}\right) \tag{4.8}$$

For unshaded walls, the load is

$$\dot{Q}_{\text{wa}} = A_{\text{wa}}\left[\bar{\alpha}_{s,\text{wa}}\left(I_r + I_{h,d} + I_{h,b}\frac{\cos i}{\sin\alpha}\right) + U_{\text{wa}}\left(T_{\text{out}} - T_{\text{in}}\right)\right] \tag{4.9}$$

For shaded walls, the load (neglecting sky diffuse and reflected radiations) is

$$\dot{Q}_{\text{wa,sh}} = A_{\text{wa,sh}}\left[U_{\text{wa}}\left(T_{\text{out}} - T_{\text{in}}\right)\right] \tag{4.10}$$

For the roof, the load is

$$\dot{Q}_{\text{rf}} = A_{\text{rf}}\left[\bar{\alpha}_{s,\text{rf}}\left(I_{h,d} + I_{h,b}\frac{\cos i}{\sin\alpha}\right) + U_{\text{rf}}\left(T_{\text{out}} - T_{\text{in}}\right)\right] \tag{4.11}$$

Sensible-cooling load due to infiltration and ventilation is

$$\dot{Q}_i = \dot{m}_a\left(h_{\text{out}} - h_{\text{in}}\right) = \dot{m}_a C_{p_a}\left(T_{\text{out}} - T_{\text{in}}\right) \tag{4.12}$$

Latent load due to infiltration and ventilation is

$$\dot{Q}_w = \dot{m}_a\left(W_{\text{out}} - W_{\text{in}}\right)\lambda_w \tag{4.13}$$

where

$\dot{Q}_{\text{wi}}$ is the heat flow through unshaded windows of area A_{wi}
$\dot{Q}_{\text{wi,sh}}$ is the heat flow through shaded windows of area $A_{\text{wi,sh}}$
$\dot{Q}_{\text{wa}}$ is the heat flow through unshaded walls of area A_{wa}
$\dot{Q}_{\text{wa,sh}}$ is the heat flow through shaded walls of area $A_{\text{wa,sh}}$
$\dot{Q}_{\text{rf}}$ is the heat flow through roof of area A_{rf}
$\dot{Q}_i$ is the heat load resulting from infiltration and ventilation
$\dot{Q}_w$ is the latent heat load
$I_{h,b}$ is the beam component of insolation on horizontal surface
$I_{h,d}$ is the diffuse component of insolation on horizontal surface
I_r is the ground-reflected component of insolation
W_{out}, W_{in} are the outside and inside humidity ratios
U_{wi}, U_{wa}, U_{rf} are the overall heat-transfer coefficients for windows, walls, and roofs, including radiation
$\dot{m}_a$ is the net infiltration and ventilation mass flow rate of dry air
C_{pa} is the specific heat of air (approximately 1.025 kJ/kg K for moist air)
T_{out} is the outside dry-bulb temperature
T_{in} is the indoor dry-bulb temperature
F_{sh} is the shading factor (1.0=unshaded, 0.0=fully shaded)
$\bar{\alpha}_{s,\text{wa}}$ is the wall solar absorptance
$\bar{\alpha}_{s,\text{rf}}$ is the roof solar absorptance
i is the solar incidence angle on walls, windows, and roof
h_{out}, h_{in} are the outside and inside air enthalpies
α is the solar altitude angle
λ_w is the latent heat of water vapor
$\bar{\tau}_{b,\text{wi}}$ is the window transmittance for beam (direct) insolation
$\bar{\tau}_{d,\text{wi}}$ is the window transmittance for diffuse insolation
$\bar{\tau}_{r,\text{wi}}$ is the window transmittance for ground-reflected insolation.

ASHRAE handbooks recommend the use of the cooling load temperature difference method. For more details of the method, refer to [13].

Example 4.5

Determine the cooling load for a building in Phoenix, AZ, assuming the north, east, and west walls are shaded and having the specifications tabulated in Table 4.6.

Solution

To determine the cooling load for the building just described, calculate the following factors in the order listed:

1. Solar altitude α at solar noon (from Equation 10.28), noting that $h_s=0$,

$$\sin\alpha = \sin\delta_s \sin L + \cos\delta_s \cos L \cos h_s = \cos(L-\delta_s) = \cos 15° = 0.966$$

Hence $\alpha=\sin^{-1}(0.966)=74.9°$

TABLE 4.6

Specifications for Example 4.5

Factor	Description or Specification
Building characteristics	
Roof	
Type of roof	Flat, shaded
Area $A_{rf,sh}$, ft^2	1,700
Walls (painted white)	
Size, north and south, ft	8×60 (two)
Size, east and west, ft	8×40 (two)
Area A_{wa}, north and south walls, ft^2	$480-A_{wi}=480-40=440$ (two)
Area A_{wa}, east and west walls, ft^2	$320-A_{wi}=320-40=280$ (two)
Absorptance $\bar{\alpha}_{s,wa}$ of white paint	0.12
Windows	
Size, north and south, ft	4×5 (two)
Size, east and west, ft	4×5 (two)
Shading factor F_{sh}	0.20
Insolation transmittance	$\bar{\tau}_{b,wi}=0.6$; $\bar{\tau}_{d,wi}=0.81$; $\bar{\tau}_{r,wi}=0.60$
Location and latitude	Phoenix, AZ; 33°N
Date	August 1
Time and local-solar-hour angle H_s	Noon; $H_s=0$
Solar declination δ_s, deg	17°–55′
Wall surface tilt from horizontal β	90°
Temperature, outside and inside, °F	$T_{out}=100$; $T_{in}=75$
Insolation I, Btu/h ft^2	$I_{h,b}=185$; $I_{h,d}=80$; $I_r=70$
U factor for walls, windows, and roof	$U_{wa}=0.19$; $U_{wi}=1.09$; $U_{rf}=0.061$
Infiltration, lbm dry air/h	Neglect
Ventilation, lbm dry air/h	Neglect
Internal loads	Neglect
Latent heat load Q_w,%	30% of wall sensible heat load[a]

[a] Approximate rule of thumb for Phoenix, AZ.

2. For a vertical wall, the wall normal is horizontal. Hence at solar noon, the incident angle, i, is equal to the solar altitude. Thus, for a vertical, south-facing surface, $i=\alpha=74.9°$ and $\cos i=0.26$
3. South-facing window load (from Equation 4.7)

$$\dot{Q}_{\text{wi}} = 40\left\{\left(0.20\times 0.6\times 185\frac{0.26}{0.966}\right)+(0.81\times 80)+(0.60\times 70)+\left[1.09(100-75)\right]\right\} = 5{,}600\ \text{Btu/h}$$

4. Shaded-window load (from Equation 4.8) Assuming the north, east, and west walls are all shaded with two 20 ft^2 windows on each wall

$$\dot{Q}_{\text{wi,sh}} = (3\times 40)\left[1.09(100-75)\right] = 3{,}270\ \text{Btu/h}$$

5. South-facing wall load (from Equation 4.9)

$$\dot{Q}_{\text{wa}} = (480-40)\left\{0.12\left[70+80+\left(185\frac{0.26}{0.966}\right)\right]+0.19(100-75)\right\} = 12{,}639\ \text{Btu/h}$$

6. Shaded-wall load (from Equation 4.10)

$$\dot{Q}_{\text{wa,sh}} = \left[(480+320+320)-(3\times 40)\right]\left[0.19(100-75)\right] = 4{,}750\ \text{Btu/h}$$

7. Roof load (from Equation 4.11)

$$\dot{Q}_{\text{rf}} = 1{,}700\left[\bar{\alpha}_{s,\text{rf}}\times 0+0.061(100-75)\right] = 2{,}600\ \text{Btu/h}$$

8. Latent heat load (30% of sensible wall load, neglecting solar radiation)

$$\dot{Q}_{w} = 0.3\left[(480+480+320+320)-(4\times 40)\right]\left[0.19(100-75)\right] = 2{,}050\ \text{Btu/h}$$

9. Infiltration load

$$\dot{Q}_{i} = 0$$

10. Total cooling load for the building described in the example

$$\dot{Q}_{\text{tot}} = \dot{Q}_{\text{wi}} + \dot{Q}_{\text{wi,sh}} + \dot{Q}_{\text{wa}} + \dot{Q}_{\text{wa,sh}} + \dot{Q}_{\text{rf}} + \dot{Q}_{w} + \dot{Q}_{i}$$

$$\dot{Q}_{\text{tot}} = 30{,}900\ \text{Btu/h}$$

Looking at the various heat loads, it is apparent that the largest heat load comes from the south-facing walls. The heat load from these walls is three times larger than that from the shaded walls, and it would therefore behoove the architect or engineer to consider the options for reducing the heat load through the southern walls. Architectural options such as overhangs could be helpful.

4.3.3 Vapor-Compression Cycle

The two principal methods of lowering air temperature for comfort cooling are refrigeration with actual removal of thermal energy from the air or evaporation cooling of the air with adiabatic vaporization of moisture into it. Refrigeration can be used under any humidity condition of entering air, whereas evaporative cooling can be used only when the entering air has a comparatively low relative humidity.

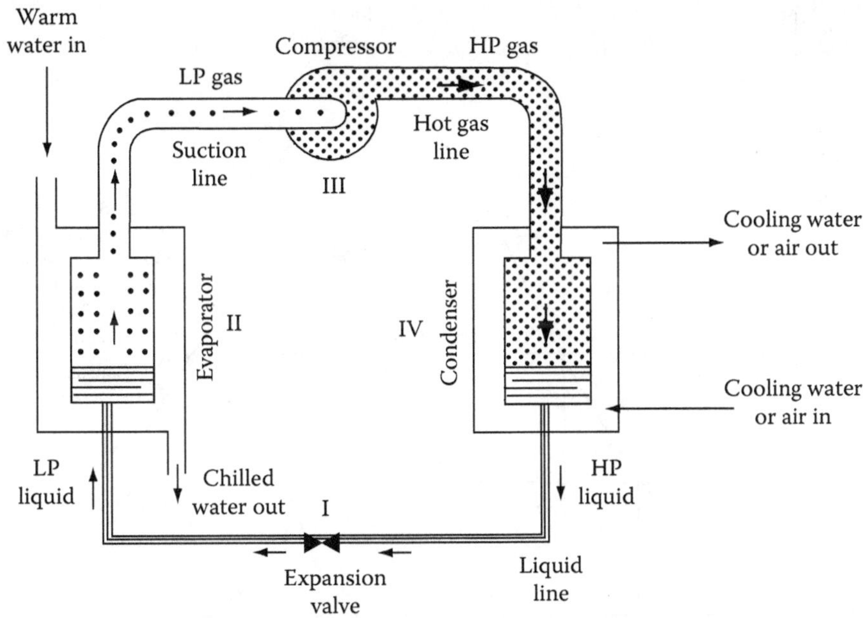

FIGURE 4.8
Schematic diagram illustrating the basic refrigeration vapor-compression cycle. (From Goswami et al. [17].)

The most widely used air-conditioning method employs a vapor-compression refrigeration cycle. Another method uses an absorption refrigeration cycle similar to that of the gas refrigerator. The vapor-compression refrigeration cycle requires energy input into the compressor, which may be provided as electricity from a photovoltaic (PV) system. Referring to Figure 4.8, the compressor raises the pressure of the refrigerant, which also increases its temperature [17]. The compressed high-temperature refrigerant vapor then transfers thermal energy via a heat exchanger to the ambient environment in the condenser, where it condenses to a high-pressure liquid at a temperature close to, but above, the environmental temperature. The liquid refrigerant is then passed through the expansion valve where its pressure is reduced, resulting in a vapor-liquid mixture at a much lower temperature. The low-temperature refrigerant is then used to cool air or water in the evaporator where the liquid refrigerant evaporates by absorbing heat from the medium being cooled. The cycle is completed by the vapor returning to the compressor. If water is cooled by the evaporator, the device is usually called a chiller. The chilled water is then used to cool the air in the building.

The operating principle of a vapor-compression refrigeration cycle can be illustrated conveniently with the aid of a pressure-enthalpy diagram, as shown in Figure 4.9. The ordinate is the pressure of the refrigerant in N/m^2 absolute, and the abscissa its enthalpy in kJ/kg. The Roman numerals in Figure 4.9 correspond to the physical locations in the schematic diagram of Figure 4.8.

Process I is a throttling process in which hot liquid refrigerant at the condensing pressure, p_c, passes through the expansion valve, where its pressure is reduced to the evaporator pressure, p_e. This is an isenthalpic (constant enthalpy) process, in which the temperature of the refrigerant decreases. In this process, some vapor is produced, and the state of the mixture of liquid refrigerant and vapor entering the evaporator is shown by point A. Since the expansion process is isenthalpic, the following relation holds:

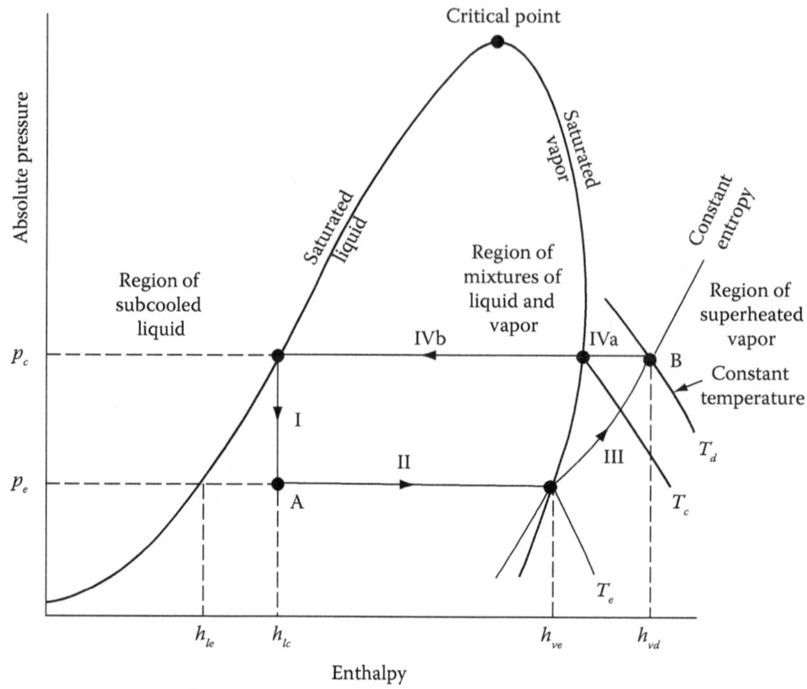

FIGURE 4.9
Simple refrigeration cycle on a pressure-enthalpy diagram. (From Goswami et al. [17].)

$$h_{ve} f + h_{le}(1 - f) = h_{lc} \tag{4.14}$$

where

f is the fraction of mass in vapor state; subscripts *v* and *l* refer to vapor and liquid states, respectively

c and *e* refer to states corresponding to condenser and evaporator pressures, respectively and

$$f = \frac{h_{lc} - h_{le}}{h_{ve} - h_{le}} \tag{4.15}$$

Process II represents the vaporization of the remaining liquid. This is the process during which heat is removed from the chiller. Thus, the specific refrigeration effect per kilogram of refrigerant q_r is

$$q_r = h_{ve} - h_{lc} \text{ in kJ/kg(Btu/lb)} \tag{4.16}$$

In the United States, it is still common practice to measure refrigeration in terms of tons. One ton is the amount of cooling produced if 1 ton of ice is melted over a period of 24 hours. Since 1 ton = 907.2 kg and the latent heat of fusion of water is 334.9 kJ/kg:

$$1 \text{ ton} = \frac{(907.2 \text{ kg}) \times (334.9 \text{ kJ/kg})}{(24 \text{ hour}) \times (3{,}600 \text{ s/h})} = 3.516 \text{ kW} = 12{,}000 \text{ Btu/h} \tag{4.17}$$

If the desired rate of refrigeration requires a heat-transfer rate of $\dot{Q}_{r'}$, the rate of mass flow of refrigerant necessary, $\dot{m}_r$, is

$$\dot{m}_r = \frac{\dot{Q}_{r'}}{(h_{ve} - h_{lc})} \tag{4.18}$$

Process III in Figure 4.9 represents the compression of refrigerant from pressure p_e to p_c. The process requires work input from an external source, which may be obtained from a solar-driven expander-turbine or a solar photovoltaic system. In general, if the heated vapor leaving the compressor is at the condition represented by point B in Figure 4.9, the work of compression W_c is

$$W_c = \dot{m}_r (h_{vd} - h_{ve}) \tag{4.19}$$

In an idealized cycle analysis, the compression process is usually assumed to be isentropic.

Process IV represents the condensation of the refrigerant. Actually, sensible heat is first removed in the subprocess IVa as the vapor is cooled at a constant pressure from T_d to T_c, and latent heat is removed at the condensation temperature, T_c, corresponding to the saturation pressure, p_c, in the condenser. The heat-transfer rate in the condenser $\dot{Q}_c$ is

$$\dot{Q}_c = \dot{m}_r (h_{vd} - h_{lc}) \tag{4.20}$$

This heat must be rejected into the environment, either to cooling water or to the atmosphere if no water is available.

The overall performance of a refrigeration machine is usually expressed as the ratio of the heat transferred in the evaporator, $\dot{Q}_r$, to the shaft work supplied to the compressor. This ratio is called the *coefficient of performance* (COP), defined by

$$\text{COP} = \frac{\dot{Q}_r}{W_c} = \frac{h_{ve} - h_{lc}}{h_{vd} - h_{ve}} \tag{4.21}$$

The highest COP for any given evaporator and condenser temperatures would be obtained if the system were operating on a reversible Carnot cycle. Under these conditions [18],

$$\text{COP(Carnot)} = \frac{T_e}{T_d - T_e} \tag{4.22}$$

The earlier cycle has been idealized. In practice, the liquid entering the expansion valve is several degrees below the condensing temperature, while the vapor entering the compressor is several degrees above the evaporation temperature. In addition, pressure drops occur in the suction, discharge, and liquid pipelines, and the compression is not truly isentropic. Finally, the work required to drive the compressor is somewhat larger than W_c provided earlier, because of frictional losses. All of these factors reduce the COP below the maximum and must be considered in a realistic engineering design.

Example 4.6

Calculate the amount of shaft work to be supplied to a 1-ton (3.52 kW) refrigeration plant operation at evaporator and condenser temperatures of 273 and 309 K, respectively, using Refrigerant 134a (R-134a) as the working fluid. The properties of Refrigerant 134a

TABLE 4.7

Properties of Refrigerant 134A for Example 4.6

Temperature (K)	Absolute Pressure (kPa)	Vapor-Specific Volume (m³/kg)	Liquid Enthalpy (kJ/kg)	Vapor Enthalpy (kJ/kg)	Vapor Entropy (kJ/kg·K)
Saturated					
273	292.8	0.0689	50.02	247.2	0.919
309	911.7	0.0223	100.25	266.4	0.9053
Superheated					
308.5	900	0.0226	—	266.18	0.9054
313	900	0.0233	—	271.3	0.9217
312.4	1,000	0.0202	-	268.0	0.9043
313	1,000	0.0203	-	268.7	0.9066

are tabulated in Table 4.7. Also calculate the COP and the mass flow rate of the refrigerant based on the ideal cycle described earlier.

Solution

From the property table, the enthalpies for process 1 are
Saturated vapor at 273 K h_{ve}=247.2 kJ/kg
Saturated liquid at 309 K h_{lc}=100.3 kJ/kg
Saturated liquid at 273 K h_{le}=50.0 kJ/kg
Therefore, from Equation 4.15:

$$f = \frac{100.3 - 50.0}{247.2 - 50.0} = 0.255$$

The mass flow rate of refrigerant, $\dot{m}_r$, is obtained from Equation 4.18 and the enthalpies given earlier, or

$$\dot{m}_r = \frac{3.52 \text{ kW}}{(247.2 - 100.3) \text{ kJ/kg}} = 0.024 \text{ kg/s}$$

The specific shaft-work input required is

$$\frac{W_c}{\dot{m}_r} = h_{vd} - h_{ve}$$

The entropy s_e of the saturated vapor entering the compressor at 273 K and 292.8 kPa is 0.919 kJ/kg K. For an isentropic compression, the vapor will have the same entropy at point B and temperature T_d. From the property table, superheated vapor at a pressure of 911.7 kPa has an entropy of 0.919 kJ/kg K at a temperature of 313 K with an enthalpy of 270.8 kJ/kg. Thus, the energy input to the working fluid by the compressor is

$$W_c = 0.024(270.8 - 247.2) = 0.566 \text{ kW}$$

Finally, the heat-transfer rate from the refrigerant to the sink, or cooling water in the condenser, is from Equation 4.20:

$$\dot{Q}_c = \dot{m}_r (h_{vd} - h_{lc}) = 0.024(270.8 - 100.3) = 4.09 \text{ kW}$$

The COP of the thermodynamic cycle is

$$COP = \frac{247.2 - 100.3}{270.8 - 247.2} = 6.2$$

whereas the Carnot COP is 273/36 or 7.6.

Air conditioners are commonly characterized using a metric related to the COP called the energy efficiency ratio, or EER. EER is the ratio of output cooling energy in Btu to the electric energy input in watt-hours. Whereas COP is dimensionless, EER has units of Btu/Wh. EER is related to COP via a simple units conversion, that is, $EER = 3.41 \times COP$. Considerable strides have been made to improve the EER and seasonal EER (SEER) of air conditioners. The latest improvements are aimed at using variable-speed compressors. Also, air-conditioning refrigerants have very high global warming potentials (GWPs). For example, the R134a (HFC 134a) used in the example problem has a 100-year GWP of 1,530, that is, it has 1,530 times the global warming potential of carbon dioxide. Just as the Montreal Protocol, originally signed in 1987, led to the phase-out of ozone-depleting refrigerants, there are now many research efforts focused on replacing today's common refrigerants with new ones that have lower GWPs.

4.3.4 Evaporative Cooling

When air with less than 100% relative humidity moves over a water surface, the evaporation of water causes both the air and the water to cool. The lowest temperature that can be reached by this direct evaporative cooling effect is the wet-bulb temperature of the air, which is directly related to the relative humidity, with lower wet-bulb temperature associated with lower relative humidity. Thus, dry air (low relative humidity) has a low wet-bulb temperature and will undergo a large temperature drop with evaporative cooling, whereas humid air (high relative humidity) can only be slightly cooled evaporatively. The wet-bulb temperature for various relative humidity and air temperature conditions can be found via the psychrometric chart available in most thermodynamic texts. Normally, an evaporative cooling process cools the air only part of the way down to the wet-bulb temperature. To achieve the maximum temperature decrease, it is necessary to have a large water surface area in contact with the air, and interior ponds and fountain sprays are often used to provide this air-water contact area.

The use of water sprays and open ponds on roofs provides cooling primarily via evaporation. The hybrid system involving a fan and wetted mat, the "swamp cooler," is by far the most widely used evaporative cooling technology. Direct, indirect, and combined evaporative cooling system design features are described in [13,18].

4.3.5 Energy Efficiency in Commercial Buildings

Commercial buildings come in a wide variety of types and encompass many activities as shown for the United States in Figure 4.10 [19]. This figure shows the total commercial building floor space, the primary energy consumption, and the energy use intensity (EUI) for various activities. Office buildings, stores, and schools are the largest primary energy users in that order, while food service, sales, and health care have the highest energy use intensities per square foot of floor area.

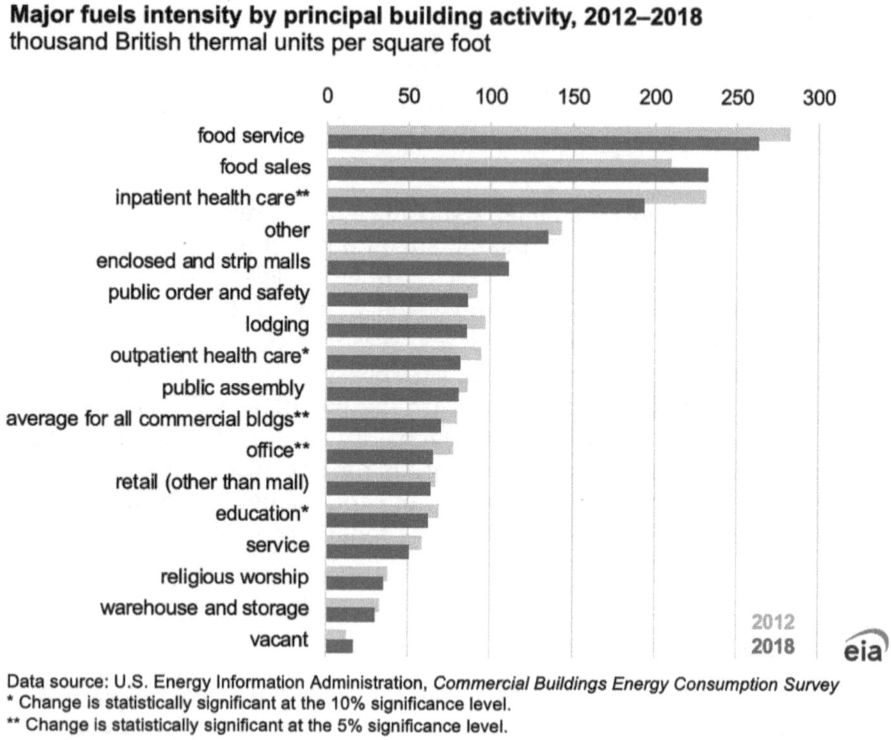

FIGURE 4.10
Fuel use by building activity. (From EIA [19].)

Commercial building energy uses in the United States by major end uses are shown in Figure 4.11 [20].

Significant gains have been made in reducing heating and lighting loads, as their total energy consumption has gone down despite the growth in building square footage.

Energy-efficient building design starts with a goal for energy use, usually expressed in terms of EUI, which is the annual energy use per square foot of total building floor area, for example, 25,000 Btu/ft^2/year. Experience has shown that aggressive energy design goals can be achieved if combined design-build teams compete to win the building contract and if the EUI goal is specifically tied to financial incentives.

In making buildings more efficient, we can focus on building types (for example, supermarkets and hospitals) or end uses (cooling and computing). For building types, the U.S. Department of Energy (DOE) has partnered with ASHRAE to develop Advanced Energy Design Guides (AEDGs). These cover various building types and provide guidelines for either a 30% or 50% energy reduction compared to code (ASHRAE 90.1) [20]. These are available online at ASHRAE [21]. Fifty percent reductions are more in line with recent practice, and these design guides are available for:

- Grocery stores
- Large hospitals
- Small to medium office buildings
- K–12 schools
- Medium- to big-box retail buildings.

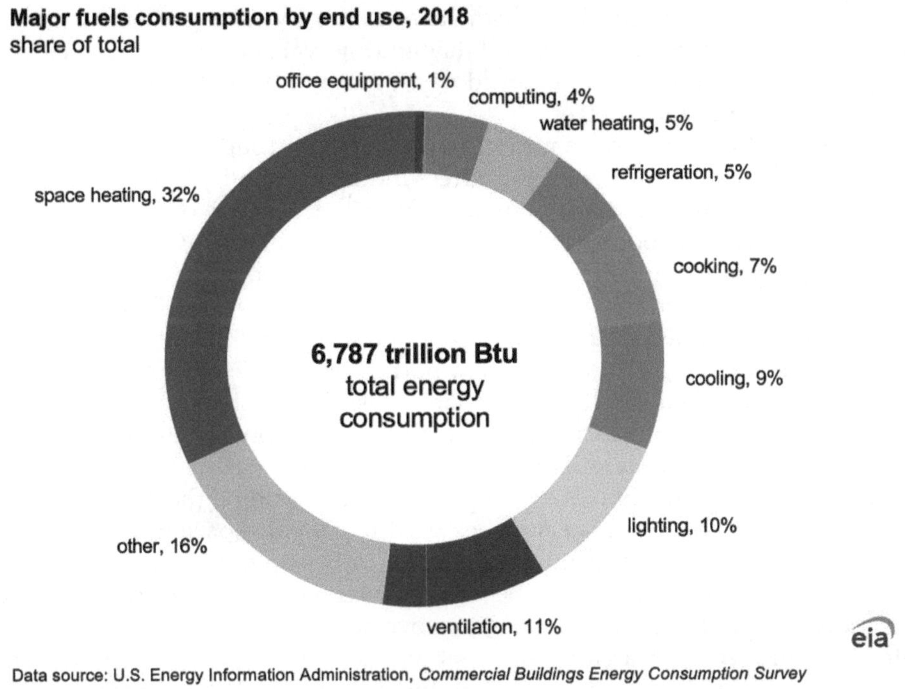

FIGURE 4.11
Energy use in U.S. commercial buildings by major end uses. (From EIA [19])

ASHRAE also now has zero-energy design guides for:

- Multifamily buildings
- K-12 schools
- Small to medium office buildings.

A variety of EEMs are considered in minimizing commercial building energy use.

Building envelope: High levels of envelope insulation, highly reflective fenestration, and air-tight construction are essential for minimizing heat gain in the summer and heat loss in the winter. The optimum amounts and orientation of glazing and shading elements are essential for minimizing summer heat gain and winter heat loss, providing adequate daylighting and allowing for some passive solar heat gain. (Passive solar heating concepts and the concept of climate-responsive design are described in Chapter 12.) Double- or triple-pane windows reduce convective heat loss. And low emittance (low-e) coatings also reduce radiative heat exchange. Low solar heat gain factors are especially important in hot, sunny climates and in buildings with high internal heat gain.

Lighting: Highly efficient fluorescent or LED lighting not only reduces the electricity needed for lighting but also reduces air-conditioning loads that result from the waste heat associated with inefficient lighting. LED lights are becoming especially popular because of their rapidly falling costs and high efficacy. (Efficacy is the radiant flux of visible light in units of lumens divided by the total energy consumed in watts. Today's LEDs have efficacies comparable to natural daylight—although daylight consumes no electric power—and are still improving.) LEDs (p-n junctions) emit light by electroluminescence, whereby electrons passing through the diode drop to a lower energy state and emit photons of light.

Automatic adjusting of lighting levels can save energy by taking advantage of natural daylight—balancing artificial lighting and daylighting. Various means, such as films and reflectors, can be used to reflect and spread daylighting to the building interior. Lighting can also be aimed at product shelves.

Ventilation: Tight construction to minimize infiltration through air leaks in the building envelope has made it more important to address indoor air quality. Commercial buildings require ventilation for the health of occupants. Ventilation is one of the largest contributors to space heating and cooling loads in commercial buildings. Controlled ventilation facilitates minimum air changes that maintain indoor air quality and meet code, thereby minimizing associated heating and cooling loads. The COVID-19 pandemic has taught us, however, that well-designed building ventilation systems and air filtering are necessary to protect occupants during an outbreak of airborne viruses. Heat recovery devices can be used to preheat or precool incoming ventilation air using the exhaust air stream, and when outdoor conditions are favorable, natural or forced ventilation of outside air can displace mechanical cooling (commonly called economizer cooling).

Heating and cooling equipment: Variable-speed heat pumps (heat pumps are essentially air conditioners that can also run backward to provide space heating) have a COP, which is the ratio of the rate of heating or cooling in Btus per hour divided by the compressor power using the same units. A heat pump uses mechanical energy (an electrically driven compressor in a refrigeration cycle) to move heat from a low-temperature source to a high-temperature sink. For an air-to-air heat pump operating in summer cooling mode, the source is the indoor air, and the sink is the outside air. When operating in winter heating mode, the source is the outside air, and the sink is the inside air. The smaller the difference between the source and sink temperatures, the higher the efficiency of the heat pump. A ground-coupled heat pump uses the deep ground as the heat source and sink instead of outside air and yields, on average, a lower delta-T and thus a higher average COP. (The ground temperature at depth is close to the average annual air temperature and so is cooler than outdoor air in summer cooling mode and warmer than outdoor air in winter heating mode.)

Research is being done on desiccant-based cooling systems that allow latent heat (moisture) to be removed from indoor air separately from sensible cooling (i.e., lowering the air temperature). In humid climates, moisture is removed by running the air conditioner evaporator coil below the indoor air dew point temperature to cause the moisture to condense out. This requires running the evaporator at a lower temperature than would be needed to sensibly cool the air and often requires "re-heat" for comfort. Both these effects reduce efficiency. In a desiccant cooling system, moisture is removed by passing indoor air through the desiccant. The dried air can then be cooled via evaporative cooling (the same mechanism by which perspiration cools the skin), which requires only a source of water and a small amount of fan power. (Studies have shown that the water used on site is similar in volume to the water used for the cooling of central thermal power plants that have traditionally provided the electricity for conventional air conditioners [22], but power plant water consumption has decreased with the growth of solar and wind power.) A separate heat source is needed, however, to "bake out" the desiccant to remove and reject the collected moisture to the outside air. In an all-electric design, a small heat pump powered by PV electricity can be used for desiccant regeneration.

Office computers and equipment: Office equipment energy use can be reduced by using central printers and copiers (private documents can be printed out via codes), use of laptops instead of desktop computers, local task LED lighting, and energy-saving power strips. Data centers are becoming major energy users. Capturing the waste heat from

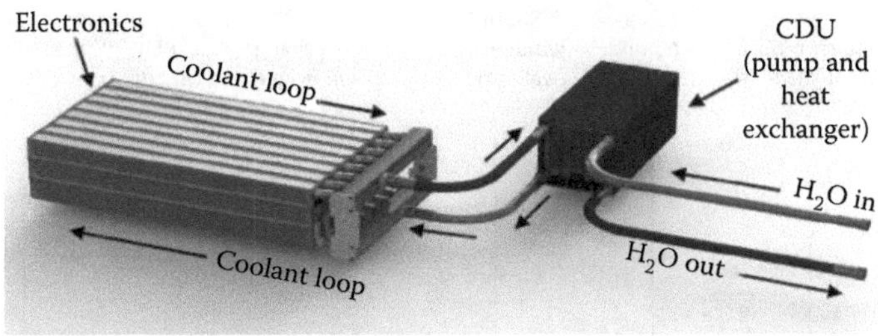

FIGURE 4.12
Product for direct liquid cooling of data center CPUs. (From NREL [23]. Courtesy of LiquidCool Solutions.)

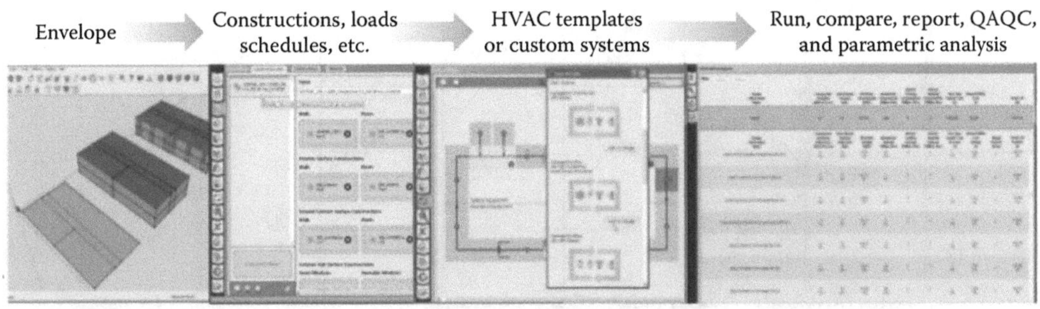

FIGURE 4.13
The OpenStudio flow process for modeling a building. (From Open Studio [24].)

servers and using it to provide building space heating can result in considerable energy savings. This has been done using both air and water cooling. There is current research underway on capturing the heat from CPUs by immersing them directly in a dielectric fluid that efficiently carries the heat away and uses it for other purposes. See, for example, Figure 4.12. There is also research to design more heat-tolerant CPUs so that the waste heat is at more useful temperatures [23].

There are clearly many design features to consider in maximizing energy efficiency. Evaluating them properly requires a detailed hour-by-hour energy analysis using typical weather data corresponding to the building site. To make this analysis widely available, DOE has developed OpenStudio [24], a suite of free and open-source software applications that facilitates building energy analysis using DOE's EnergyPlus calculation engine and a program called Radiance for daylighting calculations. Figure 4.13 shows the flow process for an OpenStudio building model. OpenStudio has a Building Component Library that allows the modeler to drag and drop in a wide range of building components and their associated performance parameters.

Once a model is created, parametric runs allow the designer to compare many different options, as shown in Figure 4.14. The five items on the left of the figure (number of stories, building aspect ratio, building rotation, window-to-wall ratio, and south overhang) are building model input parameters. The three items on the right (energy use intensity, peak electric demand, and lighting energy) are outputs of the simulation. After parametric runs are made, the user can adjust the sliders to see how the other parameters (both input design parameters and output values) change. The various gray lines represent the different runs that were made for the parametric analysis. A number of other building energy simulation programs are listed on the DOE Qualified Software List [25].

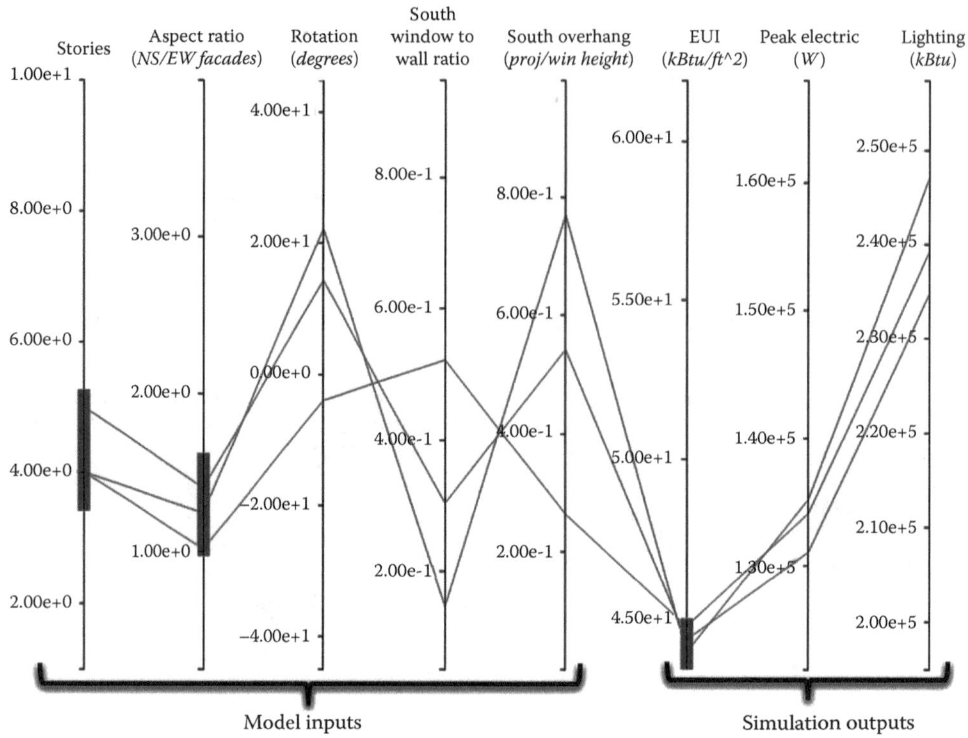

FIGURE 4.14
An example OpenStudio parallel coordinates plot that captures the results of thousands of parametric runs and allows the user to easily investigate the impact that changes in inputs and outputs have on each other by moving the sliders to vary the inputs and outputs. (Courtesy of D. Goldwasser, NREL.)

4.3.5.1 Commercial Buildings Case Studies

A common goal in commercial building efficiency is to achieve net-zero (or even positive energy production) status. Different definitions have been used for net-zero-energy buildings. According to DOE's recent report, *A Common Definition of Zero Energy Buildings,* a net-zero energy building is "an energy-efficient building where, on a source energy basis, the actual annual delivered energy is less than or equal to the on-site renewable exported energy" [26]. In other words, net zero does not imply a building that uses no energy, but rather one whose annual energy use is equal to the annual on-site production of renewable energy. Thus, net-zero buildings directly address climate change. We will consider two case studies: a large office building and a small commercial building that combines retail and some grocery.

The National Renewable Energy Laboratory Research Support Facility: The National Renewable Energy Laboratory (NREL) Research Support Facility (RSF) is the nation's largest net-zero energy office building. A key to accomplishing net-zero energy at the RSF was using a performance-based acquisition process that required a design-build team to meet a specific energy design goal, specifically an annual energy use (including the data center) of 35,000 Btus/ft^2/year. Building energy computer modeling was used to analyze the impacts of a variety of energy performance measures that were employed to achieve the design goal, including [27]:

- East–west orientation of 60-ft-deep wings with significant south-facing glass that provides passive solar heating in the winter and daylighting throughout the year, the latter aided by LightLouvers™, which reflect daylight deep into the interior
- Radiant heating/cooling in ceilings that eliminates ducting and fan power and operates at an efficient water temperature
- A dedicated outdoor air system that maintains air quality with minimal impact on heating and cooling loads; the system uses evaporative cooling when needed in hot weather
- Automated and manually operable windows for natural ventilation cooling
- High-efficiency lighting with automatic controls to augment daylighting
- Precast wall panels with 2″ of concrete for thermal mass on the inside and outside surfaces with an internal core of 4″ of rigid board insulations; the thermal mass allows the building to be cooled in the summer via purging the building with cold nighttime air
- Transpired Solarwall® for preheating building ventilation air in cold weather
- Shading devices and electrochromic glazing to reduce solar heat gain in summer
- Overhead tubular daylighting devices
- Minimal workstation plug loads at only 70 W per station.

In addition to the energy efficiency features, the RSF utilizes a total of 2.5 MW of photovoltaic modules mounted on the RSF roof, the NREL parking garage, and the visitors' parking lot canopy. A wood chip heating plant (utilizing wood chips from bark beetle kill when available) and data center waste heat augment natural gas used to provide hot water heating. Sufficient excess solar photovoltaic electricity is exported to the grid to make up for the natural gas burned on site.

Net-zero Walgreens store: Robbins et al. [28] describe the development of a net-zero-energy Walgreens store. Walgreens stores present a valuable opportunity because they combine both a retail and small supermarket space in one building and are ubiquitous with 76% of Americans living within 5 miles of a Walgreens. When Walgreens replaced an existing store in Evanston, Illinois, they decided to take advantage of the fact that their corporate engineers are located only 15 miles away and replace it with a state-of-the-art net-zero store. A challenge the engineers faced was to achieve net-zero status while maintaining more glass area than is optimum from an energy standpoint in order to make the store attractive to retail customers. The 15,000 ft^2 building features include the following [28]:

- R-30 roof and R-20 walls
- 840 roof-mounted PV modules rated at 256 kW
- South-facing rooftop clerestories that admit daylighting with automatic operable windows to allow for natural ventilation during appropriate ambient conditions
- A revolving entrance door to reduce infiltration loads
- Light-redirecting film on high west-facing glass to bounce daylight off the ceiling and deep into the building
- All-LED lights directionally aimed at retail shelves with a lighting density of only 0.89 W/ft^2 (compared to an ASHRAE Standard 90.1-2016 value for retail spaces of 1.4 W/ft^2) [21]

- A ground-source heat pump for heating and cooling using the ground under the parking lot; transcritical CO_2 is used as the working fluid, thus eliminating HCFC refrigerants that have a high global warming potential; the heat pump takes advantage of waste heat from the freezers.

As a first-of-a-kind project, Walgreens engineers learned valuable lessons. For example, while CO_2 worked well as a refrigerant, its high pressure required more expensive piping. Measurement and verification (M&V) would have been easier if it had been sufficiently considered early enough in the project and checked at each stage. The building has performed as net zero for many months but not on an annual basis. Nevertheless, its annual operating cost was 80% lower than that of the previous store located on that site, despite the fact that the new store has a larger floor area.

4.3.6 Energy Efficiency in Residential Buildings

Residential building on-site energy consumption and end uses vary by housing type (single family attached, single family detached, multifamily), vintage, and region. Details are given in the EIA Residential Energy Consumption Survey (RECS) [29]. It indicates that space heating is responsible for 42% of on-site energy use, water heating is 18%, and space cooling is 9%. The rest is split among lighting, cooking, electronics, refrigeration, and other uses.

The breakdown is different when one looks at primary energy use. Table 4.8 gives a breakdown of residential building primary (or source) energy uses [30]. Primary energy use accounts for fossil fuels used on site and the fuels used to produce electricity. While space heating represents 45% of on-site energy use (due largely to natural gas furnaces), it represents only 25% of primary energy use because of the fuels consumed in power plants to provide electricity to the buildings.

DOE evaluated energy savings potentials for both residential and commercial buildings, and these are shown in Figure 4.15 [31].

TABLE 4.8

U.S. Residential Buildings Primary Energy Use in 2016

End Use	Quads	Percent
Space heating	5.11	24.9%
Space cooling	2.31	11.2%
Water heating	2.68	13.0%
Refrigeration	1.06	5.2%
Cooking	0.57	2.8%
Clothes dryers	0.68	3.3%
Freezers	0.23	1.1%
Lighting	1.48	7.2%
Clothes washers	0.08	0.4%
Dishwashers	0.29	1.4%
Televisions and related equipment	0.98	4.8%
Computers and related equipment	0.33	1.6%
Furnace fans and boiler pumps	0.35	1.7%
Other uses	4.39	21.4%
Total	20.54	100.0%

Source: From EIA [30].

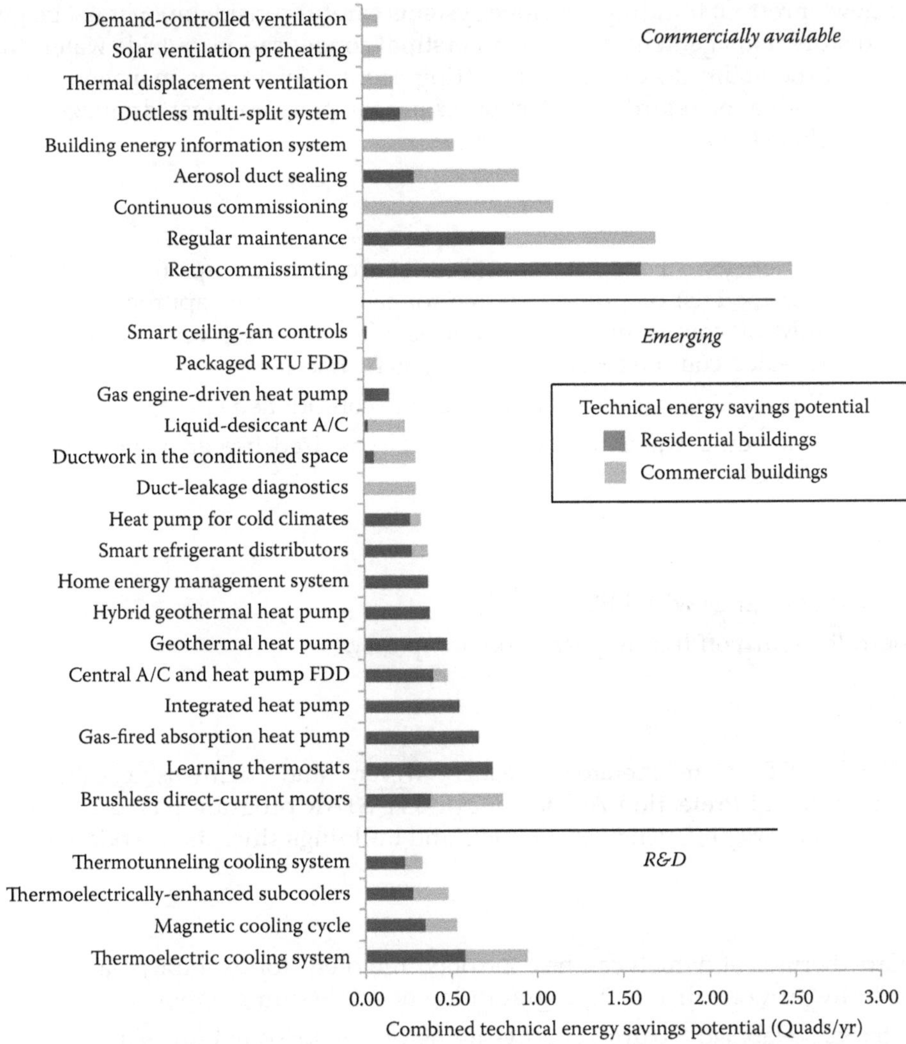

FIGURE 4.15
DOE energy savings potentials for residential and commercial buildings. Residential is shown by the dark bars and commercial by the light bars (From U.S. Department of Energy [31].)

Roughly 80% of residential building square footage and energy consumption is in single-family buildings, so the DOE's energy savings efforts are currently focused on that market. The following are example areas for energy efficiency improvements in residential buildings [31].

Building envelope:

- Roof/attic insulation
- Exterior wall insulation
- Air sealing to reduce infiltration to 3–5 ACH measured at a 50 Pa differential pressure during blower door testing; this corresponds to about 0.2–0.3 natural air changes per hour

- For new or retrofit building envelope systems careful consideration must be given to moisture management including interstitial condensation, outside water drainage, and the ability to dry after a wetting event. Moisture is managed through detailing of vapor retarders (selection of proper perm ratings), drainage planes, and weather skins.

Equipment:

- Higher efficiency air conditioners and air-source heat pumps (now available in cold-climate models) or ground source for new homes in appropriate climates (multifamily or community scale projects offer the best economics) or high-efficiency, sealed combustion condensing gas furnaces
- Duct sealing and insulation (or moving inside conditioned space)
- Ducted heat pump water heater or tankless gas-fired heater (which eliminates tank heat losses).

Lighting:

- Replace all fixtures with LED
- Controls to turn off lighting based on occupancy.

Appliances:

- ENERGY STAR refrigerator, washer, dryer, and dishwasher. (The U.S. Environmental Protection Agency's ENERGY STAR program has identified and promoted energy efficiency in products and buildings since its inception in 1992.)

Other:

- Smart thermostat (which can be controlled remotely, for example, via Wi-Fi, and generally provides different programming or self-learning options)
- WaterSense-labeled fixtures, energy-saving power strips to reduce miscellaneous electrical loads (MELs), also called plug loads
- Energy Star laptop computer.

The design of all buildings and the impacts of energy efficiency measures depend strongly on the climate region in which the building is located. This is especially true of residential buildings, which, unlike many commercial buildings, are impacted more by external weather conditions than by internal human and equipment loads. The International Energy Conservation Code (IECC) classifies 7 different climate regions in the United States as shown in the map of Figure 4.16 [32].

Researchers at NREL have performed detailed computer analysis called ResStock of the single-family residential building stock in the United States using a number of data sources [33]. They considered the impact of climate zone, building vintage, and fuel type to determine the energy savings potentials for various energy efficiency measures. Table 4.9 shows rolled-up results aggregated by region for four different energy efficiency measures:

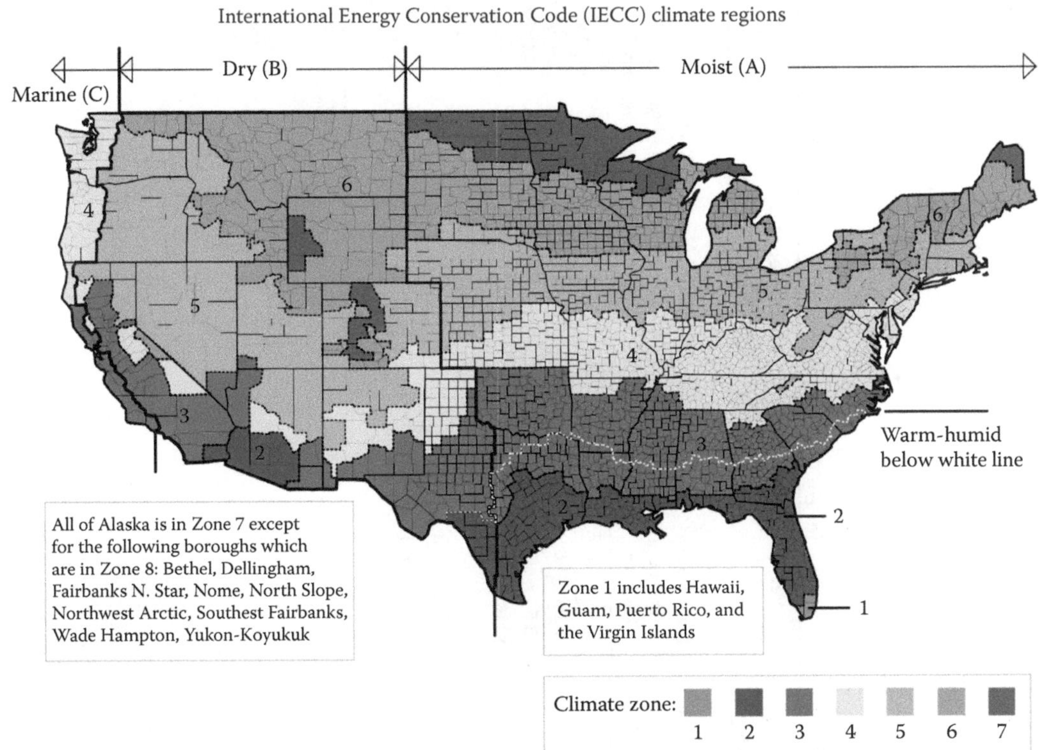

FIGURE 4.16
International Energy Conservation Code climate regions. (Republished with permission from Pacific Northwest National Laboratory [32].)

1. Replacing an electric water heater with a heat pump water heater
2. Replacing single-pane windows with double-pane low-e windows
3. Air sealing to reduce infiltration to 5 ACH as measured using a blower door test operating at an indoor-outdoor pressure differential of 50 Pa (This would be equivalent to about 0.3 natural ACH; the current International Energy Conservation Code calls for 3 or more ACH [depending on climate zone] during a blower door test at 50 Pa pressure difference)
4. Adding attic insulation to achieve a level of R-49.

The aggregated energy savings for these four measures in each region are expressed both as aggregated savings in the region in source energy quads per year and as per-house energy savings in thousands of Btus of source energy per year [33].

To analyze the value of energy efficiency measures in new and existing homes, NREL has developed a computer model called BEopt, which optimizes various combinations of measures. BEopt accepts as input a full range of energy performance specifications including window area and *R*-value, air changes, type and efficiency of space conditioning equipment, insulation levels (attic, wall, basement), type and efficiency of water heater (including solar), lighting type and efficiency, appliance efficiencies, equipment schedules, PV arrays (size, efficiency, tilt, and azimuth), and so on. Quantities such as attic or wall insulation are provided in a list of actual thicknesses that can be purchased.

TABLE 4.9

Primary (or Source) Energy Savings Potential of Four Energy Efficiency Measures for Detached Single-Family Homes in the United States

IECC Climate Zone	Baseline Source Energy Consumption [Quads/Year]	Aggregate Energy Savings [Source Quads/Year]					Per-House Energy Savings, Number of Upgraded Houses [Source MBtu/Year], [Millions]							
		Heat Pump Water Heater[a]	Windows[b]	Air Sealing[c]	Attic Insulation[d]	Total	Heat Pump Water Heater[a]		Windows[b]		Air Sealing[c]		Attic Insulation[d]	
1	0.21	0.01	0.01	0.01	0.01	0.04	17.4	0.7	21.9	0.5	17.0	0.8	6.5	0.8
2	2.2	0.12	0.09	0.11	0.06	0.38	18.0	6.9	19.4	4.9	10.5	10.3	5.9	10.4
3	3.31	0.13	0.12	0.13	0.12	0.5	17.5	7.5	15.0	8.3	7.6	17.3	6.4	18.5
4	3.4	0.12	0.09	0.24	0.1	0.55	17.0	6.9	16.4	5.5	15.0	15.7	6.3	15.3
5	4.04	0.07	0.08	0.26	0.11	0.52	16.7	4.5	17.5	4.7	15.4	16.6	6.3	16.7
6	0.89	0.02	0.02	0.04	0.01	0.09	17.3	0.9	19.5	0.9	13.6	3.0	5.7	2.6
7	0.07	0	0	0	0	0	16.3	0.1	22.8	0.1	17.1	0.2	5.8	0.2
Total	14.12	0.47	0.41	0.79	0.41	2.08		27.6		24.7		64.0		64.5

Source: From Wilson et al. [33].

Excludes Alaska and homes with propane or other atypical fuel types.

[a] Replacing electric tank with heat pump water heater.

[b] Single pane to double low-e.

[c] To 5 ACH50.

[d] To R-49.

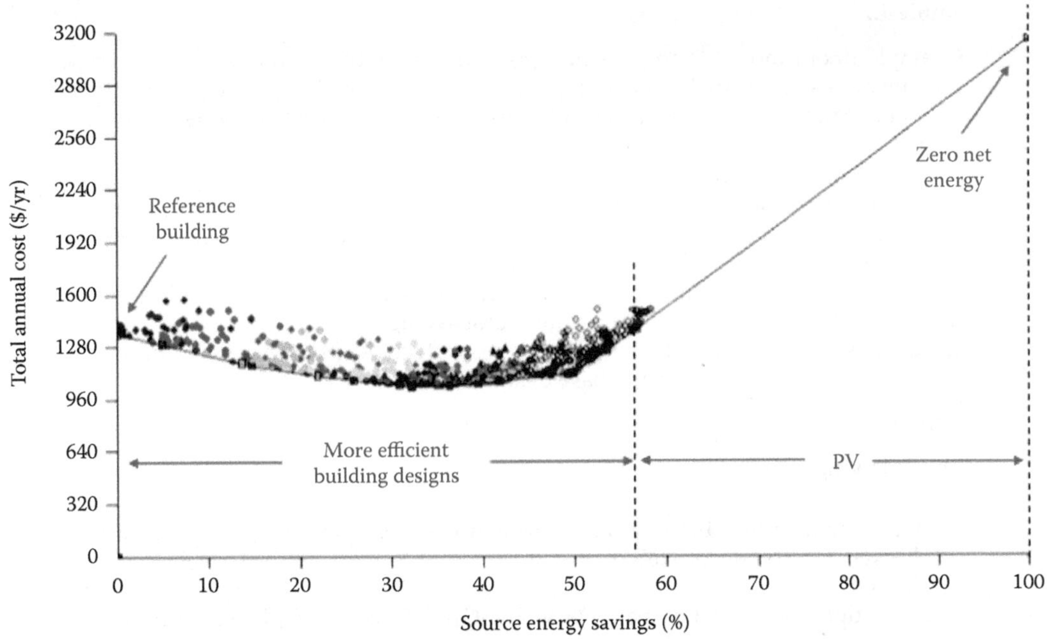

FIGURE 4.17
Example output from BEopt analysis of the cost and energy savings impact of various combinations of energy efficiency measures. (From Christensen et al. [34].)

Figure 4.17 shows an example plot of data points representing various combinations of a multitude of example EEMs for a single-family residence. The reference building is whatever you start with before you add additional energy efficiency features, for example, a new home built to code or your existing home. Different combinations of features can achieve the same savings. The *y*-axis represents total annual cost, which is the sum of energy (utility) costs and total loan costs (mortgage and any additional loan for purchasing the EEMs). As efficiency measures are added, energy savings costs exceed additional loan costs and so the total decreases until it reaches a minimum. In this example, at an energy savings of about 57%, the total cost is equal to the original total cost. This is called the neutral cost point. These savings are achievable at no net cost to the homeowner.

Achieving zero energy (assuming an all-electric home) requires the addition of PV modules, which increases the cost. However, the cost of PV has dropped rapidly. As a result, the straight PV line is moving to the left (the point at which the addition of PV is less expensive than the marginal cost of additional energy efficiency measures, which reach a point of diminishing returns) and its slope is decreasing. Given the right financial terms, the net-zero point can be reached with little or no increase in total annual cost, which is equivalent to positive cash flow for the building owner from day one.

BEopt is available as a free download at https://beopt.nrel.gov [34]. Like OpenStudio, its calculations use the EnergyPlus model.

Example 4.7

The sample BEopt model "Sample—designs and parametrics" compares a base case single-family residence to 5 energy improvement cases involving changes in different components: walls, window type, lighting, AC & furnace, and water heater. Do the following:

a. Construct a table showing for each of the five improvements the absolute and percent source energy savings and energy-related cost savings relative to the base design
b. Identify the design with the maximum energy savings
c. Identify the design with the maximum cost savings
d. Identify the designs that are not cost-effective (that is, have higher energy-related costs than the base design).

Solution:

Do the following steps:

1. After downloading BEopt, make sure that the boxes on the main screen are checked for Standard, New Construction, and Single-Family Detached and click OK.
2. In the upper left-hand corner choose File-Open Project-Sample—designs and parametrics.BEopt
3. Click the what-ifs tab in the upper left-hand corner.

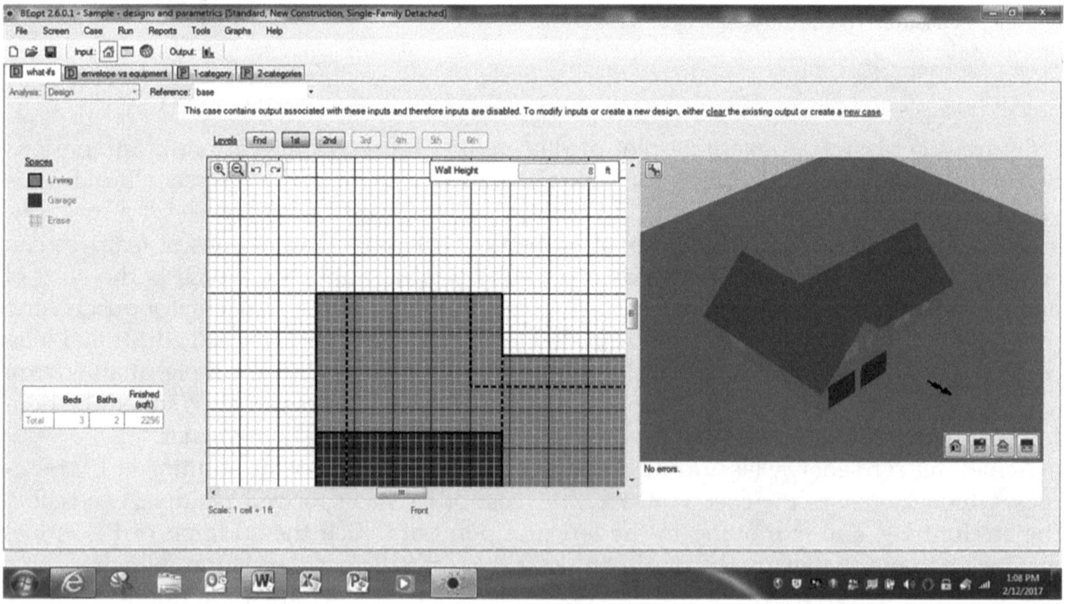

4. Click on the small bar chart icon in the Output tab at the top of the screen

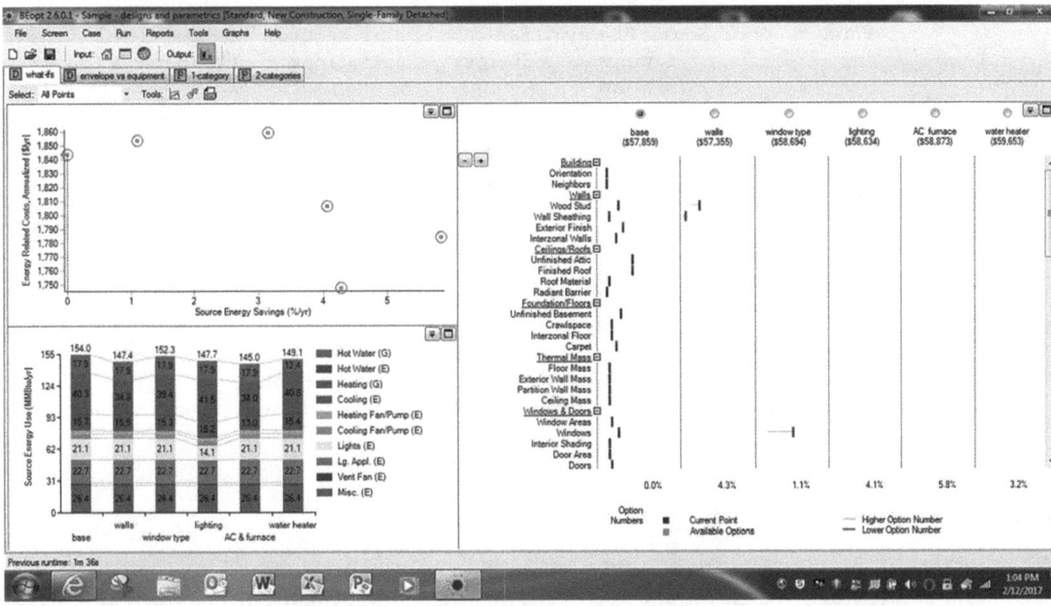

5. Click in the check boxes for the various cases modeled at the upper right of the screen. For each case, move your cursor over the blue bars to see what design changes were made. (For example, in the window type case, double-pane windows have been replaced by triple-pane windows.)
   a. To get the absolute source energy savings compared to the base case for each of the five improvements, refer to the bar chart in the lower left and subtract the source energy use in each case from the base case value. You can get the percent savings from the bottom of the case columns at the right of the screen. To get the cost savings for each case, click in the pull-down menu in the upper left next to Select. For each case selected, the energy cost in $/year is in the lower right of the screen. To get the savings for each improvement case, subtract this value from the value for the base case. Calculate the percentage savings by dividing the savings for each improvement case by the base case energy use.

      You can also find these values compiled in the BEopt Detailed Report. To download that spreadsheet, click on Reports—Detailed Report at the top of screen and choose What ifs. After you download it, open it and scroll down to the table "Output-Cost/Energy Graph Metrics" near the bottom of the spreadsheet. Note: Energy-Related Costs include utility savings and annualized investment costs for efficiency measures and future replacements. Percentages are of whole-house values, including significant quantities of conditioning energy use.

The following table results:

Point	Source Energy Savings [MMBtu/Year]	Source Energy Savings [%/Year]	Energy-Related Cost Savings, Annualized [$/Year]	Energy-Related Cost Savings %, Annualized [%/Year]
Walls	6.6	4.3%	97	5.2%
Window types	1.7	1.1%	–27	–1.4%
Lighting	6.3	4.1%	55	2.9%
AC & furnace	9.0	5.9%	61	3.3%
Water heater	4.9	3.2%	–14	–0.8%

b. The design with the maximum energy savings is the AC & furnace case
c. The design with the maximum cost savings is the walls case
d. The cases that are not cost-effective are the window type and water heater cases.

4.3.7 Zero-Carbon Urban Districts

We have described energy-efficient and zero-energy buildings. To achieve faster carbon emissions reductions, we can't just address buildings one at a time. We need to extend these concepts to a multibuilding, or district, scale. Polly et al. [35] provide details on the concept of zero-energy urban districts, some of which we will summarize here. (To be more accurate, we are calling these zero-carbon districts here.) As the authors point out, there is especially a need to use the best building energy design concepts in cities, which are growing rapidly in developing nations. Over half of the world's population currently lives in urban areas. The United Nations is projecting that this number will increase to 66% by mid-century [36]. In 2011, the organization Architecture 2030 estimated that by 2030, 900 billion ft^2 (836 million m^2) of new and rebuilt buildings would be constructed, mostly in the developing world [37]. This is equivalent to three-and-a-half times the total floor area currently existing in the United States.

The need to address climate change is recognized by mayors around the world who are already facing extreme weather events such as record heat waves and rainfall events. According to the Compact of Mayors website, as of March 2018, 7,513 cities worldwide, representing nearly 700 million people, had committed to the Global Covenant of Mayors, an agreement to measure and reduce greenhouse gas emissions [38]. In the United States, over 1,000 mayors signed the Mayors' Climate Protection Agreement and vowed to reduce carbon emissions below 1990 levels [39].

The concept of net-zero energy can be extended from buildings to a community scale. In 2009, Carlisle et al. developed definitions for net-zero energy communities [40]. More recently, DOE published source-energy-based definitions for zero-energy campuses and communities [26]. In the United States, cities offer the benefit that residents' needs can be met within walking distance. Mixed-use developments that combine homes, retail stores, restaurants, recreation, and workplaces into one community provide new energy savings opportunities. Energy can be moved from an internally load-dominated commercial building that needs cooling to a residential one that needs heating. Also, aggregating buildings makes it possible for renewable generation to be shared. And economies-of-scale may be achieved when purchasing large quantities of photovoltaic modules.

TABLE 4.10
Net-Zero Energy District Design Principles

Design Principle	Design Subprinciple
Maximize building efficiency	**Orientation:** Maximize natural daylighting, take advantage of passive solar design (see Chapter 12)
	Enclosure: efficiencies currently being implemented in net-zero-energy building industry (e.g., DOE's Zero Energy Ready Program[a]; ASHRAE's 50% Advanced Energy Design Guides[b])
	Miscellaneous electric loads: Carefully select best-in-class products; develop robust control strategies; verify with ongoing monitoring to minimize miscellaneous electric loads
	Lighting: 100% LED, controls for occupancy and daylighting variability
	Heating, ventilation, and air conditioning (HVAC): District-connected systems that maximize thermal energy recovery opportunities from low-grade heat sources across the district. For example, "ambient loops" circulate water that is close to the indoor air temperature and serve as heat source and sink for building heat pumps. Wastewater has been used in ambient loops. These loops can also be connected to the ground.
Maximize solar potential	Arrange buildings in districts to prevent building-to-building shading (shorter buildings to south, ideally)
	Orient buildings and roof slopes for maximum solar access
	Minimize other buildings systems that require roof space to provide room for PV modules
	Reserve all parking lots and garages to be shaded parking with PV
	To improve potential for off-grid resiliency, maximize rooftop solar
Maximize renewable district thermal energy	Evaluate potential for renewable thermal energy systems and waste heat recovery (e.g., ground-source district heat pump systems, industrial waste heat recovery, wastewater heat recovery)
Maximize load control	Establish controls for building and district system energy demands to accommodate the variable renewable energy supplies (e.g., PV and wind) and to support the district's interaction with the electric grid

Source: From Polly et al. [35].
[a] See DOE 2016 [45].
[b] See ASHRAE 2018 [21].

The organization EcoDistricts has published a protocol for sustainable urban development, and they state that "the district is the optimal scale to accelerate sustainability—small enough to innovate quickly and big enough to have a meaningful impact" [41]. There are a number of zero-energy districts and communities under development. For example, the University of California, Davis, has pursued a net-zero energy community in the West Village development [42]. Two Denver projects, the National Western Center and the Sun Valley Neighborhood, are briefly described in [35]. DOE's Better Buildings program ran a Zero-Energy District Accelerator to assist new projects [43]. Finally, European work on district energy systems can be found in the UN District Energy Initiative [44].

Table 4.10 from Polly et al. shows design principles for zero-energy districts [35].

4.3.8 Beneficial Building Electrification

As damage from climate change has mounted around the world, a clear strategy for addressing it has emerged. During the 10-year period from 2009 to 2019, the cost of utility-scale wind electricity dropped by 70%, and the cost of utility-scale solar dropped by about 90%. From 2015 to 2020, the cost of battery storage dropped by 75%. These dramatic reductions in the costs of carbon-free renewable electricity have spurred an international effort to replace thermal power plants fired by carbon dioxide-emitting coal and natural gas with variable solar and wind electricity, backed up by batteries. To utilize this clean electricity, there is a concurrent effort to electrify the end-use sectors of buildings, transportation, and industry. This is widely called "beneficial electrification" because it has multiple benefits: it addresses climate change and air pollution, as well as having other economic and social benefits.

Thus, to address climate change, there is a movement toward all-electric buildings. In the United States, buildings already consume about 74% of grid electricity. This is used for building cooling, lighting, and various appliances. Although some electricity is used for space heating, water heating, and cooking, the majority of energy for these needs is provided by burning fuel oil, propane, and especially natural gas on site. These fuels can be replaced by electric resistance heat. In the case of water heating and especially space heating, however, it is much more efficient to utilize heat pumps, because they can provide several kilowatt-hours of thermal energy for every kilowatt-hour of electricity consumed. There is thus a growing movement around the world to replace fossil fuel fired furnaces and hot water heaters with heat pumps. In 2023, heat pump sales in the U.S. surpassed gas furnace sales for the first time. (It is worth mentioning that some have promoted the use of green hydrogen produced from solar- and wind-powered electrolysis as a replacement for natural gas in furnaces and boilers. However, when one considers the efficiency losses associated with producing, transporting, and storing green hydrogen, as well as the COP advantage of heat pumps, using green hydrogen for heating would require on the order of 5 times more renewable electricity than would the direct use of that electricity to power heat pumps.)

In most applications, heat pumps use two common sources for building heat (which are also sinks for rejecting building heat during summer cooling): outside air and the ground. The former are in the form of either air-to-air or (less commonly) air-to-water, depending on whether air or water is used for heating and cooling inside the building. The commonly used air-to-air heat pump consists of an outdoor unit (which serves as the evaporator in heating mode and the condenser in cooling mode) and an indoor unit (which conversely serves as the condenser in heating mode and the evaporator in cooling mode). The indoor unit will often include an air handler and fan in which heat is exchanged from the indoor coil to air that is distributed through the building's ductwork. Alternatively, refrigerant can be piped to one or more individual "mini-split" units in the building that contain small condenser coils and fans. Where ductwork already exists, an air handler system is typically installed. In new buildings, mini-split units can avoid the cost of installing ductwork and allow for zone control in different rooms.

Heat pumps that use the outside air as the heating source can have seasonal average COPs in the range of 2–4 depending on the climate. Some cold-climate heat pumps today can operate at full capacity and provide a COP as high as 1.75 at an outdoor temperature of 5°F and continue to operate at temperatures as low as −13°F. For very cold climates, an auxiliary electric resistance heater (COP = 1) can be included to provide additional heat during the very coldest days.

Considerably higher efficiencies can be obtained by employing the ground as a heat source. At a certain depth that depends on location, the ground temperature will be reasonably constant year-round at close to the average annual ambient air temperature. Plastic piping carrying water or a water-glycol mixture is buried underground either in long trenches or in deep boreholes. Either of these entails significant disruption to the property and add considerable expense, so a ground-coupled system is best applied for large new buildings, home developments, and campuses, where the existing property is not disrupted and the expense of bringing trenching or drilling equipment to the site can be spread out over many trenches or boreholes.

Ground-coupled heat pumps for new buildings can have both first-cost and operating cost advantages. A study by the New Buildings Institute [46] concluded that a ground-coupled heat pump system for new buildings can be cost-effective in part because it can avoid the costs of all the natural gas infrastructure. Jobs that would ordinarily be needed for installing gas piping are replaced by similar ones needed for water piping. Once installed, the fact that in the winter ground temperature is generally much higher than outside air temperature means that the seasonal COP can be as much as two times higher than for air-source systems. Ground-coupled systems can also ameliorate another concern. As buildings in cold climates electrify, utilities will experience high electric demand on very cold days, which means more electric generating capacity must be built. Because ground-coupled systems are more efficient, they will minimize the increase in the winter peak. The non-profit organization HEET, based in Boston, is working to spread knowledge about networked ground-coupled systems [47].

Heat pump water heaters can reduce the electricity consumption for water heating in an all-electric home. These extract heat from the indoor air, which is a benefit in summer weather. In the winter the heat extracted from the indoor air must be made up by the space heating heat pump system, but that is more efficient than using an electric resistance water heater because the space heating system has a COP greater than one. Heat pump electric dryers are also available now.

Electric stoves avoid the NO_x emissions in the kitchen that occur when cooking on a gas stove, which has been shown to increase the incidence of asthma in children [48]. While electric stoves have been in widespread use for a long time, new inductance stovetops heat a pot (which must contain magnetic iron or iron-based material) directly and so provide faster heating and result in less heat loss to the kitchen.

Virtually all buildings consume some amount of operating energy. The concept of net-zero buildings was a way to allow for the continued burning of fossil fuels in buildings with the idea that solar photovoltaics would be installed on site to overproduce and export enough carbon-free energy to compensate for the on-site emissions. But to ultimately address climate change, we need to eliminate all fossil fuel emissions in buildings. As we have described in this section, the way to do that is to electrify buildings and provide that electricity with a grid decarbonized primarily with solar and wind electricity.

Efficiency continues to be important even in the case of all-electric buildings operating on a decarbonized grid. A high level of efficiency will reduce the owner's electric bills and will minimize the additional electric generating capacity that utilities must build, as well as the winter peak demands. In fact, when retrofitting existing buildings with heat pumps, bundling energy efficiency improvements can avoid an increase in heating bills that might otherwise occur in cases where a natural gas furnace is being replaced, and natural gas is much cheaper than electricity.

Buildings already consume the majority of grid electricity. By controlling when that electricity is used, buildings can play an important role in accommodating the variable nature of solar and wind through what are called demand response (DR) measures. The addition of home batteries and electric vehicles that interact with the home and grid and the use of rooftop solar photovoltaic arrays add to the ability of homes to play a key role in the concept of virtual power plants (VPPs). These concepts are covered in Chapter 5.

4.3.8.1 Networked Thermal Energy Case Study: The Whisper Valley Community

An example of an all-electric networked thermal district utilizing ground-coupled heat pumps is the award-winning 2,000-acre Whisper Valley Community in Austin, Texas [49]. It is a multi-use walkable community that consists of 7,500 all-electric homes, 2 million ft^2 of commercial space, two schools, and a 600-acre park. It utilizes an integrated energy package developed by EcoSmart and includes the use of an ambient loop connecting all homes.

The ambient loop concept was first developed by Integral Group Vancouver for the development of the Whistler Olympic Village [50]. A pump circulates water through an uninsulated pipe network buried below the frost line. The temperature of the soil is near the annual average ambient air temperature, and the water communicates with it thermally. Heat can be added to the water by buildings rejecting heat and removed by buildings needing heat. Heat pumps located at individual buildings or at other points along the ambient loop add and extract heat from the loop and can also move heat between deep geothermal wells and the circulating water.

In the Austin, TX development, the highly insulated homes are each heated and cooled with individual heat pumps (Figure 4.18) having geothermal wells approximately 350 ft deep that are connected to the ambient loop (Figure 4.19). Austin has higher summer cooling loads than winter heating loads, so to keep the ground in balance and handle peak cooling loads on the hottest days, two 250-ton cooling towers are connected to the ambient loop. Each homeowner has the option to include a 5-kW rooftop solar photovoltaic system and a home battery to operate the heat pump and energy-efficient appliances, including heat pump water heaters and inductive stovetops. The homes cost about $10,000 more than comparable homes, but owners recoup the investment through lower utility bills.

FIGURE 4.18
Geothermal heat pump piping and rooftop photovoltaic array in the Whisper Valley Community. (From [49].)

FIGURE 4.19
The ambient loop piping concept with supply and return lines for each block in the Whisper Valley Community. (From [49].)

4.3.9 Low-Carbon Building Materials

As buildings become more energy-efficient and as they electrify to take advantage of decarbonized electric grids, the carbon emissions associated with building operation will approach zero. At this point, the carbon emissions associated with building materials and energy efficiency measures (e.g., insulation) will become the dominant buildings sector contributors to climate change. Under the leadership of architect Ed Mazria, the non-profit organization Architecture 2030 has set a target of a 65% reduction in embodied emissions in new construction by 2030 and zero embodied emissions by 2040. They have developed a Carbon Smart Materials Palette to aid building designers in choosing building materials [51].

Concrete, especially the CO_2 emissions that occur in the production of Portland cement, is a major source of carbon emissions in new building construction. A variety of supplemental cementitious materials (SCMs) are being used for lower-carbon cement mixtures. Steel is also a major construction material and there are increasing efforts to use recycled steel. Structural wood such as cross-laminated timber (CLT) is increasingly being used to decrease the amount of concrete and steel in new construction. When using wood, the source of the wood is important. Architecture 2030 has recommended that builders use reclaimed or recycled wood products where possible and otherwise specify that wood be purchased from new-growth, sustainably managed forests.

It is also important to consider the embodied carbon in the materials used for energy efficiency measures. For example, blown-in materials such as cellulose and fiberglass have a lower carbon footprint than rigid and spray foam insulations, and this should be accounted for in life-cycle emissions assessments.

Architecture 2030 has summarized key design steps for minimizing embodied emissions [51]. A description of available tools for building life-cycle assessment is given on the Building Green website via a free sign-up [52].

4.4 Industrial Energy

4.4.1 Background

Industrial energy use falls into the different industry categories shown in Figure 4.20, which shows the global values in exajoules ($EJ=10^{18}J$) [53]. Note that the largest energy consumers are chemicals/petrochemicals, iron and steel, non-metallic minerals (especially cement), and nonferrous metals (e.g., aluminum).

Industries also vary in their energy intensity (the ratio of energy consumed in the industrial process to the monetary value of a product) with the cement industry having the highest energy intensity. Figure 4.21 shows relative energy intensities for different U.S. manufacturing industries [54]. Cement and lime stand out as having the greatest potential for energy efficiency improvements (e.g., with higher efficiency kilns). U.S. industry has shown a 25% decline in energy intensity since 1980 [6].

Many improvements have been made to decrease energy consumption in industrial processes. One very significant change has been a shift to the simultaneous production of electricity and process heat. This makes use of the heat that is rejected in a power cycle and is typically lost to the environment. This has been called cogeneration and is more commonly referred to today as combined heat and power, or CHP. The U.S. CHP capacity increased from 16,000 MW in 1980 to over 80,000 MW in 2013 [6]. While this means that fossil fuels have been used much more efficiently to simultaneously produce both process heat and electricity, the challenge going forward will be to transition completely off fossil fuels.

4.4.2 Improving Industrial Processes

Today, a key environmental goal in improving energy processes is to reduce greenhouse gas emissions, only one factor of which is energy efficiency. As described by the Intergovernmental Panel on Climate Change (IPCC) Working Group III, annual greenhouse gas emissions, G, can be expressed as follows [55]:

$$G=\frac{G}{E}\times\frac{E}{M}\times\frac{M}{P}\times\frac{P}{S}\times S \tag{4.23}$$

where

G/E = emissions intensity in GHG emissions per unit of energy consumed. This is reduced by switching to lower-carbon or carbon-free sources of energy.

E/M = energy-material intensity or energy consumed per unit of material used. E/M is reduced by adopting energy efficiency measures.

M/P = material-product intensity or the amount of material required to make the product. This is reduced by minimizing scrap or by recycling product.

P/S = product-service intensity or the amount of product needed to provide a service. This can be reduced by making products that last longer or that are more effective so that a smaller quantity of product is needed.

S = the demand for the service or product. This is reduced by directly reducing the product or service demand. Energy *conservation*, as opposed to *efficiency* (e.g., wearing a sweater as opposed to adding more attic insulation), is an example of direct demand reduction.

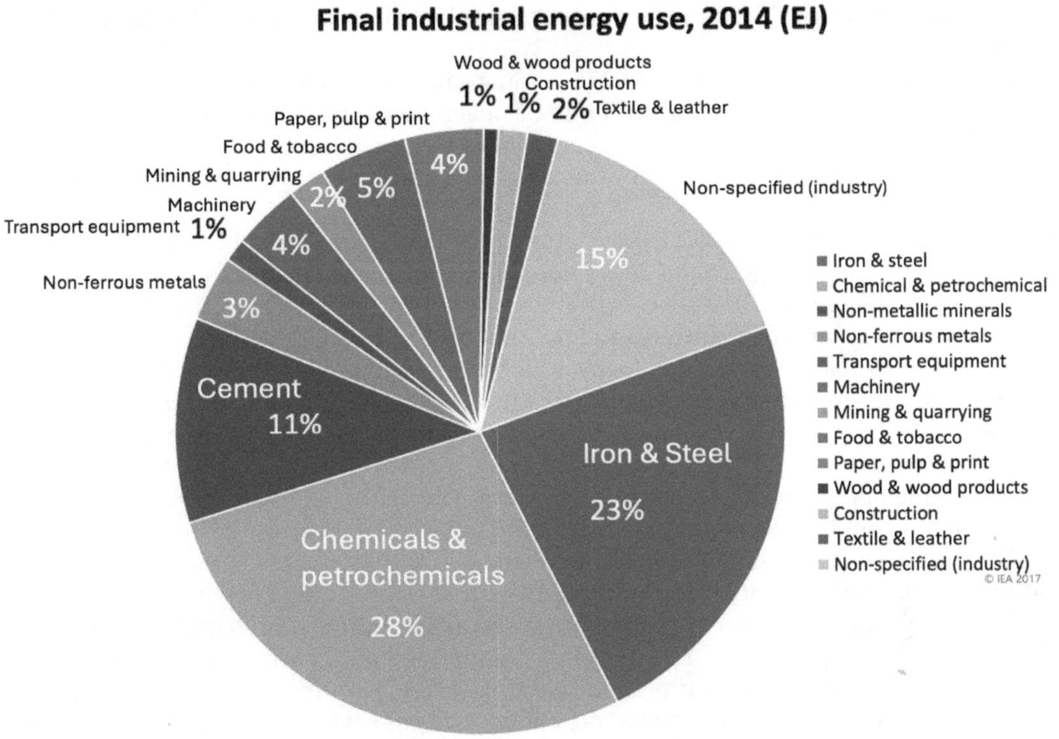

FIGURE 4.20
Energy use by different industry types. In the legend the categories are listed in a clockwise direction on the pie chart beginning with iron and steel. (From IEA [53]. 2017. Licensed under Creative Commons license CC BY 4.0.)

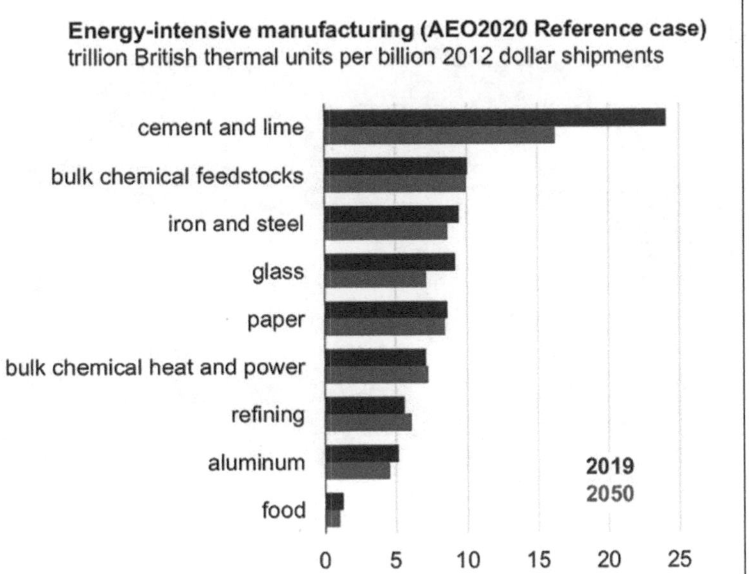

FIGURE 4.21
Energy Intensity of different manufacturing industries. (From EIA [54].)

The IPCC has examined how each of these categories can be addressed. Examples of each follow.

***G/E*:** The historical breakdown of U.S. industrial energy use is shown in Figure 4.22 [56]. Note that coal use in the United States has decreased significantly. Coal still provides about 25% of world industrial energy, however, as well as 25% from oil, 25% from natural gas, 16% from electricity and 9% from renewables. And much of the world's electricity is still provided by coal [57].

Fuel switching from oil and coal to natural gas can reduce greenhouse gas emissions provided methane leakage in the entire supply chain is kept to a very low value. Carbon capture and storage (CCS) is also being considered to limit carbon dioxide emissions. In some cases, carbon can be harnessed into a useful product. Ultimately, however, switching to renewable sources is needed to drive carbon emissions to zero. Solar thermal energy can be used for processes requiring temperatures below about 400°C. Solar and wind are increasingly being used to generate electricity, which can be used to provide very high industrial temperatures.

***E/M*:** While opportunities for improving energy efficiency will vary from one company or industry to another, studies have been done to identify the most cost-effective opportunity areas overall. Hein [58] combined energy savings potential, cost-effectiveness, and popularity of implementation to rank order the best energy efficiency measures for electricity savings. He identified the following measures in order of priority, beginning with the highest: lighting, process, refrigeration, compressed air, controls, variable frequency drives, heating, ventilation, and air conditioning (HVAC) equipment upgrades, new construction, motors, and building envelope. For natural gas savings, his recommended measures are HVAC, process, building envelope, new construction, and refrigeration. Note

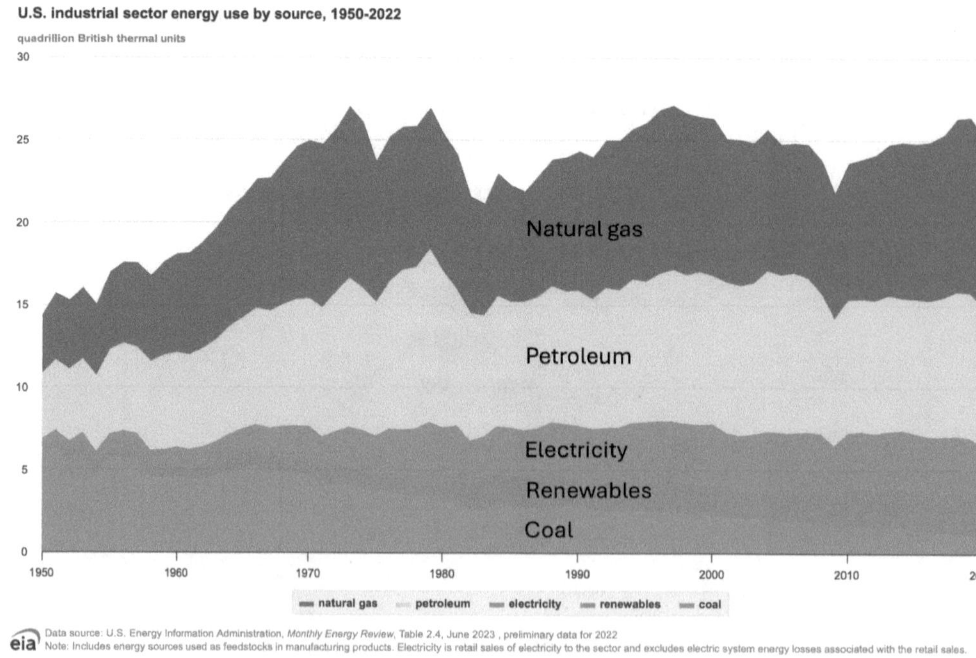

FIGURE 4.22
U.S. industrial energy use by source, 1950–2022 (quadrillion Btu) (From EIA [56].)

that many of the items in the industrial list, such as lighting, HVAC, and envelope, are also pertinent to commercial and residential buildings.

McKinsey [8] has found that most companies can reduce energy use by 10%–35%, depending on the size of the investment. In addition to the measures listed by Hein, they recommended greater use of CHP, heat recovery, and improved piping to reduce pumping power.

***M/P*:** Manufacturing processes can be improved to reduce the amount of waste. Scrap materials can be recycled. Recycled materials can, in general, be used in place of new materials. Materials that are damaged can be repurposed. In the construction of the NREL's Research Support Facility, discarded natural gas pipe was used for support structures, and bark beetle kill was used for accent wood. Improving packaging can reduce waste. For example, foam boxes have been replaced with simple paper wraps for fast food products.

***P/S*:** Making products more durable is a challenge because many products have planned obsolescence to increase sales. But durability information provided by consumer organizations can impact customer selection and provide an incentive for manufacturers to make more durable products. Making products that are higher strength and require less volume can also serve as a sales tool, for example, stronger laundry detergent.

***S*:** Reducing overall demand is perhaps the most difficult opportunity to address because there may be resistance to changes that are seen as compromising lifestyle. For example, acceptance of smaller homes would reduce the demand for building materials and switching to a vegetarian (or less meat-intensive) diet would reduce emissions associated with beef production, but many people are resistant to these changes. It is possible, however, for government or employers to support or incentivize energy-saving behavior. For example, employer support for telecommuting and video conferencing can reduce the amount of travel.

Achieving significant energy savings and lower emissions requires an energy and waste management program that is tailored to each industry. Lean energy programs are aimed at reducing the various forms of waste. McKinsey [8] has outlined eight forms of waste that industry can reduce through common sense planning and maintenance measures, as shown in Table 4.11.

TABLE 4.11

Eight Kinds of Waste

Sources of Waste	Definition	Example
Overproduction	Producing more energy than needed	Venting steam
Waiting	Energy consumption during stopped process	Laser welding line at 40% power
Transportation	Inefficient transport of energy	Steam and heat leaks in piping
Overspecification	Using more energy than necessary	Furnace operating 100°C higher than necessary
Inventory	Stored goods can waste energy	Crude steel cools in storage and has to be reheated for rolling
Rework/scrap	Need to redo a process product quality is inadequate	Re-drying polymer lines that did not get coagulated in drying process
Inefficient processes	Nonoptimal process design or operation	Excess oxygen in steam boiler
Employee potential	Failure to obtain human input	Employees not involved in developing energy-saving initiatives

Source: McKinsey & Company [8].

4.4.3 Improvements in Industrial Equipment

Opportunities for equipment improvements vary by industrial sector. Examples of improvements that can be made in the major sectors as summarized by the IPCC are given in Table 4.12 [55]. Motors, compressed air systems, and refrigeration systems are commonly used industrial equipment that have been the subject of recent energy efficiency case studies.

Motors: According to DOE, over 20% of electricity sold in the United States is for electric motors totaling 680 TWh/year. It has been estimated that electric motors consume 45% of world electricity [59]. They thus represent a major opportunity for electricity savings.

When motor speed is plotted against torque, efficiency reaches a peak at the optimum torque and drops off rapidly at lower torque. Thus, to achieve high efficiency, it is important not to oversize a motor. The National Electrical Manufacturers' Association (NEMA) ranks motor efficiencies. High-efficiency motors are given the NEMA Premium performance label. To encourage the use of more efficient motors, Minimum Energy Performance Standards have been developed around the world [60].

To further reduce motor electricity consumption, the Consortium for Energy Efficiency (CEE) Motors and Motor Systems Committee has studied the potential for electronically controlled permanent magnet AC (PMAC) motors. The use of permanent magnets eliminates rotor slip inherent in traditional AC induction motors and hence provides greater efficiency. Coupling PMAC motors with variable frequency drives (VFDs) makes them

TABLE 4.12

Example Improvements to Reduce Greenhouse Gas Emissions in Various Industries

Industry	Example Improvements
Iron and steel	• Improved heat recovery from process streams • Recycling of structural steel • Adoption of more realistic structural safety margins for buildings
Cement	• Adoption of EPA Energy Performance Indicator score to identify potential improvements • Harvesting of CO_2 • Production of higher-strength, longer-lasting concrete
Chemicals	• Generating pure waste streams to enable more recycling • Greater adoption of combined heat and power • Material-conserving plastic packaging
Pulp and paper	• Improved heat recovery in drying process • Greater paper recycling • Reduced paper weight and duplex printing
Nonferrous metals (e.g., aluminum)	• Adoption of improved electrolysis methods • Minimization and reuse of scrap
Food processing	• Less energy-intensive drying techniques, for example, mechanical dewatering • Local sourcing of food • Transition from meat to vegetarian foods
Textiles and leather	• More efficient motors and boilers
Mining	• Apply latest resource characterization methods to obtain higher quality ore and reduce amount crushing and grinding • Use of more efficient crushing technologies • Great metal recycling

Source: From Intergovernmental Panel on Climate Change (IPCC) Working Group III [55].

especially efficient for fans, pumps, and compressors supplying variable loads. Several case studies of the PMAC motor/drive combination have shown energy savings of 2%–26% compared to industry-standard induction motors having VFDs [61]. The greatest savings occurred where reduced speed operation was common. The advantage of PMAC motors was shown in bench tests of fan motors by the Sacramento Municipal Utility District (SMUD). The efficiency of the PMAC motor compared to standard and premium induction motors as functions of speed and torque is shown in Figure 4.23.

Compressed air systems: Air compressors of a variety of types are used for various processes throughout industry. The Northeast utility National Grid conducted an energy efficiency study of 41 compressed air systems in their Upstate New York and Massachusetts service area. Opportunities include leak maintenance, use of variable-speed compressors, and better controls. Hessmer et al. [62] provide details on how these opportunities can be addressed.

Refrigeration systems: Refrigeration systems consist of a compressor, a heat-rejecting condenser, expansion valve, heat-absorbing evaporator, and associated fans to move air over the evaporator and condenser. Two recent case studies aimed at improving refrigeration system efficiency applied a number of energy efficiency measures [63]. Typical measures to improve efficiency include:

- A VFD to drive the compressor
- High-efficiency fan motors with variable-speed drives
- Oversizing of condensers
- Variable condenser pressure control allowing for lower average condenser pressure reducing compressor load during winter months.

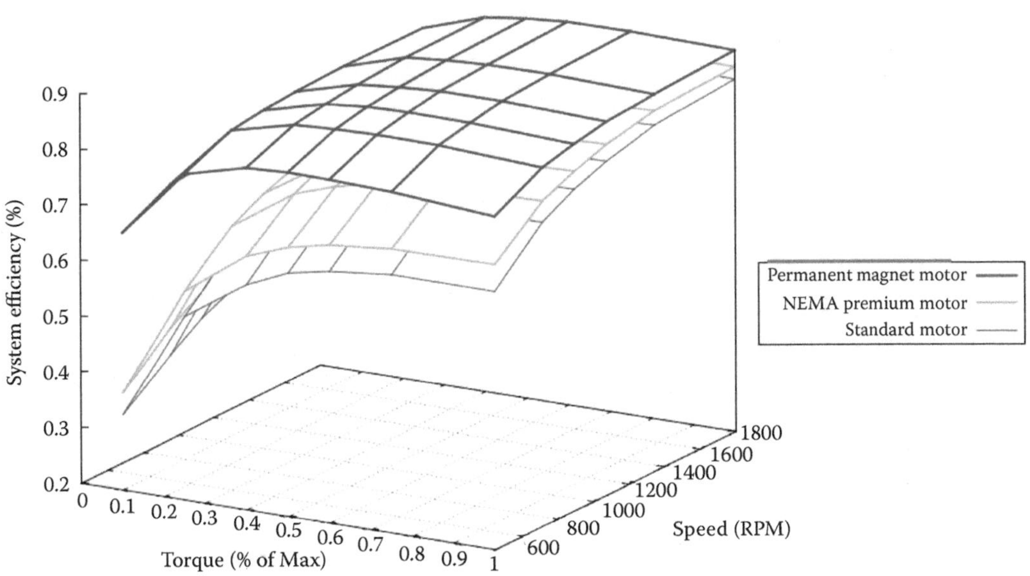

FIGURE 4.23
Efficiencies of three types of fan motors (PMAC, NEMA Premium, and standard) as functions of speed and torque. (From Larsen et al. [61].)

4.4.4 Industrial Electrification

As with the buildings sector, electrifying industries to take advantage of a decarbonizing grid is a key part of an overall climate change strategy. Fuel combustion for process heat comprises 40% of global industrial emissions, and much of this can be addressed with electricity. A particularly important opportunity lies with the fact that, depending on the industry and the particular fuel that is burned, anywhere from 30% (with coke) to over 80% (with diesel) of the energy consumed is at temperatures below 200°C, meaning that a significant portion of this could be addressed by industrial heat pumps, which would be significantly more efficient than the direct use of electricity. This is shown in Figure 4.24. [64].

Depending on the source temperature, industrial heat pumps can provide heat at temperatures up to about 160°C as shown in Figure 4.25 [65].

For process temperatures higher than the limit for heat pumps, there are a variety of ways in which electricity can be directly used as shown in Figure 4.26 [66]. Reference [67] summarizes the electrification opportunities for various industries.

There may be industrial process heating applications that don't lend themselves to direct electrification either because the proper flux profile cannot be obtained, or the conversion from fuel to electrical equipment is too expensive. In that case, electricity can be used (via electrolysis of water) to produce carbon-free green hydrogen that can then serve as a replacement for fossil fuel. The hydrogen can be delivered via pipeline or truck, or it can potentially be produced at the industrial site. (More details about hydrogen are given in Chapter 14.)

An emerging use of green hydrogen is in the processing of iron ore. Most of the steel production in the United States uses recycled steel, which can be processed in an electric arc furnace. But for the production of new steel, iron must be separated from iron ore before the iron can be sent to an electric arc furnace. This entails the reduction of iron

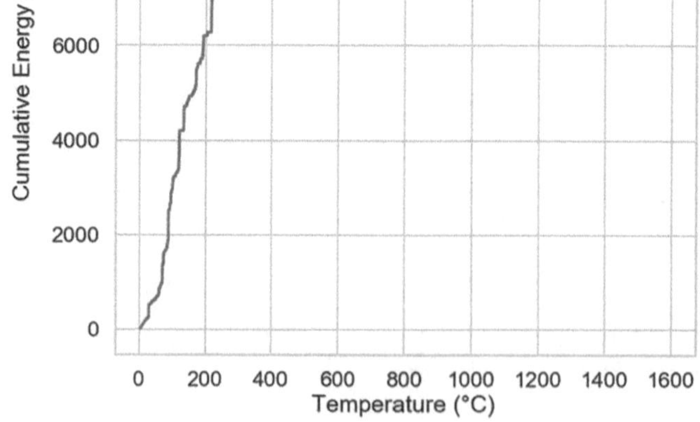

FIGURE 4.24
Cumulative industrial energy consumed vs. temperature for different fuels. (From McMillan [64].)

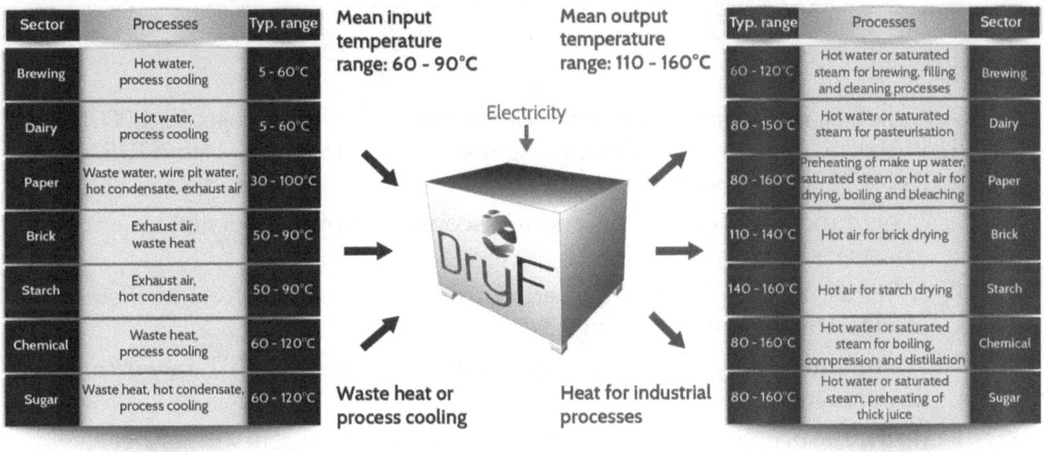

FIGURE 4.25
Source and supply temperature ranges for industrial heat pumps. (From Austrian Institute of Technology [65].)

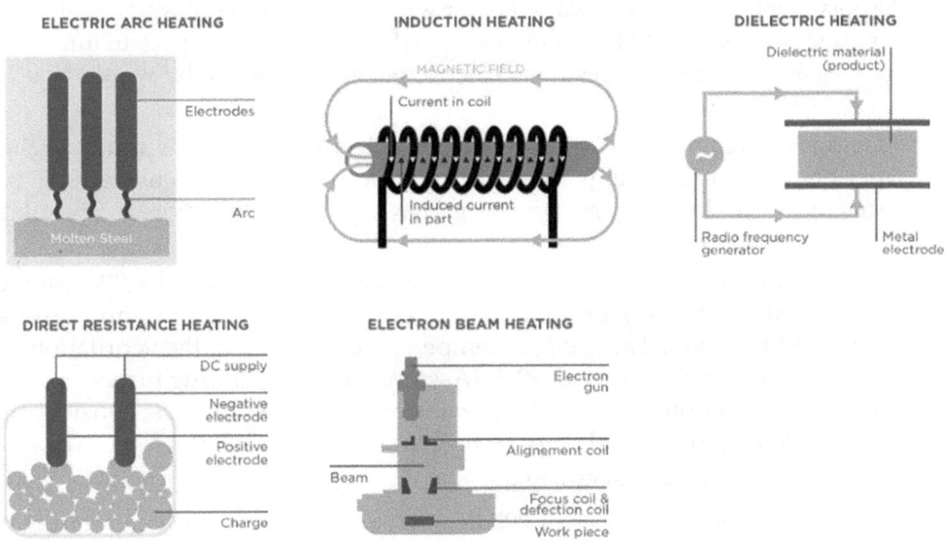

FIGURE 4.26
Different types of direct electric heating methods. (From Friedmann et al. [66].)

oxide to iron. In a basic oxygen furnace, coke (a high-carbon, coal-based fuel) is burned, and oxygen in the air combines with the carbon in coke to yield iron and carbon dioxide. A new process called direct reduced iron (DRI) burns green hydrogen in place of coke, and the oxygen combines with the hydrogen to emit steam instead of carbon dioxide.

The preceding discussion is only intended to introduce the student to industrial energy. More details on this subject can be found in Section 5 of *Accelerating the US Clean Energy Transformation: Challenges and Solutions by Sector*, by Kutscher, Logan, and Coburn, University of Colorado Boulder, December 2020 [68].

Problems

4.1 A home in a cold climate is heated with natural gas, which costs \$8/1,000 ft^3, or \$1.10/therm. It is estimated that raising the attic insulation level from R15 to R40 will cost \$1,200 and save 60 therms/year.

a. Assuming a 30-year period and a discount rate of 3%, calculate the cost of conserved energy.

b. What is the simple payback period for this investment (cost of insulation divided by the annual savings)?

c. Comment on why this payback is so long. Why might it still be a worthwhile investment?

4.2 Consider the primary energy savings for four energy efficiency measures shown in Table 4.9.

a. Air sealing shows similar per-house energy savings in climate zones 1, 4, 5, and 7 (see Figure 4.16 for climate zones) but only has significant aggregate energy savings in zones 4 and 5. Why do you think that is the case?

b. Replacing an electric hot water heater with a heat pump hot water heater has a high per-house energy savings everywhere but a low aggregate impact in zones 6 and 7 (few homes have electric water heating). Why would that be the case?

c. Upgrading single-pane windows to double-pane low-e has much higher per-house energy savings than adding attic insulation to achieve R-49, yet they have the same aggregate energy savings (0.41 Quads/year). Speculate on why that would be the case.

4.3 An auditorium at sea level is designed to hold an audience of 6,000 people. Outdoor ventilation air is provided at the current ASHRAE minimum guideline of 5 CFM/person. The outdoor temperature is 10°F and the ventilation air is exhausted at a temperature of 70°F. (Assume that the building furnace makes up for any conduction losses through walls and windows and just consider the ventilation load and internal body heat generation.)

a. Calculate the mass flow rate of air

b. Assuming a body heat rate of 225 Btu/h/person, what is the temperature of the ventilation air after it enters the room and body heat is added?

c. Assuming a ventilation heat recovery system reduces the ventilation heating by 80%, how much heat must be supplied to the incoming ventilation air?

4.4 The no-load temperature of a building with internal heat sources is given by Equation 4.3. How would this equation be modified to account for heat losses through the surface of an unheated slab, the heat losses being independent of ambient temperature?

4.5 An unheated garage is placed on the north wall of a building to act as a thermal buffer zone. If the garage has roof area A_r, window area A_{wi}, door area A_d and wall area A_{wa}, what is effective U value for the north wall of the building if its area is A_n? The garage floor is well insulated and has negligible heat loss.

Express the effective U value in terms of the U values and areas of the several garage surfaces.

4.6 What is the annual energy demand for a building in Denver, Colorado, if the peak heat load is 150,000 Btu/h based on a design temperature difference of 75°F? Internal heat sources are estimated to be 20,000 Btu/h, and the design building interior temperature is 70°F.

4.7 The table below gives the characteristics of a building in Houston, Texas. Determine the cooling load for July 30 at solar noon. Any information regarding the load not given may be assumed or neglected.

Assume solar loads only occur for the unshaded south-facing wall and window. You will use equations appearing in Section 4.3.2.

Note: Houston's latitude is 29.44°N. Equations for solar declination and altitude can be found in Chapter 10. For a vertical wall, the tilt angle is $\beta = 90°$.

Note that the angle of incidence for a south-facing vertical wall at solar noon ($a_w = a_s = 0$) is equal to the solar altitude.

Assume white paint has an absorptivity of 0.12.

Factor	Description or Specification
Roof:	
Type	Flat, shaded
Area $A_{rf,sh}$ (m^2)	250
U factor (W/m^2K)	$U_{rf} = 0.35$
Walls:	
Type	Vertical, painted white
Orientation, size ($m \times m$)	North, South, 3×10; East, West, 3×25
U factor (W/m^2K)	$U_{wa} = 1.08$
Windows:	
Orientation, area (m^2)	North, 8; South, 8; East, 20; West, 25
U factor (W/m^2K)	$U_{wi} = 6.2$
Insolation transmittance	$\bar{\tau}_{b,wi} = 0.6$; $\bar{\tau}_{d,wi} = 0.80$; $\bar{\tau}_{r,wi} = 0.55$
Temperature:	
Inside, outside (°C)	$T_{in} = 24$; $T_{out} = 37$
Insolation:	
Beam, diffuse, reflected (W/m^2)	$I_{hb} = 580$; $I_{hd} = 250$; $I_r = 200$

4.8 An air-conditioning system working in a vapor-compression cycle is used to manage the load for the building in Problem 4.7. If the high and low pressure in the cycle are 915 and 290 kPa, respectively, find (a) the flow of Refrigerant 134a (R-134a) used for the equipment and (b) the COP of the cycle. Assume the compressor is 100% efficient.

Note: Saturated and superheated R-134 properties are available in property tables.

Hints:

- For part a, look up saturated vapor and liquid properties and plug into Equation 4.18
- For part b, use Equation 4.21. You obtain h_{vd}, the enthalpy of the superheated vapor exiting the compressor, assuming an isentropic compression. You will

need to interpolate between the high and low pressure on the R-134 superheated vapor at the constant entropy.

4.9 An industrial condenser unit uses two 500 hp (373 kW) fans that operate in parallel. For 3,000 h/year, the unit operates at half flow and one fan is turned off with the other operating at full power. The plant owner is considering purchasing a variable frequency drive for each fan and running both fans at half flow instead of shutting one fan off. The efficiency of the VFD itself at half load is 98%.

a. How much electric energy is saved if VFDs allow both fans to be operated at half flow? (Note: fan power is proportional to flow rate cubed, as one can see from a simple dimensional analysis.)

b. If each VFD costs $200 per hp and the plant pays 8 ¢/kWh for electricity, what is the simple payback period?

4.10 A building owner of a small retail store wants to upgrade the 20-year-old HVAC system. The store has two 10-ton rooftop units (RTUs) with an original rated full load EER of 8.7 Btu/Wh. The owner is considering two options for replacement: two standard-efficiency RTUs (11.2 EER) or two high-efficiency RTUs (13 EER). As the design engineer, your job is to estimate the energy cost savings of the two replacement options compared to the existing RTUs.

a. Estimate the annual energy cost savings of each option.

b. Determine the simple payback period for the high-efficiency RTUs compared to the standard-efficiency RTUs with no utility incentives and with a utility rebate for using high efficiency of $350/ton. Assume:

The performance of the existing RTUs has degraded 1%/year

The cost of electricity is $0.1/kWh and there are no peak demand charges

The RTUs operate at an annual equivalent full load hours of 2,200 hours. (That is, their operation over a year is equivalent to operating a full load of 2,200 hours for the year.)

The installed cost of the RTUs is $900/ton for the standard-efficiency RTUs and $1,400/ton for the high-efficiency RTUs.

References

1. Rosenfeld, H. and Poskanzer, D., 2009. A graph is worth a thousand gigawatt-hours: How California led the United States in energy efficiency. *Innovations* 4(4), 57–79.
2. Energy Information Administration (EIA), August 2024. Monthly energy review, Tables 2.1a and b, https://www.eia.gov/totalenergy/data/monthly/#consumption.
3. EIA, 2023. International energy outlook, https://www.eia.gov/Outlooks/Ieo/pdf/IEO2023_Release_Presentation.pdf.
4. International Energy Agency (IEA), 2013. Energy efficiency market report 2013, https://iea.blob.core.windows.net/assets/f15f5fab-62cf-45f9-8016-2d1129fb8ca6/EEMR2013_free.pdf.
5. NRDC Fact Sheet, July 2013. California's energy efficiency success story: Saving billions of dollars and curbing tons of pollution, https://www.nrdc.org/sites/default/files/ca-success-story-FS.pdf.

6. Nadel, S. ACEEE, April 15, 2020. U.S. energy use held steady for 15-plus years, https://www.aceee.org/blog-post/2020/04/us-energy-use-held-steady-15-plus-years.
7. Molina, M., March 2014. The best value for America's energy dollar: A national review of the cost of utility energy efficiency programs, ACEEE Report No. U1402.
8. McKinsey & Company, March 2010. Energy efficiency: A compelling global resource, www.mckinsey.com.
9. Center for Climate and Energy Solutions, Federal vehicle standards, December 2004, https://www.c2es.org/document/comparison-of-passenger-vehicle-fuel-economy-and-ghg-emission-standards-around-the-world/.
10. Burgess, J., 2015. The second generation of strategic energy management programs, ACEEE Study on Energy Efficiency in Industry.
11. ISO 50001 – Energy Management, 2018, https://www.iso.org/iso/home/standards/management-standards/iso50001.htm.
12. International Energy Agency, https://www.iea.org/energy-system/buildings#. Last accessed January 2025.
13. ASHRAE, 2021. *ASHRAE Handbook—Fundamentals*, American Society of Heating, Refrigerating and Air-Conditioning Engineers, Atlanta, GA, https://www.ashrae.org/technical-resources/ashrae-handbook/description-2021-ashrae-handbook-fundamentals.
14. Kreith, F., and Manglik, R., 2016. *Principles of Heat Transfer*, 8th ed., Cengage Learning, Boston, MA.
15. Klein, S.A., Cooper, P.I., Freeman, T.L., Beekman, D.M., Beckman, W.A., and Duffie, J.A., 1975. A method for simulation of solar processes and its applications, *Solar Energy* 17, 29–37.
16. ASHRAE, Climatic design conditions, https://ashrae-meteo.info/v2.0/. Last accessed January 2025.
17. Goswami, Y., 2000. *Principles of Solar Engineering*, 2nd ed., Taylor & Francis, Philadelphia, PA.
18. ASHRAE, 2023. *ASHRAE Handbook—Heating, Ventilating, and Air-Conditioning Applications*, American Society of Heating, Refrigerating and Air-Conditioning Engineers, Atlanta, GA.
19. EIA, 2018. Commercial buildings energy consumption survey: Energy usage summary, https://www.eia.gov/consumption/commercial/2018.
20. ASHRAE, 2016. *ANSI/ASHRAE/IES Standard 90.1-2016—Energy Standard for Buildings Except Low-Rise Residential Buildings*, Atlanta, GA.
21. ASHRAE, 2018. Advanced energy design guides, https://www.ashrae.org/technical-resources/aedgs.
22. Torcellini, P., Long, N., and Judkoff, R., 2003. Consumptive water use for U.S. power production, https://www.nrel.gov/docs/fy04osti/33905.pdf.
23. NREL, August 2016. NREL+Liquid cool solutions, Fact Sheet, NREL/FS-5C00-66712, https://www.nrel.gov/docs/fy16osti/66712.pdf.
24. Open Studio, https://www.openstudio.net. Last accessed January 2025.
25. U.S. Department of Energy, 2025, Qualified software for calculating commercial building tax deductions, https://energy.gov/eere/buildings/qualified-software-calculating-commercial-building-tax-deductions. Last accessed January 2025.
26. U.S. Department of Energy, 2015. A common definition for zero energy buildings, DOE/EE-1247, https://energy.gov/eere/buildings/downloads/common-definition-zero-energy-buildings.
27. U.S. Department of Energy, July 2012. Research support facility, DOE/GO-102011-3311.
28. Robbins, J., Skelton, B., and Olden, R., 2015. Revealing net zero, *High Performing Buildings*, Fall 2015, https://www.hpbmagazine.org/walgreens-net-zero-store-evanston-il/.
29. EIA, 2020. Residential energy consumption survey (RECS), https://www.eia.gov/consumption/residential/.
30. EIA, 2015. Annual energy outlook, https://www.eia.gov/outlooks/aeo/data/browser/#/?id=4-AEO2015&cases=ref2015&sourcekey=0.

31. U.S. Department of Energy, October 2012. Energy savings potential and research, development & demonstration opportunities for residential building heating, ventilation, and air conditioning systems, DOE/EE-0850.
32. Pacific Northwest National Laboratory, January 2013. Building science-based climate maps, PNNL-SA-90570.
33. Wilson, E., Christensen, C., Horowitz, S., Robertson, J., and Maguire, J., January 2017. Electric end-use energy efficiency potential in the U.S. single family housing stock, NREL/TP-5500-65667.
34. Christensen, C., Anderson, R., Horowitz, S., Courtney, A., and Spencer, J., August 2006. BEopt software for building energy optimization: Features and capabilities, NREL/TP-550-39929.
35. Polly, B., Kutscher, C., Macumber, D., Schott, M., Pless, S., Livingood, B., and Geet, O.V., 2016. From zero energy buildings to zero energy districts, in *Proceedings of the ACEEE Summer Study on Energy Efficiency in Buildings ACEEE*, Washington, DC.
36. UN (United Nations), July 10, 2014. World's population increasingly urban with more than half living in urban areas, https://www.un.org/en/development/desa/news/population/world-urbanization-prospects-2014.html.
37. Architecture 2030, June 2014. Roadmap to zero emissions, https://www.architecture2030.org/wp-content/uploads/2018/12/Roadmap-to-Zero-Emissions.pdf.
38. Global Covenant of Mayors, (formerly Compact of Mayors), https://www.globalcovenantofmayors.org. Last accessed January 2025.
39. U.S. Conference of Mayors, https://www.usmayors.org/programs/mayors-climate-protection-center/. Last accessed January 2025.
40. Carlisle, N., Van Geet, O., and Pless, S., 2009. Definition of a zero net energy community, NREL/TP-7A2-46065, National Renewable Energy Laboratory, Golden, CO, https://www.nrel.gov/docs/fy10osti/46065.pdf.
41. EcoDistricts, June 2014. The EcoDistricts protocol, https://ecodistricts.org/wp-content/uploads/2013/03/EcoDistricts_Protocol_Executive_Summary_ISSUE_6.242.pdf.
42. Dakin, B., and German, A., 2014. Early performance results from a zero net energy community, in *Proceedings of the ACEEE* 2014 *Summer Study on Energy Efficiency in Buildings*, ACEEE, Washington, DC.
43. DOE Better Buildings Solutions Center, https://betterbuildingssolutioncenter.energy.gov/accelerators/zero-energy-district. Last accessed January 2025.
44. UN Environment, District energy in cities initiative, https://www.unep.org/resources/factsheet/district-energy-cities-initiative. Last accessed January 2025.
45. U.S. Department of Energy, 2016. DOE Zero Energy Ready Home (ZERG) program requirements, https://www.energy.gov/eere/buildings/doe-zero-energy-ready-home-zerh-program-requirements.
46. New Buildings Institute, Cost study of the building decarbonization code, April 2022, https://newbuildings.org/resource/cost-study-of-the-building-decarbonization-code/.
47. https://www.heet.org, Last accessed January 2025.
48. Kashtan, Y., Nicholson, M., Finnegan, C.J., Ouyang, Z., Garg, A., Lebel, E.D., Rowland, S.T., Michanowicz, D.R., Herrera, J., Nadeau, K.C., and Jackson, R.B., May 2024. Nitrogen dioxide exposure, health outcomes, and associated demographic disparities due to gas and propane combustion by U.S. stoves, *Science Advances* 10(18), 3, doi:10.1126/sciadv.adm8680.
49. https://www.whispervalleyaustin.com/news/#geothermal, Last accessed January 2025.
50. Mikler, V., June 9, 2017, District Energy 101, Integral Group, https://www.introba.com/news/district-energy-101-7-steps-success.
51. https://www.materialspalette.org/palette/, Last accessed January 2025.
52. Building Green, Embodied carbon tools: Assessing the options, https://www.buildinggreen.com/news-analysis/embodied-carbon-tools-assessing-options.
53. IEA, 2017. Global iron and steel technology roadmap, https://iea.blob.core.windows.net/assets/imports/events/245/ISTRM_Session0_IEA_201117.pdf.

54. EIA, January 2020, Annual energy outlook 2020, https://www.eia.gov/outlooks/aeo/pdf/aeo2020%20full%20report.pdf.
55. Intergovernmental Panel on Climate Change (IPCC) Working Group III. 2014. Climate change 2014: Mitigation of climate change, Chapter 10, Industry, https://www.ipcc.ch/report/ar5/wg3/.
56. EIA, Use of energy explained, Last accessed January 2025. https://www.eia.gov/energyexplained/use-of-energy/industry.php.
57. https://www.statista.com/statistics/263471/industrial-energy-consumption-worldwide/, Last accessed January 2025.
58. Hein, M., 2015. The silver bullet list: Impactful investments in energy efficiency, in *Proceedings of the ACEEE Summer Study on Energy Efficiency in Industry*, ACEEE, Washington, DC.
59. Waide, P., and Brunner, C., 2011. Energy-efficiency policy opportunities for electric motor-driven systems, International Energy Agency Working Paper.
60. Delaney, D., 2015. Global motor energy efficiency program, in *Proceedings of the ACEEE Summer Study on Energy Efficiency in Industry*, ACEEE, Washington, DC.
61. Larsen, W., et al. 2015. Exploring the customer benefits of permanent magnet motors: Test results and opportunities for next generation motor programs, in *Proceedings of the ACEEE Summer Study on Energy Efficiency in Industry*, ACEEE, Washington, DC.
62. Hessmer, C., Olmsted, J., Meserve, S., and Kondapi, R.K., 2015. Compressed air system energy efficiency upgrades implemented vs. underutilized measures, in *Proceedings of the ACEEE Summer Study on Energy Efficiency in Industry*, ACEEE, Washington, DC, https://www.aceee.org/files/proceedings/2015/data/papers/4-15.pdf.
63. Kumpanon, A., and Syed, R., 2015. Industrial refrigeration projects: Challenges and opportunities for energy efficiency, *Proceedings of the ACEEE Summer Study on Energy Efficiency in Industry*, ACEEE, Washington, DC.
64. McMillan, C., NREL, https://www.nrel.gov/analysis/solar-industrial-process-heat.html. Last accessed January 2025.
65. Austrian Institute of Technology, DryFiciency project, https://dryficiency.eu/industrial-heat-pumps/. Last accessed January 2025.
66. Friedmann, S.J., Fan, Z., and Tang, K., 2019. Low-carbon heat solutions for heavy industry: Sources, options, and costs today, Columbia Center on Global Energy Policy, Columbia University, New York, https://www.energypolicy.columbia.edu/publications/low-carbon-heat-solutions-heavy-industry-sources-options-and-costs-today/.
67. Deason, J., Wei, M., Leventis, G., Smith, S.J., and Schwartz, L.C., 2018. *Electrification of Buildings and Industry in the United States: Drivers, Barriers, Prospects, and Policy Approaches*, Lawrence Berkeley National Laboratory, Berkeley, doi:10.2172/1430688.
68. Kutscher, C.F., Logan, J.S., and Coburn, T.C., December 2020. *Accelerating the US Clean Energy Transformation: Challenges and Solutions by Sector*, University of Colorado, Boulder, https://www.colorado.edu/rasei/2020/12/01/accelerating-us-clean-energy-transformation.

5

Electricity Supply Systems

5.1 Historical Development of the U.S. Electric Power System

Access to electric power has a pervasive impact on the lives of most people around the world. The National Academy of Engineering named electrification the greatest engineering achievement of the twentieth century. The development of the electric grid began in the previous century. The first use of generated electricity was a hydropower plant completed in 1868 on the estate of English industrialist William Armstrong. The electricity was used to power early incandescent light bulbs developed by the English inventor Joseph Swan, as well as heat water and run various devices. Thomas Edison made improvements to Swan's light bulbs, and his success spurred him to build the first two public power stations: the Edison Electric Light Station in London and, a few months later, the Pearl Street Station in lower Manhattan. Both employed coal firing to power DC (direct current) electric generators.

The Edison plants generated low-voltage, high-current DC electricity, which required thick copper cables to minimize line losses and allowed for efficient transmission over only short distances on the order of a mile. Power loss in a transmission line is the product of the square of the current, I, being carried and the resistance of the cable, R (which is inversely proportional to the cross-sectional area of the cable): $P = I^2R$. To allow minimal losses when carrying electric current over large distances, it's important to minimize the current. Nikola Tesla and George Westinghouse recognized that by using alternating current (AC) instead of DC, a transformer could be used to achieve lower currents by raising the voltage. (The electric power, which can also be expressed as the product of the current and voltage, $P = VI$, stays approximately constant on each side of a transformer.) In 1885, Westinghouse's employee, William Stanley, built upon European concepts to develop the first reliable transformer that could boost AC electricity to higher voltages, thus achieving the lower currents that would allow efficient, longer distance transmission.

One of the first AC power stations, located near Telluride, Colorado, was completed in 1891. The Ames hydroelectric power plant provided 3,000 V to power an ore crusher located 2.6 miles away. Some of the benefits of AC power were demonstrated when Westinghouse and Tesla won the contract to provide electric lighting for the 1893 Chicago World's Fair using AC power. While there was no need to transmit the electricity a great distance, it successfully demonstrated to the public the safe use of AC. In 1896, Westinghouse built an 11,000-V AC transmission line to transmit electricity from a hydropower plant at Niagara Falls to Buffalo, New York, located over 20 miles away [1]. Because of its ability to easily convert to higher voltages for transmitting power over long distances, AC won out over DC. Voltages and distances kept increasing with time.

DOI: 10.1201/9781003586616-5

The electric grid today is over a century old, and modernizing the grid is a major national priority in the United States. The infrastructure needs to be upgraded to handle threats from terrorism (both physical and cyberattacks) as well as the increasing threat of extreme weather events associated with climate change (for example, storms, flooding, and heat waves). One way in which grid resilience is being improved is via the growth of microgrids, which are local electric grids containing both generators and storage that can operate in conjunction with the main grid or operate independently in an "islanding" mode when the main grid is down. Both the main grid and microgrids must be able to handle variable distributed generation sources on the grid and allow for two-way flow of both electricity and information associated with more and more devices communicating with the internet. The addition of variable sources of renewable electricity such as wind and solar photovoltaics, the interaction with battery electric vehicles, and various measures to control the demand create special challenges and opportunities that will be covered in this chapter.

5.2 Electrical Transmission

Figure 5.1 shows a schematic of a typical AC transmission system [2]. A central generating plant provides 3-phase/3-wire AC electricity at between 11,000 and 25,000 V. Single wind turbines and photovoltaic (PV) arrays typically generate AC power at relatively low levels (480–600 V) and then step up the voltage to 34,000 V before increasing the voltage to transmission levels. Power transformers at a transmission substation commonly step the voltage up to as high as 765 kV in North America and even over 1 MV internationally, proportionately decreasing the current. The low current associated with the high transmission voltage results in low transmission losses. These losses are a function of transmission line length and are on the order of 3% of the energy transmitted in the U.S. Eastern Interconnection and 5% in the Western Interconnection. After hundreds of miles of primary transmission, step-down transformers at another transmission substation reduce the voltage in the secondary transmission system to provide lower-voltage power to industrial and commercial customers, and to neighborhood transformers mounted on utility poles, which further reduce the voltage for household use.

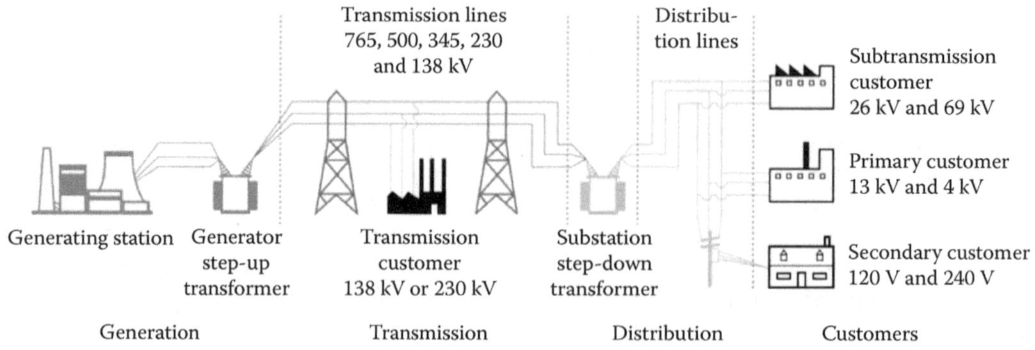

FIGURE 5.1
Diagram of AC electric transmission grid. (From North American Electric Reliability Corporation [2]. Reproduced with permission.)

At the customer level, a range of lower voltages is used depending on the customer class. Commercial and industrial customers typically use three-phase AC power. For a typical household, a three-wire, single-phase power drop provides three lines: +120 V, neutral (from the center of the utility pole transformer secondary winding), and –120 V. The 120 V household circuit is between the 120 V lines and neutral, whereas the 240 V circuit is between the +120 and –120 V lines. The household wiring system also includes a line to ground (connected to the bottom prong in a 3-prong plug). (Ground fault circuit interrupt, or GFCI, electrical outlets used in bathrooms and kitchens sense when current is escaping instead of flowing between a 120 V line and neutral and automatically break the circuit, thus preventing a current leak from passing through a homeowner to ground.)

High ambient temperatures increase the resistance of electric transmission lines. During times of high temperature, air-conditioning loads are high, meaning that the hot transmission lines carry more current and so further heat up due to I^2R heating. This can cause lines to sag due to thermal expansion, causing an electric arc to ground through trees. On August 14, 2003, a hot 345-kV transmission line in northern Ohio sagged into tree branches and caused a major cascading blackout that affected 50 million people. Thus, keeping transmission lines cleared of trees is a major maintenance activity for utilities.

AC transmission and distribution systems contain both resistive and reactive elements. In resistive elements such as resistive-wire light bulbs, the sinusoidally varying current and voltage are in phase with each other and all the power passing through the element is dissipated as heat (and light). Reactive elements such as generators and motors make use of electromagnetic interactions that can cause current and voltage to be out of phase. Some of the supplied power goes into varying electric and magnetic fields and is not lost to the environment. In the case of capacitors, a changing current builds up the voltage across the capacitor and current leads voltage by 90°. The capacitor alternately charges and discharges but delivers no net power over a cycle. In the case of an inductor, the changing voltage induces a changing current and current lags voltage. Each of these elements has a reactance measured in ohms that is analogous to a resistance. Capacitive reactance is $1/\omega C$, and inductive reactance is ωL where ω is the angular frequency, C is capacitance and L is inductance.

The real power, P, dissipated in resistors is measured in watts (W). The reactive power, Q, does no work and has units of VARs, which stands for volts-amps-reactive. The apparent power, S, in an AC circuit is VI where V is the root-mean-square voltage, and I is the root-mean-square current. It is plotted as a vector on a plot of reactive power, Q, on the y-axis (in VARs) and real power, P, on the x-axis (in watts) as shown in Figure 5.2. (The reactive component, or y-axis, is typically expressed as the imaginary axis. While reactive power is not imaginary, it does no work, and expressing it as an imaginary component of a complex number allows for electrical engineers to harness the calculational efficiency of complex numbers.)

The real power is thus the x-component of the apparent power vector or,

$P = VI \cos \phi$ expressed in watts, where ϕ is the phase angle and $\cos \phi$ is the power factor.

The reactive power is the y-component (or "imaginary" component) of the apparent power vector or,

$Q = VI \sin \phi$ expressed in VARs

The apparent power can be expressed in terms of real and reactive power as

$$S = \sqrt{P^2 + Q^2} \tag{5.1}$$

A residential utility customer pays only for real power because that is what is consumed. If loads are all resistive, $\phi = 0$ and $\cos \phi = 1$, and there is no reactive power component.

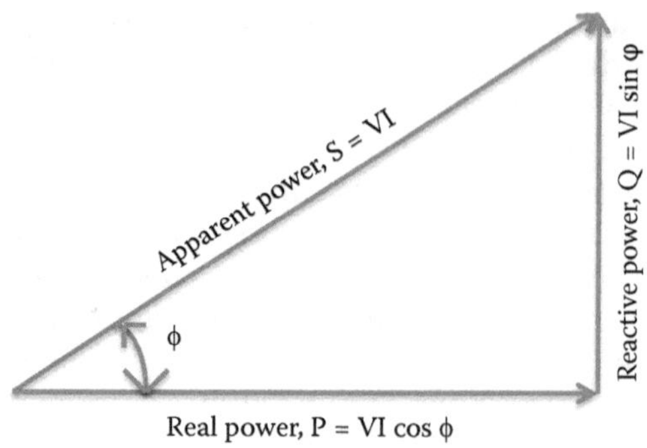

FIGURE 5.2
The power triangle showing apparent power, *S*, and its two components: real power, *P*, and reactive power, *Q*.

TABLE 5.1
Typical Baseline Costs for AC Transmission Lines

Line Voltage (kV)	Single Circuit Cost ($/Mile)	Double Circuit Cost ($/Mile)
230	959,700	1,536,400
345	1,343,800	2,150,300
500	1,919,450	3,071,750

Source: From Black and Veatch [4].

In that case apparent power=real power. If an industrial customer is using large motors or other inductive devices that cause the voltage and current to be out of phase, the power factor is less than 1. The utility needs to send current for the apparent power (both the real and reactive power components), which results in a larger current in the transmission lines and associated equipment than is needed for the real power that is consumed. Also, the additional current needed to meet reactive loads can cause transformers to overheat. For these reasons, utilities want the power factor to be close to 1. (However, some reactive power is needed to control voltage.) PV inverters provide only real power since they are compensated only for real power production. Current interconnection standards such as IEEE 1547 also do not allow the PV inverters to actively regulate the local voltage. So, the addition of PV increases the fraction of reactive power on the grid provided by the utility and decreases the local power factor of the customer. IEEE standards were revised to allow better accommodation of distributed PV into the grid by allowing for local voltage regulation from inverters [3].

As electricity from wind farms and central solar plants becomes a larger fraction of the supply, the need for new or upgraded transmission lines to transport the electricity from where the renewable resources are highest to population centers becomes an increasingly important issue.

Transmission line costs are a function of the AC voltage and the distance. Black & Veatch developed a methodology to calculate new transmission and substation capital costs for the Western Electricity Coordinating Council (WECC) [4]. Their baseline per-mile costs as a function of voltage are shown in Table 5.1.

The Black & Veatch report includes various multipliers on these baseline costs to account for different conductor types, tower types, terrain, and so on. These costs are for

TABLE 5.2

Typical Baseline Costs for HVDC Transmission Lines

Line Voltage (kV)	Bi-Pole Cost ($/Mile)
500	1,536,400
600	1,613,200

Source: From Black and Veatch [4].

above-ground transmission. Underground transmission is more aesthetic and better protected from storm damage but costs 3–10 times as much and is more expensive to repair. As we build new transmission lines, the need to secure rights-of-way and the proper permits is an important issue.

Interestingly, more than a century after Tesla and Westinghouse won the AC–DC debate over Edison on the basis of AC being better for transmission, there is increasing interest in utilizing high-voltage direct current (HVDC) for very long-range transmission. DC transmission lines do not suffer from the electromagnetic interference and capacitance/inductance effects associated with AC. In addition, long-distance HVDC transmission involves fewer conductors (two for bipolar HVDC compared to three for 3-phase AC) that can be spaced closer together, meaning less wire, less robust transmission towers, and narrower rights-of-way. Also, the constant power in DC lines is generally less than the peak power in oscillating AC lines, so the conductors do not need to be as thick. Baseline costs for HVDC transmission lines are shown in Table 5.2.

The need to go to high voltage to minimize the current and associated I^2R losses still holds, however. So HVDC transmission requires a method to provide step-up and step-down voltage. HVDC systems use power electronic rectifiers and inverters at each end to convert from AC to DC and back again to AC. Thus, the cost savings associated with the HVDC transmission lines must be weighed against the additional costs of the conversion equipment. Because the added equipment cost for HVDC transmission is fixed independent of mileage, the net advantage of DC over AC occurs for long transmission distances. A study by Argonne National Laboratory [5] concluded that HVDC transmission in the United States is cheaper than AC transmission for distances exceeding about 435 miles.

5.3 The Electric Grid and Electricity Markets

The electric grid in the United States and Canada (and a small part of western Mexico) is divided into different synchronous interconnections as shown in Figure 5.3. There are four separate interconnects: the Western Interconnection, the Eastern Interconnection, the Quebec Interconnection, and the Electric Reliability Council of Texas (ERCOT). Electricity exchanges mostly occur within these four regions, as the generators are synchronized to the same frequency within each region. To transfer electricity between these interconnects, the AC electricity must be converted to DC, sent over a DC transmission line, and then converted back to AC to be properly synchronized in the next interconnect.

In 1968, following the 1965 blackout in the Northeast, the Federal Energy Regulatory Commission (FERC) established the National Electric Reliability Corporation, later

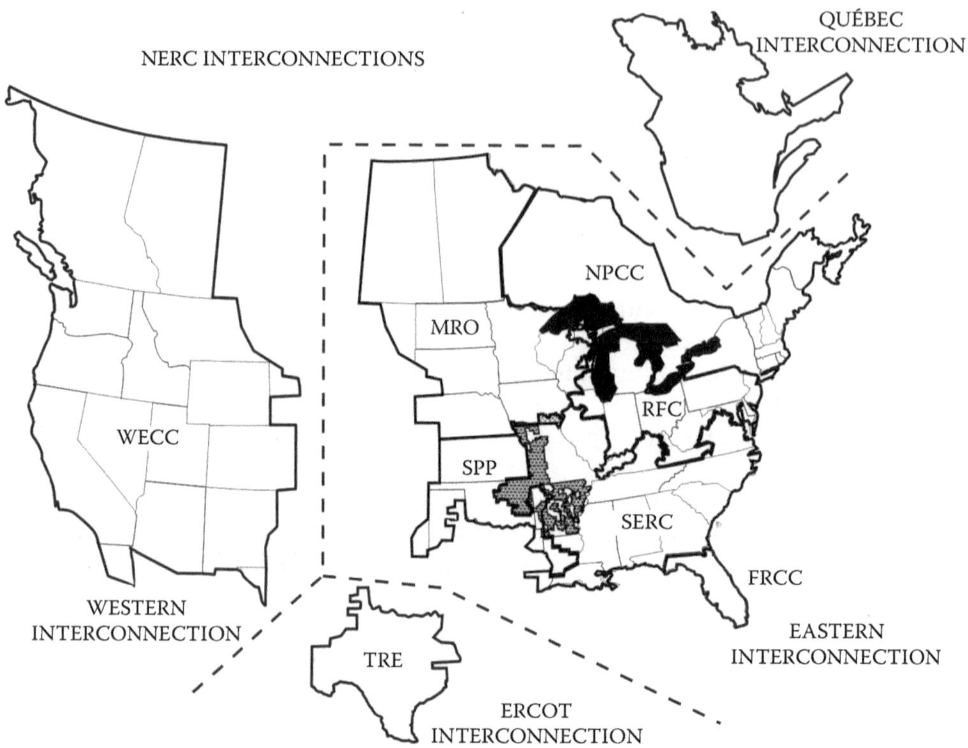

FIGURE 5.3
The four main North American interconnections and the various regions that set reliability standards. (From North American Electric Reliability Corporation [2]. Reproduced with permission.)

renamed the North American Electric Reliability Corporation (NERC). This organization subdivided the grid into various regions to ensure reliability and develop operating standards. The Eastern Interconnection was divided into six regions, whereas ERCOT and the Western Interconnection are each represented by one region. All eight regions are shown in Figure 5.3 [2]. However, in terms of how electricity is actually transmitted, the four main interconnection regions have greater relevance than the reliability regions.

The U.S. electric grid developed in piecemeal fashion over many decades and regions, and thus the electric power markets are complex and vary across the regions. They involve both private and public ownership and are regulated at both the federal and state levels. Utilities are of three types: for-profit investor-owned utilities or IOUs, and two types of nonprofit public-owned utilities: municipal utilities that are operated by a city (also called "munis") and rural electric co-ops, which are owned and operated by members in a rural region. All of these utilities fall under the umbrella of "load-serving entity" or LSE.

Customers can purchase electricity from an LSE or, in the case of very large industrial consumers (typically with electricity purchases exceeding 1 MW), they can purchase electricity directly from the electricity marketplace, with the electricity being transported by an LSE. This is shown in Figure 5.4 [6]. This shows the purchase of both generated electricity and the electricity savings resulting from demand response (DR) programs.

Most electricity consumed in the United States is purchased through an LSE. Figure 5.5 shows the flow of electricity and the flow of money for two cases: a traditional vertically

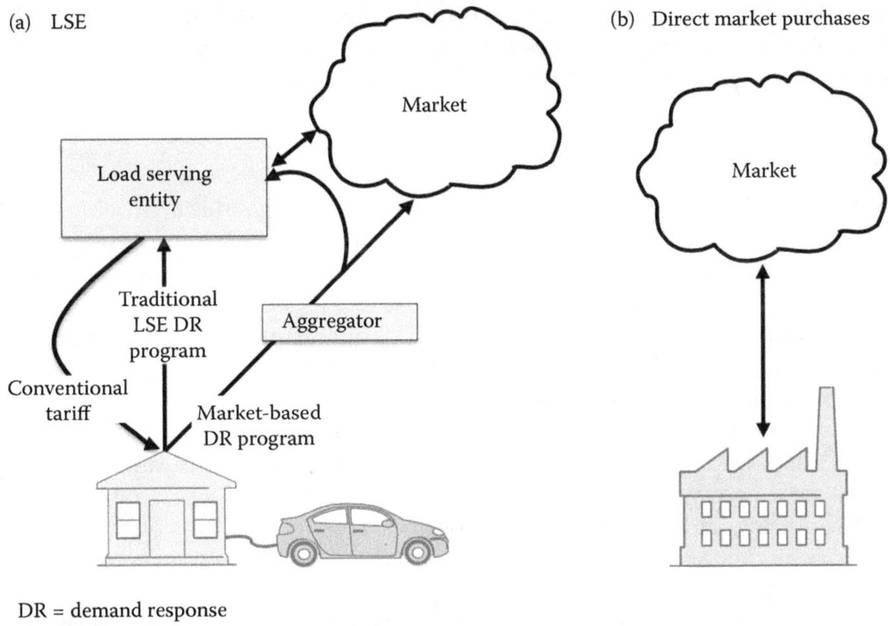

FIGURE 5.4
The purchase of electricity: (a) through a load-serving entity or (b) directly from the electricity market. (From Denholm et al. [6].)

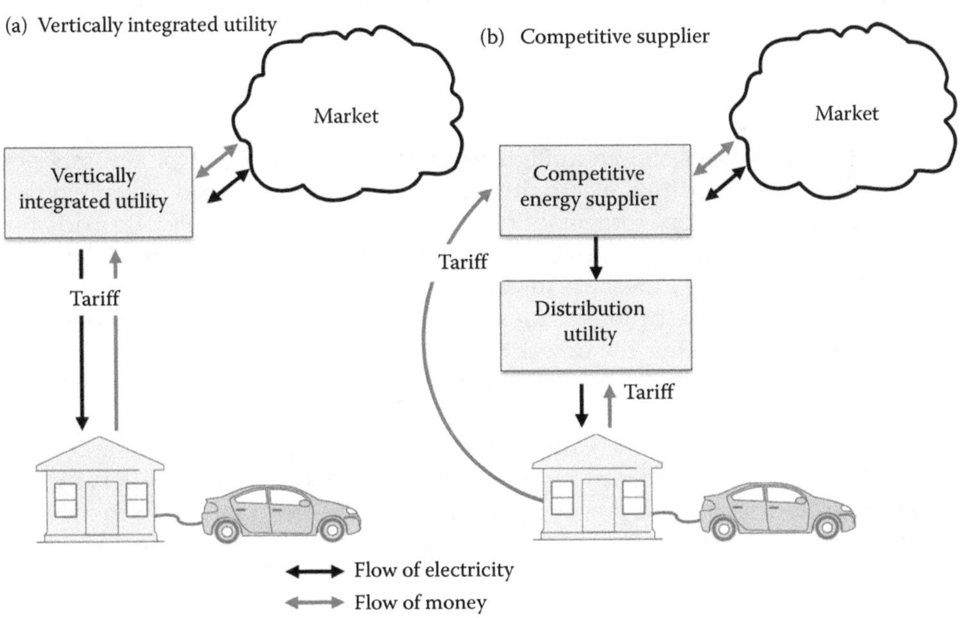

FIGURE 5.5
Electricity sales via an LSE in the case of (a) a vertically integrated utility and (b) where customer choice is allowed. (From Denholm et al. [6].)

integrated utility that controls generation, transmission, and distribution and—where customer choice is allowed—from a competitive generator. In the latter case, the customer pays both the electricity generator and the utility that provides the local distribution network.

5.3.1 Rate Structures

Residential electricity rates generally consist of a fixed charge and a variable one that is a function of the number of kilowatt-hours consumed. The fixed charge covers things like the generating and transmission equipment capital cost, conversion equipment, electric meter, the labor of the meter reader, and the cost of the record keeping and billing. The cost per kilowatt-hour consumed covers the cost of operating the generator or purchasing the power, maintaining the transmission lines, providing the various services needed to ensure reliability, and profit for the utility.

Commercial and industrial customers often have separate energy (kWh) and demand charges (kW). For example, in addition to being charged a certain number of cents per kWh for the total energy consumed in a month, there will be a separate charge for the highest kilowatt demand over a 15-minute period. This incentivizes electric customers to reduce their peak demand. High demands mean that an LSE must have more generating capacity available.

Another way to limit peak demand is to charge different prices for electric energy (kWh) at different times. Time-of-use pricing has a schedule of different charges per kWh for each time of the day and for different days of the week, usually with higher fixed energy prices during defined high-demand periods. A more accurate reflection of reality is real-time pricing, which continually changes in real time to reflect the demand on the LSE. As will be described later in this chapter, time-of-use and real-time pricing structures can encourage consumers to limit peak demand.

5.3.2 Electricity Markets

The electric energy markets shown in Figures 5.4 and 5.5 are unique for a number of reasons. Unlike other commodities, electricity cannot generally be stored cheaply on a large scale (although long-duration storage options are being explored for electric grids that are increasingly being powered by variable solar and wind energy). Electricity production constantly changes to exactly meet the instantaneous demand. Also, once electricity enters the power system, the source cannot be identified. For example, a residential consumer can opt to pay for wind power, but the power consumed cannot be directly associated with a wind farm. Finally, electric power is not directed; it simply flows down the path of least resistance.

Wholesale power markets in the United States are of two types: (1) traditional regulated markets and (2) regional markets, or regional transmission organizations (RTOs). In regulated markets, utilities own the power plants and sell the electricity to customers. Prices are set through regulatory control. Traditional markets are common in the Southwest (except California), the Southeast, and the Pacific Northwest. They include all three types of utilities: investor-owned, municipals, and rural electric co-ops. These utilities are often vertically integrated and own all three components: generation, transmission, and distribution. They include large federal systems: the Western Area Power Administration (WAPA) in the Southwest, the Tennessee Valley Authority (TVA) in the Southeast, and the Bonneville Power Administration (BPA) in the Pacific Northwest. These federal organizations work with utilities in their states to market hydropower and provide other services to multi-state regions. Utilities in traditional power markets produce their own electricity

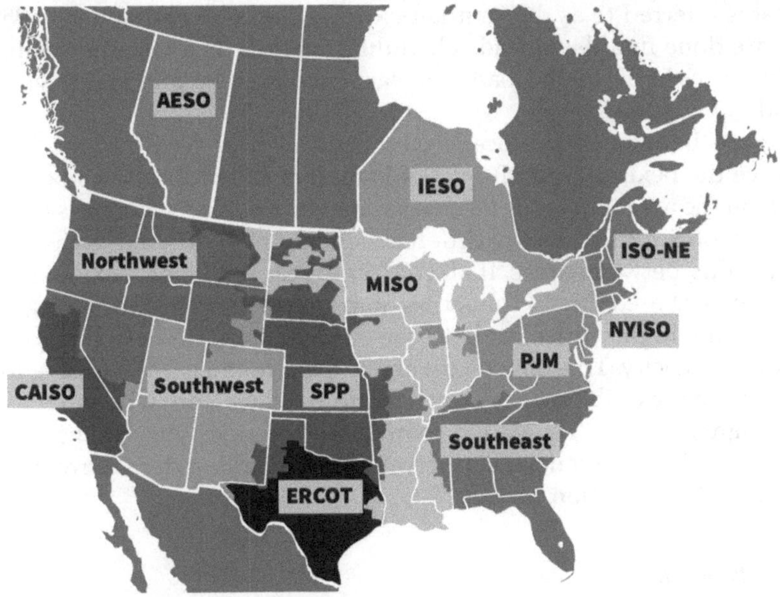

FIGURE 5.6
North American Independent System Operators (ISOs), Regional Transmission Organizations (RTOs), and Bilateral Regions. (From Federal Energy Regulatory Commission [7].)

and trade electricity with other utilities via bilateral transactions between a seller and a buyer. They determine and build the needed electric capacity, determine the sequence of electric dispatch to balance supply and demand, and plan new transmission.

In other regions of the country, power pools involving multiple utilities were developed. This allows for reserve capacity and the costs of new generating capacity to be shared among multiple entities. Approximately two-thirds of the U.S. population is served by these Regional Transmission Organizations or RTOs. Figure 5.6 is a map showing the independent system operators (ISOs) and regional transmission organizations (RTOs) in North America [7].

RTO markets are deregulated, with prices being set in competitive markets. They operate as an auction, whereby electric generating companies submit bids and consumers submit offers.

There are four different types of electricity markets:

- Energy
- Capacity
- Ancillary services
- Transmission.

These are described in the following sections.

5.3.2.1 Energy Market

The energy market sets the price of electricity on a dollar per megawatt-hour basis. There is a high demand for electricity at a low price and a low demand for electricity at a high price. At an electricity price somewhere in between, the supply matches the demand.

This is what is referred to as the "market-clearing price." Approximately 95% of energy transactions are done in a day-ahead scheduling market. The rest are real-time prices to reconcile differences between the load that was scheduled the previous day and the actual real-time load.

RTOs use a mechanism called locational marginal price (LMP) to determine the cost needed to serve the next (marginal) MW of load. It is called locational because the price depends both on the area where the electricity is produced and where it is consumed. This cost accounts not only for the costs of different generators in different areas but also for the transmission ability between areas. If all consumers cannot be provided with the electricity from the cheapest area due to transmission congestion, areas importing cheaper electricity must pay the higher price for both the cheaper imported electricity and the more expensive local electricity. This added cost above the amount actually going to the generators is called the congestion cost. It is collected to pay investors who purchased financial transmission rights as a hedge against transmission congestion. Financial transmission rights are an example of a financial derivative investment that derives from the electricity commodity market (see Section 5.3.2.4).

5.3.2.2 Capacity Market

Capacity is the continuous electric power in megawatts (MW) that LSEs can transmit to their customers. Capacity resources are of several types [8]:

- Electric generation (both existing and planned) either generated by the LSE or purchased via a bilateral agreement or on the spot market
- Demand resources, including interruptible loads
- Energy efficiency measures
- Transmission upgrades.

RTOs have various means for attracting new capacity investment and ensuring sufficient future capacity. In the United States, four RTOs operate a capacity market: PJM, New York ISO, ISO New England, and Midcontinent ISO. These markets assign the needed capacity to meet forecast demand, provide the income to support existing generators, and ensure sufficient reserve margins. Different types of auctions are used to allow generators to bid on providing future capacity. A price cap is used to prevent excessive prices.

RTOs that do not have a capacity market have programs designed to ensure that sufficient capacity is developed. For example, the California ISO, CAISO, uses a program established by the California Public Utilities Commission to ensure adequate generation. Load-serving entities are required to contract a year ahead for capacity services that can provide a margin of 15% above the predicted demand [8].

The capacity market has gone through significant changes in recent years. The low natural gas prices since the latter half of 2008 have shifted new generation from base load coal plants to natural gas plants with peaking capability. There is also a shift from new generation to demand resources and efficiency measures, thus reducing the need for new generation. In recent years, over two-thirds of U.S. electric capacity additions have been in wind and solar resources, a trend that is likely to continue due to the decreasing costs of these technologies, state renewable energy requirements, and federal tax incentives. At the time of this writing in late 2024, the sum of global solar and wind generation was about to surpass hydropower, and the sum of all renewable electricity sources was projected to soon surpass coal generation [9].

5.3.2.3 Ancillary Services Market

Ancillary services are various services that ensure that electricity is reliably provided to the load. There are several types of ancillary services including frequency regulation, operating reserve, voltage control, and black start service. Each of these is described below.

5.3.2.3.1 Frequency Regulation

When demand suddenly exceeds supply, as when a generator goes down, the operating generators will slow, dropping the frequency below the system operating frequency (60 Hz in the United States, 50 Hz in Europe). The opposite situation occurs if demand suddenly drops below the supply. To maintain the operating frequency within a ±0.5 Hz band, an automatic control signal is continuously provided to special frequency regulation generators to increase or decrease power (referred to as "regulation up" or "regulation down") to raise or lower the frequency, respectively. Generators that can ramp quickly and provide fine-tuning will bid into a special regulation market, and RTOs will award contracts based on the best offer prices and the availabilities of the units. Frequency regulation can also be provided by controlling demand. For example, air conditioner compressors could be temporarily shut off to provide regulation up. It has also been suggested that electric vehicles hooked into the grid could be programmed to adjust the demand in response to grid frequency changes.

5.3.2.3.2 Operating Reserves

Each RTO will require that there be an operating reserve of contracted power generation assets to handle contingencies. For example, the New York ISO requires that there be an operating reserve capable of providing power equal to 150% of the largest generator on the grid to allow for that generator to unexpectedly shut down and still have some contingency. There are different types of operating reserve: spinning and non-spinning. Spinning reserves are up and running and synchronized to the grid but are providing less than their rated capacity. They can ramp up within a 10–30-minute time window. Non-spinning reserves are not operating but can start up and synchronize quickly and reach full output in 10–30 minutes. As is the case for other ancillary services, RTOs accept bids for operating reserves for the different parts of their operating territory.

5.3.2.3.3 Voltage Control

For a stable grid both the real power must be controlled (to match supply and demand and thus keep frequency steady) and the reactive power must be controlled. The reactive power has a strong impact on the current in transmission lines and hence the voltage drop in those lines, which is proportional to the total power being delivered. Thus, to ensure that a steady voltage within defined limits (e.g., ±6%) is being delivered to the load, a utility must control the reactive power.

AC transmission and distribution lines also contain both inductance and capacitance and so also affect reactive power. Because most reactive industrial loads are inductive, a utility can bring the voltage and current back into phase by using capacitance. Then current can flow back and forth between capacitors and inductors. In terms of reactive power, a synchronous generator acts like a motor in reverse and so also has inductive reactance but of an opposite sign. A utility can control the amount of generator reactive power by adjusting the amount of current through the rotor windings.

Voltage control is thus an ancillary service that involves controlling the reactive power. This service is provided by online generators (via current control), capacitor banks, and devices that have the ability to consume reactive power. Companies providing these voltage control services are paid a monthly service fee.

5.3.2.3.4 *Black Start Service*

In the case of grid electrical failure, some facilities must be available to start up and generate electricity that is synchronized to the grid (e.g., at ±0.5 Hz) without the benefit of an outside electricity supply. Natural gas combustion turbines and hydropower turbines are examples of systems that can start during a blackout. Units that can provide this service are paid regardless of whether or not they are used because of the reliability they represent.

5.3.2.4 Financial Transmission Rights Market

If a consumer must pay a higher price for electricity because a transmission line is congested (at its limit) and transmission from a lower-cost generator is thus limited, the additional cost is paid to holders of financial transmission rights (FTRs). (Payments are not made to transmission owners, as this would provide an incentive to create congestion.) FTRs are essentially insurance contracts that provide a hedge against transmission line congestion. They are bought and sold via auctions on various time scales ranging from monthly to long term.

5.4 Generation Resource Planning and Dispatch

Utilities have two different types of decisions to make about what types of generation to deploy: decisions about building or retiring generating facilities (capacity planning) and decisions about what units to operate at a particular point in time (unit commitment and dispatch). In adding new capacity to the grid, a utility must consider the total capital and operating costs of plants [10]. When considering which units to commit on a particular day and what amount of electricity to generate from them, operating costs, especially fuel costs, are most important. In regulated markets, the electric power utility makes the short-term dispatch decisions based on forecast loads and the utility's knowledge of the operating costs of their generating units, with those units having the lowest operating costs deployed first, provided all operating constraints on the system are met. In RTO markets, participating generators submit bids for power production, and the ISO accepts bids in the order of the lowest to the highest cost. In either system, to the extent they are available, "non-dispatchable" wind and solar resources, which have no fuel costs, are the first to be deployed.

As a first cut in capacity planning decisions, various plant options can be compared in terms of the levelized cost of energy, or LCOE, which was introduced in Chapter 2. In this chapter, we consider simplified estimates of LCOE for coal- and gas-fired generators to illustrate the importance of the capacity factor in determining a plant's average cost of producing electricity. Assuming that a constant amount of electricity is generated every year, and that fuel costs and O&M costs are constant in real terms, the real LCOE in units of \$/kWh is estimated by:

$$\text{LCOE} = \text{Fixed Cost} + \text{Variable O \& M cost} + \text{Fuel cost} \times \text{Heat Rate} \tag{5.2}$$

where

$$\text{Fixed Cost} = \frac{\text{Capital Cost} \times \text{Capital Recovery Factor} + \text{Fixed Annual O \& M Cost}}{\text{Capacity Factor} \times 8{,}760}$$

TABLE 5.3

Example Values for Calculating LCOE

Plant Type	Capital Cost ($/kW)[a]	Heat Rate (Btu/kWh)[b]	Fuel Cost ($/$10^6$Btu)[c]	Variable O&M ($/MWh)[d]	Fixed O&M ($/kW/Year)[d]
Ultra supercritical coal	3,636	8,800	2.16	4.6	42.1
Combined-cycle gas	978	6,600	2.40	3.50	11
Gas turbine	1,101	10,000	2.40	3.5	17.5

[a,b,c] EIA [11].
[d] EIA [12].

and

$$\text{Capital Recovery Factor} = \text{CRF} = \frac{i(1+i)^t}{(1+i)^t - 1} \tag{2.19}$$

The capital cost is the plant construction cost divided by the rated plant capacity and has units of \$/kW. The Fixed Annual O&M Cost depends on plant size and type and has units of \$/kW/year. The variable O&M cost is proportional to the amount of energy produced and has units of \$/kWh. The Heat Rate is the average amount of thermal energy the plant must receive to produce a kilowatt-hour of electricity (i.e., it is inversely proportional to plant efficiency) and has units of Btu/kWh. The fuel cost is then given as \$/Btu. In Equation 2.19, i is the real discount rate and t is the plant lifetime.

The numerator of the Fixed Cost equation consists of annualized costs, so the 8,760 h/year converts that to an hourly cost. The capacity factor is the ratio of the amount of electrical energy produced over a year divided by the electrical energy that would be produced if the plant operated at its rated plant capacity for every hour of the year. Thus, a capacity factor of 1 would give the minimum levelized cost of electricity. If the plant operated at an average capacity factor of 0.5, its electricity would cost twice as much per unit of fixed costs.

Note that for most renewable energy plants, such as solar, wind, and geothermal, the fuel cost is zero. (Biomass plants do incur a cost for the biomass feedstock.) Thus, these plants are not susceptible to future increases in fuel costs as can occur for natural gas, for example.

Table 5.3 gives some example values for the different parameters for conventional plants [11,12]. (These values change frequently. See the EIA website for the latest values.)

Example 5.1

In comparing different plants, the LCOE is determined based on a plant's expected capacity factor. The LCOE decreases with increasing capacity factor because more annual energy is produced per investment dollar. Consider the case for three common fossil fuel plants: natural gas combustion turbine, gas combined cycle, and coal. Using the equation for LCOE, calculate the LCOE for each power plant type as a function of capacity factor and plot these on the same graph. For plot a, use a high natural gas cost of \$7 per million Btu. For plot b, use the latest natural gas cost (as of 2016) from Table 5.3. Use the cost of coal and other relevant costs from Table 5.3. In determining the capital recovery factor, assume a plant lifetime of 40 years and a discount rate of 4%.

Calculating the capital recovery factor based on a 40-year lifetime and a 4% discount rate:

$$\text{CRF} = \frac{i(1+i)^t}{(1+i)^t - 1}$$

$$\text{CRF} = \frac{0.04(1+0.04)^{40}}{(1+0.04)^{40} - 1} = 0.05052$$

We then calculate the LCOE for a range of capacity factor values using this capital recovery factor and the values from Table 5.3 and repeat the calculations substituting the higher natural gas cost. So, for example, for a capacity factor of 10%, the fixed cost of the coal plant would be

$$\text{Fixed cost} = \frac{\$3{,}636 \times 0.05052 + \$42.1/\text{kW/year}}{0.1 \times 8{,}760 \text{ hours}} = \$0.26/\text{kWh}$$

and (converting variable O&M cost from \$/MWh to \$/kWh)

$$\text{LCOE} = \$0.26\,/\,\text{kWh} + \$0.0046\,/\,\text{kWh} + \$2.16\,/\,\text{MMBtu} \times \left(1 \text{ MMBtu}\,/\,10^6 \text{ Btu}\right) \times 8{,}800 \text{ Btu/kWh}$$

$$= \$0.283\,/\,\text{kWh}$$

The results for a range of capacity factors are plotted below. Note that at the high gas prices, which have often existed, coal plants are the least cost option at high capacity factors. Coal plants have often been run as base load plants with capacity factors over 60%. However, with low natural gas prices, coal plants are not cost-effective even at the highest capacity factors. Typically having very low capacity factors, simple combustion turbines have traditionally been used as "peakers" to provide electric power during peak summer hours when air-conditioning loads are high. The latest EIA cost numbers show that even at low capacity factors, combined-cycle plants have a lower levelized cost of electricity than combustion turbines. This is because their high efficiency results in a lower capital cost as expressed in dollars per kilowatt. Because combustion turbines have a lower absolute capital cost, however, they require a smaller initial investment.

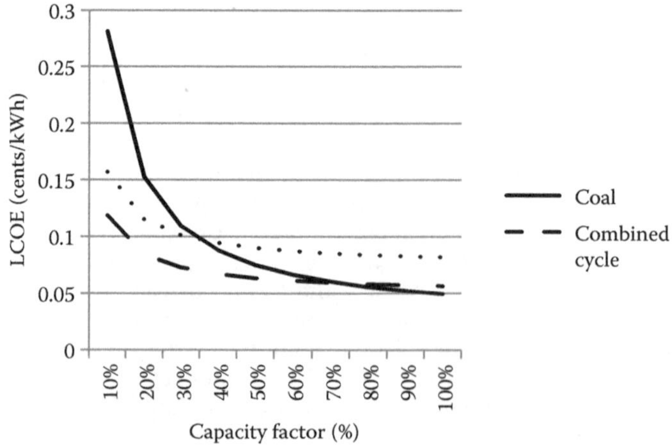

(a) High natural gas costs.

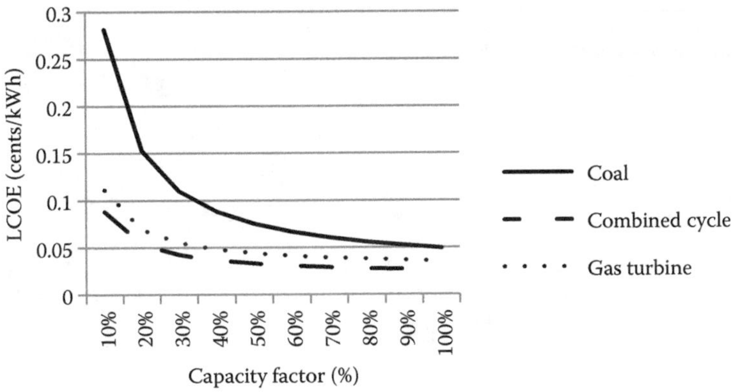

(b) Low natural gas costs.

Actual recent average capacity factors for all the operating electric generating plants in the United States in 2023 are shown in Table 5.4 [13,14]. Note that these are averages across all plants and account for actual operation. So, for example, utilities shifted to burning more natural gas, which left some coal plants idle, thus decreasing the average nationwide coal capacity factor. It is important to also note that the spread of values can be large. For example, a particular plant's capacity factor will be reduced if it experiences operating issues resulting in shutdowns.

The previous equation for LCOE provides a simple way to compare different plant options when planning for capacity additions and retirements. In reality, depreciation; pollution control requirements; tax considerations including government incentives; energy (kWh) vs. peak demand (kW) electricity rates; and loan parameters such as loan period, debt service coverage ratio, and so on, require a more sophisticated calculation. In addition, state renewable portfolio standards may require a utility to deploy more renewable energy. Thus, utilities use sophisticated financial models to compare alternatives for new generation. Any price on carbon dioxide emissions, as currently exists in California, for example, will skew the results more toward renewable options. Finally, when analyzing the addition of a new power plant, a utility needs to understand how that plant will integrate with other plants

TABLE 5.4

Average Capacity Factors of U.S. Electric Generating Plants in 2023

Plant Type	Capacity Factor (%)
Coal[a]	42
Nuclear[a]	93
Gas turbine[a]	14
Gas combined cycle[a]	59
Large hydropower[b]	33–66
Biomass combustion[b]	60
Geothermal[b]	80–90
Land-based wind[b]	26–50
Offshore wind[b]	27–47
Utility-scale solar PV plus battery[b]	21–33
Concentrating solar power (incl. thermal storage)[b]	51–67

[a] EIA [13].
[b] NREL [14].

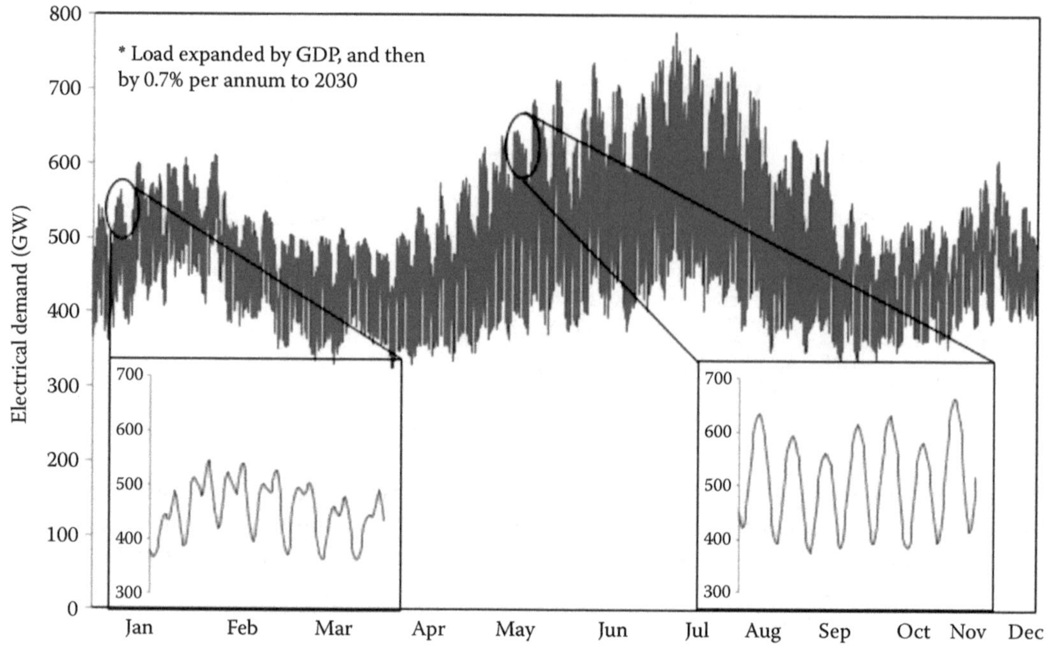

FIGURE 5.7
Example plot of U.S. electric load throughout the year. On typical winter days, both a morning and early evening peak are common, whereas summer days typically have a late afternoon peak. (From Clack [15].)

on the grid. In the case of adding wind and PV on the grid, it requires a detailed resource assessment including hour-ahead and day-ahead forecast modeling.

Once generating plants are online, the question turns to which plants will be operated at a given time. The first step, "unit commitment," involves deciding which plants to turn on or off, and is usually taken about a day in advance. The second "dispatch" step determines the output level of the committed units and requires many short-term adjustments to balance variations in load. Unit commitment and dispatch considerations include a unit's operating costs and physical constraints such as start-up times and the rates at which operating levels can be ramped up or down. An example plot of the total U.S. electric load throughout the year is shown in Figure 5.7 [15]. Not shown on this graph is that about a 15% margin of capacity above the peak must be available on a long-term basis to handle unexpected peaks or loss of capacity.

Figure 5.8 [16] shows an example normalized load duration curve for conventional electric generating plants broken into different segments. The bottom area is the region where base load plants operate. These have traditionally been coal, nuclear power plants, and hydropower because they have typically been the cheapest to operate, so it makes sense to use them for more hours. The peak load is provided by combustion gas turbines. They are the most expensive to operate due to lower efficiency, but they have the greatest ability to follow rapidly changing loads. In between these two are natural gas combined-cycle plants, which are very efficient and have some load-following capability. (Because of the low natural gas prices since 2015, natural gas plants have been replacing coal plants even as base load.) For each plant type, its shaded area represents the total power it delivers over a year. If one extends the area to the right as a horizontal rectangle, the capacity factor is the ratio of the shaded area in the horizontal rectangle to the total area of that rectangle.

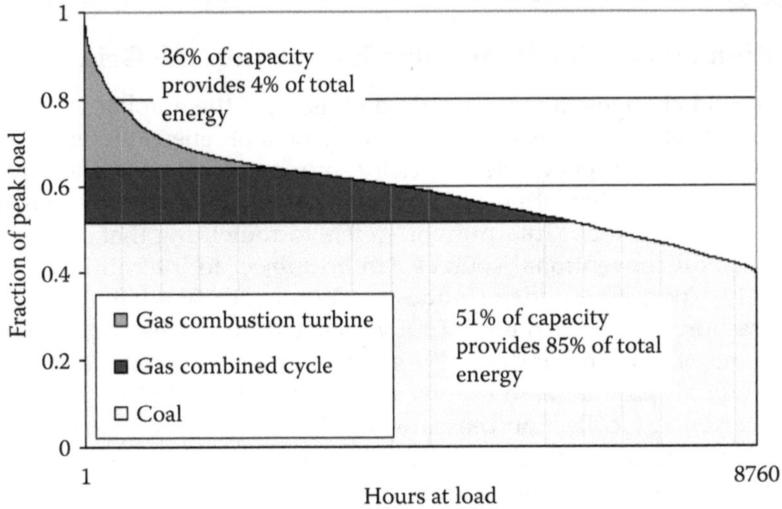

FIGURE 5.8
Example traditional load duration curve showing how many hours per year the load is at a given level and the different dispatch components. The peak load occurs for only a small number of hours per year, whereas the load is at or above a base load value for all 8,760 h/year. (From F. Kreith lecture notes [16].)

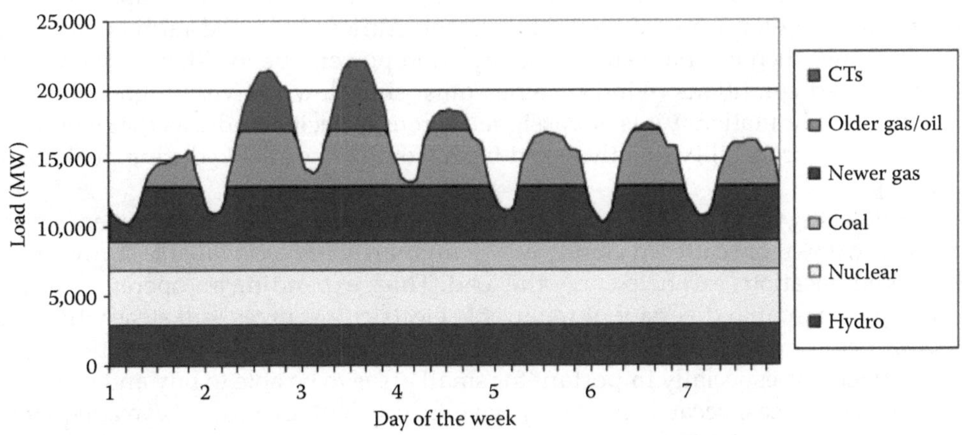

FIGURE 5.9
Example conventional dispatch stack. (From MCEN Sustainable Energy [17].)

Of course, a utility operator must dispatch plants throughout the day as the load changes. Based on operating costs, the utility operator might typically dispatch available hydropower, nuclear, and coal plants in that order to provide the base load. After that would come the most modern and efficient combined-cycle natural gas plants, followed by older combined-cycle plants. Finally, to meet those peak loads that occur on certain days, the more expensive to operate combustion gas turbines would be used to follow the load. Figure 5.9 shows how different conventional generators might be dispatched throughout a week [17].

5.5 Integration of Variable Renewable Energy into the Grid

As the electric grid changes to reduce carbon emissions, the emphasis is on energy efficiency to reduce the electricity need and the deployment of renewable energy sources. The two most rapidly growing renewable electricity sources are solar PV and wind. These are sometimes referred to as intermittent resources. However, many experts consider the word "variable" to be more correct. "Intermittent" refers to something that can turn on and off suddenly. In fact, all conventional sources can be subject to intermittency because they will occasionally experience sudden shutdowns. Wind and PV are more correctly considered variable resources because their output varies with time when operating as opposed to turning on and off. When wind and PV systems are spread out over a wide area, it is unlikely they will all provide zero output at any given time, but their output will vary with time. (Solar energy collection ceases when the sun sets, of course, but utility-scale solar power plants are increasingly incorporating storage: battery storage for PV plants and thermal storage for concentrating solar power plants.)

Utilities assign a term called "capacity credit" (different from "capacity factor") to account for the probability that an electric generation technology will be available during peak load periods. The capacity credit of a wind farm or PV installation is the ratio of the capacity (in megawatts) of the thermal plant that is displaced to the rated output (in megawatts) of the wind or PV, usually expressed as a percentage. It indicates the "firm capacity" that the renewable energy system can supply. Because utility operators have a wide range of different combinations of supplies for meeting electricity demand, capacity credit is viewed from a long-term probabilistic standpoint. Although a wind farm could conceivably have a time when it is not generating any wind power, due to either very low or very high wind speed conditions (wind turbines must shut down at very high wind speeds due to structural limitations), it is not assigned a zero-capacity credit, as that would mean that there is zero probability that the wind farm would be available during peak demand periods [18].

Renewable energy in any one location can be extremely variable. However, while the wind may die down or scattered clouds may temporarily reduce available sunlight for PV systems, other locations can help carry the load. Thus, expanding an operator's region to achieve a greater spatial diversity of renewable electricity sources will steady the electric output. Allowing increased interchange of electricity between different utilities provides the same effect. It is especially important for small areas to be able to buy and sell electricity from adjacent areas, because they may have insufficient area to achieve adequate spatial diversity. Figure 5.10 [19] shows how access to greater numbers of wind turbines and PV plants will smooth the total aggregated output.

The dispatch of a power system becomes more complex when renewable sources are brought to bear. Figure 5.11 shows the dispatch of generators on a summer day when more and more PV is added to the grid [20]. This shows the daily load profiles in California for four cases: no PV, 2% PV, 6% PV, and 10% PV. On hot, sunny afternoons, PV output is high and so it coincides well with afternoon air-conditioning loads. Thus, adding 2% PV to the grid reduces the peak load that the utility must address with natural gas combustion turbines, which are the highest-cost electricity producers. Note also, however, that the addition of PV has increased the ramp rate for the conventional power that must be dispatched as the solar resource declines at the end of the day.

As the PV penetration increases to 6%, it continues to cut into the peak, displacing combustion turbines. As shown by the dashed black line, however, as the PV gets to 10%

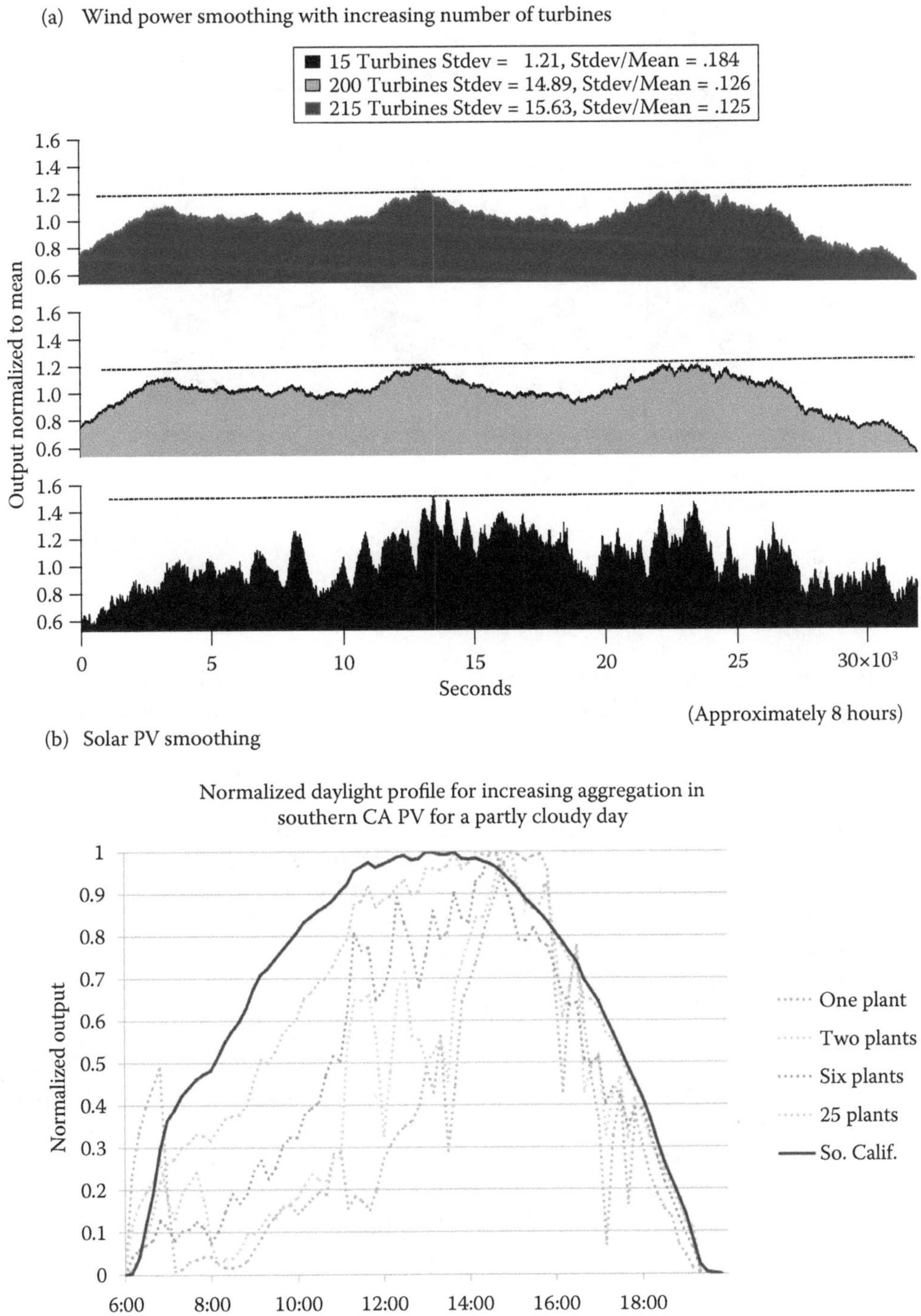

FIGURE 5.10
Smoothing of variable renewable electricity by drawing upon multiple locations: (a) effect of increasing number of wind turbines (NREL wind plant data), and (b) effect of increasing number of PV plants on a partly cloudy day. (From Lew et al. [19].)

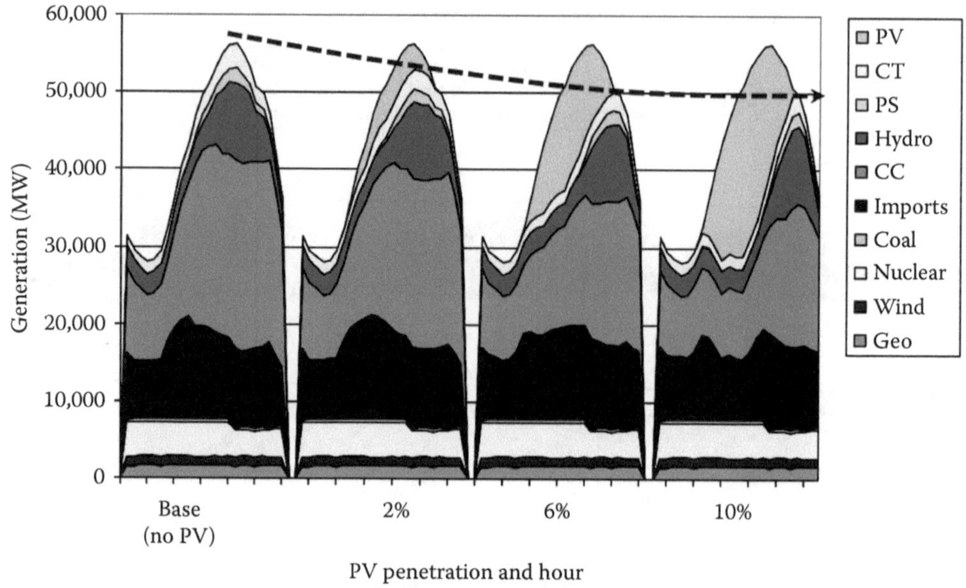

FIGURE 5.11
Impact of adding additional PV to the grid on a typical summer day. (From Denholm and Mehos [20].)

penetration, we are reaching a point of diminishing returns in reducing the peak load. The peak that the utility must match with conventional energy has shifted to later in the afternoon and evening, at which point the sun is setting and increased PV cannot continue to reduce the peak. Adding more PV will continue to displace conventional fuel on a summer day and thus reduce carbon emissions; however, it will not continue to reduce the peak.

The situation in the swing seasons is different. Figure 5.12 shows the situation for a sunny April day [20]. In this case, PV output is similar to what it is on a summer day. (The number of sunlight hours will be similar to the extent that the spring and summer days are on opposite sides of the summer solstice. If the array is at a tilt angle close to the local latitude, the average incident angle in the spring may be better than that in the summer.) The peak is now in the early evening when people get home from work and so PV doesn't impact the peak. Also, the total load that the utility must meet is lower. In this case, PV quickly reduces the need for generation from other plants. Note that at just 6% penetration, the wind output is impacted. This shows that the wind power must be curtailed (typically by adjusting the pitch of the turbine blades) so that the electric power supplied matches the lower demand. That's not good, because it means the cost of the wind farm is not being fully amortized, making wind power effectively more expensive. At 10% penetration, the PV is providing so much electricity on this spring day that the base load coal and nuclear plants need to be curtailed, and these plants are not typically operated at partial loads. The reality in this case is that the PV itself will likely be curtailed.

Curtailment of renewable sources already occurs at times on the electric grid. For example, CAISO has to curtail wind and solar for short periods, especially in the late winter and early spring. Curtailing wind and PV is undesirable because it increases the effective cost of these sources, and it reduces the carbon dioxide emissions reduction benefits of renewable energy. In fact, there is no real physical limit to how much variable renewable energy can be put on the grid. But at high penetration values, curtailment causes renewables to be very expensive unless storage or other measures are taken to increase grid flexibility.

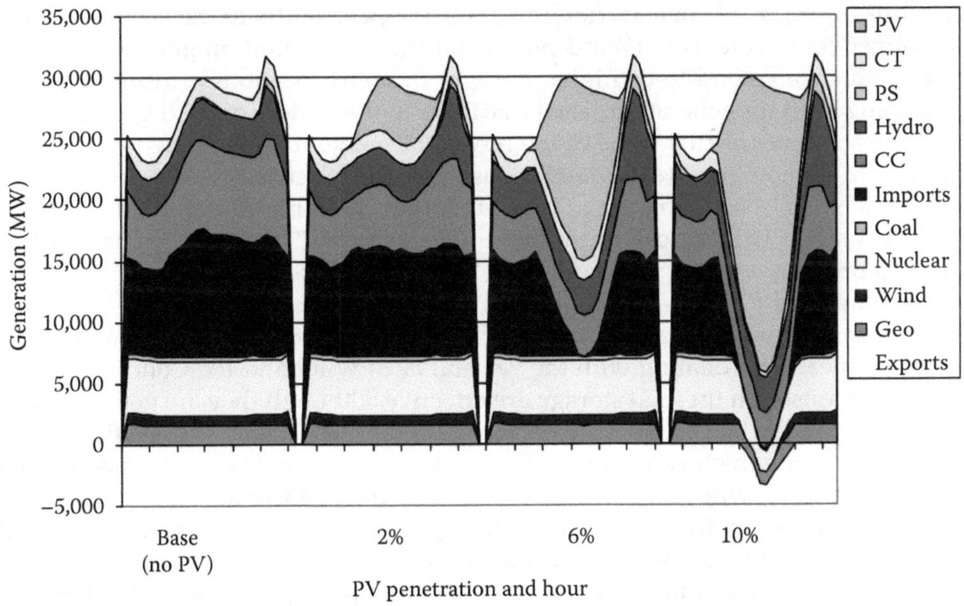

FIGURE 5.12
Impact of adding additional PV to the grid on a typical spring day. (The negative portion of the graph corresponds to exported electricity.) (From Denholm and Mehos [20].)

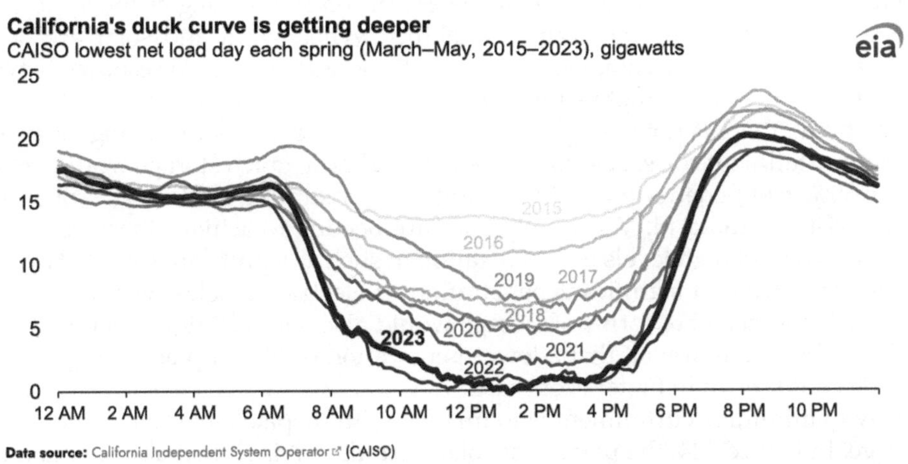

FIGURE 5.13
Utility "duck curves" showing the lowest net load day in California for each year from 2015 to 2023. As solar deployment as grown, the net load (remaining after renewable energy) has shrunk. When it dips below zero, the solar energy exceeds the needed load and must be curtailed. (From Energy Information Administration [21].)

Figure 5.13 shows another way to look at this spring generation problem [21]. In 2013, the California Independent System Operator (CAISO) projected the impact in the spring of adding more and more photovoltaic generation to the California grid each year. Figure 5.13 shows the actual family of net load curves in the spring (indicating the remaining load the utility must meet after the renewable supply) for each year from 2015 to 2023, with each progressive year showing additional photovoltaic penetration. As with the previous figure, as PV penetration increases the peak shifts to the evening, when PV can no longer contribute to it. (Wind power, on the other hand, might reduce the peak because wind speeds tend to be higher at night in many locations.) Adding PV reduces the remaining need for generation. The large belly in the middle of the day when solar PV output is greatest means that some of the renewable generation will need to be curtailed if the curve dips below zero. (Alternatively, usage of this electricity can be increased, such as for electric vehicle (EV) charging.) Note, too, that the ramp rate in the late afternoon becomes very steep. Utilities call these curves "duck curves" because they have the shape of a duck in profile.

Because wind and PV are variable sources, it is sometimes argued that a large amount of electrical storage must be added to the grid to accommodate them. However, utilities have various means for dealing with the variability of wind and PV. Concentrating solar power (CSP) plants with thermal storage are effective, although they are generally limited to desert regions like the U.S. Southwest. Many regions, especially the Pacific Northwest, utilize hydropower, which can provide low-cost base load and yet can also adjust very rapidly to follow a varying load. Denmark produces over 50% of its electricity from wind power and purchases hydropower from Norway when wind speeds drop. Some utilities use pumped-hydro storage, pumping water from a reservoir at low elevation to one at a higher elevation at times of low load and then meeting peak power needs by flowing that water downhill through a turbine. And, of course, because the U.S. electric grid is so interconnected, electricity is continually bought and sold between different operating regions.

To accommodate renewable supply shortages, operators can plan on greater use of natural gas combustion turbines, or they can run combined-cycle gas plants at part load. To effectively address climate change, however, carbon-free strategies must increasingly be tapped to accommodate renewable variability. When wind and solar are deployed together, they are generally complementary. That's because the wind resource is typically greater in the winter and at night—the opposite of the solar resource.

At some point, there is not enough flexibility in the grid and some storage is needed to avoid large curtailment costs. A study [22] examined how much storage would be needed to achieve 50% grid penetration of PV in California, assuming another 25% of renewable resources exists on the grid. Various changes are needed to achieve these very high PV penetrations to avoid high levels of PV curtailment such as significant electricity exchange with adjacent states and significant penetration of electric vehicles with vehicle-to-grid capability. (Of course, other forms of storage would also support higher penetration.) The study also indicated, however, that a lower penetration of PV coupled with greater wind penetration could result in better economics.

One way to minimize curtailment is to mix PV with dispatchable CSP containing thermal storage. In Figure 5.14, the percent of solar curtailment is shown as a function of total solar penetration on the grid for three cases: PV with a limited amount of grid flexibility, PV with enhanced flexibility, and a combination of PV and CSP with thermal energy storage. The enhanced flexibility includes variable output from conventional generators, some form of storage (e.g., stationary battery storage or pumped hydro), greater geographic diversity, and some DR. The thermal storage of the CSP plants provides additional flexibility regarding when solar electricity is dispatched. This allows greater grid penetration of solar electricity and reduces the amount of solar curtailment. Because of the large decreases that have occurred in the costs of PV and battery storage, utilities today are now building PV plants with integrated battery storage as opposed to higher cost CSP with thermal storage. As of July 2024, the United States has 20.7 GW of utility-scale battery power capacity, almost all of which has been added since 2020 [23].

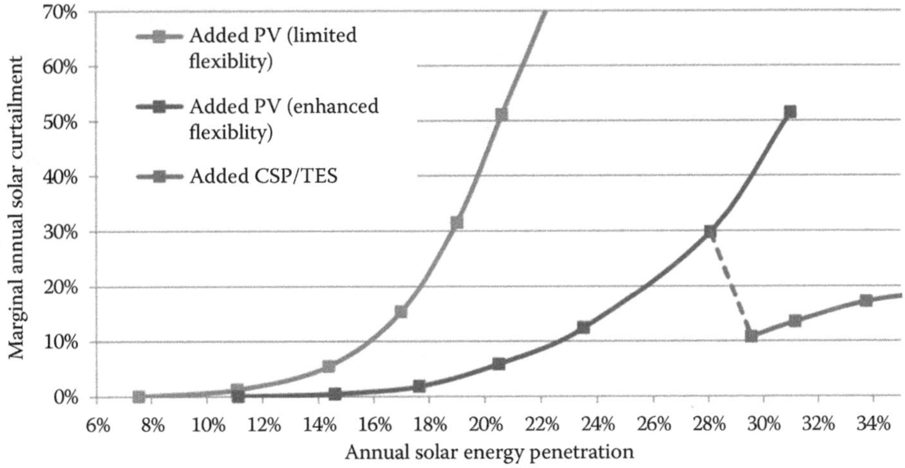

FIGURE 5.14
Comparison of PV only with two assumptions regarding the amount of grid flexibility to the same amount of PV plus CSP with thermal storage (that provides an amount of energy equal to the PV) over a range of different solar energy penetrations. (From Denholm et al. [22].)

Numerous utilities have conducted grid integration studies to determine the additional costs of integrating wind energy into the grid. Although variable renewable energy generators save fuel costs (they have zero fuel costs, although they do incur some operating costs), because their resource varies, utilities have calculated what additional costs they might incur, and these are called integration costs. An example of an integration cost is the cost associated with having spinning reserve available for when renewable output drops. These studies have generally shown that even at penetrations as high as 25%–30% of peak power, the integrating renewables adds no more than about a half-cent per kilowatt-hour to the cost of generating electricity.

Figure 5.15 shows the results of an Arizona Public Service study on grid integration costs for varying degrees of wind energy penetration on their grid [24]. The figure shows that at 10% energy penetration, the additional cost is only $4/MWh or 0.4 ¢/kWh. Much of the cost is associated with the uncertainty in day-ahead and hour-ahead forecasting of the wind resource. Accurate forecasting can reduce the amount of spinning reserve that a load-serving entity must have and so is a major area of research.

To plan dispatch, RTOs use mathematical models to forecast the load in their service regions. Forecasts are generally done hour ahead, day ahead, week ahead, and longer term. Weather prediction plays the most important role in short-term forecasts. Temperature and humidity are key parameters, temperature being important for both winter heating and summer air-conditioning loads, humidity being a factor in the latter. Temperature is also important in determining the performance of thermal power plants. The higher the outdoor temperature, the less efficient the heat rejection and thus the lower the cycle efficiency. This is especially the case for air-cooled power plants. The accuracy of weather forecasts results in an uncertainty of 1%–3% in day-ahead load forecasts.

Unlike fossil and nuclear fuels, renewable resources (wind and solar) also require energy resource forecasting in addition to weather forecasting. A utility's ability to utilize renewable energy depends a great deal on how well the availability of the renewable resources can be forecast an hour ahead, a day ahead, and a week ahead. The U.S. National Oceanic and Atmospheric Administration (NOAA) provides the public with information

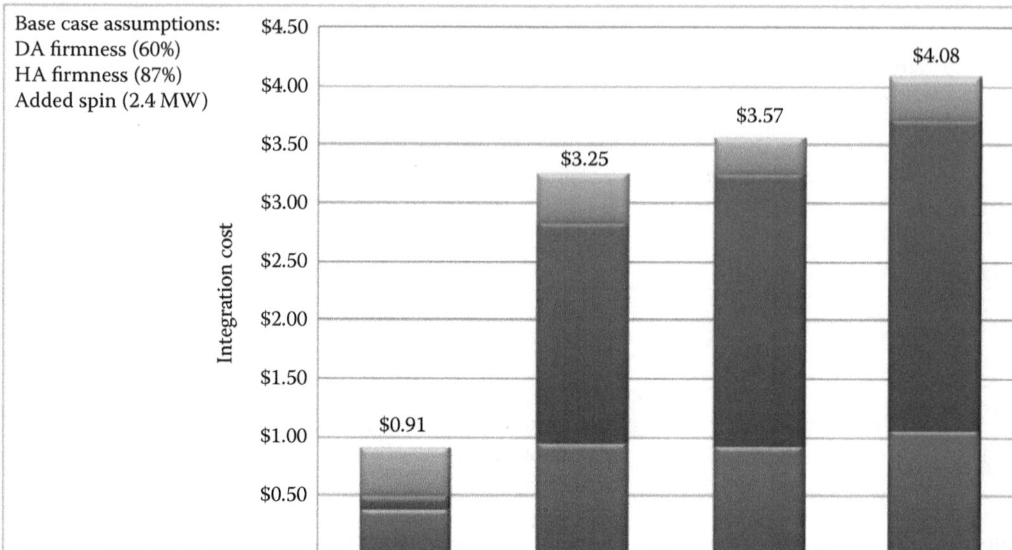

Wind energy penetration	1%	4%	7%	10%
Within-hour regulating	$0.40	$0.41	$0.31	$0.37
Hour-ahead uncertainty	$0.11	$1.88	$2.32	$2.65
Day-ahead uncertainty	$0.39	$0.95	$0.93	$1.06

FIGURE 5.15
Arizona Public Service study of the costs (in $/MWh) of integrating wind power as a function of grid penetration and forecasting ability. (From Acker [24].)

about wind and solar energy resources on these short time scales through the National Weather Service. Wind energy forecasting began in the 1980s and has generally focused on day-ahead forecasting. More recent efforts have looked at much shorter time horizons, from minutes to 6 hours ahead. Forecasts are made of wind speed, direction, and ramp rates (the expected time rate of change of wind speed).

A collaborative effort by NOAA, the U.S. Department of Energy, universities, and private companies developed improved renewable energy forecasts across a range of time scales, thus making the job of integrating wind power easier and less costly. Forecasting uses a combination of improved numerical weather prediction (NWP) models and enhanced measurement networks. Because wind turbine hub heights keep increasing to reach higher wind speeds, obtaining data at these greater heights and developing models to extrapolate measurements to greater heights have received increasing attention. A research area of special interest has been the prediction of low-level jets that periodically occur in the southern Great Plains of the United States, which is an area with a large wind resource. These jets can greatly increase wind turbine output but also result in a significant vertical wind speed gradient that impacts blade performance [25].

Forecasting solar energy is different from wind forecasting in that it is heavily dependent on cloud cover and movement but can also depend on fog and snow. Solar forecasting employs different techniques depending on the forecast time horizon. Long-term forecasts are accomplished via climate models. Numerical weather prediction models can be used for time horizons shorter than 2 weeks down to as short as an hour, but they lack accuracy below 6 hours because of insufficient resolution to predict individual clouds. For time horizons shorter than 6 hours, forecasts utilize satellite imagery and/or sequences of photographic images of the whole sky. Statistical learning methods can blend results from different models and use the success of previous forecasts to reduce model error. Individual rooftop, or behind-the-meter (BTM) systems, can be very dependent on local conditions, but, as is the case with the smoothing of the wind resource, aggregation of installations over a wide region reduces overall forecasting error.

Although numerous utilities have done renewable energy integration cost studies like the one shown in Figure 5.15, the accuracy of these studies has been called into question. For one thing, whenever any new generator—renewable or conventional—is brought onto the grid, it can negatively impact the operation of the existing generators. And no two plants are alike. Utilities send out automatic generation control (AGC) signals to adjust the outputs of different generators to meet the demand. One coal plant may be able to follow the signal closely, whereas another may not. So, integration costs are not just associated with renewable energy technologies and should be considered to be associated with each individual plant on the grid rather than being viewed by technology type.

While utilities have often assigned integration costs only to variable renewables, they accrue from any changes in the grid. Thus, the true cost of integrating renewables should be evaluated plant-by-plant and should be considered in the same way as any other changes in the grid mix. Finally, from a societal standpoint, the true grid integration costs for renewables should account for the societal savings associated with the elimination of emissions, including carbon dioxide emissions. A comprehensive scientific study of the social cost of carbon dioxide arrived at a value of $185 per metric ton in 2020 dollars at a near-term real discount rate of 2% [26].

It is important to remember that non-variable sources of renewable energy can also provide power to the electric grid. Combining variable wind and PV power with existing fast-ramping hydropower, steady biomass power, and geothermal power can allow for still higher total renewable energy penetration before one needs to build advanced electrical storage systems such as stationary batteries, compressed air energy storage, or additional pumped-hydro storage. And, as we have seen, the thermal storage in CSP plants can also reduce PV curtailment costs. Arizona Public Service built the 250-MW Solana parabolic trough plant with 6 hours of molten salt thermal energy storage that can provide electricity for Phoenix in the evening and morning peak demand periods.

In the United States, non-hydro renewables provided 15.7% of electrical energy or 654 TWh in 2023 [27]. In the world, non-hydro renewable energy represents 16.0% of world's electrical energy or 4,703 TWh in 2023 [28]. Analysts who have studied how to approach 100% penetration of renewable energy on the electric grid have found that economically achieving the last 10%–20% presents a special challenge [29]. First, there can be relatively rare periods of a week or more when weather conditions cause both solar and wind resources to be low. The Germans refer to this condition as "dunkelflaute," or dark doldrums. Second, there is a difference between spring and summer that we saw earlier; namely, there is a considerable increase in total load from spring to summer but not a commensurate increase in the renewable resource. Figure 5.16 shows this situation for ERCOT. While the peak electricity demand occurs in July and August, peak renewables (wind plus solar) occur in April and May. If one deploys sufficient wind and solar to meet the summer load, they would have to be curtailed in the spring. Such over-deployment can be a solution if the cost of the renewable energy is low enough. Potential long-duration storage options include biofuels, green hydrogen, and high-temperature thermal storage.

Critics of renewable electricity often point to the variability of renewables (the sun doesn't always shine, and the wind doesn't always blow). They argue that renewables are

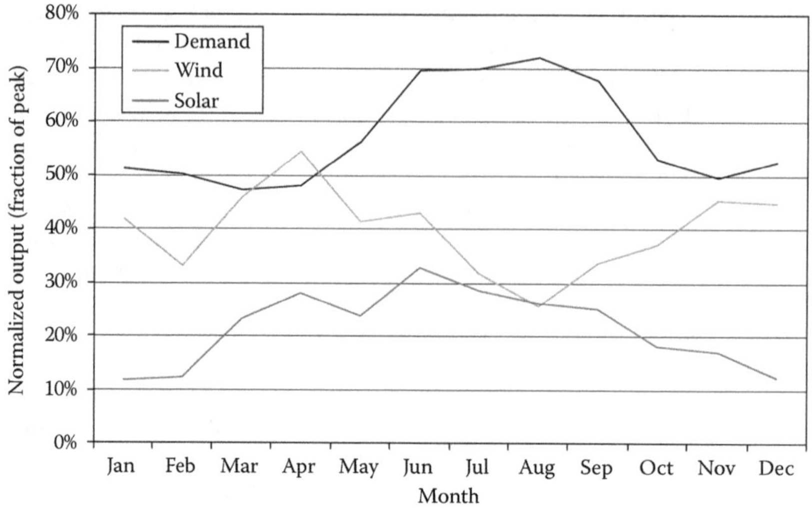

FIGURE 5.16
ERCOT profiles of monthly demand and wind and solar outputs (in that order from top to bottom). (From Denholm and Hand [29].)

not compatible with a grid that has to provide reliable electricity 99.97% of the time (which corresponds to one day of outage every 10 years). However, this doesn't mean that each power plant must operate with that reliability. Indeed, none do. Rather, it means that the fleet as a whole must meet that standard via a combination of all the electricity sources that grid operators have at their disposal. As we have seen, utilities have a wide range of options available to meet the load requirements. Even base load power does not have to be met by the traditional combination of coal and nuclear plants running at full power 24 h/day. Steady base load power output can be comprised of a changing mix of many renewable sources—variable sources like wind and PV and steady sources like geothermal, hydropower, and biomass—that work with storage to meet the demand.

Despite the options that utilities have to integrate renewables, various challenges associated with transitioning to utilities have to be addressed on a variety of levels: building, substation, generating plant, and overall utility. Kroposki [30] has summarized the challenges and potential solutions. These are shown in Table 5.5.

Utilities today are facing multiple challenges: the need to modernize the infrastructure of a century-old grid, the threat of both physical and cyberattacks, the growth of data centers, and the impact of millions of rooftop solar energy systems adding power to the grid. Many states have net metering laws that, with various stipulations, pay rooftop PV owners retail rates for the power they send back to the grid. Thus, PV owners are essentially using the grid as back-up power for free. Utilities still have to cover the cost of maintaining transmission lines and providing reserve power. They must also deal with power quality issues that the PV power injection can present. While net metering laws helped boost sales of rooftop PV, they are now being modified or eliminated in many states. To make use of solar energy collected when home usage is low, we can expect greater use of home batteries and the batteries in EVs. As suggested in Table 5.5, utilities will have to adopt new business models such as charging for services instead of merely selling electricity as a commodity.

Hawaii is at the leading edge of many of these issues. Because fossil fuels must be shipped in, Hawaii has very high electricity rates and so is aggressively adopting renewable energy. California has also been a leader in renewable energy adoption and has recently increased

TABLE 5.5

Summary of Potential Challenges and Solutions for Transitioning to a Grid with a Large Penetration of Variable Renewable Sources

Challenge	Solution
Building	
Addition of PV array to building causes outgoing power to exceed electrical power rating	Upgrade building service to accommodate maximum expected outgoing power
Overcurrent protection inadequate	Wiring and circuit breakers may need to be upgraded to accommodate higher peak current
Increased voltage and cycling caused by PV power injection to grid impacts voltage regulation	Advanced inverters at building and improved voltage regulation at substation
PV supply of 100% real power to grid decreases fraction of reactive power and increases cycling of reactive power controls	Advanced inverters can supply some reactive power if needed
Substation	
Substation designed only to provide power out experiences injection of additional power from renewable energy	Reconfigure substations to route power from circuits with excess RE generation to circuits with load to serve
Network protectors prevent inflow of RE power to substation	Advanced inverters can limit output to grid needs or on-site battery storage can accommodate excess power
Generating Stations	
Additional spinning reserve needed to accommodate sudden drops in RE consumes fuel, and the additional runtime increases O&M costs	Improved RE forecasting, adoption of demand response measures and addition of grid electric storage will reduce needed spinning reserve
Generators must maintain frequency synchronization in face of varying RE power injected into grid	Improved generator controls may be needed
Utility/Grid Level	
Tripping of PV inverters during a low-voltage condition could exacerbate the problem	New inverter standards will allow inverters to ride though low-voltage and low-frequency conditions and will allow some reactive power to be provided for voltage control
Marginal cost of solar and wind is essentially zero and their energy supply suppresses wholesale price and sales of fossil energy, the main income source utilities have to maintain the electric grid	Instead of charging for kilowatt-hour sales, a new utility business model can be based on a service charge for providing reliable, high-quality power that allows distributed energy consumers to use the grid as a storage/back-up device

Source: From Kroposki [30].

investments in utility-scale battery storage. The rest of the United States is closely watching how Hawaii and California are managing this twenty-first-century energy transition.

5.6 Demand Response and Transactional Controls

Most of the previous discussion has involved the ability of utilities to draw on different sources of electricity supply to meet the changing electric demand. Utilities also have demand-side management programs that are focused on the demand side. These programs have traditionally involved energy efficiency measures that tend to reduce total demand and demand response (or DR) measures, which focus more on shifting demand and shaving peak demand.

Reducing peak demand reduces the cost of transmission and distribution. As more variable renewables enter the grid (at the time of this writing EIA projected that in 2024 75% of new primary U.S. generating capacity would be solar and 17% would be wind), utilities have become increasingly interested in changing the load profile to accommodate those renewable supplies. DR thus involves both demand reduction and reshaping the load profile. It can also provide ancillary services. Kirby [31] describes the following types of DR:

1. **Price response:** Time-of-use pricing (prices set according to a time schedule) and real-time pricing (prices adjust with supply and demand) can shape the demand profile.
2. **Peak shaving:** Programs focus on reducing peak demand on the highest load days/hours.
3. **Reliability response:** This involves rapid change of demand to accommodate power system events like a generator dropping off-line.
4. **Regulation response:** In this case, demand is adjusted to follow the minute-to-minute changes in supply. This and item 3 can have financial value as ancillary services.

The electric grid was developed for one-way transmission of electric power from generating stations to homes and businesses. Local meters on homes have typically recorded electric usage with meter readers coming around periodically to read them. In the past, if part of the grid went down due to a line breakage in an ice storm or a transformer failure, the only way a utility had of knowing the specific location of a power failure was through phone calls from affected customers. However, there is now a growing movement to deploy smarter grids that allow two-way communications. According to the U.S. Energy Information Administration, U.S. electric utilities had installed (as of the end of 2022) about 119 million smart meters, or 72% of total meter installations) as part of a growing advanced metering infrastructure (AMI), with about 88% of the smart meters for residential customers [32]. These meters record electricity usage at a minimum of hourly intervals and provide this data to the utility and the customer at least once a day. Some of the meters can transmit instantaneous data. They can also monitor and report on power flow in both directions, identify power quality issues and automatically report on outages and restoration status, as well as identify power tampering and energy diversion attempts.

Utilities influence demand by sending price signals. Electricity pricing plans can separately price energy sales in cents per kWh of electrical energy and peak demand charges in $/kW in order to incentivize lower power consumption during peak periods. Utilities often sell specially discounted electricity to industrial customers who agree to have their power curtailed during peak demand periods. In the United States, buildings consume approximately 75% of grid electricity, so measures that control the shape of the building load profile can allow easier penetration of wind and PV on the grid. For example, some utilities have programs for residential customers whereby homeowners can receive a discount by agreeing to allow their air conditioner compressors to be shut off for short periods of time during summer heat waves. Electric hot water heaters and a building's thermal mass can be

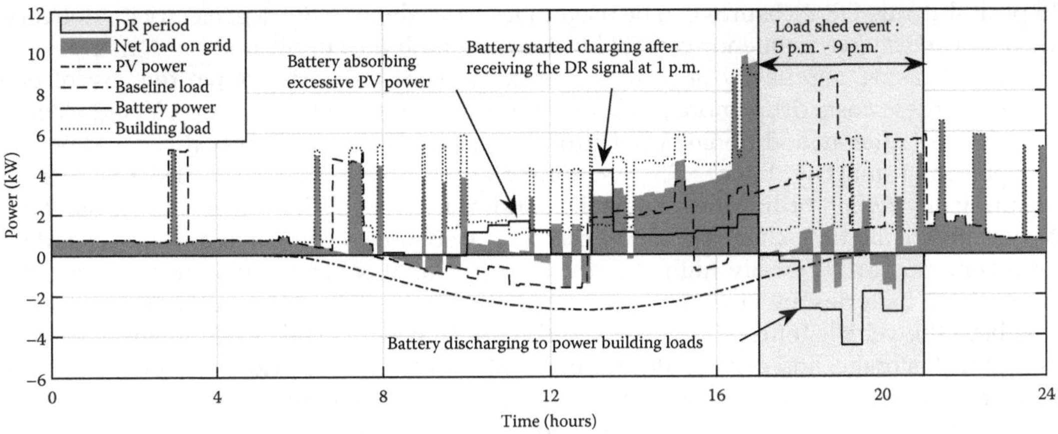

FIGURE 5.17
An example of using demand response measures, including PV-charged battery storage, to address an anticipated summer evening load shed event in the Pacific Northwest. (From Jin et al. [34].)

employed as thermal storage. The defrost operations of supermarket refrigerator cases can be run in such a way as to shift demand times and minimize peak demand charges.

Building equipment manufacturers are developing smart appliances that can time their operation to coincide with lower utility electric rates. So-called "transactional controls" will allow various devices to operate in a manner consistent with getting the best time-of-day or real-time electric prices. For example, a home energy management system could control the various appliances and serve as the point of contact with the utility. (In such a system the utility would monitor the total load profile for the house, avoiding the need for details of how the electricity is used by each appliance, and thus preserving customer privacy.) Beyond buildings, other uses of grid electricity offer opportunities for demand shifting. The water supply and treatment industry consumes a lot of electricity for pumping water, and controlling the times for pumping offers a good opportunity for demand control [33].

Figure 5.17 is a modeled example that shows how a home energy management system responds to a signal received from the utility at 1 p.m. indicating that a load shed event will occur between 5 and 9 p.m. [34]. To prepare for the event, the building is precooled, and the battery is operated in charging mode only, drawing electricity from rooftop PV and the grid. Then at 5 p.m., the battery is discharged, and appliance loads are curtailed. During the 4-hour load shed event, 15.66 kWh of electricity were saved. This savings is composed of 7.90 kWh from the fully discharged battery and 7.76 kWh of reduced building loads. Note in Figure 5.17 how the shaded area representing the net load on the grid is greatly reduced during the 4-hour load shed period.

Of course, as the electric grid becomes ever more digitally connected, everything will interact. Actions designed to minimize peak demands will themselves shift that demand. So, for example, if the electricity rate is expected to increase at 3 p.m. on a hot, summer afternoon, smart air conditioners may turn on at full power at 2 p.m. to precool buildings to allow the compressors to shut down at 3 p.m. But if there are enough smart air conditioners doing this at the same time, the utility peak load will actually just shift an hour earlier instead of decreasing in magnitude. Algorithms will be needed to prevent this. In addition to peak shifting, DR measures can be used to improve the reliability of electricity delivery. Utilities can achieve frequency control by adjusting both the supply and the demand.

The increasing popularity of battery electric vehicles will drive down the cost of batteries. As these costs drop, more affordable home batteries tied to rooftop PV and electric vehicles will offer an additional opportunity to control demand in a way that could be less complex than smart appliances [35]. As more solar has been added to the grid, the number of sunny days when utilities must curtail solar has increased. Changing time-of-use rates to incentivize daytime charging of EVs can reduce solar curtailment and decrease carbon emissions that are typically high at night. Transactional controls can determine at what times those batteries are charged and even when they sell electricity back to the grid so that both the vehicle/building owner and the utility benefit. The development of electric equipment, storage, and control algorithms to allow the electric grid to operate in an efficient manner that minimizes costs and carbon emissions is an exciting and emerging field.

Rooftop solar, electric vehicle batteries, hot water heater tanks, smart building controls, and various flexible loads in residential and commercial buildings and industry can all be considered examples of distributed energy resources, or DERs. When these resources are harnessed together to control electricity demand and provide grid services, they are called Virtual Power Plants, or VPPs. While VPPs are quite different from central utility plants, they are playing an increasingly important role in contributing to and controlling grid electricity in the twenty-first-century electric grid.

5.7 Expanded Transmission Capability

In 2023, 55 miles of new high-voltage transmission lines were built in the United States [36], but much more will be needed. After many years of limited annual growth, electricity demand is suddenly growing rapidly. This is a result of the growth of data centers to handle both artificial intelligence and cryptocurrency mining, as well as the electrification of buildings, transportation, and industry to take advantage of the growth of solar and wind [9]. More electricity means that more transmission capacity will be needed. Also, as the percent of renewable electricity increases, batteries and other storage options (see Chapter 14), as well as increased demand flexibility, can all firm the variable electricity supply, but an expanded transmission network can also help by harnessing solar and wind from a large region.

The United States extends east-west over four time zones. This means that peak electricity demand and peak solar occur at different times in different parts of the country. For example, when people get home from work in the densely populated cities of the Northeast, there is still solar energy available in the desert Southwest. To take advantage of this, there has long been an interest in building a "super grid" consisting of HVDC transmission lines that connect the three different interconnects. NREL did a modeling study (called the "Seam Study") that demonstrated the advantage of that type of approach, as shown in Figure 5.18 [37]. The ability to transmit electricity on HVDC lines between the Western Interconnection and the Eastern Interconnection, shown by the thick line segments, significantly improves the reliability and flexibility of a renewables-based electric grid and lowers the total cost.

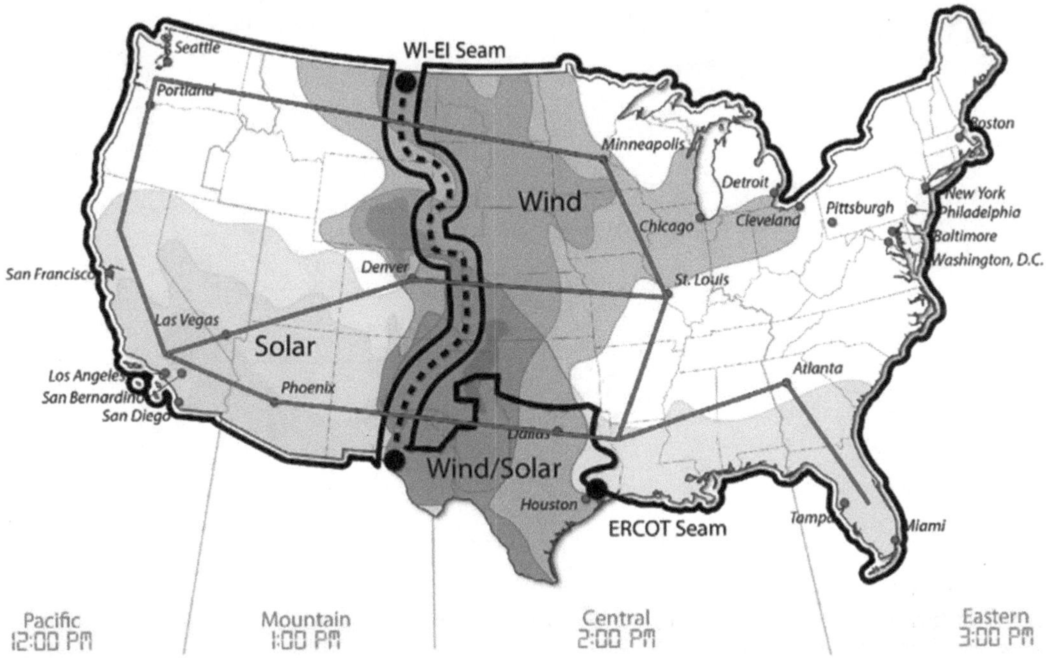

FIGURE 5.18
HVDC interconnection map from NREL Seam Study. The thick line segments show potential HVDC lines that can transmit electricity across the "seams" between the three interconnections. (From NREL [37].)

Berkeley Lab conducted a study showing historical values of additional transmission capacity, indicating where the greatest value would have been obtained [38]. This is shown in Figure 5.19. The number associated with each transmission line is the average absolute value of the difference in the marginal wholesale price of electricity between the two locations at each end of the line. This represents the profit that the transmission line owner or representative would make by purchasing electricity at the lower price end and selling it at the higher price end. Prices at each location can vary, so a profit can be made selling in a different direction at different times; hence the absolute value of each difference is summed in determining the mean value for the study period.

The numbers on the right side of the legend bar at the right of the map indicate the dollar value of the annual electricity transmitted per GW of line size (in million \$/GW) corresponding to the marginal \$/MWh value for the given line $\left(\text{converted by multiplying } \frac{\$}{\text{MWh}} \text{ by } \frac{1{,}000\ \text{MW}}{1\ \text{GW}} * \frac{8{,}760\ \text{h}}{\text{year}}\right)$. From a practical standpoint, the Berkeley Lab authors estimate that the actual values of \$/GW would be on the order of 10%–30% less than the numbers shown for a 1-GW interregional line.

A renewable electricity grows, interconnecting new solar and wind plants with the grid has become a major bottleneck. As a result, there are many examples throughout the U.S. where coal plants are being replaced by solar farms and battery storage facilities to take advantage of the fact that the coal plants are already wired to the grid [39]. Even if the site

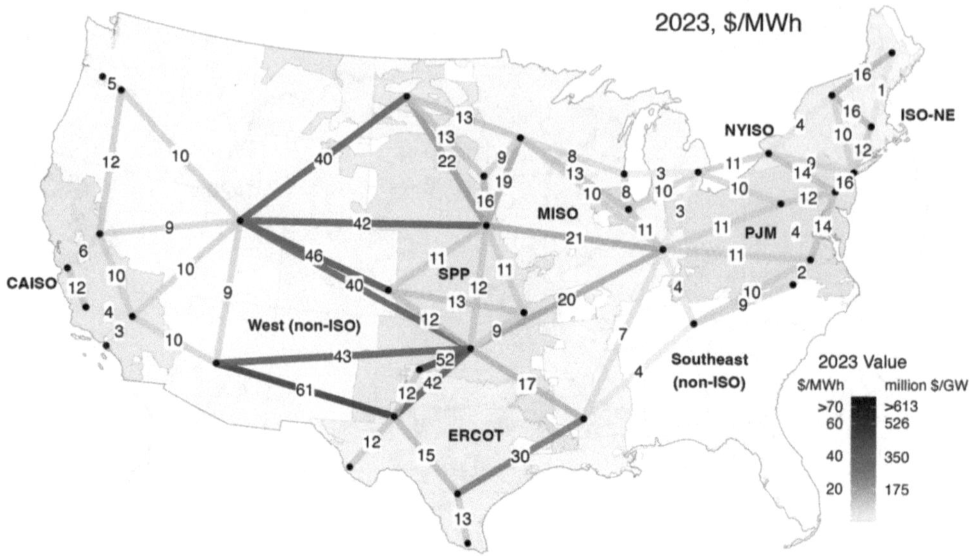

FIGURE 5.19
Map of transmission monetary values in $/MWh showing high values for connections between the three interconnections. (From Berkeley Lab [38].)

might not be ideal from a solar resource standpoint, the fact that lines already exist to connect that power to the grid offers great speed and financial advantages.

A challenge of adding new transmission pathways is obtaining rights-of-way. One way to address this is to utilize existing rights-of-way along highways, railroads, and natural gas lines. Another is to increase transmission along existing transmission corridors. There are various ways existing lines can be upgraded to carry more electricity. These are referred to as grid-enhancing technologies, or GETs, and include the following:

1. **Dynamic line ratings:** Transmission lines have limits on how much current they can carry to avoid overheating and sagging. These limits have traditionally been based on the worst weather conditions. Greater transmission capability can be obtained by varying the line rating based on weather conditions. For example, on cold, windy days lines can safely carry more current. Taking advantage of this will become important as buildings electrify and have peak heating loads on cold days.
2. **Power flow control:** This involves the use of hardware devices and controls that allow system operators to route electricity through the most efficient lines at any given time and avoid congestion.
3. **Line upgrades:** This can mean adding additional lines, increasing line voltage, or even replacing AC lines with HVDC lines. It can also mean replacing existing lines with lines of higher conductance, referred to as reconductoring. Aluminum Conductor Composite Cores (ACCCs) can operate at higher temperatures and double the current carrying capacity [40].

As electric grids approach 100% renewable, a combination of storage, expanded transmission, and demand flexibility will all play important roles in ensuring that electricity supplies continue to have high reliability.

Problems

5.1 An induction motor has a power factor of 0.82 and draws 800 W of apparent power. Calculate the reactive power and the real power and show these on a power triangle with the correct angles indicated. What is the phase angle difference between the current and the voltage?

5.2 A utility is planning to build a 500-mile-long 500-kV, 3,000-MW transmission line and is considering both double circuit AC and bi-pole DC lines.

a. Assuming the lines are going over flat land, using the baseline transmission costs in Tables 5.1 and 5.2 and assuming the DC transmission line would require a DC to AC conversion station above and beyond the AC case with a cost of $460,708,000, calculate the difference in cost between the two transmission methods.

b. Assuming the transmission lines are going over rolling terrain and the terrain cost multiplier for the transmission lines is 1.4, calculate the savings due to DC transmission.

5.3 Consider three ways that a residential air conditioner can be operated as shown in the figure, which plots home air conditioner electric power consumption (compressor power plus constant fan power) versus hour of the day.

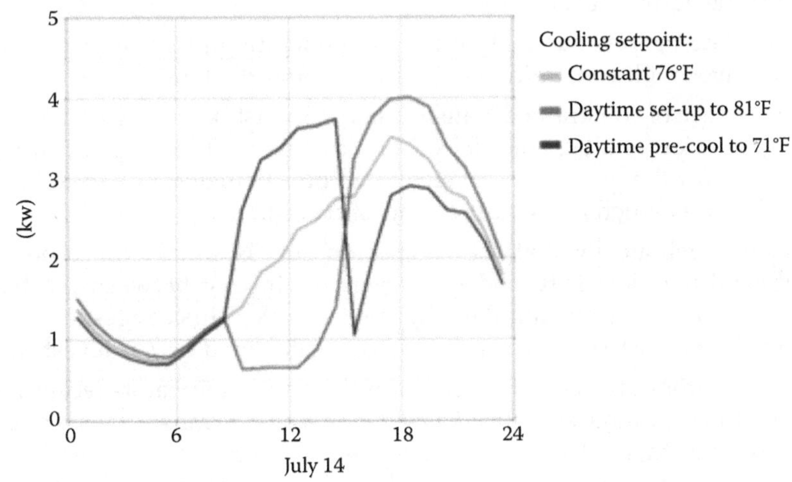

The three options are:

1. Setting the house programmable thermostat to 76°F all day.
2. Setting the house thermostat to 81°F until 2 p.m. and then 76°F.
3. Setting the house thermostat to 71°F until 3 p.m. and then 76°F.

Option 2, setting the thermostat up to 81°F, might typically be considered the energy-conserving option. However, consider the duck curves of Figure 5.13 and explain why option 3, which takes advantage of the thermal mass of the house, could be a better option for purchasing electric power in the case of large PV penetration scenarios. Why is the 81°F setting the worst one in this case?

5.4 As in the case of Example 5.1, consider a utility trying to decide whether to invest in a coal plant. It is anticipated that the coal plant will operate at a capacity factor of 60%. What is the minimum natural gas price at which the coal plant would have a lower levelized cost of electricity than a combined-cycle natural gas plant?

5.5 Consider the generation and load profiles in Figure 5.16. To meet the summer load with only wind and solar energy, these generators would have to be large enough that they would not be fully utilized in other months. Propose different means for meeting the summer load with all renewable energy without having to curtail it in other seasons. Consider not only energy supply alternatives but also how to control the demand.

5.6 Consider a case in which generating plants are located in two different areas. Area 1 has a plant that can generate 500 MW and sells it at a cost of $30/MWh. Area 2 has a plant that can generate 250 MW and must sell it at a higher price of $40/MWh. A customer in area 2 needs to purchase 300 MWh over a 1-hour period. Assuming there is no transmission constraint between the two areas, the customer in Area 2 would purchase the full 300 MWh from Area 1 and pay $9,000. Now consider the case where the transmission line is limited to a power flow of 250 MW. Purchasing the full 300 MWh over the 1-hour period from Area 1 is not possible because the transmission line is congested. In such a case Area 1 would provide 250 MWh (the limit of the transmission line) and Area 2 would provide the additional 50 MWh.

a. Calculate the revenue each of the two generating plants in Area 1 and Area 2 receives with constrained transmission and the total.

b. In the case of constrained transmission, the customer in Area 2 must pay the higher price for all its electricity. Compare the total cost paid by the customer to the cost the generators receive that you calculated in part a. What is the difference, which is called the congestion cost?

c. Show algebraically that the congestion cost is simply the product of the transmitted electricity and the difference in cost between the two areas. Verify this using the numbers in this example. This congestion cost is paid to holders of FTRs. (The congestion cost calculated here is an estimate.)

In this example, the customer pays more than the generators receive. But if the customer had purchased an FTR as a hedge against transmission congestion, it could recover that extra expense. The cost of the FTR will depend on auction prices. The customer has a choice to take a chance that congestion will not occur or purchase an FTR as insurance. FTRs are an example of where a primary commodity (kilowatt-hours of purchased electricity) generates secondary financial markets that address risk. FTRs are just one example of the various electricity markets described in Section 5.3.

5.7 Market-clearing price problem. In an energy market, an RTO or ISO (independent system operator) receives both bids to sell certain amounts of electricity at specific prices from electric generators (the supply) and offers to purchase specific amounts of electricity at various prices from large consumers (the demand). The RTO/ISO must match the supply and demand. It does so by essentially plotting the prices of supply and demand versus the aggregated electricity quantity. It builds up the supply by starting with the lowest generator cost offers and it builds up the demand by starting with the highest consumer offers. When the decreasing supply price and the increasing demand price match (i.e., when the supply and demand curves intersect), that is called the market-clearing price (MCP) and the aggregated electric quantity at that price is called the market-clearing quantity (MCQ).

Consider two generators, Generator 1 and Generator 2, and two consumers, Consumer 1 and Consumer 2, that bid, respectively, quantities of electricity to sell and buy, at various bid prices. These are shown in the tables below.

Bids to Sell Electricity from Two Generators (Supply Values):

Generator 1

Price ($/MWh)	Quantity (MWh)
10	200
40	150
60	50

Generator 2

Price ($/MWh)	Quantity (MWh)
20	100
80	60

Bids to Purchase Electricity from Two Consumers (Demand Values):

Consumer 1

Price ($/MWh)	Quantity ($/MWh)
50	100
30	150

Consumer 2

Price ($/MWh)	Quantity (MWh)
80	50
40	100
20	200

To determine the MCP, first stack up the five supply quantities in order of increasing bid starting with the lowest bid and stack up the five demand quantities in order of decreasing bid starting with the highest bid and put them in the tables below.

Aggregated Supply Quantities:

Price ($/MWh)	Generator Company (1 or 2)	Quantity (MWh)	Aggregated Quantity Range (MWh)
10			
20			
40			
60			
80			

Aggregated Demand Quantities:

Price ($/MWh)	Consumer (1 or 2)	Quantity (MWh)	Aggregated Range
80			
50			
40			
30			
20			

Now on a plot of price vs. aggregated quantity, plot in stair step fashion the decreasing demand and increasing supply curves. What is the MCP and at what MCQ does it occur?

5.8 Frequency regulation service problem. If total generation exceeds demand, frequency will increase above the operating value (60 Hz in the United States). If demand exceeds generation the frequency will decrease. An RTO/ISO must thus have controllable power to provide frequency regulation, which is considered one of the ancillary services. In a regulation market, generators submit bids to an RTO/ISO for the maximum number of megawatts of variable power (regulation capability) they are willing to provide for a particular hour at a particular price. The RTO/ISO then stacks up the various offers to determine the price it will pay.

For the given hour, a generator can bid a block of electricity both into the regulation market and into the energy market. To purchase the regulation, the RTO must pay not only the regulation bid price but, for each MWh of electricity purchased for regulation, it must pay what the generator lost as a result of not selling that MWh into the main energy market. The latter is called the lost opportunity cost, or LOC. It is the difference between the market energy price and the energy bid. (One could think of it as the net profit the generator would have received above its bid price.)

An RTO wants to purchase 20 MW of frequency regulation for a 1-hour period. The table below shows six different generators, the regulation capability each is bidding (in megawatts) into 1 hour, the regulation bid, the energy bid for that hour, the market price for electricity, and the LOC.

Generator	Regulation Capability (MW)	Regulation Bid ($/MWh)	Energy Bid ($/MWh)	Market Price ($/MWh)	LOC ($/MWh)	Total Regulation Cost ($/MWh)
1	10	12	20	40	20	
2	8	7	10	40	30	
3	12	3	20	40	20	
4	6	6	40	40	0	
5	14	5	15	40	25	
6	3	4	15	40	25	

Calculate the total regulation cost in $/MWh for each generator. Then, in the next table, order the generators from the lowest to the highest regulation cost.

Generator	Regulation Capability (MW)	Aggregated Capability (MW)	Regulation Cost ($/MWh)

The MCP is the minimum price needed to meet a sufficient number of bids to provide the power it must purchase for regulation, in this case 20 MW. The 20 MW will be purchased from multiple generators, namely all those who bid to sell regulation power at or below the MCP. They will all be paid the regulation MCP. Consider that there will be some uncertainty regarding the needed regulation. In this example, if you were the RTO executive, would you consider increasing the MCP a bit above the value you obtained? Why?

5.9 LCOE for wind and solar.

Consider the following costs for combined-cycle natural gas and utility-scale wind and photovoltaics systems:

Plant Type	Capital Cost ($/kW)[1]	Heat Rate (Btu/kWh)[2]	Fuel Cost ($/$10^6$Btu)[3]	Variable O&M ($/MWh)[4]	Fixed O&M ($/kW/year)[4]	Capacity Factor
Combined Cycle	978	6,600	2.40	3.5	11.0	56.3
Wind	1,877	NA	0	0	39.7	32.5
Solar PV	2,671	NA	0	0	23.4	28.6%

Assume a 30-year lifetime and a 4% discount rate, and neglect energy performance degradation with time.

a. Calculate the LCOE for each of the three technologies.

b. At what natural gas cost would solar PV have a lower LCOE than the combined cycle?

c. At what natural gas cost would wind have a lower LCOE than the combined cycle?

5.10 Impact of penetration on wind capacity credit and cost. The figure below shows how wind capacity credit as a percentage of installed wind capacity in Germany decreases with deployed capacity. In Great Britain, wind capacity credit in gigawatts is approximately equal to the square root of the installed capacity in gigawatts.

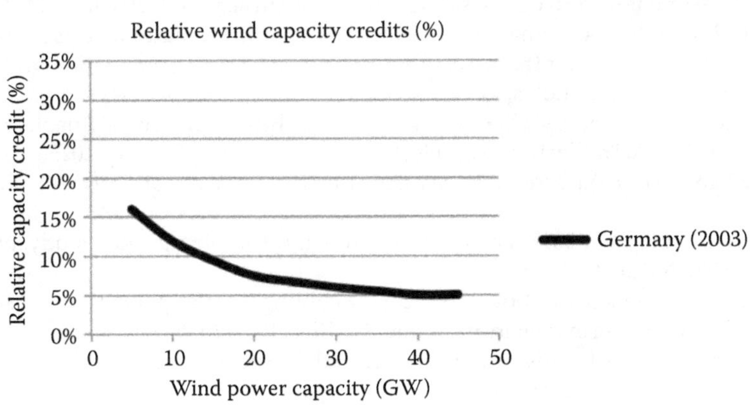

a. Add a plot of relative capacity credit vs. installed capacity for Great Britain on this plot.

b. Explain why the curve for Great Britain is higher than the one for Germany. (Hint: consider the geographic orientation of each country.)

References

1. Brown, M.H., and Sedano, R.P., 2004. Electricity transmission: A primer, National Council on Electric Policy, https://www.energy.gov/oe/articles/electricity-transmission-primer.
2. North American Electric Reliability Corporation, March 2023. Understanding the grid, https://www.nerc.com/news/Documents/March%202023%20Understanding%20the%20Grid.pdf.
3. Basso, T., 2014. IEEE 1547 and 2030 standards for distributed energy resources interconnection and interoperability with the electricity grid, NREL/TP-5D00–63157, NREL, Golden, CO.
4. Black & Veatch, 2014. Capital costs for transmission and substations: Updated recommendations for WECC transmission expansion planning, https://pdf4pro.com/view/capital-costs-for-transmission-and-1efa33.html.
5. Argonne National Laboratory, 2007. The design, construction, and operation of long-distance high-voltage electric transmission technologies, ANL/EVS/TM/08-4, Department of Energy National Laboratory, Argonne, IL.
6. Denholm, P., Eichman, J., and Markel, T., 2015. Summary of market opportunities for electric vehicles and dispatchable load in electrolyzers, NREL/TP-6A20-64172.
7. Federal Energy Regulatory Commission, 2024. Energy primer: A handbook of energy market basics, https://www.ferc.gov/media/energy-primer-handbook-energy-market-basics.
8. National Energy Technology Laboratory (NETL), 2019. Power market primers, DOE/NETL-2019/2059, https://www.osti.gov/servlets/purl/1556069.
9. International Energy Agency, July 2024. Electricity mid-year update, https://www.iea.org/reports/electricity-mid-year-update-july-2024.
10. NREL, LCOE calculator documentation, https://www.nrel.gov/analysis/tech-lcoe-documentation.html. Last accessed January 2025.
11. U.S. Energy Information Administration, November 2016. Capital cost estimates for utility scale electricity generating plants, https://www.eia.gov/analysis/studies/powerplants/capitalcost/pdf/capcost_assumption.pdf.
12. U.S. Energy Information Administration, May 2016. Monthly energy review, https://www.eia.gov/TOTALENERGY/data/monthly/archive/00351605.pdf
13. EIA, Electric power monthly, 2025, gas turbine: https://www.eia.gov/electricity/monthly/epm_table_grapher.php?t=epmt_6_07_a, nuclear: https://www.eia.gov/electricity/monthly/epm_table_grapher.php?t=epmt_6_07_b
14. NREL, 2024. Annual technology baseline, https://atb.nrel.gov/electricity/2024/index.
15. Clack, C.T., 2015. Decarbonizing the U.S. electric sector, presentation, https://www.slideshare.net/sercuser/decarbonizing-the-us-energy-sector-a-study-in-optimization-and-tradeoffs.
16. F. Kreith personal lecture notes.
17. MCEN Sustainable Energy, Electricity systems, https://mcensustainableenergy.pbworks.com/w/page/20638049/Electricity%20Systems. Last accessed January 2025.
18. Boyle, G., 2009. *Renewable Electricity and the Grid: The Challenge of Variability*, Earthscan, New York, NY.
19. Lew, D., et al., September 2013. The western wind and solar integration study, phase 2, NREL/TP-5500–55588, NREL, Golden, CO.
20. Denholm, P., and Mehos, M., November 2011. Enabling greater penetration of solar power via the use of CSP with thermal energy storage, NREL/TP-6A20–52978, NREL, Golden, CO.
21. Energy Information Administration, June 21, 2023. Today in energy, https://www.eia.gov/todayinenergy/detail.php?id=56880.
22. Denholm, P., Clark, K., and O'Connell, M., 2016. On the path to SunShot: Emerging issues and challenges in integrating high levels of solar into the electrical generation and transmission system, NREL/TP-6A20–65800, https://www.nrel.gov/docs/fy16osti/65800.pdf.
23. Energy Information Administration, September 5, 2024. Today in energy, https://www.eia.gov/todayinenergy/detail.php?id=63025.

24. Acker, T., September 2007. Final report: Arizona public service wind integration cost impact study, Northern Arizona University.
25. Wilczak, J., Finley, C., Freedman, J., Cline, J., Bianco, L., Olson, J., and Djalalova, I., October 2015. The wind forecast improvement project, (WFIP), *Bulletin of the American Meteorological Society* 96, 1699–1718.
26. Rennert, K., Errickson, F., Prest, B.C., Rennels, L., Newell, R.G., Pizer, W., Kingdon, C., Wingenroth, J., Cooke, R., Parthum, B., Smith, D., Cromar, K., Diaz, D., Moore, F.C., Müller, U.K., Plevin, R.J., Raftery, A.E., Ševčíková, H., Sheets, H., Stock, J.H., Tan, T., Watson, M., Wong, T.E., and Anthoff, D., October 2022. Comprehensive evidence implies a higher social cost of CO_2, *Nature* 610, 687–692, https://www.nature.com/articles/s41586-022-05224-9.
27. US EIA, Frequently asked questions, "What is U.S. electricity generation by energy source?," https://www.eia.gov/tools/faqs/faq.php?id=427&t=3. Last accessed January 2025.
28. Our World in Data, January 2024, Electricity Mix, https://ourworldindata.org/electricity-mix.
29. Denholm, P., and Hand, M., 2011. Grid flexibility and storage required to achieve very high penetration of variable renewable electricity, *Energy Policy* 39(3), 1817–1830.
30. Kroposki, B., 2016. Energy systems integration, in: Walker, A., Scheib, J., Turchi, C.S., Robicaud, R., Tomberlin, G., Burman, K., Hillesheim, M., Kroposki, B., and Qu, M. eds., *Integration of Renewable Energy Systems*, ASME Press, New York, pp. 121–136.
31. Kirby, B., Introduction to Bulk Power Systems, course notes.
32. EIA, Frequently asked questions, "How many smart meters are installed in the United States, and who has them?,"https://www.eia.gov/tools/faqs/faq.cfm?id=108&t=3. Last accessed January 2025.
33. Sparn, B., and Hunsberger, R., November 2015. Opportunities and challenges for water and wastewater industries to provide exchangeable services, Technical Report, NREL/TP-5500–63931.
34. Jin, X., Baker, K., Christensen, D., and Isley, S., 2018. Foresee™: A user-centric home energy management system for energy efficiency and demand response, *Applied Energy* 205, 1583–1595, NREL/JA-5500–69073.
35. Jin, X., and Meintz, A., July 2015. Challenges and opportunities for transactive control of electric vehicle supply equipment: A reference guide, Technical Report, NREL/TP-5500–64007.
36. Howland, E., July 30, 2024. 55 miles of high-voltage transmission lines added in 2023 as US buildout slows: Report, *Utility Dive*, https://www.utilitydive.com/news/high-voltage-transmission-buildout-aceg-grid-strategies-report/722725/.
37. NREL, October 18, 2021. Where the east meets the west: Interconnections Seam Study shows value in joining U.S. transmission grids, https://www.nrel.gov/news/program/2021/where-the-east-meets-the-west-interconnections-seam-study.html.
38. Berkeley Lab, July 17, 2024. Transmission value in 2023: Market data shows the value of transmission remained high in certain locations despite overall low wholesale electricity prices, https://emp.lbl.gov/news/transmission-value-2023-market-data-shows-value-transmission-remained-high-certain.
39. Shao, E., July 15, 2022. In a twist, old coal plants help deliver renewable power, Here's how, *New York Times*, https://www.nytimes.com/2022/07/15/climate/coal-plants-renewable-energy.html.
40. Fairley, P., September 25, 2024. A faster, cheaper way to double line capacity. *IEEE Spectrum*, https://spectrum.ieee.org/grid-enhancing-technologies.

6

Fossil Fuels

According to the International Energy Agency, global primary energy production in 2021 totaled 553 million terajoules (TJ) (14,700 Mtoe) [1]. Oil accounted for 29.5% of the total, coal and peat accounted for 27.2%, natural gas for 23.6%, and nuclear for 5.0%. Only 14.7% of primary energy production came from sources that are or can be renewable: hydropower, biofuels, waste, geothermal, solar, and wind. As discussed in Chapter 2, society depends heavily on fossil fuels because they are convenient and "cheap," from a market perspective. However, the broader lens of sustainability requires that we reexamine this assumption.

In the United States, coal overtook wood as the largest energy source around 1890, with oil and gas overtaking coal around 1960. Combined, coal, petroleum, and natural gas supplied approximately 80% of primary energy to meet U.S. energy consumption in 2023 [2]. According to U.S. Energy Information Administration (EIA) business-as-usual projections, fossil fuels will still supply 68% of U.S. primary energy consumption in 2040 [2]. EIA projects that U.S. reliance on coal will decline sharply over this period, but natural gas and oil use drop more slowly. Given the dominance of fossil fuel and nuclear energy in current supplies, it is important to understand the consequences of continued reliance on these resources, as a prerequisite to examining opportunities to quickly reduce our dependence on them.

6.1 Fossil Fuel Resources and Extraction

6.1.1 Coal

Coal is the most abundant fossil fuel and is widely distributed around the world. Historically, coal has been used for space heating, as fuel for ships and trains, and to provide process heat for steel and other industries. Today, 92% of coal consumption in the United States is for electricity production (see http://www.eia.gov/coal for trends).

Coal was formed largely on land in areas of swamps and bogs where vegetation decay was slower than the sedimentation rate. High-energy-content coal was lithified by being buried by other sediments and exposed to elevated geothermal temperature and pressure. The conditions for coal formation changed over tens of millions of years, so coal is bedded in other geologic sediment layers. Thick coal seams less than 200 ft below the surface are normally extracted in open cast mines. The coal is blasted and fractured in strips, hence the term strip-mining. In the United States, more than 65% of coal is surface mined. Underground coal mines have typically been developed up to 1,500 ft deep, but some mines in China reach depths of 3,900 ft. Spoil refers to the overburden, top soil and sedimentary layers between coal seams that must be removed and stored. The overburden rate refers to the depth of overburden compared to the coal seam depth. Typical overburden rates exceed 10:1 (10 ft of soil for 1 ft of coal). Blasting and moving of coal can produce dust that is carried by the wind and deposited in surrounding areas. Sediment run-off of

DOI: 10.1201/9781003586616-6

coal fines can be a serious water pollution issue. Within the United States, coal is typically transported by rail. Large mines in Wyoming and Montana supply coal to eastern power plants, with transport distances in excess of 1,500 miles.

Coal is not a single substance, but has differing composition and is classified by its carbon, volatiles, and heat content. Coal types reflect the degree of chemical processing that has occurred within the coal seam over millions of years at elevated temperatures and pressures. The maturation of coal from the source wood follows the sequence:

$$\text{wood} \rightarrow \text{peat} \rightarrow \text{lignite} \rightarrow \text{sub-bituminous} \rightarrow \text{bituminous} \rightarrow \text{anthracite}$$

The overall effect of the chemical processing is to remove oxygen and hydrogen, so the most "mature" coal types, anthracite and bituminous, have a greater percentage of carbon and a higher heat content than the less mature coals. The highest grade of coal is coking coal that is used in making steel. Anthracite, also known as hard coal, has the highest energy content (31 MJ/kg). Brown coal or lignite can have an energy content not much better than wood (14–18 MJ/kg) [3]. Coal composition is characterized by ultimate analysis, which gives the elemental composition of the material on a dry mass basis. DeNevers [4] lists the typical mass-based composition of subbituminous coal from the Powder River Basin: 72.5% C, 6.1% H, 17.2% O, 0.7% N, 0.4% S, and 3.1% ash. In addition, coal contains trace amounts of many elements. Some of these elements, like mercury, are of serious concern for air pollution. Coal from the Powder River Basin is relatively low in sulfur compared to coals from other source regions in the United States, which can have sulfur levels as high as 4% [3].

Coal is crushed before use as a fuel. Coal must be burned with excess combustion air in a high-temperature combustion chamber with a long fuel residence time. All coal has a mineral content that will remain in the combustion chamber as slag and also be carried out in the combustion gas stream in small particles as fly ash. Fly ash includes silica and other minerals, along with other compounds that are formed or liberated during combustion that condense into the solid phase. Most of the fly ash can be collected in fabric filters or by electrostatic collection onto charged plates called electrostatic precipitation. Both slag and fly ash pose storage and disposal issues. Some fly ash can be used in concrete. As discussed below, coal combustion causes air pollution with serious health and environmental effects.

Coal abundance is categorized at the broadest level by *resources*, defined as the quantity of naturally occurring concentrations or deposits of coal in the Earth's crust, in such forms and amounts that economic extraction is currently or potentially feasible. Resources may be either known or undiscovered. Coal *reserves* are more narrowly defined, as quantities of unextracted coal that comprise the demonstrated base for future production, including both proved and probable reserves (http://www.eia.gov/tools/glossary). Coal reserves are the subset of resources that could be *economically* extracted at the time the determination is made, considering legal and technological constraints, but not requiring that mining facilities be in place to do so. By definition, reserves are relatively well known at a given time, but change over time due to past production and due to changing technology or economics.

The depth of the seam, the amount of overburden, and the quality of the coal limit coal reserves. The United States has the largest coal reserves of any country [5], but China is currently producing the most coal annually. Worldwide, approximately 168 million TJ of coal were produced in 2021, with 52.8% of the production in China, 7.2% in the United States, and 7.6% in India [1]. Total coal reserves for the United States are estimated at about 260 billion short tons [5]. Within the United States, the states of Wyoming, West Virginia and Kentucky are the largest coal producers [6]. Ninety percent of Wyoming's coal comes from surface mines. Production in West Virginia and Kentucky is more evenly split between underground and surface mines [7].

Coal mining has historically been a hazardous occupation, although deaths and injuries have declined dramatically in some countries, due to improved regulations and safer technology. The U.S. Mine Safety and Health Administration reports that in the United States in the first half of the twentieth century, there were 452 coalmine disasters that each resulted in five or more deaths. Since 1950, there have been 53 such disasters. In 2010, an underground mine explosion at the Upper Big Branch Mine-South, in Raleigh County, West Virginia, resulted in 29 deaths [8]. Total coal mining fatalities in the United States dropped from 242 in 1978, the year after the Federal Mine Safety and Health Act was passed, to 20 in 2013. As reported in 2015, the U.S. coal mining fatality rate had averaged approximately 23 deaths per 100,000 workers over the previous decade [9]. For comparison, the overall occupational fatality rate in the United States was 3.3 per 100,000 workers in 2013; the rate in the construction industry was 9.7 per 100,000 [10]. Coal mine safety conditions have been much worse in other countries, including China, Turkey, and Ukraine. The International Energy Agency reported in 2009 that between 1985 and 2005, from 5,000 to 7,000 deaths per year occurred in coal mining operations in China [11]. Due in part to the consolidation of the Chinese mining industry and to a greater focus on safety, the coal mine fatality rate subsequently dropped, with about 2,000 deaths reported in 2012 [12].

Both surface and underground coal mining can deleteriously affect the environment where mining occurs. Underground mines can lead to subsidence. Mining wastes containing pyrite (FeS_2) cause problems with acid mine drainage. Active strip mines can disturb large areas with excavations, roads, and displaced overburden and soil material (Figure 6.1). To address these disturbances, the Surface Mining Control and Reclamation Act of 1977 requires that mine sites be restored to their previous condition or a condition that supports higher uses. However, in Appalachia, coal mining on steep terrain is done using the mountaintop mining/valley fill method, also known as mountaintop removal. Blasting the mountaintop generates large volumes of spoil that cannot be returned to the original location and instead is placed in adjacent valleys. This process permanently reconfigures the terrain, sometimes covering streams and wetlands.

FIGURE 6.1
Surface mining activity in northwestern Colorado. (Photo by J. Milford.)

Methane gas is also formed in coal beds as the sediments are heated by geothermal energy. This gas poses a serious hazard for coal miners. To control this hazard, most underground mines vent methane to the atmosphere, although the methane can be captured and used for energy. Methane can also be released from abandoned mines through fissures, vents, or boreholes. These abandoned mines may also represent opportunities for methane recovery. Methane emissions from coal mines also need to be minimized to mitigate climate change impacts, because methane is a potent greenhouse gas (GHG).

6.1.2 Natural Gas

Natural gas is an important fossil fuel for industrial process heat, heating buildings, and electricity production. Natural gas is also used as a chemical feedstock, primarily for the production of ammonia, methanol, and hydrogen. In 2023 in the United States, a total of about 35.5 million TJ was consumed, with about 40% used for power generation, 26% in industry, and 24% in the residential and commercial sectors [13]. Globally over the past three decades, natural gas has increased its share of primary energy from 19% of primary energy supplies in 1990, to 24% of supplies in 2021 [1].

Natural gas is comprised of methane and other light hydrocarbons that are produced from the decomposition of organic matter. Thermogenic natural gas was formed over millions of years from buried organic matter processed at high temperature and pressure. Methane is also formed biogenically, as microorganisms decompose waste in anaerobic environments, including wetlands and landfills. Large deposits of natural gas have conventionally been found in relatively porous rock underneath an impervious cap that traps the gas in place. Unconventional deposits are now being developed in tighter source rock, with hydraulic fracturing used to stimulate gas flow.

As delivered to end-users, natural gas is mainly methane, CH_4. Natural gas wells produce mixtures of methane with water, CO_2, hydrogen sulfide (H_2S), N_2, and other volatile light hydrocarbons including ethane, propane, butane, and pentanes. Water and natural gas liquids (the volatile hydrocarbons) are typically removed near the wellhead, using separators and dehydrators. If present in sufficient quantity, the light hydrocarbons are captured and sold as fuel or chemical feedstock. Gas processing plants remove CO_2, H_2S, and N_2 as unwanted contaminants. Natural gas is commonly transported in pipelines. The United States has more than 300,000 miles of natural gas transmission pipelines that carry gas long distances from supply regions to end-use markets. Pipeline diameters and pressures are stepped down from the transmission system to the distribution system, as gas is conveyed to commercial buildings and residences. International trade in natural gas is much more expensive if it must be liquefied and then transported by ship.

Similar to coal, natural gas and crude oil resources are estimated as the total quantities that might be produced at some time in the future. Portions of the overall resource can be further classified as follows:

1. Total remaining oil and gas in place
2. Technically recoverable resources, which can be produced based on current technology, industry practice, and geologic knowledge
3. Economically recoverable resources, the portion of technically recoverable resources that can be profitably produced, considering both the costs of production and current prices
4. Proved reserves, which are the volumes of oil and natural gas that geologic and engineering data demonstrate with reasonable certainty to be recoverable in future years from known reservoirs under existing economic and operating conditions [14].

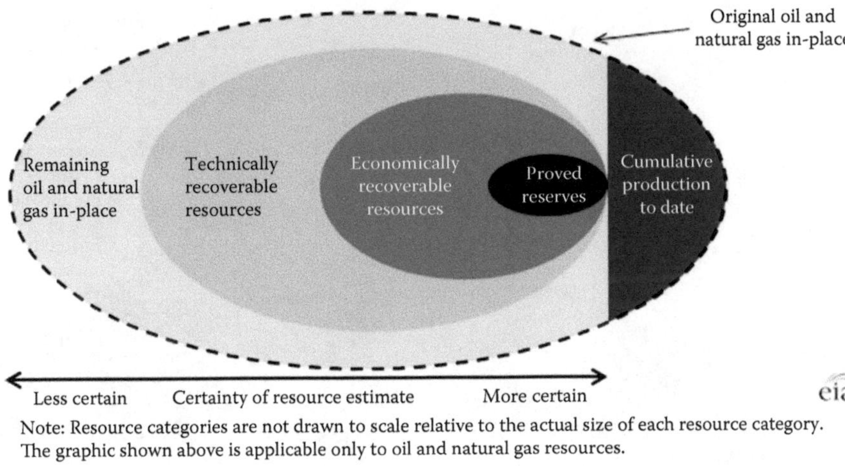

FIGURE 6.2
Schematic representation of oil and natural gas resource categorizations (not to scale). (From EIA [14].)

Figure 6.2 provides a schematic diagram of the relationship between these categories, showing how the degree of uncertainty in the resource estimate increases from the most restrictive category of proved reserves to the most expansive category of oil and gas in place.

For natural gas in the United States, the technically recoverable resource estimate in 2021 was about 3,000 trillion cubic feet [15], corresponding to more than 80 years' worth of production at the current consumption rate. Reflecting the rapid advance in technology for recovering natural gas from shale and other tight formations, the 2021 estimate of technically recoverable natural gas was more than double the 2004 estimate [16].

From 2008 to 2021, natural gas production in the United States increased from 20 to 38 trillion cubic feet per year, mostly as a result of horizontal drilling in tight formations that were previously inaccessible [13]. "Unconventional" natural gas formations include shale rock, tight "sands" (sandstone or carbonaceous rock formations), and deep coal seams from which natural gas can be recovered. Natural gas is produced from shale and other tight formations by hydraulic fracturing (fracking) and horizontal drilling (Figure 6.3 [17]). Experts have known for a long time that natural gas and oil deposits exist in deep shale formations, but until recent decades, these resources were not thought to be economically recoverable. Fracking is the process of creating fissures in underground formations to allow natural gas (or oil) to flow more easily. To create the fractures, fluid is pumped under high pressure into the shale formation. The fluid used is approximately 99% water, with sand added to prop open the fractures, and a small amount of chemical additives. With directional drilling, well bores can be extended horizontally to increase the exposure to the gas formation.

The rapid growth of unconventional natural gas production in the United States has lowered natural gas prices and helped shift electricity generation from coal-fired power plants to cleaner, natural gas-fired generation. However, the expansion of natural gas and oil production due to hydraulic fracturing and directional drilling has raised significant environmental concerns. These concerns include contamination of drinking water with methane, spills of toxic drilling fluids, and flowback of produced water with natural salts, heavy metals, hydrocarbons, and radioactive materials [18,19]. There have also been concerns that in some areas, hydraulic fracturing or associated water reinjection can stimulate earthquakes [19,20]. Earthquakes linked to the hydraulic fracturing process itself have been

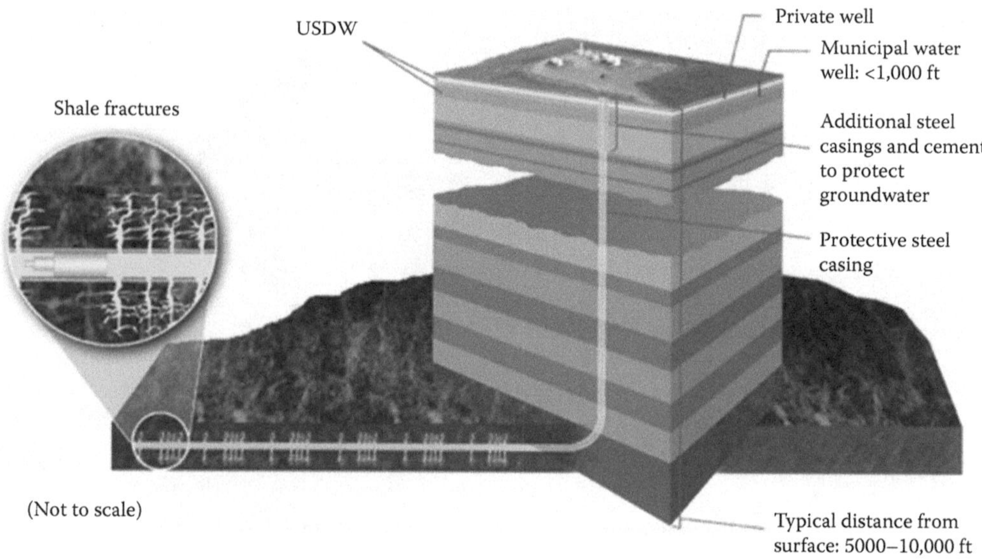

FIGURE 6.3
Schematic of a hydraulically fractured shale gas system (not to scale). (From NETL [17].)

below the threshold associated with structural damage; however, larger earthquakes have apparently been caused by the disposal of oil and gas-associated wastewater in deep strata or basement formations [21]. The U.S. Environmental Protection Agency estimates that natural gas and petroleum production account for approximately 30% of U.S. emissions of methane, a potent greenhouse gas [22]. Natural gas production operations also release volatile organic compounds and oxides of nitrogen, which contribute to ground-level ozone formation as well as hazardous air pollutants such as benzene and formaldehyde [23]. Although some of the environmental impacts of unconventional gas production can be mitigated through careful engineering, operations, and regulatory oversight [18], in some areas, environmental regulators have been left scrambling to catch up with the rapid growth of the activity.

6.1.3 Petroleum

Petroleum refers to crude oil and natural gas liquids. Petroleum is refined by high-temperature catalytic reactions and separation into a wide range of liquid fuels and tar or bitumen. Combustion of liquid fuels requires atomization into a hot combustion chamber, mixing with air and sufficient residence time to vaporize the liquid and completely burn all of the different chemical components in the fuel. Light fuels such as kerosene are used in jet engines and cookstoves. Petroleum and diesel fuel can produce unburned volatile organic compounds (VOCs), soot and CO if engines are slightly out of tune, not enough air is supplied, or if the mixing and flame residence times are not sufficient. Heavy oil, used engine oil, and diesel are used in boilers with air blast or steam blast atomization.

Petroleum reserves were formed 50–200 million years ago by the accumulation of biological sediments in coastal waters where low oxygen content slowed decay. These areas with hydrocarbon sediments were then buried under further sedimentation and heated by geothermal energy, causing chemical breakdown of the organic matter. If the matrix rock was porous enough, the lighter organic material could become mobile, and due to its lower density, could migrate upward. Where non-porous sediment layers were folded to form anticlines or where faults caused cap rock traps, the organic material could accumulate into an oil field. If there were no cap rock structure, then the oil, gas, or CO_2 would percolate up through the soils and be released on the surface.

The productivity of an oil field depends on how porous the resource rock is and how light the hydrocarbons are. Hydrocarbons that have been subjected to the highest temperature for the longest time may be decomposed to CO_2, and the lightest energy deposits of natural gas and oil normally have a significant fraction of CO_2 in the resource that must be separated in production prior to compression. A "sweet" gas or petroleum resource refers to a low CO_2 content.

Crude oil can be refined into a wide range of products, including pure chemicals (e.g., ethylene or propylene) and chemical mixtures. Most crude oil is refined to produce either gasoline or distillate fuel oil, which includes diesel. Gasoline is a mixture of hydrocarbons, comprised of molecules with carbon numbers in the range of C3–C12. The mixture includes straight chain, branched, and cycloalkanes; aromatics; and alkenes. Constituents of diesel fuel are mostly saturated hydrocarbons in the range from C13–C17. Heavy fuel oil is comprised of molecules in the C20–C45 range. Crude oil often contains sulfur, which is removed or partially removed in the refining process.

As noted above, proved reserves are amounts of oil and gas that have been demonstrated with reasonable certainty to be recoverable under existing economic and operating conditions. Worldwide, proved oil reserves are estimated at over 1.6 trillion barrels. Venezuela and Saudi Arabia lead the world in proved reserves, with 302 and 266 billion barrels, respectively [24]. Notably, proved reserves (along with current production numbers) now include significant quantities of oil resources that would have been considered uneconomical to produce two decades ago—from the Canadian tar sands, Venezuelan heavy oil belt, and U.S. shale oil formations. Proved reserves are often much lower than estimates of technically recoverable resources. For example, in the United States, proved reserves at the beginning of 2021 were estimated at 38.2 billion barrels of oil, while technically recoverable oil resources (excluding oil shale) were estimated at 359.5 billion barrels [25]. In 2023, the United States produced about 4.7 billion barrels of oil, leading the world in production [26].

The "low-hanging fruit" of petroleum resources is the light sweet crude oil found in highly porous source rock under massive anticline features near the land surface in the Middle East. When drilling first breaches the cap rock of a high-quality oil field, the oil and gas are at very high pressure and can be produced without pumping. Later in the production cycle, oil wells lose pressure, start to produce water mixed with oil, and require enhanced recovery by injection of natural gas, steam, or CO_2 to continue production. Recent growth in petroleum production in the United States has been spurred by the development of light oil found in tight source rock, which is commonly referred to as shale oil. The resource is developed using similar hydraulic fracturing and horizontal drilling techniques to those used in shale gas production. The lowest grade of petroleum resource currently being exploited is the heavy, viscous hydrocarbons found in the Athabasca "tar" sands of Canada, and the Orinoco oil belt in Venezuela. These deposits are "heavy" because they have not been thermally broken down. Production in the tar sands requires either steam injection to reduce the viscosity of the hydrocarbons in place before they can be pumped to the surface, or open-pit mining followed by processing of the extracted tar sands with hot water to separate the sand and the bitumen. Bitumen is subsequently upgraded to produce synthetic crude oil.

Oil shale is a form of petroleum resource that hasn't yet been tapped in significant quantities. Oil shale contains kerogen, a solid hydrocarbon substance, which can be converted into liquid oil through thermochemical processing. Possible approaches for producing oil shale include mining the shale, then crushing and heating it in a retort system to release the hydrocarbons, or heating it in place. In either case, the process would require significant energy, water, and land disturbance.

In addition to increasingly exploiting heavier, more viscous petroleum resources or resources in less porous deposits, oil and gas development continues moving into deeper formations and more remote locations. The first offshore oil well was completed in 1938, a mile-and-a-half off the coast of Louisiana, in 14 ft of water [27]. The push into the Gulf of Mexico accelerated in the 1950s, with the introduction of mobile drilling units. By 1957, nearly 450 production platforms were operating off the coasts of Texas and Louisiana, in water depths of up to 100 ft. Advances across numerous technological fronts, from seismology to stronger steel to digital computing, pushed the production limit to 350-ft depths by the late 1960s. In 1979, Shell Oil started production in a Gulf of Mexico oil field under 1,000 ft of water. Shell's production pushed on to 3,000-ft water depths in the early 1990s, and then on to astonishing 10,000-ft depths by the early 2000s [27].

In the 1960s, the physical challenges of the offshore environment and pressures to keep costs down compromised safety in the U.S. oil industry, earning offshore oil work a reputation as a high-risk occupation. In 1969, a blowout on an oil platform in the Santa Barbara Channel led to the release of an estimated 80,000–100,000 barrels of oil and a slick that covered 800 square miles [27]. Today, that accident is widely credited with helping launch the U.S. environmental movement. Two more disastrous blowouts, including one that killed four people, occurred the following year in the Gulf of Mexico. In the wake of these accidents, the U.S. government finally stepped up its regulation of the industry, and the industry took on voluntary efforts to improve its safety practices. However, accidents and oil spills from wells, pipelines, and tankers remain all too common.

In the United States, two catastrophic oil spills are of particular note. On March 24, 1989, the oil tanker Exxon Valdez struck a reef in Prince William Sound in Alaska. More than 11 million gallons of oil were dumped into the pristine environment, impacting over 1,300 miles of coastline in Alaska. Scientists are still studying the long-term impacts of the disaster on fish, marine mammals, and birds. There are still pockets of crude oil in some locations and damage to the ecosystem continues to be observed. In the wake of the Exxon Valdez disaster, the U.S. Congress enacted the Oil Pollution Prevention, Response, Liability, and Compensation Act of 1990, which included requirements for double-hulled oil tankers operating in U.S. waters.

On April 20, 2010, the $350 million Deepwater Horizon oil drilling rig exploded in the Gulf of Mexico, killing 11 workers. The blowout of BP's Macondo well released more than 4 million barrels of crude oil into the Gulf over the course of nearly 5 months [28]. The cause of the BP disaster was not the result of a single human error, but rather of systemic negligence of engineering practice and operations in the oil industry [27]. In July 2015, the U.S. Department of Justice and the states of Alabama, Florida, Louisiana, Mississippi, and Texas announced they had reached a settlement agreement with BP for the company to pay approximately $18.7 billion in fines and compensation for damage from the spill [29]. The case tragically reiterated the need for the oil industry to adhere to the highest standards of operations and safety, including redundant control and containment systems at all points in the supply process.

6.2 Fossil Fuel Combustion and Energy Conversion Technologies

6.2.1 Heat of Combustion

Neglecting minor constituents of fuel and air and treating air as 21% O_2 and 79% N_2 by volume, the stoichiometric combustion of hydrocarbon fuels in the air can be represented as follows:

$$C_xH_y + (x + y/4)O_2 + 3.76(x + y/4)N_2 \rightarrow xCO_2 + y/2H_2O + 3.76(x + y/4)N_2 \tag{6.1}$$

where x is the number of carbon atoms in the hydrocarbon molecule and y is the number of hydrogen atoms. For example, the reaction for the stoichiometric combustion of gasoline in air can be approximated as follows:

$$C_8H_{16} + 12O_2 + 45.1N_2 \rightarrow 8CO_2 + 8H_2O + 45.1N_2 \tag{6.2}$$

The heat of combustion, or "heating value" of a fuel, is the energy released per unit mass of fuel when it undergoes complete combustion in air. The heat of combustion is numerically equal (but opposite in sign) to the enthalpy of combustion of a fuel, which is the difference between the standard state enthalpy of the products and reactants for combustion of a stoichiometric mixture:

$$\Delta h_C = q_{out} - \Delta h_R = -(h_{products} - h_{reactants}) \tag{6.3}$$

In Equation 6.3, h is the specific enthalpy of the product or reactant mixture at standard state conditions. The combustion process is assumed to be complete. From the first law of thermodynamics, the difference in standard state enthalpies is equal to the heat given off by the reactions, q_{out}, again on a mass basis.

The higher heating value, HHV, is the heat of combustion calculated by assuming that all of the water in the products has condensed to liquid. In contrast, the lower heating value, LHV, is the heat of combustion calculated by assuming that none of the water condenses. The LHV gives a more realistic estimate of heating value for applications where it is not practical to recover the latent heat of condensation, such as in a motor vehicle engine. Heating values are given in Table 6.1 for methane, typical gasoline, and several coal types. The difference between LHV and HHV can be significant, demonstrating the value for energy recovery and efficiency of adding a condensing system where practical. For example, modern gas furnaces for residential applications include a condensing stage to significantly increase their efficiency.

TABLE 6.1

Energy Content of Various Fossil Fuels

Fuel	Heat Content (MJ/kg)
Methane	55 (HHV)
Anthracite coal	~32 (HHV)
Subbituminous coal (Powder River Basin)	~20 (HHV)
Diesel	~45
Gasoline	~45

6.2.2 Fossil Fuel Use for Heat

In 2023, the industrial sector accounted for 33% of energy (primary energy and end-use electricity) consumed in the United States, and 32% of direct natural gas consumption. The sector mainly uses natural gas for process heat, and secondarily for space heating. The commercial and residential sectors together accounted for 37% of energy use, and 23% of direct natural gas consumption [30]. Most of the remaining natural gas (40%) is used to produce electricity. The commercial and residential sectors use primary energy mainly for space heating, and secondarily for hot water. In 2020, about half of U.S. residences met their space heating needs mainly with natural gas, 36% with electricity, and the balance with fuel oil, propane, or wood [31].

Home furnaces and boilers are rated by Annual Fuel Use Efficiency (AFUE), which is the ratio of the annual heat output compared to the total annual fossil fuel energy they consume. The metric does not account for heat losses in ducts or piping. The U.S. Department of Energy requires a minimum AFUE for residential gas furnaces of 80%. High-efficiency systems can achieve 90%–98.5% AFUE using a combination of electronic ignition and airflow control, sealed combustion, and a secondary heat exchanger to condense flue gases [32].

Fossil fuels are important fuels for manufacturing and processing plants. Much of the fuel is burned to produce high temperatures for material processing, including for the production of metals, glass, cement, and fertilizer. Once the process is completed, the product must then be cooled. The industrial process to remove this heat and exhaust it to the environment produces a large thermal resource. Some industrial facilities can take advantage of cogeneration opportunities to extract process heat and electric power from the same energy source, an approach known as combined heat and power (CHP). Steam cycle "topping" cogeneration plants route part of the steam from the turbine to use as process heat, without routing it through a condenser. In a waste-heat-to-power system, the flue gases from fuel combustion are run through a heat exchanger to vaporize the working fluid in an organic Rankine cycle (ORC) power generation plant before the heat is exhausted to the environment. The ORC uses a working fluid with a lower boiling point than water, enabling its use at relatively low temperatures. This is known as a bottoming cycle CHP. This same technology is already used in geothermal power generation. While the thermal science and the plant technology are mature, developing these waste heat resources requires site-specific analysis and development efforts. These waste heat streams are part of existing production plants and the particular details of the extraction and power plant design and construction must be determined for those specific site conditions. On the other hand, development of industrial waste heat does not require new land or new water resources because these are already being used by the industry for cooling. Most of these industries have sufficient on-site power demand that the power generated can be directly used and does not need to be tied into the grid. Waste heat power plants would not need new environmental assessments and would probably have little public opposition, as they would be built on existing industrial sites. Due to the relatively low temperatures involved with ORC, the plant efficiencies are often less than half those of coal-fired steam power plants. However, the power is produced at the site where it can be consumed, and it does not need new transport of fuels or wastes. The U.S. Environmental Protection Agency (EPA) has estimated the recoverable waste heat available at temperatures above ~230°C in the U.S. could support over 7,600 MW of electrical generating capacity. There is additional potential from waste heat at lower temperatures that could be used in ORC systems, but it was not quantified in EPA's assessment [33].

6.2.3 Electricity Generation from Pulverized Coal

The dominant configuration for existing coal-fired power plants is pulverized coal (PC), where coal is milled to a fine powder and mixed with preheated air before being introduced to the burners. A simplified schematic diagram for a PC power plant is shown in Figure 6.4. PC plants operate on a steam Rankine cycle, shown schematically in Figure 6.5. In this cycle, coal combustion in the boiler (process 1–2) heats water in the tubes that form the boiler walls, producing superheated steam. The superheated steam exits the boiler

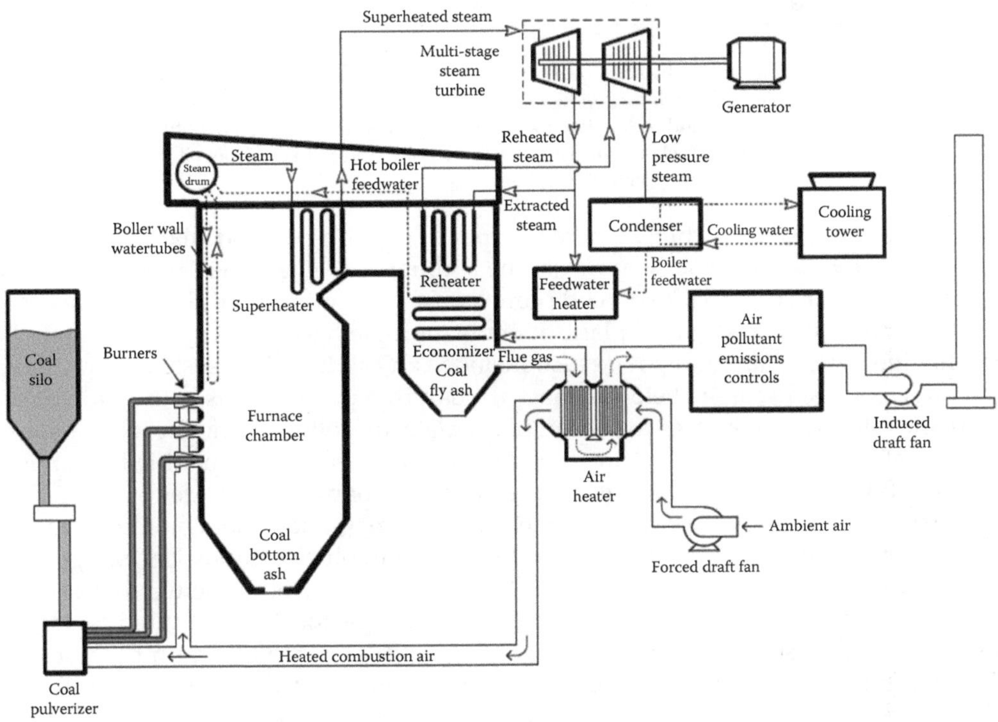

FIGURE 6.4
Simplified schematic diagram for a typical pulverized coal power plant. (From EPA [34].)

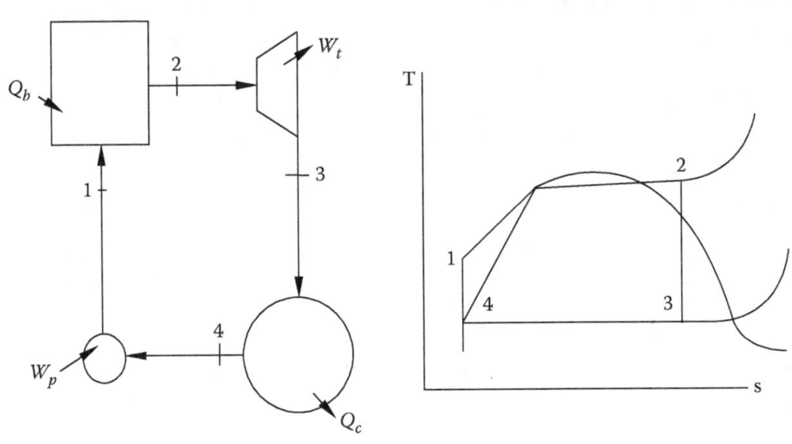

FIGURE 6.5
Ideal Rankine cycle. (From Wong [35].)

TABLE 6.2

Average Efficiencies for Fossil and Nuclear Power Plants in the United States in 2022

Fuel Type	Heat Rate (Btu/kWh)	Efficiency (%)
Coal[a]	10,689	32.6
Petroleum[b]	11,116	31.8
Natural gas	7,740	44.1
Nuclear	10,448	32.7

Source: EIA [36].

[a] Coal includes anthracite, bituminous, subbituminous, lignite, waste coal, and synthetic coal.

[b] Petroleum includes distillate fuel oil (all diesel and No. 1 and No. 2 fuel oils), residual fuel oil (No. 5 and No. 6 fuel oils) and bunker C fuel oil, jet fuel, kerosene, petroleum coke, and waste oil.

and enters a multi-stage turbine, where it expands, rotating a shaft that is connected to a generator (process 2–3). The steam exits the turbine and enters the condenser, where it cools and condenses (process 3–4). Feed pumps return the water to the boiler (process 4–1) to continue the cycle. Modern PC plants include multiple heat exchangers to recover heat from the flue gases. Plant equipment also includes an ash handling system, and multiple control devices. As discussed below, almost all coal-fired power plants in the United States have fabric filters or bag houses to control particulate air pollution and most have flue-gas desulfurization units to limit sulfur dioxide emissions as well as modified combustion systems or flue-gas control devices to reduce emissions of nitrogen oxides.

The product of combustion efficiency, thermal efficiency, and generator efficiency determines the overall efficiency of a coal-fired power plant. Since combustion and generator efficiencies are relatively close to 100%, the thermal efficiency of the actual Rankine cycle implemented at the plant will be the most influential factor. In the United States, heat rate is commonly used as an alternative way to express efficiency, in units of Btu or kJ of heat input per kWh of electricity output. The convention in the United States is to use HHV for power plant efficiency and heat rate calculations. Efficiency values also depend on whether the gross or net output from the plant is considered. Gross output is the total power produced by the facility. Net output is the gross output minus the parasitic load of electricity required to operate the plant, including requirements for pumps, fans, coal and ash handling and pollution control equipment. Operating flue-gas desulfurization equipment, for example, commonly requires 1%–2% of plant output [34]. Table 6.2 shows average heat rate and efficiencies for fossil and nuclear power plants in the United States in 2022, calculated by the U.S. EIA based on net electricity output [36].

Example 6.1

Consider an ideal Rankine cycle that operates with steam as the working fluid. The steam leaves the boiler and enters the turbine at 600°C and 5 MPa. It enters the condenser at a pressure of 10 kPa. The working fluid enters the pump at 10 kPa as saturated liquid and exits at 5 MPa. The cycle produces 150 MW net power output. Find (a) the thermal efficiency of the cycle and (b) the mass flow rate of steam.

Solution:

Figure 6.5 shows a schematic of the cycle and diagrams each process as a function of temperature, T, and entropy, s, indicating which processes are assumed to be isentropic. The T-s diagram also shows how the processes lay out with respect to the vapor dome, indicating the phase of the working fluid as saturated liquid at the pump inlet (state 4), liquid–vapor mixture in the boiler (1–2) and condenser (3–4) and superheated vapor at the turbine inlet (state 2). The cycle is analyzed by applying the first law of thermodynamics to the sequence of four steady flow processes that take the working fluid through the cycle from states 1 through 4. Each thermodynamic state is fixed by specifying two independent properties.

a. The overall thermal efficiency is given by

$$\eta_{\text{th}} = \frac{\dot{W}_{\text{net}}}{\dot{Q}_{\text{boiler}}} = \frac{\dot{m}_{\text{steam}}\left((h_2 - h_3) - (h_1 - h_4)\right)}{\dot{m}_{\text{steam}}\left(h_2 - h_1\right)}$$

where

$\dot{W}_{\text{net}}$ = net rate of work output from the cycle

$\dot{Q}_{\text{boiler}}$ = net rate of heat input to the cycle

$\dot{m}_{\text{steam}}$ = mass flow rate of working fluid

h_i = specific enthalpy of state i.

In the following, P_i = pressure at state i, v_i = specific volume at state i, s_i = entropy at state i, and T_i = temperature at state i.

State 4 is the inlet to the pump and exit from the condenser. At state 4, P_4 = 10 kPa and the water is saturated liquid. From the steam tables [35], h_4 = 191.81 kJ/kg; v_4 = 0.001010 m^3/kg.

State 1 is the exit from the pump and inlet to the boiler. P_1 is given as 5 MPa. The pump work is

$$w_{\text{pump,in}} = v_4\left(P_1 - P_4\right)$$

So, from the first law of thermodynamics:

$$h_1 = h_4 + v_4\left(P_1 - P_4\right) = 196.85\,\frac{\text{kJ}}{\text{kg}}$$

State 2 is the boiler outlet and turbine inlet, at which point the steam is superheated with P_2 = 5 MPa and T_2 = 600°C. From the steam tables, h_2 = 3,666.9 kJ/kg and s_2 = 7.2605 kJ/kg-K.

State 3 is the turbine exit and condenser inlet. Assuming the turbine is isentropic, $s_3 = s_2$. P_3 = 10 kPa, as given. Since the entropy at state 3 is known, using the limiting entropies from the steam tables for pure liquid and pure vapor at P_3 = 10 kPa, the quality at that state is found to be 88.2% vapor. The corresponding enthalpy is h_3 = 2,301.64 kJ/kg.

The overall cycle efficiency is then

$$\eta_{\text{th}} = \frac{\left((h_2 - h_3) - (h_1 - h_4)\right)}{\left(h_2 - h_1\right)} = \frac{\left((3{,}666.9 - 2{,}301.64) - (196.85 - 191.81)\right)}{(3{,}666.9 - 196.85)} = 0.392$$

b. The mass flow rate of steam can be calculated from the overall power output of the cycle:

$$\dot{m}_{\text{steam}} = \frac{150 \times 10^3\,\dfrac{\text{kJ}}{\text{s}}}{\left((3{,}666.9 - 2{,}301.64) - (196.85 - 191.81)\right)\dfrac{\text{kJ}}{\text{kg}}} = 110.3\,\frac{\text{kg}}{\text{s}}$$

Some efficiency improvements at existing coal-fired power plants are likely possible, even without major changes in plant configuration. In 2014, in preparing to propose CO_2 emissions regulations for existing coal-fired power plants in the United States, EPA estimated that on average, existing coal-fired power plants in the United States could reduce their heat rates by about 6% using established, cost-effective measures [37]. Opportunities for improving efficiencies included improved boiler combustion control, optimizing heat exchangers, upgrading motors, pumps, fans and drives to high-efficiency equipment, condenser and turbine upgrades, and improving water quality to reduce scale build-up in boiler tubes.

In new coal-fired power plants, higher efficiencies can be achieved through supercritical or ultra-supercritical operation of the steam cycle. These terms refer to the critical point of water, corresponding to a temperature of 647 K and pressure of 22.1 MPa. While thermal design efficiencies are estimated to range from 33% to 37% for new subcritical PC plants, supercritical operation at ~850 K and 24 MPa increases the design efficiency range to 39%–40%, and ultra-supercritical operation increases the range to 42%–45%. These conditions require upgraded boilers and turbines that can withstand high temperatures and pressures. Hundreds of supercritical plants have been built worldwide.

Coal is on the decline in the United States and in many other countries around the world. Due to retirements, U.S. coal plant capacity shrank by 39% and electricity generation from coal declined by 45% between 2012 and 2022. More retirements and no new additions are expected in the U.S. in the coming years [38].

6.2.4 Electricity Generation from Natural Gas

The most efficient natural gas power plants are combined cycle units, which use a gas power (Brayton) cycle topping a steam turbine (Rankine) cycle. The gas power cycle typically operates with turbine inlet temperatures of about 1,700 K, with turbine outlet temperatures of over 770 K. This allows the exhaust from the gas turbine to be used as an energy source for the steam cycle. New natural gas combined cycle (NGCC) power plants advertise design thermal efficiencies of up to 60%. However, if reported on the basis of net electricity output and using the HHV for natural gas, operating efficiencies above 50% would reflect good performance. In the United States, the overall average efficiency of power production with natural gas (Table 6.2) is also lowered by the use of relatively inefficient combustion turbines that operate on a simple Brayton cycle, as well as older NGCC units that are less efficient than the state-of-the-art models. Despite their lower efficiency, the simple cycle combustion turbines are utilized as peaking units, as they are relatively easy to ramp up and down.

Currently in the United States, as coal-fired power generation capacity is declining, natural gas capacity is increasing. On average, the cost of generation from a new NGCC unit is expected to be less than that from a new coal-fired power plant. U.S. natural gas plant capacity grew by 16% and electricity generation from natural gas increased by 38% between 2012 and 2022. For natural gas power plants, near-term capacity additions are anticipated to roughly balance retirements [38].

6.2.5 Integrated Gasification Combined Cycle

Integrated gasification combined cycle (IGCC) power plants (Figure 6.6) are designed to take advantage of the efficiency gains of combining gas and steam cycles but starting from a solid fuel (coal or biomass) [39]. To accomplish this, the solid fuel is first converted

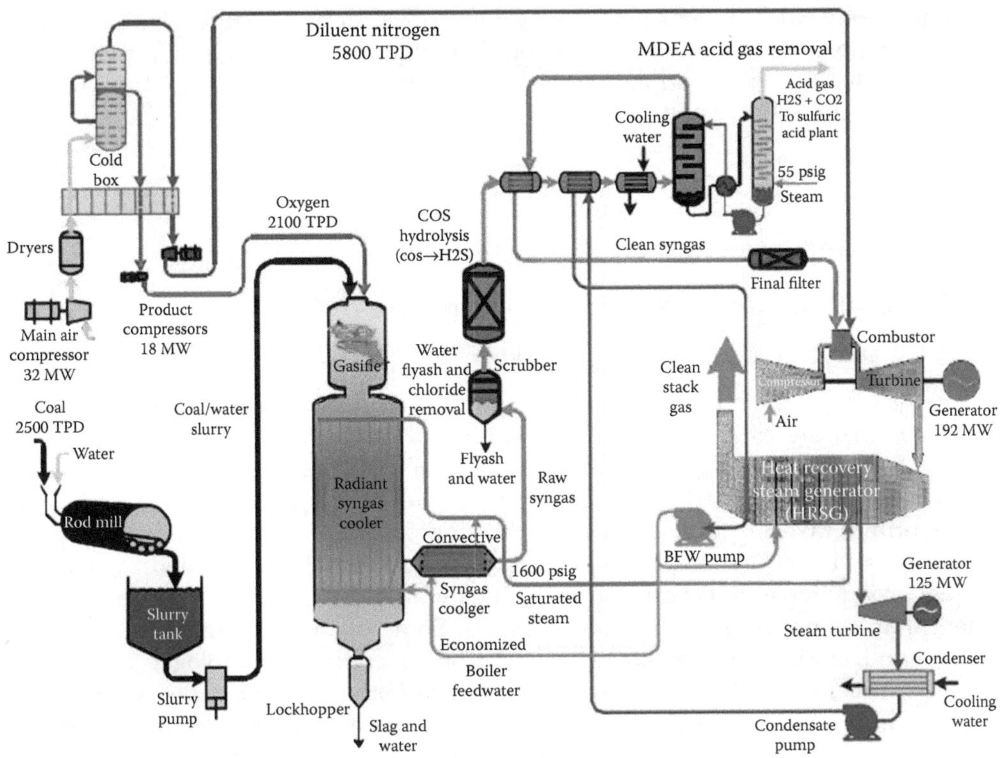

FIGURE 6.6
Simplified process flow diagram for the Tampa Electric IGCC plant. (From NETL [39].)

to synthesis gas, which is mostly CO and H_2, by reacting the fuel with air or oxygen at high temperature and pressure. Reacting with oxygen requires an air separation unit, but avoids diluting the synthesis gas with nitrogen, thus, reducing the costs of gas cleanup. The syngas is cleaned to remove particulate matter, sulfur, and mercury. In some designs, the CO in the syngas is reacted with water in a water-gas shift reactor to produce H_2 and CO_2. The CO_2 can then be captured pre-combustion, when it is present at relatively high concentration and can be captured much more easily than would be the case if the CO_2 were formed from CO in the combustion process. A 2007 MIT study estimated that without CO_2 capture, overall operating efficiencies of IGCC plants processing bituminous coal would approach 40% (based on HHV), on par with a supercritical PC plant [3].

Several IGCC plants have been constructed and operated around the world, benefiting from government subsidies as demonstration projects. A number of them have now been converted, suspended operations, or retired. Three of the longest running IGCC plants were the Wabash River Coal Gasification Repowering Project near West Terre Haute, Indiana, a 265 MW plant that started up in 1995 and was retired in 2016; the Tampa Electric IGCC project in Polk County, Florida, a 250 MW plant that started up in 1996 but in 2024 had the gasification process on standby; and the Elcogas IGCC plant in Puertollano, Spain, a 300 MW plant that started up in 1997 and was retired in 2016. A 582 MW IGCC plant that was scheduled to open in 2016 in Kemper County, Mississippi, was to be the first IGCC plant in the United States to capture and sequester carbon. However, the gasifier section was not put into operation and was removed from the facility in 2021 [40].

Costs of IGCC are currently not competitive with NGCC or new coal-fired units. However, fees or restrictions on carbon emissions could change the economics, as pre-combustion capture in an IGCC configuration is easier and less expensive than post-combustion capture in a conventional PC plant.

6.3 Air Pollution from Fossil Fuel Combustion

6.3.1 Local and Regional Scale Air Pollution

Historically, fossil fuels have been recognized for providing "cheap" energy and delivering services that are essential to the functioning of modern society, starting with heat, light, and transportation. However, the use of fossil fuels carries tremendous costs that are not covered in the price we pay for electric power or for gasoline at the pump. These costs sometimes take the form of economically inefficient government subsidies that distort markets and encourage excess consumption. Additionally, unwanted byproducts of fossil fuel production and use can cause significant damage to human health and the environment that are not reflected in market prices. Some of this damage can be prevented through good engineering practices, application of control technology, and remedial actions. However, fossil fuels are extracted and exploited in some areas where requirements to minimize environmental harm are not in place or are not effectively enforced. And even when protective regulations and safety systems are in place, accidental spills and releases can still occur. Most critically, there is currently no established solution to the problem of global warming caused by fossil fuel combustion.

Emissions of air pollutants, including greenhouse gases and other climate-forcing aerosols, are arguably the chief environmental concern associated with fossil fuel use. Since the 1970s, significant progress has been made in developing and implementing control technology to reduce emissions of "conventional" air pollutants from fossil fuel combustion systems, covering the range of applications from residential heating to motor vehicle engines to utility-scale electricity generation. The pollutants that have been widely addressed include particulate matter, carbon monoxide, sulfur oxides, VOCs, and nitrogen oxides. Nevertheless, significant challenges in reducing emissions of these pollutants remain, especially with older equipment and facilities in less-developed countries. Emissions of mercury from coal-fired power plants are an ongoing global concern. Every state in the United States has issued health advisories banning fishing and consumption of fish due to toxic mercury exposure in at least one lake or river. Mercury bioaccumulates in the fatty tissues and oils of all animals and is never broken down or dispersed. Thus, release of mercury into the atmosphere from coal combustion is a serious sustainability issue for humans, agriculture, and wild animals, particularly aquatic predator species such as tuna, seals, and dolphins. The U.S. EPA adopted stringent limits on mercury emissions from coal-fired power plants in 2012, with compliance required by 2015. However, because atmospheric mercury has a global distribution, health and environmental concerns will persist until action to limit emissions is taken internationally.

Anenberg et al. [41] estimated that the increase in anthropogenic air pollution emissions that occurred from preindustrial times up to the year 2000 accounted for 0.7 ± 0.3 million deaths per year from respiratory disease attributable to long-term exposure to ozone air pollution. They also found that the increase in exposure to fine particulate air pollution (airborne particles with aerodynamic diameter less than 2.5 μm, known as $PM_{2.5}$) accounted for 3.5±0.9 million deaths per year from cardiopulmonary disease and 220,000±80,000 deaths from lung cancer. The estimated number of excess deaths was greatest in densely populated parts of Europe and Asia, where emissions were also high. Mortality rates (deaths per million people) were also high in the eastern United States.

Anenberg et al. [41] arrived at these estimates by using a global chemistry and transport model to estimate how the increase in anthropogenic emissions since preindustrial times had changed ozone and $PM_{2.5}$ concentrations, then applying concentration-response functions for the specified disease endpoints to estimate the number of excess premature deaths. The concentration-response functions are well established and supported by a very large body of literature on air pollution health effects. Fossil fuel combustion is the dominant source of the anthropogenic emissions contributing to these deaths.

More recently, the Global Burden of Disease project estimated that long-term (chronic) exposure to ambient $PM_{2.5}$ and ozone air pollution accounted for 4.5 million premature deaths a year around the world in 2019, with $PM_{2.5}$ contributing the largest share [42]. For that same timeframe, McDuffie and colleagues [43] estimated that more than 1 million premature deaths per year from $PM_{2.5}$ exposure could have been avoided by eliminating fossil fuel combustion.

Ground-level ozone is formed in the atmosphere through reactions of VOCs and nitrogen oxides (NO_x) in the presence of sunlight. Major anthropogenic sources of VOC include evaporation of liquid fuels and incomplete combustion of liquid and solid fuels. VOCs are also emitted by vegetation, a source category that is considered biogenic, not anthropogenic. Nitrogen oxides are emitted from high-temperature combustion sources that convert N_2 in combustion air to NO and NO_2. Sources of NO_x thus include coal and gas-fired power plants, motor vehicles with internal combustion engines, aircraft, burning natural gas for industrial process heat, and fuel combustion for cooking and space heating.

Some $PM_{2.5}$ is directly emitted to the atmosphere from abrasive processes including manufacturing, agricultural activities, and road-use wear of tires and vehicles, and from incomplete combustion of liquid or solid fuels, including gasoline, diesel, fuel oil, coal, and biomass. Additional $PM_{2.5}$ is formed in the atmosphere from chemical reactions involving sulfur oxides, nitrogen oxides, ammonia, and carbonaceous compounds. Anthropogenic emissions of sulfur oxides are mainly due to combustion of sulfur-containing fuels, with coal combustion being the largest source.

In many countries around the world, air pollutant emissions from fossil fuel combustion have been regulated with increasing stringency since the 1970s. In the United States, for example, the series of regulations imposed since the 1970 Clean Air Act Amendments were enacted led to sharp reductions in emissions of SO_2 and NO_x, among other pollutants. U.S. EPA estimates that anthropogenic SO_2 emissions in the United States were reduced from over 28 million metric tons per year in 1970 to under 2 million metric tons per year in 2023, a 95% decline. Anthropogenic NO_x emissions were reduced by 74% over the same period, from 24 million to 6 million metric tons per year [44]. Reductions in emissions of sulfur dioxide resulted mainly from switching from high- to low-sulfur coal, installing flue-gas desulfurization (FGD) emissions controls, and coal plant retirements. NO_x emissions have been reduced mainly by the use of advanced post-combustion controls on coal and natural gas-fired power plants, coal plant retirements, and catalytic control devices on cars and trucks. Many of these measures were driven by regulations developed under the Clean Air Act.

Since the mid-2000s, China has made significant progress in reducing the emissions rates of sulfur oxides from its expanding fleet of coal-fired power plants [45]. In 2000, only about 1% of Chinese coal-fired power plant capacity was equipped with FGD equipment. By 2010, FGD penetration was reported to have increased to over 80% of capacity. McDuffie et al. [46] estimate that anthropogenic SO_2 emissions in China peaked at about 31 million metric tons per year in 2004, but had been reduced to about 12 million metric tons per year in 2017. China's NOx emissions started declining after 2012, but were still the highest in the world in 2017, at 24 million metric tons per year. In India, emissions of SO_2 have increased along with energy consumption, from about 4 million metric tons per year in 1970 to about 10 million metric tons per year in 2017. Anthropogenic NO_x emissions in India are estimated to have increased from about 2 million to about 9 million metric tons over the same period [46].

6.3.2 Greenhouse Gas Emissions and Climate Change

For more than 100 years, scientists have understood that carbon dioxide absorbs heat and acts as a greenhouse gas. The concentration of CO_2 in the atmosphere is now greater than 420 ppm, up from a preindustrial level of about 280 ppm. Along with changes in other greenhouse gases and aerosols, the increase in atmospheric CO_2 levels has shifted long-standing climate patterns across the globe. The scientists who observe the effects of climate disruption including loss of polar ice, altered growing seasons, more frequent heat waves and extreme weather and accelerated species loss are calling for immediate and deep reductions in CO_2 emissions so that climate systems do not become catastrophically destabilized. The Intergovernmental Panel on Climate Change (IPCC) series of assessments describe the risks and changes that are likely to happen if the production and use of fossil fuels are not quickly curtailed. A recent IPCC assessment states that society needs to at least reach net zero CO_2 emissions by mid-century and sharply reduce emissions of other GHG to keep the global temperature increase below catastrophic levels [47].

CO_2 and other long-lived greenhouse gases accumulate in the atmosphere, so some global warming is already locked in for decades to come. The issue with continued reliance on fossil fuels is how much worse it will get. The main human CO_2 additions to the atmosphere are from burning fossil fuels and from manufacture of cement, plus forest clearance and soil disturbance. From 1750 to 2019, fossil fuel combustion and cement manufacture are estimated to have released 445±20 metric Gigatons of carbon (Gt C) into the atmosphere. Deforestation and land use change are estimated to have released an additional 240±70 Gt C. Of the total released, scientists estimate that 285±5 Gt C has been retained in the atmosphere. The ocean has absorbed 170±20 Gt C, which is causing acidification and harming coral reefs and other marine systems. Forests and other terrestrial ecosystems have absorbed the remaining 230±60 Gt C [48]. The net release to the atmosphere has led to the current CO_2 atmospheric concentration, which has not existed for over 3 million years.

Production of fossil fuels also contributes to emissions of methane, a potent greenhouse gas with 30 or more times as much radiative forcing effect per unit mass as CO_2. Methane is estimated to account for nearly 30% of the rise in global temperatures that has occurred since the Industrial Revolution. Critically, methane has an atmospheric lifetime on the order of a decade, so reducing its emissions could lead to a relatively rapid climate benefit. Globally, fossil fuel production and use (oil, natural gas, and coal) are estimated to contribute about 37% of anthropogenic methane emissions [49]. While the agricultural sector contributes more methane emissions, the fossil fuel sector offers more near-term opportunities to significantly cut emissions by eliminating flaring and venting and improved practices to quickly find and fix methane leaks.

Black carbon (BC) aerosol is another potent climate-forcing agent that is emitted from fossil fuel combustion. BC absorbs solar radiation at all wavelengths and also contributes to warming by reducing the surface albedo of snow and ice. BC is also a significant component of $PM_{2.5}$ air pollution, discussed above in the context of its health effects. In contrast to BC, some other $PM_{2.5}$ constituents, such as the sulfate aerosol that is produced from SO_2 oxidation, have a cooling effect on climate, because they tend to scatter sunlight (direct effect) and to enhance the cooling properties of clouds (indirect effect). While the overall effect of aerosols on climate change has more uncertainty, it is clear that reducing BC emissions would have benefits for mitigating climate change and reducing health effects. Furthermore, BC has a short atmospheric lifetime (days to weeks) so the atmospheric response to reductions in BC emissions will be faster than the response to reductions in long-lived greenhouse gases. In the United States, about 50% of BC emissions are estimated to be due to fossil fuel combustion, primarily from the use of diesel. The balance is mainly due to open biomass burning [50]. Worldwide, open biomass burning and residential fuel combustion are estimated to contribute about two-thirds of BC emissions. Emissions from diesel vehicles in the United States have been declining due to increasingly stringent control requirements [50].

Table 6.3 shows estimates of greenhouse gas emissions per unit energy for coal, natural gas, and a variety of crude oil products [51–54]. In each case, the table shows both the direct combustion emissions and estimates of the "upstream" emissions associated with fossil fuel extraction and processing. Upstream greenhouse gas emissions for natural gas are mainly due to methane venting and leaks, which can be further reduced with good engineering practice and regulations. Upstream emissions are also a significant fraction of the total for gasoline and diesel from conventionally produced crude oil, due mainly to emissions from oil refining. The upstream contribution increases as petroleum production shifts to unconventional resources such as the tar sands and especially oil shale. Total emissions per unit energy for gasoline could double if oil shale is used.

TABLE 6.3

Estimated Greenhouse Gas Emissions Factors for Fossil Fuels (g CO_2-equivalent/MJ)

	Gasoline from Conventional Crude Oil [51]	Gasoline from Tar Sands [51]		Gasoline from Oil Shale [51]	
		Low	High	Low	High
Upstream	20.5	34.1	57.9	47.7	183.3
Combustion	73.7	73.7	73.7	73.7	73.7
Total	94.2	107.8	131.6	121	256.7
	Diesel from Conventional Crude Oil [51]	**Natural Gas [52,53]**		**Coal [53,54]**	
Upstream	16.1	12.7		5.3	
Combustion	77.4	51.1		89.0	
Total	93.5	63.8		94.2	

Example 6.2

Estimate the direct combustion emissions of CO_2 per kWh of electricity produced for an average natural gas plant in the United States, assuming the natural gas is 100% methane, stoichiometric combustion (Equation 6.1), and using the average heat rate from Table 6.2.

Solution:

The average heat rate for natural gas plants is 7,740 Btu/kWh and the heat content of methane is 55.5 MJ/kg. The molecular weight of CO_2 is 44 g/mol; that of methane is 16 g/mol. Hence:

$$\frac{7{,}740\left(\frac{\text{Btu}}{\text{kWh}}\right)\cdot 1{,}055\left(\frac{\text{J}}{\text{Btu}}\right)}{55.5\cdot 10^{6}\left(\frac{\text{J}}{\text{kgCH}_4}\right)} = 0.147\,\frac{\text{kgCH}_4}{\text{kWh}}$$

$$0.147\,\frac{\text{kgCH}_4}{\text{kWh}}\cdot\frac{1\,\text{molCO}_2}{\text{molCH}_4}\cdot\frac{44\,\text{gCO}_2/\text{molCO}_2}{16\,\text{gCH}_4/\text{molCH}_4} = 0.405\,\frac{\text{kgCO}_2}{\text{kWh}}$$

As discussed elsewhere in this book, CO_2 emissions associated with the energy system can be addressed by reducing demand for energy or shifting to low-carbon or carbon-free energy sources. Emissions of methane from energy production can be reduced significantly by avoiding venting, by detecting and fixing leaks, and by upgrading equipment. Although methane is a valuable product, it is often vented off when it is produced in association with oil in areas that lack natural gas pipelines. It is also deliberately vented when wells are cleaned out during completions, liquids unloading, and workovers. The U.S. EPA and individual state regulatory agencies have required the oil and gas industry to reduce or eliminate these emissions. In addition, it is technically feasible to capture CO_2 from large combustion sources and sequester it underground. However, this approach has relatively high costs and energy penalties and limited applicability. The technology to capture and sequester carbon from motor vehicles operating on gasoline or diesel does not exist.

6.3.3 Carbon Capture and Sequestration

Prospects for carbon capture and geological sequestration (CCS) of CO_2 remain uncertain, due to relatively high costs and limited experience with application of the technology in the electricity sector. However, some analysts think application of CCS could be an important element of strategies to reduce CO_2 emissions, particularly in China and other countries where abundant resources and past investments have created strong inertia to move away from the use of coal.

Geologic sequestration means capturing CO_2 from a combustion system, then storing it deep underground in structures that will trap the gas. Geological CCS could have application for concentrated CO_2 emissions streams from cement manufacturing, fertilizer production, and electricity generation from coal, natural gas, or biomass. We discuss the main capture alternatives that have been examined for electricity generation from coal, but similar technologies apply to the other source types.

Methods for capturing CO_2 from air-blown pulverized coal-fired power plants are established, but expensive. Post-combustion CO_2 is present in flue gases from coal-fired power plants at concentrations of about 15% (v). These relatively low concentrations make CO_2 gas separation challenging. CO_2 capture in a relatively dilute flue-gas stream would likely be accomplished in a packed absorption tower using chemical absorption with an aqueous amine solution. This approach has been used for decades to separate CO_2 (and H_2S) from natural gas in the oil and gas industry, commonly with monoethanolamine as the solvent. As a second step in the process, CO_2 would be stripped from the solution by using low-pressure steam to elevate the temperature, regenerating the absorbent for future use. In an aqueous solution, the amines are weak bases that can react with weak acids (such as CO_2) to form water-soluble salts. Finally, the CO_2 is cooled, dried, and compressed for transport and storage. MIT estimates that a 500 MW subcritical PC plant would require a 37% increase in heat input to maintain level output if an amine-based CO_2 capture system were added [3]. The solvent regeneration step and CO_2 compression step would require most of the extra energy required for the capture system.

An alternative approach for CO_2 capture from PC plants involves combusting the coal in oxygen, rather than air. Doing so results in flue gases that are mainly CO_2 and water, without dilution by N_2. Following flue-gas cleanup, CO_2 can be captured directly by compression. However, this oxy-fuel combustion and capture approach requires an air separation unit, which itself imposes a significant energy penalty on plant operations. The MIT study estimates that the energy efficiency of a new supercritical PC plant would be reduced from 38.5% to 30.6% if an oxy-fuel CO_2 capture system were included, due primarily to energy requirements for air separation and CO_2 compression [3].

The last approach that is widely discussed for CO_2 capture from coal-fired power plants is adding CO_2 capture to an IGCC system. In this case, three extra operations have to be added to a standard IGCC configuration. The first is a shift reactor, in which CO from the syngas is reacted with steam over a catalyst, producing CO_2 and H_2O. This allows the CO_2 to be produced prior to the combustion stage so that it isn't diluted with nitrogen. The shift reactor is followed by a CO_2 separation unit. In this case, the CO_2 is more concentrated than it would be in post-combustion flue gases, so it can be separated from the gas stream using a weakly binding physical solvent such as Selexol. The solvent can be regenerated by dropping the pressure, requiring much less energy than needed for amine regeneration in the post-combustion case. After separation, the fuel gas sent to the turbine is mainly H_2, which requires turbine modifications due to its high flame temperature. As with the other capture methods, the last step is compression and drying of the CO_2 to prepare it for transport and storage.

Geologic sequestration essentially means putting CO_2 gas into geologic structures with porous reservoir rocks and non-porous cap rock structures that will trap and permanently hold the gas. The types of formations with these characteristics that we know the most about are depleted oil and gas reservoirs. In enhanced oil recovery processes, CO_2 and methane produced with the oil are compressed and reinjected into the oil field to keep the reservoir pressure up and keep the oil production going. However, there have been a limited number of studies of the long-term storage in such formations. As well, the enhanced oil production may not be consistent with the need for sharp cuts in economy-wide CO_2 emissions. Another challenge with geologic sequestration is that the locations of the possible depleted oil and gas fields are not often near coal-fired power plants. Saline aquifers have been proposed for geologic storage as these formations are more widespread and not considered to be of use as groundwater. The technology to compress CO_2 and transport it via pipelines is known, but the economics are challenging.

The Boundary Dam project in Saskatchewan, Canada, was the first power plant to operate with full-scale carbon capture when it started up in October 2014. This 110 MW (net) unit is designed to use post-combustion amine scrubbing to capture 90% of the CO_2 produced by the facility, with the CO_2 piped to a nearby oil field for use in enhanced oil recovery. In a January 2024 status update, the facility reported CCS plant availability of about 80% and average emissions of 380 metric tons CO_2 per GWh of electricity produced [55]. The reported CO_2 emissions rate is about 60% lower than typical of a coal-fired power plant without CCS.

6.3.4 Leaving It in the Ground

According to the IPCC working group on mitigation of climate change, in order to have a better than even chance of limiting global warming to 2°C above the temperature in preindustrial times, the release of carbon from fossil fuels between 2020 and 2050 would need to be limited to about 1,150 Gt CO_2. The cap corresponding to a 1.5°C limit is about 500 Gt [56]. Even the 2°C carbon limit is an order of magnitude less than the carbon contained in global fossil fuel resources, which is estimated to be equivalent to roughly 11,000 Gt CO_2 [57]. Thus the pressing question society faces right now about fossil fuels is not whether we're running out of them, but how much more we can afford to use if we want to avoid catastrophic climate change.

In its Sixth Assessment, the IPCC found that in modeled emissions scenarios that limited warming to 1.5°C, global GHG emissions needed to peak by around 2025 and then start a sharp decline with deep cuts by 2050 [56]. Following such a trajectory will entail a fast transition of the world's energy system, including rapidly:

- Increasing energy efficiency in the industrial, buildings, and transportation sectors and reducing demand through recycling and circular economy approaches and improved urban design
- Accelerating research, development and commercialization efforts to decarbonize manufacture of cement, steel and other emissions intensive materials
- Eliminating methane emissions from oil and gas production and coal mining and then phasing down demand for these commodities
- Scaling up low- or zero carbon electricity generation and storage technologies while retiring fossil fuel power plants
- Electrifying end uses, including residential, commercial and industrial heating as well as transportation
- Advancing low or zero emissions fuel and propulsion systems for ships and airplanes

The IPCC report cautions against continuing to install fossil fuel infrastructure unless carbon capture and sequestration technology is available to abate emissions [56].

As discussed in Chapter 1, in Paris, France, on December 12, 2015, more than 190 countries reached an agreement to keep the global temperature increase this century to less than 2°C above preindustrial levels, with the further aim of limiting the increase to below 1.5°C. The Paris Agreement entered into force on October 5, 2016, when it exceeded the threshold of ratification by at least 55 countries representing 55% of global emissions. Parties to the Paris Agreement are obligated to pursue their best efforts to reduce emissions of greenhouse gases and other climate-forcing agents. Individual countries submit "Nationally Determined Contributions" (NDCs) that establish their commitments to the global community. The agreement requires countries to report regularly on their progress and requires the international community to periodically reassess progress and goals. A third round of NDCs is due in 2025. The Conference of Parties (COP) held in December

2023 assessed parties' progress and called for doubling energy efficiency and tripling the amount of renewable energy installed worldwide by 2030. At the COP meeting, 200 countries committed to transition away from fossil fuels [58]. According to the Executive Secretary of the conference, that action marks "the beginning of the end" [59].

Problems

6.1 (Thermodynamics review problem) Examine the effect of steam temperature and pressure on the efficiency of an ideal Rankine cycle. (a) Compute the thermal efficiency of a cycle in which superheated steam enters the turbine at 3.5 MPa and 400°C and is condensed at a pressure of 10 kPa. (b) Compute the efficiency of a cycle in which the steam enters the turbine at 12.5 MPa and 500°C and is again condensed at 10 kPa. In both cases, assume the compression and expansion are isentropic, and assume the steam exits the condenser as saturated liquid at 10 kPa.

6.2 (Thermodynamics review problem) Consider the effect on an ideal Rankine cycle of using a two-stage turbine with steam reheating in between the stages. Compute the thermal efficiency of a reheat cycle for the following conditions. Steam enters the first stage turbine at 12.5 MPa and 500°C, and then is reheated back to 500°C at a pressure of 2 MPa, before entering the second stage turbine. The steam is condensed at 10 kPa and exits the condenser as a saturated liquid. Assume the compression and expansion processes are isentropic.

6.3 (Thermodynamics review problem) An ideal Brayton cycle operates with isentropic compression and expansion, with air entering the compressor at 300 K and leaving the combustion chamber at 1,000 K. The pressure ratio is 7:1. Use property data for air and compute the thermal efficiency of the cycle.

6.4 (Thermodynamics review problem) Determine the thermal efficiency and net power output of an ideal combined cycle power plant, using property data for air in the top cycle and steam properties for the bottom cycle. In the top cycle, air enters the compressor at 300 K at a rate of 300 kg/s and is heated to 1,200 K in the combustion chamber. The top cycle has a pressure ratio of 12. The steam in the bottom cycle is heated to 400°C at 8 MPa in a heat exchanger, and after leaving the turbine is condensed at 10 kPa. The combustion gases leave the heat exchanger at 400 K. Assume that all compression and expansion processes are isentropic.

6.5 Compute the levelized cost of electricity for a natural gas combustion turbine power plant, assuming the thermal efficiency is 40%, combustor efficiency is 95% and generator efficiency is 95%. The plant is rated at 150 MW and has a capacity factor of 50%. The capital equipment cost is $165 million and plant lifetime is 20 years. Plant financing is arranged at a nominal interest rate of 8% over a 20-year term. Natural gas is supplied to the plant during the first year at a price of $5/1,000 ft^3, with the price expected to escalate at a rate of 2% per year. Other operation and maintenance (O&M) costs are $10 million per year in the first year, escalating at 2% per year. Neglect tax considerations. Show the calculations and results for the following steps needed to reach the final answer:

a. Calculate the annual natural gas requirement for the facility, in units of thousand cubic feet. The average heat content of natural gas is 1.032 MMBtu/Mcf. (Mcf = 1,000 ft^3.)

b. Calculate the annual cost of the natural gas required in the initial year.

c. Calculate the annual payment required to pay back the plant financing in equal payments over 20 years.

d. Calculate the net present value of the 20-year stream of total costs using a nominal discount rate of 10%.

e. Calculate the LCOE.

6.6 A coal-fired power plant with a steam cycle efficiency of 36% produces 1.8 billion kWh/year of electricity. Assume the generator is 97% efficient, and the boiler is 95% efficient. Assume the coal has a heating value of 29 MJ/kg and is 75% carbon by mass. Determine:

a. The total coal energy content required per year

b. The mass of coal required per year in metric tons

c. The mass of CO_2 emitted per year from the plant

d. The mass of CO_2 emitted per MWh produced in pounds per MWh

6.7 In an April 2023 report, the U.S. EIA estimated that the average levelized cost of electricity for conventional combined cycle natural gas plants entering service in 2028 would be about $43/MWh (2022 US$) and that for offshore wind would be about $100/MWh (2022 US$) [60]. Based on these levelized costs, estimate the emissions fee in $ per metric ton of CO_2-e that would be required to incentivize shifting from NGCC electricity generation to offshore wind generation. Assume fees will be imposed on life cycle greenhouse gas emissions of ~600 g CO_2-e/kWh for NGCC and 20 g CO_2-e/kWh for offshore wind. How does your estimated emissions fee compare to the U.S. government's social cost of carbon for 2030, as presented in Table 2.10?

References

1. IEA, 2023. Energy statistics data browser, International Energy Agency, https://www.iea.org/data-and-statistics.
2. EIA, 2023. *Annual Energy Outlook 2023*, U.S. Energy Information Administration, Washington, DC, https://www.eia.gov/outlooks/aeo/.
3. Ansolabehere, S., Beer, J., Deutch, J., Ellerman, A.D., Friedman, J., Herzog, H., Jacoby, H., Joskow, P., McRae, G., Lester, R., Moniz, E., and Steinfeld, E., 2007. *The Future of Coal: Options for a Carbon-Constrained World*, Massachusetts Institute of Technology, Cambridge, MA.
4. De Nevers, N., 2000. *Air Pollution Control Engineering*, McGraw-Hill, New York.
5. EIA, 2015. International energy statistics, U.S. Energy Information Administration, https://www.eia.gov/beta/international/.
6. EIA, 2015. Top coal producing states, 2013, U.S. Energy Information Administration, https://www.eia.gov/energyexplained/index.cfm?page=coal_home.
7. NRC, 2010. *Hidden Costs of Energy: Unpriced Consequences of Energy Production and Use*, National Research Council Committee on Health, Environmental, and Other External Costs and Benefits of Energy Production and Consumption, Washington, DC.

8. MSHA, 2015. Historical data on mine disasters in the United States, Mine Safety and Health Administration, https://www.msha.gov/MSHAINFO/FactSheets/MSHAFCT8.HTM.
9. MSHA, 2015. Mine safety and health at a glance, Mine Safety and Health Administration, https://www.msha.gov/MSHAINFO/FactSheets/MSHAFCT10.asp.
10. BLS, 2015. Census of fatal occupational injuries charts, Bureau of Labor Statistics, https://www.bls.gov/iif/oshcfoi1.htm.
11. IEA, 2009. *Cleaner Coal in China*, International Energy Agency, Paris, France.
12. Areddy, J.T., 2012. China coal sector has safety setback, *The Wall Street Journal.*
13. EIA, 2024. Natural gas summary, consumption, U.S. Energy Information Administration, https://www.eia.gov/naturalgas.
14. EIA, 2015. Oil and natural gas resource categories reflect varying degrees of certainty, *Today in Energy*, U.S. Energy Information Administration, https://www.eia.gov/todayinenergy/detail.cfm?id=17151.
15. EIA, 2023. *Assumptions to Annual Energy Outlook 2023, Oil and Gas Supply Module*, U.S. Energy Information Administration, Washington, DC.
16. EIA, 2006. *Assumptions to Annual Energy Outlook 2006, Oil and Gas Supply Module*, U.S. Energy Information Administration, Washington, DC.
17. NETL. 2015. On-site research, National Energy Technology Laboratory, U.S. Department of Energy, https://www.netl.doe.gov/research/on-site-research/research-portfolio/oil--gas-research/ucr.
18. Small, M.J., Stern, P.C., Bomberg E., Christopherson, S.M., Goldstein, B.D., Israel, A.L., Jackson, R.B., Krupnick, A., Mauter, M.S., Nash, J., and North, D.W., 2014. Risks and risk governance in unconventional shale gas development. *Environmental Science & Technology* 48, 8289–8297.
19. Jackson, R.B., Vengosh, A., Carey, J.W., Davies, R.J., Darrah, T.H., O'Sullivan, F., and Pétron, G., 2014. The environmental costs and benefits of fracking, *Annual Review of Environment and Resources* 39, 327–362.
20. Petersen, M.D., Mueller, C.S., Moschetti, M.P., Hoover, S.M., Rubinstein, J.L., Llenos, A.L., Michael, A.J., Ellsworth, W.L., McGarr, A.F., Holland, A.A., and Anderson, J.G., 2015. *Incorporating Induced Seismicity in the 2014 United States National Seismic Hazard Model—Results of 2014 Workshop and Sensitivity Studies*, U.S. Geological Survey, Reston, VA.
21. Ellsworth, W.L., 2013. Injection-induced earthquakes, *Science* 341(6142), 1225942.
22. EPA, 2024. *Inventory of U.S. Greenhouse Gas Emissions and Sinks: 1990–2022*, EPA 430R-24004, U.S. Environmental Protection Agency, https://www.epa.gov/ghgemissions/inventory-us-greenhouse-gas-emissions-and-sinks-1990-2022.
23. Moore, C., Zielinska, B., Pétron, G., and Jackson, R.B., 2014. Air impacts of increased natural gas acquisition, processing, and use: A critical review, *Environmental Science & Technology* 48(15), 8349–8359.
24. CIA, 2021. Country comparisons: Crude oil-proved reserves, The World FactBook (2021 Archive), U.S. Central Intelligence Agency, https://www.cia.gov/the-world-factbook/about/archives/2021/field/crude-oil-proved-reserves/country-comparison.
25. EIA, 2023. Assumptions to the annual energy outlook 2023: Oil and gas supply module, U.S. Energy Information Administration, https://www.eia.gov/outlooks/aeo/assumptions/pdf/OGSM_Assumptions.pdf.
26. EIA, 2024. Petroleum and other liquids, U.S. Energy Information Administration, https://www.eia.gov/petroleum/.
27. U.S. GPO, 2011. *Deep Water: The Gulf Oil Disaster and the Future of Offshore Drilling*, National Commission on the BP Deepwater Horizon Oil Spill and Offshore Drilling, U.S. Government Printing Office, Washington, DC.
28. Robertson, C., and Krause, C., August 2, 2010. Gulf spill is largest of its kind, scientists say, *New York Times.*
29. Robertson, C., Perez-Pena, R., and Krause, C., July 2, 2015. BP to pay $18.7 billion for deep water horizon spill, *New York Times.*
30. EIA, 2024. Monthly energy review, U.S. Energy Information Administration, https://www.eia.gov/totalenergy/data/monthly/#consumption, Tables 2.1a, 2.1b, and 4.3.

31. EIA, 2023. RECS survey data, Residential Energy Consumption Survey (RECS), U.S. Energy Information Administration, https://www.eia.gov/consumption/residential/index.php.
32. DOE, 2015. Furnaces and boilers, U.S. Department of Energy, https://energy.gov/energysaver/articles/furnaces-and-boilers.
33. EPA, 2022. Waste heat to power systems, U.S. Environmental Protection Agency, https://www.epa.gov/sites/production/files/2015-07/documents/waste_heat_to_power_systems.pdf.
34. EPA, 2010. *Available and Emerging Technologies for Reducing Greenhouse Gas Emissions from Coal-Fired Electric Generating Units*, U.S. Environmental Protection Agency, Office of Air and Radiation, Research Triangle Park, NC.
35. Wong, K.V., 2012. *Thermodynamics for Engineers*, 2nd ed., CRC Press, Boca Raton, FL.
36. EIA, 2023. Average operating heat rate for selected energy sources, Electric Power Annual, U.S. Energy Information Administration, https://www.eia.gov/electricity/annual/.
37. EPA, 2014. *Carbon Pollution Emission Guidelines for Existing Stationary Sources: Electric Utility Generating Units; Proposed Rule*, U.S. Environmental Protection Agency, 79 FR 34830, Government Printing Office, Washington, DC.
38. EIA, 2023. Electric power annual, Tables 3.1.A, 4.3, and 4.5, U.S. Energy Information Administration, https://www.eia.gov/electricity/annual/.
39. NETL, 2015. Simplified process flow diagram for the Tampa Electric IGCC plant, National Energy Technology Laboratory, U.S. Department of Energy, https://www.netl.doe.gov/research/coal/energy-systems/gasification/gasifipedia/igcc-config.
40. Swartz, K., 2021. The Kemper project just collapsed: What it signifies for CCS, *E&E News*, October 24.
41. Anenberg, S.C., Horowitz, L.W., Tong, D.Q., and West, J.J., 2010. An estimate of the global burden of anthropogenic ozone and fine particulate matter on premature human mortality using atmospheric modeling, *Environmental Health Perspectives* 118(9), 1189–1195.
42. Murray, C., Aravkin, A., Zheng, P., Abbafati, C., Abbas, K., Abbasi-Kangevari, M., Abd-Allah F, Abdelalim A, Abdollahi M, Abdollahpour I, Abegaz KH. 2020. Global burden of 87 risk factors in 204 countries and territories, 1990–2019: A systematic analysis for the Global Burden of Disease Study 2019, *The Lancet* 396(10258), 1223–1249, doi:10.1016/S0140-6736(20)30752-2.
43. McDuffie, E., Martin, R., Yin, H., and Brauer, M., 2020. *Global Burden of Disease from Major Air Pollution Sources (GBD MAPS): A Global Approach*, Health Effects Institute, Boston, MA.
44. EPA, 2024. National Emissions Inventory (NEI) air pollutant emissions trends data, U.S, Environmental Protection Agency, https://www.epa.gov/air-emissions-inventories/air-pollutant-emissions-trends-data.
45. Lu, Z., Zhang, Q., and Streets, D.G., 2011. Sulfur dioxide and primary carbonaceous aerosol emissions in China and India, 1996–2010, *Atmospheric Chemistry and Physics* 11, 9839–9864.
46. McDuffie, E., Smith, S.J., O'Rourke, P., Tibrewal, K., Venkataraman, C., Marais, E.A., Zheng, B., Crippa, M., Brauer, M., and Martin, R.V., 2020. A global anthropogenic emission inventory of atmospheric pollutants from sector- and fuel-specific sources (1970–2017): An application of the Community Emissions Data System (CEDS), *Earth Syst. Sci. Data*, 12, 3413–3422, doi:10.5194/essd-12-3413-2020.
47. IPCC, 2023. Summary for policymakers, in: Lee, H. and Romero, J. eds., *Climate Change 2023: Synthesis Report. Contribution of Working Groups I, II, and III to the Sixth Assessment Report of the Intergovernmental Panel on Climate Change* [Core writing team H. Lee and J. Romero (eds)], IPCC, Geneva, Switzerland, pp. 1–34, doi:10.59327/IPCC/AR6-9789291691647.001.
48. IPCC, 2021. Canadell, J.G., Monteiro, P.M.S., Costa, M.H., Cotrim da Cunha, L., Cox, P.M., Eliseev, A.V., Henson, S., Ishii, M., Jaccard, S., Koven, C., Lohila, A., Patra, P.K., Piao, S., Rogelj, J., Syampungani, S., Zaehle, S. and K. Zickfeld, Global carbon and other biogeochemical cycles and feedbacks, Chapter 5, Table 5.1, in: Masson-Delmotte, V., Zhai, P., Pirani, A., Connors, S.L., Péan, C., Berger, S., Caud, N., Chen, Y., Goldfarb, L., Gomis, M.I., Huang, M., Leitzell, K., Lonnoy, E., Matthews, J.B.R., Maycock, T.K., Waterfield, T., Yelekc, O., Yu, R. and Zhou, B. eds., *Climate Change 2021: The Physical Science Basis. Contribution of Working Group I to the Sixth Assessment Report of the Intergovernmental Panel on Climate Change*, Cambridge University Press, Cambridge, UK and New York, 2391 pp., doi:10.1017/9781009157896.

49. IEA, 2023. *Global Methane Tracker 2023*, International Energy Agency, Paris, https://www.iea.org/reports/global-methane-tracker-2023, Licence: CC BY 4.0.
50. EPA, 2012. Report to Congress on black carbon, U.S. Environmental Protection Agency, https://www3.epa.gov/airquality/blackcarbon/2012report/fullreport.pdf.
51. Brandt, A.R., and Farrell, A.E., 2007. Scraping the bottom of the barrel: Greenhouse gas emission consequences of a transition to low-quality and synthetic petroleum resources, *Climatic Change* 84, 241–263.
52. Weber, C.L., and Clavin, C., 2012. Life cycle carbon footprint of shale gas: Review of evidence and implications. *Environmental Science & Technology* 46(11), 5688–5695.
53. EPA, 2010. Air markets program data for 2010. U.S. Environmental Protection Agency, https://ampd.ea.gov/ampd/%3E.
54. Jaramillo, P., Griffin, W.M., and Matthews, H.S., 2007. Comparative life-cycle air emissions of coal, domestic natural gas, LNG, and SNG for electricity generation, *Environmental Science & Technology* 41(17), 6290–6296.
55. SaskPower, 2024. BD3 status update: Q4 2023, January 16, https://www.saskpower.com/about-us/our-company/blog/2024/bd3-status-update-q4-2023.
56. IPCC, 2022. Summary for Policymakers, in: Shukla, P.R., Skea, J., Reisinger, A., Slade, R., Fradera, R., Pathak, M., Al Khourdajie, A., Belkacemi, M., van Diemen, R., Hasija, A., Lisboa, G., Luz, S., Malley, J., McCollum, D., Some, S. and Vyas, P. eds., *Climate Change 2022: Mitigation of Climate Change. Contribution of Working Group III to the Sixth Assessment Report of the Intergovernmental Panel on Climate Change*, Cambridge University Press, Cambridge, UK and New York, doi:10.1017/9781009157926.001.
57. McGlade, C., and Ekins, P., 2015. The geographical distribution of fossil fuels unused when limiting global warming to 2°C, *Nature* 517, 187–190.
58. Plumer, B., and Bearak, M., 2023. In a first, nations at climate summit agree to move away from fossil fuels, *The New York Times*, December 13, 2023.
59. Stiell, S., 2023. Speech at the closing of COP 28, Dubai, United Arab Emirates, December 13, 2023, https://unfccc.int/news/we-didn-t-turn-the-page-on-the-fossil-fuel-era-but-this-outcome-is-the-beginning-of-the-end-un.
60. EIA, 2023. Levelized costs new generation resources in the Annual Energy Outlook 2023, U.S. Energy Information Administration, April, https://www.eia.gov/outlooks/aeo/electricity_generation/pdf/AEO2023_LCOE_report.pdf.

7

Nuclear Power

7.1 Introduction

Nuclear power is not considered a sustainable energy technology because it relies on mining a finite source of nuclear fuel; however, it is a major contributor to global electricity, accounting for approximately 20% of electricity generation in the United States alone. Additionally, because decarbonization is a driving force behind the energy transition and nuclear power plants do not emit carbon dioxide, nuclear power plays an important role in the ongoing energy discussion. This chapter provides an overview of this electricity-generating technology.

The basic process used in today's nuclear power plants is the fission of heavy element nuclei. In this process, a carefully controlled fission reaction generates heat to produce steam to power conventional turbines for electricity generation. An alternative nuclear reaction, fusion, is the same process by which the sun's radiation is generated. Its mechanism has been under investigation for a long time but is not likely to reach commercial feasibility for some years to come. This chapter will therefore concentrate on power derived from fission, a commercially available means of generating power that has been developed and employed in many countries since the 1950s. In this type of nuclear reaction, a power source called a nuclear reactor captures the energy from the fission process as heat, which is then transferred to a working fluid that can be used to generate electricity, similar to the Rankine cycle in a coal-fired power plant. The primary fission fuel currently used in a majority of nuclear power plants is Uranium-235 (^{235}U)—an isotope of the element Uranium-238 (^{238}U). Uranium is a widely available element in the Earth's crust. It occurs naturally as an ore containing about 99% ^{238}U and about 0.7% of the isotope ^{235}U in all natural uranium deposits. When ^{235}U is fissioned in a reactor, it provides about one megawatt day of energy for each gram of ^{235}U that is fissioned.

To understand the nuclear power option, it is necessary to consider the entire nuclear fuel cycle, not just the part that occurs in the reactor. The cycle begins with the mining of uranium—an ore that is processed or enriched to extract the isotope ^{235}U, which is used today in the majority of electricity-producing reactors.

The enrichment process is the key to the use of uranium as a fuel in a power plant to generate heat for electricity generation. Once enriched, the uranium is formed into a suitable fuel for power plants, usually small pellets that can be formed into fuel rods. After the fission process is complete, the spent fuel can either be reprocessed or stored for eventual disposal. Radioactive waste is generated in all nuclear power cycles and must be disposed of safely. The storage and transportation of fissionable material are important parts of the fuel cycle and constitute critical steps susceptible to terrorist action.

DOI: 10.1201/9781003586616-7

The first nuclear reactor was Chicago Pile-1 (CP-1), developed under the direction of Enrico Fermi at the University of Chicago. It achieved criticality (continuous fission reaction) on December 2, 1942 [1]. The application of fissionable material for power production began at the end of World War II, when engineers and scientists who developed the atomic bomb realized that a controlled nuclear chain reaction could be used as a source of heat for generating electricity. Electricity production from nuclear energy began in 1954. That year, the first nuclear power plant generating heat from uranium was connected to the grid in the former USSR [2]. The world's first large-scale nuclear power plant was built in Pennsylvania and began operation in 1957. It was a pressurized-water reactor (PWR) capable of producing 231 MW of thermal energy and 68 MW of electric power. The PWR remains the most widely used nuclear power plant design.

There is worldwide debate on the contribution that nuclear energy will make toward the energy future. That contribution depends on the availability and cost of renewable resources. In countries such as Finland, South Korea, or Japan that do not have plentiful sustainable resources, nuclear power will likely be one of the main contributors of energy and power. The second largest operating nuclear power plant in the world, after the Taishan plant in China (1,750 gross MW), is the Olkiluoto power plant in Finland (1,660 gross MW) [3]. Japan restarted its nuclear power program after careful safety reevaluation following the Fukushima nuclear disasters in 2011 [4]. On the other hand, countries such as the United States or Brazil, which have plentiful renewable resources, are likely to rely primarily on these sources in the future, with nuclear power potentially continuing to provide some base load power.

Nuclear power plants are similar to coal-fired or natural gas power plants, except for the energy source, which is uranium. In 2024, there were 415 reactor units in operation worldwide [5], contributing about 9% of the world's electricity production [6]. The majority of reactors are located in the United States (94 reactor units), China (56 units), France (56 units), Russia (36 units), and South Korea (26 units) [5] as shown in Figure 7.1.

There are 65 reactors under construction (totaling 66,100 MW) mostly in China and India, as shown in Figure 7.2 [7].

Although the United States is the largest producer of nuclear power, it experienced a two-decade period without the startup of any new nuclear power plants until the Tennessee Valley Authority's startup of the 1,150 MW Watts Bar Unit 2. When this unit was connected to the power grid on June 3, 2016, it became the first nuclear reactor unit in the United States to come online since 1996, when Watts Bar Unit 1 began operations. This is shown in the photo below (Figure 7.3) [8]. Watts Bar Unit 2 was the first nuclear plant in the United States to meet new regulations from the U.S. Nuclear Regulatory Commission (NRC) that were established after the 2011 earthquake and tsunami that damaged the Fukushima Daiichi Nuclear Plant in Japan.

In 2013, construction began on four other U.S. nuclear reactors, all 1,100-MW Westinghouse AP1000 PWRs: V.C. Summer Units 2 and 3 by South Carolina Electric & Gas and Vogtle Units 3 and 4 in Georgia by Southern Company. The two V.C. Summer reactors were ultimately canceled due to repeated construction delays, cost overruns, and charges of fraud related to false reports of progress [9]. After also experiencing long delays and cost overruns, Vogtle Units 3 and 4 were finally completed and began commercial operation on July 31, 2023 (the first new nuclear startup in the United States since Watts Bar 2 in 2016) and April 29, 2024, respectively. However, the total cost was over $35 billion ($16,000/kW),

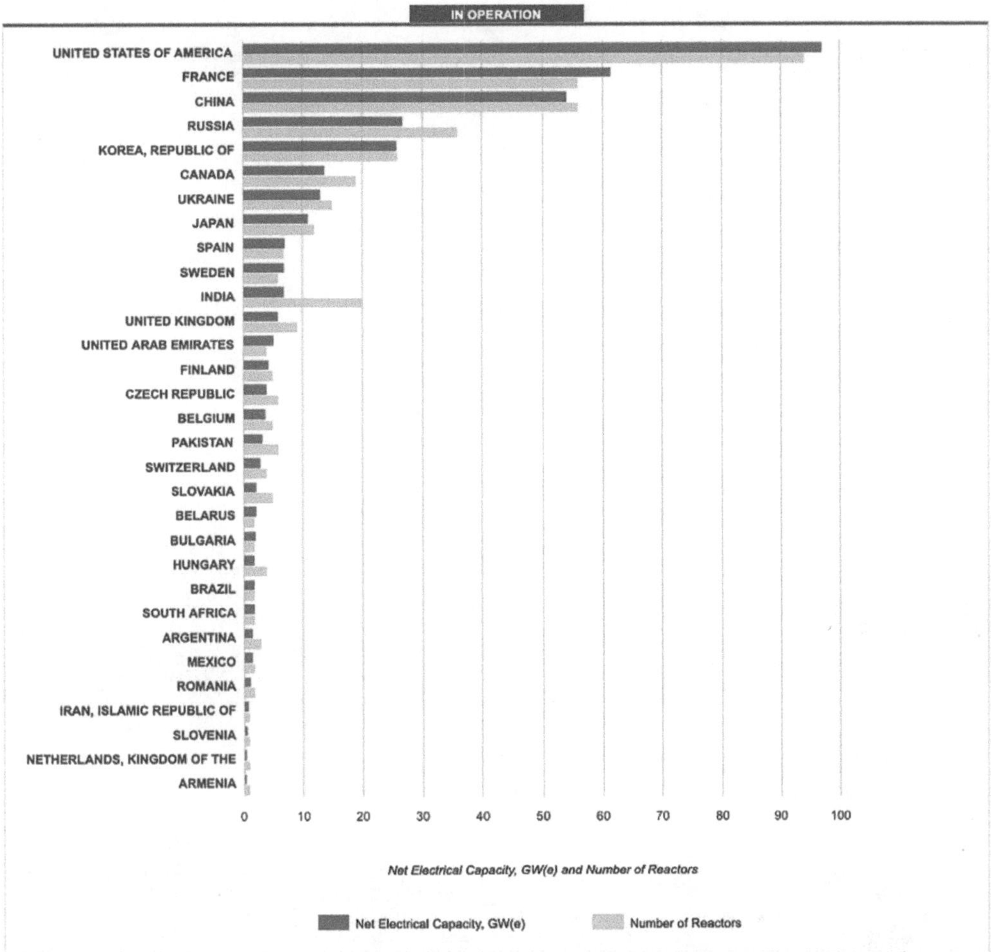

FIGURE 7.1
Number of reactors in operation, worldwide, as of November 2024. (© IAEA [5].)

or approximately 2.5 times the original certified construction cost of $14 billion, resulting, according to consumer organizations, in an estimated average $35/month electricity bill increase for Georgia Power ratepayers [10].

It is important to note that most of the reactors in operation today are more than 35 years old. Many of them are currently operating under extensions of their original lifetime limits. The age distribution of operating reactors is shown in Figure 7.4 [11].

As a result of the age of these reactors, within the next 10 or 20 years, at least 150 will have to be shut down and replaced. Thus, at least 150 new reactors will need to be built before nuclear energy can make a net increase in its contribution to global needs. The cost of construction and the initial energy investment to replace these old reactors are enormous. Because of today's high cost of new nuclear reactors in the United States (see Section 7.8), replacing the old reactors would require a huge initial investment.

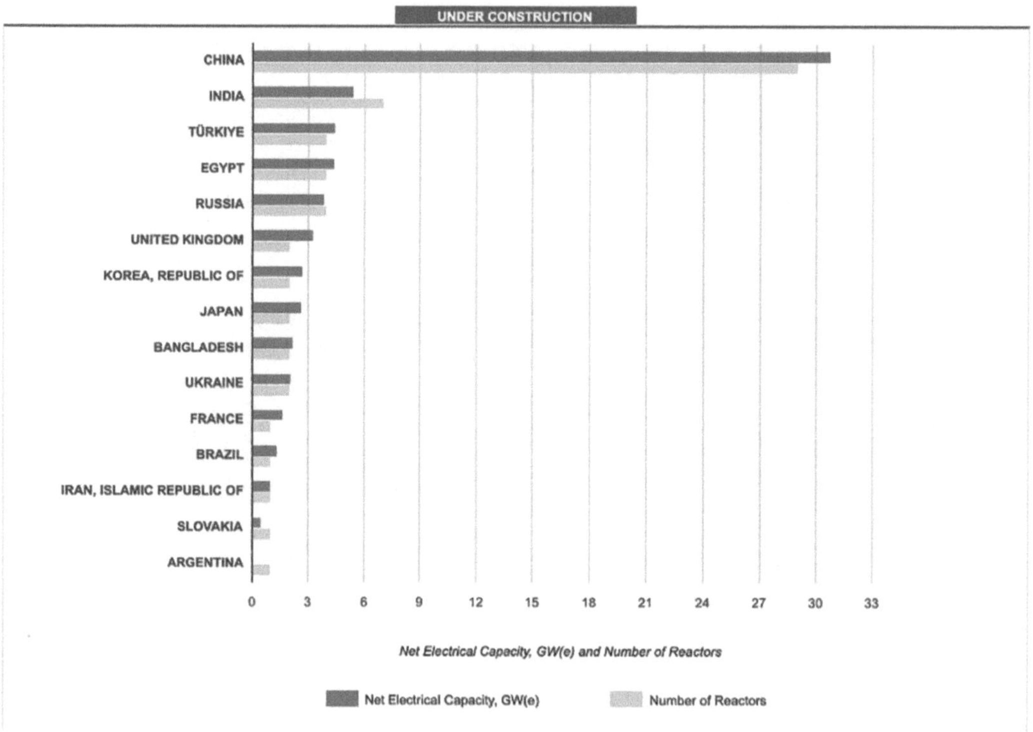

FIGURE 7.2
Number of reactors under construction, as of November 2024. (© IAEA [7].)

FIGURE 7.3
Watts Bar Nuclear Generating Site. (Republished with permission from the Tennessee Valley Authority.)

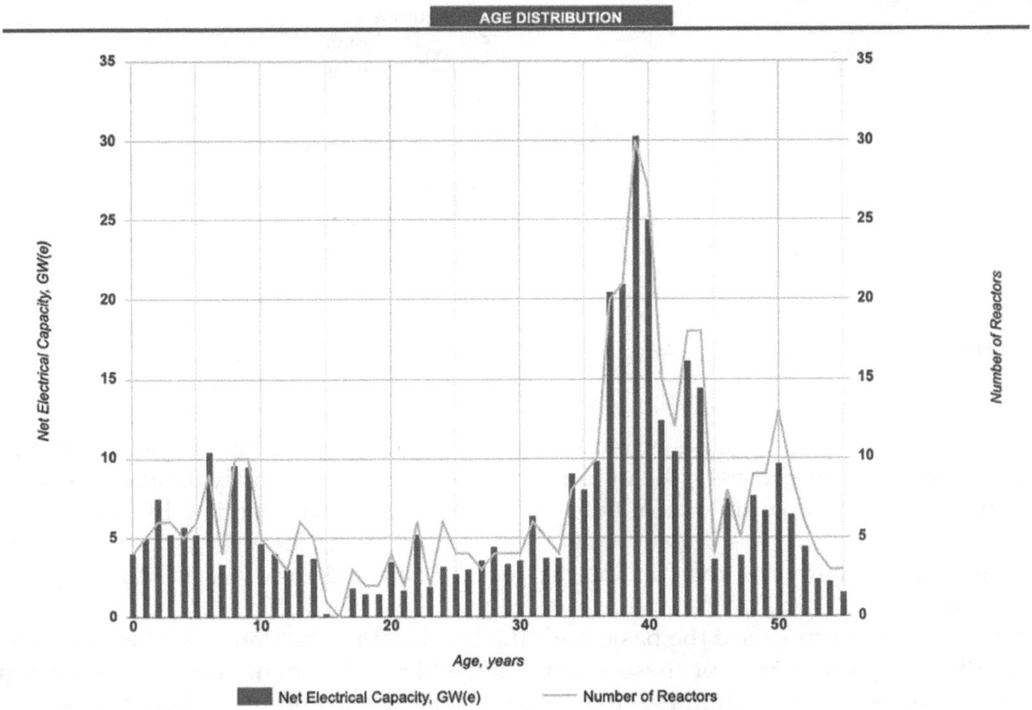

FIGURE 7.4
Number of nuclear reactors worldwide by age, as of November 2024. (© IAEA [11].)

7.2 Fission Mechanism

Whereas a conventional power station generates electricity by using the energy released from burning fossil fuels such as coal, a nuclear reactor converts the energy released by controlled nuclear fission into thermal energy for subsequent conversion into electric power. The fission mechanism involves absorption of a neutron by a large fissile atomic nucleus such as Uranium-235 or Plutonium-239. In the fission process, the heavy nucleus splits upon impact with a neutron into two or more lighter nuclei, called fission products, releasing kinetic energy, gamma rays, and free neutrons. Some of these neutrons may subsequently be absorbed by other fissile atoms and trigger additional fission events. In fission reactors, the release of these neutrons results in a chain reaction.

To sustain this chain reaction and prevent a dangerous runaway, it is controlled (in a conventional "thermal" reactor) by a combination of a moderator that slows the neutrons to increase their capture from one fuel rod to another and thus enhance fission, and control rods, which move in and out of the reactor core. The control rods are made of boron or other materials that absorb neutrons and thus reduce fission. In a nuclear reactor, safety is provided by automatic and manual systems to shut down the fission reaction when monitoring devices indicate a potentially unsafe condition. Commonly used moderators include regular or light water (used in three-quarters of the existing reactors), solid graphite, or heavy water. Water also serves as a coolant and heat transfer medium. Breeder reactors, which operate with fast neutrons and do not benefit from moderation, have used liquid metals as a coolant, but there are only two fast breeder reactors currently operating in the world, both in Russia, and they utilize liquid sodium as the coolant.

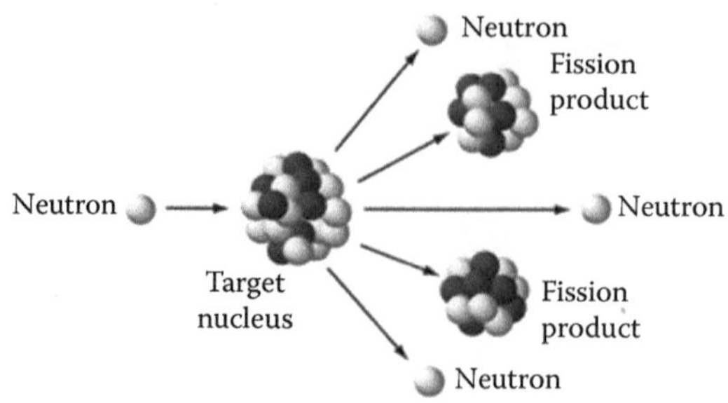

FIGURE 7.5
Typical fission reaction. (From Atomic Archive [12].)

The reason nuclear energy is attractive, from technical and economic perspectives, is because in a nuclear process, 1 kg of uranium releases approximately 3 million times as much energy as a kilogram of coal burned in conventional combustion. In other words, a nuclear reaction releases 7×10^{13} J/kg of ^{235}U compared to 2.4×10^{7} J/kg in the burning of coal. Furthermore, the fuel can be reprocessed and used for additional power production.

Protons and neutrons are the basic building blocks of nuclear matter. The sum of the individual masses of each of these subatomic particles that make up a given element will yield a greater mass than the mass of the nucleus. The energy difference is known as the nuclear binding energy, which is the energy necessary to maintain the position of the protons and neutrons. The energy released during fission is the source of nuclear power.

In the fission process, a nucleus splits into two fragments called the fission products, as shown in Figure 7.5 [12]. Each of these is about equal to half the original mass. In addition, two or more neutrons are emitted. The sum of the masses of the fission products is slightly less than the original mass, as a result of the release of the binding energy. This difference in mass (about 0.1% of the original mass) has been converted into energy in accordance with Einstein's famous equation, which states that energy and mass are equivalent:

$$E = mc^2 \tag{7.1}$$

where c is the speed of light.

When the emitted neutrons collide with neighboring nuclei, and induce additional fission, a self-sustaining series of nuclear fission reactions, known as a chain reaction, can occur.

The fission of ^{235}U releases two or three neutrons. If these neutrons are absorbed by other ^{235}U nuclei, these neutrons induce more fission reactions, and the rate of these reactions increases geometrically, as shown in Figure 7.6 [13].

An example of one of the fission processes that occurs is as follows [14]:

$$^{235}_{92}\mathrm{U} + ^{1}_{0}\mathrm{n} \rightarrow ^{90}_{36}\mathrm{Kr} + ^{144}_{56}\mathrm{Ba} + 2^{1}_{0}\mathrm{n} + 200\ \mathrm{MeV} \tag{7.2}$$

It has been found experimentally that some minimum mass of a fissile isotope, called the critical mass, is necessary to sustain such a chain reaction. The amount of mass required to be critical depends on the purity of the material and its shape.

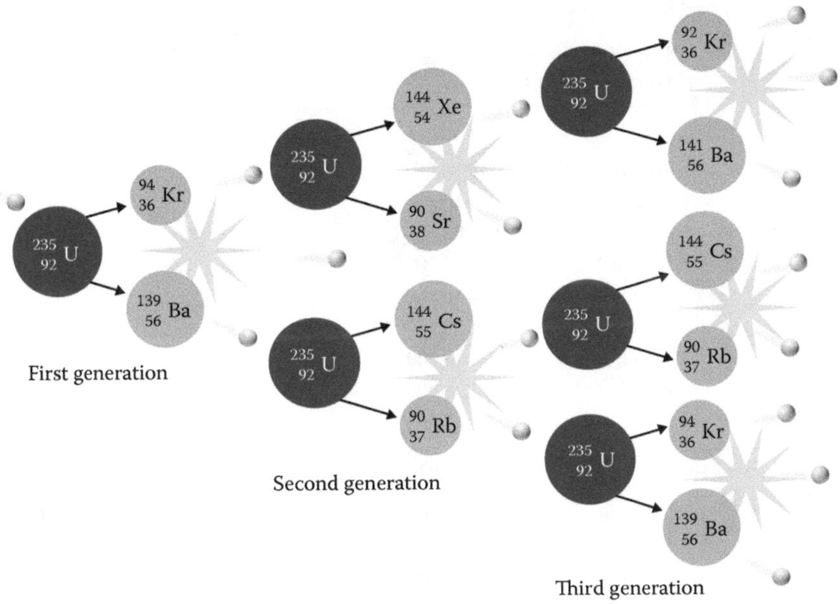

FIGURE 7.6
Schematic of a typical chain reaction. (From Shutterstock Image No. 73672591. [13].)

7.3 Available Nuclear Resources

7.3.1 Uranium Resources

Uranium is a naturally occurring element in the Earth's crust. Uranium mines operate in 20 countries, but over 97% of the world's production comes from mines in eight countries: Kazakhstan, Canada, Namibia, Australia, Uzbekistan, Russia, Niger, and China, with Kazakhstan accounting for 44% of the total [15].

In conventional mines, the ore is passed through a mill where it is crushed. In the next step, it is ground in water to produce a slurry of fine ore particles. The slurry is leached with sulfuric acid to dissolve the uranium oxide. This leaves the remaining rock and other material undissolved as mine tailings. Recently, however, a different method has been used in mining, called *in situ* leaching. In this process, the mining is accomplished without any major disturbance of the environment. First, groundwater with a lot of oxygen is injected into the ore. The water is then circulated through the mine, and uranium is dissolved. The solution with its dissolved uranium is then pumped to the surface for further processing. Both mining methods produce a liquid, which is filtered for the uranium and subsequently separated by means of ion exchange. The filtered and dried product is uranium oxide, which is then placed into drums and sealed. Since the concentrate has a bright yellow color, the product is usually called "yellow cake."

Available resources of uranium are divided into two groups: reasonably assured resources (RAR) and inferred resources (IR). RAR are the uranium in known deposits of sufficient size, grade, and configuration that it could be recovered within reasonable production costs with currently available mining and processing technology. IR are inferred

TABLE 7.1

Reasonably Assured Resources of Uranium (Plus Inferred Resources), 2021

	Tonnes U	Percentage of World
Australia	1,684,100	28%
Kazakhstan	815,200	13%
Canada	588,500	10%
Russia	480,900	8%
Namibia	470,100	8%
South Africa	320,900	5%
Niger	311,100	5%
Brazil	276,800	5%
China	223,900	4%
Mongolia	144,600	2%
Uzbekistan	131,300	2%
Ukraine	107,200	2%
Botswana	87,200	1%
United States	59,400	1%
Tanzania	58,200	1%
Jordan	52,500	1%
Other	266,600	5%
World total	6,078,500	

Source: From World Nuclear Association [15].

resources whose availability is uncertain. The categories are further divided into various cost classes according to estimated extraction costs. The division of these classes into below $50/kg, below $80/kg, and below $130/kg is widely accepted. According to some experts, RAR for less than $80/kg uranium are called proven resources. It is estimated that approximately 2.6 million tons of this type are available globally. When used in light-water reactors, this corresponds to an energy equivalent of 28 billion tons of hard coal. Table 7.1 shows the deposits of uranium and the various countries in which they are located [15]. If mining costs up to $130/kg are included, the global uranium sources are increased to 3.3 million tons. The total, irrespective of costs, may be as high as 15 million tons. The amount of energy that can be extracted from these uranium resources depends upon whether or not spent fuel from light-water reactors can be reprocessed. Reprocessing all of the spent fuels from light-water reactors and recycling the uranium and plutonium in mixed oxide (MOX) fuel to operate light-water reactors with 30% MOX and 70% uranium would lead to a cumulative savings of some 600,000 tons of natural uranium in the next 40 years. However, current law in the United States forbids reprocessing because it produces plutonium which could be used to make a nuclear bomb.

7.3.2 Plutonium

According to the World Nuclear Association, over one-third of the energy produced in nuclear power plants comes from plutonium. Plutonium is created as a by-product in a uranium type of nuclear reactor. Plutonium is recovered by reprocessing the spent uranium reactor fuel and is recycled as mixed oxide fuel usually called MOX. When a typical

1,000 MW_e nuclear power reactor is operating, it contains within its uranium fuel load several hundred kilograms of plutonium [16]. Plutonium is a heavy metal and alpha emitter. It is not particularly dangerous as an external radiation source because large alpha particles cannot penetrate skin. However, it is dangerous if inhaled as a dust or injected.

There are in practice two different kinds of plutonium: reactor-grade and weapons-grade. The first is recovered as a by-product of typical fuel from a nuclear reactor after the fuel has been irradiated or burned for about 3 years. Weapons-grade plutonium requires a specialized plutonium production reactor and cannot be produced in a commercial-type of electric power-producing uranium reactor. Plutonium is a valuable energy source when integrated into the nuclear fuel cycle. In a conventional nuclear reactor, 1 kg of the isotope ^{239}Pu can produce sufficient heat to generate nearly 8 million kWh of electricity.

Plutonium has a number of isotopes differing in the number of neutrons in the nucleus. All of these isotopes are fissionable with fast neutrons, but only two of them are fissile with slow neutrons: ^{239}Pu and ^{241}Pu. ^{239}Pu plays a major role in a conventional light-water power reactor. It is formed by neutron capture in ^{238}U followed by beta decay. It yields about the same amount of energy as the fission of ^{235}U. A 1,000 MW_e light-water reactor produces about 25 tons of used fuel per year containing as much as 290 kg of plutonium. It can be used as a direct substitute for ^{235}U with ^{239}Pu being the main fissile part and ^{241}Pu contributing a minor amount of energy. In order to extract plutonium for recycling, the used fuel is reprocessed. The recovered plutonium oxide is then mixed with uranium oxide to produce MOX that contains about 8% ^{239}Pu.

Although not significant in the world's electricity production, the decay heat of the isotope ^{238}Pu can be used as an electricity source. It produces about 0.56 W/g and is used in some cardiac pacemakers, space satellites, and navigation beacons. Plutonium used in Radioisotope Thermoelectric Generators (RTGs) has also powered some 30 U.S. space vehicles and enabled spacecraft to send back photos of distant planets [17].

7.4 Reactor Types

7.4.1 Pressurized-Water Reactors (PWRs)

The basic components of a PWR generating electricity are shown in Figure 7.7 [18]. A PWR consists of a reactor core contained within a pressure vessel and is cooled by light water under high pressure. The core of the plant contains uranium dioxide fuel pellets enclosed in zircaloy rods that are held together in fuel assemblies. Depending upon the reactor size, there are 200–300 rods in an assembly, and 100–200 fuel assemblies in the reactor core. The rods are arranged vertically and contain as much as 80–100 tons of enriched uranium. The rods are spaced apart in such a way that the water surrounding the fuel rods provides the optimum amount of moderation of neutrons to maximize capture cross-section and power output, i.e., the fuel rods are arranged in the most reactive configuration.

The working fluid in a PWR is pressurized water at 315°C circulated through the steam generators. The steam generator is a typical tubular heat exchanger with the hot water circulating inside the tube. The steam generator separates the reactor cooling water, which is radioactive, from the steam that is taken outside the nuclear reactor to a conventional turbine. This turbine generates electricity. The pressure vessel containing the reactor core and the steam generator is located inside the containment structure, which is designed to

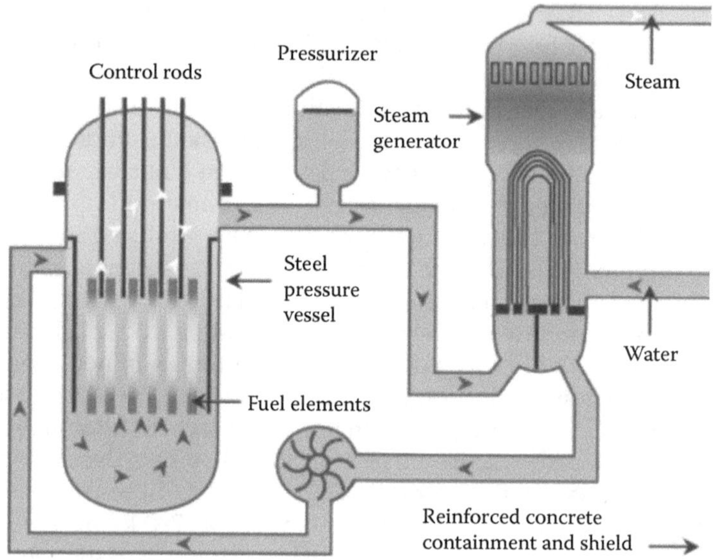

FIGURE 7.7
Schematic of a typical PWR power plant. (From World Nuclear Association [18].)

isolate the radioactive part of the system from the turbine and condenser. The part of the system outside the reactor is an ordinary Rankine cycle steam power plant isolated from the reactor containing all the radioactive elements in the cycle. The condenser cooling water is usually circulated to a cooling tower, where it is cooled by evaporation. In places where water is scarce, the condenser can also be cooled by air, but this type of design is less efficient. The PWR design is widely considered to be the safest type of nuclear power plant and constitutes the largest number of reactors used to generate electricity worldwide. Almost all of the nuclear power plants built since 2008 and under construction are of the PWR type, with sizes ranging up to 1,600 MW_e capacity.

7.4.2 Boiling Water Reactors (BWRs)

A boiling water reactor schematic, as shown in Figure 7.8, consists of a reactor core located inside a reactor vessel that is cooled by circulating light water [18]. The BWRs range in size from 400 to 1,200 MW_e. BWRs are the second largest reactor type used for generating electricity. In a BWR, the cooling water is heated to 285°C in the reactor vessel, and the resulting steam is sent directly to a turbine outside the pressurized container. There is no intermediate loop in a BWR, as there is in the PWR, and the system is therefore somewhat simpler than the PWR because it does not need an intermediate heat exchanger. The steam from the turbine is condensed in a heat exchanger using outside cooling water and then returned into the reactor vessel. The system operates also on a Rankine cycle with the heat source inside the steel pressure vessel, as shown. BWRs depend on careful control of the boiling process to avoid any pockets of steam forming inside the pressurized containers.

7.4.3 Heavy Water-Cooled and -Moderated Reactor

Heavy water-moderated reactors, called the CANDU reactors, were designed and built in Canada. An advantage of the CANDU reactor is that it uses less expensive unenriched

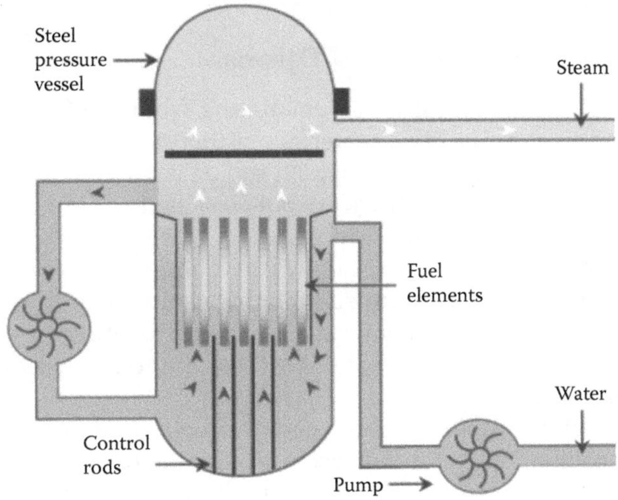

FIGURE 7.8
Schematic of a typical BWR power plant. (From World Nuclear Association [18].)

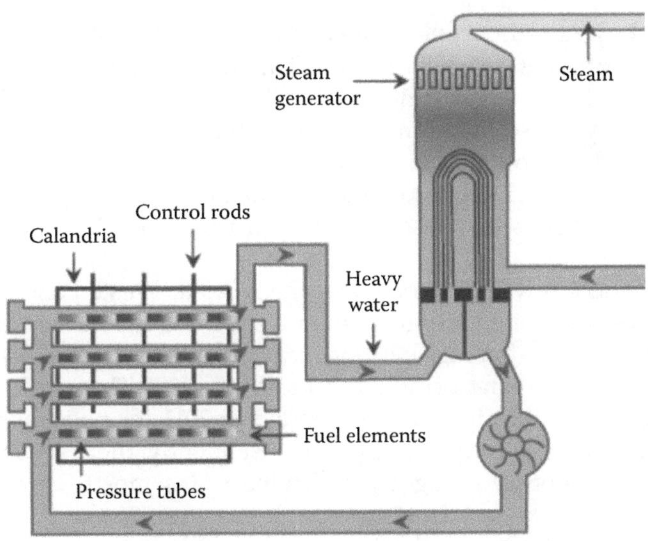

FIGURE 7.9
Schematic of a typical CANDU reactor power station. (From World Nuclear Association [18].)

uranium, which is held in zirconium alloy cans loaded into horizontal zirconium alloy tubes. Heavy water serves as the moderator and extracts heat from the fuel as it is pumped through the tubes under high pressure to prevent boiling. The water is then passed through a steam generator where it produces steam from ordinary light water. This design requires additional moderation, which is achieved by immersing the tubes in an unpressurized container called a calandria, a device containing more heavy water. By inserting or withdrawing cadmium rods from the calandria, the nuclear reaction is controlled. The entire assembly is placed inside a concrete shield and containment vessel, as shown in Figure 7.9 [18].

7.5 Nuclear Waste Management and Disposal

There are four types of radioactive waste emanating from the fuel cycle. The first type is the naturally occurring radioactive materials in the uranium ore during mining. The decay of these materials produces radon, which can be contained by covering the tailings with clay. Technically, these mine tailings are not classified as nuclear waste.

Low-level waste (LLW) is generated by the nuclear fuel cycle as well as by medical and industrial uses. The radioactivity from such wastes comes from affected paper, clothing, and soils. They are short-lived and are suitable for shallow-land burial. The volume of these materials can be reduced by compacting and incinerating in specialized facilities prior to their disposal. LLWs make up about 90% of the total nuclear waste but contain only about 1% of its radioactivity.

Intermediate-level wastes (ILW) are generated during the operation of reactors as well as in the reprocessing of spent fuel and the decommissioning of nuclear power plants. These materials contain higher amounts of radioactivity and are generally made up of resins, chemical sludges, fuel cladding, and contaminated material from decommissioning. Some of these wastes are processed before disposal by solidifying them in concrete. ILW make up about 7% of the volume and have about 4% of the radioactivity of the total radioactive wastes.

The fourth and largest radioactive waste component is high-level waste (HLW) generated during the operation of a nuclear reactor. This waste consists of fission products and transuranic elements generated during fission. This material is not only highly radioactive but also thermally hot so that it must be both shielded and cooled. It contains 95% of the radioactivity produced by nuclear power generation. HLW will eventually have to be disposed in deep geological facilities to allow its decomposition safely over time. Several countries have selected sites for these facilities. At this time, the United States does not have an HLW depository.

In 1987, Yucca Mountain was designated by the Nuclear Waste Policy Act Amendments to be the long-term storage site for high-level radioactive waste in the United States. Yucca Mountain is a deep geological storage location located on federal land adjacent to the Nevada Nuclear Test Site, about 130 km northwest of Las Vegas [19]. The project was approved in 2002 by the U.S. Congress, but funding was ended in 2011 for political reasons according to the Government Accountability Office [20]. Consequently, there are many thousands of tons of spent fuel in temporary storage in the United States, mostly at reactor sites.

In countries where spent fuel is reprocessed, the uranium and plutonium are first removed during reprocessing, and then the much smaller volume of remaining HLW is solidified using a vitrification process. In this process, the fission products are mixed in glass materials, vitrified in stainless steel canisters, and finally stored in shielded facilities for later permanent disposal. The disposal process is still one of the unresolved issues regarding the wide-scale use of nuclear power. Information about current waste management processes adopted by countries throughout the world can be found in [21].

Radioactive wastes are produced during any reactor fuel cycle. The cost of managing these wastes must be included in the economics and are a part of the electricity costs. Radioactive materials decay with time, and each radioactive isotope has a half-life, which is the time it takes for half of the material to decay away. Eventually, these materials decay to a stable, nonradioactive form, but this may take a very long time. Hence, the process of managing radioactive waste requires the protection of people from the harmful effects of radiation. The long-lived radioactive materials emit alpha and beta particles. It is relatively easy to shield humans from this type of radiation, but ingesting alpha or beta radiation can be harmful.

7.6 Spent Fuel Storage and Reprocessing

Spent fuel is routinely discharged from operating nuclear reactors. As it is discharged, it is moved to specially designed storage pools of ordinary water, which are an integral part of a nuclear power plant. These storage pools hold fuel from many years of operation, and the heat produced from the spent fuel is removed by circulating cooling water through the pool. The spent fuel capacity of many older reactors is reaching its capacity of on-site storage. At this point, the fuel must be transferred to dry storage, which takes place in large metal or concrete storage facilities. These dry storage facilities can be passively cooled by circulating air around them, but eventually, a final storage for radioactive waste is required.

Reprocessing is the recovery of plutonium from the spent fuel of a nuclear reactor. It extends the amount of energy retrievable from the original uranium by approximately 30%–40%. The process currently used for reprocessing is called PUREX. To reprocess the spent fuel from light-water reactors, the spent fuel is first dissolved in nitric acid, and the acid solution is then mixed with an organic solvent consisting of tributyl phosphate in a concrete container. The uranium and plutonium are extracted in this organic phase, and the fission products remain in the aqueous phase. Additional processing permits the separation of uranium and plutonium. PUREX has been used successfully for many years by all the countries that reprocess spent fuel.

7.7 Nuclear Power Plant Accidents

The first serious commercial nuclear reactor accident occurred at Three Mile Island Unit 2, near Middletown, Pennsylvania, in 1979. The accident was the worst in the history of the U.S. commercial power industry. Mechanical problems and human error caused loss of coolant water, which led to a partial nuclear core meltdown and the release of radioactive gases. Orders for new nuclear power plants were canceled after the accident, and public sentiment shifted to concern about safety and issues of nuclear waste disposal.

On April 26, 1986, a catastrophic accident at the Chernobyl Nuclear Power Plant, Ukraine, involved a massive steam explosion and fire that released large amounts of radioactive contamination across the USSR and Europe. Three hundred fifty thousand people were evacuated and permanently resettled from the most highly contaminated areas of Belarus and Ukraine. The accident killed 31 plant workers and caused a further 64 deaths from radiation, primarily affecting emergency and plant workers. Over 300,000 people were relocated, and an estimated 400,000 children developed thyroid cancers [22]. The Chernobyl disaster was the most devastating nuclear accident in terms of released radioactivity, about an order of magnitude worse than Fukushima.

On March 11, 2011, a massive 9.0 magnitude earthquake struck 109 miles (175 km) from Fukushima, Japan, in the Pacific Ocean. The tsunami that followed breached the safety wall that had been built to protect the nuclear reactors at the Fukushima Daiichi Nuclear Power Plant and flooded the pumps necessary for cooling the reactor cores. The 4.7 GW_e Fukushima Daiichi Nuclear Plant had six reactor units designed by General Electric [23]. All of the reactors came online between 1970 and 1979, which made them all more than 30 years old. The reactors were of the boiling water type. The number one reactor was slated to be shut down in 2011 but was granted a 10-year extension by Japanese regulators in 2009, although risks associated with the reactor had been previously identified.

Three of the units at Fukushima Daiichi Nuclear Power Plant had been shut down prior to the earthquake for maintenance, and the other three reactors shut down automatically after the earthquake, with the remaining decay heat of the fuel rods being cooled using power supplied by emergency generators. The 5.7 m tsunami wall was breached by waves reaching heights of 14 m. The tsunami not only destroyed the primary cooling system but also disabled the emergency diesel generators that are required to provide power to the pumps used to cool the reactor core in case of primary system failure. In the following 3 weeks, there was evidence of partial nuclear meltdowns in Units 1, 2, and 3, visible explosions believed to be caused by hydrogen gas in Units 1 and 3, and a suspected explosion in Unit 2 that may have damaged the primary containment vessel. Also, due to a loss of water in the pools used to store the spent fuel rods for three of the reactors, some of the rods were uncovered and heated beyond the safe limit.

Although the events leading up to the nuclear disaster at Fukushima were not the direct result of shortcomings in the nuclear plant operations, they did highlight the vulnerability of nuclear power plants to natural disasters. In response to concerns in the United States about the safety of its own nuclear power plants, the NRC prepared a report outlining the changes in safety precautions recommended for nuclear power plants [23]. The NRC recommended that nuclear facility operating licenses should be denied unless operators can demonstrate the ability to deal with a complete loss of power for 8 hours using backup generators, and be able to provide cooling to the radioactive core and spent fuel rods in a water pool for at least 72 hours. The pump failures at Fukushima also led to tightly packed spent fuel rods stored in cooling water ponds catching fire, believed to be the result of oxidation of the fuel rod zirconium cladding, which produced hydrogen. Spent fuel rods generate heat for many years after they are removed from the reactor. This is an important concern for overall safety, as the rods in Japan were even less tightly packed than the fuel rods in cooling ponds in the United States. As a precaution to a cooling water accident in the United States, the recommendations from the NRC include requirements for dry storage of the fuel rods after no more than 5 years in a cooling pond. All of these measures will make nuclear reactors safer but also increase the cost of nuclear power in the United States.

7.8 Current Status and Cost of Nuclear Technology

The ability of nuclear reactors to produce electricity without generating greenhouse gases has revitalized interest in nuclear power. The fact that nuclear power plants operate at high capacity factors, 24 h/day make them potentially attractive for meeting the high electricity demands of data centers. Many lessons have been learned from the operation of current power plants that have improved the safety of new designs, and there is continuing research and development for a new generation of nuclear reactors.

Capital cost estimates by Lazard for new U.S. nuclear plants [24] range from $8,765 to $14,400/kW. (Recall that Vogtle Units 3 and 4 cost approximately $16,000/kW.) The levelized cost of electricity from new nuclear plants is estimated to be in the range of $142–222/MWh. These costs are on the order of three times the current costs of onshore wind and utility-scale solar photovoltaic electricity, which are still dropping. The most recent Lazard LCOE reports have attempted to add in the costs of storage needed to firm the costs of wind and solar power in various regional transmission organizations. Even when that is done, the LCOE of nuclear is on the order of 50% higher than for solar and wind [24].

7.9 Next-Generation Nuclear Technologies

The ability of nuclear power to provide reliable low-carbon electricity has spurred an interest in developing new reactor designs. The development of nuclear power has progressed through four general phases. The initial development of prototype reactor designs (Generation I) occurred in the 1950s and 1960s. Development and deployment of large commercial plants (Generation II) took place in the 1970s and 1980s, when over 100 nuclear power plants were built and deployed in the United States. The development of advanced light-water reactors (Generation III) is an ongoing effort to overcome the high costs of the large (over 1,000 MW) nuclear reactors in operation. The fourth phase of nuclear reactor development, known as Generation IV technologies, consists of a number of ideas proposed for more economical and safer nuclear power designs.

Generation II nuclear reactors have effectively demonstrated the viability of nuclear power all over the world. But future development and deployment in the United States faces several challenges that must be overcome for nuclear power to grow significantly. These challenges include (1) public concerns about safety, (2) high capital cost and complications associated with the construction of new plants, (3) the long time required for construction and licensing, (4) concerns about disposal of nuclear waste, and (5) potential for terrorists acquiring material for nuclear weapons. However, since nuclear power plants do not emit pollutants that cause global climate change or local air pollution, their increased use in the future is considered by many to be an important part of a low-carbon energy strategy.

Generation III's key objective is to use modern manufacturing technology with advanced safety systems to develop improved light-water reactors. Some designs are smaller than 300 MW in size, which are classified as small modular reactors, or SMRs. The reasons offered for developing SMRs include the following. (1) Smaller reactors can be produced in a factory and moved as a single unit to be installed. This would shorten the time of construction and potentially reduce the cost. (2) If larger power stations were needed in the future, they could be made up of multiple small units in accordance with demand. Current large-scale units often produce more electricity than needed nearby and require long transmission lines. (3) Smaller reactors can serve remote areas and thus reduce transmission costs. (4) Smaller reactors can be operated with passive safety systems, which would avoid the possibility of the kind of accidents that concern the public, for example, the accident that occurred at Three Mile Island.

According to IAEA there are more than 80 small modular reactor (SMR) designs being developed around the world [25]. Many utilize Generation IV advanced reactor technologies (see below). The thermodynamic power cycle of the SMR plants is essentially the same as that described previously in this chapter for large-scale light-water reactors. However, many of the detailed features of the SMRs are proprietary. There are two floating 35 MW_e SMRs that began commercial operation in Russia in May 2020. These have reportedly had capital costs of $10,500/kW [26]. The CAREM-25 25 MW_e net SMR in Argentina began construction in February 2014 [27]. It is reported to be 85% complete, and the latest target year for completion was 2028, but it has been delayed again due to a lack of sufficient government funds [28]. Its estimated capital cost was $10,500/kW [26], but the current delay will likely increase that cost.

The U.S. NRC shows on its website that it is involved in only one SMR licensing review and it is for the NuScale design consisting of multiple 77 MW_e modules [29]. It is an integral pressurized light-water reactor based on the MASLWR (Multi-Application Small Light-Water Reactor) design developed at Oregon State University in the early 2000s. It is a natural circulation reactor with the core and helical coil steam generator located in a common reactor vessel placed in a cylindrical steel container. The reactor vessel/containment module is submerged in water in the reactor building safety pool. Thus, in case of an accident, the safety cooling system operates by natural convection without the need of electric power. The entire reactor building is located below ground with the reactor building designed to hold up to 12 SMR modules.

The First Commercial six-module NuScale project was planned to be built on the site of Idaho National Laboratory. It would be owned by the Utah Associated Municipal Power System and operated by Energy Northwest. It was canceled in November 2023, however, due to cost overruns [30]. The project had a target of delivering 40 years of electricity at a cost of $55/MWh, but the cost estimate had risen to $89/MWh after accounting for a $30/MWh benefit from the Inflation Reduction Act [30]. Because NuScale was the only SMR design with NRC approval, the collapse of this project was a major blow to the industry. However, NuScale is developing other potential projects including a power plant in Romania [31]. Westinghouse is also marketing its AP300 SMR concept, which is a scaled-down, 300 MW_e version of its commercial AP1000 reactor [32].

Recently, momentum for nuclear generation has picked up again due to growing data center needs for electricity as well as commitments to net zero made by technology companies. The planned restart of Three Mile Island Unit 1 by Microsoft [33], power purchase agreements by Amazon for traditional and small modular reactors [34], and Google's agreement with Kairos Power LLC [35] build on declarations made by 25 countries at COP28 to triple nuclear capacity by 2050 to achieve global climate goals.

By virtue of their smaller size, SMRs lower the total project cost of nuclear. To compete in the electricity market however, a lower LCOE is needed. An SMR could conceivably lower the cost per megawatt by virtue of achieving factory production of a standardized design. But some market success would first be needed to justify mass production. A challenge for SMRs is that they defy economy of scale principles.

Every SMR requires a power block to convert the heat to electricity, and power blocks are well studied and modeled. In estimating the costs of power block components (boilers, condensers, pumps, piping, heat exchangers, turbine) the six-tenths rule is commonly used. This says that the cost of a component varies as the size of the component raised to the six-tenths power (the exponent varies with component type, but 0.6 is an average value) [36]. Thus, for example, if the size of the component is reduced to one-quarter the original size, the cost is reduced not to 25% of the original cost but to only $(0.25)^{0.6}$ or 44% of the original cost. The various reactor components, like the pressure vessel and pumps, would also follow a power law cost model. One would thus expect the capital cost (in \$/MW) to be higher for a 300-MW SMR than for a 1,100-MW full-size reactor. As we have seen, conventional 1,100 MW reactors suffer from cost overruns and construction delays. Both of these are likely to be worse, at least at first, when trying to construct a new reactor design.

In addition to the expected higher capital cost per MW (and higher LCOE) for SMRs compared to large conventional light-water reactors (LWRs), one study concluded that SMRs will have fuel costs that are 15%–70% higher [37]. Recently older conventional reactors have shut down because their operating costs were too high to compete in today's electricity markets. So higher fuel costs could exacerbate this problem for SMRs.

The Generation IV International Forum (GIF) is an international effort to share R&D of Generation IV reactors. The GIF was initiated by the U.S. Department of Energy and chartered in 2001, with participation from 13 members: the United States, Argentina, Brazil, Canada, China, France, Japan, Russia, South Korea, South Africa, Switzerland, and the United Kingdom, along with the EU (Euratom) [38]. The goal of GIF is to identify the most promising reactor concepts, referred to as the Generation IV technology road map. This effort was further supported by the U.S. Department of Energy, which announced \$82 million funding for commercializing advanced reactors in June 2016.

Approximately 100 Generation IV candidate technologies were initially evaluated by GIF, and seven reactor systems are now undergoing further evaluation and development (there are six general reactor designs but two types of molten salt reactors). These are listed in Figure 7.10 [38]. More details about Generation IV reactor concepts can be found in [38].

The Generation IV designs called fast reactors are breeders. A breeder can produce enough fuel for its own operation and generate additional fissile material for other reactors. This is achieved primarily by the conversion of the non-fissionable ^{238}U isotope into fissionable plutonium. This process makes much more efficient use of the energy contents of the natural uranium as mined. Current reactors "burn" only ^{235}U, which as mentioned previously, constitutes only 0.7% of natural uranium ore. The main drawback of the breeder reactor is that it requires fast neutrons and therefore, ordinary water, which acts to slow down the neutrons in a reactor, cannot be used as the reactor coolant.

The basic reaction through which a breeder converts some of the normally non-fissionable isotope of ^{238}U into the fissionable isotope ^{239}P occurs in the following steps:

$$^{238}_{92}U + ^{1}_{0}n \rightarrow ^{239}_{92}U \xrightarrow{\beta} ^{239}_{93}Np \xrightarrow{\beta} ^{239}_{94}Pu \tag{7.3}$$

The neutrons captured by the ^{238}U form ^{239}U, which has a half-life of 23 minutes. The ^{239}U decays into Neptunium-239 by beta decay with a half-life of 2.4 days. The ^{239}Np decays into ^{239}Pu with a half-life of 24,400 years.

The above reaction also occurs in thermal reactors, but is very inefficient due to their design. A typical 1,000 MW_e nuclear reactor produces about 200 kg of plutonium per year. This is insufficient to replace the fissile material destroyed, whereas breeders are constructed to maximize the production of plutonium. Breeders use high-energy neutrons, so-called "fast neutrons," because these fast neutrons have a better chance of being captured by ordinary ^{238}U than by the isotope ^{235}U. When fission is produced with fast neutrons it creates, on average, 2.9 neutrons compared to 2.4 produced per fission event.

Generation IV reactor designs under development by GIF

	Neutron spectrum (fast/thermal)	Coolant	Temperature (°C)	Pressure*	Fuel	Fuel cycle	Size (MWe)	Use
Gas-cooled fast reactors	fast	helium	850	high	U-238 +	closed, on site	1200	electricity & hydrogen
Lead-cooled fast reactors	fast	lead or Pb-Bi	480-570	low	U-238 +	closed, regional	20-180** 300-1200 600-1000	electricity & hydrogen
Molten salt fast reactors	fast	fluoride salts	700-800	low	UF in salt	closed	1000	electricity & hydrogen
Molten salt reactor - **advanced high-temperature reactors**	thermal	fluoride salts	750-1000		UO_2 particles in prism	open	1000-1500	hydrogen
Sodium-cooled fast reactors	fast	sodium	500-550	low	U-238 & MOX	closed	50-150 600-1500	electricity
Supercritical water-cooled reactors	thermal or fast	water	510-625	very high	UO_2	open (thermal) closed (fast)	300-700 1000-1500	electricity
Very high temperature gas reactors	thermal	helium	900-1000	high	UO_2 prism or pebbles	open	250-300	hydrogen & electricity

* high = 7-15 MPa
\+ = with some U-235 or Pu-239
** 'battery' model with long cassette core life (15-20 yr) or replaceable reactor module.

FIGURE 7.10
Generation IV reactor designs under development. (From World Nuclear Association [38].)

The breeder design can use 60% of the energy content in natural uranium, compared to 1% or 2% in the light-water reactor. Thus, the breeder has a much higher energy return on energy investment than current LWR designs.

Breeder reactors commercialized in France in the 1970s used liquid sodium as the reactor coolant, and it remains the preferred coolant choice. Sodium slows the fission neutrons less than ordinary water or gas. The disadvantage of sodium is that the metal reacts violently with water and can burn spontaneously in air at the temperature of the breeder. Thus, special efforts must be made to prevent leakage of sodium from the cooling system.

One Gen IV reactor that has received considerable press coverage is the TerraPower Natrium reactor for which Bill Gates has provided considerable funding. This is a 345-MW_e liquid sodium-cooled fast reactor with a demonstration project scheduled for construction to begin in Kemmerer, Wyoming in 2025 and the goal of providing power to PacifiCorp's electric grid in 2030. A schematic of this type of reactor is shown in Figure 7.11 [39]. The project will include molten salt thermal energy storage with the capacity to provide 500 MW of power for a period of more than five and a half hours when electricity prices are highest [40].

In summary, nuclear engineers have proposed many different new concepts for nuclear power plants, and many countries have built and are experimenting with small versions of these ideas. However, none of them has yet reached the point of commercialization.

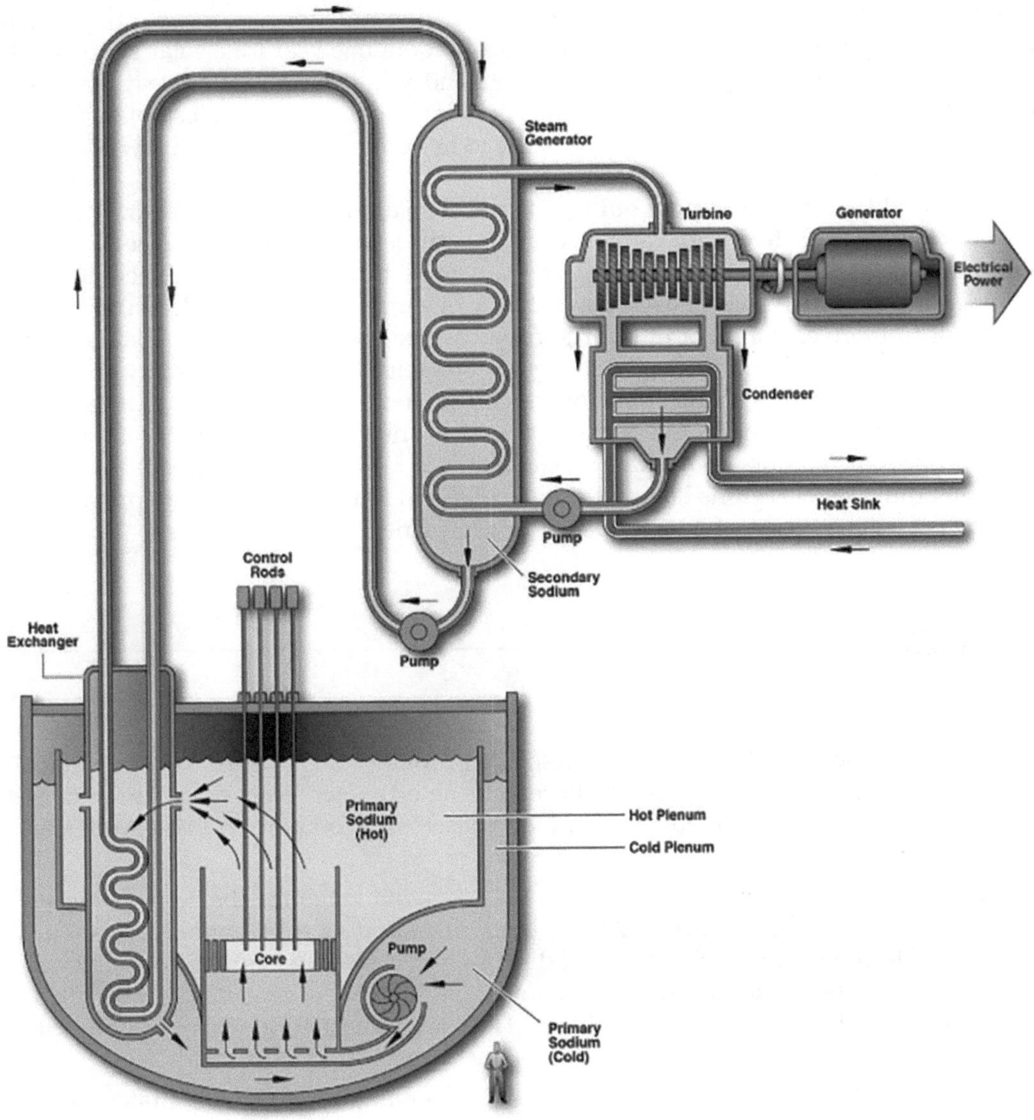

FIGURE 7.11
Sodium-cooled fast reactor. (From DOE [39].)

7.10 The Future of Nuclear Power

Nuclear power has already played an important role in reducing carbon emissions by providing reliable, low-carbon electricity at high capacity factors. The key challenge that nuclear power faces going forward is the competition from very low-cost solar, wind, and batteries. As we have seen, nuclear power plants have very high capital costs. They also have long construction times. The average construction time for new nuclear plants is on the order of eight years compared to an average of about one year for a solar PV plant and six months for a wind farm. Investing in solar and wind instead of nuclear could save a large amount of carbon emissions during the eight years that a nuclear plant is being constructed. And once a nuclear plant is ready to be connected to the grid it will compete against still lower renewable energy costs.

Also, a grid that is increasingly built on variable solar and wind needs varying power to firm it. Because of their high capital cost, U.S. utilities are reluctant to throttle nuclear power plants and reduce their revenue stream (and throttling also reduces efficiency and increases maintenance costs). The addition of a new baseload nuclear power plant can result in more curtailment of solar and wind. The addition of thermal storage, as is being done with the Natrium reactor, addresses this issue but adds complexity and cost. It remains unclear whether SMRs or any of the Generation III or IV concepts will reduce the capital cost per MW or the LCOE of nuclear power. The growth of 24 h/day data centers might prove to be an opportunity for new reactors or for the extended operation of older reactors. Countries or regions having limited solar and wind resources may prove to provide the best opportunities for new nuclear power plants.

Discussion Questions

1. In the United States, the Price-Anderson Act provides taxpayer-funded insurance in the event of nuclear accidents. Research the estimated costs of the three most recent major nuclear accidents. How do these costs compare to Price-Anderson coverage?
2. Long-term storage of nuclear waste remains a challenge for the nuclear industry. Yucca Mountain in Nevada has been proposed as a potential storage site. Research the history of this site and discuss why it has not been utilized. Do you believe it offers a good long-term solution? What other measures can be taken to address the nuclear waste problem? How are other nations dealing with this issue?
3. There are 94 nuclear reactor units currently operating in the United States. Some environmentalists want them shut down. Determine how much carbon dioxide emissions these plants save and how much renewable energy (assume half solar photovoltaics and half wind power) would need to be deployed to achieve an equivalent amount of emissions savings and how much this would cost.
4. New nuclear power plants in the United States have proven to be expensive. Research the most recent cost of a new plant in the United States as well as new plants in China and India. Explain why these costs differ and what might be done to reduce U.S. plant costs.
5. The proliferation of nuclear weapons material has been a long-standing concern with nuclear power. Discuss how some new reactor designs address this issue.
6. Research and describe new passive reactor safety mechanisms.

Acknowledgment

The lead author of this chapter (Frank Kreith) thanks his colleague Ken Kok for providing significant information about many aspects of nuclear energy in his handbook *Nuclear Engineering Handbook, Second Edition*. CRC Press, 2017.

References

1. DOE, The Manhattan Project, An interactive history, https://www.osti.gov/opennet/manhattan-project-history. Last accessed January 2025.
2. Kok, K.D., 2017. *Nuclear Engineering Handbook*, 2nd ed., CRC Press, Boca Raton, FL.
3. Statista, 2023, Leading nuclear power reactors in operation worldwide as of December 2023, but gross capacity, https://www.statista.com/statistics/273111/largest-nuclear-power-reactors-worldwide-by-capacity/.
4. ABC, August 12, 2016. Japan reactor restarts, despite protests, boosting Tokyo's nuclear push, *ABC News*, https://www.abc.net.au/news/2016-08-12/japan-reactor-restarts-in-post-fukushima-nuclear-push/7729892.
5. IAEA Operational reactors by country, Vienna, Austria, https://www.iaea.org/PRIS/WorldStatistics/OperationalReactorsByCountry.aspx. , Last accessed January 2025.
6. World Nuclear Association, Nuclear power in the world today, https://world-nuclear.org/information-library/current-and-future-generation/nuclear-power-in-the-world-today. Last accessed January 2025.
7. IAEA, Under construction reactors by country, Vienna, Austria, https://www.iaea.org/PRIS/WorldStatistics/UnderConstructionReactorsByCountry.aspx. Last accessed January 2025.
8. EIA, 2016. Today in energy, https://www.eia.gov/todayinenergy/detail.php?id=26652.
9. Crees, A., December 4, 2018, The failed V.C. Summer nuclear project: A timeline, https://www.chooseenergy.com/news/article/failed-v-c-summer-nuclear-project-timeline/#:~:text=Summer%20Nuclear%20Generating%20Station%20owners,ultimately%20abandoned%20in%20July%202017.
10. Energy & Climate, Press & Media, May 29, 2024. https://www.gcvoters.org/blog/2024/05/29/report-new-nuclear-reactors-to-cost-georgia-ratepayers-extra-420-annually-on-average/.
11. IAEA, Power reactor information system, Age distribution, Vienna, Austria, https://pris.iaea.org/PRIS/WorldStatistics/OperationalByAge.aspx. Last accessed January 2025.
12. Atomic Archive, Nuclear fission: Basics, https://www.atomicarchive.com/science/fission/index.html. Last accessed January 2025
13. LibreTexts, https://chem.libretexts.org/Bookshelves/Physical_and_Theoretical_Chemistry_Textbook_Maps/Supplemental_Modules_(Physical_and_Theoretical_Chemistry)/Nuclear_Chemistry/Fission_and_Fusion/Fission_and_Fusion, Last accessed January 2025
14. World Nuclear Association, January 10, 2025. Physics of uranium and nuclear energy, https://world-nuclear.org/information-library/nuclear-fuel-cycle/introduction/physics-of-nuclear-energy
15. World Nuclear Association, May 16, 2024, World Uranium Mining Production, https://world-nuclear.org/information-library/nuclear-fuel-cycle/mining-of-uranium/world-uranium-mining-production.

16. World Nuclear Association, August 16, 2023, Plutonium, https://world-nuclear.org/information-library/nuclear-fuel-cycle/fuel-recycling/plutonium.
17. World Nuclear Association, March 5, 2025, Nuclear reactors and radioisotopes for space, https://world-nuclear.org/information-library/non-power-nuclear-applications/transport/nuclear-reactors-for-space.
18. World Nuclear Association, March 21, 2025, Nuclear power reactors, https://world-nuclear.org/information-library/nuclear-fuel-cycle/nuclear-power-reactors/nuclear-power-reactors.
19. DOE, March 2004. Nuclear waste policy act as amended, https://www.energy.gov/downloads/nuclear-waste-policy-act.
20. Northey, H., May 10, 2011. GAO: Death of Yucca Mountain caused by political maneuvering, New York Times, https://archive.nytimes.com/www.nytimes.com/gwire/2011/05/10/10greenwire-gao-death-of-yucca-mountain-caused-by-politica-36298.html?pagewanted=all.
21. The World Nuclear Waste Report, 2019. https://worldnuclearwastereport.org.
22. IAEA, 2006. The Chernobyl Forum 2003–2005, Chernobyl's legacy: Health, environmental, and socio-economic impacts and recommendations to the governments of Belarus, the Russian Federation and Ukraine, Austria, https://www.iaea.org/sites/default/files/chernobyl.pdf.
23. U.S. Nuclear Regulatory Commission, July 12, 2011. Recommendations for enhancing reactor safety in the 21st century: The near-term task force review of insights from the Fukushima Dai-Ichi accident, https://www.nrc.gov/docs/ML1118/ML111861807.pdf.
24. Lazard, June 2024. Levelized cost of energy analysis, V.17.0, https://www.lazard.com/media/xemfey0k/lazards-lcoeplus-june-2024-_vf.pdf.
25. International Atomic Energy Agency (IAEA), 2022. Advances in small modular reactor technology developments, https://aris.iaea.org/Publications/SMR_booklet_2022.pdf.
26. Makhijani, A., and Ramana, M.V., March 25, 2021. Why small modular nuclear reactors won't help counter the climate crisis, Environmental Working Group, https://www.ewg.org/news-insights/news/why-small-modular-nuclear-reactors-wont-help-counter-climate-crisis.
27. World Nuclear Association, https://world-nuclear.org/nuclear-reactor-database/details/carem25.
28. Nuclear Engineering International, September 17, 2024, Argentina's CAREM-25 SMR faces setbacks, https://www.neimagazine.com/news/argentinas-carem-25-smr-faces-setbacks/.
29. U.S. NRC, Design Certification – NuScale US600, https://www.nrc.gov/reactors/new-reactors/advanced/who-were-working-with/past-license-activities/nuscale.html. Last accessed January 2025.
30. Bright, Z., November 9, 2023. NuScale cancels first-of-a-kind nuclear project as costs surge, *E&E News*, https://www.eenews.net/articles/nuscale-cancels-first-of-a-kind-nuclear-project-as-costs-surge/.
31. Skidmore, Z., November 12, 2024, NuScale bullish on SMR nuclear prospects despite regulatory setbacks, Data Center Dynamics, https://www.datacenterdynamics.com/en/news/nuscale-bullish-on-smr-nuclear-prospects-despite-regulatory-setbacks/.
32. Clarion Energy Content Directors, February 8, 2024. Westinghouse signs agreement to deploy AP300 SMR fleet in the U.K, *Power Engineering*, https://www.power-eng.com/nuclear/new-projects-nuclear/westinghouse-signs-agreement-to-deploy-ap300-smr-fleet-in-the-u-k/.
33. Constellation Energy, September 20, 2024. Constellation to launch clean energy center, restoring jobs and carbon-free power to the grid, https://www.constellationenergy.com/newsroom/2024/Constellation-to-Launch-Crane-Clean-Energy-Center-Restoring-Jobs-and-Carbon-Free-Power-to-The-Grid.html.
34. Amazon, October 16, 2024. Amazon signs agreements for innovative nuclear energy projects to address growing energy demands, https://www.aboutamazon.com/news/sustainability/amazon-nuclear-small-modular-reactor-net-carbon-zero.
35. Terrell, M. Google, October 14, 2024. New nuclear clean energy agreement with Kairos Power, https://blog.google/outreach-initiatives/sustainability/google-kairos-power-nuclear-energy-agreement/.

36. Whitesides, R., 2012. Process equipment cost estimating by ratio and proportion, https://www.pdhonline.com/courses/g127/g127content.pdf.
37. Pannier, C., and Skoda, R., 2014. Comparison of small modular reactor and large nuclear reactor fuel cost, *Energy and Power Engineering* 6(5), 82–94, https://www.scirp.org/html/2-6201621_45669.htm.
38. World Nuclear Association, April 30, 2024, Generation IV Nuclear Reactors, https://world-nuclear.org/information-library/nuclear-fuel-cycle/nuclear-power-reactors/generation-iv-nuclear-reactors.
39. DOE,2021.Sodium-cooledfastreactor,https://www.energy.gov/ne/articles/3-advanced-reactor-systems-watch-2030.
40. Martucci, B., June 12, 2024. TerraPower begins construction at 345-MW advanced reactor site in Wyoming, Utility Dive, https://www.utilitydive.com/news/terrapower-smr-advanced-nuclear-reactor-bill-gates/718722/.

8

Wind Energy

... As yet, the wind is an untamed, and unharnessed force; and quite possibly one of the greatest discoveries hereafter to be made, will be the taming, and harnessing of the wind.

Abraham Lincoln
Discoveries and Inventions (1858)

8.1 Introduction

Wind power has led the way in the transition from fossil fuels to renewable electricity. According to the International Energy Agency, 2.1 million GWh of wind energy was generated globally in 2022, a remarkable 68-fold increase over the 31,000 GWh generated from wind in 2000 [1]. In the United States, electricity from wind grew from 6,000 GWh in 2000 to 434,000 GWh in 2022, a 72-fold increase [2]. In 2022, Texas led the United States with 120,000 GWh of electricity from wind, followed by Iowa with 41,900 GWh [2].

Turbine designs continue to advance to allow higher hub heights, accessing greater wind resources. Improvements in turbine control systems, wind forecasts, and grid integration are supporting increased grid penetration levels. Although well behind land-based wind in total capacity, offshore installations are expanding rapidly, led by China, the United Kingdom, and Germany [3].

People have harnessed wind energy for millennia. Wind energy was used to propel boats along the Nile River more than 5,000 years ago. Around the beginning of the Common Era, simple windmills were used in China for pumping water. At about the same time, vertical-axis windmills with woven reed sails were used to grind grain in the Middle East. In the seventh century, a vertical-axis windmill was used in Sistan, Afghanistan, to process grain and pump water. People in the Middle East used windmills for food production in the eleventh century; the crusaders brought this idea back to Europe where it spread rapidly. The Dutch refined windmill technology and adapted it for draining lakes and marshes in the Rhine River Delta. American colonists used windmills to grind wheat and corn and to cut wood at sawmills. Across the United States, windmills were used extensively for pumping water on farms and ranches in the nineteenth and early twentieth centuries.

Professor James Blyth built the first wind turbine (WT) for electricity production in Scotland, successfully generating electricity in 1887 [4,5]. Soon afterward, Charles F. Brush built the world's first automatically operated WT in Cleveland, Ohio, in the winter of 1887–1888 [6]. Like the configuration that dominates today, Brush's turbine was a horizontal-axis wind turbine (HAWT), in which the axis of the rotor was horizontal to the ground. The first megawatt-sized WT to supply electricity through the electric grid was built in 1941 on a hill called Grandpa's Knob, in Vermont (Figure 8.1). Palmer C. Putnam designed the 1.25 MW turbine, which had two blades encompassing a 53 m rotational diameter.

DOI: 10.1201/9781003586616-8

FIGURE 8.1
Smith Putnam wind turbine on Grandpa's Knob in Vermont, circa 1941. The Smith-Putnam machine was rated at 1.25 MW in winds of about 30 mph. It was removed from service in 1945. (https://commons.wikipedia.org/wiki/File:Wind_turbine_1941.jpg.)

The turbine operated intermittently for 4 years until one of the blades failed [7,8]. Interest in wind-electric turbines continued into the 1950s, led by Danish engineers, but efforts there were subsequently sidelined by the availability of cheap oil and low energy prices [9]. The picture changed in the 1970s, when oil shortages and interest in alternative energy sources brought WT development to the fore all over the world. For example, Danish manufacturers Kuriant, Vestas, Nordtank, and Bonus began producing 20–30 kW machines, while NASA, at the Lewis (now Glenn) Research Center in Cleveland, Ohio, developed a

FIGURE 8.2
The 2.5 MW Liberty wind turbine (Clipper Windpower) near Medicine Bow, Wyoming. (Courtesy of DOE/NREL, Golden, CO. With permission.)

government research and development program producing the first of the "Mod" series of wind turbines ranging from 0.2 to 2 MW [9].

Much work has occurred over the last few decades leading to optimization of performance and cost in WT design. Currently, the exclusive commercial WT design (e.g., 1 MW and larger) is a HAWT, in which the axis of the rotor is horizontal to the ground, as shown in the photograph in Figure 8.2 and schematically in Figure 8.3. In contrast, vertical-axis wind turbines (VAWTs) have blades that spin about a vertical axis, as shown in Figure 8.4. The biggest advantage of this design is that it does not have to be pointed in the direction of the wind. However, HAWTs are generally more efficient. Readers interested in more information on a variety of alternative WT concepts are directed to V. Nelson and K. Starcher's *Wind Energy* [11].

Figure 8.5 shows the installed onshore wind capacity in the year 2023 in the top ten countries with wind energy generation. The leader in installed wind capacity is China, which had a capacity of more than 400 GW in 2023 [12]. China also leads in offshore installations, accounting for half of the global total of 75 GW in place in 2023 [12].

8.2 Environmental Impact

Early in its history, the wind industry faced challenges in maintaining reliability as some turbines experienced structural fatigue problems or broke under severe conditions. In addition, the industry has historically faced concerns about aesthetics, noise, and impacts

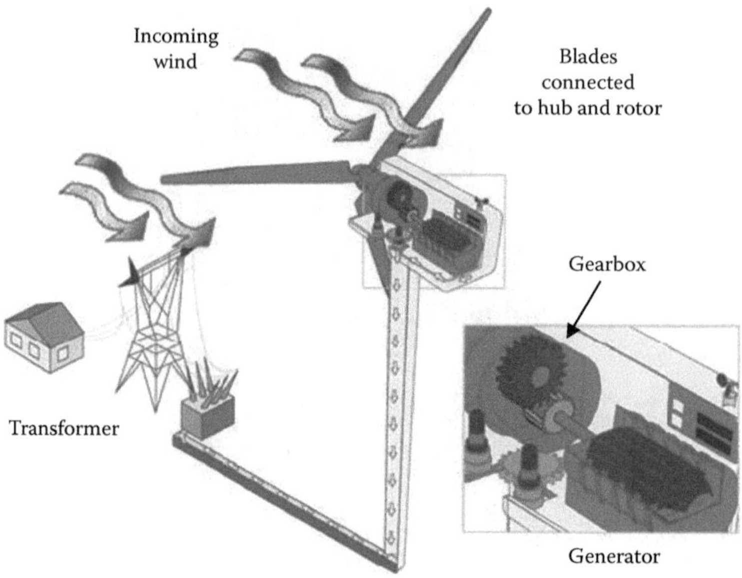

FIGURE 8.3
Schematic of typical wind turbine layout. (From Department of Energy [10]. With permission.)

FIGURE 8.4
An example of a Darrieus VAWT in Quebec, Canada.

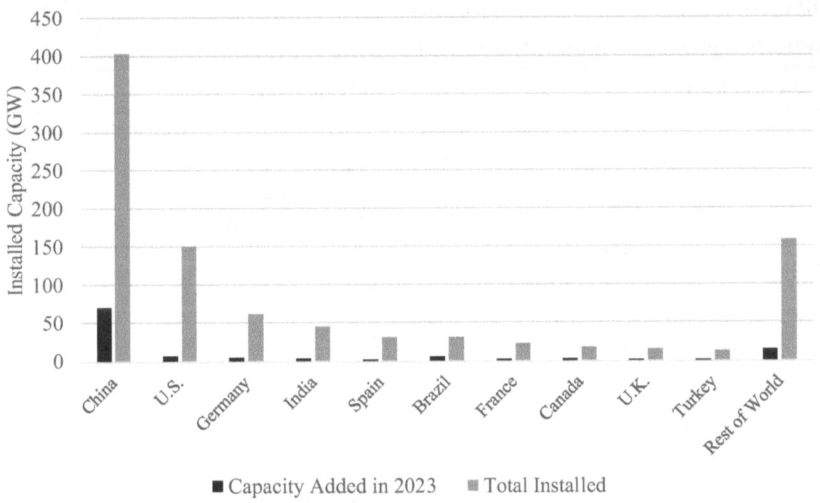

FIGURE 8.5
Installed global land-based wind power capacity in 2023. (Data from Global Wind Energy Council [12].)

on birds, bats, and wildlife. Significant improvements have been made on all these fronts as the industry has developed, but these concerns still warrant attention to minimize impacts. On the other hand, WTs have significant advantages over other electricity-generating technologies in life-cycle emissions and water use.

8.2.1 Noise and Visual Impact

Noise issues related to wind energy have been reported since the 1980s. Problems were initially found with the mechanical design as well as blade design and operation. For example, the 2-MW MOD-1 turbine developed by NASA in the late 1970s was a two-bladed turbine (61 m rotor diameter) [13]. It operated with the turbine blades downstream of a lattice tower structure. Low-frequency noise was produced as the turbine blades moved in and out of the "tower shadow" caused by wind flowing around and through the lattice tower. Noise of significant magnitude was produced, causing complaints from local residents.

Today, most large commercial WTs have the blades operating upstream of the tower and use a monopole tower. Modern commercial WTs have noise levels of approximately 35–45 dB at a distance of 300 m [14]. For comparison, typical noise or loudness levels in everyday life are given in Table 8.1 [15]. As seen in Table 8.1, the modern turbine (at distances of 750–1,000 ft) has a noise level comparable to a library or an average home (40–50 dB). When locating turbines, the distance between residences and WTs needs to be considered. Small WTs can actually have higher noise levels due to higher rates of rotation.

The sound typically heard from a WT comes from the blades. Careful aerodynamic design of the blades, as well as understanding the maximum allowable rotational speed of the blades, has reduced the aerodynamic noise level from previous levels. Similarly, mechanical components have been redesigned to virtually eliminate mechanical noise, leaving blade noise as the predominant noise source, albeit at a low noise level [16]. Wind farm developers are typically required to address noise concerns as part of local permitting and should demonstrate that they meet restrictions on sound levels, including through the use of detailed site-specific acoustic modeling [14].

TABLE 8.1

Loudness Levels of Common Noises [15]

Noise Level (dB)	Equivalent Activity
120	Jet plane takeoff, amplified rock music at 4–6 ft, car stereo, band practice
110	Rock music, model airplane
100	Snowmobile, chain saw, pneumatic drill
90	Lawnmower, shop tools, truck traffic, subway
80	Alarm clock, busy street
70	Vacuum cleaner
60	Conversation, dishwasher
50	Average home
40	Quiet library
30	Quiet bedroom

In Europe, a moving "shadow flicker" created by sunlight shining through rotating WT blades on residences has occurred due to placing WTs too close to the residences. The resulting flicker can be very annoying. Proper planning and accounting for latitude allow the necessary distance between turbine and residence to be calculated so no turbine shadow falls on the residence [17]. Minimum setbacks from residences and other occupied facilities should also consider the noise level of the operating turbine. The turbine should be located a sufficient distance away (based on sun angle) to eliminate any shadow flicker or noise concerns.

8.2.2 Life-Cycle Greenhouse Gas Emissions, Land, and Water Use

As discussed in previous chapters, an important consideration for any technology is the greenhouse gas (GHG) emissions for the total life cycle. Each phase of the life cycle, including resource extraction, facility construction, facility operation, and end-of-life disposition, must be evaluated to quantify and compare emissions from different technologies. As shown in Chapter 3 (Figure 3.3), even when the operations needed to manufacture WTs and build wind farms are included in a greenhouse gas emission analysis, the use of wind power compared to coal- or natural gas-fired electricity generation provides significant reductions. In a review of more than 40 studies, Nugent and Sovacool [18] found a median estimate of GHG emissions from wind generation of 12 g CO_2-e/kWh, and a mean estimate of 34 g CO_2-e/kWh. For comparison, life-cycle emissions from coal-fired electricity generation are about 1000 g CO_2-e/kWh [19]. Compared to thermal power plants, which require significant water for cooling, wind also has the advantage of very low life-cycle water use.

Other environmental impacts from WTs do exist. For example, construction of a wind farm could cause erosion due to the construction and use of roads needed to build the farm (truck and crane traffic, electrical line/equipment installation, etc.), but this impact can be mitigated using standard environmental reclamation techniques. Land occupancy and corresponding disruption of wildlife habitat or interference with other land use are also considerations. Denholm et al. [20] estimated that wind farms in the United States occupy a perimeter area that averages about 35 ha/MW capacity, although much of that area may be available for other uses. They estimate the area directly occupied by turbine pads, access roads, service buildings, and other infrastructure is about 0.7 ha/MW during construction and 0.3 ha/MW during the plant's operating life.

8.2.3 Bird and Bat Fatalities from Onshore Wind Turbines

Expanding wind energy conversion is critical for addressing the threat that climate disruption poses to both people and wildlife. That aim will be advanced by responsible development of wind resources that address concerns about wildlife habitat loss and bird and bat fatalities from collisions with WTs and associated infrastructure. In North America and elsewhere in the world, bird populations have declined sharply over the past half century, raising significant concern [21]. The reasons are not fully understood, but habitat loss and disturbance, intensification of agriculture and urbanization (in the surveyed areas and in other areas where birds migrate) are likely the main drivers. Acute fatalities from domestic and feral cats and collisions with transmission lines and other infrastructure, buildings, windows, and vehicles add to the toll [22,23]. WTs account for a tiny fraction of acute fatalities compared to cats, collisions with buildings and other anthropogenic causes [22,23], but nevertheless their impacts should be minimized through responsible siting and operations.

Based on numerous monitoring studies, scientists estimate that bird fatalities from wind turbine collisions average three to six birds per MW per year, which adds up to about 400,000–900,000 fatalities per year based on current installed capacity [23]. These numbers are two to four orders of magnitude lower than those for other causes of acute fatalities. For most passerine species (perching birds), fatalities from wind turbine collisions are not expected to be significant at the population level. However, wind turbine collisions account for a higher portion of raptor deaths from human causes and may have population-level impacts [23]. Bat fatalities with wind turbines are estimated to average four to seven deaths per MW per year, totaling 600,000 to 1 million deaths per year [23]. For bats, the population-level significance of wind turbine collisions remains unclear.

Climate change deeply threatens bird and bat populations and diversity. Wind energy that is sited and operated responsibly to protect birds, bats and other wildlife is an important part of the clean electricity transition needed to slow that overriding threat. The U.S. Fish and Wildlife Service (USFWS) has published guidelines for onshore wind developments, covering recommended practices for site screening and pre-construction field studies to identify and avoid critical habitat, post-construction monitoring of impacts, and decommissioning and site restoration [24]. The USFWS guidelines highlight the point that in addition to wind turbines, associated roads, fencing and especially power lines can also disturb, displace or kill birds, bats and other wildlife. Siting turbines away from topographic and landscape features that attract raptors and bats is a critical step [23,24]. Turbines can be curtailed during anticipated migration periods, or in real-time when automated monitoring shows raptors or bats approaching [23]. Curtailing rotation of turbines at low wind speeds can sharply reduce collisions with minimal impact on overall power output. Ultrasonic acoustic deterrent devices have been shown to deter some bat species [23,24].

8.2.4 Impacts of Offshore Wind Development on Marine Life

Offshore wind energy holds great promise for delivering low-carbon electricity close to densely populated coastal areas. As with onshore wind, however, potential impacts on local ecosystems and marine life need to be addressed in siting, site assessment, construction, operation and decommissioning of these facilities. Although offshore wind development has taken off very recently in China and to date has been quite limited in North America, offshore wind has been exploited in the North Sea for more than 20 years. Experience there, together with modeling studies, analogies to comparable disturbances from other settings, and laboratory tests, offers insights into what impacts might be expected and how they can best be managed as the offshore wind industry expands.

A recent review prepared for the U.S. Congress identifies hydrodynamic (mixing and stratification) changes from turbine wakes, noise from acoustic underwater surveys and construction, entanglement risks from undersea cables and moorings, electromagnetic fields from underwater transmission lines, water pollution from spills, and collision with turbine blades and construction and maintenance vessels as disruptions from offshore wind development that warrant attention [25]. While many potential impacts are negative, some species of invertebrates and fish may benefit from the fixed foundations of offshore turbines acting as artificial reefs. Noise impacts on marine mammals appear to mainly disrupt behavior rather than cause acute injury. Noise can also disrupt behavior of fish and invertebrates. Some seabirds and bats that migrate over water may face collision risks with turbines, while marine mammals could collide with undersea structures and marine vessels. Most of these impacts, if they occur, are expected to be limited to the immediate vicinity of the infrastructure [25]. Whether or not they are significant will depend on the location and configuration of the development and the marine life that uses the area. Contrary to some claims, no connections have been established between humpback whale mortalities and offshore wind development. Rather, whale deaths have been attributed to ship collisions and entanglement by fishing gear [26].

Opportunities to minimize the impacts of offshore wind development on marine life and ecosystems include siting facilities away from sensitive areas, measures to reduce noise generation or dampen sound propagation, work restrictions during migration, mating or hatching periods, and temporary turbine curtailment if protected animals come inside a safety zone [25]. In the United States, leasing and permitting for offshore wind development in federal waters is led by the Bureau of Ocean Energy Management (BOEM) under the Department of Interior. Before projects can proceed, BOEM must complete environmental impact assessments in consultation with the federal agencies that oversee the Endangered Species Act, the Marine Mammals Protection Act, and the Migratory Bird Treaty. State approvals may also be required for infrastructure in state waters or coastal areas. These assessments, which often require multiple years to complete, provide a vital opportunity for public input and typically require impact monitoring, reporting and mitigation measures. As decades of experience with onshore energy development have shown, careful monitoring and adaptive management of impacts will be essential as new offshore wind projects advance through site surveys, construction, and operating phases.

8.2.5 Turbine Blade Waste

Waste from large wind turbine blades has been a growing issue. Blades have typically been made from a combination of fiberglass and epoxy resin, which is not easily recycled. As a result, at the end of an approximate 20-year life, turbine blades have typically been chopped up and disposed of in landfills. Various efforts are underway to reuse old turbine blade material (such as for cement filler) and to develop recyclable or biodegradable turbine blade materials [27].

8.3 Power and Energy of Wind

Power and energy are common terms that are often misused. Power is the rate at which energy is used or produced. Light bulbs use power; WTs generate power. Units of power are expressed in work (energy) over an amount of time:

$$\text{Power} = \frac{\text{Work}}{\text{Time}} = \frac{\text{Energy}}{\text{Time}} = \frac{N \cdot m}{s} = \frac{J}{s} = W \tag{8.1}$$

WTs are rated in terms of power. Large commercial onshore wind turbines have power ratings in the 1–5 MW range, whereas a WT used for a single home is likely rated in the 3–15 kW range.

Energy, on the other hand, is a measure of how much work can be done by a force. Units of energy are joules, J, the product of force and distance:

$$\text{Energy} = N \cdot m = J \tag{8.2}$$

The link between power and energy comes from the rate of energy usage. In other words, energy equals the product of power and time, or power is energy divided by time. If a 100-W light bulb is turned on for 24 hours, the energy usage is 100 W × 24 hour = 2.4 kWh. While turbines are denoted by their rated power production (e.g., 100 kW, 2.5 MW), we pay for electricity based on how much energy we use (e.g., $0.06/kWh).

To determine the power contained in the wind moving toward a WT, consider the kinetic energy, KE, of a mass of air, m (in kg), moving at speed, U (in m/s):

$$\text{KE} = \frac{1}{2} mU^2 \tag{8.3}$$

$$\text{Unit check}: \text{kg}\left(\frac{\text{m}}{\text{s}}\right)^2 = \frac{\text{kg} \cdot \text{m}^2}{\text{s}^2} = \left(\frac{\text{kg} \cdot \text{m}}{\text{s}^2}\right) \cdot m = N \cdot m = J$$

Now imagine a horizontal cylinder of air of area A (in m^2) and velocity U, moving toward a WT, as seen in Figure 8.6. Note that the area, A, corresponds to the area swept out by the rotating WT blades. The mass m of the air column is volume V times the density, where ρ represents the air density in kg/m^3. Mass can then also be expressed as follows:

$$m = \rho V = \rho A l = \rho A U t \tag{8.4}$$

where

- ρ is the air density, kg/m^3
- V is the volume, m^3
- A is the area, m^2
- l is the length, m
- U is the wind speed, m/s
- t is the time, s.

The length of the column of air, l, is equal to the distance the wind travels in a given time interval, t, and is found by multiplying the wind speed, U, by the time.

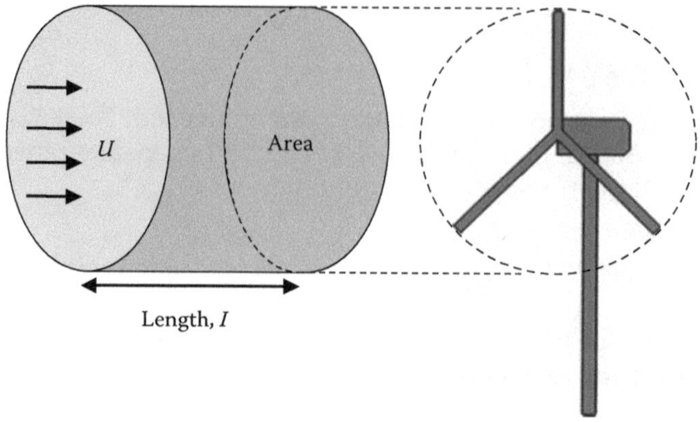

FIGURE 8.6
Incoming air column and wind turbine.

Combining Equation 8.3 with Equation 8.4, the kinetic energy becomes

$$\text{KE} = \frac{1}{2}mU^2 = \frac{1}{2}(\rho AUt)U^2 = \frac{1}{2}\rho AtU^3 \tag{8.5}$$

Dividing the KE by time yields the power of a moving column of air:

$$\text{Power} = \frac{\text{KE}}{t} = \frac{1}{2}\rho AU^3 \tag{8.6}$$

In other words, the power in a column of wind is linearly proportional to the air density and area of the air column. Of even more significance, the power is also proportional to the cube of the wind speed. We will see later that a key reason WTs have grown in height over time is to place the WT rotor blades higher in the atmosphere where the wind speed is greater, thus harnessing a much higher kinetic energy output.

To increase power output, the area of the WT blades can be modified to affect the power output. Since the WT power output is linearly proportional to the area, a large-scale commercial WT designer would want to increase the blade length to increase the power output. Since the circular area swept out by the turbine blades is proportional to the blade radius squared (R^2), turbine power is also proportional to the blade radius squared (R^2):

$$\text{Power} = \frac{1}{2}\rho AU^3 = \frac{1}{2}\rho\pi R^2U^3 \tag{8.7}$$

In summary, longer blade lengths and taller turbines (providing access to higher wind speeds) are options for increasing WT power output. However, this performance increase comes at a price. Longer blades need more sophisticated manufacturing techniques to build blades strong enough to withstand wind gusts and extreme wind speeds (in excess of 100 mph). Taller towers must be strong enough to withstand the forces transmitted from the blades, weight of the nacelle, and other parts while still being cost-effective. Transporting turbine blades from manufacturing facilities to installation sites also becomes more challenging and more costly.

Some applications, for example, remote homes off the electrical grid, require more modest amounts of power. The design challenges here are just as significant: structures (blades, towers, etc.) must be strong enough to withstand wind forces, but also must be light, cost-effective, and reliable.

8.4 Power Coefficient

A key performance measure in WT design is the power coefficient, C_p, which is a ratio of the aerodynamic power extracted by the WT, P_{aero}, divided by the total power in the wind, P_{wind}, and is calculated as follows:

$$C_p = \frac{P_{aero}}{P_{wind}} = \frac{P_{aero}}{1/2\rho AU^3} = \frac{P_{aero}}{1/2\rho\pi R^2U^3} \tag{8.8}$$

In Equation 8.8, $P_{aero} = Q \times \Omega$, where Q is the aerodynamic torque and Ω is the rotational rotor speed.

The theoretical maximum turbine efficiency was derived by Albert Betz in 1920 [28], as well as others [29], and is

$$C_p = \frac{16}{27} \approx 0.59 = \text{Betz limit} \tag{8.9}$$

In addition, C_p is a function of how fast the tips of the WT blades are rotating compared to the incoming wind. The ratio of the blade's rotational speed divided by the incoming wind speed is known as the tip-speed ratio (TSR):

$$\text{Turbine tip speed ratio} = \frac{\Omega R}{U} \tag{8.10}$$

where

Ω is the turbine blade rotational speed, rad/s
R is the rotor radius, m
U is the wind speed, m/s.

In Figure 8.7 [30], the power coefficient is plotted as a function of TSR for different WT configurations. Several key points can be seen in Figure 8.7, starting from the top. The highest line shows how the theoretical maximum power coefficient increases with increasing TSR. The Betz limit value of 0.59 is approached asymptotically as the TSR increases toward a value of seven (TSR = 7). The next highest curve denotes the performance of today's modern high-speed horizontal-axis two- or three-bladed turbines. The maximum theoretical power coefficient for these turbines is approximately $C_p = 0.5$, whereas the Darrieus vertical-axis rotor has a maximum C_p of approximately 0.40–0.45. Thus, the modern two- and three-bladed HAWTs are approximately 10% more efficient than a Darrieus (VAWT) machine and have the potential to make 10% more energy.

Not only are the large commercial WTs configured horizontally, but they are also predominantly three-bladed upwind (blades rotate in front of the tower). Why are HAWTs mainly upwind? As described earlier with noise issues, the blades on a downwind turbine experience cyclic loading (alternating higher forces and lower forces) due to the blades passing behind the tower, which may cause fatigue failure of the blades as well as noise.

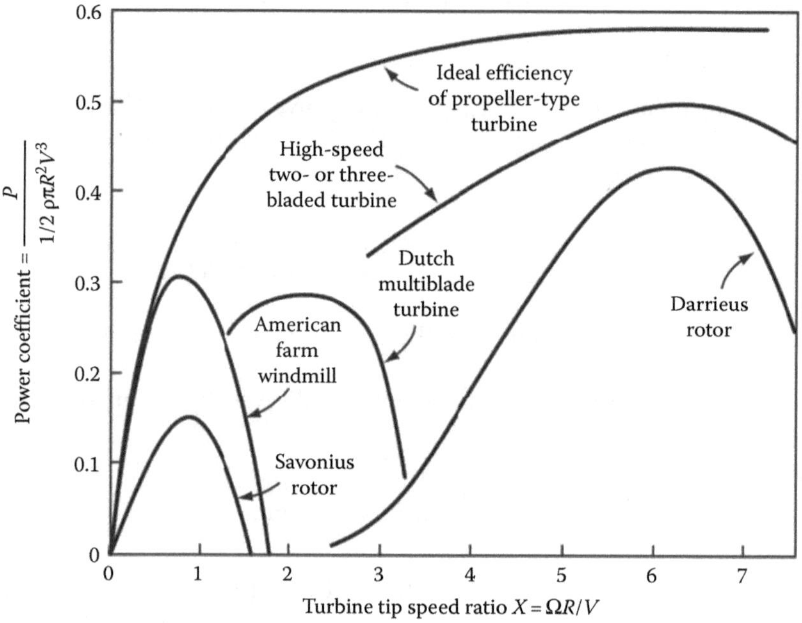

FIGURE 8.7
Various turbine power coefficients versus TSR. (Note: The figure uses the symbol V for wind speed.) (From Wilson and Lissaman [30]. With permission.)

8.5 Aerodynamics

Examining the aerodynamics of WT blades helps explain why various WT configurations have such different power coefficients, C_p [31]. Figure 8.8 shows a WT blade rotating (in the plane of the paper) in a clockwise direction. The atmospheric wind, U_w, is blowing toward the blade (i.e., into the plane of the paper). A section view of the blade is then taken near the tip of the blade looking in toward the root of the blade (Figure 8.8). The cross-sectional view of the blade airfoil is shown in the right-hand side of the figure. The two airflows acting on the blade are the incoming wind speed, U_w, which is perpendicular to the plane of the rotating blades, and the rotational wind speed, $U_{Rot} = \Omega R$, created by the turbine blade rotating at an angular speed, Ω, at the radius of the cross section, R. Note that the side of the airfoil facing the wind (designated the front side) is relatively flat compared to the curved rear side of the airfoil. The leading edge of the airfoil hits the airflow arising from the blade rotation causing the airflow to separate into two streams over the sides of the airfoil. Finally, the trailing edge is where the airflow over the two sides of the airfoil is rejoined.

In Figure 8.9, a blade cross section near the tip of the blade is shown with the resultant forces created by the passing wind, the lift force, F_{Lift}, and the drag force, F_{Drag}. The straight line connecting the front (leading) and rear (trailing) edges of the airfoil is known as the chord line, and the distance between the two points is designated as the chord, c. Also shown are the resultant velocity component, U_{Res}, between the wind speed, U_w, and the wind arising from the blade's rotational speed, U_{Rot}. Finally, the angle between the resultant wind speed, U_{Res}, and the chord line, c, is the angle of attack, α.

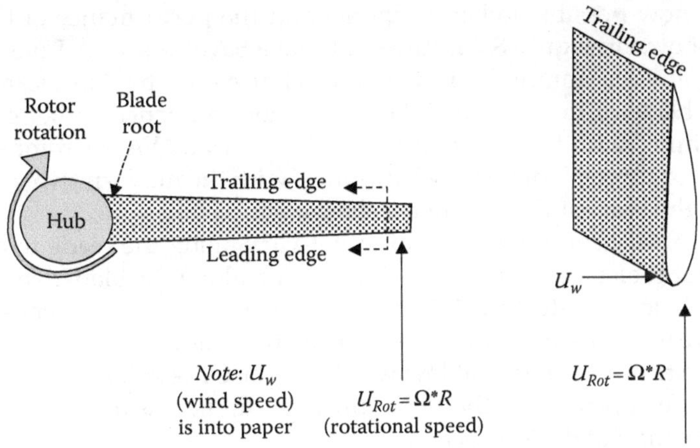

FIGURE 8.8
Rotor blade diagram.

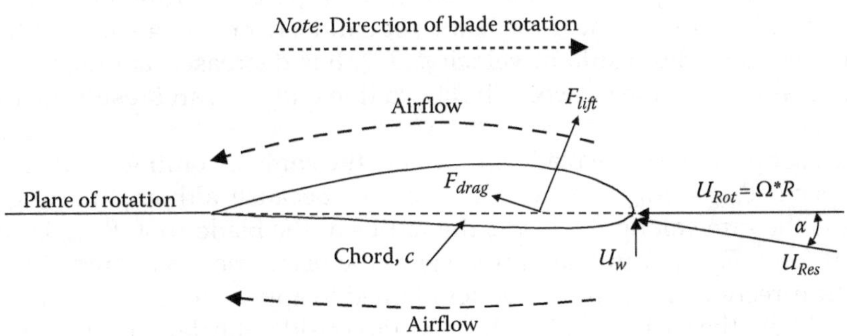

FIGURE 8.9
Lift and drag forces on a blade cross section near the tip.

Via Bernoulli's law, Equation 8.11, the velocity is highest across the curved (upper) surface of any airfoil (e.g., airplane wing). In addition, Bernoulli's law tells us that the static pressure is the lowest where the velocity is the highest. In the case of a WT blade, the pressure is lower over the rear (i.e., upper) surface of the airfoil compared to the front (i.e., lower) airfoil surface.

$$p + \frac{1}{2}\rho U^2 = \text{const.} \tag{8.11}$$

In summary, the lowest pressure is near the leading edge ("nose") of the airfoil, where the wind velocity is the highest, and over the rear (or upper) curved surface. This results in a net force, F_{Lift}, on the airfoil, causing it to be drawn forward, in the direction of the "nose" of the airfoil and hence rotating. By convention, the resultant forces on the airfoil are drawn perpendicular, F_{Lift}, and parallel, F_{Drag}, to the resultant velocity, U_{Res}. The higher the lift force (and conversely the lower the drag force) along a WT blade, the greater the rotational force, which means the WT is capturing more of the wind's energy and converting it to electricity. In other words, the higher the lift-to-drag ratio (*L*/*D*) is, the more efficient the blade will be at capturing the wind's energy.

Note that it is now possible to better understand the performance of the different WT configurations shown in Figure 8.7. It turns out that a Savonius VAWT has a poor *L/D* ratio and thus a very low maximum C_p, as shown in Figure 8.7. The American farm windmill and Dutch multiblade turbines have higher *L/D* ratios and hence higher maximum C_p values. Lastly, machines with high *L/D* ratios are the Darrieus VAWT rotor and the modern high-speed two- or three-bladed HAWT turbines, with the modern two- and three-bladed turbines having the highest C_p, as noted earlier.

WT blades have very complicated shapes that vary along the blade to account, in part, for the fact that local blade speed varies with position along the blade. The shape variation is needed not only to help start the WT rotating but also to operate very efficiently as the wind speed increases. The airfoil is twisted near the blade root to provide the required torque to start the turbine rotating at low wind speeds. Once rotating, the optimized blade profiles in the outer portions of the blade provide the necessary aerodynamic forces to keep the blade rotating at design speeds.

Recall that Figure 8.9 illustrates a cross section of a blade in the outer portion of the blade. Figure 8.10 illustrates the lift and drag forces on a blade cross section near the root of the blade. Note that the blade is twisted, and the chord line now forms an angle θ with the plane of rotation of the blade. This twist angle, θ, is a function of turbine blade radius and has a larger value near the blade root and a decreasing value with increasing blade radius. Because the resultant velocity, U_{Res}, has decreased approaching the blade root, this increases the angle between the chord line and U_{Res}, and results in an increased angle of attack.

While the root portion of the blade encounters the same incoming wind speed, U_w, the rotational speed, $U_{Rot} = \Omega R_{Root}$, is lower. This occurs because, although the angular rate of rotation, Ω, is the same along the blade, the radius at the blade root, R_{Root}, is smaller. The resulting lift force, F_{Lift}, per unit area is lower at the blade root compared to the blade tip. However, the direction of the force is better aligned to apply a torque to begin rotating the blades. In addition, the surface area of the blade is considerably larger at the root. This provides sufficient area for the lower magnitude aerodynamic forces to act over, resulting in enough torque to start rotation at low wind speeds. Finally, a larger surface area is needed to accommodate the necessary structural strength to handle the cantilevered loads arising from wind forces on the blade, weight of the blade, and so on.

Figure 8.11 shows the vector diagrams from Figures 8.9 and 8.10 side by side. Assuming a constant wind speed at both the blade root and tip cross sections, we will consider the case where the root cross section is located at a radial location one-fifth of the distance along the blade to the blade tip. This last assumption means that the rotation speed at the

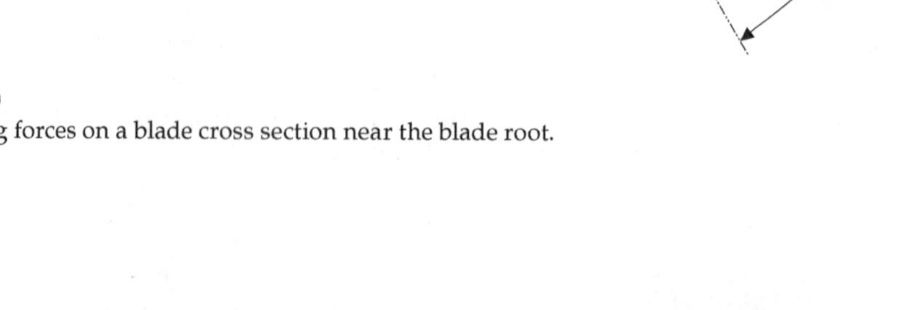

FIGURE 8.10
Lift and drag forces on a blade cross section near the blade root.

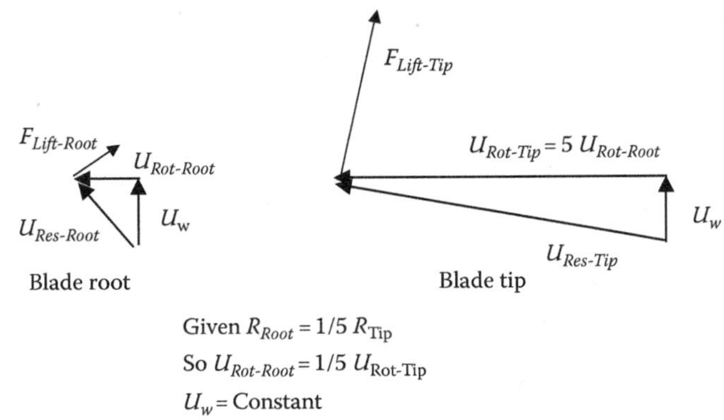

FIGURE 8.11
Comparison of velocity vector diagrams for blade root and tip cross sections.

blade root, $U_{Rot\text{-}Root}$, is one-fifth the speed at the blade tip, $U_{Rot\text{-}Tip}$. Note that $U_{Rot\text{-}Root}$ and $U_{Rot\text{-}Tip}$ are drawn to scale. The resultant velocity vector at the tip is much larger due to the contribution of the rotational wind speed. This results in a larger lift force per unit area at the tip. At the blade root, the lift force is better aligned to provide a start-up torque. While the blade root lift force is lower per unit area, it is applied over a large surface area. Interestingly, at the tip, the contribution of the wind speed to the resultant velocity is small compared to the contribution from the rotational wind speed when the turbine is in operational conditions.

8.6 Wind Characteristics

8.6.1 Wind Generation

Now we will turn our attention to the atmospheric wind: generation, distribution, characteristics, and so on. The motion of air, otherwise known as wind, is caused by uneven heating of the Earth by solar radiation and the rotation of the Earth. The predominant wind direction varies with latitude and altitude above the Earth's surface, as shown in Figure 8.12 [31–33]. Energy is transferred from the equator to the poles via wind. As the wind moves, the Earth's rotation influences the wind direction in two ways. First, Coriolis forces accelerate a particle of air to the right in the Northern Hemisphere (and to the left in the Southern Hemisphere). Second, each air particle has an angular momentum from west to east due to the Earth's rotational direction. As the air particle moves toward the pole, the conservation of angular momentum requires its west-to-east velocity component to increase. This effect is pronounced in the mid-latitudes, in the region called the westerlies, near the ground. In general, the westerlies are the best-suited region for wind energy applications. In addition, the polar jet stream (7–12 km or 20,000–40,000 ft above sea level) can have a profound impact on the westerlies, moving them north or south, resulting in lower or higher energy production from year to year.

Of course, the most critical part of the atmosphere for a WT is relatively close to the ground. The atmospheric boundary layer (ABL) is the lowest part of the atmosphere and varies from the ground up to 500–2,000 m in height depending on the time of day,

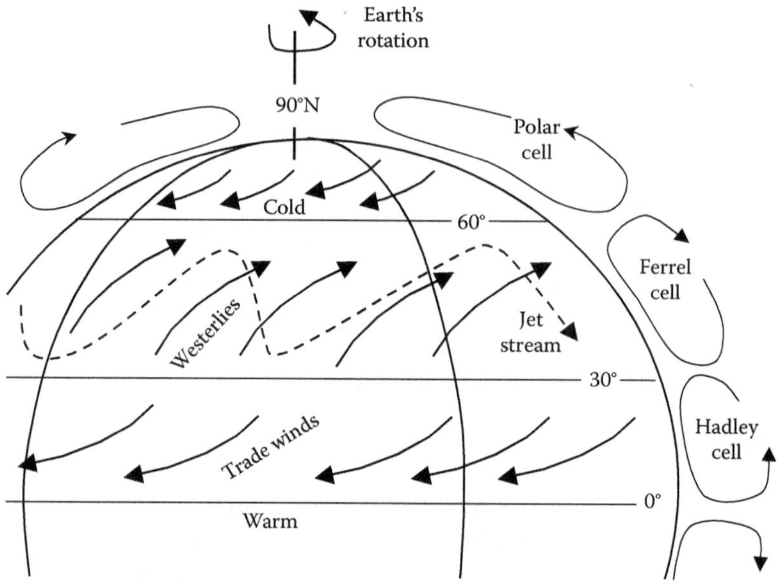

FIGURE 8.12
Wind flow in the Northern Hemisphere.

temperature, and pressure. In the ABL, the mean wind speed varies from zero at or near the ground up to the free stream velocity of the atmosphere at the top of the ABL. In addition, the velocity field exhibits turbulent fluctuations. In other words, each of the three velocity components for an air particle has a mean as well as a fluctuating component. The turbulent velocity component can be quite large and by nature is time-varying. This results in unanticipated fluctuating forces (high velocity causes high blade loading), which can lead to fatigue failures of WT components such as the blades, gearboxes, and tower. The turbulent portion of the flow field is typically described statistically. Methods of accurately simulating turbulent flow fields for a WT design are an active field of research [34].

8.6.2 Distribution of Wind

Now we transition from a global view of wind to a local area view of the wind resource. The wind resource is the mean wind speed as well as how many hours per year the wind blows at a given speed. A wind farm developer needs high confidence that the wind speeds at a given location are "high" and consistent from year to year. How much of a concern is this? If the average wind speed is only 5% lower than predicted, the maximum possible power will be 14% lower than anticipated! High confidence in the wind resource allows the developer to estimate the probable energy output and compare that to the investment needed to lay out a wind site and purchase or finance, install, and operate the wind farm (consisting of tens or even hundreds of turbines).

Wind resource maps have been developed for the entire world. For example, a U.S. wind resource map is shown in Figure 8.13 [35]. Similar maps are available for other countries. Figure 8.13 maps the estimated annual average wind speed 80 m above the ground. The 80 m height is a relevant starting point for current commercial installations, but taller turbines are increasingly used. A color version of the map in Figure 8.13 as well as wind maps

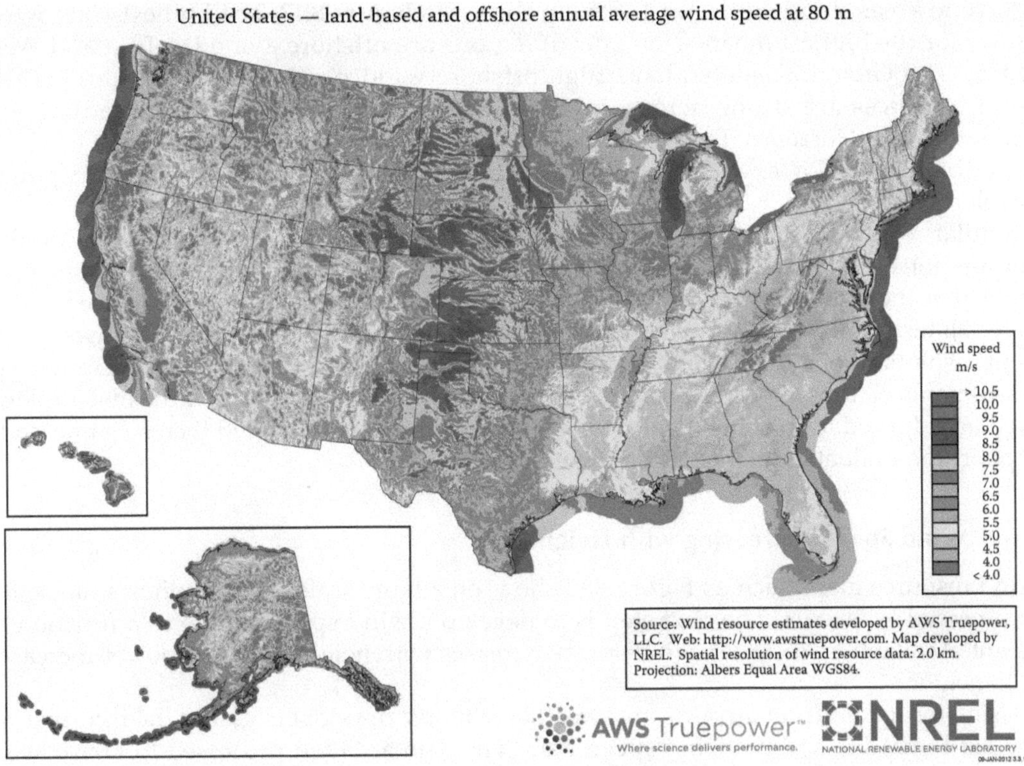

FIGURE 8.13
Wind resource map of the United States. This map shows the annual average wind speed estimates at a height of 80 m. The estimates are model-derived. Color maps for the United States for a range of heights are available at windexchange.energy.gov. (Figure Courtesy of U.S. Department of Energy, National Renewable Energy Laboratory, Washington, DC.)

TABLE 8.2
Rating Wind Energy Resource Potential

Wind Power Class	Resource Potential	Wind Power Density at 50 m (W/m²)	Mean Wind Speed[a] at 50 m (m/s)	Mean Wind Speed[a] at 50 m (mph)
1	Poor	0–200	0.0–5.9	0.0–13.2
2	Marginal	200–300	5.9–6.7	13.2–15.0
3	Fair	300–400	6.7–7.4	15.0–16.6
4	Good	400–500	7.4–7.9	16.6–17.7
5	Excellent	500–600	7.9–8.4	17.7–18.8
6	Outstanding	600–800	8.4–9.3	18.8–20.8
7	Superb	>800	>9.3	>20.8

Source: Enlarged from Wind Powering America [37].
[a] Wind speeds are based on a Weibull k value of 2.0 (Rayleigh).

for 100 m heights and maps at a higher spatial resolution are available from NREL [36]. Table 8.2 illustrates the relationship between wind power density (in W/m² swept area) and mean wind speed (in m/s) [37].

Several areas stand out in the U.S. map shown in Figure 8.13. The highest-class wind power (or the highest mean wind speed) regions are offshore along the East and West coasts. The Great Lakes also have high offshore wind power resources. Land-based wind resources are strong across the middle of the country from the Canadian border south into Northern Texas. Correspondingly, states in the middle of the country, including Texas, Iowa, and Wyoming, have been leading locations for wind power development.

Similar regions of high annual wind power can be found in many areas of the country but are not visible at national-scale resolution. This highlights a key point, not surprisingly, that the wind resource is highly dependent on local conditions such as vegetation (forests), terrain (hills, mountains), and water (ponds, rivers, lakes, oceans). Given sufficient resolution, one can search for and decide on a site to locate a WT. Screening-level assessments of annual power output of a turbine may be done using routine meteorological data, adjusted for turbine height. More detailed information would then be gathered to support the evaluation of a proposed wind project.

8.6.3 Wind Speed Increasing with Height

Wind resource maps such as Figure 8.13 show, on a large scale, where winds suitable for commercial WTs exist. The next step is to develop a wind speed profile as a function of height. As mentioned earlier, wind speed increases with height and thus power increases with height.

Using local data from airports and weather stations, developers gather the information needed to get site-specific wind speed data. The data are then processed to arrive at an equation expressing the wind speed as a function of height. Two common approaches are the log law and the power law.

8.6.4 Log Law Wind Speed Profile

The log law wind speed profile has a physical basis. It is assumed that the shear stress in the wind is constant with height, the pressure gradient near the Earth's surface is small, and the Earth's rotation is ignored. The height of this constant shear stress layer is 100–200 m and is known as the "surface layer" of the ABL (see, for example, [38]). For a smooth surface, the mixing length, l, of the turbulent wind is given as

$$l = \kappa z \tag{8.12}$$

where

κ is von Karman's constant ($\kappa = 0.4$)

z is the height above the ground.

The velocity profile as a function of height can be found to be

$$U(z) = \frac{U^*}{\kappa} \ln\left(\frac{z}{z_0}\right) \tag{8.13}$$

where

$U(z)$ is the wind speed at a height, z

U^* is the friction velocity arising from shear stress

z_0 is the surface roughness length.

Values for z_0 are shown in Table 8.3.

TABLE 8.3

Surface Roughness for Various Outdoor Surfaces

Terrain Type	Roughness Length, z_0 (m)
Cities, forests	0.7
Suburbs, wooded countryside	0.3
Countryside with trees and hedges	0.1
Open farmland	0.03
Flat grassy plains	0.01
Flat desert, rough sea	0.001

Source: Burton et al. [39]; Manwell et al. [40].

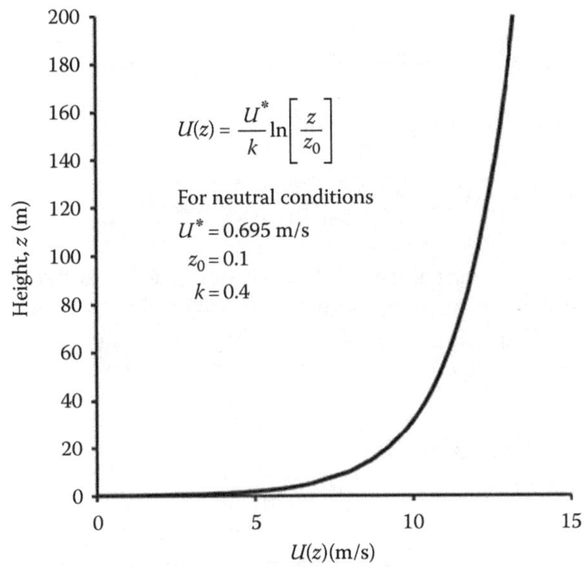

FIGURE 8.14
Log law velocity distribution.

There are two unknowns in this equation: U^* and z_0. Substituting typical values for U^* and z_0, the shape of the resulting atmospheric velocity profile is illustrated in Figure 8.14 for a neutrally stable atmosphere. One sees a rapid exponential increase in wind speed through the first 20 m or so of the ABL and then a more gradual growth approaching a steady-state value above 200 m.

There are three stability states of the atmosphere: neutral, stable, and unstable [39]. Neutral stability occurs when there is no change in temperature with height through the ABL due to adiabatic cooling. This state is typically transitional between unstable and stable periods. A stable atmosphere occurs when the air temperature is cold on the ground and increases with altitude. This condition is associated with high wind speeds and results in a steep increase in the wind speed with height (i.e., large wind shear). An unstable atmosphere occurs when the air temperature near the ground is warmer than the temperature at higher altitudes. The air on the ground rises while the colder air above sinks, thereby creating turbulent wind conditions.

If one wants to determine an unknown wind speed at a given height, it can be estimated from a known wind speed at another height using Equation 8.13. This approach is especially valuable because wind speed data are typically known at heights of 50 m or less (reference height); however, turbines operate at much higher heights where wind speed data are unknown. Using this method, wind speeds can be estimated at hub height from data available at lower heights.

So, taking Equation 8.13 at a height z, for the wind speed, $U(z)$, and then dividing it by the same equation for a reference height, z_r at the reference wind speed, $U(z_r)$, eliminates the unknown friction velocity, U^*, and yields

$$\frac{U(z)}{U(z_r)} = \frac{\ln\left(\frac{z}{z_0}\right)}{\ln\left(\frac{z_r}{z_0}\right)} \tag{8.14}$$

This equation now has only one unknown, z_0, the roughness length, which is estimated from Table 8.3. Consider a meteorological tower taking measurements at 40 m ($z_r = 40$ m), measuring a wind speed of $U(z_r) = 7.2$ m/s at z_r, located in a wooded countryside, $z_0 = 0.3$, then the wind speed, $U(z)$, at a hub height of $z = 95$ m is

$$U(z) = 7.2 \cdot \frac{\ln(95/0.3)}{\ln(40/0.3)} = 8.5\ \text{m/s} \tag{8.15}$$

In other words, the wind speed is predicted to increase by approximately 18% from 40 to 95 m if the atmospheric assumptions in the model hold. In this case, the wind power would increase by $(8.5/7.2)^3 = 1.65$ or 65%. Height does indeed matter when it comes to wind turbines.

8.6.5 Power Law Wind Speed Profile

Another model commonly used in engineering wind studies is the power law profile. It is used to estimate wind speeds given a reference height, z_r, to some desired height, z. The power law profile is used when one does not know the surface roughness, z_0, and/or stability information (i.e., earlier defined friction velocity, U^*) is unknown. The power law profile is given as

$$U(z) = U(z_r) \cdot \left[\frac{z}{z_r}\right]^{\alpha} \tag{8.16}$$

where α is the power law or shear exponent.

Typical values for α are

$$\text{For neutral flows}: \alpha \cong \frac{1}{7} = 0.143$$

$$\text{For unstable flows}: \alpha < 0.1$$

$$\text{For strongly stable flows}: \alpha > 0.3$$

A comparison between the log law and power law wind speed profiles is shown in Figure 8.15. This figure shows that the wind speed profile of the power law profile for a

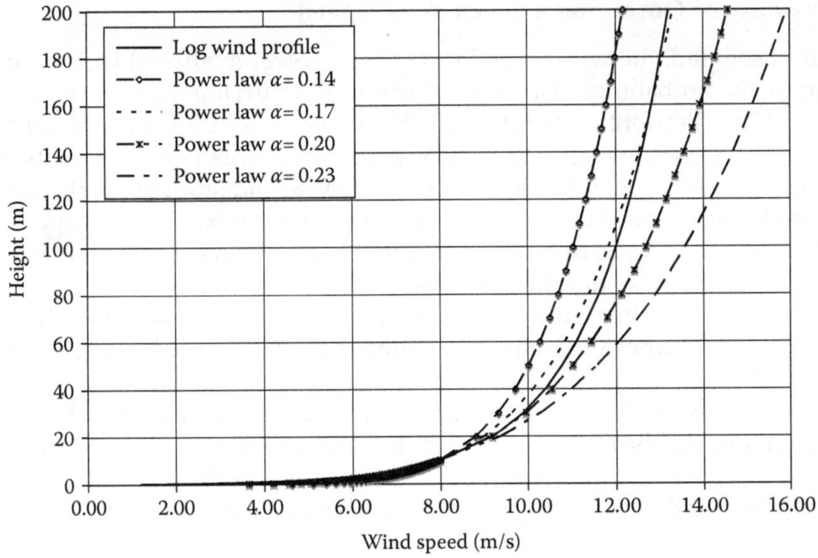

FIGURE 8.15
Log law and power law velocity profile comparison, with $U(z_r) = 8$ m/s at $z_r = 10$ m.

TABLE 8.4
Variation in Wind Speed with α at a Height of 100 m

α	Wind Speed $U(z)$ (m/s)
0.14	11.0
0.23	13.6

value of $\alpha = 0.17$ is similar in shape to the log law wind profile plotted earlier for neutral atmospheric conditions. In any case, it should be noted that the value of α is highly dependent on temperature, weather, geographic location (elevation and terrain), and time (of day and season). The power law profile is commonly used in engineering studies due to its simplicity.

Table 8.4 shows the calculated wind speeds for two different values of α at a height of 100 m, which is a typical hub height for modern commercial WTs. The change in predicted power based on these wind speeds is the ratio $(13.6/11)^3 = 1.89$. This indicates that the predicted power can nearly double at 100 m if the value of α increases from 0.14 to 0.23. This has large economic implications. In other words, determining the correct value of α for the power law profile (or using an entirely different method to determine wind speed as a function of height) for a given location and time period is critical when making choices about where to build wind farms. The best option is to gather long-term meteorological data at hub height in a sufficient number of locations at a potential WT site (if doing so is affordable). Methods have been developed, and continue to be developed, to choose the best value of α given a set of wind data and environmental wind factors.

8.6.6 Probability of Observing a Given Wind Speed

As has been discussed, the wind speed (and hence power available) varies. The next step is to determine the probability of a given wind speed occurring over a desired time period using a probability distribution function (PDF). A common PDF is the normal or Gaussian PDF, which can be seen in Figure 8.16. Many natural events can be described by a normal PDF, for example, the height distribution of 20-year-old students in the United States. When graphed, this Gaussian PDF yields the commonly known "bell-shaped" curve. For all PDFs, the vertical axis is the observed frequency of a given event within a certain range or magnitude, for example, number of students between the heights of 5′0″ and 5′1″. The horizontal axis contains the range of all events, for example, 4′0″ to 6′10″. The peak of any Gaussian PDF curve (where the curve is denoted by $f(x)$) is the mean magnitude of all occurring events.

The area under any section of a PDF curve corresponds to the probability, P, of a range of events occurring. As shown in Figure 8.16, the area represents the probability of having an event occur with a magnitude between a and b. The probability, P, of a range of events occurring is obtained by integrating the PDF, $f(x)$, between the range of the events, a and b.

$$P = \int_a^b f(x)dx \tag{8.17}$$

It should also be noted that integrating the PDF between the limits of $-\infty$ and ∞ yields a value of 1. In other words, the probability of having an event occur of any magnitude between the limits of all events is 100%.

$$P = \int_{-\infty}^{\infty} f(x)dx = 1 \tag{8.18a}$$

Two commonly used PDFs, Weibull and Rayleigh, are used to describe the probability of observing a given wind speed. These distributions are used because they more closely

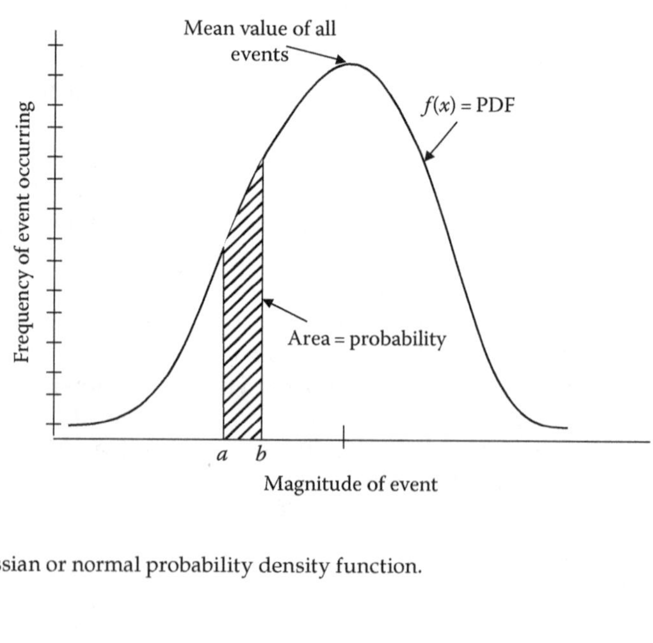

FIGURE 8.16
Illustration of a Gaussian or normal probability density function.

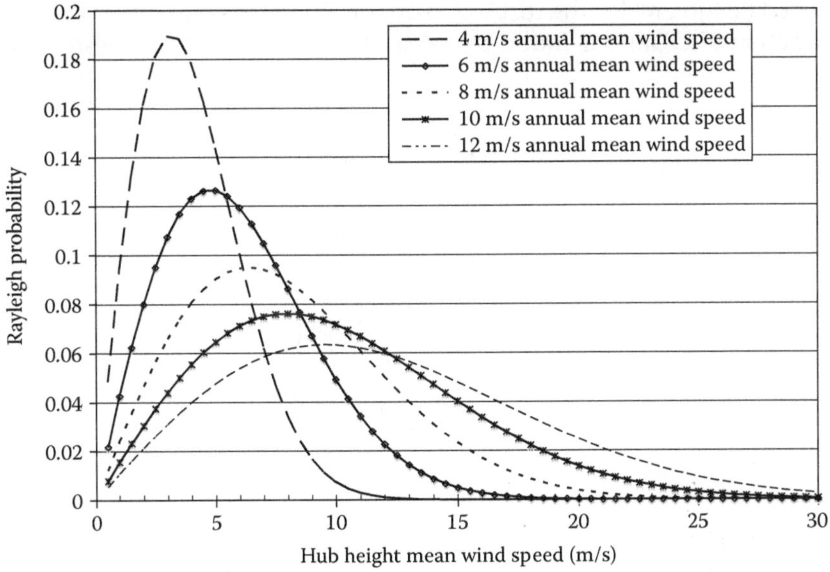

FIGURE 8.17
Rayleigh wind distributions.

represent the distribution of wind speeds at a given location. These PDFs are not symmetric, as is the Gaussian distribution (see Figure 8.16), but are skewed to the "left" (see Figure 8.17).

A wind speed distribution skewed to the "left" means that the probability of seeing a lower wind speed is higher than the probability of seeing a higher wind speed (i.e., there is a higher probability of a 3 m/s wind speed than a 25 m/s wind speed at a typical WT site). These PDFs are typically determined at a desired height, for example, hub height. These probability distributions are, of course, only defined for values greater than zero. Integrating the PDF will provide the probability, $P(U)$ of wind speed, U, occurring between two wind speeds, U_a and U_b:

$$P = \int_{U_a}^{U_b} f(U)\,dU \tag{8.18b}$$

Several other important statistical relationships can be found given a wind speed PDF, $f(U)$, including

$$\text{Mean wind speed:} \quad \bar{U} = \int_0^\infty U f(U)\,dU \tag{8.19}$$

$$\text{Standard deviation:} \quad \sigma = \sqrt{\int_0^\infty \left(U - \bar{U}\right)^2 f(U)\,dU} \tag{8.20}$$

Lastly, one is likely to want to know the probability (or the percent of time) that the wind will be equal to or less than a given wind speed. For example, what percent of a year will the wind speed be too low for the turbine to operate? This is known as the cumulative distribution function, $C(U)$, and is found by integrating from 0 m/s to a desired wind speed, U m/s:

$$C(U) = \int_0^U f(U)\,dU \tag{8.21}$$

The Weibull PDF and cumulative distribution function are given by

$$f(U) = \left(\frac{k}{c}\right)\left(\frac{U}{c}\right)^{k-1} \exp\left[-\left(\frac{U}{c}\right)^k\right] \tag{8.22}$$

$$C(U) = 1 - \exp\left[-\left(\frac{U}{c}\right)^k\right] \tag{8.23}$$

where
  k is the shape parameter
  c is the scale parameter.

Both k and c are functions of the mean wind speed, U, and the standard deviation of the wind speed, σ. The Rayleigh PDF and cumulative distribution function are given by

$$f(U) = \frac{\pi}{2}\left(\frac{U}{\bar{U}^2}\right)\exp\left[-\frac{\pi}{4}\left(\frac{U}{\bar{U}}\right)^2\right] \tag{8.24}$$

$$C(U) = 1 - \exp\left[-\frac{\pi}{4}\left(\frac{U}{\bar{U}}\right)^2\right] \tag{8.25}$$

The Rayleigh PDF only depends on one parameter, the mean wind speed, $\bar{U}$, making it easier to use (but not necessarily more accurate!) Note that the Rayleigh distribution is a special case of the Weibull distribution when $k = 2$ and $c = \bar{U}\left(\frac{4}{\pi}\right)^{\frac{1}{2}}$. An example of the Rayleigh distribution for different annual mean wind speeds is shown in Figure 8.17. This figure shows that as the mean speed increases, there is a larger probability of having higher wind speeds at a given site, and hence the annual power production has the potential to be higher.

Example 8.1: Rayleigh Wind Speed Distribution Calculations

An analysis of wind speed data (10-minute interval average, taken over a 1-year period) has yielded an average speed of $\bar{U} = 6$ m/s for a potential WT site. It has been determined that a Rayleigh wind speed distribution gives a good fit to the wind data. Recall that the Rayleigh PDF is given by Equation 8.24.

A. Estimate the number of hours per year that the wind speed will be between $U_A = 9.5$ and $U_B = 10.5$ m/s during the year.

   To do this, we will find the probability the wind speed will be in a given range during the year, $P(U_A \le U \le U_B)$, and multiply the probability by the number of hours in a year:

$$P(U_A \le U \le U_B) = \int_{U_A}^{U_B} f(U)dU = \int_0^{U_B} f(U)dU - \int_0^{U_A} f(U)dU = C(U_B) - C(U_A)$$

For a Rayleigh distribution $C(U) = 1 - \exp\left[-\frac{\pi}{4}\left(\frac{U}{\bar{U}}\right)^2\right]$

So,

$$C(U_B) - C(U_A) = \left[1 - \exp\left[-\frac{\pi}{4}\left(\frac{U_B}{\bar{U}}\right)^2\right]\right] - \left[1 - \exp\left[-\frac{\pi}{4}\left(\frac{U_A}{\bar{U}}\right)^2\right]\right] = P(U_A \le U \le U_B)$$

Substituting $\bar{U} = 6$ m/s, $U_B = 10.5$ m/s, and $U_A = 9.5$ m/s yields

$$P(U_A \le U \le U_B) = 0.049$$

In other words, the wind speed will be between U_B and U_A m/s 4.9% of the year. Thus, the total number of hours the wind speed is between U_B and U_A in a year is

$$\frac{\#\,\text{h}}{\text{year}} = 0.0494 \cdot \frac{24\ \text{h}}{\text{day}} \cdot \frac{365\ \text{day}}{\text{year}} = 432\ \text{h/year}$$

B. Estimate the number of hours per year that the wind speed is above 16 m/s. This problem is the same as part A with the wind speed ranges being

$$U_B = \infty \quad \text{and} \quad U_A = 16\ \text{m/s}.$$

$$P(16 \le U \le \infty) = \int_{16}^{\infty} P(U)dU = C(\infty) - C(16)$$

$$C(\infty) - C(16) = [1 - \exp[-\infty]] - \left[1 - \exp\left[-\frac{\pi}{4}\left(\frac{U_A}{\bar{U}}\right)^2\right]\right] = \exp\left[-\frac{\pi}{4}\left(\frac{U_A}{\bar{U}}\right)^2\right] = 0.0038$$

Thus, the total number of hours in a year the wind speed is greater than 16 m/s (i.e., between $U_A = 16$ m/s and $U_B = \infty$) is

$$\frac{\#\,\text{h}}{\text{year}} = 0.0038 \cdot \frac{24\ \text{h}}{\text{day}} \cdot \frac{365\ \text{day}}{\text{year}} = 33\ \text{h/year}$$

8.7 Turbine Performance

Given a WT site, selected by examining long-term wind speed data, a turbine can be picked and/or designed for the site. To do this, detailed engineering work has to be conducted to design the blades, tower, electrical systems, and so on. In this section, we build on Section 8.5 and focus on understanding the basics of blade aerodynamics to design an efficient WT. Aerodynamics is the study of bodies moving through gases. Its use allows the blade and tower shape to be optimized as well as optimizing blade rotational speed. The goal is

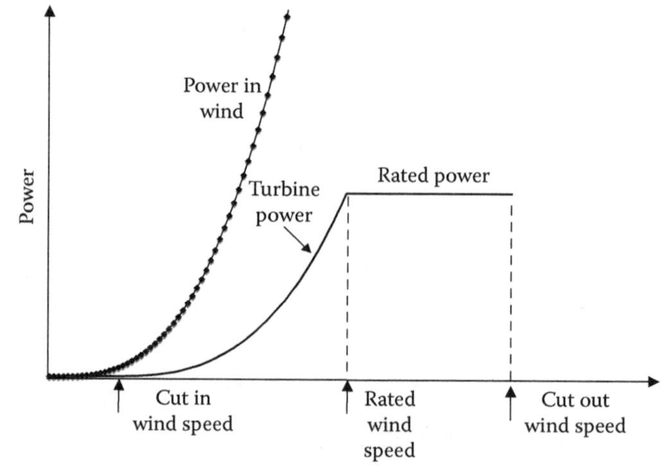

FIGURE 8.18
Wind power and turbine power curves versus wind speed.

to maximize the extraction of power from the incoming wind stream and minimize the effect of wind loads (both steady and unsteady).

The efficiency of a WT is the power it extracts compared to the power contained in the wind flowing through the area of the blades. Figure 8.18 contains a schematic comparing these quantities. Similar power curves are published for commercial turbines based on detailed engineering analysis. Note that the power contained in the wind is always greater than the power extracted by the turbine.

The turbine power curve contains several important points. The turbine does not start producing power until the wind speed reaches the cut-in wind speed. This is the lowest wind speed in which the rotor's aerodynamic power is greater than the sum of equipment loads and mechanical and electrical inefficiencies. The cut-out wind speed is the speed at which additional energy production is less than the cost of structural strength to withstand incremental fatigue damage. Lastly, the rated wind speed is the speed at which the turbine is operating at or near its peak efficiency. The goal is to operate the WT near or at its maximum efficiency.

Earlier, we saw that the standard measure of a WT's efficiency is known as the power coefficient, C_p, and is defined as

$$C_p = \frac{P_{\text{aero}}}{P_{\text{wind}}} = \frac{Q\Omega}{1/2\rho AU^3} \tag{8.26}$$

where

Q is the aerodynamic torque
Ω is the rotor rotational speed.

A homework problem will show that the maximum theoretical limit of C_p is given by the Betz limit, $C_{p,\max} = C_{p,\text{Betz}} = 16/27 \approx 0.5926$. In other words, the maximum power a WT should be able to extract from the wind is approximately 59% of the power crossing the area swept by the WT blades.

Another heavily used design metric is the TSR, which was defined in Equation 8.10. Design guidelines currently suggest limiting the tip speed to 90 m/s to minimize noise issues.

Example 8.2: Tip-Speed Ratio Calculation

What is the TSR for a 90-m-diameter turbine rotating at 15 rpm at a wind speed of 10 m/s?

$$R = \frac{D}{2} = 45 \text{ m} \quad \text{and} \quad U_{\text{wind}} = 10 \text{ m/s}$$

$$\Omega = 15 \text{ rpm} = 15 \frac{\text{rev}}{\text{min}} \cdot 2\pi \frac{\text{rad}}{\text{rev}} \cdot \frac{\text{min}}{60 \text{ s}} = 1.57 \text{ rad/s}$$

$$\text{TSR} = \frac{\Omega R}{U_{\text{wind}}} = \frac{1.57 \text{ rad/s} \cdot 45 \text{ m}}{10 \text{ m/ s}} = 7.07$$

Actuator disk theory allows one to find simple relationships for power and thrust (the force of the wind on the turbine). Another more comprehensive theory beyond the scope of this chapter allows one to design the blade shape to optimize power and thrust. This theory, called blade element momentum (BEM) theory, is covered in detail in many of the comprehensive wind energy textbooks (see Burton et al. [39] and Hau [31]).

The first assumption in actuator disk theory is that the area swept by the rotor blades is a permeable disk, shown in Figure 8.19, with an area swept out by the turbine blades. As the wind flows through the disk, momentum (i.e., power and thrust) is extracted.

$$U_{\text{wind}} > U_{\text{disk}} > U_{\text{wake}} \tag{8.27}$$

From the conservation of mass, one can see that as the velocity decreases, the affected area of the column of wind increases (i.e., the mass flow rate remains constant).

Note that the area of the wind column increases after passing through the actuator disk.

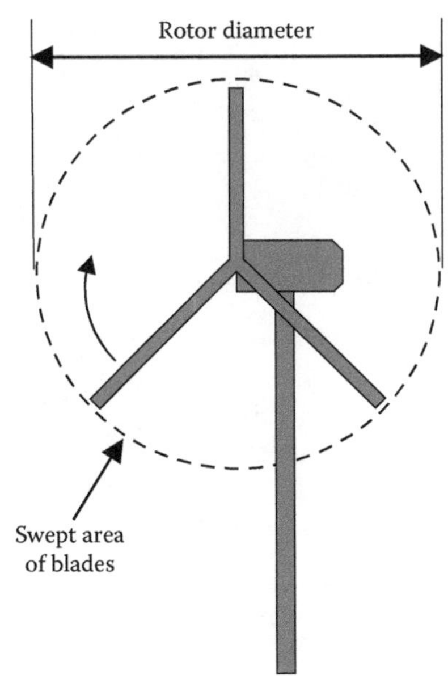

FIGURE 8.19
Actuator disk.

Additional assumptions are used to develop this theory. Key assumptions include the following:

- Homogeneous, incompressible, inviscid, and steady-state flow.
- Static pressure far upstream and downstream are equivalent.
- Velocity on either side of the actuator disk is the same.
- Infinite number of blades that are infinitesimally thin.
- Uniform thrust over actuator disk.
- Nonrotating wake.
- Velocity across the disk is constant.

Bernoulli's theory is valid:

$$P + \frac{1}{2}\rho U^2 = \text{constant} \tag{8.28}$$

More complicated theories such as BEM ease some of these restrictions.

Next, a momentum balance will be made on the actuator disk (see Figures 8.20 and 8.21) to find the power coefficient, C_p. The change in momentum from upwind to the wake is caused by the thrust on the disk aligned with the wind direction, which is the sum of the forces on the disk:

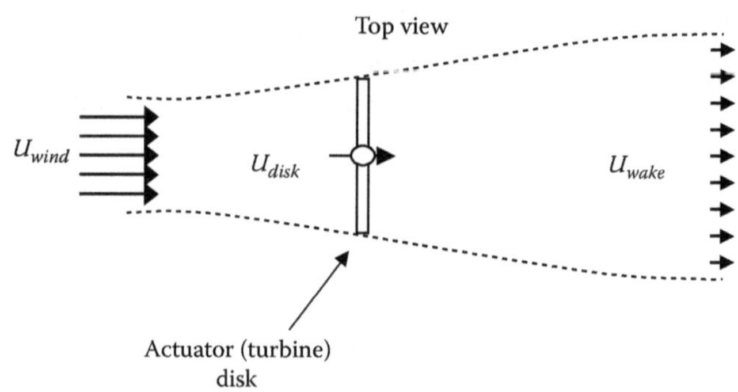

FIGURE 8.20
Actuator disk (top view), showing incoming wind passing through the actuator disk. (Note: Wind speed is denoted by arrow length.)

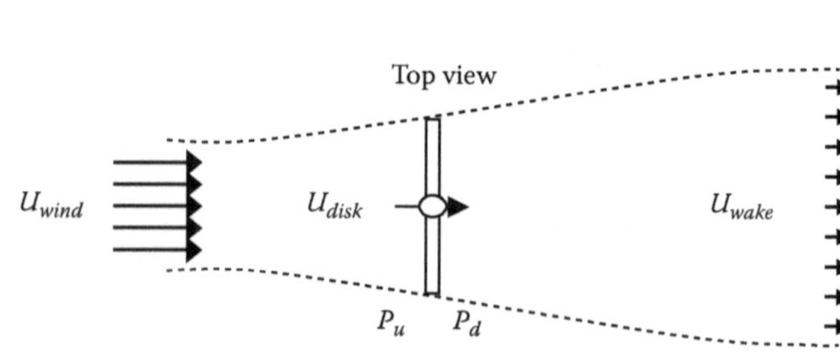

FIGURE 8.21
Actuator disk (top view) with quantities needed for momentum balance. (Note: P_u is the static pressure immediately upstream of the actuator disk; P_d is the static pressure immediately downstream of the disk.)

$$\sum F_{\text{disk}} = (\rho A U_{\text{disk}})(U_{\text{wind}} - U_{\text{wake}}) \tag{8.29}$$

For a purely axial force, the sum of forces can also be expressed in terms of the upstream to downstream pressure difference:

$$\sum F_{\text{disk}} = A(P_u - P_d) \tag{8.30}$$

The axial induction factor, a, expresses the reduction in wind speed at the disk compared to the free stream wind speed:

$$a = \frac{U_{\text{wind}} - U_{\text{disk}}}{U_{\text{wind}}} \tag{8.31}$$

or

$$U_{\text{disk}} = (1-a)U_{\text{wind}} \tag{8.32}$$

Combining Bernoulli's equation with the change in momentum, one can show

$$U_{\text{wake}} = (1-2a)U_{\text{wind}} \tag{8.33}$$

To find the power coefficient, C_p, we first find the aerodynamic power P_{aero} as the product of the thrust on the disk and the wind velocity at the disk, substituting in Equations 8.29 and 8.32:

$$P_{\text{aero}} = \sum F_{\text{disk}} U_{\text{disk}} = \left(2\rho A U_{\text{wind}}^2 a(1-a)\right)U_{\text{disk}} = \left(2\rho A U_{\text{wind}}^2 a(1-a)\right)U_{\text{wind}}\,(1-a)$$

Simplifying yields

$$P_{\text{aero}} = 2\rho A U_{\text{wind}}^3 a(1-a)^2 \tag{8.34}$$

The expression for P_{aero} can then be substituted into the relationship for C_p (Equation 8.8):

$$C_p = \frac{P_{\text{aero}}}{P_{\text{wind}}} = \frac{2\rho A U_{\text{wind}}^3 a(1-a)^2}{1/2\rho A U_{\text{wind}}^3} \tag{8.35}$$

which simplifies to

$$C_p = \frac{P_{\text{aero}}}{P_{\text{wind}}} = 4a(1-a)^2 \tag{8.36}$$

It is left as a homework problem to find the Betz limit of $C_{p\max} = 16/27 = 0.59$ from this relationship.

Similarly, a thrust coefficient, C_T, can be found that characterizes the axial thrust on the actuator disk. Summing forces on the disk yields the thrust, T:

$$T = \sum F_{\text{disk}} \tag{8.37}$$

The thrust can be related to the upstream wind speed using Equations 8.29, 8.32, and 8.33, and simplifying to

$$T = 2\rho A U_{\text{wind}}^2 a(1-a) \tag{8.38}$$

Nondimensionalizing the thrust by the power in the wind, P_{wind} yields the thrust coefficient:

$$C_T = \frac{2\rho A U_{\text{wind}}^2 a(1-a)}{1/2\rho A U_{\text{wind}}^2} \tag{8.39}$$

Simplifying Equation 8.39 yields

$$C_T = 4a(1-a) \tag{8.40}$$

The power and thrust coefficients are plotted in Figure 8.22 as functions of the axial induction factor, a. Note that the maximum power and thrust occur at different axial induction factors. The maximum power output of $C_{p\max} = C_{p\text{Betz}} = 0.59$ occurs at $a = 1/3$, whereas the maximum nondimensional thrust of 1.0 occurs at a value of $a = 0.5$. When power is a maximum at $a = 1/3$, $C_T = 8/9$.

Also note that the axial induction factor theory is not valid for $a > 0.5$. Equation 8.33 shows that the speed of the wake, U_{wake}, behind the turbine decreases as the axial induction factor increases from 0.0 to 0.5. At $a = 0.5$, the wake speed equals zero. For values of $a > 0.5$, the wake speed would be negative, violating the assumptions of this simple theory.

Actuator disk theory provides a theoretical limit for the maximum power coefficient of a WT, the Betz limit. In practice, C_p is a function of many variables including wind speed, pitch angle, and TSR. Values of C_p are obtained through experiments or numerical modeling. Figure 8.23 illustrates the wake tip vortex produced by a rotating WT located in a wind tunnel.

With C_p determined for a particular turbine design through testing or modeling, the aerodynamic power of the wind is determined by rearranging

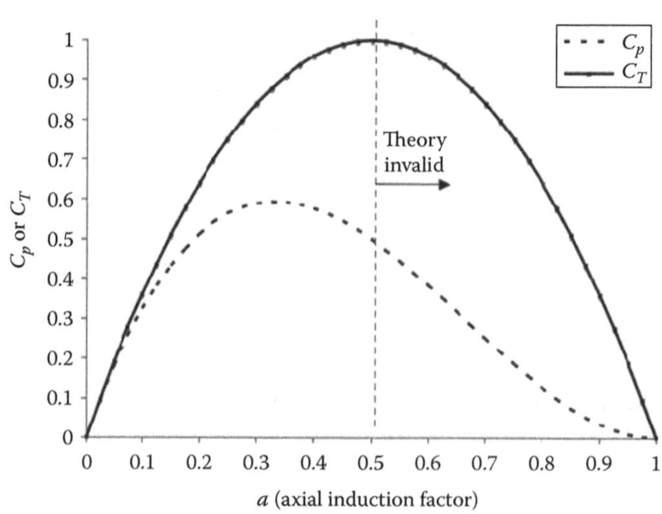

FIGURE 8.22
Power and thrust coefficients versus induction factor showing maximum power coefficient, $C_{p,\max}$.

FIGURE 8.23
Flow visualization tests conducted in the NASA Ames wind tunnel using smoke emitted from the tips of the turbine helped researchers determine the extent of the wake under a controlled set of conditions. (Courtesy of DOE/NREL, Golden, CO.)

$$C_p = \frac{P_{\text{aero}}}{P_{\text{wind}}} \tag{8.41}$$

to yield

$$P_{\text{aero}} = C_p P_{\text{wind}} = \frac{1}{2} C_p \rho A U_{\text{wind}}^3 \tag{8.42}$$

Given C_p and the expected probability of wind distribution, the annual energy production can be estimated.

Example 8.3: Determine Annual Energy Production

Given a WT with the following parameters, calculate the annual energy production: $C_p = C_{p\max} = 0.48$ (C_p value assumed maximum and constant for all wind speeds).

Rated power = 5,000 kW = 5.0 MW
Annual average wind speed = 8.0 m/s
Rated wind speed = 13.0 m/s
Cut-out wind speed = 25 m/s
Rotor radius = 48.0 m
Area = 7,238.2 m^2
Air density = 1.20 kg/m^3

The annual energy production for all wind speeds will be illustrated by first calculating the annual energy production at a single wind speed. Then, the methodology will be applied to all wind speeds to arrive at the annual energy production for all wind speeds.

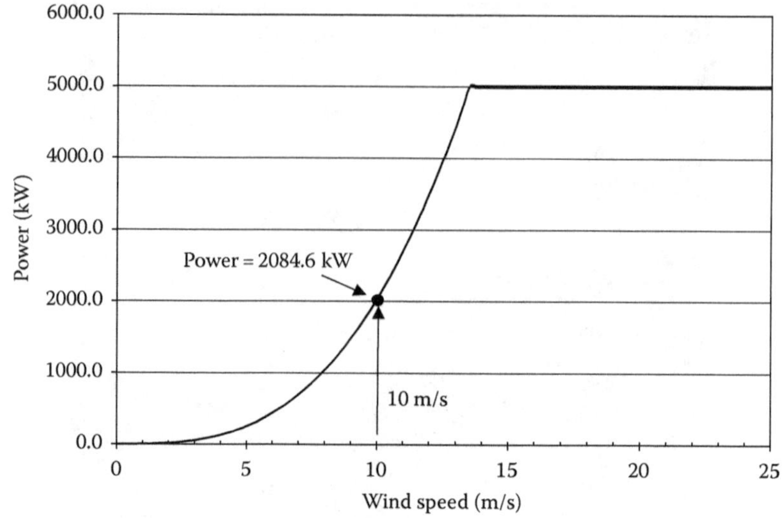

FIGURE 8.24
Power versus wind speed for a 5,000-kW turbine with $C_p = C_{pmax} = 0.48$.

First, determine the power production at a single wind speed, say 10 m/s. Then, a Rayleigh distribution will be calculated to determine the number of hours the wind speed is at 10 m/s. The power and number of hours are then multiplied to determine the yearly energy production for a 10 m/s wind speed.

Power production: For a wind speed of 10 m/s, the power produced for the values shown earlier is

$$P_{aero} = \frac{1}{2} C_P \rho A U_{wind}^3 = \frac{1}{2}\left(0.48 \cdot 1.2 \cdot 7{,}238.2 \cdot 10^3\right) = 2{,}084.6 \text{ kW}$$

A plot of the power produced for all wind speeds is shown in Figure 8.24. Note that above the rated wind speed, from 13 m/s up to the cut-out wind speed, the power production is constant at the rated level of 5,000 kW.

Number of hours at 10 m/s: Assuming a Rayleigh PDF with an average wind speed of 8 m/s, the annual probability of a 10 m/s wind, calculated for the interval from 9.5 to 10.5 m/s, is

$$C(10.5) - C(9.5) = \left[1 - \exp\left[-\frac{\pi}{4}\left(\frac{10.5}{8}\right)^2\right]\right] - \left[1 - \exp\left[-\frac{\pi}{4}\left(\frac{9.5}{8}\right)^2\right]\right] = 0.0719 = 7.19\%$$

Thus, in a given year, the number of hours the wind will blow at 10 m/s at this site is the probability times the number of hours in a year.

$$0.0719 \cdot \frac{24 \text{ h}}{\text{day}} \cdot \frac{365 \text{ day}}{\text{year}} = 629.8 \text{ hours}$$

A plot of the number of hours the wind blows at a given wind speed for all wind speeds is shown in Figure 8.25.

Finally, the annual energy production for this WT, at this site, at a wind speed of 10 m/s is found by multiplying the power production by the number of hours the power is produced, in this case:

Annual energy production at 10 m/s = 2,084.6 kW · 629.8 hours = 1,312.9 MWh

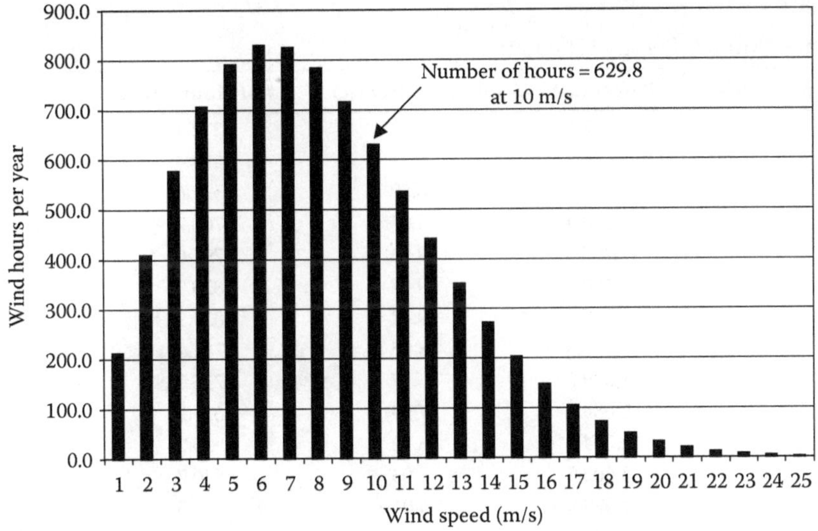

FIGURE 8.25
Number of hours at a given wind speed for a Rayleigh distribution with an annual average wind speed of 8.0 m/s.

Similarly, this is done for all wind speeds up to cut out. The energy production from all wind speeds is then summed to determine the annual energy production estimate. A summary table of the calculations is shown in Table 8.5.

Examination of Table 8.5 shows that the annual energy produced at this site is calculated to be 13,873 MWh. If electricity costs 10 ¢/kWh ($0.10/kWh), the annual value of the energy produced in this example would be

$$\text{Annual value} = 13{,}873{,}070\ \text{kWh} \cdot \frac{\$0.10}{\text{kWh}} = \$1.4\ \text{million}$$

Now the rationale to build wind farms comprised of dozens of WTs with a similar energy (cash) output can be seen. The wind is literally producing money!

As noted above and shown in Table 8.5, the power produced above 13 m/s (rated wind speed) is a constant value of 5,000 kW. Control schemes are typically implemented to limit the maximum amount of power produced by a turbine to the rated turbine power (5 MW in this case) at the rated wind speed (13 m/s). Limiting the power production is important from both a structural standpoint and an electrical generation standpoint. In both cases, WT components must be sized to operate under maximum defined loads and thus cannot handle larger loads.

Another measure of a WT's performance is the capacity factor. Because the wind never blows continuously, a turbine will never be able to operate steadily at full power. The capacity factor is the ratio of the energy produced to the theoretical maximum the turbine could produce (i.e., rated power production, 24 hours a day, 7 days a week for a year):

$$\text{Capacity factor} = \frac{\text{Energy generated}}{\text{Rated power} \times (\#\ \text{h/year})} = \frac{\text{Energy generated}}{\text{Rated power} \times 8{,}760\ \text{hours}}$$

For this example, the capacity factor is found to be

$$\text{Capacity factor} = \frac{13{,}880\ \text{MWh}}{5\ \text{MW} \times 8760\ \text{hours}} = 31.7\%$$

TABLE 8.5

Power and Annual Energy Calculation

Wind Speed (m/s)	Power (kW)	Probability (%)	# of Hours/Year	Energy (kWh)
1	2.1	2.42	212.4	442.7
2	16.7	4.67	409.4	6,827.6
3	56.3	6.59	577.6	32,507.8
4	133.4	8.07	706.7	94,283.4
5	260.6	9.03	791.0	206,114.7
6	450.3	9.47	829.3	373,429.3
7	715.0	9.42	824.9	589,808.3
8	1,067.3	8.95	784.2	837,017.8
9	1,519.7	8.17	716.1	1,088,282.6
10	2,084.6	7.19	630.2	1,313,739.7
11	2,774.6	6.12	535.7	1,486,478.3
12	3,602.2	5.03	440.7	1,587,567.2
13	4,579.9	4.01	351.3	1,608,943.1
14	5,000.0	3.10	271.6	1,358,130.2
15	5,000.0	2.33	203.9	1,019,404.1
16	5,000.0	1.70	148.7	743,288.9
17	5,000.0	1.20	105.4	526,756.5
18	5,000.0	0.83	72.6	362,992.5
19	5,000.0	0.56	48.7	243,323.3
20	5,000.0	0.36	31.7	158,710.6
21	5,000.0	0.23	20.2	100,758.5
22	5,000.0	0.14	12.5	62,274.8
23	5,000.0	0.09	7.5	37,478.8
24	5,000.0	0.05	4.4	21,967.4
25	5,000.0	0.03	2.5	12,541.7
		Total: 99.76		**Total: 13,873,070.1 kWh**

For comparison, typical WT capacity factors are 20%–40% [41].

Finally, the calculations shown in Table 8.5 would typically be done at wind speed increments of 0.10 or 0.25 m/s. Note that slight differences in values between the spreadsheet table and the example calculations are due to round-off errors in the example calculations.

8.7.1 Control Schemes

As a comparison, we will briefly examine the major control schemes currently used in large commercial WTs. In the first case, called constant-speed/variable-pitch (CSVP) the rotor speed is held fixed or constant while the blade pitch is variable. In the second case, called variable-speed/variable-pitch (VSVP), both the rotor speed and the blade pitch are variable. The CSVP turbine has a simpler mechanical design, the control scheme is easier to implement, and it also offers advantages by simplifying the electrical power generation. The VSVP control scheme is more complicated, both mechanically and electrically. Is the more complicated VSVP turbine worth it from an energy production standpoint?

Figure 8.26 illustrates the power output versus wind speed for both WT control methods. The operating parameters cited in Example 8.3 were also used for Figure 8.26. In addition, the results in Example 8.3 are the same as shown here for the VSVP turbine.

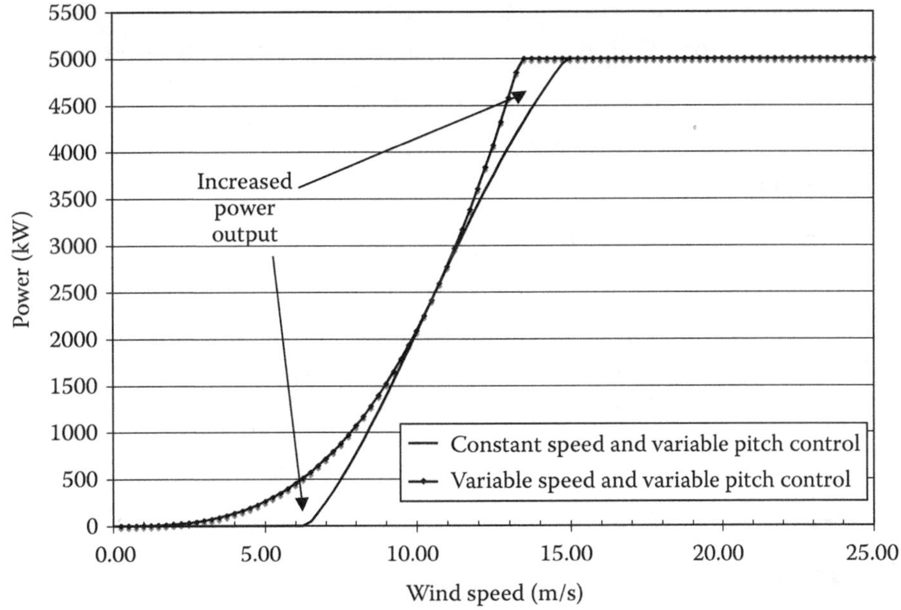

FIGURE 8.26
Power output for constant-speed/variable-pitch and variable-speed and pitch-controlled schemes.

The key point is that the VSVP turbine has a higher power output. Two regions can be seen where the VSVP turbine power output is higher than the CSVP turbine output. This increased power output will lead to a higher annual energy production and capacity factor. This is because of the ability to operate the VSVP turbine at $C_p = C_{p\max}$ at all wind conditions. The CSVP runs at a variable, C_p. Because the turbine cannot vary rotor speed, it cannot run at $C_p = C_{p\max}$ at all conditions.

Note that both turbines eventually reach a rated power of 5 MW. However, the CSVP reaches the 5 MW turbine rating at a higher wind speed, 15 m/s, whereas the VSVP turbine reaches 5 MW rated power at 13.5 m/s. This is because the CSVP turbine is operating at a lower C_p and thus needs a higher wind speed to reach the higher 5 MW power rating. For the purposes of this illustration, C_p for the CSVP was assumed to be a function of TSR (ΩR/(wind speed)) and a fixed pitch angle of 0°:

$$C_p(\lambda, \beta) = 0.5176\left(\frac{116}{\lambda_i} - 0.4\beta - 5\right)e^{-21/\lambda_i} + 0.0068\lambda \tag{8.43}$$

where

$$\lambda = \text{Tip speed ratio} = \frac{\Omega R}{\text{Wind speed}}$$

$$\frac{1}{\lambda_i} = \frac{1}{\lambda + 0.08\beta} - \frac{0.035}{\beta^3 + 1}$$

β is the blade pitch angle.

Examining Table 8.6, we can see that the VSVP-controlled turbine produces over 15% more energy (i.e., 15% more yearly revenue), and thus has a capacity factor 15% higher than

TABLE 8.6

Energy Production and Capacity Factor for VSVP and CSVP Wind Turbines

	CSVP	VSVP	Percent Increase from CSVP to VSVP (%)
Energy (MWh)	12,000	13,880	15.3
Capacity factor	27.5%	31.7%	15.3

the CSVP turbine. The goal of WT manufacturers is to design and build turbines, such as the VSVP turbine, that produce more power while minimizing the cost to design and build these turbines, to achieve the highest level of performance possible.

8.8 Wind Farms

A wind farm is a group of WTs producing electricity in a single location (see Figure 8.27). To have a successful wind farm, selecting a proper location is key [42,43]. The most important item is selecting a site with the desired wind resource. The winds must be consistent and have a sufficient speed to generate the desired energy. Wind resource maps (see Section 8.6.2) are a typical starting point. Meteorological data from "met" towers erected by companies interested in a site give a detailed view of the wind resource at a specific site.

Potential impacts of wind farms also need to be considered. As noted earlier, bird/bat activity and avian migration need to be considered to prevent negative effects on the environment. Noise effects can also have an impact on the area local to a proposed wind farm site. Other land-use questions are also studied, for example: Will the turbines create interference for local air traffic? These additional factors must be weighed in addition to the wind resource when selecting a WT site.

It should be noted that wind farms can be reliably located in extreme freezing climates. In February 2021 portions of Texas experienced an extreme winter storm with record low temperatures. Severe power outages occurred that were primarily related to frozen infrastructure at natural gas facilities. However, WTs in West Texas also experienced failures due to the freezing weather. That is because those turbines had been installed without cold weather packages that are routinely used in colder climates. These protections include de-icing equipment, component heating systems, and special turbine blade coatings [44].

Besides wind resource and environmental effects, other factors that must be considered include land access (lease or purchase, zoning, and permits must be obtained) and sufficient capital to fund the project (approximately $1 million/MW of installed generating capacity) [42]. The wind farm would ideally be located near existing transmission lines to limit the cost (up to $1 million/mile) of installing large lengths of new transmission lines that drive the balance of system (BOS) costs up. In addition, there must be a market available and an agreement in place to buy the electric power produced. Assuming that other factors are favorable, based on wind resource and turbine costs, the wind farm developer selects a WT model. Additional steps include checking with the manufacturer for availability and considering a service provider for annual maintenance.

FIGURE 8.27
Wind farm located in Tehachapi, California, operational since the early 1980s. (Courtesy Ian Kluft.)

Another consideration is the issue of variability. The power output from a single WT can vary because the wind speed can fluctuate. On the other hand, a wind farm, with WTs distributed over a large, geographically diverse area, will have less chance of the wind speed being low for all WTs in the wind farm. A study by Archer and Jacobson [45] indicates that when multiple wind farms are interconnected together in an array through the electric transmission grid, the correlation between the wind speeds at the sites decreases, thus producing steadier output, as was shown in Chapter 5. In addition, the benefits of interconnecting wind farms continued to grow as the number of connected wind farms increased.

On the other hand, distributing WTs in a wind farm over a large geographic area drives up BOS costs due to the increased cost of electrical interconnects and the potential need for new transmission lines. Wind farms are typically arranged with a rectangular grid pattern of turbines with each successive row staggered, as shown in Figure 8.28. Turbines are staggered by a specified number of rotor diameters apart, both side to side (crosswind spacing) and between rows (downwind spacing) of turbines. This distance varies with the wind farm owner, site conditions, and wind resource. If only BOS costs are considered, the turbines should be placed closer together. From a power output standpoint, if 100 turbines rated at 2 MW each are to be placed at this site, the total power output could be anticipated to be 200 MW. Unfortunately, this is not the case.

Consider a typical wind farm development scenario. Once transmission, land-use issues, and a power purchase agreement are in place, the wind farm development can proceed. The wind resource estimates from the met tower for a potential site are anticipated to be sufficient for power production; however, wind farms do not always produce the expected power. Several factors could be responsible for this unexpected reduction in power output.

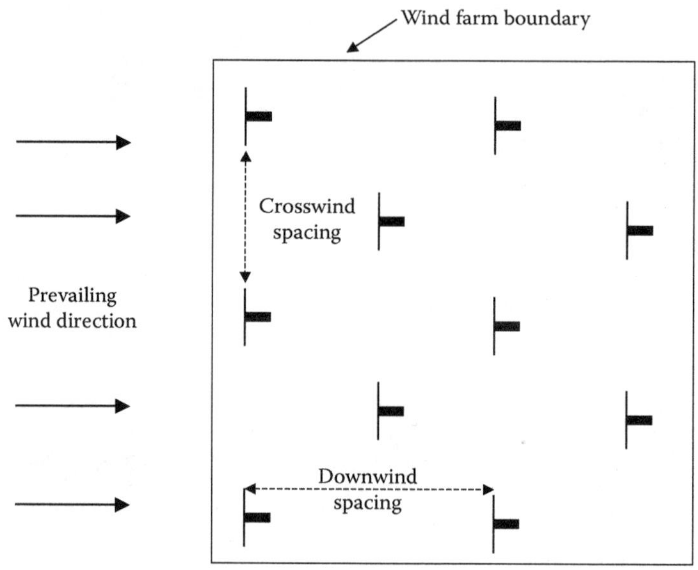

FIGURE 8.28
Schematic (top view) of typical wind farm layout.

For example, the met tower used to collect data to make the wind resource predictions can only measure wind at one location. The wind resource could change throughout the site depending on terrain such as ridges and hills or on surface features like a forest. As turbines are built farther away from the met tower, the resource prediction becomes less accurate.

Another suspect for producing less-than-expected power is wake effects or array effects. Depending on the locations of the turbines positioned on the site and the direction of the wind, some turbines can be located downwind of other turbines. As a turbine extracts energy from the wind, the flow downstream becomes slower and more turbulent. This slower, more turbulent flow persisting downwind of a turbine is known as a wake. Turbines built in the wake of other turbines have a diminished wind resource and do not produce the power levels of the upwind turbines. This will result in the production of less energy, which means a reduced monetary return for investors. Downstream of a WT (perhaps 15 rotor diameters or much more depending on conditions), the wind speed in the wake will begin to approach the free stream velocity, thereby allowing higher energy production by downstream turbines. Unfortunately, increasing the spacing between turbines increases BOS costs (e.g., more construction and materials), resulting in a lower return on investment. Several projects studying wake effects strive to quantify wake losses to better design wind farms.

Wake modeling is an active area of research and detailed treatment is beyond the scope of this textbook. However, a 2009 study by Stovall [46] illustrates the significance of wake interaction effects. Stovall used computational fluid dynamics and large-eddy simulations to examine the overlying boundary layer effects on wake propagation. Two turbines, with 100 m rotor diameters, are modeled as being inline. The incoming wind speed to the upstream turbine is approximately 8–10 m/s. The objective is to determine the effect the wake has on the power output of the downstream turbine compared to power output of the upstream turbine. In this study, the turbines have a downwind spacing of 9 rotor diameters. The power coefficient for the turbines in this simulation is approximately $C_p = 0.50$.

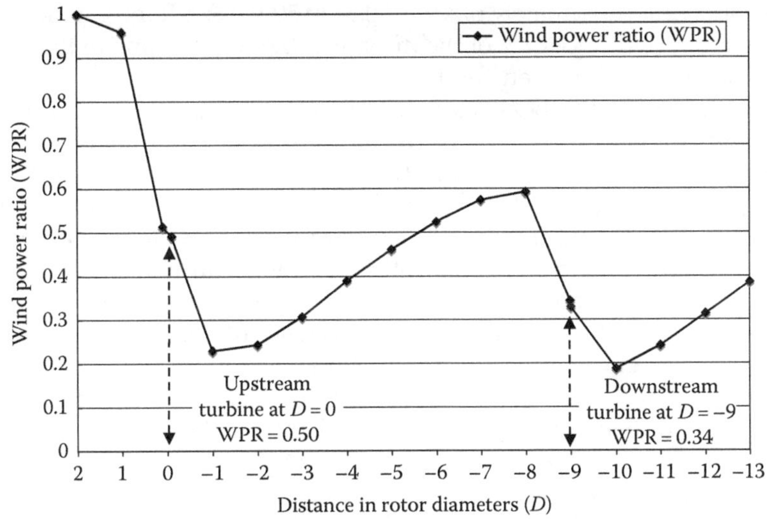

FIGURE 8.29
Wind power wake reduction: Local wind power normalized by wind power level at an inlet. (From Stovall [46]. With permission.)

Prior to modeling the power output of the WTs, a simulation of the ABL is created. Once the ABL simulation is validated, turbines are imposed into the domain with the turbine rotors modeled via actuator disks. The actuator disks impose a pressure drop on the flow field proportional to the power extracted by a turbine.

Results from the numerical simulation show the diminished wind resource caused by the wake-forming downstream of a WT. Figure 8.29 illustrates the power in the wind calculated as the area integral of velocity cubed (Equation 8.6). Distances between the turbines are shown in rotor diameters (*D*). The inlet of the computational domain used in this numerical simulation is located at $D = +2$. The first turbine is located at $D = 0$, whereas the second turbine is located 9 rotor diameters downstream of the first turbine at $D = -9$. All wind power is expressed as a wind power ratio (WPR) of the wind power at a given location divided by the inlet wind power (located at $D = +2$). This results in a WPR = 1.0 at the inlet ($D = +2$).

Notice that the nondimensional wind power extracted by the first turbine is half of the wind power at the inlet, yielding a WPR = 0.50, as would be expected because the power coefficient is $C_p = 0.50$.

Downstream of the first turbine, the wake is represented by a further reduction in wind power due to turbulent losses. As distance increases downstream of the first turbine, the wake begins to recover (*D* increases from −1 to −8) as surrounding higher speed (i.e., higher energy) wind mixes with the wake. The power in the wake, however, does not achieve free stream power levels prior to reaching the second turbine.

Because the incoming power of the wind is lower immediately upstream of the second turbine, the power output of the second turbine should be lower than the power output of the first turbine. Indeed, the WPR of the second turbine in this simulation is WPR = 0.34, indicating that the second turbine was only able to extract 34% of the energy available in the incoming stream at $D = +2$. The wake from the first turbine adversely impacts the power output of the second turbine. Another way to view this is that the ratio of the power extracted by a second turbine (located 9*D* downstream of the first turbine in this study) to the first turbine is approximately 67% (i.e., 0.34/0.50 = 0.67). In economic terms, the second turbine will only produce 67% of the income of the first turbine in this scenario.

Increasing the spacing between the turbines in the downwind direction reduces the wake effect but increases the BOS costs. A trade-off between these two competing economic factors remains an active area of research. Although the results from the Stovall study match up with experimental data from wind farms, the simulation addresses a relatively simple configuration. The computational resources required for modeling an entire wind farm (dozens of turbines) are significant and remain an area of research for government and industry to develop for optimizing wind farm layouts [47,48]. Significant work remains to develop a robust, accurate, and fast optimization tool for wind farm wake effects. With such a tool, BOS costs can be optimized simultaneously with turbine location resulting in maximum energy (income) production.

Because the wakes generated by upstream turbines decrease the performance of downstream turbines, it has been found that computer-controlled steering of the orientation to the wind of individual turbines in a manner that slightly decreases the output of those turbines but reduces the wake effects on downstream turbines can increase the total electricity output of an entire wind farm. This is a classic example of the benefits of optimizing the performance of an overall system as opposed to its individual components. An NREL modeling study found that wake steering can increase annual electricity production by between 0.4% and 1.7% with even higher gains in annual revenue [49]. Simley et al. [50] describe experimental results for two wind turbines in a commercial wind farm. At the time of this writing, wake steering has not yet been applied at any scale in operating wind farms.

8.9 Offshore Wind Energy

Offshore wind farms (see Figure 8.30) address limits of land-based wind farms such as transmission line access and capacity [51], and space for WTs (more of an issue in higher-density population areas such as Europe), while providing access to a vast and steady wind resource. Significant development remains for wind farms in water deeper than 30 m. BOS costs can double in an offshore installation compared to land-based installation. Still, the complexity of offshore technology development is more than offset by the need and desire to harness the higher wind resource available offshore (ocean as well as large inland lakes).

Offshore wind energy has a number of benefits including:

Better wind resources:

- Reduced turbulence resulting in steadier winds
- Higher wind speed resulting in higher energy production
- Higher-capacity factors resulting in better electric load matching.

Closer proximity to electric load centers resulting in:

- Lower transmission costs and constraints
- Ability to serve high-cost regions.

FIGURE 8.30
Offshore wind farm. (Courtesy of Hans Hillewaert.)

Avoids land-based size limits:

Shipping—no land-based roadway limits
Assembly—sea-based cranes are taller
Larger machines are more economical.

These benefits are significant throughout the world. There is over 75,000 MW of installed offshore wind power capacity in the world; approximately half of this is off the coast of China [12]. The rapid increase in offshore wind capacity through 2023 is shown in Figure 8.31 [12,52]. The United States sharply lags Europe and Asia in offshore wind development. The first offshore wind farm in the United States was installed in 2016 — a $300 million, 30-MW project off the coast of Rhode Island (Block Island Wind Farm) [53]. The United States had 174 MW of operational offshore wind power as of May 2024, with three new projects totaling 4,100 MW of capacity under construction [54]. The cost of offshore wind power has come down and is projected to continue to drop as the industry expands. The mid-range levelized cost of energy (LCOE) estimate for fixed-bottom offshore turbines exploiting wind resources like those being developed off the U.S. Atlantic Coast is projected to decline from $110/MWh in 2022 to $70/MWh a decade later [55]. These estimates are based on 12 MW turbines with 216 m rotor diameter and hub heights of 137 m.

Most of the offshore wind capacity installed or under construction around the world is sited at locations with an ocean depth that is less than 40 m and distances to shore of less than 50 km [56]. However, new projects are advancing beyond these envelopes.

Of the 48 contiguous United States, 28 have a coastal boundary, and these 28 states use three-quarters of the electricity in the United States [57]. Because these coastal load centers are typically located hundreds of miles from land-based wind energy sources, offshore wind power provides a nearby source of energy. Many of these coastal states have relatively modest onshore wind resources, but they have significant offshore resources.

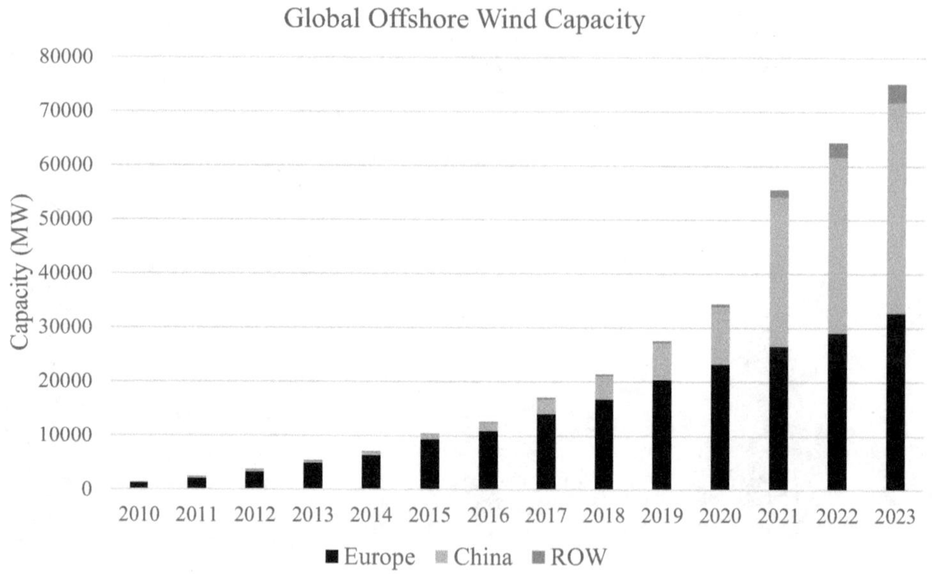

FIGURE 8.31
Global cumulative offshore wind capacity from 2010 through 2023. (Data from Global World Energy Council [12] and Global Offshore Wind Report [52].)

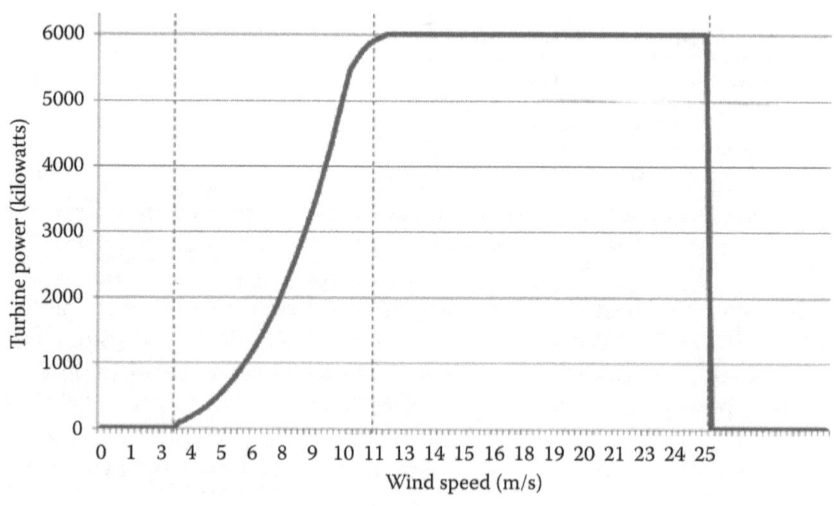

FIGURE 8.32
Power curve for a 6 MW wind turbine assumed to be representative of currently available offshore technology. (From Musial et al. [58].)

A 2016 NREL study [58] assessed the potential available offshore wind power as a function of depth and coastal location for the United States. As shown in Figure 8.13, the United States has significant wind resources located off the Atlantic and Pacific Coasts, in the Gulf of Mexico and Great Lakes, and surrounding Hawaii. The study estimated resources assuming a 100 m hub height, with technical potential constrained by currently available technology. Figure 8.32 [58] shows the power curve for a 6 MW offshore turbine that was used as the basis for the assessment. The estimates of technical potential also exclude identified areas with economic or environmental conflicts, although more detailed assessments

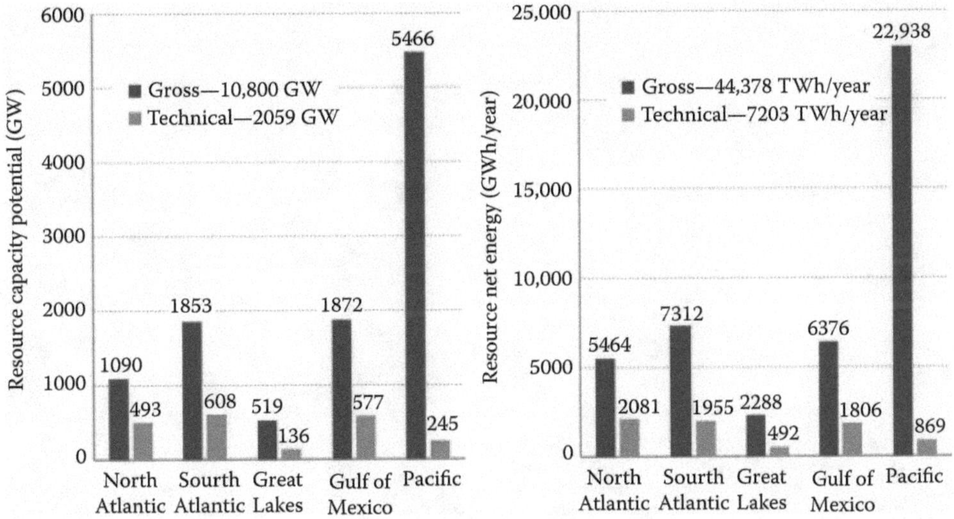

FIGURE 8.33
Capacity (left) and net energy (right) offshore gross resource (dark shading) and final net technical (light shading) potential estimates for five U.S. offshore wind resource regions. (From Musial et al. [58].)

would be required before development could move forward at specific sites. Losses from wake effects and electrical systems were accounted for in determining net technical potential. Figure 8.33 [58] shows the resulting estimated gross and technical potential for offshore wind by U.S. region. Of the 7,203 TWh/year of technical potential, 22% is at <30 m depth and an additional 20% at depths between 30 and 60 m. The availability of shallow water sites is more limited in some areas such as California, where only 4% of technical potential is at a depth of <60 m. Regarding distance offshore, 7% of the U.S. technical potential is located <3 nautical miles (5.6 km) offshore, 13% between 3 and 12 nautical miles (5.6 and 22.2 km), and 41% between 12 and 50 nautical miles (22.2 and 92.6 km) [58].

Design of substructures to support wind turbines in offshore environments has been a key technical challenge for this industry. Figure 8.34 illustrates six different turbine substructures [59]. The fixed-bottom substructures are used in relatively shallow waters, at depths of up to 60 m. This technology is relatively well established, although innovative designs are still required to accommodate seabed conditions at specific locations. Significant technology challenges exist to develop floating WTs that could be deployed for waters deeper than 60 m. According to the Global Wind Energy Council, in 2022 there were only five operating wind farms with a total of 32 turbines using floating substructures, although more were under development [52]. NREL's 2024 Annual Technology Baseline report anticipates floating WT technology will not be commercially feasible in the United States before 2030 [55]. Modeling a floating WT is a challenging task because it must include effects from wind and wave interaction with the turbine and floating substructure, a nonstationary and rotating turbine, and a control system, which monitors and controls the turbine for these effects.

In addition to design improvements and cost reductions for wind turbine and substructure technology, opportunities for offshore wind energy would benefit from improved engineering solutions for installation of equipment in offshore environments and for approaches to streamline operation and maintenance (O&M) activities. Expanding offshore wind development also demands improved transmission infrastructure, both offshore and onshore. European countries, led by Denmark, are pursuing the development of

FIGURE 8.34
Six different substructures for offshore wind turbines. The three on the left are fixed-bottom substructures (monopole, jacket, and inward battered guide); the three on the right are floating substructures (semisubmersible, tension leg platform, and spar). Illustration by Josh Bauer, NREL. (From Gilman et al. [59].)

"energy islands," which could host transmission infrastructure and O&M operations for a network of offshore wind farms sending power to multiple destinations. These islands might also house facilities for using wind-generated electricity to produce green hydrogen or other fuels that could be shipped or piped to shore.

Another need is for improved resource and site characterization information for offshore locations. As discussed in Section 8.2, concerns have been raised about the potential impacts of offshore wind installations on migratory birds, marine mammals, and fisheries. Biological surveys have been conducted to characterize the distribution of wildlife in relation to potential offshore wind locations, but empirical information on actual impacts or mitigation methods is limited for this nascent technology. There is also need for better information on impacts of offshore wind farms on human activities and other technologies, including fishing and potential interference with radar systems for air traffic control, weather, and navigation [59].

8.10 System Advisor Model for Wind Farm Analysis

As introduced in Chapter 1, the System Advisor Model (SAM) is a numerical performance and economic model designed by NREL to make performance predictions and estimate financial metrics for stand-alone or grid-connected electric power projects in the distributed and central generation markets. For wind projects, the model calculates the cost of generating electricity based on information about the project's desired location and corresponding wind resource, WT type, installation configuration, capital and operating costs, and various financial parameters such as interest rates, tax treatment, and incentives. The performance model makes hour-by-hour calculations for the power output of wind systems generating a set of 8,760 hourly values that represent the system's electricity production during a specified year.

Running SAM for wind energy requires wind resource information for the project location. Hourly wind resource files for a few dozen representative locations come packaged with the model. Alternatively, the user can download wind resource files from NREL's WIND Toolkit or specify the parameters of a Weibull distribution to approximate the wind conditions. Users can select from a library of commercial WT models with associated power curves or specify turbine characteristics from which the power curve can be estimated. Both onshore and offshore installations can be evaluated. The model comes with a comprehensive set of default input data covering financial and performance variables, which are frequently updated based on NREL's Annual Technology Baseline assessments (atb.nrel.gov). The analyst is responsible for modifying the input data as appropriate for the project being evaluated.

To utilize SAM, begin by downloading and installing NREL's free software at https://sam.nrel.gov. From the SAM home page, you can create a new utility wind energy project as follows. The following instructions are for the "SAM 2023.12.17" release.

Example 8.4: SAM Method to Determine LCOE

Using NREL's SAM software, model a 50 MW utility wind farm using turbines with a nameplate capacity of 2,000 kW. Assume a 25-year lifetime, a 12% internal rate of return (IRR) target to be met in year 15, an inflation rate of 2%, a real discount rate of 7%, and a 2.75 ¢/kWh production incentive credit. Run the model for two different locations: southern Wyoming and central Iowa. For each location, find the project's capacity factor, LCOE, Power Purchase Agreement (PPA) price, and Net Present Value (NPV).

Solution:

SAM allows users to modify turbine cost and performance specifications, financing, government incentives, wind farm layout, etc., allowing them to create a model as simple or complex as needed. For this example, we will only be modifying the model to fit the assumptions in the problem statement while leaving all other options as the defaults.

Start up the SAM program and start a new project, selecting "Wind," "Power Purchase Agreement," and "Single Owner." In a PPA project, the generating plant will be connected to the grid and earn revenue through electricity sales at a negotiated price that may escalate over time. A "Single Owner" project is one that is owned, built, and operated by a single entity.

With the type of project selected, you will then see a menu on the left side with links to parameter sets for "Wind Resource," "Wind Turbine," "Wind Farm," etc. The "Wind Resource File" page is open first, with options to choose a representative file or download a file from the WIND Toolkit.

The first wind farm location can be selected through the "Wind Resource File" drop-down menu on the "Wind Resource" tab. Start by selecting the "Southern Wyoming flatlands" file from the library. Run the model for this location after setting up the rest of the model parameters. For central Iowa, we will download a file from the NREL WIND Toolkit. We will return to that after completing the run for the southern Wyoming location.

Next, select the "Wind Turbine" tab and select the "Vestas V80–2.0" turbine, which has a rated output of 2,000 kW and 80 m rotor diameter. Figure 8.35a shows the power curve from SAM for this turbine. Take note of the default hub height and the cut-in, rated, and cut-out wind speeds for this turbine.

Now select the "Wind Farm" tab. At the top of the page, choose the option "specify desired farm size." Set the farm size to 50,000 kW. Note the resulting number of turbines (25) and system nameplate capacity (50 MW). When the "desired farm size" option is chosen, SAM determines a wind farm layout, which is shown in the grayed-out "Turbine Layout" section of this tab (Turbines per row = 5, number of rows = 5, turbine spacing = 8 rotor diameters, row spacing = 8 rotor diameters, offset for rows = 0 diameters, offset type = every other row, row orientation = 0 degrees), see Figure 8.35b. If the option to "Specify number of turbines" is selected instead, the user can define their own layout. The "Wind Farm" tab also allows the user to select an internal wake model or set a constant loss percentage due to internal wake effects. This example uses the "Simple Wake Model" incorporated into SAM's wind model.

The "Losses" tab (Figure 8.35c) allows SAM users to specify the degree to which turbine and wind farm performance fall short of ideal production, including losses due to wake effects, downtime, electricity line losses, and sub-optimal turbine performance. Note but do not change the default values.

Next, on the "Installation Costs" tab, note the turbine cost, BOS costs, and O&M costs, but do not modify them. The costs represented in SAM 2023.12.17 are shown in Figure 8.35d. SAM specifies costs in year 1 dollars, corresponding to the first year of power production. (Pre-production costs are all counted in year 0.) Turbine costs include the rotor, drivetrain, nacelle and tower for each turbine, expressed on a per kW basis. BOS costs for onshore wind include site preparation and foundations; connection, transformer, and other electrical substation equipment; and permitting, engineering, management and construction labor costs. We are using the defaults for this example, but SAM also provides BOS models for onshore and offshore installations that allow for detailed user input.

The "Operating Costs" tab (Figure 8.35e) is next. The default value for this case is specified as a fixed cost per kW-yr of capacity. Multiplying this value by the total wind farm capacity gives the annual operating costs. Alternatively, users can specify fixed annual costs or variable costs based on generation (assuming maintenance costs scale with the amount of electricity produced). The "value/sched" switch allows the option of modifying operating costs over time, for example to allow for extensive maintenance in a particular year. An annual decline in capacity can be specified in the "AC Degradation" tab. For this example, we will leave it at 0% degradation.

SAM can be run in two solution modes. In the first mode, a PPA electricity price is supplied to the model and the internal rate of return is calculated based on that price. In the second mode, the project developer's target IRR and year the target should be achieved are specified and SAM calculates the power price that would deliver that IRR. The choice of mode is entered on the "Revenue" tab (Figure 8.35f). For this example, select the "Specify IRR target" option as the solution mode, and set the IRR target to 12% and the target year to 15. Next, on the "Financial Parameters" tab, set the "Analysis period" to 25 years, the "Inflation rate" to 2%/year, and the real discount rate to 7%. Note but don't change the state and federal tax rates, sales tax, insurance costs, and salvage values. Keep all other values on this tab as defaults. A screenshot of the upper portion of this tab from SAM is shown in Figure 8.35g.

Select the "Incentives" tab from the left menu. Set the "Federal Investment Tax Credit" to "0" if it is not at that value. Likewise, set the "Federal Production Tax" credit to $0.0275/kWh with a 10-year term, escalating at 2.5% per year, if those values are not already being used. Leave all other values on this tab as defaults. The relevant portion of this screen is shown in Figure 8.35h. Review the values shown on the Depreciation tab. These values determine the schedule for deducting the reduction in value of capital equipment from taxable income, as described in Chapter 2. The 5-year and 15-year MACRS (Modified Accelerated Cost Recovery System) options are depreciation schedules defined in the U.S. tax code, with set percentages of equipment value to be taken in each year. SAM's "help" utility in the upper right corner provides more explanation.

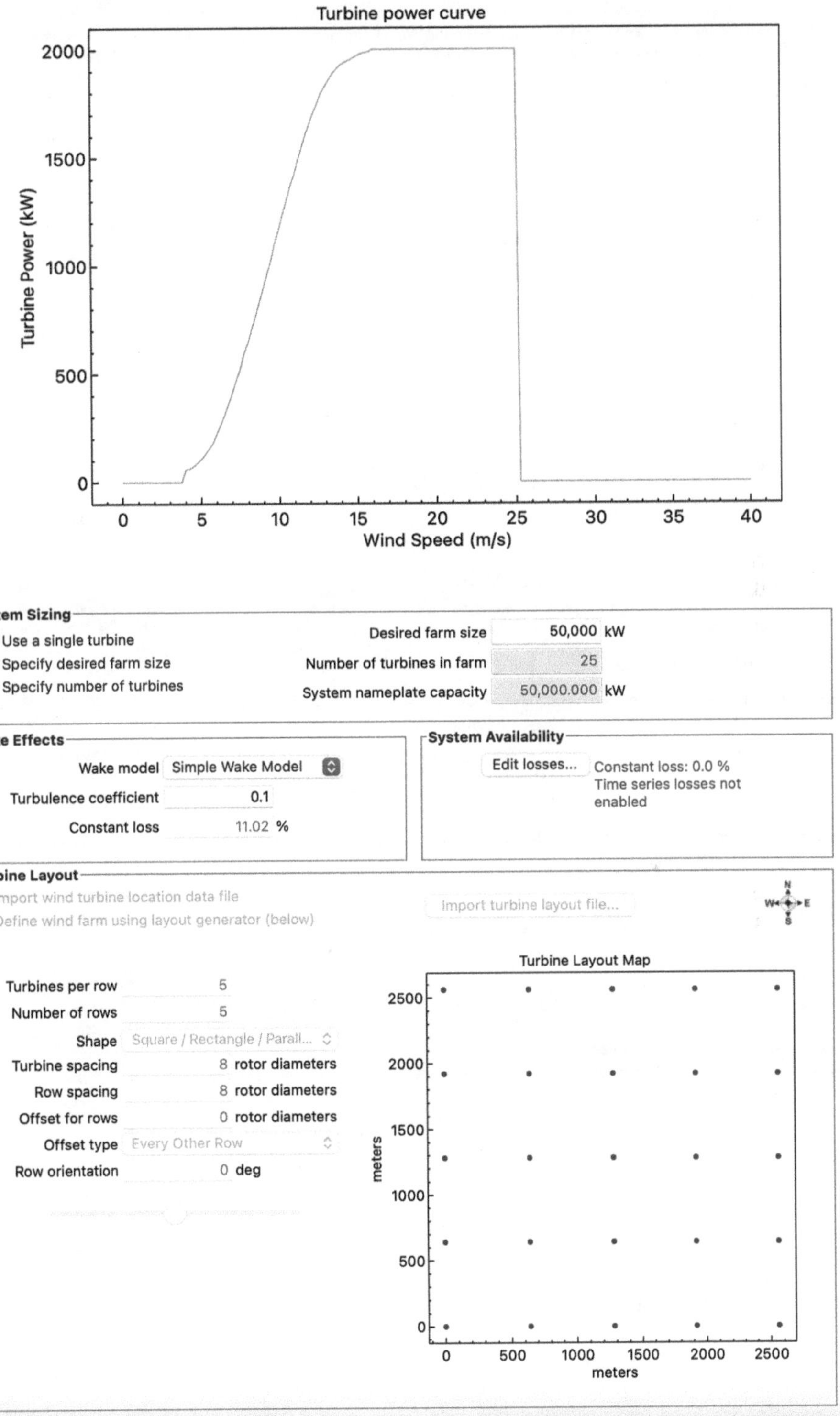

FIGURE 8.35
(a) Power curve for the Vestas V80–2.0 wind turbine from SAM. (b) Wind farm parameters for the 50 MW example.

(Continued)

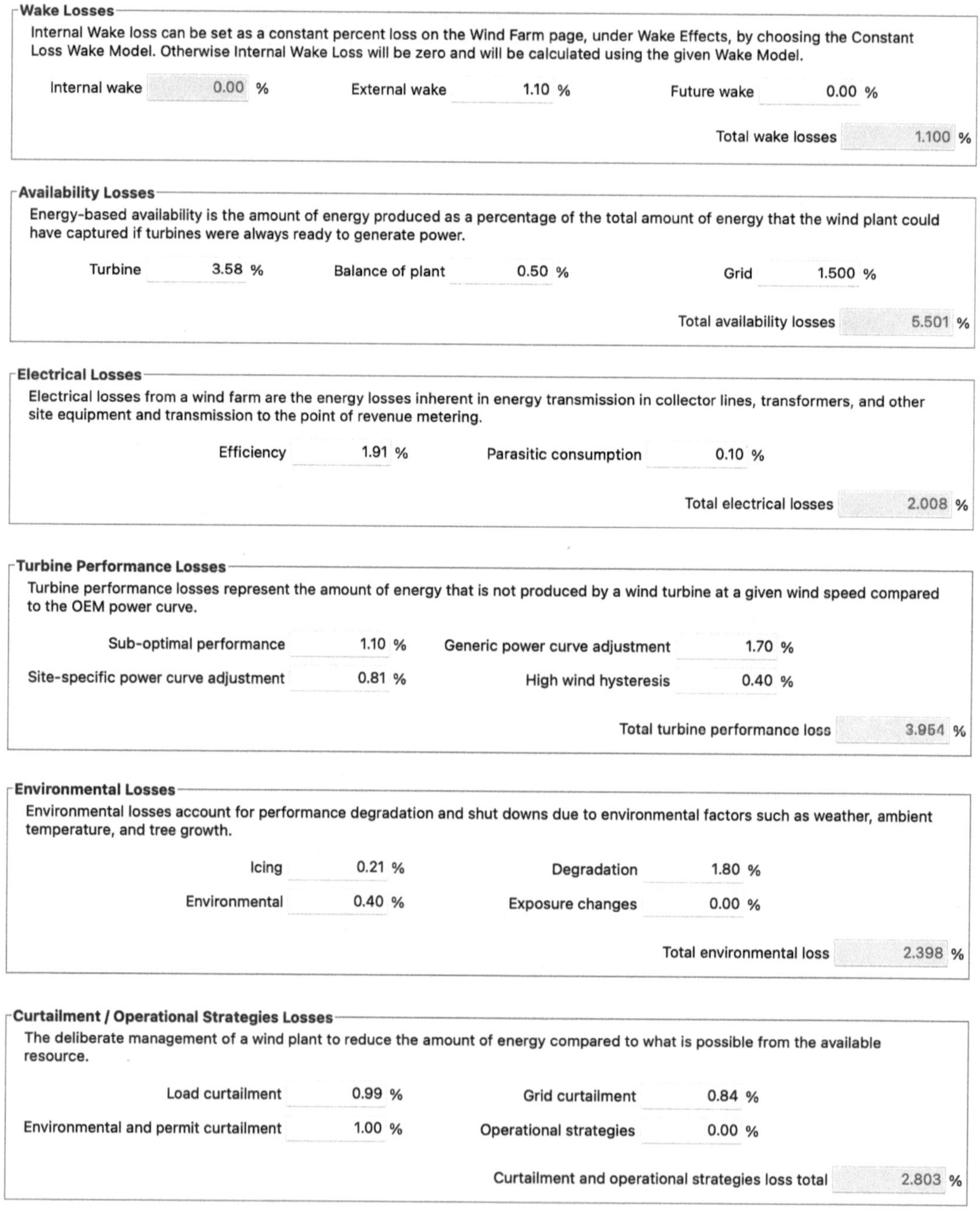

FIGURE 8.35 (*Continued*)
(c) Wind farm electricity production losses from SAM.

(*Continued*)

Lastly, on the Electricity Purchases tab, select "Use retail electricity rate(s)." This is required when running in the IRR target mode, as SAM will be solving for the PPA price that meets the IRR target, so it isn't known in advance. For the southern Wyoming site, use the "search for rates" function with zip code 82005 for Cheyenne. SAM returns the rate schedules for Cheyenne Light, Fuel, and Power. Choose the Commercial Service rate schedule.

Run the simulation by pressing "Simulate" in the lower left corner and review the summary output, hourly time series plots for wind speed and power generation, daily profile plots of wind speed and power generation, and monthly statistics for the same two variables. Figure 8.35i shows the monthly energy production results for the southern Wyoming site. The corresponding capacity factor is 40.8%. Figure 8.35j shows the summary

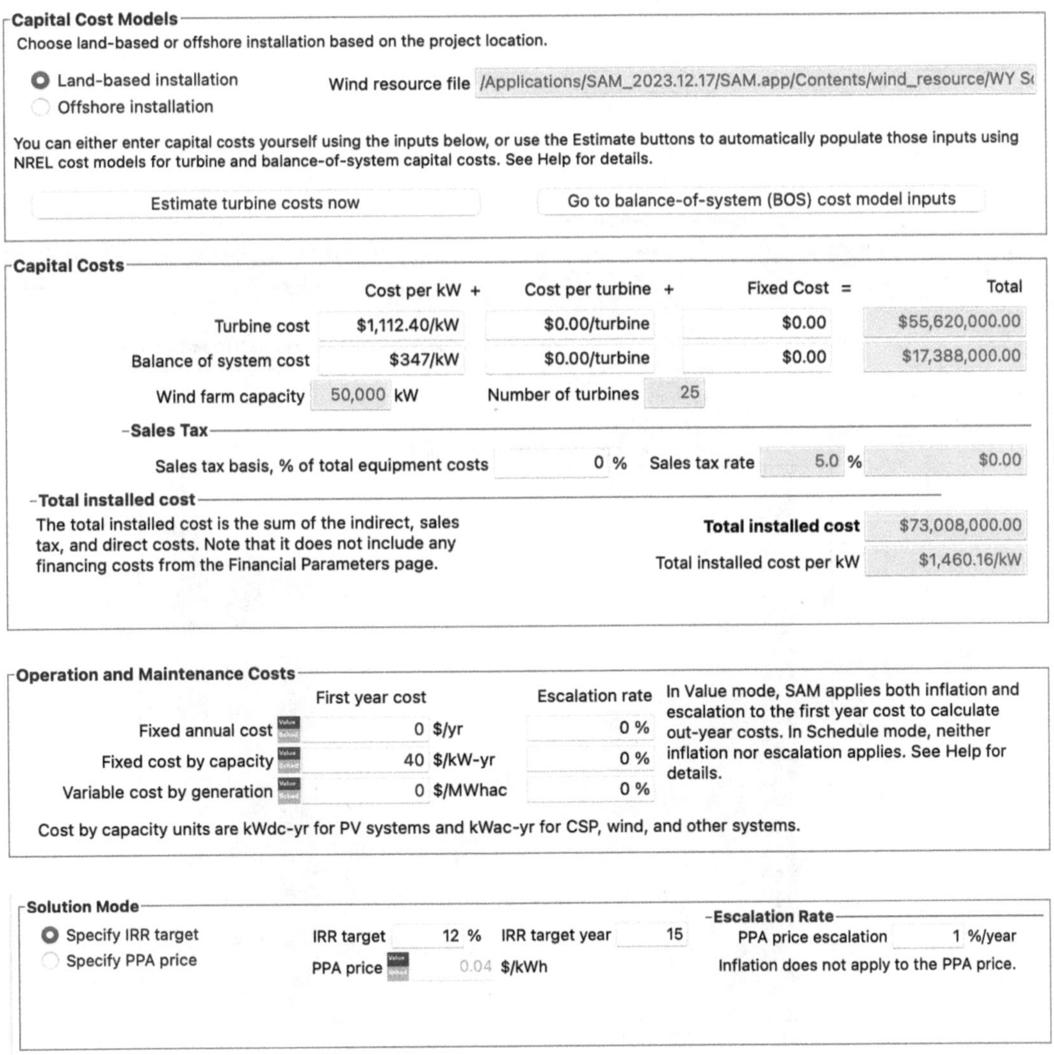

FIGURE 8.35 ***(Continued)***
(d) Equipment costs for the 50 MW wind farm with Vestas V80–2.0 turbines from SAM. (e) Operation and maintenance costs for the 50 MW wind farm from SAM. (f) Revenue tab from SAM showing selection of IRR target solution mode.

(Continued)

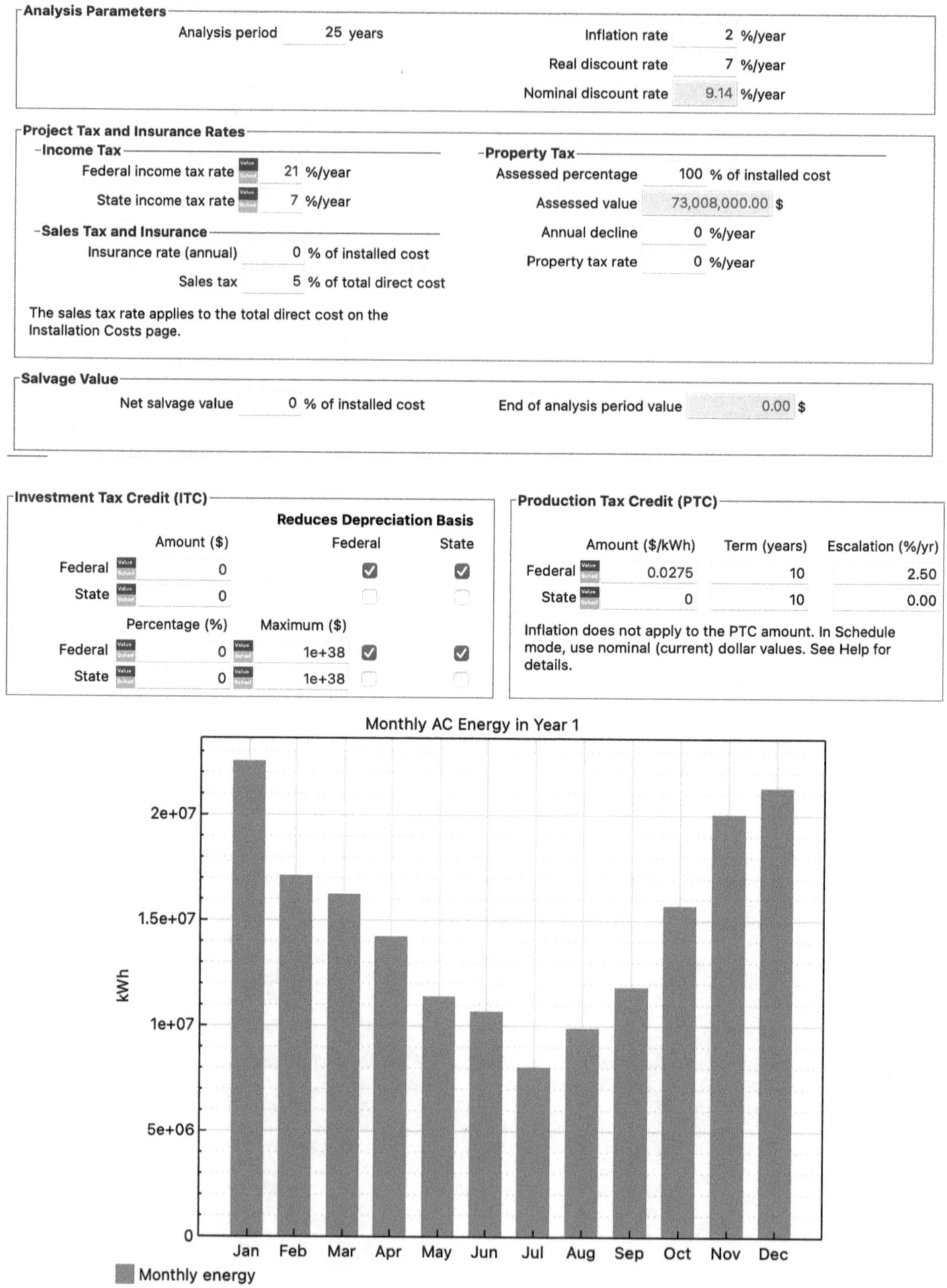

FIGURE 8.35 (*Continued*)
(g) Financial model parameters for the 50 MW wind farm example. (h) Federal tax incentives for the 50 MW wind farm example. (i) Monthly electricity production from the 50 MW wind farm in southern Wyoming.

(*Continued*)

Metric	Value
Annual AC energy in Year 1	178,869,792 kWh
Capacity	50,000 kW
Capacity factor in Year 1	40.8%
PPA price in Year 1	3.35 ¢/kWh
PPA price escalation	1.00 %/year
LPPA Levelized PPA price nominal	3.63 ¢/kWh
LPPA Levelized PPA price real	3.03 ¢/kWh
LCOE Levelized cost of energy nominal	3.07 ¢/kWh
LCOE Levelized cost of energy real	2.56 ¢/kWh
NPV Net present value	$9,702,325
IRR Internal rate of return	12.00 %
Year IRR is achieved	15
IRR at end of project	13.07 %
Net capital cost	$78,166,376
Equity	$48,543,912
Size of debt	$29,622,462
Debt percent	37.90%

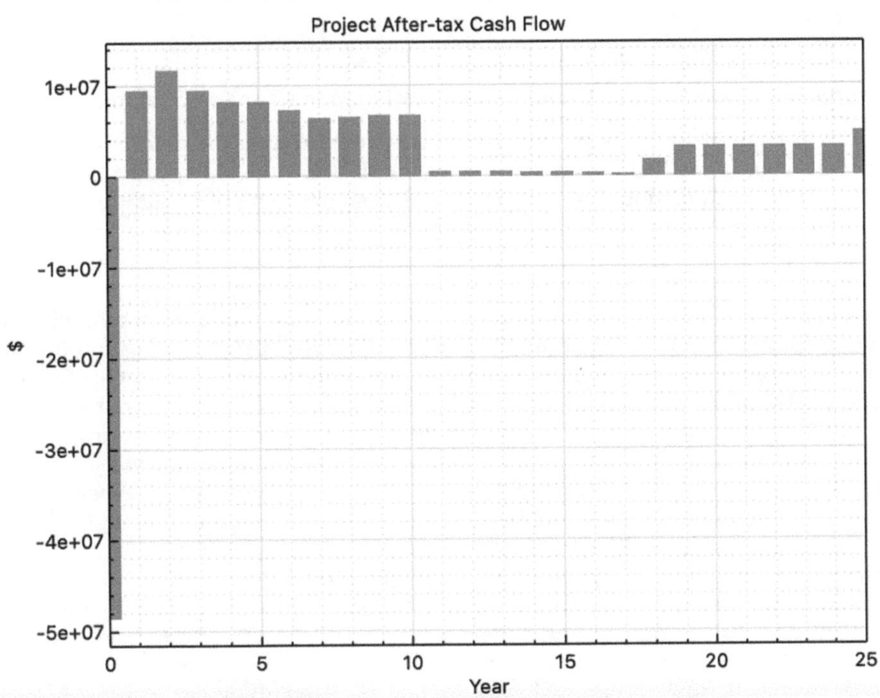

FIGURE 8.35 ***(Continued)***
(j) Summary financial performance results and (k) yearly cash flow for the 50 MW wind farm in southern Wyoming.

financial performance results and yearly cash flow, with a real LCOE =2.56 ¢/kWh, real levelized PPA price = 3.03 ¢/kWh, and Net Present Value = $9.7 million. The yearly cash flow graph (Figure 8.35k) is also of interest, showing the large outlay of investment costs in year 0, followed by 10 years of strong positive returns due to electricity sales, production tax credit revenue, and reduced tax liability from equipment depreciation. Cash flow drops after 10 years when the production tax credit (PTC) payments end and picks up again in year 18 after the project debt is paid off.

In addition to the preloaded wind resource files, SAM allows users to download wind resource data from NREL's WIND Toolkit for locations in the continental United States. To run SAM for a site in central Iowa, return to the Wind Resource page and select the Download option in the second section of the screen. Enter "Des Moines IA" as the location, 2014 as the wind data resource year, and 80 m as the wind measurement height. If the download is successful, click "OK." The system will download a resource file for Des Moines at 41.59°N latitude and 93.62°W longitude. In the next section of the tab, check the box to "Use a wind resource file stored on your computer." One other change is needed to run the new case. On the Electricity Purchases tab, use zip code 50302 for Des Moines, and select MidAmerican Energy Company as the utility. Then select the Rate LS—Large Electric Service rate schedule. We are keeping everything else the same in this example, so with those changes made, run the new simulation.

The SAM results for the Des Moines location indicate that its wind resource is not quite as good as that in southern Wyoming, giving a capacity factor of 23.8%. The real LCOE for the new location is 5.74 ¢/kWh, and the levelized PPA price is 6.5 ¢/kWh. With that PPA price, the 12% IRR target is met in year 15, and the NPV for the project is $9.9 million. The financial parameters indicate that wind farms at both locations would be viable, but the Des Moines location would need to secure a higher price for its electricity.

Example 8.5: Using SAM to Examine the Effect of Wind Turbine Spacing

Using NREL's SAM software and the parameters from Example 8.4, determine the effect on annual energy production, capacity factor, and LCOE if the WT spacing is cut in half to reduce land acquisition costs at the "Wind Resource" site of "Southern Wyoming flatlands."

Solution:

If you are not continuing from Example 8.4, set up the SAM model as described in that example for the "Southern Wyoming flatlands" "Wind Resource." If you are continuing from Example 8.4, make sure the desired "Wind Resource" location is set to "Southern Wyoming flatlands" by choosing that file from the library. Now the desired wind farm layout will be set. In Example 8.4, under the "Wind Farm" tab, the option "specify desired farm size" was chosen and the "Desired farm size" was specified as 50,000 kW. This choice resulted in the following default settings: (see Figure 8.38b, under "Turbine Layout"): turbines per row = 5, number of rows = 5, shape = Square/Rectangle/Parallel, turbine spacing = 8 rotor diameters, row spacing = 8 rotor diameters, offset for rows = 0 diameters, offset type = every other row, row orientation = 0°.

For this example, under the "Wind Farm" tab, choose the option "Specify number of turbines" and reduce the spacing between turbines by setting the following options under "Turbine Layout": turbines per row = 5, number of rows = 5, turbine spacing = 4 rotor diameters, row spacing = 4 rotor diameters. Keep all other default values (see Figure 8.36a). The wind farm now requires approximately 25% of the area of the wind farm in Example 8.4, which would reduce the land acquisition cost.

Run the simulation. The results for the reduced turbine spacing are shown in Figure 8.36b.

The annual energy production for the reduced turbine spacing in year 1 is 172.0 million kWh, down from 178.9 million kWh in the baseline case, resulting 3.9% less income. The capacity factor for the reduced turbine spacing is reduced to 39.3%, from 40.8% for the default spacing. The corresponding LCOE values (real) are 2.74 with reduced spacing and 2.56 ¢/kWh for the original layout.

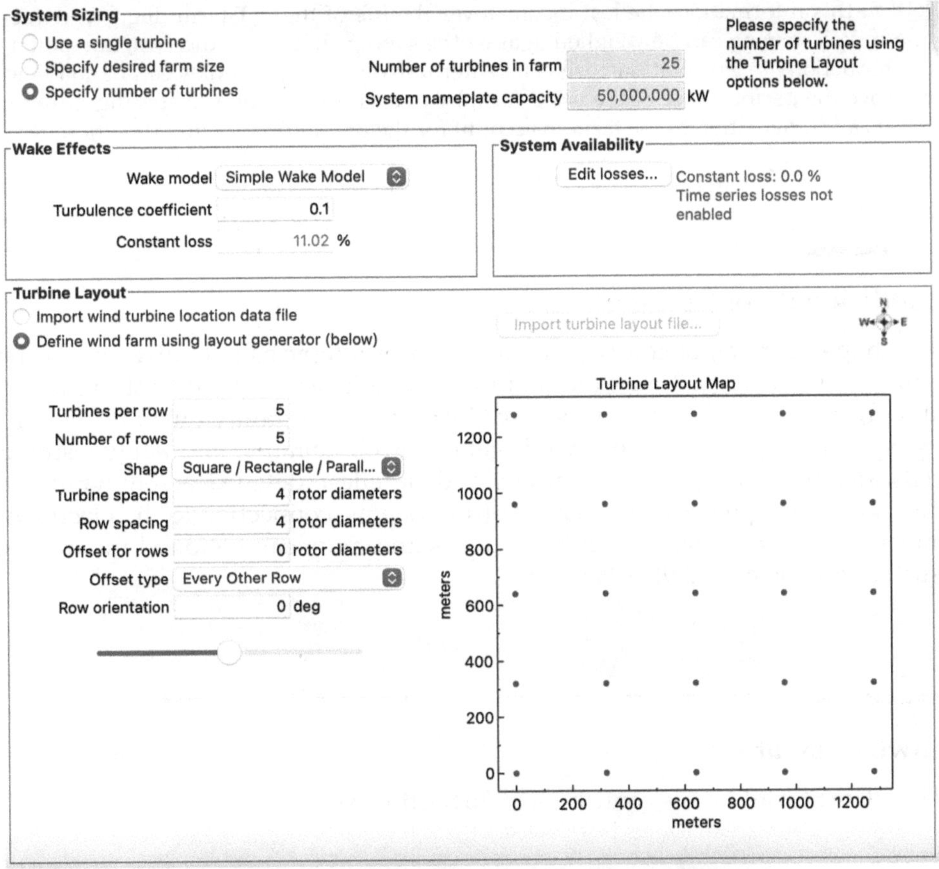

Metric	Value
Annual AC energy in Year 1	172,022,880 kWh
Capacity	50,000 kW
Capacity factor in Year 1	39.3%
PPA price in Year 1	3.57 ¢/kWh
PPA price escalation	1.00 %/year
LPPA Levelized PPA price nominal	3.86 ¢/kWh
LPPA Levelized PPA price real	3.22 ¢/kWh
LCOE Levelized cost of energy nominal	3.28 ¢/kWh
LCOE Levelized cost of energy real	2.74 ¢/kWh
NPV Net present value	$9,718,905
IRR Internal rate of return	12.00 %
Year IRR is achieved	15
IRR at end of project	13.13 %
Net capital cost	$78,247,576
Equity	$47,526,796
Size of debt	$30,720,776
Debt percent	39.26%

FIGURE 8.36
(a) Wind farm parameters with reduced turbine spacing for the 50 MW wind farm example. (b) Summary of financial performance results for the 50 MW wind farm with reduced turbine spacing in southern Wyoming.

With this information, the lost income over the life of the WT farm due to a reduction in WT spacing can be weighed against the savings in initial land cost, any potential reductions in installation costs, etc. Alternatively, similar studies can be made to improve the performance of the wind farm by determining optimum spacing, turbine offsets and orientation of the farm to account for the predominant wind direction.

8.11 Additional Topics for Study

A single chapter cannot, of course, completely cover a topic as rich, diverse, and rapidly changing as wind energy. Topics omitted for brevity include (but are not limited to) statistical representation of the wind resource including uncertainty quantification, remote sensing, systems engineering and wind plant control, vibration and fatigue, strength of materials, manufacturing (blades, towers, and nacelles), detailed aerodynamic design, gearbox, generators, power converters, control systems, connection to the electrical grid, and transmission. Discussion of all these topics and more can be found in wind energy texts such as Burton et al. [39] or Hau [31].

Acknowledgment

Thanks to Tim Stovall for his contributions to Section 8.8.

Problems

8.1 While completing this homework, imagine the study space you are using employs a space heater with a rating of 30 kW, a refrigerator (for snacks) that is rated for 5 kW, and a light fixture with two 100 W bulbs, all operating at a constant rate for the 1.5 hours it takes you to finish.

a. What is the total power consumed at any given moment?

b. How much energy do you consume to complete the task?

c. Assume your local power company, XEnergy, charges 6.5 ¢/unit of energy in part b, how much did it cost you to complete the assignment?

8.2 A residential-size WT has a diameter of 1 m. Due to the size, its efficiency is only 25% of the Betz limit. What is the expected power output at a wind speed of 20 mph? Assume the air density is 1.225 kg/m^3.

8.3 You have assembled a modern three-bladed upwind WT with a rotor radius of 5 m at sea level where the density is 1.225 kg/m^3.

a. If the wind is blowing at an average speed of 10 m/s, what is the power of the wind traveling through the swept area of the rotor?

b. What is the average actual power produced by a turbine of this type at this wind speed? (**Hint:** Find C_p.)

Hint: Find the maximum power coefficient, C_p, based on the following equation for axial induction:

$$C_p = 4a(1-a)^2$$

8.4 An investor comes to you and tells you that a company wants to invest in a new wind turbine design with the following performance characteristics:
112 m rotor diameter
13 m/s wind speed
2.1 MW
1.223 kg/m^3 air density.

a. What is the claimed rotor power coefficient, C_p, for this turbine design?

b1. Do you encourage the investor to invest the company's $1 million? Answer in complete sentence(s).

b2. Why? Answer in complete sentence(s).

8.5 **Weibull distribution:** From an analysis of wind speed data (hourly interval average, taken over a 1 year period), the Weibull parameters are determined to be $c = 6$ m/s and $k = 1.8$ at a potential WT site.
The Weibull PDF is given by

$$p(U) = \left(\frac{k}{c}\right)\left(\frac{U}{c}\right)^{k-1} \exp\left[-\left(\frac{U}{c}\right)^{k}\right]$$

Estimate the number of hours per year that the wind speed will be between $U_A = 6.5$ and $U_B = 7.5$ m/s.

Estimate the number of hours per year that the wind speed will be above 16 m/s.

8.6 Given a WT with the following parameters, calculate the annual energy production assuming a Rayleigh PDF:

$C_p = C_{pmax} = 0.40$ (C_p value assumed maximum and constant for all wind speeds)

Rated power = 3,000 kW = 3.0 MW

Annual average wind speed = 7.0 m/s

Rated wind speed = 12.5 m/s

Cut-out wind speed = 25 m/s

Rotor radius = 45.0 m

Area = 6,361.73 m^2

Air density = 1.20 kg/m^3.

Wind Speed (m/s)	Power (kW)	Probability (%)	# of Hours/Year	Energy (kWh)
1	1.5			
2	12.2			
3	41.2			

(Continued)

(*Continued*)

Wind Speed (m/s)	Power (kW)	Probability (%)	# of Hours/Year	Energy (kWh)
4	97.7			
5	190.9			
6	329.8			
7	523.7			
8	781.7			
9	1,113.0			
10	1,526.8			
11	2,032.2			
12	2,638.3			
13	3,000.0			
14	3,000.0			
15	3,000.0			
16	3,000.0			
17	3,000.0			
18	3,000.0			
19	3,000.0			
20	3,000.0			
21	3,000.0			
22	3,000.0			
23	3,000.0			
24	3,000.0			
25	3,000.0			

8.7 a. What is the levelized cost of energy for a wind turbine with the following assumptions:

5 MW rating

25-year turbine life

31% capacity factor

$1,250/kW turbine cost

$91,000 O&M cost/year (constant dollars)

BOS costs of $1,999,999

Equipment replacement costs of $135,000/year (constant dollars)

Discount rate of 9% to cover financing and expected return on investment

b. If a gearbox replacement was required every 5 years at a constant dollar cost of $600,000, what would the levelized cost of energy be?

8.8 Power deficit inside wind farms—Barthelmie et al. [60]—present results of modeling the behavior of WT wakes to improve power output predictions. The following figure shows the predicted power deficit in the Danish offshore wind farm Horns Rev.

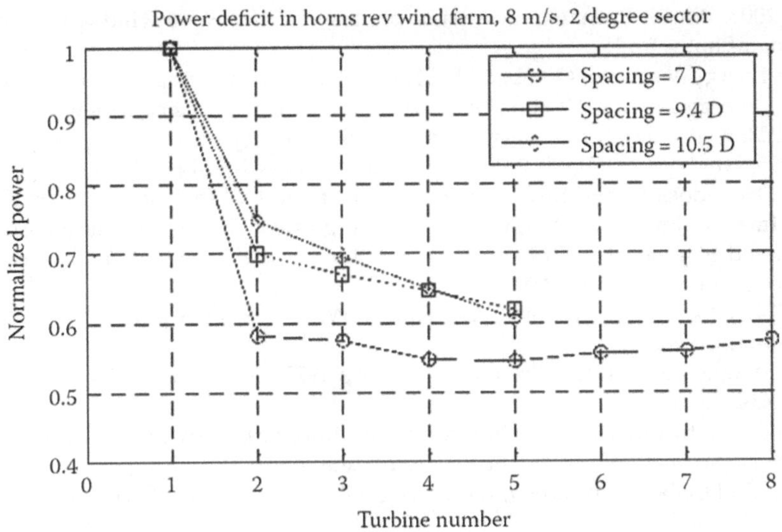

As an example, if two 2.3 MW turbines are spaced 9.4D apart directly in line with the wind, what is the total power produced?

Front turbine, normalized power is 1.0, power = 1.0 × 2.3 MW = 2.3 MW

Second turbine is "waked," normalized power is 0.7, waked power = 0.7 × 2.3 MW = 1.61 MW

$$\text{Total power} = 2.3 + 1.61\ \text{MW} = 3.91\ \text{MW}$$

a. What is the power of three turbines, 9.4D apart, directly in line with the wind?

b. What is the power of three turbines, 9.4D apart, perpendicular with the wind?

References

1. IEA, 2024. Electricity generation by source, energy statistics data browser, International Energy Agency, https://www.iea.org/data-and-statistics/data-tools/energy-statistics-data-browser.
2. EIA, 2024. Net generation, renewable sources: Total – All sectors, electric power monthly, Table 1.1A, U.S. Energy Information Administration, September, https://www.eia.gov/electricity/monthly/.
3. GWEC, 2023. Global offshore wind report 2023, Global Wind Energy Council, https://gwec.net/gwecs-global-offshore-wind-report-2023/.
4. Nebraska Wind and Solar, 2008. History of wind power, https://nebraskawindandsolar.com/history.aspx.
5. Price, T.J., 2005. James Blyth—Britain's first modern wind power pioneer, *Wind Engineering* 29(3), 191–200.
6. Danish Wind Industry Association (DWIA), 1997. A wind energy pioneer: Charles F. Brush, https://drømstørre.dk/wp-content/wind/miller/windpower%20web/en/pictures/brush.htm.

7. Gipe, P., 2003. Smith-Putnam industrial photos, https://www.wind-works.org/photos/Smith-PutnamPhotos.html.
8. Renewable Energy Vermont, October 24, 2013. The story of Grandpa's Knob: How Vermont made wind energy history, *Vermont Biz*, https://www.vermontbiz.com/news/october/story-grandpa%E2%80%99s-knob-how-vermont-made-wind-energy-history.
9. Gipe, P., 1995. *Wind Energy Comes of Age*, John Wiley & Sons, Inc., New York.
10. EERE, 2010. How does a wind turbine work? Office of Energy Efficiency & Renewable Energy, U.S. Department of Energy, https://www1.eere.energy.gov/wind/wind_animation.html.
11. Nelson, V., and Starcher, K., 2018. *Wind Energy: Renewable Energy and the Environment*, CRC Press, Taylor and Francis, Boca Raton, FL.
12. GWEC, 2024. Global wind report 2024, Global Wind Energy Council, https://gwec.net/global-wind-report-2024/.
13. Winds of Change, 2006. American federal projects 1975–1985, https://www.windsofchange.dk/WOC-usastat.php.
14. EERE, 2024. Wind turbine sound, wind exchange, Office of Energy Efficiency and Renewable Energy, U.S. Department of Energy, https://windexchange.energy.gov/projects/sound.
15. Sengpiel Audio. Decibel table, https://www.sengpielaudio.com/TableOfSoundPressureLevels.htm.
16. AWEA, 2010. Wind power myths vs. facts, American Wind Energy Association, https://www.awea.org/pubs/factsheets/MythsvsFacts-FactSheet.pdf.
17. DWIA, 1997. Shadow casting from wind turbines, Danish Wind Industry Association, https://drømstørre.dk/wp-content/wind/miller/windpower%20web/en/tour/env/shadow/index.htm.
18. Nugent, D., and Sovacool, B.K., 2014. Assessing the lifecycle greenhouse gas emissions from solar PV and wind energy: A critical meta survey, *Energy Policy* 65, 229–244.
19. Sathaye, J., Lucon, O., Rahman, A., Christensen, J., Denton, F., Fujino, J., Heath, G., Kadner, S., Mirza, M., Rudnick, H., Schlaepfer, A., and Shmakin, A., 2011. Renewable energy in the context of sustainable development, in: Edenhofer, O., Pichs-Madruga, R., Sokona, Y., Seyboth, K., Matschoss, P., Kadner, S., Zwickel, T., Eickemeier, P., Hansen, G., Schlömer, S., and von Stechow, C. eds., *IPCC Special Report on Renewable Energy Sources and Climate Change Mitigation*, Chapter 9, pp. 707–790. Cambridge University Press, Cambridge, UK.
20. Denholm, P., Hand, M., Jackson, M., and Ong, S., August 2009. *Land-Use Requirements of Modern Wind Power Plants in the US*, NREL/TP-6A2–45834, National Renewable Energy Laboratory, Golden, CO.
21. Rosenberg, K.V., Dokter, A.M., Blancher, P.J., Sauer, J.R., Smith, A.C., Smith, P.A., Stanton, J.C., Panjabi, A., Helft, L., Parr, M., and Marra, P.P., 2019. Decline of the North American avifauna, *Science*, 366, 120–124, doi:10.1126/science.aaw1313.
22. Erickson, W.P., Wolfe, M.M., Bay, K.J., Johnson, D.H., and Gehring, J.L., 2014. A comprehensive analysis of small-passerine fatalities from collision with turbines at wind energy facilities, *PLoS ONE* 9(9), e107491, doi:10.1371/journal.pone.0107491.
23. AWWI, 2021. Wind energy interactions with wildlife and their habitats: A summary of research results and priority questions, American Wind Wildlife Institute, https://rewi.org/resources/summary-of-wind-power-interactions-with-wildlife/.
24. U.S. FWS, 2012. Land-based wind energy guidelines, U.S. Fish and Wildlife Service, https://www.fws.gov/sites/default/files/documents/land-based-wind-energy-guidelines.pdf.
25. Keating-Bitonti, C., Marshak, A.R., Comay, L.B., Clark, C.E., Ward, E.H., and Sheikh, P.A. 2024. Potential impacts of offshore wind on the marine ecosystem and associated species: Background and issues for Congress, R47894, Congressional Research Service, Washington DC.
26. Axelrod, J., 2024. The science is clear: Offshore wind isn't what's killing whales, *Scientific American*, June 26, 2024, https://www.scientificamerican.com/article/whales-are-dying-but-not-from-offshore-wind/.
27. Kim, M. 2024. Turbine blades have piled up in landfills. A solution may be coming, *New York Times*, August 30, 2024.

28. Betz, A., 1920. Das maximum der theoretisch möglichen ausnützung des windes durch windmotoren, *Zeitschrift für das gesamte Turbinenwesen* 26, 307–309.
29. van Kuik, G.A.M., 2007. The Lanchester–Betz–Joukowsky limit, *Wind Energy* 10, 289–291.
30. Wilson, R.E., and Lissaman, P.B.S., 1974. *Applied Aerodynamics of Wind Power Machines*, Oregon State University, Corvallis, OR, https://ir.library.oregonstate.edu/jspui/handle/1957/8140.
31. Hau, E., 2006. *Wind Turbines—Fundamentals, Technologies, Application, Economics*, Springer, Berlin, Germany.
32. Frost, W.D., and Asphiden, D., 1994. Characteristics of the wind, in: Spera, D.A. ed., *Wind Turbine Technology*, ASME Press, New York, pp. 467–542.
33. Pidwirny, M., 2006. *Fundamentals of Physical Geography*, 2nd ed., Chap. p. 7, Fundamentals eBook. https://www.physicalgeography.net/fundamentals/7p.html.
34. Kelley, N.D., and Jonkman, B.J., 2007. Overview of the TurbSim stochastic inflow turbulence simulator version 1.21, National Renewable Energy Laboratory, Technical Report, NREL/TP-500–41137, Golden, CO, https://wind.nrel.gov/designcodes/preprocessors/turbsim/TurbSimOverview.pdf.
35. NREL, 2017. Wind maps, U.S. 80 m wind resource maps, National Renewable Energy Laboratory, https://www.nrel.gov/gis/wind.html.
36. NREL, 2024. Wind resource maps and data, National Renewable Energy Laboratory, https://www.nrel.gov/gis/wind-resource-maps.html.
37. Wind Powering America, 2009. Wind resource maps, https://www.windpoweringamerica.gov/wind_maps.asp.
38. Holton, J., 2004. The planetary boundary layer, in: *Dynamic Meteorology*, International Geophysics Series, vol. 88. Chap. 5, Elsevier Academic Press, Burlington, MA, pp. 116–139.
39. Burton, T., Sharpe, D., Jenkins, N., and Bossanyi, E., 2001. Wind Energy Handbook, Wiley, Chichester, UK.
40. Manwell, J.F., McGowan, J.G., and Rogers, A.L., 2002. *Wind Energy Explained, Theory, Design and Application*, Wiley, Chichester, UK.
41. Renewable Energy Research Laboratory, 2006. Capacity factor, intermittency, and what happens when the wind doesn't blow, U. Massachusetts at Amherst, https://www.ceere.org/rerl/about_wind/RERL_Fact_Sheet_2a_Capacity_Factor.pdf.
42. AWEA, 2010. 10 steps in building a wind farm, American Wind Energy Association, https://www.awea.org/pubs/factsheets/10stwf_fs.pdf.
43. DOE, 2008. 20% Wind energy by 2030; increasing wind energy's contribution to U.S. electricity supply, https://www.nrel.gov/docs/fy08osti/41869.pdf.
44. Carpenter, S., 2021. Why wind turbines in cold climates don't freeze: De-icing and carbon fiber, *Forbes*, February 16. https://www.forbes.com/sites/scottcarpenter/2021/02/16/why-wind-turbines-in-cold-climates-dont-freeze-de-icing-and-carbon-fiber/.
45. Archer, C.L., and Jacobson, M.Z., 2007. Supplying baseload power and reducing transmission requirements by interconnecting wind farms, *Journal of Applied Meteorology and Climatology* 46(11), 1701–1717.
46. Stovall, T.D., 2009. Simulations of wind turbine wake interactions in OpenFOAM, MS thesis, University of Colorado, Boulder, CO.
47. Riso, D.T.U., 2003. *Wind Atlas Analysis and Application Program (WAsP)*, DTU Wind Energy, Roskilde, Denmark, www.wasp.dk.
48. NREL, 2017. Wind-plant integrated system design and engineering model (WISDEM), National Renewable Energy Laboratory, https://nwtc.nrel.gov/WISDEM.
49. Simley, E., Millstein, D., Jeong, S., and Fleming, P., 2021. The value of wake steering wind farm flow control in US energy markets, *Wind Energy Science* 9, 219–234, doi:10.5194/wes-9-219-2024.
50. Simley, E., Fleming, P., Girard, N., Alloin, L., Godefroy, E., and Duc, T., 2021. Results from a wake steering experiment at a commercial wind plant: Investigating the wind speed dependence of wake-steering performance, *Wind Energy Science* 6, 1427–1453, doi:10.5194/wes-6-1427-2021.

51. Piwko, R., Osborn, D., Gramlich, R., Jordan, G., Hawkins, D., and Porter, K., 2005. Wind energy delivery issues: Transmission planning and competitive electricity market operation, *IEEE Power and Energy Magazine* 3(6), 47–56.
52. GWEC, 2023. Global offshore wind report 2023, Global Wind Energy Council, https://gwec.net/gwecs-global-offshore-wind-report-2023/.
53. Rocky Mountain PBS, 2016. U.S. builds first offshore wind farm, https://www.pbs.org/newshour/bb/u-s-builds-first-offshore-wind-plant.
54. McCoy, A., Musial, W., Hammond, R., Hernando, D. M., Duffy, P., Beiter, P., Pérez, P., Baranowski, R., Reber, G., and Spitsen, P., 2024. Offshore wind market report, 2024 Edition, NREL/TP-5000–90525, National Renewable Energy Laboratory, Golden, CO, https://www.nrel.gov/docs/fy24osti/90525.pdf.
55. NREL, 2024. Annual technology baseline: Offshore wind, National Renewable Energy Laboratory, https://atb.nrel.gov/electricity/2024/offshore_wind.
56. Smith, A., Stehly, T., and Musial, W., September 2015. 2014–2015 Offshore wind technologies market report, NREL/TP-5000–64283, National Renewable Energy Laboratory, Golden, CO, https://www.nrel.gov/docs/fy15osti/64283.pdf.
57. EIA, 2006. Electricity. Spread sheet state sales, U.S. Energy Information Administration, www.eia.doe.gov/cneaf/electricity/epa/sales_state.xls.
58. Musial, W., Heimiller, D., Beiter, P., Scott, G., and Draxl, C., September, 2016. 2016 Offshore wind energy resource assessment for the United States, NREL/TP-5000–66599, National Renewable Energy Laboratory, Golden, CO, https://www.nrel.gov/docs/fy16osti/66599.pdf.
59. Gilman, P., Maurer, B., Feinberg, L., Duerr, A., Peterson, L., Musial, W., Beiter, P., Golladay, J., Stromberg, J., Johnson, I., Boren, D., and Moore, A., 2016. National offshore wind strategy: Facilitating the development of the offshore wind industry in the United States, U.S. Department of Energy and U.S. Department of the Interior, https://energy.gov/sites/prod/files/2016/09/f33/National-Offshore-Wind-Strategy-report-09082016.pdf.
60. Barthelmie, R.J., Rathmann, O., Frandsen, S.T., Hansen, K., Politis, E., Prospathopoulos, J., Rados, K., Cabezón, D., Schlez, W., Phillips, J., Neubert, A., Schepers, J.G., and van der Pijl, S.P., 2007. Modelling and measurements of wakes in large wind farms, *Journal of Physics Conference Series* 75, 012049, *The Science of Making Torque from Wind*, IOP Publishing.

9

Capturing Solar Energy through Biomass

Biorenewable resources, sometimes referred to as biomass, are organic materials of recent biological origin [1]. This definition is deliberately broad with the intent of only excluding fossil fuel resources from the wide variety of organic materials that arise from the biotic environment.

We have the technology to convert biorenewable resources into a variety of gaseous and liquid fuels using both thermal and biological processes. Which process and product we choose depends upon the nature of available biomass feedstock and the target market. Gaseous bioenergy products include biogas from anaerobic digestion and producer gas or synthetic gas (syngas) from thermal gasification; and hydrogen, methane, ammonia, and dimethyl ether (DME) from the upgrading of the primary products of anaerobic digestion or gasification. Liquid bioenergy products are conveniently classified according to the intermediate substrate derived from physically, chemically, or thermally processing biomass: carbohydrates, triglycerides, syngas, and bio-oil/bio-crude.

The majority of carbohydrate conversion processes focus on ethanol production, although butanol, isoprenes, furans, and even alkanes are also possibilities. Similarly, triglycerides can be transesterified to methyl esters (biodiesel) or hydroprocessed into alkanes. Researchers are investigating promising processes to convert syngas into a variety of liquid fuels: Fischer–Tropsch liquids, methanol, ethanol, and methanol-to-gasoline; and to upgrade bio-oil or bio-crude into a range of hydrocarbons such as gasoline, diesel fuel, or aviation (jet) fuel. Finally, biomass can be used to generate electricity, which offers an alternative approach to providing energy for transportation.

This chapter introduces biomass feedstocks, availability, and conversion to valuable energy products. It also includes sections on pressing topics such as land use, environmental impact, and energy efficiency.

9.1 Biomass Production and Land Use

Biomass is a term that encompasses a wide range of materials. Scientists generally classify biorenewable resources as either wastes or dedicated energy crops. A waste is a material that has been traditionally discarded because it has no apparent value or represents a nuisance or even a pollutant to the local environment. Dedicated energy crops are plants grown specifically to produce bio-based products, that is, for purposes other than food or feed. These resources are typically the product of converting solar energy into organic compounds. The properties of these organic compounds, and the means by which they are formed, are subject to numerous factors.

This section describes the different types of biomass and their properties, the use of land for crop production, and the principles of solar energy conversion. These are important considerations in order to understand the potential of biorenewable resources for capturing solar energy.

DOI: 10.1201/9781003586616-9

9.1.1 Waste Materials

Categories of waste materials that qualify as biorenewable resources include municipal solid wastes (MSWs), agricultural and forest residues and their by-products, and manure. MSWs are whatever is thrown out in the garbage and clearly include materials that do not qualify as biorenewable resources, such as glass, metal, and plastics. MSW includes food processing waste that is the effluent from a wide variety of industries ranging from breakfast cereal manufacturers to alcohol breweries. Another category of waste product is agricultural residues. Agricultural residues (see Figure 9.1) are simply that part of a crop discarded by farmers after harvest such as corn stover (husks and stalks), rice hulls, wheat straw, and bagasse (fibrous material remaining after the milling of sugar cane). The recent concentration of animals into giant livestock facilities has led to calls to treat animal wastes in a manner similar to that for human wastes. Table 9.1 shows the potential quantities of agricultural and forest residue available in the United States [2].

Waste materials share few common traits other than the difficulty of characterizing them because of their variable and complex composition. Thus, waste biomass presents special problems to engineers who are tasked with converting this sometimes unpredictable feedstock into reliable power or high-quality fuels and chemicals. The major virtue of waste materials is their low cost. By definition, waste materials have little apparent economic value and can often be acquired for little more than the cost of transporting the material from its point of origin to a processing plant. In fact, it is possible to acquire wastes at a negative cost because of the rising costs for solid waste disposal and sewer discharges and restrictions on landfilling certain kinds of wastes; that is, a biorenewable resource processing plant is paid by a company seeking to dispose of a waste stream. For this reason, many of the most economically attractive opportunities in biorenewable resources involve waste feedstocks.

(a) (b) (c)

FIGURE 9.1
Corn fields during the early growing season (a), middle growing season (b), and baled corn stover after harvest (c).

TABLE 9.1

Potential Agricultural, Forest, and Process Waste Supply in the United States

	Annual Biomass Supply (Million Dry Mg/Year)
Logging and other residue	58
Fuel treatments	54
Urban wood residues	43
Wood processing residues	64
Pulping liquor	67
Fuelwood	47
Crop residues	405
Process residues	79

Source: From Perlack et al. [2]. With permission.

Clearly, a waste material that can be used as feedstock for an energy conversion process is no longer a waste material. As demand for these new-found feedstocks increases, those that generate it come to view themselves as suppliers and may demand payment for the one-time waste: a negative feedstock cost becomes a positive cost. Such a situation developed in the California biomass power industry during the 1980s [3]. Concerns about air pollution in California led to restrictions on open field burning of agricultural residues, a practice designed to control infestations of pests. With no means for getting rid of these residues, an enormous reserve of biomass feedstocks materialized. These feedstocks were so inexpensive that independent power producers recognized that even small, inefficient power plants using these materials as fuel would be profitable. A number of plants were constructed and operated on agricultural residues. Eventually, the feedstock producers had plant operators bidding up the cost of their once nuisance waste material. In the end, many of these plants were closed because of the escalating cost of fuel.

9.1.2 Energy Crops

Energy crops are defined as plants grown specifically as an energy resource. We should note that firewood obtained from cutting down an old-growth forest does not constitute an energy crop. An energy crop is planted and harvested periodically. Harvesting may occur on an annual basis, as with sugar beets or switchgrass, or on a 5–7-year cycle, as with certain strains of fast-growing trees such as hybrid poplar or willow. The cycle of planting and harvesting over a relatively short time period assures that the resource is used in a sustainable fashion; that is, the resource will be available for future generations.

Energy crops contain significant quantities of one or more of four important energy-rich components: oils, sugars, starches, and lignocellulose (fiber). Farmers historically cultivated crops rich in the first three components for food and feed: oils from soybeans and nuts; sugars from sugar beets, sorghum, and sugar cane; and starches from corn and cereal crops. Oil, sugars, and starches are easily metabolized. On the other hand, humans find it hard to digest lignocellulose. Certain domesticated animals with specialized digestive tracts are able to break down the polymeric structure of lignocellulose and use it as an energy source. From this discussion, it might appear that the best strategy for developing biomass resources is to grow crops rich in oils, sugars, and starches. However, even for "oil crops" or "starch crops," the largest single constituent is invariably lignocellulose

TABLE 9.2

Typical Woody Biomass Compositions

Component	Weight (%)
Cellulose	44±6
Hemicellulose	28±4
Lignin	20±5

(Table 9.2), which is the structural (fibrous) material of the plant: stems, leaves, and roots. If we harvest oils, sugars, and starches and leave the lignocellulose behind as an agricultural residue rather than used as fuel, we will waste the greatest portion of the biomass crop.

Research has shown that energy yields (J/km^2/year) are usually greatest for plants that are mostly "roots and stems"; in other words, plant resources are directed toward the manufacture of lignocellulose rather than oils, sugars, and starches. As a result, there has been a bias toward the development of energy crops that focus on lignocellulosic biomass, which is reflected in the discussion that follows.

Dedicated energy crops are typically high-fiber crops grown specifically for their high productivity of holocellulose (cellulose and hemicellulose). Harvesting may occur on an annual basis, as with switchgrass, or on a 5–7-year cycle, as with certain strains of fast-growing trees such as hybrid poplar. Lignocellulosic crops are conveniently divided into herbaceous energy crops (HECs) and short rotation woody crops (SRWCs) [4].

Herbaceous crops are plants that have little or no woody tissue. The above-ground growth of these plants usually lives for only a single growing season. However, herbaceous crops include both annuals and perennials. Annuals die at the end of a growing season and must be replanted in the spring. Perennials die back each year in temperate climates but reestablish themselves each spring from rootstock. Both annual and perennial HECs are harvested on at least an annual basis, if not more frequently, with yields averaging 550–1,100 Mg/km^2/year, with maximum yields between 2,000 and 2,500 Mg/km^2/year in temperate regions [4]. As with trees, yields can be much higher in tropical and subtropical regions.

Herbaceous crops more closely resemble hardwoods in their chemical properties than they do softwoods. Their low lignin content makes them relatively easy to delignify, which improves accessibility of the carbohydrate in the lignocellulose. The hemicellulose contains mostly xylan, which is highly susceptible to hydrolysis, compared to the cellulose. As a result, microbes can easily degrade agricultural residues, destroying their processing potential in a matter of days if exposed to the elements. Herbaceous crops have relatively high silica content compared to woody crops, which can present problems during processing.

SRWC is used to describe woody biomass that is fast growing and suitable for use in dedicated feedstock supply systems. Desirable SRWC candidates display rapid juvenile growth, wide site adaptability, and pest and disease resistance. Woody crops grown on a sustainable basis are harvested on a rotation of 3–10 years. Annual SRWC yields range between 500 and 2,400 Mg/km^2/year.

Woody crops include hardwoods and softwoods. Hardwoods are trees classified as angiosperms, which are also known as flowering plants. Examples include willow, oak, and poplar. Hardwoods can be regrown from stumps, a process known as coppicing, which reduces their production costs compared to softwoods. Advantages of hardwoods in processing include high density for many species; relative ease of delignification and accessibility of wood carbohydrates; the presence of hemicellulose high in xylan, which can be removed relatively easily; low content of ash, particularly silica, compared to softwoods and herbaceous crops; and high acetyl content compared to most softwoods and herbaceous crops, which is an advantage in the recovery of acetic acid.

Softwoods are trees classified as gymnosperms, which encompass most trees known as evergreens. Examples include pine, spruce, and cedar. Softwoods are generally fast growing, but their carbohydrate is not as accessible for chemical processing as the carbohydrates in hardwood. Since softwoods have considerable value as construction lumber and pulpwood, they are more readily available as waste material in the form of logging and manufacturing residues compared to hardwoods. Logging residues, consisting of a high proportion of branches and tops, contain considerable high-density compression wood, which is not easily delignified. Therefore, logging residues are more suitable as boiler fuel or other thermochemical treatments than as feedstock for chemical or enzymatic processing.

9.1.3 Algae

Algae is a broad term that encompasses several eukaryotic organisms. Eukaryotic organisms are characterized by complex structures enclosed within their cell membranes. Although algae do not share many of the structures that define terrestrial biomass, they are capable of photosynthesis and capturing carbon. Algae's affinity to convert CO_2 into lipids has drawn academic and industrial attention as a means to simultaneously lower carbon emissions and produce biofuels.

Algal biomass uses CO_2 as its carbon source and sunlight as its energy source. About 1.8 kg of CO_2 is fixed for every kg of algal biomass, which contains up to 50% carbon by dry weight. Controlled production of renewable fuels from algae has been proposed in either raceway ponds or photobioreactors. Raceway ponds consist of open, shallow recirculation channels with mechanical flow control and surfaces that enhance light retention. Raceway ponds are inexpensive, but relatively inefficient when compared to photobioreactors. There are various photobioreactor designs with the common goal of maintaining a monoculture of algae that is efficiently exposed to sunlight and carbon dioxide. A common design employs arrays of tubes arranged vertically to minimize land use and oriented north–south to maximize light exposure.

Given that algae do not require fresh water or fertile soils, waste lands have been suggested as potential locations to grow algae. One suggestion is to build algae ponds in the desert Southwest United States, where inexpensive flat land, abundant sunlight, water from alkaline aquifers, and CO_2 from power plants could be combined to generate renewable fuels. Algae's potential for yields of 1.12–9.40 million liters of oil/km^2/year promises significant reductions in the land footprint required to produce biofuels.

9.1.4 Land Use for Biomass Production

Global land use is broadly defined by five categories: pasture, crop, forest, urban, and abandoned. Pasture is land devoted primarily to animal grazing; crop lands are areas actively cultivated for food production; forest land contains primarily large trees; urban areas are heavily populated regions; and abandoned lands are territories that formerly fit one of the previous categories but are no longer employed for human activities. Humans, because of population migrations or land-use change, alter the portions of land devoted to each of these categories over time.

TABLE 9.3

Nominal Annual Yields of Biomass Crops

Biomass Crop	Geographical Location	Annual Yield (Mg/km^2)
Corn: grain	North America	700
Corn: cobs	North America	130
Corn: stover	North America	840
Jerusalem artichoke: tuber	North America	4,500
Jerusalem artichoke: sugar	North America	640
Sugar cane: crop	Hawaii	5,500
Sugar cane: sugar	Hawaii	720
Sugar cane: bagasse (dry)	Hawaii	720
Sweet sorghum: crop	Midwest U.S.	3,800
Sweet sorghum: sugar	Midwest U.S.	530
Sweet sorghum: fiber (dry)	Midwest U.S.	490
Switchgrass	North America	1,400
Hybrid poplar	North America	1,400
Wheat: grain	Canada	220
Wheat: straw	Canada	600

Source: From Wayman and Parekh [6]. With permission.

Researchers estimate that 14.5 and 33.2 million km^2 of global land area were devoted to crops and pasture respectively in 2000 [5]. These land-use groups can coexist within the same region. For example, the U.S. Midwest and parts of the Southeast include regions with more than 70% of the land devoted to crops, and the western sides of the Midwest and Southern U.S. states have a high concentration of land for pasture.

Most North American farmers devote their production to a small selection of crops depending on socioeconomic factors. Table 9.3 shows a sample of biomass crops grown in various geographical regions and their annual yields [6]. Crops such as corn and sugar cane can serve both food and energy needs due to their high yields of sugar-rich biomass and biomass residue (stover and bagasse respectively).

We can estimate the amount of biomass available in a given region by assuming nominal values for crop productivity and available land-use data using the following equation:

$$\text{Total Biomass}\left[\frac{\text{kg}}{\text{year}}\right] = f \times \text{Crop}_{\text{yield}}\left[\frac{\text{kg}}{\text{km}^2 \times \text{year}}\right] \times \text{Land}_{\text{area}}\left[\text{km}^2\right] \tag{9.1}$$

In Equation 9.1, f is a factor that accounts for crop rotations, farmer participation, and land conservation among other considerations that restrict the land use. As an example, Iowa has a total land area of 144,700 km^2 that is predominantly covered by corn and soybeans. In 2010, farmers planted 37.5% of Iowa land with corn netting an average yield of 165 bushels per acre (1,035 Mg/km^2). Thus, the total amount of corn grown in Iowa that year was 56.2 million Mg.

Farmers and seed companies have managed to increase crop yields every year for the past couple of decades. Crop yield increases follow the exponential growth formula:

$$\text{Crop}_{\text{yield}}(t)\left[\frac{\text{kg}}{\text{km}^2}\right] = \text{Crop}_{\text{yield},0}e^{kt} \tag{9.2}$$

where k is the growth rate and t is the period of time since the initial value, $Crop_{yield,0}$. Iowa corn yields increased from an average of 1,162 Mg/km² in 1990 to 1,533 Mg/km² in 2010. This improvement represents a 1.32% annual growth rate as shown in Equation 9.3:

$$k = \frac{\ln(1,533/1,162)}{21} = 0.0132 \tag{9.3}$$

The U.S. Department of Agriculture (USDA) maintains a comprehensive database of agricultural statistics (available online for free at http://quickstats.nass.usda.gov). The data span several years and include county-level data for crops, demographics, economics, animals and products, and environmental impacts.

Example 9.1: Biomass Collection and Transportation Distance Calculations

A facility would like to collect 10,000 metric tons of corn stover per day for 330 days/year into the outskirts of Springfield, Illinois. We can estimate how far, on average, corn stover would need to travel if it is all delivered by truck.

Springfield is centrally located in Sangamon County and surrounded by the counties listed in Table 9.4 [7]. The total amount of stover produced in the area is 6,306,000 metric tons/year or enough to supply 19,110 metric tons/day to the facility for 330 working days/year. However, let us assume that only 70% of the stover is sustainably recoverable (f=0.7). The distance between county seats and Springfield is the travel distance estimated from online mapping tools.

1. The facility has decided to collect the stover from farm gates. They approximated the transport distance by assuming that the facility would need to collect stover from a circular area surrounding the facility. The total area for the seven closest counties is approximately 10,860 km². The average distance for transport within a circle with this total area is

$$\tau \times \frac{2}{3} \times \sqrt{\frac{\text{Area}}{\pi}} = 50.95 \text{ km} \tag{9.4}$$

where τ is known as the tortuosity factor—a way to account for windings and road degradation in transport networks. Here we assumed that τ is 1.3, and it varies between 1 (straight line) and 1.5 (square grid network).

TABLE 9.4

Counties Surrounding Springfield, IL, and Their County Seats, Distance to Springfield, and 2010 Corn Stover Yields

County	County Seat	Distance to Springfield, IL (km)	Stover Production (Mg/year)	Planted Area (km²)	County Area (km²)
Sangamon	Springfield	0	969,347	1,000	2,271
Menard	Petersburg	38.3	310,667	340	816
Christian	Taylorville	43.1	913,248	927	1,854
Cass	Virginia	54.4	351,039	376	995
Morgan	Jacksonville	56.8	580,518	631	1,481
Logan	Lincoln	61	838,452	858	1,603
Montgomery	Hillsboro	78.2	746,398	761	1,839
Macon	Decatur	80.5	749,017	708	1,515
Macoupin	Carlinville	83.5	847,035	919	2,248

Source: United States Department of Agriculture [7].

2. Alternatively, to save costs, the facility has convinced farmers to deliver stover to satellite depots in their respective county seats. For this case, the average stover transport distance for the facility is the weighted average of stover transport distance to Springfield (Equation 9.5) for the amount required (Equation 9.6):

$$\text{Distance}(\text{ave. km}) = \frac{\sum_{i=0}^{n} D_i \times S_i}{\sum_{i=0}^{n} S_i} \quad \forall i \in \text{County Seats} \tag{9.5}$$

$$\sum_{i=0}^{n} S_i \leq 10{,}000 \text{ tons/day} \times 330 \text{ days} \tag{9.6}$$

In Equation 9.5, D_i is the distance to Springfield for county i and S_i is the county's stover production. For this case, the seven nearest counties (Sangamon, Menard, Christian, Cass, Morgan, Logan, and Montgomery) can sustainably provide 3,300,000 metric tons/year to the facility. The average distance that stover would ship from the county seat depots to Springfield is about 45.2 km.

Every year, 5.6 million exajoules (EJ) of solar radiation impacts the top of our planet's atmosphere. This is enough energy to supply the world's energy consumption for several thousand years as the world consumes about 570 EJ/year. Unfortunately, this energy is very diffuse and difficult to convert efficiently. A majority of the energy is unreachable on land surfaces, and only a minuscule amount is converted into biomass.

Only 70% of solar radiation makes it through the planet's atmosphere, and the remaining 30% is absorbed, reradiated, or reflected in the atmosphere—the Earth's surface area is approximately 29.2% land. Furthermore, biomass covers approximately 21% of the Earth's surface. This leaves 6.1% of the solar energy incident on the Earth's surface available for crops.

Biomass captures a small amount of the incident solar irradiation energy via photosynthesis. Equation 9.7 can summarize the overall reaction:

$$6CO_2 + 6H_2O + (\text{sunlight}) \rightarrow C_6H_{12}O_6 + 6O_2 \tag{9.7}$$

The efficiency depends on the plant's carbon fixation pathway. Plants employ carbon fixation as a biochemical mechanism to capture carbon from airborne CO_2 and convert it to organic forms. Scientists categorize carbon fixation pathways after the carbon chain length of the first carbohydrate formed. C3 and C4 carbon fixation pathways proceed from initial molecules with three and four carbon atoms respectively. The vast majority of plants employ C3 carbon fixation whereas about 3% of known species employ the C4 pathway. Corn, sugar cane, and sorghum are common C4 plants.

Theoretical estimates for solar energy to biomass conversion are 4.6% and 6% for C3 and C4 plants respectively. C3 plants consume energy during carbohydrate synthesis, photorespiration, and respiration. C4 plants employ more energy for carbohydrate synthesis and respiration, but do not suffer from photorespiration energy penalties resulting in higher overall efficiencies. Table 9.5 summarizes each step involved in converting solar energy at the plant's surface into chemical energy (biomass) [8].

In practice, the most efficient conversion measured in C4 plants is 3.7% of the incident solar energy, and C3 plants managed 2.4% conversion efficiency.

TABLE 9.5

Photosynthesis Steps and Efficiencies

Photosynthesis Step	% Total Energy	
Incident solar energy (on leaf surface)	100%	
Energy in photosynthetically active spectrum	48.7%	
Absorbed energy	43.8%	
Photochemically converted energy	37.2%	
Carbon fixation pathway	C3	C4
Energy in synthesized carbohydrates	12.6%	8.5%
Energy available after photorespiration	6.5%	8.5%
Energy available after respiration	4.6%	6.0%

Source: From Zhu et al. [8]. With permission.

Corn grain and soybeans are C4 and C3 plants grown throughout the U.S. Midwest, where global irradiation on the horizontal surface averages around 4 kWh/m²/day. We can estimate the efficiency of solar energy collection in this region. First, we estimate the amount of solar irradiation in the area:

$$\text{Iowa irradiation energy} = 4\,\frac{\frac{\text{kWh}}{\text{m}^2}}{\text{day}} \times 365\,\frac{\text{days}}{\text{year}} \times 3,600\,\frac{\text{s}}{\text{h}} \times 1,000\,\frac{\text{J}/\text{s}}{\text{kW}} \tag{9.8}$$

$$= 5.25 \times 10^9\,\frac{\text{MJ}}{\text{km}^2\text{year}}$$

Second, we estimate how much biomass grows on a plot of land. Total biomass includes above- and below-ground material. The ratio of below-ground biomass to above ground is known as the root-to-shoot ratio:

$$\frac{\text{Root}}{\text{Shoot}} = \frac{\text{root + exudates}}{\text{crowne or leaves or branches + trunk or stalk}} \tag{9.9}$$

Corn has a root-to-shoot ratio of about 0.55. Therefore, the total above- and below-ground biomass in land growing corn includes corn grain (700 Mg/km²), corn cobs (130 Mg/km²), corn stover (840 Mg/km²), and root and exudates (918 Mg/km²) which sums to 2,588 Mg/km². Based on an assumed energy value of 17.58 MJ/kg of corn [9], the total corn energy is

$$\text{Corn Energy}\left[\frac{\text{MJ}}{\text{km}^2}\right] = 17.58\left[\frac{\text{MJ}}{\text{kg}}\right] \times 2,588,000\left[\frac{\text{kg}}{\text{km}^2}\right] = 45.6 \times 10^6\left[\frac{\text{MJ}}{\text{km}^2}\right] \tag{9.10}$$

Based on these estimates, corn has an annual solar energy conversion efficiency of 0.87%. Farmers plant corn seeds between April and June, and they harvest in October. Thus, corn has a growth period of about 180 days yielding a 1.76% growing season efficiency.

Although biomass has low solar energy conversion efficiency, it requires minimal inputs or maintenance. This low efficiency is sufficient for nature's purposes, but scientists are investigating how to increase it. We could dramatically increase the size of our biomass resource base with incremental improvements in biomass efficiency yielding more food and energy for human use.

TABLE 9.6
Physical and Thermochemical Properties of Selected Biomass

	Feedstock	Corn Stover	Herbaceous Crop	Woody Crop
Organic composition (wt.%)	Cellulose	53	45	50
	Hemicellulose	15	30	23
	Lignin	16	15	22
	Other	16	10	5
Elemental analysis (dry wt.%)	C	44	47	48
	H	5.6	5.8	5.9
	O	43	42	44
	N	0.6	0.7	0.5
	Ash	6.8	4.5	1.6
Proximate analysis (dry wt-%)	Volatile matter	75	81	82
	Fixed C	19	15	16
	Ash	6	4	1.3
HHV (MJ/kg)		17.7	18.7	19.4
Bulk density (kg/m^3)		160–300	160–300	280–480
Yield (Mg/ha)		8,400	14,000	14,000

9.1.5 Important Properties of Biomass

Scientists need information about plant composition, heating value, bulk density, and production yields to evaluate the potential resource of biomass feedstock. Studies report composition in terms of organic components, proximate analysis, or ultimate analysis. Table 9.6 shows properties of representative fibrous biomass feedstock. Compilations of relevant biomass properties are found in Brown [1]. Excellent online databases of biomass properties include the U.S. Department of Energy Biomass Feedstock Composition and Property Database and the ECN Phyllis database.

Organic component reports describe types and quantities of biomass compounds including proteins, oils, sugars, starches, and lignocellulose (fiber). These components vary widely among plant parts. For example, corn grain is mostly starch (72 wt.%) with a relatively small amount of fiber (13 wt.%) while corn stover, that part of the crop left on the field, is mostly fiber (84 wt.%) with very little starch content. Often the fiber, which is a polymeric composite of cellulose, hemicellulose, and lignin, is reported in terms of these three constituents. Biochemical and thermochemical process design requires knowledge of the organic components because they influence the final product composition.

Proximate analysis is important in developing thermochemical conversion processes for biomass. Proximate analysis reports the yields (% mass basis) of various products obtained upon heating the material under controlled conditions; these products include moisture, volatile matter, fixed carbon, and ash. Since moisture content of biomass is so variable and can be easily determined by gravimetric methods (weighing, heating at 100°C, and reweighing), the proximate analysis of biomass is commonly reported on a dry basis. Volatile matter is that fraction of biomass that decomposes and escapes as gases upon heating a sample at moderate temperatures (about 400°C) in an inert (nonoxidizing) environment. Knowledge of volatile matter is important in designing burners and gasifiers for biomass. The remaining fraction is a mixture of solid carbon (fixed carbon) and mineral matter (ash), which can be distinguished by further heating the sample in the presence of oxygen: the carbon is converted to carbon dioxide leaving only the ash.

Ultimate analysis is simply the (major) elemental composition of the biomass on a gravimetric basis: carbon, hydrogen, oxygen, nitrogen, sulfur, and chlorine along with moisture and ash. Sometimes this information is presented on a dry, ash-free (daf) basis. Compared to fossil fuels, biomass has relatively high oxygen content (typically 40–45 wt.%), which detracts from its heating value and represents new challenges in converting these compounds into substitutes for the hydrocarbons that currently dominate our economy. In many instances, a generic molecular formula based on 1 mol of carbon is convenient for performing mass balances on a process. For example, cellulose and starch have the generic molecular formula $CH_{1.7}O_{0.83}$, hemicellulose can be represented by $CH_{1.6}O_{0.8}$, and wood is $CH_{1.4}O_{0.66}$ [10].

Efficient processing facilities keep close track of energy use. They usually employ heating value as a measure of energy content. Heating value is the net enthalpy released upon the reaction of a particular fuel with oxygen under isothermal conditions (the starting and ending temperatures are the same). If water vapor formed during reaction condenses at the end of the process, the latent enthalpy of condensation contributes to what is known as the higher heating value (HHV). Otherwise, the latent enthalpy does not contribute and the lower heating value (LHV) prevails. These measurements are typically performed in a bomb calorimeter and yield the HHV for the fuel. Heating values of biomass are conveniently estimated from the percent of carbon in the biomass on a dry basis using the empirical relationship [10]:

$$\text{HHV}(\text{dry})\left[\frac{\text{MJ}}{\text{kg}}\right] = 0.4571(\%\text{C on dry basis}) - 2.70 \tag{9.11}$$

For estimating purposes, assuming an HHV of 18 MJ/kg is a good approximation for many kinds of (dry) biomass. Enthalpies of formation are very useful in thermodynamic calculations, but these data are rarely tabulated for biomass because of the wide variability in its composition. However, if HHV for a biomass fuel has been determined in a bomb calorimeter, its enthalpy of formation can be determined by summing the enthalpies of formation of the products of combustion and subtracting the HHV from this sum.

Biomass transport properties such as bulk density and volumetric energy content are important for transport and storage applications. Bulk density is determined by weighing a known volume of biomass that is packed or baled in the form anticipated for its transportation or use. Clearly, solid logs will have higher bulk density than the same wood chipped. Bulk density will be an important determinant of transportation costs and the size of fuel storage and handling equipment. Volumetric energy content is also important. Volumetric energy content, which is simply the enthalpy content of fuel per unit volume, is calculated by multiplying the HHV of a fuel by its bulk density.

Biomass energy projects need knowledge of appropriate biomass properties to be successful. These include knowledge of the types and quantities of biomass grown in a given area, which depend on many factors including plant variety, crop management (fertilization and pest control), soil type, landscape, climate, weather, and water drainage. Thus, each project will require site-specific information obtained through discussions with state extension agents and local agronomists in combination with field trials in advance of detailed manufacturing plant design.

Example 9.2: Enthalpy of Formation Calculations for Switchgrass

Assume that a sample of switchgrass has an elemental analysis that gives a generic molecular formula of $CH_{1.4}O_{0.8}$ (molecular weight of 26.2 kg/kmol) and its HHV is measured to be 18.1 MJ/kg. The complete combustion of 1 kmol of switchgrass can be represented by

$$CH_{1.4}O_{0.8} + 0.95O_2 \rightarrow CO_2 + 0.7H_2O\ (\text{liquid}) \tag{9.12}$$

The enthalpy of reaction, ΔH_R, is calculated from the various enthalpies of formation, h_f° (note: molecular elements like O_2 have an h_f° equal to zero):

$$\begin{aligned}\Delta H_R &= h_{fCO_2}^\circ + 0.7h_{f\,H_2O(\text{liquid})}^\circ - \left(h_{fCH_{1.4}O_{0.8}}^\circ + 0.95h_{fO_2}^\circ\right) = -18.1\left[\frac{\text{MJ}}{\text{kg}}\right]\times 26.2\left[\frac{\text{kg}}{\text{kmol}}\right]\\ &= -393.5\left[\frac{\text{MJ}}{\text{kmol}}\right] + 0.7\times\left(-285.8\left[\frac{\text{MJ}}{\text{kmol}}\right]\right) - \left(h_{f\,CH_{1.4}O_{0.8}}^\circ + 0\right)\end{aligned} \tag{9.13}$$

Solving for the enthalpy of formation for $CH_{1.4}O_{0.8}$ yields

$$h_{f\,CH_{1.4}O_{0.8}}^\circ = -119.3\left[\frac{\text{MJ}}{\text{kmol}}\right] \tag{9.14}$$

This value can be used in various thermodynamic calculations for this switchgrass sample, whether gasification to hydrogen or hydrolysis to fermentable sugars. Heat of formation data are available from most engineering thermodynamic textbooks such as Moran et al. [11].

9.2 Biomass Process Economics and Technology

There are many ways to convert biomass into valuable chemicals, fuels, and electricity. This section will discuss biomass conversion technologies for gaseous and liquid fuel production as well as power generation. Gaseous fuels include biogas from anaerobic digestion and synthetic gas from gasification, and liquid fuels encompass a wide range of products such as ethanol, gasoline, diesel, and Fischer–Tropsch liquids. Combinations of these technologies can improve the overall conversion efficiency, and we will show how to compare these technologies from an energy efficiency perspective.

9.2.1 Biomass Process Economics

Conversion costs are a major factor that determines whether a technology becomes a commercial success. Conversion costs can be grouped into three major categories: capital, operating, and feedstock costs.

Capital costs include equipment purchasing and installation costs as well as interest on financial loans. They are difficult to estimate. The accuracy for most analyses ranges between +100% and –30% (meaning that the project could be twice as expensive or 30% cheaper). The accuracy can be improved with detailed engineering analysis and/or cost databases.

Operating costs consist of all expenses required to maintain the operation of a given facility such as labor, maintenance, and utilities. Operating costs include both fixed and variable cost components. Fixed costs, like labor and maintenance, are independent of the plant throughput capacity. On the other hand, variable costs are typically related to the amount of feedstock input. For example, a facility would have higher heating costs from drying increasing quantities of biomass.

Finally, feedstock costs are the costs of purchasing and delivering biomass to the facility. Facilities may choose to pay a fixed, plant-gate cost in which case the supplier carries the burden of delivering the feedstock. However, delivery costs are an important factor in how much a biorefinery pays for biomass.

The summation of these cost components allows us to estimate the cost of a biomass product. These are typically reported on a normalized basis such as Equation 9.15:

$$\text{Fuel Cost}\left[\frac{\$}{\text{unit output}}\right]=\frac{\text{Capital}+\text{Operating}+\text{Feedstock}}{\text{Product Output}} \tag{9.15}$$

In Equation 9.15, costs and output would typically be expressed on an annual basis, with capital costs annualized using a capital recovery factor (Equation 2.19). Fuel costs typically depend on the scale of operation. Large-scale facilities can optimize their operations to take advantage of economies-of-scale. The impact of economies-of-scale is greatest for capital costs, and research has shown a strong relationship between the plant capacity and unit capital costs. This relationship is captured by the power law:

$$\text{Capital Cost}=\text{Capital Cost}_0\times\left(\frac{\text{Capacity}}{\text{Capacity}_0}\right)^n \tag{9.16}$$

where

n is the scale factor

the index 0 indicates the base or known value (cost or capacity).

Typical values for the scale factor are 0.63 for biochemical facilities and 0.7 for thermochemical plants. The higher value is due to the larger capital investment required for most thermochemical projects. Economies-of-scale allow us to estimate the costs of scaling up a facility. For example, if a 40-million gallon per year (MMGPY) ethanol plant constructed in 2010 cost $46.7 million, the equivalent 200-MMGPY facility built in 2010 would have cost:

$$\text{Capital Cost}_{200\text{ MMGPY}}=\$46.7\times10^6\times\left(\frac{200\times10^6}{40\times10^6}\right)^{0.63}=\$128.7\text{million} \tag{9.17}$$

In this case, per unit capital costs decrease from $1.17/gallon of annual capacity to $0.64. This shows the potential cost reductions from scaling-up operations. In reality, these benefits are not fully realized because of diseconomies of scale. In other words, additional considerations become important at different plant capacities. However, several industries have successfully scaled refineries from a few thousand barrels per day of capacity to over 400,000 barrels/day and benefit from dramatic reductions in per unit costs.

Most operating costs scale linearly with capacity ($m=1$). However, large facilities can also reduce their per unit operating costs with labor management and equipment maintenance strategies. Thus, operating costs are sometimes scaled with a 0.9 scale factor.

Feedstock costs for biorefineries are notable for their diseconomies of scale—their per unit costs increase with capacity. Unlike fossil fuels that travel from a single source (coal mine or oil well) to a facility, biomass must be collected from a wide area. The larger the facility, the further away it must seek feedstock. This relationship can be captured simply using a scale factor greater than 1 ($p \approx 1.5$).

From these relationships, we can roughly estimate the costs of scaling-up a known technology using Equation 9.18:

$$\text{Fuel Cost}_M = C_0 \times \left(\frac{\text{Capacity}}{\text{Capacity}_0}\right)^n + O_0 \times \left(\frac{\text{Capacity}}{\text{Capacity}_0}\right)^m + F_0 \times \left(\frac{\text{Capacity}}{\text{Capacity}_0}\right)^p \quad (9.18)$$

where C, O, and F stand for capital, operating, and feedstock costs, respectively. Per unit costs in units of \$/gal can be calculated by dividing total costs by the biorefinery output capacity.

As discussed in Chapter 2, inflation should be taken into account when comparing estimates published in different years. There are several indices available to account for inflation: consumer price index (CPI), chemical engineering plant cost index (CEPCI), and the Marshall and Swift index. The CPI is maintained by the U.S. government Bureau of Labor Statistics and is a general measure of the time value of money. The other indices are more relevant for industrial estimates but are not publicly available. The CPI is available at https://www.bls.gov/data/inflation_calculator.htm. Table 9.7 shows select CPI values for 2010–2023, extracted from the longer time series of CPI values provided in Table 2.1. The U.S. CPI averages a 3.5% annual growth rate, but in 1917, the CPI increased by more than 20%. Practitioners should consider the inflation basis year to account for unusual economic variations.

Adjusting for inflation, the \$46.7 million corn ethanol facility estimate, developed in 2010, would be more expensive. Based on the annual average CPI value for the years 2010 and 2023, the cost would be

$$\$46.7 \times 10^6 \times \frac{304.702}{218.056} = \$65.3 \times 10^6 \quad (9.19)$$

The previous approach is suitable for quick comparisons of biorefineries operating at different scales. However, it would not help us determine the costs for a new facility. To determine the costs of a new biorefinery, we would need detailed process and cost data of the unit operations. These data can be acquired from process modeling tools and cost databases such as Aspen Plus™ and Aspen Process Economic Analyzer™.

The following sections include brief descriptions of mature biorefinery technologies and their costs. This information can help us conduct comparisons such as the one described here.

TABLE 9.7

Consumer Price Index Factors for 2010–2023

Year	CPI Factor
2010	218.056
2015	237.017
2020	258.811
2023	304.702

9.2.2 Conversion of Biomass to Gaseous Fuels

There are two main approaches to converting biomass into gaseous fuels: anaerobic digestion and gasification. Both processes generate a combustible gas suitable for power generation applications. Anaerobic digestion gas is commonly known as biogas, and biomass gasification products are called synthetic gas (syngas) or producer gas depending on the gas composition. The key difference between biogas and syngas is that syngas is better suited for liquid fuel production. Therefore, biomass gasification is a pathway to both power generation and liquid fuel synthesis.

9.2.2.1 Biomass to Biogas

Anaerobic digestion is the decomposition of waste to gaseous products by bacteria in an oxygen-free environment. Bacteria decompose organic wastes into a mixture of methane, carbon dioxide, and trace gases. Although bacteria can convert a wide range of waste materials, their productivity is sensitive to feedstock properties such as moisture and ash content. Anaerobic digestion yields vary from 23 to over 250 m^3/Mg (wet basis) with a methane content of 55%–75% by volume. The thermodynamic efficiency of this process is about 60%.

Biogas with a high methane content can substitute for natural gas in many applications. Clean biogas is suitable for engine generator sets, small gas turbines, and some kinds of fuel cells. Biogas contaminants, sulfur in particular, must be removed before use in some equipment.

The biological processes within an anaerobic digester that lead to biogas, as summarized in Figure 9.2, are relatively complicated [12]. They involve a series of steps through

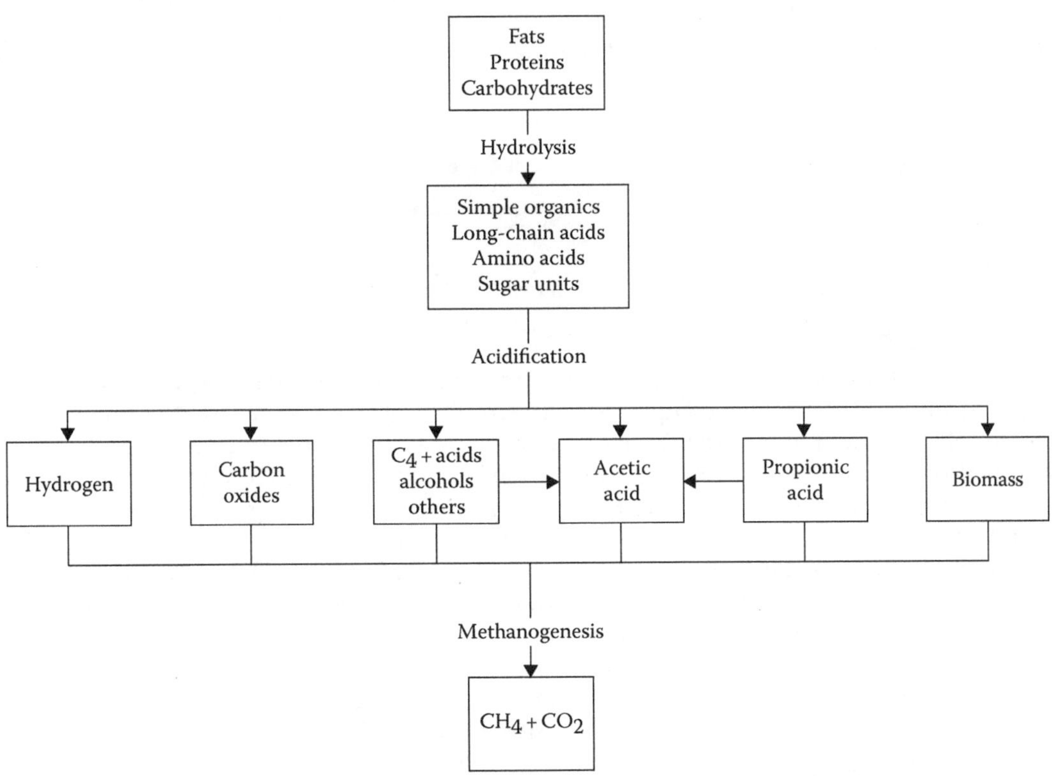

FIGURE 9.2
Microbial phases in anaerobic digestion. (Adapted from Klass [12].)

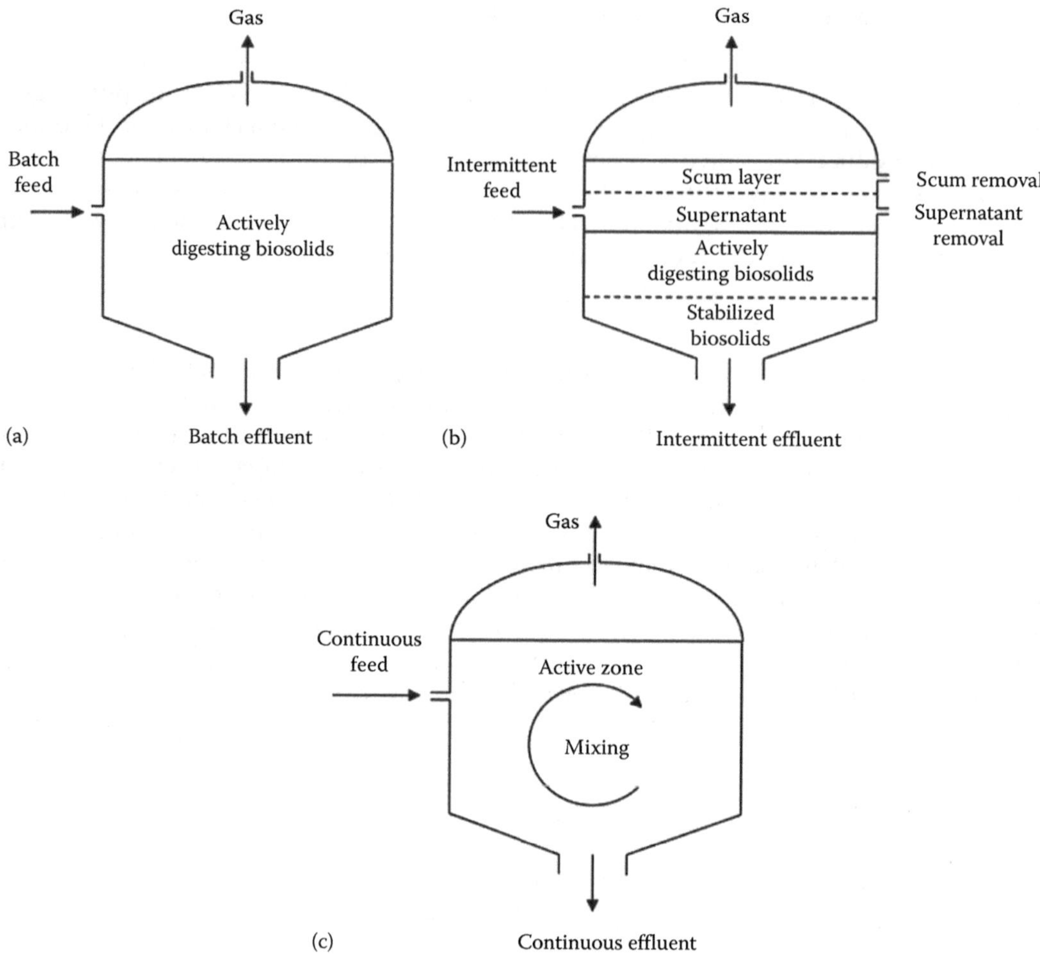

FIGURE 9.3
Types of anaerobic digesters. (a) Batch fed, (b) intermittently fed, and (c) continuously stirred and fed.

which several bacteria species break down proteins, carbohydrates, and fats into simple acids, alcohols, and gaseous compounds. Anaerobic digester systems, on the other hand, are simple as shown in Figure 9.3. These systems are typically operated as either batch or steady flow units although semi-batch operation is also an option. The effluents from these systems are biogas and sludge. The biogas may undergo scrubbing to remove hydrogen sulfide that would otherwise yield the pollutant sulfur dioxide.

Anaerobic digester designs include batch, plug-flow, and continuously stirred tank, upflow, and two-tank reactors. The batch reactor is a single vessel design in which all steps of the digestion process take place. Advanced reactors aim to improve waste contact with active bacteria and/or to separate and control the environments for acid-forming and methane-forming bacteria. Modifying multiple variables can optimize anaerobic digestion: waste pretreatment, heating, mixing, nutrient addition, specialized bacteria addition, and pH among others.

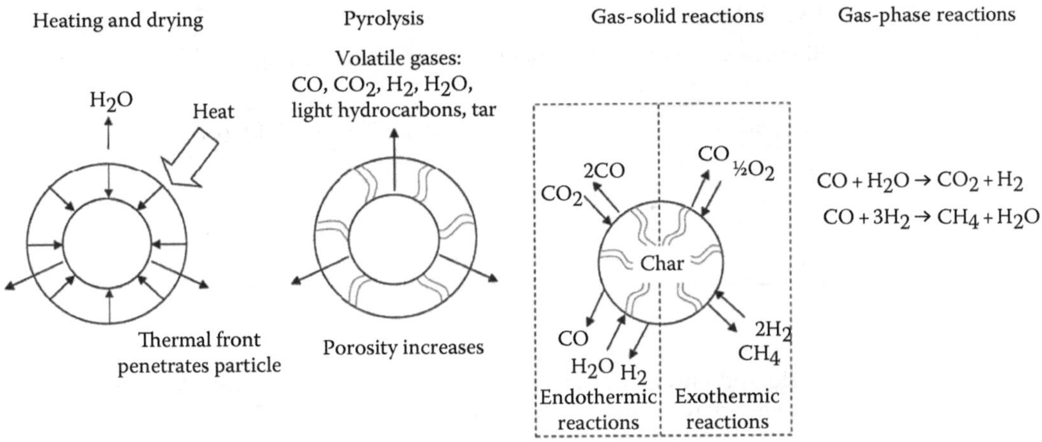

FIGURE 9.4
Processes of thermal gasification.

9.2.2.2 Biomass to Synthetic Gas

Gasification is the high-temperature (750°C–850°C) conversion of solid, carbonaceous fuels into flammable gas mixtures consisting of carbon monoxide (CO), hydrogen (H_2), methane (CH_4), nitrogen (N_2), carbon dioxide (CO_2), and smaller quantities of higher hydrocarbons. When the gas mixture has a high nitrogen content, it is known as producer gas; otherwise, it is known as synthetic gas. The overall process is endothermic and requires either the simultaneous burning of part of the fuel or the delivery of an external heat source to drive the process [13]. Gasification converts over 90% of biomass feed into gas with the remaining material ending as solid char and ash, or viscous tar in some applications. Typical biomass-to-syngas efficiencies range between 70% and 90%.

Figure 9.4 illustrates the steps involved in gasification: heating and drying, pyrolysis, gas–solid reactions that consume char, and gas-phase reactions that adjust the final composition. Solid particles are initially heated and dried at temperatures below 100°C. Pyrolysis takes place as the particle temperature reaches 600°C; pyrolysis produces the intermediate products of char, gases, and condensable vapors (including water, methanol, acids, and heavy hydrocarbons). At high enough temperatures (above 1,000°C), it is possible to convert over 95% of the feed carbon with char-consuming reactions. Gas-phase reactions achieve equilibrium in high-temperature gasification; however, gasification is sometimes conducted at low temperatures (approximately 850°C) yielding nonequilibrium products (mostly tars and light hydrocarbons).

Example 9.3: The Role of Carbon Conversion in Gasification

Carbon conversion is a key measure of gasification efficiency. Commercial high-temperature gasifiers can achieve carbon conversions of over 97%. Low-temperature gasifiers commonly employed with biomass operate at lower carbon conversion rates of 90%–95%. Solid residue from the gasifier contains most of the unconverted carbon otherwise known as char. Even though char material yields are low (10 wt.%), char energy content can be high. This example illustrates how much energy can remain unconverted in the relatively small fractions of char and tar.

TABLE 9.8

Biomass Gasification Product Distribution and Energy Content for Example 9.3

Compound	Yield (wt.%)	Energy Content (MJ/kg)
Biomass	—	17.4
Syngas	81	17.1
Char	12	22.7
Tar	7	12.1

Based on the yield and energy content values in Table 9.8, we can calculate the percentage of biomass energy remaining as char and tar:

$$\mathrm{Char_{Energy}}\left[\frac{\mathrm{MJ}}{\mathrm{MJ_{Biomass}}}\right]=\frac{0.12\left[\mathrm{kg/kg}\right]\times 22.7\left[\mathrm{MJ/kg}\right]}{17.4\left[\mathrm{MJ/kg}\right]}=15.6\% \quad (9.20)$$

$$\mathrm{Tar_{Energy}}\left[\frac{\mathrm{MJ}}{\mathrm{MJ_{Biomass}}}\right]=\frac{0.07\left[\mathrm{kg/kg}\right]\times 12.1\left[\mathrm{MJ/kg}\right]}{17.4\left[\mathrm{MJ/kg}\right]}=4.87\% \quad (9.21)$$

These results show that 20.5% of the biomass energy remains unconverted as char and tar after low-temperature gasification. Higher conversion rates are possible by recirculating the char into the gasifier, and some of the tar energy can be recovered by combustion downstream from the gasifier.

Gasification requires an oxygen source for exothermic processes that help with drying, pyrolysis, and endothermic gas-phase reactions. Air, pure oxygen, and steam are the main oxygen sources employed in gasification. Air gasification dilutes the produced gas with nitrogen lowering the product's energy density. Thus, there are advantages in using oxygen or steam despite the additional costs. Oxygen requirement is an important process parameter that can be estimated using combustion equations. The stoichiometric requirement for complete combustion can be calculated using the combustion formula:

$$C_aH_bO_cN_dS_e + fO_2 + 3.76f\ N_2 \rightarrow aCO_2 + \frac{b}{2}H_2O + eSO_2 + \left(3.76f + \frac{d}{2}\right)N_2 \quad (9.22)$$

For full combustion, the moles of oxygen required are f:

$$f = a + \frac{b}{4} + e - \frac{c}{2} \quad (9.23)$$

For example, the moles of oxygen required to fully combust coal are

$$\text{Coal composition} = CH_{0.739}O_{0.091}N_{0.014}S_{0.003} \quad (9.24)$$

$$f = 1 + \frac{0.739}{4} + 0.003 - \frac{0.091}{2} = 1.14 \text{ mol of } O_2 \quad (9.25)$$

We can calculate the stoichiometric oxygen requirement on a mass basis using

$$\frac{1.14 \text{ mol of } O_2}{\text{mol of Coal}} \times \frac{32.0 \text{ g of } O_2}{\text{mol of } O_2} \times \frac{1 \text{ mol of Coal}}{14.49 \text{ g of Coal}} = 2.52 \frac{\text{g of } O_2}{\text{g of Coal}} \tag{9.26}$$

Therefore, a kilogram of coal with this composition consumes 2.52 kg of oxygen for full combustion. Note that it is common practice to burn fuel with about 25% excess oxygen than the stoichiometric amount. We can convert the oxygen requirement into an equivalent amount of air. Since air is 23.2 wt.% oxygen, the total amount of air needed for complete combustion of this fuel is 10.87 kg (oxygen required/0.232). However, gasification requires that oxygen levels remain below 25% of the stoichiometric amount for full combustion. Therefore, gasification of this coal feedstock would require less than 0.63 kg of oxygen (2.72 kg of air) per kilogram of coal.

There are four main types of gasifiers: updraft (countercurrent), downdraft (cocurrent), fluidized bed, and entrained flow [14]. These are illustrated in Figure 9.5, and their performance characteristics are summarized in Table 9.9. The most common types of gasifiers are variations of the entrained flow and fluidized bed designs [15].

The entrained flow gasifier is a highly efficient reactor. This reactor was developed for steam-oxygen gasification of coal at temperatures of 1,200°C–1,500°C. These high temperatures assure excellent char conversion (approaching 99%) and low tar production and convert the ash to molten slag, which drains from the bottom of the reactor. The technology

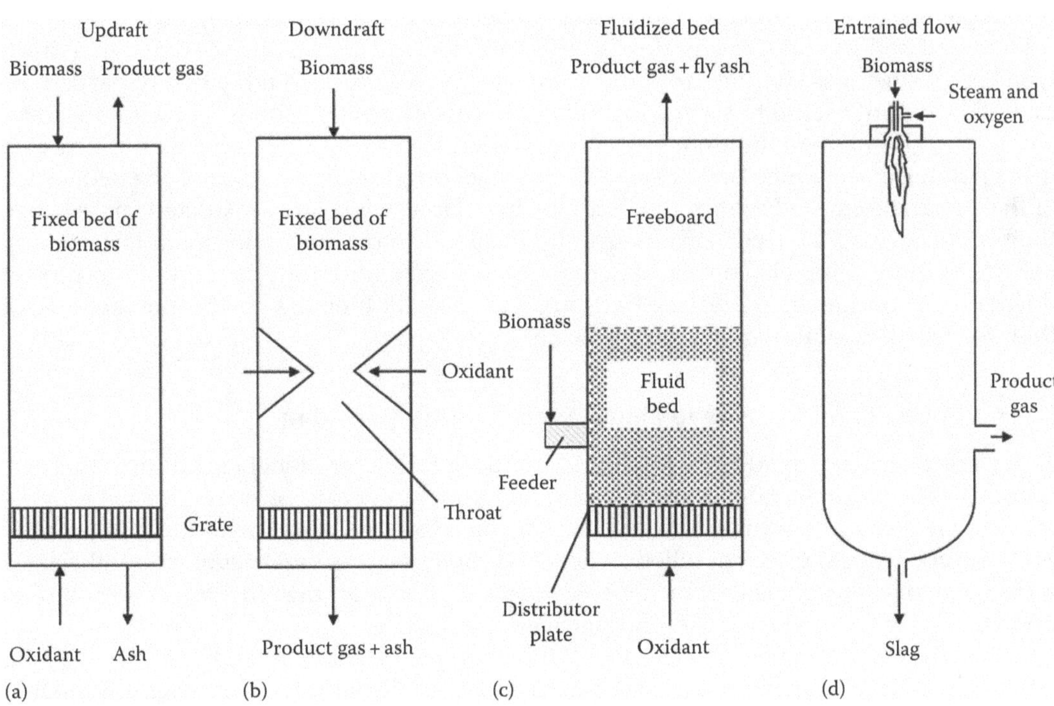

FIGURE 9.5
Common types of biomass gasifiers: (a) updraft, (b) downdraft, (c) fluidized bed, and (d) entrained flow.

TABLE 9.9

Producer Gas Composition from Various Kinds of Gasifiers

	Gaseous Constituents (vol.% Dry)						Gas Quality	
Gasifier Type	H_2	CO	CO_2	CH_4	N_2	**HHV (MJ/m³)**	**Tars**	**Dust**
Air-blown updraft	11	24	9	3	53	5.5	High (≈10 g/m³)	Low
Air-blown downdraft	17	21	13	1	48	5.7	Low (≈1 g/m³)	Medium
Air-blown fluidized bed	9	14	20	7	50	5.4	Medium (≈10 g/m³)	High
Oxygen-blown downdraft	32	48	15	2	3	10.4	Low (≈1 g/m³)	Low
Indirectly heated fluidized bed	31	48	0	21	0	17.4	Medium (≈10 g/m³)	High

is attractive for advanced coal power plants but has not been widely explored for biomass because of the expense of finely dividing biomass, the difficulty of reaching high temperatures with biomass, and the presence of alkali in biomass, which leads to severe ash sintering. Despite these difficulties, entrained flow reactors are being considered for gasification of pretreated biomass (such as torrefied biomass and pyrolysis liquids) [16].

Torrefied biomass is the product of heating biomass to between 200°C and 350°C—a process known as torrefaction. At these temperatures, biomass undergoes drying and minor loss of volatile content. The remaining product has a closer resemblance to coal: high energy density and improved grindability. It is also hydrophobic, which improves its storage properties by reducing water-related degradation. Torrefaction has attracted recent interest as an approach to introduce biomass into coal-based facilities.

Fluidized bed gasifiers are markedly different from entrained flow designs. In a fluidized bed gasifier, gas streams from the bottom of the reactor through a bed of particulate material (typically sand) to form a turbulent mixture of gas and solids. Feed is added at a small enough rate to maintain a low concentration in the bed. Typically, a fluidized gasifier operates in the range of 700°C–850°C. By injecting fuel in the base of the bed, much of the tar can be cracked within the fluidized bed. However, a large insulated space above the bed, known as the freeboard, is usually included to promote additional tar cracking as well as more complete conversion of char. Further conditioning is required to protect downstream equipment. Fluidized beds are attractive for biomass gasification because of their feedstock flexibility and ability to scale.

9.2.3 Conversion of Biomass to Liquid Fuels for Transportation

There are two major platforms for liquid biofuel production: biochemical and thermochemical. The biochemical platform employs microorganisms to convert primarily sugars into mostly alcohols, and it is best known for the corn- or sugarcane-to-ethanol process. The thermochemical platform relies on heat and catalysts (precious metals) to synthesize a wide range of compounds that include alcohols and hydrocarbons; this technology builds upon research from the coal, gas, and petroleum industries.

This section will discuss how to convert biomass into transportation fuels (biofuels) via the biochemical and thermochemical platforms. We will discuss the overall efficiencies, the various products generated, and overall process costs.

9.2.3.1 Corn Ethanol

Ethanol is the dominant biofuel with global production exceeding 100 billion L. The majority of this ethanol comes from Brazilian sugarcane and U.S. corn grain. Microbes convert

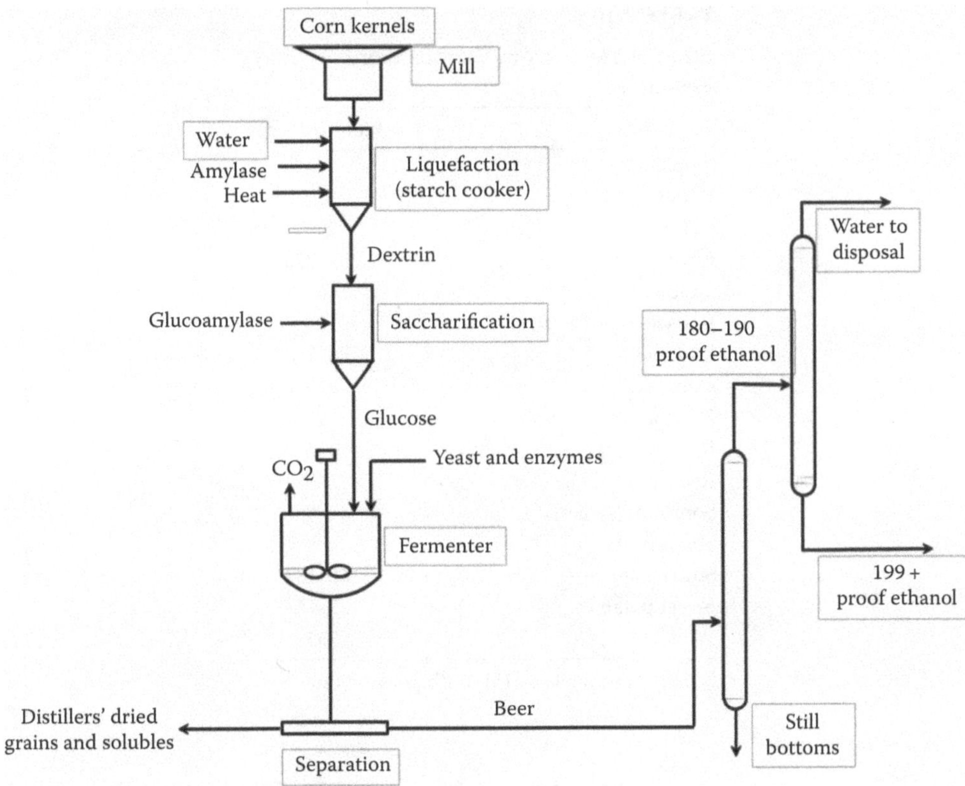

FIGURE 9.6
Dry-grind corn to ethanol.

these sugar-rich crops into alcohols with relative ease and at low cost. Two major types of corn-to-ethanol facilities have been operated in the United States: dry-grind and wet-milling.

Dry-grind plants, shown schematically in Figure 9.6, grind the whole kernel while wet-milling plants soak the grain with water and acid to separate the corn germ, fiber, gluten, and starch components before mechanical grinding [17]. The capital investment for dry-grind is less than that for a comparably sized wet-milling plant. However, the higher value of its by-products, greater product flexibility, and simpler ethanol production can make a wet-milling plant a more profitable investment.

Dry-grind ethanol production takes place in four major steps: pretreatment, cooking, fermentation, and distillation. Pretreatment consists of grinding the corn kernel into flour "meal," which is mixed with water, enzymes, and ammonia. This mixture ("mash") is then "cooked" to reduce bacteria levels. After cooling, the mash is sent to the fermenter where it remains for 40 hours or more. The beer resulting from fermentation consists of a mixture of ethanol and stillage. Energy-intensive distillation of the beer is necessary to separate the stillage and water content from the ethanol and achieve maximum concentrations of 95%. This is followed by further purification to 100% ethanol using molecular sieves.

A modern dry-grind plant will produce over 2.7 gal of ethanol per bushel of corn processed. Yields of coproducts per bushel of corn are 7.7–8.2 kg (17–18 lb) of dry distiller's dried grains with solubles and 7.3–7.7 kg (16–17 lb) of carbon dioxide evolved from fermentation. The former is sold as animal feed while the latter can be sold to the carbonated beverage industry and there are proposals to sequester it in saline aquifers to accomplish

TABLE 9.10

Ethanol Yields from Various Biorenewable Resources

Feedstock	Yield (L/Mg)
Apples	64
Barley	330
Cellulose	259
Corn	355–370
Grapes	63
Jerusalem artichoke	83
Molasses	280–288
Oats	265
Potatoes	96
Rice (rough)	332
Rye	329
Sorghum (sweet)	44–86
Sugar beets	88
Sugarcane	160–187
Sweet potatoes	125–143
Wheat	355

Source: From Klass [15]. With permission.

net carbon dioxide removal from the atmosphere. As a rule of thumb, the three products are produced in approximately equal weight per bushel. Table 9.10 shows ethanol yields for various feedstocks [15].

In its 2022 Annual Technology Baseline for transportation fuels, the National Renewable Energy Laboratory reported U.S. average wholesale prices for fuel ethanol of $1.39 per gallon of ethanol or $2.14 per gasoline gallon equivalent (2022 dollars) (https://atb.nrel.gov/transportation/2022/ethanol).

9.2.3.2 Cellulosic Ethanol

Much of the carbohydrate in plant materials is structural polysaccharides, providing shape and strength to the plant. This structural material, known as lignocellulose, is a composite of cellulose fibers embedded in a cross-linked lignin–hemicellulose matrix [19]. Depolymerization to basic plant components is difficult because lignocellulose is resistant to both chemical and biological attacks [20]. However, depolymerization is necessary for microbes to efficiently convert cellulosic biomass into alcohols.

Cellulose to ethanol consists of four steps: pretreatment, enzymatic hydrolysis, fermentation, and distillation [17]. Of these, pretreatment is the costliest step, accounting for about 33% of the total processing costs [21,22]. An important goal of all pretreatments is to increase the surface area of lignocellulosic material, making the polysaccharides more susceptible to hydrolysis. Thus, comminution, or size reduction, is an integral part of all pretreatments.

Enzymatic hydrolysis was developed to better utilize both cellulose and hemicellulose from lignocellulosic materials. Three basic methods for hydrolyzing structural polysaccharides in plant cell walls to fermentable sugars are available: concentrated acid hydrolysis, dilute acid hydrolysis, and enzymatic hydrolysis [6,20]. The two acid processes hydrolyze both hemicellulose and cellulose with very little pretreatment beyond comminution of the lignocellulosic material to particles of about 1 mm size. The enzymatic process must be preceded by extensive pretreatment to separate the cellulose, hemicellulose, and lignin fractions.

Although thermodynamic efficiencies for conversion of carbohydrates to ethanol can be calculated, it is more typical to report the volumetric yield of ethanol per unit mass of feedstock. The yield of ethanol from energy crops varies considerably. Among sugar crops, sweet sorghum yields 80 L/metric ton, sugar beets yield 90–100 L/ton, and sugar cane yields 75 L/ton. Among starch and inulin crops, the ethanol yield is 350–400 L/ton of corn, 400 L/ton of wheat, and 90 L/ton of Jerusalem artichoke. Among lignocellulosic crops, the potential ethanol yield is 400 L/ton of hybrid poplar, 450 L/ton for corn stover, 510 L/ton for corn cobs, and 490 L/ton for wheat straw.

In its 2022 Annual Technology Baseline, NREL estimated costs for cellulosic ethanol at $4.00 per gasoline gallon equivalent (2022 dollars) based on high volume production from corn stover (https://atb.nrel.gov/transportation/2022/ethanol).

Example 9.4: Cellulosic Ethanol Energy Return on Investment Calculations

There is growing interest in calculating the Energy Return on Investment (EROI) of biofuel technologies and ethanol in particular [23]. The process of estimating EROI can be summarized into three steps: identifying system inputs and outputs, calculating energy transfer quantities, and determining the EROI following a systematic approach.

1. **System identification:** The cellulosic ethanol system involves inputs of fuel, electricity, fertilizer, and herbicides. The primary output is ethanol, but excess heat can be considered a coproduct. Energy inputs/costs can be distributed among coproducts using different methods.
2. **Energy transfer:** The life cycle analysis literature includes reports of energy input/output in biofuel systems. Hall et al. [23] reported the values shown in Table 9.11 for cellulosic ethanol production. These values are based on converting switchgrass to ethanol, in an optimistic scenario with high switchgrass yields combined with low fertilizer input. This process employs a portion of the switchgrass feed to provide process heat and electricity.

TABLE 9.11

Advanced Cellulosic Ethanol Energy Input/Output

Input	Input/Output Value (MJ/L Ethanol)
Agriculture: fuel	0.19
Agriculture: electricity	0.00
Fertilizer	0.33
Pesticides/herbicides	0.10
Feedstock transport	0.29
Biorefinery: fuel	0
Biorefinery: electricity	0
Ethanol distribution	≈0.00
Biorefinery: coproducts	0
Total direct	0.91
Indirect	0.13
Total input	1.04

Source: From Hall et al. [23]. With permission.

3. **EROI calculations:** In addition to ethanol, the biorefinery supporting Table 9.11 generates 4.79 MJ/L of excess electricity, which displaces three times the amount of energy in the form of fuel input (based on fuel-to-power efficiency of ≈33%). Thus, the authors concluded that cellulosic ethanol has an energy output of 37.10 MJ/L, including 14.4 MJ/L of electricity. The net EROI for the advanced cellulosic ethanol is 37.10 MJ/L/1.04 MJ/L = 35.7.

EROI calculations are subject to numerous assumptions and methodologies. Differences in energy input/output values and calculation methods explain some of the large differences in reported EROI estimates. For ethanol, coproduct allocations and assumptions about process energy sources and use of residual energy are important factors that need to be clearly specified.

9.2.3.3 Biomass Fermentation to Alternative Fuels

Ethanol has several limitations as a transportation fuel, including its affinity for water, which prevents it from being fully compatible with the existing fuel infrastructure, and its low volumetric heating value, which is only two-thirds that of gasoline. For this reason, fermentations that produce metabolites other than ethanol have been proposed. Alternative fermentations could produce hydrophobic molecules that are less oxidized than ethanol including higher alcohols (most prominently butanol), fatty acids, fatty alcohols, esters, alkanes, alkenes, and isoprenes.

Alternative biochemical biofuels could address the energetic and fuel compatibility challenges faced by ethanol. Metabolic engineering, reactor design, and hybrid approaches are some of the approaches employed by researchers to identify and improve microbes' selectivity and performance for biofuel production.

9.2.3.4 Biomass to Fischer–Tropsch Liquids

Fischer–Tropsch liquids from biomass have antecedents in the coal-to-liquids industry. Germany extensively developed the Fischer–Tropsch process during World War II when it was denied access to petroleum-rich regions of the world. Likewise, when South Africa faced a world oil embargo during its era of apartheid, it employed Fischer–Tropsch technology to sustain its national economy. A comprehensive bibliography of Fischer–Tropsch literature can be found on the internet [24].

Fischer–Tropsch catalysis produces a large variety of hydrocarbons including light hydrocarbon gases, paraffinic waxes, and alcohols according to the generalized reaction shown in Equation 9.27:

$$\begin{aligned} n\mathrm{CO} + (2n+1)\mathrm{H_2} &\rightarrow \mathrm{C}_n\mathrm{H}_{2n+2} + n\mathrm{H_2O} \\ n\mathrm{CO} + 2n\mathrm{H_2} &\rightarrow \mathrm{C}_n\mathrm{H}_{2n} + n\mathrm{H_2O} \end{aligned} \tag{9.27}$$

Fischer–Tropsch liquids composition depends on the process selectivity. Process selectivity is affected by various factors including catalyst and feed gas properties. The Anderson–Schulz–Flory (ASF) distribution (Equation 9.28) describes the probability of hydrocarbon chain growth where the molar yield for a carbon chain can be calculated using the following equation [25]:

$$C_n = \alpha^{n-1}(1-\alpha) \tag{9.28}$$

where α is the chain growth probability of a hydrocarbon of length n. Light hydrocarbons (mostly methane) can be fed into a gas turbine to provide power. Fischer–Tropsch liquids can be separated into various products in a process similar to petroleum distillation. Product distribution is a function of temperature, pressure, feed gas composition (H_2/CO), catalyst type, and composition [26]. Depending on the types and quantities of Fischer–Tropsch products desired, either low-(200°C–240°C) or high-temperature (300°C–350°C) synthesis at pressures ranging between 10 and 40 bar is used. For example, high gasoline yield can be achieved using high process temperatures and an iron catalyst. Fischer–Tropsch synthesis requires careful control of the H_2/CO ratio to satisfy the stoichiometry of the synthesis reactions as well as avoid deposition of carbon on the catalysts (coking). The optimal H_2/CO ratio for the production of naphtha and diesel range fuels sold in North American markets is 2:1.

Swanson et al. developed an analysis of Fischer–Tropsch liquid fuels from biomass [27]. Their estimates of a 2,000 Mg/day corn stover facility found capital costs of $498–$606 million (2007 dollars) with minimum fuel selling prices of $4.27 and $4.83 per gallon of gasoline equivalent depending on whether the process was based on a fluid bed or entrained flow gasifier.

9.2.3.5 Biomass Pyrolysis Oil to Gasoline and Diesel

Pyrolysis is the thermal decomposition of organic compounds in the absence of oxygen [28]. The resulting product streams depend on the rate and duration of heating. Liquid yields exceeding 70% are possible under conditions of fast pyrolysis, which is characterized by rapid heating rates (up to 1,000°C/s), moderate reactor temperatures (450°C–600°C), short vapor residence times (<0.5 seconds), and rapid cooling at the end of the process. Rapid cooling is essential if high-molecular-weight liquids are to be condensed rather than further decomposed to low-molecular-weight gases.

Pyrolysis liquid, also known as bio-oil, is a low-viscosity, dark-brown fluid with up to 15%–20% water, which contrasts with the black, tarry liquid resulting from slow pyrolysis [29]. Fast pyrolysis liquid is a complicated mixture of organic compounds arising from thermal degradation of carbohydrate and lignin polymers in biomass [30]. The liquid is highly oxygenated, approximating the elemental composition of the feedstock, which makes it highly unstable. The HHV of pyrolysis liquids ranges between 17 and 20 MJ/kg with liquid densities of about 1280 kg/m^3. Assuming the conversion of 72% of the biomass feedstock to liquid on a weight basis, yields of pyrolysis oil are about 135 gal/ton.

Example 9.5: Energy Content in Pyrolysis Oils

Producing bio-oils with a high energy content is one of the main challenges for biomass pyrolysis. Bio-oil material yields can exceed 70 wt.% but suffer from high moisture content. In this example, we calculate the energy conversion efficiency for a typical pyrolysis experiment.

Based on yield and energy density values from Table 9.12, we can estimate the biomass to bio-oil energy efficiency using Equation 9.29:

$$\text{Efficiency}_{\text{Bio-oil}} = \frac{0.65\left[\text{kg/kg}\right]\times 16.5\left[\text{MJ/kg}\right]}{15.8\left[\text{MJ/kg}\right]} = 68.0\% \tag{9.29}$$

TABLE 9.12
Pyrolysis Mass and Energy Yield Values for Example 9.5

Component	Yield (wt.%)	Energy Content (MJ/kg)
Biomass	—	15.8
Pyrolysis oil	65	16.5
Pyrolysis gas	21	5.7
Pyrolysis char	14	27.5

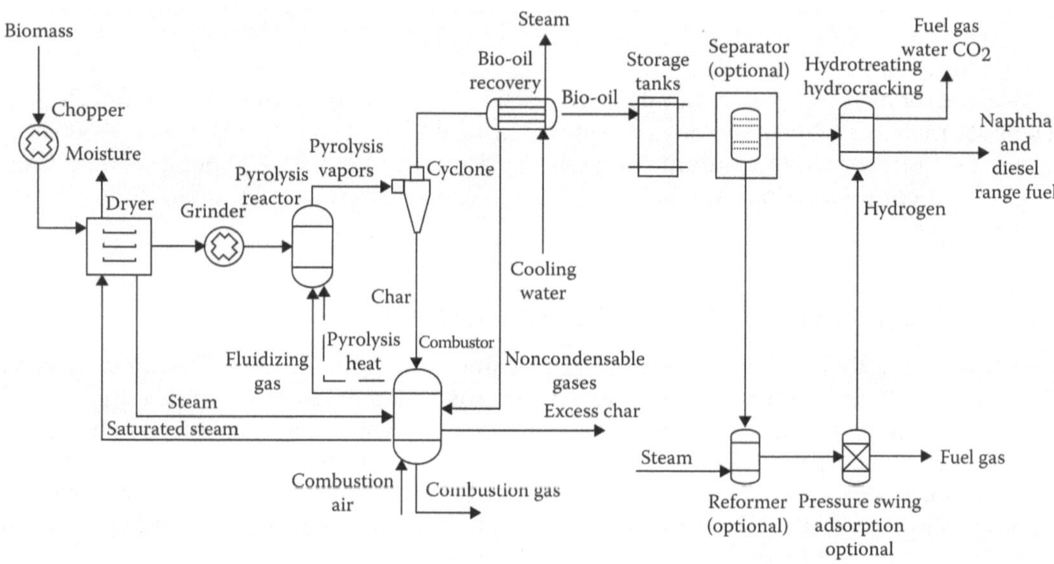

FIGURE 9.7
Biomass fast pyrolysis and bio-oil hydrotreating/hydrocracking to gasoline and diesel.

Production of pyrolysis oils and its coproducts involves several steps [31] that are illustrated in Figure 9.7. Lignocellulosic feedstock, such as wood or agricultural residues, is milled to a fine powder to promote rapid reaction. The particles are augured into the pyrolysis reactor where they are rapidly heated and converted into condensable vapors, liquid aerosols, noncondensable gases, and charcoal. These products are transported out of the reactor into a cyclone operating above the condensation point of pyrolysis vapors where the charcoal is removed. Vapors and gases are transported to a quench vessel where a spray of pyrolysis liquid cools vapors sufficiently for them to condense. The noncondensable gases, which include flammable carbon monoxide, hydrogen, and methane, are burned in air to provide heat for the pyrolysis reactor. The condensable liquids consist of a mixture of hundreds of organic compounds commonly known as bio-oil.

Bio-oil can be upgraded using conventional oil refinery processes. The most common of these processes are hydrotreatment and hydrocracking. Hydrotreatment can remove most of the bio-oil impurities such as nitrogen, alkali metals, and oxygen carried over from the original biomass. The purpose of hydrocracking is to break down heavy hydrocarbons with long carbon chain lengths into compounds within the naphtha and diesel range (7–20 carbon atoms). Bio-oil contains organic compounds with molecular weights in the hundreds. These compounds can be cracked into lighter hydrocarbons increasing the yield of naphtha and diesel range fuels.

Gasoline and diesel from corn stover fast pyrolysis followed by bio-oil upgrading could cost between $2.00 and $3.10 per gasoline gallon equivalent (~2010 dollars) for a 2,000 Mg/day biorefinery [31]. This facility would generate 35.4 million gallons/year.

9.2.3.6 Compressed Gases as Transportation Fuel

The ideal transportation fuel is a stable liquid at ambient temperature and pressure that can be readily vaporized and burned within an engine. However, some gaseous compounds are converted to transportation fuels by increasing their density through compression. Among these gaseous transportation fuels are hydrogen, methane, ammonia, and dimethyl ether (DME).

Hydrogen can be manufactured from syngas via the water–gas shift reaction. This moderately exothermic reaction is best performed at relatively low temperatures in one or more stages with the aid of catalysts. Biomass-to-hydrogen processes face the same fuel delivery challenges as hydrogen from fossil sources in addition to the increased costs associated with using biomass.

Methane can be the main product of gasification under conditions known as hydrogasification [13]:

$$\begin{aligned} C + 2H_2 &\rightarrow CH_4 \\ CO + 3H_2 &\rightarrow CH_4 + H_2O \end{aligned} \tag{9.30}$$

Although methane is more easily pressurized or liquefied than hydrogen, its density is still too low to be an attractive transportation fuel except in some urban mass transit applications [32].

Ammonia is produced by the Haber process at 200 bar and 500°C [33]:

$$N_2 + 3H_2 \rightarrow 2NH_3 \tag{9.31}$$

As a widely employed agricultural fertilizer, the United States already has in place production, storage, and distribution infrastructure for its use.

DME, like liquefied petroleum gas, is a nontoxic, flammable gas at ambient conditions that is easily stored as a liquid under modest pressures [34]. It can be produced from syngas and can substitute for diesel after minor engine modifications.

9.2.3.7 Modern Concepts in Biofuel Conversion

There are a wide range of technologies under development that could have a dramatic impact on how we convert biomass to transportation fuels. These technologies include syngas to alcohols, lipid to fuels, hydrothermal processing (HTP), catalytic methylated furan synthesis, and compressed gases.

Syngas affords the possibility of generating many different types of alcohols. Efforts in Germany during World War II to develop alternative motor fuels discovered that iron-based catalysts could yield appreciable quantities of water-soluble alcohols from syngas, especially ethanol [35]. The overall conversion can be described by Equation 9.32:

$$2CO + 4H_2 \rightarrow CH_3CH_2OH + H_2O \tag{9.32}$$

These early efforts yielded liquids containing as much as 45%–60% alcohols of which 60%–70% was ethanol. Working at pressures of around 50 bar and temperatures in the range of 220°C–370°C, researchers have developed catalysts with selectivity to alcohols of over 95%, but production of pure ethanol has been elusive. Researchers have proposed several alternative approaches: direct carbonylation of methanol and syngas fermentation.

Methanol, which can be synthesized from syngas, is readily converted into ethanol with direct carbonylation (Equation 9.33). Direct carbonylation of methanol has the advantage of yielding ethanol without coproduct water, which would eliminate energy-intensive distillations. However, the cost-effectiveness of this approach to ethanol synthesis has not been proven:

$$CH_3OH + 2CO + H_2 \rightarrow CH_3CH_2OH + CO_2 \tag{9.33}$$

Methanol can not only be converted into ethanol, but it can also be converted to alcohols with a wide range of carbon chain lengths. The overall reaction (Equation 9.34) can be described by the conversion of CO and H into a mixture of alcohols and water as the main by-product:

$$nCO + 2nH_2 \rightarrow C_nH_{2n+1}OH + (n-1)H_2O \tag{9.34}$$

where n ranges between 1 and 8 [36]. Methanol synthesis is favored at low temperatures and high pressures, and higher alcohols are produced as the temperature is increased. Process conditions for high-temperature, high-pressure synthesis catalysts range between 300°C and 425°C, and 12.5 and 30 MPa using modified methanol catalysts.

In 2007, NREL researchers estimated capital costs for mixed-alcohol synthesis of $137 million and operating costs of $1.01 per gallon for a 72.6-MMGPY mixed-alcohol biomass plant [37].

Syngas fermentation is a hybrid thermochemical and biochemical biofuel synthesis approach. In this process, microorganisms consume the CO, CO_2, and H_2 in syngas to form a variety of products including carboxylic acids, alcohols, and esters [38]. The advantages of this approach include ambient temperature and pressure operating conditions, and improved resistance to contaminants such as sulfur. The disadvantages include product inhibition and mass transfer limitations.

Lipids are a large group of hydrophobic, fat-soluble compounds produced by plants and animals for high-density energy storage. Triglycerides of fatty acids, commonly known as fats and oils depending on their melting points, are among the most familiar form of lipids and have been widely used in recent years for the production of diesel fuel substitutes. The solution to this problem is to convert the triglycerides into methyl esters or ethyl esters of the fatty acids, known as biodiesel, and the by-product 1,2,3-propanetriol (glycerol).

A wide variety of plant species produce triglycerides in commercially significant quantities, most of it occurring in seeds [18]. Average oil yields range from 15,000 L/km^2 for cottonseed to 81,400 L/km^2 for peanut oil although intensive cultivation might double these numbers. Soybeans are responsible for more than 50% of world production of oilseed, representing 48–82 million bbl/year. The average oil yield for soybeans is 38,300 L/km^2 [18]. A higher yielding crop is oil palm, already grown in plantations for vegetable oil production [39].

Although oil palm yields are 10 times higher than soybeans, environmentalists are concerned that its cultivation for fuel production has resulted in large-scale rainforest destruction. However, several oilseed crops have been identified that could be grown on waste land or even saline soils, which reduces concerns about competition for food crops and rainforest destruction. These alternative oilseed crops include jatropha, Chinese tallow tree, and salicornia [40,41].

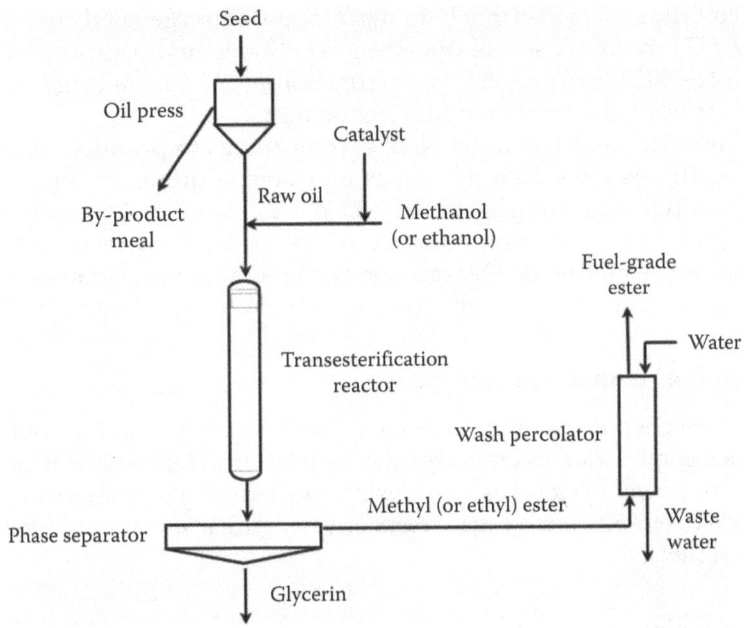

FIGURE 9.8
Conversion of triglycerides to methyl (or ethyl) esters and glycerol (biodiesel synthesis).

Triglycerides can be converted into transportation fuels via transesterification [42]. The process, described in Figure 9.8, has been commercialized for the production of biodiesel. Although biodiesel can substitute for diesel fuel, it has some shortcomings. Fatty acid methyl esters found in biodiesel are subject to microbial or oxidative attack, making them unsuitable in applications requiring long-term fuel storage. Low-temperature performance of biodiesel is sometimes problematic.

For these reasons, hydrogenation is being evaluated as a replacement for transesterification in the production of lipid-based biofuels [42]. Hydrogenation includes a number of reactions: large molecules are broken into smaller molecules; carbon–carbon double bonds are converted into single bonds; molecular structures are rearranged; and undesirable atoms such as sulfur, nitrogen, and oxygen are removed from the hydrogenated compounds. In the case of lipids, hydrogenation yields alkanes, which are highly desirable fuel molecules.

Finally, lipid-rich microalgae might be grown in brackish water or even in seawater [43]. Algae can produce as much as 60% of their body weight as lipids when deprived of key nutrients such as silicon for diatoms or nitrogen for green algae. They employ relatively low substrate concentrations, on the order of 10–40 g/L. However, algae require the proper combination of brackish water, CO_2, and sunlight, which could limit the number of sites where this process would be profitable. In a 2010 study, Lundquist et al. [44] estimated that biofuels from algae would require break-even oil prices of $332/barrel to be commercially competitive.

Researchers have pursued hydrothermal biomass processing (HTP) as an alternative to producing a variety of biofuels [45]. HTP describes the thermal treatment of wet biomass to produce carbohydrate, liquid hydrocarbons, or gaseous products depending upon the reaction conditions [46,47]. HTP liquids are commonly referred to as bio-crude. Bio-crude contains a wide range of organic compounds including hydrocarbons. Unlike pyrolysis oil, bio-crude contains a much smaller amount of oxygenated organic compounds. Unfortunately, HTP requires severe operating conditions: temperatures of 200°C–600°C and pressures of 5–40 MPa to prevent water from boiling. The metallurgy required for this process has so far limited adoption of this technology.

Destructive distillation of wood is one of several routes to produce methylated furans [48]. Methylated furans have heating values and octane numbers comparable to gasoline making them potential transportation fuel [49]. 2,5-Dimethyl furan in particular has received interest because new catalytic synthesis routes from sugars have been developed [50,51]. Furfural can be methylated by reaction with methanol over zeolite catalyst to yield methyl furan, dimethyl furan, trimethyl furan, and tetramethyl furan [49,52].

9.2.4 Conversion of Biomass to Electricity

We can convert the chemical energy in biomass into electric power by a number of different routes. Direct combustion of biomass releases heat that can be used in Stirling engines or Rankine steam power cycles. Fast pyrolysis and thermal gasification, described in earlier sections, yield bio-oil and syngas, respectively, which are suitable for firing in gas turbines or even fuel cells.

9.2.4.1 Direct Combustion

Combustion is the rapid oxidation of fuel to obtain energy in the form of heat. Since biomass fuels are primarily composed of carbon, hydrogen, and oxygen, the main oxidation products are carbon dioxide and water although fuel-bound nitrogen and sulfur can be significant sources of sulfur oxide and nitrogen oxide emissions. Depending on the heating value and moisture content of the fuel, the amount of air used to burn the fuel, and the construction of the furnace, flame temperatures can exceed 1,650°C.

Solid-fuel combustion consists of four steps, illustrated in Figure 9.9: heating and drying, pyrolysis, flaming combustion, and char combustion [53]. Heating and drying of the fuel particle is normally not accompanied by chemical reaction. Water is driven from the fuel particle as the thermal front advances into the interior of the particle. As long as water remains, the temperature of the particle does not rise high enough to initiate pyrolysis, the second step in solid-fuel combustion.

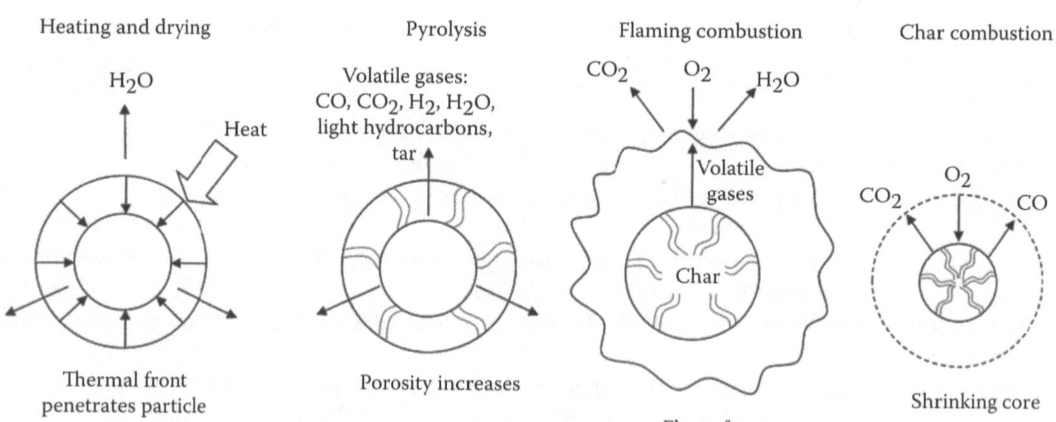

FIGURE 9.9
Processes of solid-fuel combustion.

Pyrolysis is a complex series of thermally driven chemical reactions that decompose organic compounds in the fuel [54]. Pyrolysis proceeds at relatively low temperatures, which depend on the type of plant material. Hemicellulose begins to pyrolyze at temperatures between 150°C and 300°C, cellulose pyrolyzes between 275°C and 350°C, and lignin pyrolysis is initiated between 250°C and 500°C.

The resulting decomposition yields a large variety of volatile organic and inorganic compounds, the types and amounts dependent on the fuel and the heating rate of the fuel. Pyrolysis products include carbon monoxide (CO), carbon dioxide (CO_2), methane (CH_4), and high-molecular-weight compounds that condense to a tarry liquid if cooled before they are able to burn. Fine droplets of these condensable compounds represent much of the smoke associated with smoldering fires. Pyrolysis follows the thermal front through the particle, releasing volatile compounds and leaving behind pores that penetrate to the surface of the particle.

Both the volatile gases and the char resulting from pyrolysis can be oxidized if sufficient oxygen is available to them. Oxidation of the volatile gases above the solid fuel results in flaming combustion. The ultimate products of volatile combustion are CO_2 and H_2O, although a variety of intermediate chemical compounds can exist in the flame, including CO, condensable organic compounds, and long chains of carbon known as soot. Indeed, hot glowing soot is responsible for the familiar orange flame color of wood fires.

Combustion intermediates will be consumed in the flame if sufficient temperature, turbulence, and time are allowed. High combustion temperature assures that chemical reactions will proceed at high rates. Turbulent or vigorous mixing of air with the fuel makes certain that every fuel molecule comes into contact with oxygen molecules. Long residence times for fuel in a combustor allow the fuel to be completely consumed. In the absence of good combustion conditions, a variety of noxious organic compounds can survive the combustion process including CO, soot, polycyclic aromatic hydrocarbons, and the particularly toxic families of chlorinated hydrocarbons known as furans and dioxins. In some cases, a poorly operated combustor can produce pollutants from relatively benign fuel molecules.

Both CO and CO_2 can form at or near the surface of burning char [53]:

$$\begin{aligned} &C+\frac{1}{2}O_2 \rightarrow CO \\ &CO+\frac{1}{2}O_2 \rightarrow CO_2 \end{aligned} \tag{9.35}$$

These gases escape the immediate vicinity of the char particle where CO is oxidized to CO_2 if sufficient oxygen and temperature are available; otherwise, they appear in the flue gas as a pollutant.

The next step in the combustion of solid fuels is solid–gas reactions of char, also known as glowing combustion, familiar as red-hot embers in a fire. Char is primarily carbon with a small amount of mineral matter interspersed. Char oxidation is controlled by mass transfer of oxygen to the char surface rather than by chemical kinetics, which is very fast at the elevated temperatures of combustion. Depending on the porosity and reactivity of the char and the combustion temperature, oxygen may react with char at the surface of the particle or it may penetrate into the pores before oxidizing char inside the particle. The former situation results in a steadily shrinking core of char whereas the latter situation produces a constant-diameter particle of increasing porosity.

9.2.4.2 Combustion Equipment

A combustor is a device that converts the chemical energy of fuels into high-temperature exhaust gases. Heat from the high-temperature gases can be employed in a variety of applications, including space heating, drying, and power generation. However, with the exception of kilns used by the cement industry, most solid-fuel combustors today are designed to produce either low-pressure steam for process heat or high-pressure steam for power generation. Combustors integrated with steam-raising equipment are called boilers. In some boiler designs, distinct sections exist for combustion, high-temperature heat transfer, and moderate-temperature heat transfer: these are called the furnace, radiative, and convective sections of the boiler, respectively. In other designs, no clear separation between the processes of combustion and heat transfer exists.

Solid-fuel combustors, illustrated in Figure 9.10, can generally be categorized as grate-fired systems, suspension burners, or fluidized beds [55]. Grate-fired systems were the first burner systems to be developed, evolving during the late nineteenth century and early twentieth century into a variety of automated systems. The most common system is the spreader-stoker, consisting of a fuel feeder that mechanically or pneumatically flings fuel onto a moving grate where the fuel burns. Much of the ash falls off the end of the moving grate, although some fly ash appears in the flue gas. Grate systems rarely achieve combustion efficiencies exceeding 90%.

Suspension burners suspend the fuel as fine powder in a stream of vertically rising air. The fuel burns in a fireball and radiates heat to tubes that contain water to be converted into steam. Suspension burners, also known as pulverized coal boilers, have dominated coal combustion in the U.S. power industry since World War II because of their high volumetric

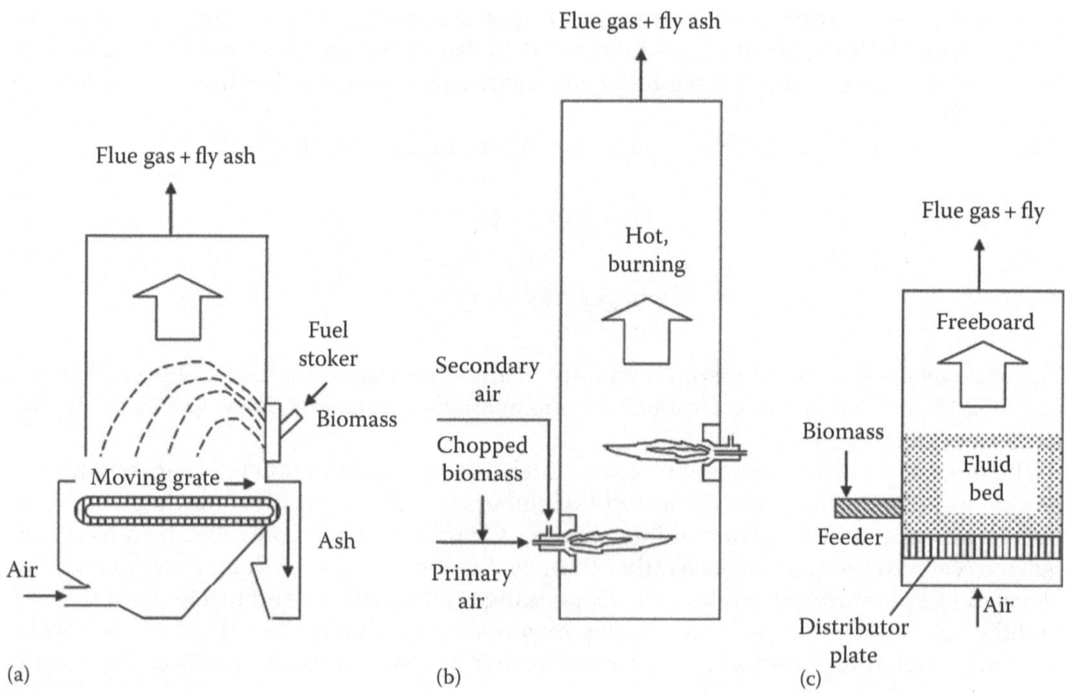

FIGURE 9.10
Common types of combustors: (a) grate-fired, (b) suspension, and (c) fluidized.

heat release rates and their ability to achieve high combustion efficiencies, often exceeding 99%. However, they are not well suited to burning coarse particles of biomass fuel, and they are notorious generators of nitrogen oxides. Biomass is fed from a bunker through pulverizers designed to reduce fuel particle size enough to burn in suspension. The fuel particles are suspended in the primary airflow and fed to the furnace section of the boiler through burner ports where it burns as a rising fireball. Secondary air injected into the boiler helps complete the combustion process. Heat is absorbed by steam tubes arrayed in banks of heat exchangers (waterwall, superheaters, and economizer) before exiting through a bag house designed to capture ash released from the fuel. Steam produced in the boiler is part of a Rankine power cycle.

Fluidized bed combustors are a more recent innovation in boiler design. Air injected into the bottom of the boiler suspends a bed of sand or other granular refractory material producing a turbulent mixture of air and sand. The high rates of heat and mass transfer in this environment are ideal for efficiently burning a variety of fuels. Furthermore, the large thermal mass of the sand bed allows the unit to be operated as low as 850°C, which lowers the emission of nitrogen compounds. A commercial market for fluidized bed boilers developed during the 1980s, especially for industrial applications.

9.2.4.3 ***Biomass Cofiring***

As an alternative to completely replacing coal with biomass fuel in a boiler, mixtures of biomass and coal can be burned together in a process known as cofiring [56]. Cofiring offers several advantages for industrial boilers. Industries that generate large quantities of biomass wastes, such as lumber mills or pulp and paper companies, can use cofiring as an alternative to costly landfilling of wastes.

The best wood-fired power plants, which are typically 20–100 MW_e in capacity, have heat rates exceeding 12,500 Btu/kWh. In contrast, large, coal-fired power plants have heat rates of only 10,250 Btu/kWh. The relatively low thermodynamic efficiency of steam power plants at the sizes of relevance to biomass power systems may ultimately limit the use of direct combustion to convert biomass fuels to useful energy.

9.2.5 Fossil and Biomass Fuel Properties

Fossil-based transportation fuels include gasoline, diesel, and jet fuel. Gasoline is intended for spark-ignition (Otto cycle) engines; thus, it is relatively volatile but resistant to autoignition during compression. Diesel fuel is intended for use in compression-ignition (diesel cycle) engines; thus, it is less volatile compared to gasoline and more susceptible to autoignition during compression. Jet fuel is designed for use in gas turbine (Brayton cycle) engines, which are not limited by autoignition characteristics but otherwise have very strict fuel specifications for reasons of safety and engine durability (aviation for example). Biomass can be converted into gasoline, diesel, and jet fuel as well as alternative fuels such as alcohols, DME, and Fischer–Tropsch liquids.

Combinations of biochemical and thermochemical technologies could generate biofuels that meet or even exceed the combustion properties of conventional fuels. Table 9.13 [57–60] shows key properties of various fossil- and biomass-derived transportation fuels. These properties help identify biofuel substitutes for conventional fuels.

Transportation fuels are characterized by several properties based on broadly recognized standards and their combustion behavior. Specific gravity is a measure of the fuel's density. Kinematic viscosity describes the fuel's ability to flow at a given temperature—a

TABLE 9.13

Comparison of Ignition and Combustion Properties of Transportation Fuels

	Fossil Fuel-Derived		Biomass-Derived						
Fuel Type	**Gasoline**	**No. 2 Diesel Fuel**	**Methanol**	**Ethanol**	**Methyl Ester (From Soybean Oil)**	**Fischer–Tropsch A**	**Hydrogen**	**Methane**	**Dimethyl Ether**
Specific gravity[a]	0.72–0.78	0.85	0.796	0.794	0.886	0.770	0.071 (liq)	0.422 (liq)	0.660
Kinematic viscosity at 20°C–25°C (mm^2/s)	0.8	2.5	0.75	1.51	3.9	2.08	105	16.5	0.227
Boiling point range (°C)	30–225	210–235	65	78	339	164–352	−253	−162	−24.9
Flashpoint (°C)	−43	52	11	13	188	58.5	−184	—	
Autoignition temperature (°C)	370	254	464	423	—	—	566–582	540	235
Octane no. (research)	91–100	—	109	109	—	—	>130	>120	—
Octane no. (motor)	82–92	—	89	90	—	—	—	—	—
Cetane no.	<15	37–56	<15	<15	55	74.6	—	—	>55
Heat of vaporization (kJ/kg)	380	375	1,185	920	—	—	447	509	402
Lower heating value (MJ/kg)	43.5	45	20.1	27	37	43.9	120	49.5	28.88

[a] Measured at 16°C except for liquefied gases, which are saturated liquids at their respective boiling points.

high fuel viscosity could be detrimental to fuel delivery to an engine. Boiling point range and flash point temperature are important because they are key factors during engine start-up. Flash point is the lowest temperature at which enough fuel vaporizes and mixes with air to form a combustible gas. Below this temperature, the fuel would fail to ignite even when exposed to a spark. Flash point differs slightly from the autoignition temperature: autoignition occurs when the fuel combusts without a spark or ignition source.

The octane number is an important transportation fuel figure of merit. The octane number indicates the tendency of a fuel to undergo premature detonation within the combustion cylinder of an internal combustion engine. The higher the octane number, the less likely a fuel will detonate until exposed to an ignition source (electrical spark). Premature detonation is responsible for the phenomenon known as engine knock, which reduces fuel economy and can damage an engine. Various systems of octane rating have been developed, including research octane and motor octane numbers. Federal regulation in the United States requires gasoline sold commercially to be rated using an average of the research and motor octane numbers. Gasoline rated as "regular" has a commercial octane number of about 87 while premium grade is 93.

These properties specify the potential uses for any type of transportation fuels. Therefore, biofuel development requires careful consideration of these properties. An extended discussion of this topic is beyond the scope of this chapter, but readers are encouraged to consult engine and combustion texts on this matter.

9.3 Use of Biomass in Developing Communities

The discussion above has focused on the conversion of biomass to liquid or gaseous biofuels, or biomass combustion to produce electricity. Until recently, however, by far the largest use of biomass energy worldwide has been direct combustion for cooking and residential heating, often in open fires or rudimentary stoves. The International Energy Agency reports that in 2022, about 9% of primary energy use worldwide was bioenergy or waste. Over one-third of this was for traditional household cooking and heating applications, mainly in low-income countries [61]. In many parts of the world, this household use of biomass is unsustainable, due to growing populations or energy demand putting increasing pressure on limited biomass resources, heavy time burdens involved in gathering fuels, and the very heavy health burdens associated with chronic exposure to biomass smoke. In 2016, the World Health Organization (WHO) estimated that roughly 3 billion people worldwide relied on burning crop residue, wood, dung, or charcoal to meet household energy needs [62]. The WHO characterizes household air pollution from solid fuels combustion as the "most important environmental health risk factor" worldwide. These burdens were recognized by the United Nations (UN) in the sustainable development goals the organization adopted in 2015. The UN's Sustainable Development Goal #7 is to provide universal access to "affordable, reliable, sustainable, modern energy by 2030" (sustainabledevelopment.UN.org). The goal focuses mainly on access to electricity and clean cooking fuel and technology. Efforts to reduce reliance on inefficient solid-fuel stoves have been underway for several decades, but with inadequate progress. The UN's 2024 report on progress toward achieving the sustainable development goals estimates that in 2022, 2.1 billion people still relied on polluting fuels and technologies for cooking [63]. While access to clean cooking fuels and technologies has improved substantially in Asian

FIGURE 9.11
Traditional three-stone fire set-up for biomass cooking in Ghana (left); Philips forced-draft gasifier stove designed as a high-efficiency, low-emissions replacement and sold internationally (center); and a Gyapa wood stove that is locally made in Ghana and features a ceramic insert to retain heat. (Photos Courtesy of REACCTING [Research of Emissions, Air quality, Climate, and Cooking Technologies in Northern Ghana].)

countries over the past decade, the population without clean cooking in Sub-Saharan Africa has increased over the same period. Major international initiatives that are currently underway to convert households to cleaner fuels include the United Nations-led Sustainable Energy for All (SE4All) program (http://www.se4all.org); the Global Alliance for Clean Cookstoves, hosted by the UN Foundation (http://cleancookstoves.org); and the Climate and Clean Air Coalition based out of the United Nations Environment Program (http://www.ccacoalition.org).

Use of solid biomass for cooking is especially common in Africa and Central and Southern Asia [63]. WHO's 2016 study estimated that in Africa, excluding the Mediterranean region, 85% of household cooking was done over biomass fires or stoves [62]. In many households, cooking is done over traditional three-stone fires such as the set-up shown in Figure 9.11. Use of biomass or other self-gathered solid fuels is most prevalent in rural areas but is not limited to them. In colder climates in some low-income communities, solid fuels including biomass and coal have been widely used to provide heat. Less information is currently available on household combustion of solid fuels for heating than for cooking. While surveys have been conducted in a few low-income countries, more information is needed to understand the prevalence of in-home solid-fuel combustion for heating [62].

WHO estimates that household air pollution from solid-fuel combustion caused more than 4 million premature deaths in 2012 [62]. The diseases linked to these deaths include chronic obstructive pulmonary disease and lung cancer in adults, especially women, and pneumonia in children under the age of 5. Chronic exposure to smoke from household fuel combustion is associated with a wide range of other diseases, including cataracts and blindness [64].

Uncontrolled incomplete combustion of biomass fuels emits hundreds of pollutants, including carbon monoxide, oxygenated organics, free radicals, and hydrocarbons. Studies of exposure to household air pollution from incomplete combustion commonly use carbon monoxide and particulate matter (PM) as primary exposure metrics. More precisely, scientists and regulators widely reference the mass concentration of inhalable particulate matter less than 10 microns in aerodynamic diameter (PM_{10}) or the concentration of fine particulate matter less than 2.5 μm in diameter ($PM_{2.5}$) as proxies for the mixture of chemicals found in condensed phase air pollution. The $PM_{2.5}$ size range is considered more dangerous because the smaller particles can penetrate deep into the alveolar region of the lungs. To protect human health, WHO has set an air quality guideline for $PM_{2.5}$ concentrations to be kept below 5 μg/m^3 on an annual average basis and below 15 μg/m^3 over a 24-hour averaging period [65]. Demonstrating the severity of the problem, some studies of household pollution in kitchen areas where unvented solid-fuel combustion was taking place have reported average $PM_{2.5}$ concentrations above 500 μg/m^3 [66].

In addition to producing pollution that harms human health, household solid-fuel combustion is estimated to produce 25% of global emissions of black carbon, which is the light-absorbing component of PM [62]. Black carbon, along with methane, is one of the main short-lived climate-forcing pollutants. Black carbon has an atmospheric lifetime on the order of a week—not decades or centuries like CO_2 or N_2O—so reducing BC presents a good near-term opportunity for mitigating climate change. Lacey et al. [67] used a global atmospheric chemistry and transport model linked to a reduced-form radiative forcing model to estimate the country-specific net surface temperature response to changing out solid-fuel cookstoves. Their analysis considered not only BC, but also methane and organic carbon (OC) from incomplete combustion, net CO_2 emissions from unsustainable biomass harvesting, and sulfate aerosol formed from SO_2 emissions from coal. Whether the cook stove interventions in each country would have a net warming or net cooling effect was not clear in advance, because OC and sulfate aerosols are reflective and have a cooling effect in the atmosphere. Nevertheless, Lacey et al. found that for almost all the countries they examined, changing out cook stoves would have a beneficial climate effect (reducing global warming) in addition to delivering significant health benefits [67].

Sustainably harvested biomass is considered to be a renewable energy resource. On the other hand, biomass used for household fuel combustion in much of the world is not sustainably managed. Deforestation or forest degradation is a centuries-old problem that is partially attributable to biomass fuel use. In a 2015 global assessment, Bailis et al. [68] estimated that 22%–30% of wood fuel supplies in tropical countries are harvested unsustainably. The African countries of Eritrea, Ethiopia, Kenya, Uganda, Rwanda, and Burundi, and the Asian countries of Pakistan, Nepal, Bhutan, Indonesia, and Bangladesh all had significant areas where more than half of wood fuel was not sustainably harvested [68]. In other areas, high rates of deforestation due to agricultural expansion constrain wood fuel availability. Significant environmental consequences of deforestation include habitat loss, soil erosion, and flooding, as well as net CO_2 emissions from reduced carbon sequestration in the biosphere. Moreover, in many areas today people who gather their own biomass for cooking or heating are relegated to walking long distances and spending hours a day to gather fuel because nearby supplies have been depleted. The WHO has found that some children spend more than 30 hours a week gathering biomass [62]. In many countries, the task of gathering biomass falls primarily to girls and is a significant impediment to their gaining an education [62].

Options for addressing the problems caused by household combustion of biofuels and other solid fuels include switching to cleaner heat sources, such as electricity, pipeline natural gas, biogas, or liquefied petroleum gas; use of improved biomass stoves with higher combustion efficiency and lower emissions, such as those shown in Figure 9.11; and/or use of stoves with chimneys to move emissions away from occupied areas. Reducing the cost of these alternatives is a key challenge, as many people who use biofuels in a rudimentary stove or fire do so because they cannot afford alternatives. Rural and remote communities often lack the infrastructure to transmit electricity or supply gas by pipeline. Programs to develop and deploy more efficient biomass stoves have often been stymied because new stoves did not meet local expectations for cooking time, fuel costs or fuel preparation requirements, ease of installation and repair, robustness, and durability. More efficient stove designs that use fans may not work in areas lacking access to electric power. Studies that have tried to distribute improved stoves have often found that emissions reductions, efficiency, and performance are variable and that improvements seen in the lab are not replicated in actual use. Recipients of new stoves have often been found to use them to supplement rather than to replace their traditional stoves, a practice known as "stacking." In colder climates, another issue is that heat given off by cooking stoves is relied upon for space heating, so replacement stoves not designed for that dual purpose are unsatisfactory. In a review of studies designed to assess the effectiveness of low-emission stoves, Bruce et al. [69] found that providing these stoves was often not enough to achieve air quality guidelines, due to stove stacking, the presence of other combustion sources in the home, and contributions from outdoor sources. They recommend taking a multi-pronged approach to reducing household air pollution that goes beyond stove replacement and places greater emphasis on improving access to clean fuels.

9.4 Conclusions

In developed countries, economic and political concerns have renewed interest in biomass development as a domestic source of clean and renewable fuels and chemicals. Initial efforts have unfortunately competed with food crops for low-cost sugars. A sustainable biomass industry will have to provide food to the world's growing population before supporting future energy needs. Thus, researchers are investigating how to convert crop residues and dedicated energy crops into valuable fuels and chemicals. These feedstocks could provide a bridge from first- to second-generation biofuels and beyond.

First-generation biofuels based on grain or sugar cane ethanol and soy diesel will continue to be important sources of alternative transportation fuels for several years because significant investment has been made in their infrastructure and due to continuing political support for government incentives. However, significant expansion of the biofuel industry will require alternatives to corn and soybean crops as feedstocks. These possibilities include both lignocellulose and lipids from alternative crops.

Each of the biofuel and bioenergy options considered in this chapter has its advantages and disadvantages in terms of market, opportunities, environmental benefits, and process economics. Accordingly, no single biofuel is likely to dominate transportation in the same manner as currently exists for petroleum-based gasoline and diesel. Niche markets and regional considerations will heavily influence the best choice of fuel for specific applications. Finally, increased research efforts could raise new possibilities that change the outlook for biofuels.

Although biochemical processing dominates current-generation biofuels, thermochemical processing is likely to play an important role in advanced biofuels because of its technical maturity and potential to efficiently produce hydrocarbon fuels. However, it is too early to designate winners and losers among the many biofuel options. What is certain is that biofuels will evolve as a result of several forces including new technologies, market demand, environmental concerns, and government policies.

Problems

9.1 Wisconsin produced 13,260,000 bushels (25.4 kg/bushel) of oats on 195,000 acres (4046.85 m^2/acre) of land. What is the corresponding yield of oat residue in metric tons per km^2 assuming a residue to grain ratio of 2:1?

9.2 The U.S. total land area is almost 9,307,800 km^2. Thirty percent of the land is forestland. How much above-ground biomass is available in the U.S. forestland assuming a 15,240 Mg/km^2 biomass density? How much land would we need to grow an equivalent amount of biomass with fast-growing trees (hybrid poplar), or energy crops (switchgrass)?

9.3 A common assumption for biorefinery capacities is 2,000 Mg/day of feedstock input capacity. What is the land area required to grow this quantity of biomass from corn stover with land availability (f) of 40%?

9.4 Given the enthalpy of formation for food waste to be −394.7 MJ/kmol, use the combustion reaction from Equation 9.11 to calculate its HHV (assume that the molecular formula is the same as switchgrass).

9.5 Solar irradiation in parts of California exceeds 9 kWh/m^2/day (32 million MJ/km^2/day). What are theoretical maximum C3 and C4 biomass energy yields assuming a 150-day crop growth period? In Florida, solar irradiation is about 3.5 kWh/m^2/day; what would the theoretical maximum yield be if the crop had a maturity period of 180 days?

9.6 A gasifier can convert 1 kg of wood (19.5 MJ/kg) into 1.5 m^3 of syngas (5.2 MJ/m^3). If we define energy conversion efficiency as the ratio of syngas energy to biomass energy, what is the energy conversion efficiency of wood to syngas?

9.7 A lactating cow can produce more than 60 kg of manure (about 7 MJ/kg on a dry basis) per day. What is the energy efficiency of converting cow manure to biogas if it yields 25 m^3/Mg with the methane content of 55% (35 MJ/m^3)?

9.8 Gasification uses an oxygen-to-fuel ratio that is 25% of the required oxygen for complete combustion. What is the mass of oxygen required to gasify 1 kg of wood ($CH_{1.4}O_{0.66}$)?

9.9 Steam gasification employs steam as the oxygen carrier. How much steam would be needed to gasify 1 kg of wood ($CH_{1.4}O_{0.66}$) if the oxygen-to-fuel ratio is 25% of the oxygen required for complete combustion?

9.10 How much air is needed to fully combust fuel oil ($CH_{1.67}O_{0.006}N_{0.005}S_{0.002}$) if 25% excess air is required?

9.11 Calculate the energy conversion efficiency of biomass to gasification products assuming the yields and energy contents from Table 9.8.

9.12 Assuming a chain growth probability of 0.8, calculate the mole yields for hydrocarbons with 1–24 carbon atoms using the ASF distribution. How does this distribution compare to that of a 0.9 chain growth probability? Graph the results (yield vs. chain length) to show the comparison.

9.13 Calculate the heat of reaction for ethanol (heat of formation −277.7 MJ/kmol) synthesis with the mixed-alcohol process and with the methanol (heat of formation −238.4 MJ/kmol) conversion process.

9.14 Assuming nominal annual yields of 700,000, 5,500,000, and 3,800,000 kg/km^2 for corn grain, sugar cane, and sweet sorghum respectively, which conversion pathway is the most productive use of land? Use conversion yield values given in the ethanol section and units of L/km^2.

9.15 A group proposes a 2,000 barrel-per-day ethanol output biorefinery using Jerusalem artichokes. What is the daily feedstock requirement for this biorefinery? What is the minimum amount of land required to grow the feedstock (f=100%)?

9.16 If the yield of bio-oil to liquid fuels (gasoline- and diesel-like fuels) is 42 wt.%, what is the overall energy efficiency? Use the values in Table 9.12 for energy content and product yields. First, calculate the biomass to liquid fuel yield based on the biomass to bio-oil and bio-oil to liquid fuel yields. Finally, assume a liquid fuel energy density of 41.2 MJ/kg and calculate the energy efficiency.

9.17 Using an ethanol fuel content of 21.2 MJ/L and Tank to Wheels (TTW) efficiency of 3.08 MJ/km, calculate the theoretical travel distance in units of km per km^2 of land for an E-85 vehicle powered by corn grain ethanol. How does this number compare to sugar cane ethanol and cellulosic ethanol? Employ appropriate assumptions for crop yields and process conversion efficiency.

References

1. Brown, R.C., 2003. *Biorenewable Resources: Engineering New Products from Agriculture*, Iowa State Press, A Blackwell Publishing Company, Ames, IA, pp. 59–75.
2. Perlack, R., Wright, L., Turhollow, A., Graham, R., Stokes, B., and Erbach, D., 2005. Biomass as feedstock for a bioenergy and bioproducts industry: The technical feasibility of a billion-ton annual supply, Technical Report A357634, Oak Ridge National Laboratory.
3. Morris, G., 2003. The status of biomass power generation in California, July 31, 2003, Technical Report NREL/SR-510–35114, National Renewable Energy Laboratory.
4. Wright, L.L., and Hohenstein, W.G., 1994. Dedicated feedstock supply systems: Their current status in the USA, *Biomass and Bioenergy* 6(3), 159–241.
5. Field, C.B., Campbell, J.E., and Lobell, D.B., 2008. Biomass energy: The scale of the potential resource, *Trends in Ecology & Evolution* 23(2), 65–72.
6. Wayman, M., and Parekh, S.R., 1990. *Biotechnology of Biomass Conversion: Fuels and Chemicals from Renewable Resources*, Open University Press, Philadelphia, PA.
7. United States Department of Agriculture, 2012. Agricultural statistics database, July 2012, https://quickstats.nass.usda.gov.
8. Zhu, X.G., Long, S.P., and Ort, D.R., 2008. What is the maximum efficiency with which photosynthesis can convert solar energy into biomass? *Current Opinion in Biotechnology* 19(2), 153–159.
9. Jenkins, B.M., Baxter, L.L., and Miles, T.R., 1998. Combustion properties of biomass, *Fuel Processing Technology* 54(1–3), 17–46.
10. Klass, D.L., 1998. *Biomass for Renewable Energy, Fuels, and Chemicals*, Academic Press, San Diego, CA, pp. 72–90.
11. Moran, M.J., Shapiro, H.N., Boettner, D.D., and Bailey, M., 2010. *Fundamentals of Engineering Thermodynamics*, Wiley, New York.
12. Klass, D.L., 1998. *Biomass for Renewable Energy, Fuels, and Chemicals*, Academic Press, San Diego, CA, p. 356.
13. Reed, T.B., 1981. *Biomass Gasification: Principles and Technology*, Noyes Data Corp., Park Ridge, NJ.

14. Bridgwater, A.V., 1995. The technical and economic feasibility of biomass gasification for power generation, *Fuel* 74(5), 631–653.
15. Klass, D.L., 1998. *Biomass for Renewable Energy, Fuels, and Chemicals*, Academic Press, San Diego, CA, p. 416.
16. Henrich, E., and Weirich, F., 2004. Pressurized entrained flow gasifiers for biomass, *Environmental Engineering Science* 21(1), 53–64.
17. Bailey, B.K., 1996. Performance of ethanol as a transportation fuel, in: Wyman, C. ed., *Handbook on Bioethanol*, vol. 1, Taylor and Francis, Washington, DC, pp. 37–58.
18. Kwiatkowski, J.R., McAloon, A.J., Taylor, F., and Johnston, D.B., 2006. Modeling the process and costs of fuel ethanol production by the corn dry-grind process, *Industrial Crops & Products* 23(3), 288–296.
19. Rowell, R.M., Young, R.A., and Rowell, J.K., 1997. *Paper and Composites from Agro-Based Resources*, CRC Press, Boca Raton, FL.
20. Schell, D.J., McMillian, J.D., and Philippidis, G.P., 1992. Ethanol from lignocellulosic biomass, *Advances in Solar Energy* 7, 373–448.
21. Lynd, L.R., 1996. Overview and evaluation of fuel ethanol from cellulosic biomass: Technology, economics, the environment and policy, *Annual Reviews in Energy and the Environment* 21(1), 403–465.
22. Aden, A., Ruth, M., Ibsen, K., Jechura, J., Neeves, K., Sheehan, J., Wallace, B., Montague, L., and Slayton, A., 2002. Lignocellulosic biomass to ethanol process design and economics utilizing co-current dilute acid prehydrolysis and enzymatic hydrolysis for corn stover, Technical Report NREL//TP-510–32438, National Renewable Energy Laboratory.
23. Hall, C.A.S., Dale, B.E., and Pimentel, D., 2011. Seeking to understand the reasons for different energy return on investment (eroi) estimates for biofuels, *Sustainability* 3(12), 2413–2432.
24. Anon., 2005. Fischer–Tropsch archive, https://www.fischer-tropsch.org.
25. Schulz, H., 1999. Short history and present trends of Fischer–Tropsch synthesis, *Applied Catalysis A, General* 186(1–2), 3–12.
26. Tijmensen, M.J.A., Faaij, A.P.C., Hamelinck, C.N., and van Hardeveld, M.R.M., 2002. Exploration of the possibilities for production of Fischer Tropsch liquids and power via biomass gasification, *Biomass and Bioenergy* 23(2), 129–152.
27. Swanson, R.M., Platon, A., Satrio, J.A., and Brown, R.C., 2010. Techno-economic analysis of biomass-to-liquids production based on gasification, *Fuel* 89, S11–S19.
28. Bridgwater, A.V., and Peacocke, G.V.C., 2000. Fast pyrolysis processes for biomass, *Renewable and Sustainable Energy Reviews* 4(1), 1–73.
29. Czernik, S., and Bridgwater, A.V., 2004. Overview of applications of biomass fast pyrolysis oil, *Energy and Fuels* 18(2), 590–598.
30. Piskorz, J., and Scott, D.S., 1987. The composition of oils obtained by the fast pyrolysis of different woods, American Chemical Society, Division of Fuel Chemistry, *Preprint Paper*, 32(2).
31. Wright, M.M., Daugaard, D.E., Satrio, J.A., and Brown, R.C., 2010. Techno-economic analysis of biomass fast pyrolysis to transportation fuels, *Fuel* 89(Supplement 1(0)), S2–S10, ISSN 0016–2361, https://www.sciencedirect.com/science/article/pii/S0016236110003765.
32. Vieira de Carvalho, A.J., 1982. Natural gas and other alternative fuels for transportation purposes, *Energy* 10(2), 187–215.
33. Satterfield, C.N., 1991. *Heterogeneous Catalysis in Industrial Practice*, McGraw-Hill Book Co., New York.
34. Sorenson, S.C., 2001. Dimethyl ether in diesel engines: Progress and perspectives, *Journal of Engineering for Gas Turbines and Power* 123, 652.
35. Klass, D.L., 1998. *Biomass for Renewable Energy, Fuels, and Chemicals*, Academic Press, San Diego, CA, pp. 274–276.
36. Forzatti, P., Tronconi, E., and Pasquon, I., 1991. Higher alcohol synthesis, *Catalysis Reviews* 33(1), 109–168.
37. Phillips, S., Aden, A., Jechura, J., Dayton, D., and Eggeman, T., 2007. Thermochemical ethanol via indirect gasification and mixed alcohol synthesis of lignocellulosic biomass, Technical Report NREL//TP-510–41168, National Renewable Energy Laboratory.

38. Brown, R.C., 2007. Hybrid thermochemical/biological processing, *Applied Biochemistry and Biotechnology* 137(1), 947–956.
39. Openshaw, K., 2000. A review of *Jatropha curcas:* An oil plant of unfulfilled promise, *Biomass and Bioenergy* 19(1), 1–15.
40. Glenn, E.P., O'Leary, J.W., Watson, M.C., Thompson, T.L., and Kuehl, R.O., 1991. *Salicornia bigelovii* torr.: An oilseed halophyte for seawater irrigation, *Science* 251(4997), 1065–1067.
41. Kumar, A., and Sharma, S., 2008. An evaluation of multipurpose oil seed crop for industrial uses (*Jatropha curcas* l.): A review, *Industrial Crops & Products* 28(1), 1–10.
42. Long, F., Liu, W., Jiang, X., Zhai, Q., Cao, X., Jiang, J., and Xu, J. 2021. State-of-the-art technologies for biofuel production from triglycerides: A review, *Renewable and Sustainable Energy Reviews*, 148, 111269, doi:10.1016/j.rser.2021.111269.
43. Klass, D.L., 1998e. *Biomass for Renewable Energy, Fuels, and Chemicals*, Academic Press, San Diego, CA, pp. 341–344.
44. Lundquist, T.J., Woertz, I.C., Quinn, N.W.T., and Benemann, J.R., 2010. A realistic technology and engineering assessment of algae biofuel production, Energy Biosciences Institute, Berkeley, CA, p. 1.
45. Peterson, A.A., Vogel, F., Lachance, R.P., Fröling, M., Antal Jr., M.J., and Tester, J.W., 2008. Thermochemical biofuel production in hydrothermal media: A review of sub-and supercritical water technologies, *Energy & Environmental Science* 1(1), 32–65.
46. Elliott, D.C., Beckman, D., Bridgwater, A.V., Diebold, J.P., Gevert, S.B., and Solantausta, Y., 1991. Developments in direct thermochemical liquefaction of biomass: 1983–1990, *Energy & Fuels* 5(3), 399–410.
47. Elliott, D.C., Neuenschwander, G.G., Hart, T.R., Butner, R.S., Zacher, A.H., Engelhard, M.H., Young, J.S., and McCready, D.E., 2004. Chemical processing in high-pressure aqueous environments. 7. Process development for catalytic gasification of wet biomass feedstocks, *Industrial and Engineering Chemistry Research* 43(9), 1999–2004.
48. Moreau, C., Naceur Belgacem, M., and Gandini, A., 2004. Recent catalytic advances in the chemistry of substituted furans from carbohydrates and in the ensuing polymers, *Topics in Catalysis* 27(1–4), 11–30.
49. Carlson, T.R., Vispute, T.P., and Huber, G.W., 2008. Green gasoline by catalytic fast pyrolysis of solid biomass derived compounds, *ChemSusChem* 1(5), 397.
50. Roman-Leshkov, Y., Barrett, C.J., Liu, Z.Y., and Dumesic, J.A., 2007. Production of dimethylfuran for liquid fuels from biomass-derived carbohydrates, *Nature* 447, 982–985.
51. Zhao, H., Holladay, J.E., Brown, H., and Zhang, Z.C., 2007. Metal chlorides in ionic liquid solvents convert sugars to 5-hydroxymethylfurfural, *Science* 316(5831), 1597.
52. Diebold, J.P., and Evans, R.J., 1988. U.S. Patent No. 4, 764, 627, U.S. Patent and Trademark Office, Washington, DC.
53. Tillman, D.A., 1991. *The Combustion of Solid Fuels and Wastes*, Academic Press, San Diego, CA.
54. Bain, R.L., Overend, R.P., and Craig, K.R., 1998. Biomass-fired power generation, *Fuel Processing Technology* 54(1–3), 1–16.
55. Kumar, J.V., and Pratt, B.C., 1996. Compositional analysis of some renewable biofuels, *American Laboratory* 28(8), 15–20.
56. Sami, M., Annamalai, K., and Wooldridge, M., 2001. Cofiring of coal and biomass fuel blends, *Progress in Energy and Combustion Science* 27(2), 171–214.
57. Borman, G.L., and Ragland, K.W., 1998. *Combustion Engineering*, McGraw-Hill Company, New York.
58. National Renewable Energy Laboratory, Alternative fuels, general table of fuel properties, https://www.eere.energy.gov/afdc/altfuel/fuelsdo5(p)roperties.html.
59. MacKenzie, J.J., and Avery, W.H., 1996. Ammonia fuel: The key to hydrogen-based transportation, in: *Energy Conversion Engineering Conference, IECEC 96*, vol. 3, Washington, DC.
60. Teng, H., McCandless, J.C., and Schneyer, J.B., 2001. Thermochemical characteristics of dimethyl ether-an alternative fuel for compression-ignition engines [r], SAE technical paper, 2001-01-0154.

61. International Energy Agency, 2023. Energy statistics data browser, Total Energy Supply by Source, World, 1990–2022. www.iea.org.
62. World Health Organization, 2016. *Burning Opportunity: Clean Household Energy for Health, Sustainable Development, and Wellbeing of Women and Children*, WHO, Geneva, Switzerland.
63. United Nations, 2024. The sustainable development goals report 2024, https://unstats.un.org/sdgs/report/2024/.
64. Pokhrel, A.K., Smith, K.R., Khalakdina, A., Deuja, A., and Bates, M.N., 2005. Case-control study of indoor cooking smoke exposure and cataract in Nepal and India, *International Journal of Epidemiology* 34, 702–708.
65. World Health Organization, 2021. *WHO Global Air Quality Guidelines: Particulate Matter ($PM_{2.5}$ and PM_{10}), Ozone, Nitrogen Dioxide, Sulfur Dioxide and Carbon Monoxide*, WHO, Geneva.
66. Clark, M.L., Peel, J.L., Balakrishnan, K., Breysse, P.N., Chillrud, S.N., Naeher, L.P., Rodes, C.E., Vette, A.F., and Balbus, J.M., 2013. Health and household air pollution from solid fuel use: The need for improved exposure assessment, *Environmental Health Perspectives* 121, 1120–1128.
67. Lacey, F.G., Henze, D.K., Lee, C.J., van Donkelaar, A., and Martin, R.V., 2017. Transient climate and ambient health impacts due to national solid fuel cookstove emissions, *Proceedings of the National Academy of Sciences of the United States of America* 114, 1269–1274, doi:10.1073/pnas.1612430114.
68. Bailis, R., Drigo, R., Ghilardi, A., and Masera, O., 2015. The carbon footprint of traditional woodfuels, *Nature Climate Change* 5, 266–272.
69. Bruce, N., Pope, D., Rehfuess, E., Balakrishnan, K., Adair-Rohani, H., and Dora, C., 2015. WHO indoor air quality guidelines on household fuel combustion: Strategy implications of new evidence on interventions and exposure-risk functions, *Atmospheric Environment* 106, 451–457.

10

Fundamentals of Solar Radiation

Having been admonished by the Holy Office [the Inquisition] entirely to abandon the false option that the Sun was the center of the universe and immovable, and that the Earth was not the center of the same and that it moved … I abjure …

Galileo

10.1 Physics of the Sun and Its Energy Transport

The nature of energy generation in the Sun is still an unanswered question. Spectral measurements have confirmed the presence of nearly all the known elements in the Sun. However, 80% of the Sun is hydrogen and 19% helium. Therefore, the remaining 100-plus observed elements make up only a tiny fraction of the composition of the Sun. It is generally accepted that a hydrogen-to-helium thermonuclear reaction is the source of the Sun's energy. Yet because such a reaction has not been duplicated in the laboratory, it is unclear precisely what the reaction mechanism is, what role the turbulent flows in the Sun play, and how solar prominences and sunspots are created.

The nature of the energy creation process is of no importance to terrestrial users of the Sun's radiation. Of interest is the amount of energy, its spectral and temporal distribution, and its variation with time of day and year. These matters are the main subject of this chapter.

The Sun is a 13.9×10^5 km diameter sphere comprised of many layers of gases, which are progressively hotter toward its center. The outermost layer, from which energy is radiated into the solar system, is approximately at an equivalent black-body temperature of 5,760 K. The center of the Sun, however, may be at 20×10^6 K. The rate of energy emission from the Sun is 3.8×10^{23} kW, which results from the conversion of 4.3×10^9 g/s of mass to energy. Of this total, only a tiny fraction, approximately 1.7×10^{14} kW, is intercepted by the Earth, which is located about 150 million km from the Sun (Figure 10.1).

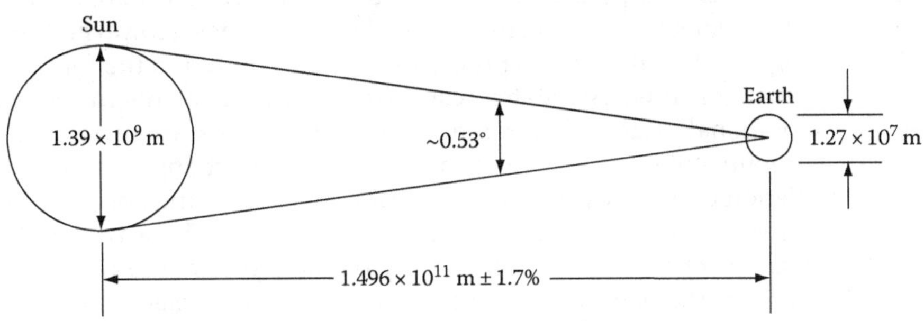

FIGURE 10.1
Relationship between the Sun and the Earth.

DOI: 10.1201/9781003586616-10

Solar energy is the world's most abundant permanent source of energy. The amount of solar energy intercepted by the planet Earth is 5,000 times greater than the sum of all other inputs (terrestrial, nuclear, geothermal, and gravitational energies, and lunar gravitational energy). Of this amount, 30% is reflected to space, 47% is converted to low-temperature heat and reradiated to space, and 23% powers the evaporation/precipitation cycle of the biosphere. Less than 0.5% is represented in the kinetic energy of the wind and waves and in photosynthetic storage in plants.

Total terrestrial radiation is only about one-third of the extraterrestrial total during a year, and 70% of that falls on the oceans. However, the remaining 1.5×10^{17} kWh that falls on land is a prodigious amount of energy—about 6,000 times the total energy usage of the United States in 2000. Only a small fraction of this total can be used because of physical and socioeconomic constraints, as described in Chapter 1.

10.2 Thermal Radiation Fundamentals

The material presented in this section has been selected from textbooks on heat transfer and radiation [1–4]. It provides the background needed to understand the nature of solar radiation for the engineering analysis of solar energy systems.

To begin with, all radiation travels at the speed of light, which is equal to the product of the wavelength and the frequency of radiation. The speed of light in a medium equals the speed of light in a vacuum divided by the refractive index of the medium through which it travels:

$$c = \lambda v = \frac{c_0}{n} \tag{10.1}$$

where

λ is the wavelength (m)
v is the frequency (s^{-1})
c is the speed of light in a medium (m/s)
c_0 is the speed of light in a vacuum (m/s)
n is the index of refraction of the medium.

Thermal radiation is one kind of electromagnetic energy, and all bodies emit thermal radiation by virtue of their temperature. When a body is heated, its atoms, molecules, or electrons are raised to higher levels of activity called excited states. However, they tend to return to lower energy states, and in this process, energy is emitted in the form of electromagnetic waves. Changes in energy states result from rearrangements in the electronic, rotational, and vibrational states of atoms and molecules. Since these rearrangements involve different amounts of energy changes and these energy changes are related to the frequency, the radiation emitted by a body is distributed over a range of wavelengths. A portion of the electromagnetic spectrum is shown in Figure 10.2. The wavelengths associated with the various mechanisms are not sharply defined; thermal radiation is usually considered to fall within the band from about 0.1 to 100 μm, whereas solar radiation has most of its energy between 0.1 and 3 μm.

For some problems in solar energy engineering, the classical electromagnetic wave theory is not suitable. In such cases, for example, in photovoltaic or photochemical processes,

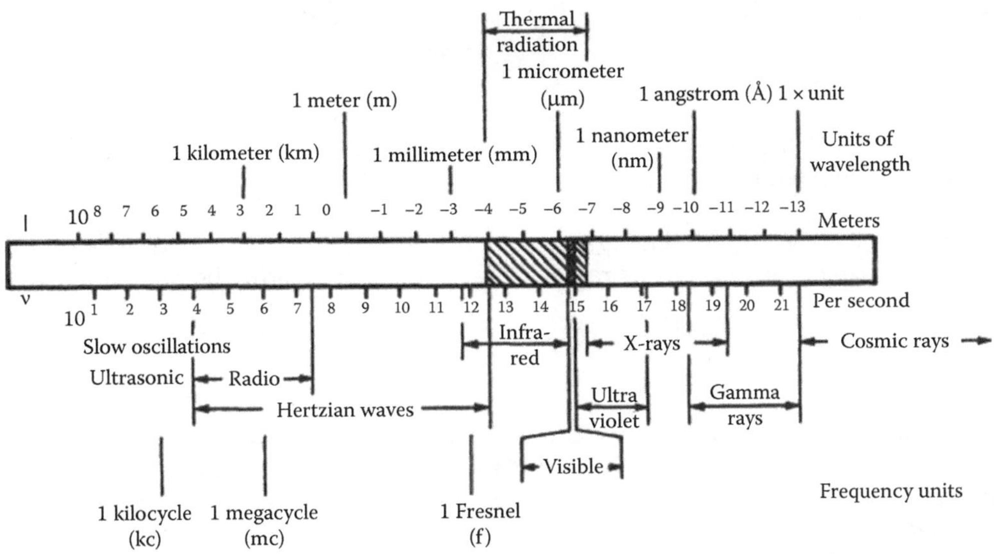

FIGURE 10.2
Electromagnetic radiation spectrum.

it is necessary to treat the energy transport from the point of view of quantum mechanics. In this view, energy is transported by particles or *photons*, which are treated as energy units or quanta rather than waves. The energy of a photon, E_p, of frequency ν_p is

$$E_p = h\nu_p \tag{10.2}$$

where h is Planck's constant (6.625×10^{-34} J·s).

10.2.1 Black-Body Radiation

The energy density of the radiation emitted at a given wavelength (monochromatic) by a perfect radiator, usually called a black body, is given according to the following relation:

$$E_{b\lambda} = \frac{C_1}{\left(e^{C_2/\lambda T} - 1\right)\lambda^5 n^2} \tag{10.3}$$

where

$C_1 = 3.74 \times 10^8$ W μm^4/m^2

$C_2 = 1.44 \times 10^4$ μm K

n = refractive index of the medium = 1.0 for vacuum; n is taken to be approximately equal to 1 for air.

The quantity $E_{b\lambda}$ has the units of W/m^2 μm and is called the monochromatic emissive power of a black body, defined as the energy emitted by a perfect radiator per unit wavelength at the specified wavelength per unit area and per unit time at the temperature T.

The total energy emitted by a black body, E_b, can be obtained by integration over all wavelengths. This yields the Stefan–Boltzmann law:

$$E_b = \int_0^\infty E_{b\lambda} d\lambda = \sigma T^4 \quad (10.4)$$

where
σ is the Stefan–Boltzmann constant = 5.67×10^{-8} W/m²K⁴
T is the absolute temperature (K).

The concept of a black body, although no such body actually exists in nature, is very convenient in engineering because its radiation properties can readily be related to those of real bodies.

10.2.2 Thermal Radiation Functions

Engineering calculations of radiative transfer are facilitated by the use of radiation function tables, which present the results of Planck's law in a more convenient form than Equation 10.3. A plot of the monochromatic emissive power of a black body as a function of wavelength as the temperature is increased is given in Figure 10.3. The emissive power shows a maximum at a particular wavelength.

These peaks, or inflection points, are uniquely related to the body temperature. By differentiating Planck's distribution law (Equation 10.3) and equating to zero, the wavelength corresponding to the maximum value of $E_{b\lambda}$ can be shown to occur when

$$\lambda_{\max} T = 2,897.8\ \mu\text{mK}\left(5,215.6\ \mu\text{mR}\right) \quad (10.5)$$

Frequently, one needs to know the amount of energy emitted by a black body within a specified range of wavelengths. This type of calculation can be performed easily with the

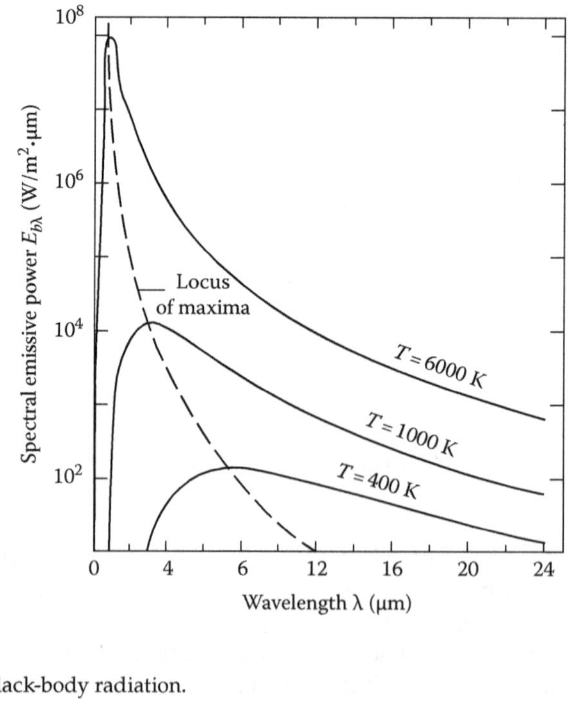

FIGURE 10.3
Spectral distribution of black-body radiation.

aid of the radiation functions mentioned previously. To construct the appropriate radiation functions in dimensionless form, note that the ratio of the black-body radiation emitted between 0 and λ and between 0 and ∞ can be made a function of the single variable (λT) by using Equation 10.3 as shown (for $n = 1$) in the following equation:

$$\frac{E_{b,0-\lambda}}{E_{b,0-\infty}} = \frac{\int_0^{\lambda} E_{b\lambda} d\lambda}{\sigma T^4} = \int_0^{\lambda T} \frac{C_1 d(\lambda T)}{\sigma(\lambda T)^5 \left(e^{C_2/\lambda T} - 1\right)} \tag{10.6}$$

The earlier relation is plotted in Figure 10.4, and the results are also shown in tabular form in Table 10.1.

In Table 10.1, the first column is the ratio of λ to λ_{max} from Equation 10.5, and the third column the ratio of $E_{b,0-\lambda}$ to σT^4 from Equation 10.6. For use on a computer, Equation 10.6 can be approximated by the following polynomials:

$$v \le 2: \frac{E_{b,0-\lambda}}{\sigma T^4} = \frac{15}{\pi^4} \sum_{m=1,2,\ldots} \frac{E^{-mv}}{m^4} \{[(mv+3)mv+6]mv+6\} \tag{10.7a}$$

and

$$v \ge 2: \frac{E_{b,0-\lambda}}{\sigma T^4} = \frac{15}{1-\pi^4} v^3 \left(\frac{1}{3} - \frac{v}{8} - \frac{v^2}{60} - \frac{v^4}{5,040} + \frac{v^6}{272,160} - \frac{v^8}{13,305,600}\right) \tag{10.7b}$$

where $v = C_2/\lambda T$.

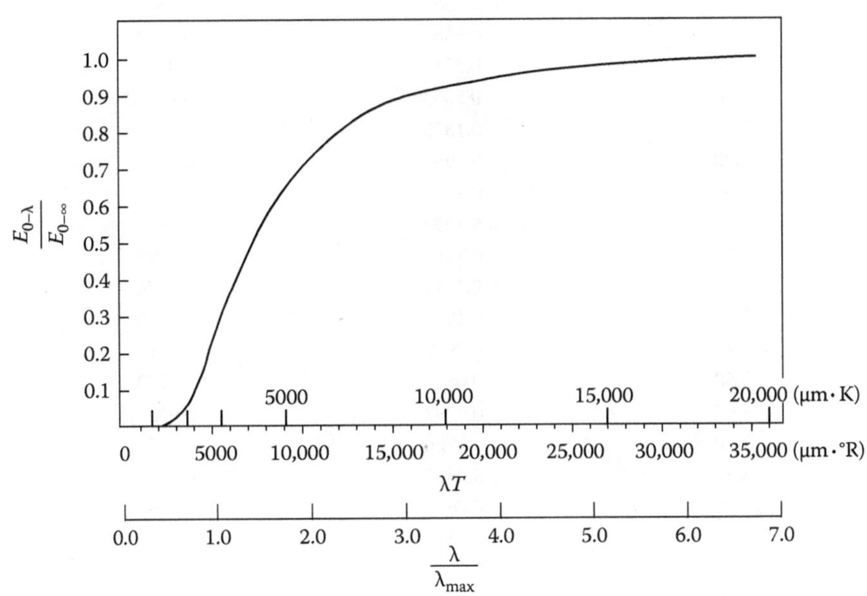

FIGURE 10.4
Fraction of total emissive power in spectral region between $\lambda = 0$ and λ as a function of λT and λ/λ_{max}.

TABLE 10.1

Thermal Radiation Functions[a]

λ/λ_{max}	$E_{b\lambda}/E_{b\lambda,max}$	$E_{b\lambda,0-\lambda}/\sigma T^4$
0.00	0.0000	0.0000
0.20	0.0000	0.0000
0.30	0.0038	0.0001
0.40	0.0565	0.0015
0.50	0.2217	0.0101
0.60	0.4664	0.0325
0.70	0.7042	0.0712
0.80	0.8776	0.1236
0.90	0.9725	0.1849
1.00	1.0000	0.2501
1.10	0.9791	0.3153
1.20	0.9277	0.3782
1.30	0.8600	0.4370
1.40	0.7854	0.4911
1.50	0.7103	0.5403
1.60	0.6382	0.5846
1.70	0.5710	0.6243
1.80	0.5098	0.6598
1.90	0.4546	0.6915
2.00	0.4054	0.7197
2.10	0.3616	0.7449
2.20	0.3229	0.7674
2.30	0.2887	0.7875
2.40	0.2585	0.8054
2.50	0.2318	0.8215
2.60	0.2083	0.8360
2.70	0.1875	0.8490
2.80	0.1691	0.8607
2.90	0.1528	0.8713
3.00	0.1384	0.8809
3.10	0.1255	0.8895
3.20	0.1141	0.8974
3.30	0.1038	0.9045
3.40	0.0947	0.9111
3.50	0.0865	0.9170
3.60	0.0792	0.9225
3.70	0.0726	0.9275
3.80	0.0667	0.9320
3.90	0.0613	0.9362
4.00	0.0565	0.9401
4.20	0.0482	0.9470
4.40	0.0413	0.9528
4.60	0.0356	0.9579
4.80	0.0308	0.9622
5.00	0.0268	0.9660

(Continued)

TABLE 10.1 (*Continued*)

Thermal Radiation Functions[a]

λ/λ_{max}	$E_{b\lambda}/E_{b\lambda,max}$	$E_{b\lambda,0-\lambda}/\sigma T^4$
6.00	0.0142	0.9790
7.00	0.0082	0.9861
8.00	0.0050	0.9904
9.00	0.0033	0.9930
10.00	0.0022	0.9948
20.00	0.0002	0.9993
40.00	0.0000	0.9999
50.00	0.0000	1.0000

[a] λ = wavelength in µm.
λ_{max} = wavelength at $E_{b\lambda,max}$ in µm = 2,898/T.
$E_{b\lambda}$ = monochromatic emissive power in W/m²·µm.
= $374.15 \times 10^6/\lambda^5[\exp(14{,}387.9/\lambda T) - 1]$,
$E_{b\lambda,max}$ = maximum monochromatic emissive power in W/m²·µm
= $12.865 \times 10^{-12}T^5$,

$$E_{b\lambda,0-\lambda} = \int_0^{\lambda} E_{b\lambda} d\lambda'$$

$\sigma T^4 = E_{b\lambda,0-\infty} = 5.670 \times 10^{-8} T^4$ W/m²
T = absolute temperature in K.

10.2.3 Intensity of Radiation and Shape Factor

The emissive power of a surface gives the total radiation emitted in all directions. To determine the radiation emitted in a given direction, we must define another quantity, the radiation intensity I. As given in Equation 10.8, this quantity is defined as the radiant energy, dE, passing through an imaginary plane in space per unit area, dA', per unit time and per unit solid angle, $d\omega$, perpendicular to the plane, as shown in Figure 10.5. I is defined by the relation

$$I = \lim_{dA' \to 0\ d\omega \to 0} = \frac{dE}{dA' d\omega} \tag{10.8}$$

Radiation intensity has both magnitude and direction. It can be related to the radiation flux, defined as the radiant energy passing through an imaginary plane per unit area per unit time in all directions.

Note that, whereas for the intensity, the area dA' is perpendicular to the direction of the radiation, for the flux, the area dA is at the base in the center of a hemisphere through which all of the radiation passes. Recalling that the definition for the solid angle between dA' and dA is $d\omega = dA'/r^2$, the radiation flux q_r emanating from dA can be obtained by integrating the intensity over the hemisphere. As shown in Figure 10.5, the unit projected area for I is $dA \cos\theta$ and the differential area dA' on the hemisphere is $r^2 \sin\theta\, d\theta\, d\phi$; thus,

$$q_r = \int_0^{2\pi} \int_0^{\frac{\pi}{2}} I\cos\theta\sin\theta d\theta d\phi \tag{10.9}$$

If the area dA is located on a surface, the emissive power E can also be obtained from Equation 10.9. For the special case of a diffuse surface, for which I is the same in all directions, Equation 10.9 gives

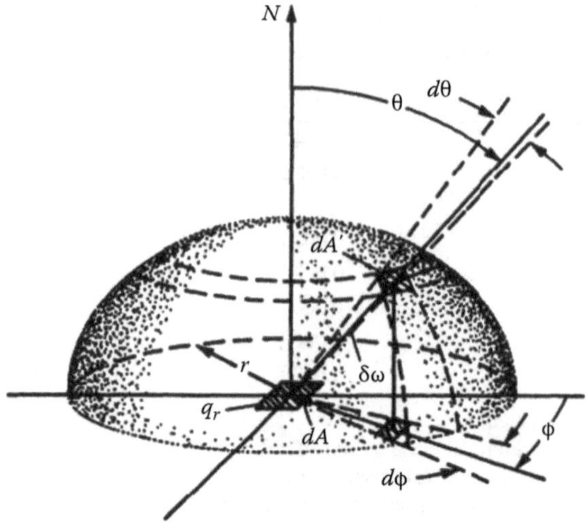

FIGURE 10.5
Schematic diagram illustrating radiation intensity and flux. *N* is the normal to the surface, *θ* is the angle between the beam direction and the surface normal, and *φ* is the angle of the projection of the beam onto the surface in the plane of the surface.

$$q_r = \pi I \tag{10.10}$$

Since all black surfaces are diffuse,

$$E_b = \pi I_b \tag{10.11}$$

Equation 10.11 can, of course, also be written for monochromatic radiation as

$$E_{b\lambda} = \pi I_{b\lambda} \tag{10.12}$$

In the evaluation of the rate of radiation heat transfer between two surfaces, not only their temperatures and their radiation properties but also their geometric configurations and relationships play a part. The influence of geometry in radiation heat transfer can be expressed in terms of the *radiation shape factor* between any two surfaces 1 and 2 defined as follows:

F_{1-2} = fraction of radiation leaving surface 1 that reaches surface 2

F_{2-1} = fraction of radiation leaving surface 2 that reaches surface 1

In general, F_{m-n} = fraction of radiation leaving surface "*m*" that reaches surface "*n*." If both surfaces are black, the energy leaving surface "*m*" and arriving at surface "*n*" is $E_{bm}A_mF_{m-n}$ and the energy leaving surface "*n*" and arriving at "*m*" is $E_{bn}A_nF_{n-m}$. If both surfaces absorb all the incident energy, the net rate of exchange $q_{m\Leftrightarrow n}$ will be

$$q_{m\Leftrightarrow n} = E_{bm}A_mF_{m-n} - E_{bn}A_nF_{n-m} \tag{10.13}$$

If both surfaces are at the same temperature, $E_{bm} = E_{bn}$ and the net exchange is zero: $q_{m\Leftrightarrow n} = 0$. This shows that the geometric radiation shape factor must obey the reciprocity relation

$$A_m F_{m-n} = A_n F_{n-m} \tag{10.14}$$

The net rate of heat transfer can therefore be written in two equivalent forms:

$$q_{m\Leftrightarrow n} = A_m F_{m-n}\left(E_{bm} - E_{bn}\right) = A_n F_{n-m}\left(E_{bm} - E_{bn}\right) \tag{10.15}$$

The evaluation of geometric shape factors is in general quite involved. For a majority of solar energy applications, however, only a few special cases are of interest. One of these is a small convex object of area A_1 surrounded by a large enclosure A_2. Since all radiation leaving A_1 is intercepted by A_2, $F_{1-2} = 1$ and $F_{2-1} = A_1/A_2$.

Another case is the exchange of radiation between two large parallel surfaces. If the two surfaces are near each other, almost all of the radiation leaving A_1 reaches A_2 and vice versa. Thus, $F_{1-2} = F_{2-1} = 1.0$, according to the definition of the shape factor. A third case of importance is the exchange between a small surface ΔA_1 and a portion of space A_2, for example, the exchange between a flat-plate solar collector tilted at an angle β from the horizontal and the sky it can see. For this situation, we refer to the definition of radiation flux (see Figure 10.5). The portion of the radiation emitted by ΔA_1 that is intercepted by the surrounding hemisphere depends on the angle of tilt. When the surface is horizontal, $F_{1-2} = 1$; when it is vertical, $F_{1-2} = 1/2$ ($\beta = 90°$). For intermediate values, it can be shown that [4]

$$F_{1-2} = \frac{1}{2}\left(1 + \cos\beta\right) = \cos^2\left(\frac{\beta}{2}\right) \tag{10.16}$$

If the diffuse sky radiation is uniformly distributed and assumed to be black, then a small black area A_1 receives radiation at the rate

$$A_1 F_{1-\text{sky}} E_{\text{sky}} = \frac{A_1}{2}\left(1 + \cos\beta\right)\sigma T_{\text{sky}}^4 \tag{10.17}$$

whereas the net radiation heat transfer is given by

$$q_{\text{sky}\Leftrightarrow 1} = A_1 F_{1-\text{sky}}\sigma\left(T_{\text{sky}}^4 - T_1^4\right) \tag{10.18}$$

If the receiving area is gray with an absorptance $\bar{\alpha}$ equal to the emittance $\bar{\varepsilon}$, the net exchange is given by

$$q_{\text{sky}\Rightarrow 1} = A_1 F_{1-\text{sky}}\bar{\alpha}\sigma\left(T_{\text{sky}}^4 - T_1^4\right) \tag{10.19}$$

10.2.4 Transmission of Radiation through a Medium

When radiation passes through a transparent medium such as glass or the atmosphere, the decrease in intensity can be described by Beer's law, which assumes that the attenuation is proportional to the local intensity in the medium. If $I_\lambda(x)$ is monochromatic intensity after the radiation has traveled a distance x, the law is expressed by the equation

$$-dI_\lambda(x) = I_\lambda(x) K_\lambda dx \tag{10.20}$$

where K_λ is the monochromatic extinction coefficient assumed to be a constant of the medium. If the transparent medium is a slab of thickness L and the intensity at $x = 0$ is designated by the symbol $I_{\lambda,0}$, the monochromatic transmittance τ_λ is equal to the ratio of the intensity at $x = L$ to $I_{\lambda,0}$. An expression for $I_\lambda(L)$ can be obtained by integrating Equation 10.20 between 0 and L, which gives

$$\ln\frac{I_\lambda(L)}{I_{\lambda,0}} = -K_\lambda L \quad \text{or} \quad I_\lambda(L) = I_{\lambda,0}e^{-K_\lambda L} \tag{10.21}$$

Then,

$$\tau_\lambda = \frac{I_\lambda(L)}{I_{\lambda,0}} = e^{-K_\lambda L} \tag{10.22}$$

The extinction coefficient K_λ is a complex property of the medium since it combines the effects of absorption, emission, and scattering by the molecules and particles that make up the medium. Fortunately, for materials such as glass and plastics with known compositions, this coefficient can be determined accurately. Transmission of radiation through such materials is discussed further in Chapter 12. In this chapter, we are concerned about the transmission of solar radiation through the atmosphere. The atmosphere consists of the molecules of gases in it, such as N_2, O_2, CO_2, and H_2O, and aerosols such as dust particles, air pollutants, water droplets, and ice crystals. The extinction processes of the atmosphere consist of (1) absorption and emission by the molecules and aerosols, (2) scattering by the molecules, and (3) scattering by aerosols. Estimates of atmospheric transmittance can be made using an empirical approach [5] to develop an average extinction coefficient for use in Equation 10.22. More details on the transmission of solar energy through the atmosphere are covered in Section 10.5.1.

10.3 Sun–Earth Geometric Relationship

Figure 10.6 shows the annual orbit of the Earth around the Sun. The distance between the Earth and the Sun changes throughout the year, the minimum being 1.471×10^{11} m at winter solstice (December 21) and the maximum being 1.521×10^{11} m at summer solstice (June 21). The year-round average Earth–Sun distance is 1.496×10^{11} m. The amount of solar radiation intercepted by the Earth, therefore, varies throughout the year, the maximum being on December 21 and the minimum on June 21.

The axis of the Earth's daily rotation is at an angle of 23.45° to the axis of its ecliptic orbital plane. This tilt is the major cause of the seasonal variation of the solar radiation available at any location on the Earth. The angle between the Earth–Sun line (through their centers) and the plane through the equator is called the *solar declination*, δ_s. The declination varies between –23.45° on December 21 and +23.45° on June 21. Stated another way, the declination has the same numerical value as the latitude at which the Sun is directly overhead at solar noon on a given day. The Tropics of Cancer (23.45°N) and Capricorn (23.45°S) are at the extreme latitudes where the Sun is overhead at least once a year, as shown in Figure 10.6. The Arctic and Antarctic circles are defined as those latitudes above which the Sun does not rise above the horizon plane at least once per year. They are located, respectively, at 66½°N and 66½°S. Declinations north of the equator (summer in the northern

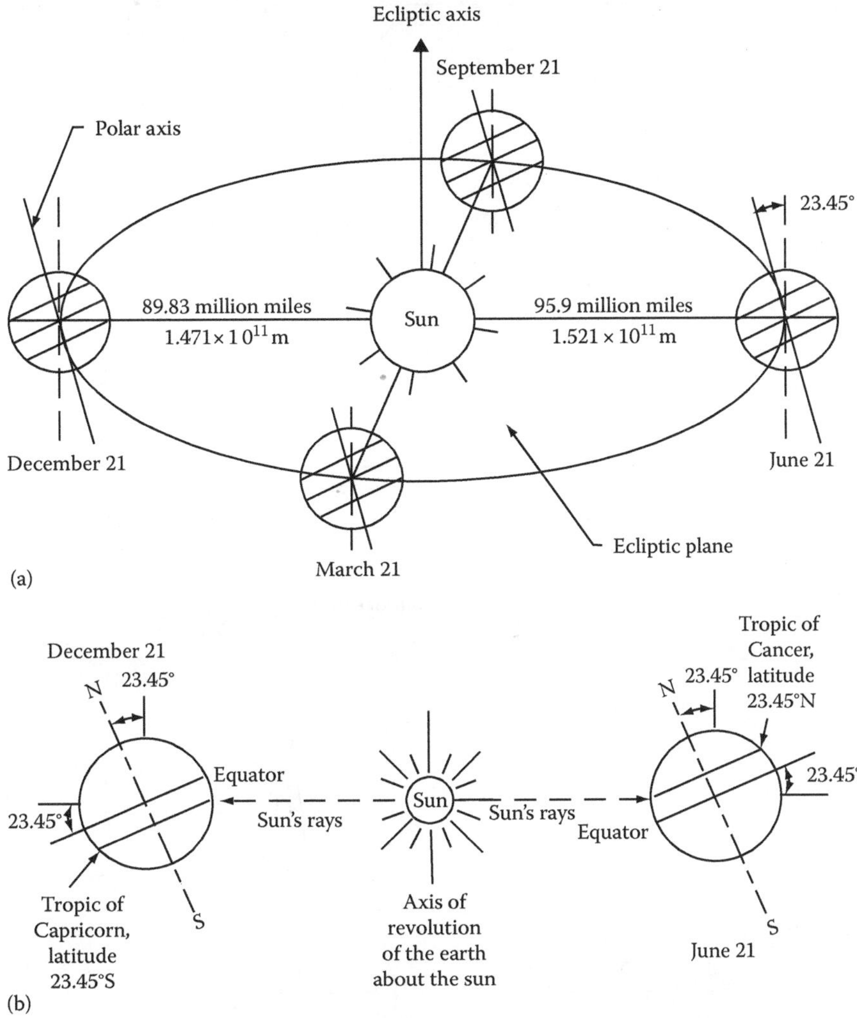

FIGURE 10.6
(a) Motion of the Earth about the Sun. (b) Location of tropics. Note that the Sun is so far from the Earth that all the rays of the Sun may usually be considered as parallel to one another when they reach the Earth (but there are exceptions, as in the case of concentrating optics).

hemisphere) are positive; those south, negative. The solar declination may be estimated by the following relation[1]:

$$\delta_s = 23.45^{\circ} \sin\left[\frac{360(284+n)}{365}\right]^{\circ} \tag{10.23}$$

where n is the day number during a year with January 1 being $n = 1$. Approximate values of declination may also be obtained from Figure 10.7. For most calculations, the declination may be considered constant during any given day.

[1] A more accurate relation is $\sin \delta_s = \sin(23.45^{\circ}) \sin[360(284 + n)/365]^{\circ}$. Because the error is small, Equation 10.23 is generally used.

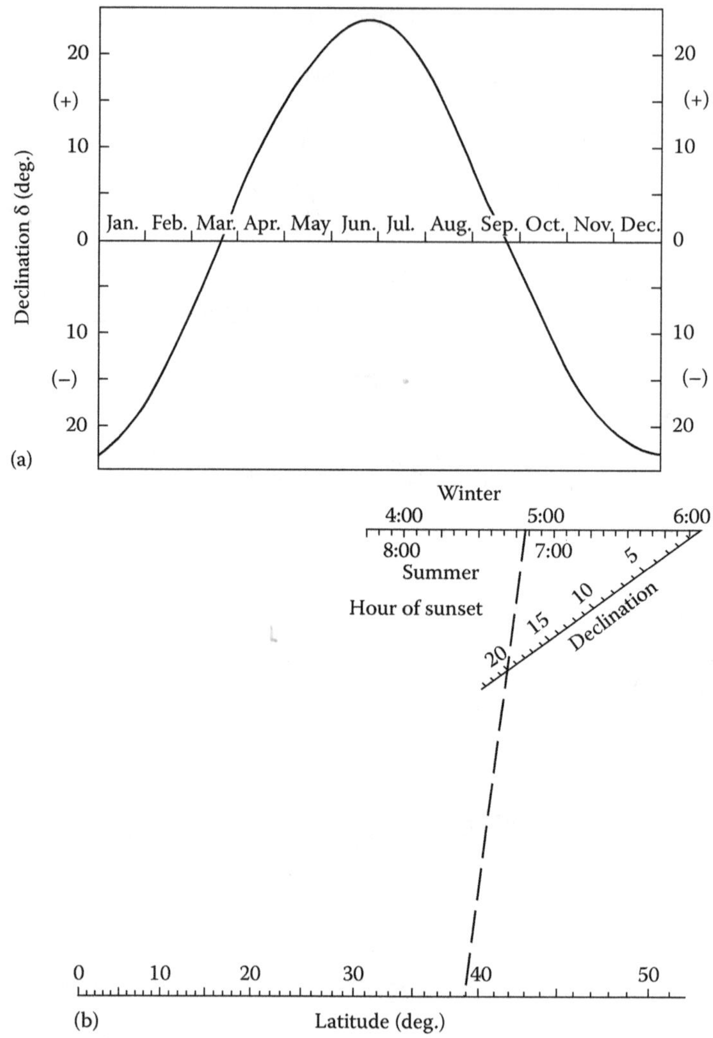

FIGURE 10.7
(a) Graph to determine the solar declination. (b) Sunset nomograph example. Example (b) shows determination of sunset time for summer (7:08 p.m.) and winter (4:52 p.m.) when the latitude is 39°N and the solar declination angle is 20°.

For the purposes of this book, the Ptolemaic view of the Sun's motion provides a simplification to the analysis that follows. It is convenient to assume the Earth to be fixed and to describe the Sun's apparent motion in a coordinate system fixed to the Earth with its origin at the site of interest. Figure 10.8 shows an apparent path of the Sun to an observer. The position of the Sun can be described at any time by two angles, the altitude and azimuth angles, as shown in Figure 10.8. The *solar altitude angle, α,* is the angle between a line collinear with the Sun's rays and the horizontal plane. The *solar azimuth angle, a_s,* is the angle between a north-south line and the projection of the site-to-Sun line on a horizontal plane. The sign convention used for azimuth angle is based on the angle becoming increasingly positive in a clockwise direction. In some conventions, 0° is taken as due north; in others, as in this book (with the exception of Sun-path diagrams described later), 0° is taken as due

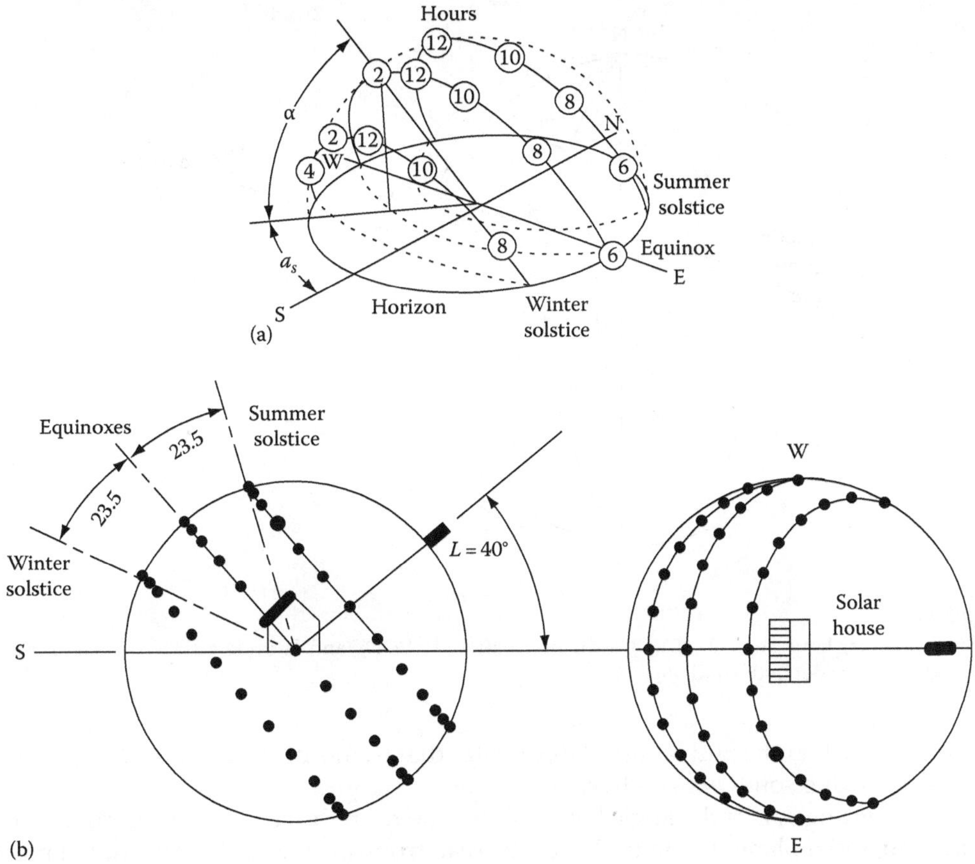

FIGURE 10.8
Sun paths for the summer solstice (6/21), the equinoxes (3/21 and 9/21), and the winter solstice (12/21) for a site at 40°N: (a) isometric view and (b) elevation and plan views.

south. Thus, in our definition, the azimuth angle is positive west of due south and negative east of due south. The *solar zenith angle, z,* is the angle between the site to Sun line and the vertical at the site:

$$z = 90° - \alpha \tag{10.24}$$

The solar altitude and azimuth angles are not fundamental angles. Hence, they must be related to the fundamental angular quantities *hour angle, latitude,* and *declination.* The three angles are shown in Figure 10.9 [6]. The solar hour angle h_s is based on the nominal time of 24 hours required for the Sun to move 360° around the Earth or 15°/h. Therefore, h_s, in degrees, can be expressed as a function of the hours or minutes, respectively, from solar noon as follows:

$$\begin{aligned} h_s &= \frac{15°}{\text{hour}} \times \text{hours from local solar noon} \\ &= \frac{\text{Minutes from local solar noon}}{4\ \text{min / degree}} \end{aligned} \tag{10.25}$$

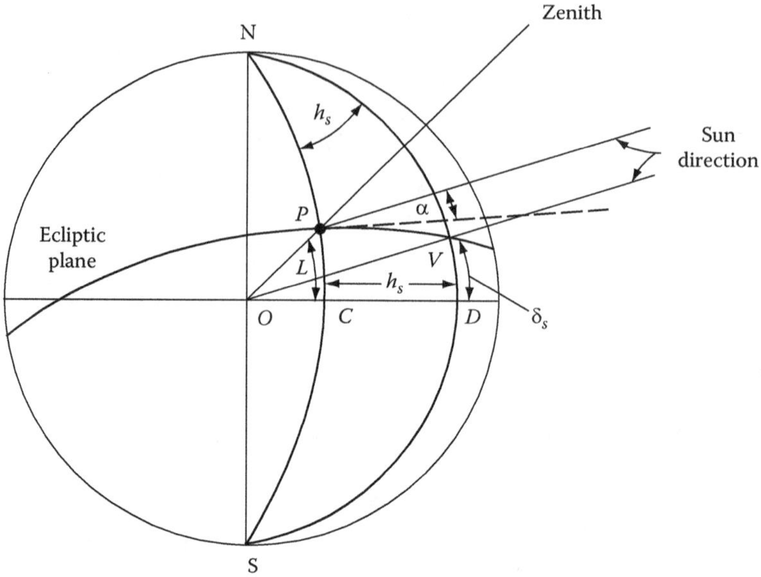

FIGURE 10.9
Definition of solar hour angle h_s (CND), solar declination δ_s (VOD), and latitude L (POC); p, site of interest. (Modified from Kreider and Kreith [6].)

Again, values when the Sun is east of due south, that is, morning values, are negative, and values west of due south are positive.

The latitude angle, L, is the angle between the line from the center of the Earth to the site and the equatorial plane. The latitude may be read from an atlas and is considered positive north of the equator and negative south of the equator.

10.3.1 Solar Time and Angles

The Sun angles are obtained from the local solar time, which differs from the local standard time (LST). The relationship between the local solar time and the LST is

$$\text{Solar time} = \text{LST} + \text{ET} + (l_{\text{st}} - l_{\text{local}}) \cdot 4 \text{ min / degree} \tag{10.26}$$

where

ET is the equation of time, which is a correction factor that accounts for the irregularity of the speed of Earth's motion around the Sun

l_{st} is the standard time meridian

l_{local} is the local longitude.

ET may be calculated from the following empirical equation:

$$\text{ET(minutes)} = 9.87 \sin 2B - 7.53 \cos B - 1.5 \sin B \tag{10.27}$$

where $B = (360(n - 81)/365)°$.

The solar altitude angle, α, can be found from the application of the law of cosines to the geometry of Figure 10.9 and can be simplified as

$$\sin \alpha = \sin L \sin \delta_s + \cos L \cos \delta_s \cos h_s \tag{10.28}$$

Using a similar technique, the solar azimuth angle, a_s, can be found as

$$\sin a_s = \frac{\cos\delta_s \sin h_s}{\cos\alpha} \tag{10.29}$$

where
α is the altitude of the Sun (angular elevation above the horizon)
L is the latitude of the observer
δ_s is the declination of the Sun
h_s is the hour angle of the Sun (angular distance from the meridian of the observer)
a_s is the azimuth of the Sun (measured west from due south).

At local solar noon, $h_s = 0$; therefore, $\alpha = 90 - |L - \delta_s|$ and $a_s = 0$.

In calculating the solar azimuth angle from Equation 10.29, a problem occurs whenever the absolute value of a_s is greater than 90°. A computational device usually calculates the angle as less than 90° since $\sin a_s = \sin(180 - a_s)$. The problem can be solved in the following way:

For $L > \delta_s$, the solar times when the Sun is due east (t_E) or due west (t_W) can be calculated by t_E or t_W = 12:00 Noon ± $\cos^{-1}[\tan \delta_s/\tan_L]°(15°/\text{hour})$ (−for t_E, + for t_W).

For solar times earlier than t_E or later than t_W, the Sun would be north (south in the southern hemisphere) of the east–west line and the absolute value of a_s would be greater than 90°. Then, a_s may be calculated as $a_s = \pm(180° - |a_s|)$.

For $L \le \delta_s$, the Sun remains north (south in the southern hemisphere) of the east–west line and the true value of a_s is greater than 90°.

Sunrise and *sunset* times can be estimated by finding the hour angle for $\alpha = 0$. Substituting $\alpha = 0$ in Equation 10.28 gives the hour angles for sunrise (h_{sr}) and sunset (h_{ss}) as

$$h_{ss} \quad \text{or} \quad h_{sr} = \pm\cos^{-1}\left[-\tan L \cdot \tan\delta_s\right] \tag{10.30}$$

It should be emphasized that Equation 10.30 is based on the center of the Sun at the horizon. In practice, sunrise and sunset are defined as the times when the upper limb of the Sun is on the horizon. Because the radius of the solar disk in the sky is 16′ where the prime indicates minutes, the sunrise would occur when $\alpha = -16'$. Also, at lower solar elevations, the Sun will appear on the horizon when it is actually 34′ below the horizon. Therefore, for apparent sunrise or sunset, $\alpha = -50'$.

Example 10.1

Find the solar altitude and azimuth angles at solar noon in Gainesville, Florida, on February 1. Also find the sunrise and sunset times in Gainesville on that day.

Solution:

For Gainesville,

$$\text{Latitude } L = 29° + 41'\text{N} \quad \text{or} \quad 29.7°\text{N}$$

$$\text{Longitude } l_{\text{local}} = 82° + 16'\text{W} \quad \text{or} \quad 82.3°\text{W}$$

On February 1, day number, $n = 32$. Therefore, from Equation 10.23, the declination is

$$\delta_s = 23.45 \sin\left[\frac{360(284+32)}{365}\right]^\circ = -17.5^\circ$$

At solar noon, $h_s = 0$. Therefore,

$$\sin\alpha = \cos L \cos\delta_s \cos h_s + \sin L \sin\delta_s$$
$$= \cos(29.7^\circ)\cos(-17.5^\circ)\cos(0) + \sin(29.7^\circ)\sin(-17.5^\circ)$$

or $\alpha = 42.8^\circ$

$$\sin a_s = \frac{\cos(-17.5^\circ)\sin(0)}{\cos(42.8^\circ)} = 0$$

or $a_s = 0$.

At solar noon, α can also be found as

$$\alpha = 90 - |L - \delta_s|^\circ = 90 - |29.7 + 17.5| = 42.8^\circ$$

$$h_{ss} \quad \text{or} \quad h_{sr} = \pm\cos^{-1}[-\tan L \cdot \tan\delta_s]$$
$$= \pm\cos^{-1}[-\tan(29.7^\circ)\cdot\tan(17.5^\circ)] = \pm 79.65^\circ$$

$$\text{Time from solar noon} = \pm(79.65^\circ)\left(4\,\frac{\text{min}}{\text{degree}}\right) = \pm 319 \text{ minutes} \quad \text{or} \quad \pm(5 \text{ hours } 19 \text{ minutes})$$

$$\text{Sunrise time} = 12:00 \text{ noon} - (5 \text{ hours } 19 \text{ minutes}) = 6 \text{ hours } 41 \text{ minutes a.m. (solar time)}$$

$$\text{Sunset time} = 12:00 \text{ noon} + (5 \text{ hours } 19 \text{ minutes}) = 5 \text{ hours } 19 \text{ minutes p.m. (solar time)}$$

To convert these times to local times, we need to find ET:

$$\text{ET} = (9.87 \sin 2B - 7.53 \cos B - 1.5 \sin B) \text{ minutes}$$

$$B = \frac{360}{364}(n - 81) = \frac{360}{364}(32 - 81) = -48.46^\circ$$

Therefore,

$$\text{ET} = -13.67 \text{ minutes}$$

$$\text{LST} = \text{solar time} - \text{ET} - 4(l_{st} - l_{local})$$

Gainesville, Florida, is in the eastern standard time (EST) zone, where l_{st} = 75°W. Therefore,

$$\begin{aligned} \text{LST} &= \text{Solar time} - (-13.67 \text{ minutes}) - 4(75 - 82.27) \text{ minutes} \\ &= \text{Solar time} + 42.75 \text{ minutes} \end{aligned}$$

Therefore,

$$\text{Sunrise time} = 6:41 \text{ a.m.} + 43 \text{ minutes} = 7{:}24 \text{ a.m. EST},$$

and

$$\text{Sunset time} = 5:19 \text{ p.m.} + 43 \text{ minutes} = 6{:}02 \text{ p.m. EST}$$

Note: Since the sunrise and sunset times are calculated when the center of the Sun is at the horizon, they differ from the apparent times. If we use $\alpha = -50'$, the apparent sunrise and sunset times would be 7:20 a.m. EST and 6:06 p.m. EST, respectively.

Knowledge of the solar angles is helpful in the design of passive solar buildings, especially the placement of windows for solar access and the roof overhang for shading the walls and windows at certain times of the year. The following example illustrates this point.

Example 10.2

Find the roof overhang L of a south-facing window of height H = 1 m, such that the window is completely shaded at solar noon on April 1 and not shaded at all at noon on November 1. Assume that the roof extends far beyond the window on either side. Location: Gainesville, Florida. Also find the overhang if S = 1.3 m.

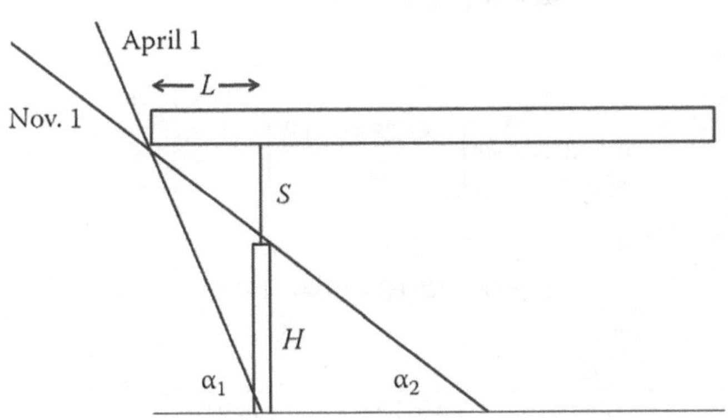

Solution:

From the geometry of the figure, we can derive the following two equations with four unknowns, L, S, α_1, and α_2:

$$L = \frac{S}{\tan \alpha_2}$$

$$L = \frac{(S + H)}{\tan \alpha_1}$$

From the following figure, it can be determined that $\alpha + |\text{Lat.} - \delta_s| = 90°$ from simple geometry. Therefore, α_1 and α_2 are dependent on the seasonal declination angle δ_s and the latitude of Gainesville, Florida, which is 29.68° from Example 10.1.

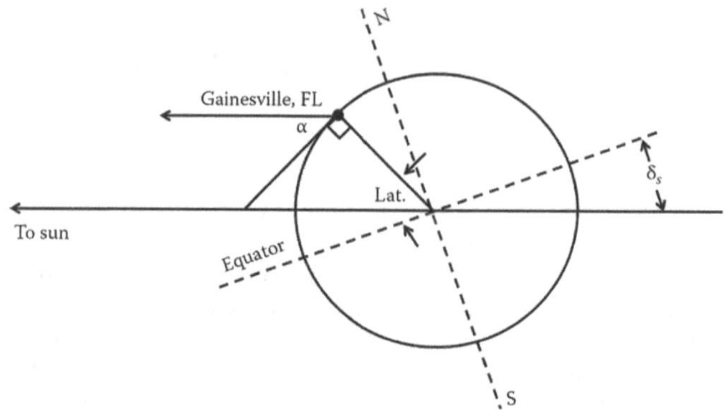

To find the declination angle δ_s on April 1, $n = 91$, we use Equation 10.23:

$$n = 91$$

$$\delta_s = 23.45° \sin\left[\frac{360(284+91)}{365}\right] = 4.02°$$

Therefore, at solar noon,

$$\alpha_1 = 90 - |29.68 - 4.02| = 64.34°$$

On November 1, $n = 305$ and

$$\delta_s = 23.45° \sin\left[\frac{360(284+305)}{365}\right]^{\circ} = -15.4°$$

Therefore,

$$\alpha_2 = 90 - |29.68 + 15.4| = 44.9°$$

$$L = \frac{H}{\tan\alpha_1 - \tan\alpha_2} = \frac{1}{\tan(64.34°) - \tan(44.9°)} = 0.92 \text{ m}$$

$$S = L\tan\alpha_2 = 0.92 \times \tan(44.9°) = 0.92 \text{ m}$$

If $S = 1.3$ m, then

$$L = \frac{1.3}{\tan 44.9°} = 1.3 \text{ m}$$

Also,

$$L = \frac{2.3}{\tan 64.34°} = 1.1 \text{ m}$$

Therefore, 1.1 m $\leq L \leq$ 1.3 m.

10.3.2 Sun-Path Diagram

The projection of the Sun's path on the horizontal plane is called a *Sun-path diagram*. Such diagrams are very useful in determining shading phenomena associated with solar collectors, windows, and shading devices. As shown earlier, the solar angles (α, a_s) depend upon the hour angle, declination, and latitude. Since only two of these variables can be plotted on a two-dimensional graph, the usual method is to prepare a different Sun-path diagram for each latitude with variations of hour angle and declination shown for a full year. Sun-path diagrams for different latitudes are shown in Figure 10.10 [7]. Note that these charts define 0° azimuth as due north instead of the due south definition we have used elsewhere in this chapter. Thus due south is an azimuth angle of 180°. Positive changes in azimuth still occur in the clockwise direction.

From Equations 10.28 and 10.29, it can be seen that the altitude and azimuth of the Sun are functions of the latitude of the observer, the time of day (hour angle), and the date (declination).

Figure 10.10a through f provides a series of charts for six different latitudes in 5° increments ranging from 25°N to 50°N giving the altitude and azimuth of the Sun as a function of the true solar time and the declination of the Sun in a form originally suggested by Hand. Linear interpolation for intermediate latitudes will give results within the accuracy to which the charts can be read.

On these charts, a point corresponding to the projected position of the Sun is determined from the heavy lines corresponding to declination and solar time.

To find the solar altitude and azimuth,

1. Select the chart or charts appropriate to the latitude.
2. Find the solar declination δ corresponding to the date.
3. Determine the *true solar time* as follows:
    a. To the *LST* (zone time), add 4° for each degree of longitude the station is east of the standard meridian or subtract 4° for each degree west of the standard meridian to get the *local mean solar time*.
    b. To the *local mean solar time*, add algebraically the equation of time; the sum is the required *true solar time*.
4. Read the required altitude and azimuth at the point determined by the declination and the true solar time. Interpolate linearly between two charts for intermediate latitudes.

It should be emphasized that the solar altitude determined from these charts is the true geometric position of the center of the Sun. At low solar elevations, terrestrial refraction may considerably alter the apparent position of the Sun. Under average atmospheric refraction, the Sun will appear on the horizon when it actually is about 34′ below the horizon; the effect of refraction decreases rapidly with increasing solar elevation. Since sunset or sunrise is defined as the time when the upper limb of the Sun appears on the horizon, and the semidiameter of the Sun is 16′, sunset or sunrise occurs under average atmospheric refraction when the Sun is 50′ below the horizon. In polar regions especially, unusual atmospheric refraction can make considerable variation in the time of sunset or sunrise.

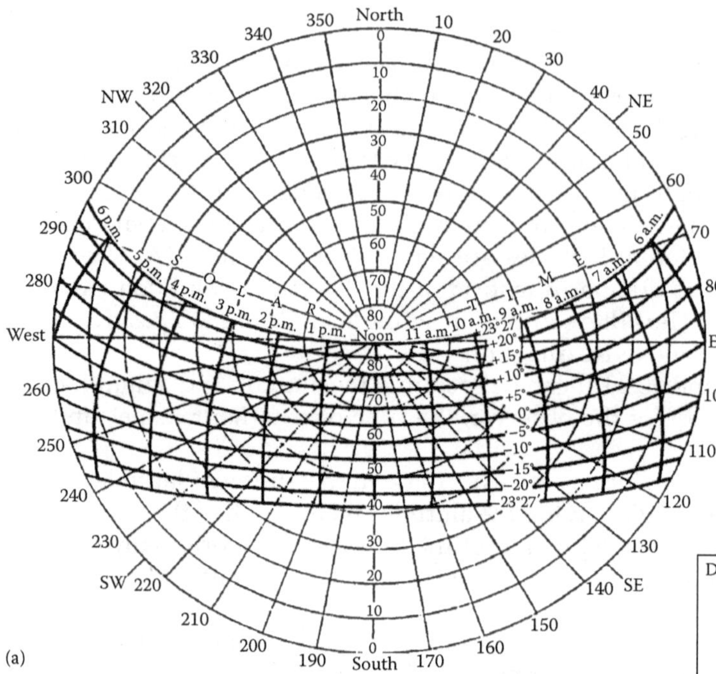

(a)

Declination	Approx. dates
+23° 27′	June 22
+20°	May 21, July 24
+15°	May 1, Aug. 12
+10°	Apr. 16, Aug. 28
+5°	Apr. 3, Sept. 10
0°	Mar. 21, Sept. 23
−5°	Mar. 8, Oct. 6
−10°	Feb. 23, Oct. 20
−15°	Feb. 9, Nov. 3
−20°	Jan. 21, Nov. 22
−23° 27′	Dec. 22

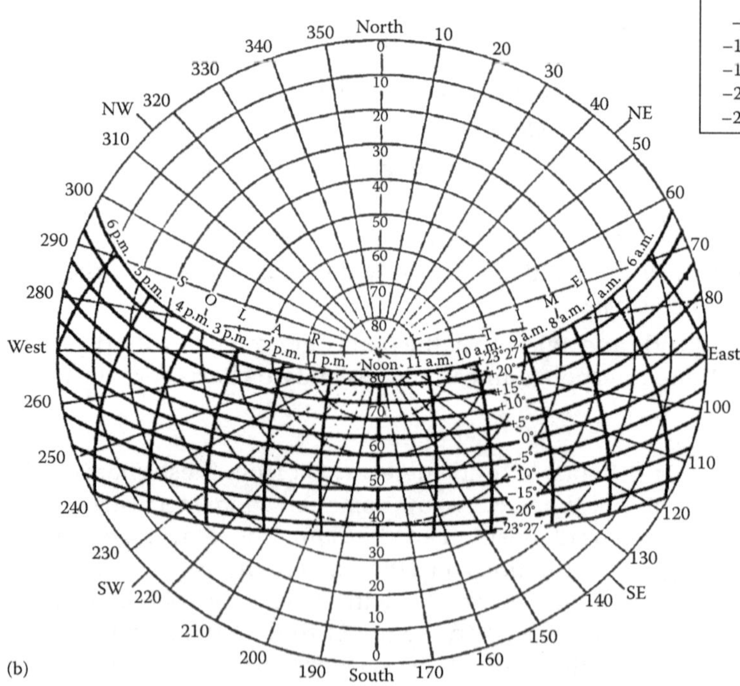

(b)

FIGURE 10.10
Description of method for calculating true solar time, together with accompanying meteorological charts, for computing solar altitude and azimuth angles: (a) chart, 25°N latitude; (b) chart, 30°N latitude; (c) chart, 35°N latitude; (d) chart, 40°N latitude; (e) chart, 45°N latitude; and (f) chart, 50°N latitude. (Reproduced from List [7].)

(Continued)

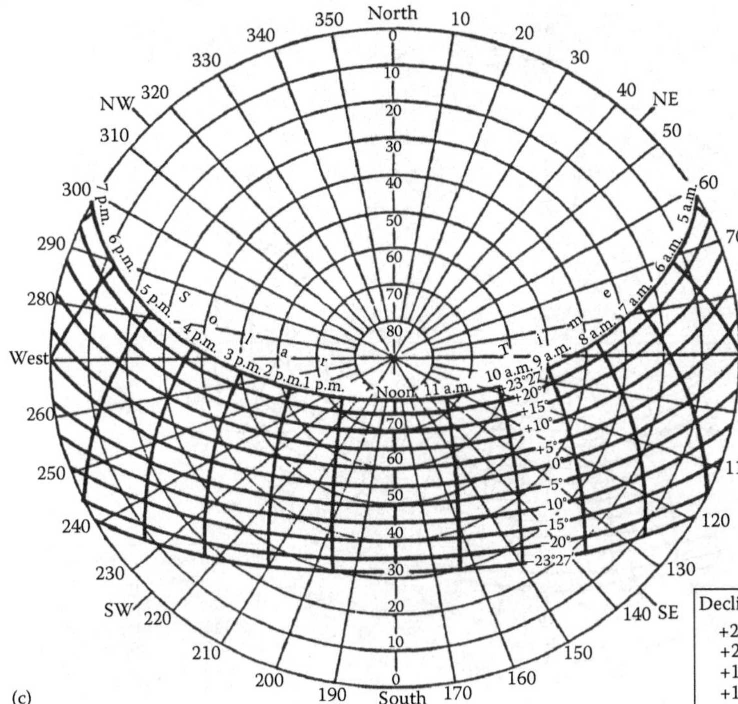

(c)

Declination	Approx. dates
+23° 27′	June 22
+20°	May 21, July 24
+15°	May 1, Aug. 12
+10°	Apr. 16, Aug. 28
+5°	Apr. 3, Sept. 10
0°	Mar. 21, Sept. 23
–5°	Mar. 8, Oct. 6
–10°	Feb. 23, Oct. 20
–15°	Feb. 9, Nov. 3
–20°	Jan. 21, Nov. 22
–23° 27′	Dec. 22

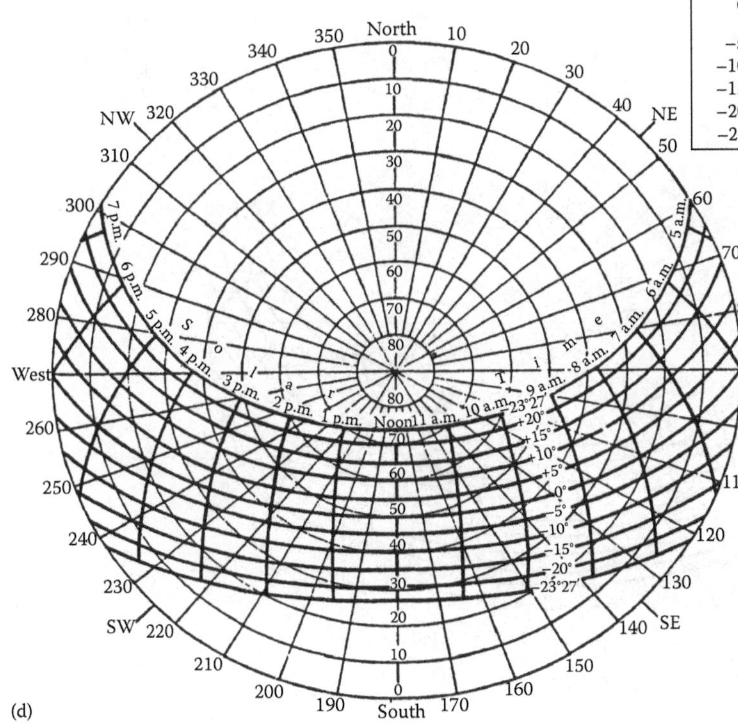

(d)

FIGURE 10.10 ***(Continued)***
Description of method for calculating true solar time, together with accompanying meteorological charts, for computing solar altitude and azimuth angles: (a) chart, 25°N latitude; (b) chart, 30°N latitude; (c) chart, 35°N latitude; (d) chart, 40°N latitude; (e) chart, 45°N latitude; and (f) chart, 50°N latitude. (Reproduced from List [7].)

(Continued)

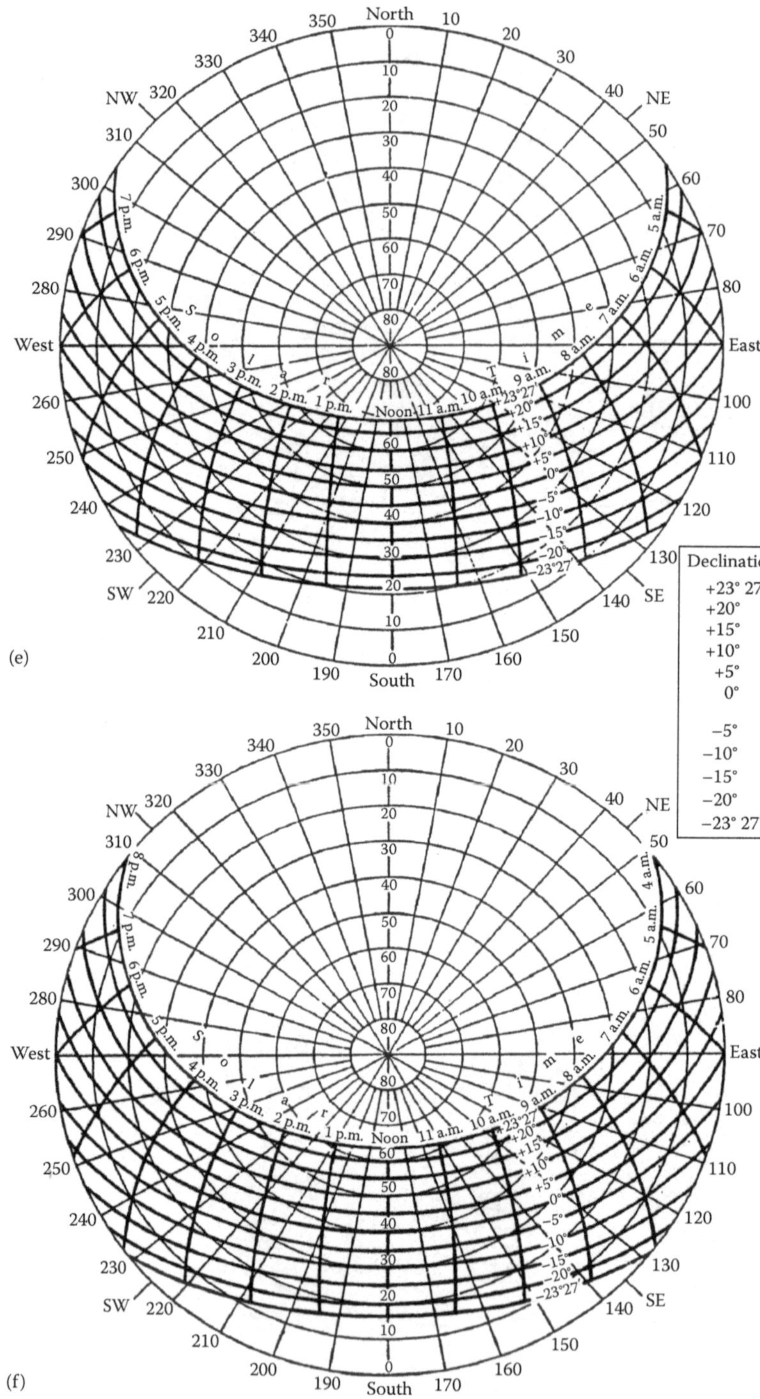

Declination	Approx. dates
+23° 27′	June 22
+20°	May 21, July 24
+15°	May 1, Aug. 12
+10°	Apr. 16, Aug. 28
+5°	Apr. 3, Sept. 10
0°	Mar. 21, Sept. 23
−5°	Mar. 8, Oct. 6
−10°	Feb. 23, Oct. 20
−15°	Feb. 9, Nov. 3
−20°	Jan. 21, Nov. 22
−23° 27′	Dec. 22

FIGURE 10.10 (*Continued*)
Description of method for calculating true solar time, together with accompanying meteorological charts, for computing solar altitude and azimuth angles: (a) chart, 25°N latitude; (b) chart, 30°N latitude; (c) chart, 35°N latitude; (d) chart, 40°N latitude; (e) chart, 45°N latitude; and (f) chart, 50°N latitude. (Reproduced from List [7].)

To compute solar altitude and azimuth for southern latitudes, change the sign of the solar declination and proceed as earlier. The resulting azimuths will indicate angular distance from *south* (measured eastward) rather than from north.

Sun-path diagrams for a given latitude are used by entering them with appropriate values of declination δ_s and hour angle h_s. The point at the intersection of the corresponding δ_s and h_s lines represents the instantaneous location of the Sun. The solar altitude can then be read from the concentric circles in the diagram, and the azimuth, from the scale around the circumference of the diagram. Sun path diagrams can be found on the internet [8].

Example 10.3

Using Figure 10.10b, determine the solar altitude and azimuth for March 8 at 10 a.m. for a site at 30°N. Compare the results to those calculated from the basic equations (Equations 10.28 and 10.29).

Solution:

On March 8, the solar declination is −5°; therefore the −5° Sun path is used. The intersection of the 10 a.m. line and the −5° declination line in the diagram represents the Sun's location.

The Sun's position lies midway between the 40° and 50° altitude circles, say at 45°, and midway between the 140° and 130° azimuth lines, or about 135°. But that is based on using the convention of 0° azimuth being due north. Converting to our convention of 0° azimuth being due south, we subtract 180° to get an azimuth of −45°. So $\alpha \cong 45°$ and $a_s \cong -45°$. Equations 10.28 and 10.29 give precise values for α and a_s:

$$\sin\alpha = \sin(30°)\sin(-5°) + \cos(30°)\cos(-5°)\cos(-30°)$$

$$\alpha = 44.7°$$

$$\sin a_s = \frac{\cos(-5°)\sin(-30°)}{\cos(44.7°)}$$

$$a_s = -44.5°$$

Therefore, the calculated values are within ±0.5° (1%) of those read from the Sun-path diagram.

10.3.3 Shadow-Angle Protractor

The shadow-angle protractor used in shading calculations is a plot of solar altitude angles, projected onto a given plane, versus solar azimuth angle. The projected altitude angle is usually called the *profile angle* γ. It is termed as the angle between the normal to a surface and the projection of the Sun's rays on a vertical plane normal to the same surface. The profile angle is shown in Figure 10.11a with the corresponding solar altitude angle. The profile angle, which is always used in sizing shading devices, is given by

$$\tan\gamma = \sec a \tan\alpha \tag{10.31}$$

where a is the solar azimuth angle with respect to the wall normal.

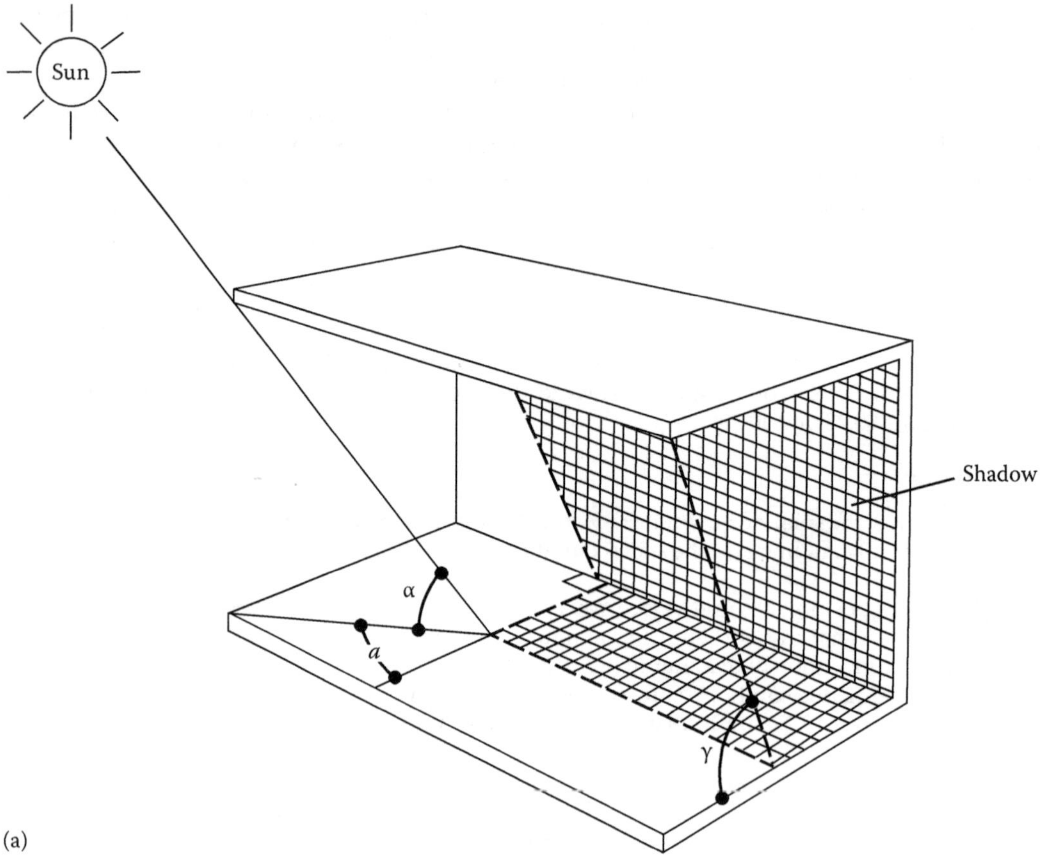

(a)

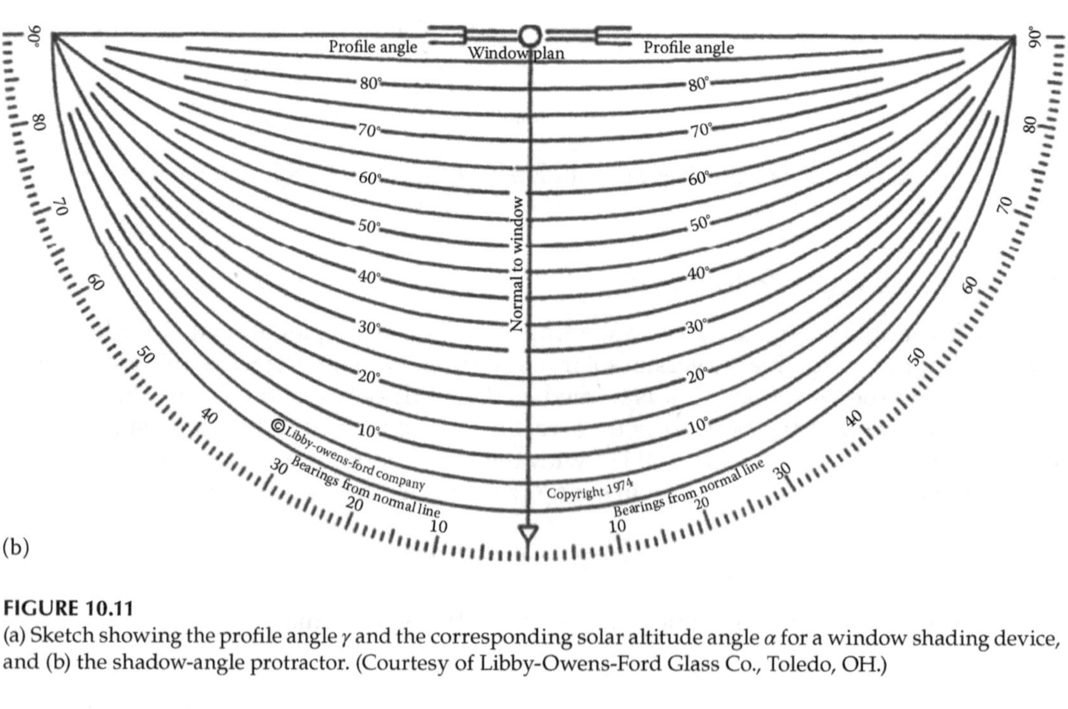

(b)

FIGURE 10.11
(a) Sketch showing the profile angle γ and the corresponding solar altitude angle α for a window shading device, and (b) the shadow-angle protractor. (Courtesy of Libby-Owens-Ford Glass Co., Toledo, OH.)

Figure 10.11b shows the shadow-angle protractor to the same scale as the Sun-path diagrams in Figure 10.10d. It is used by plotting the limiting values of profile angle γ and azimuth angle a, which will start to cause shading of a particular point. The shadow-angle protractor is usually traced onto a transparent sheet so that the shadow map constructed on it can be placed over the pertinent Sun-path diagram to indicate the times of day and months of the year during which shading will take place. The use of the shadow-angle protractor is best illustrated by an example.

Example 10.4

A solar building with a south-facing collector is sited to the north–northwest of an existing building. Prepare a shadow map showing what months of the year and what part of the day point C at the base of the solar collector will be shaded. Plan and elevation views are shown in Figure 10.12. Latitude = 40°N.

Solution:

The limiting profile angle for shading is 40° and the limiting azimuth angles are −45° and +10°, as shown in Figure 10.12. These values are plotted on the shadow-angle protractor (Figure 10.13a). The shadow map, when superimposed on the Sun-path diagram

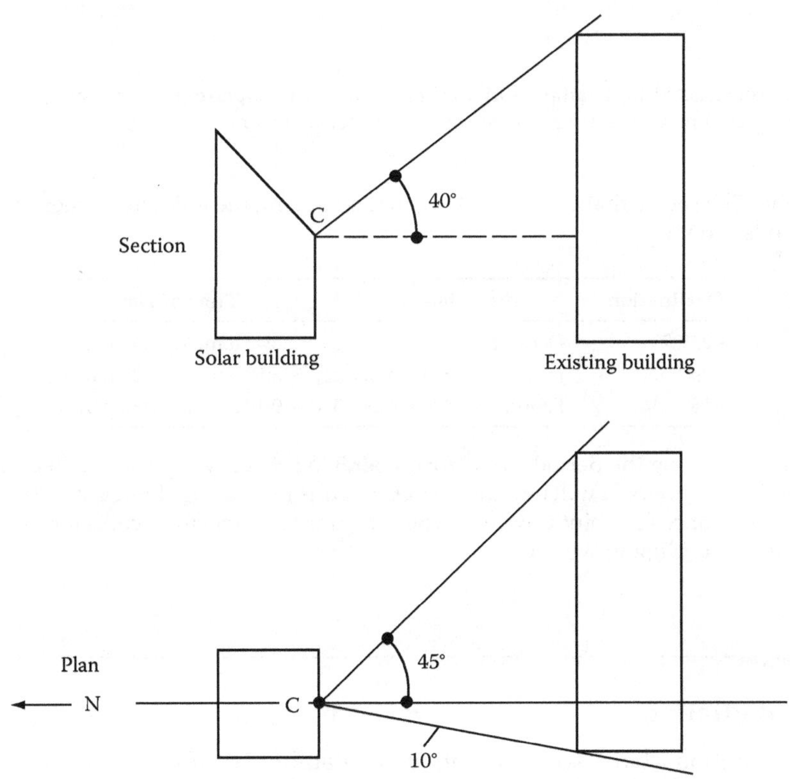

FIGURE 10.12
Plan and elevation views of proposed solar building and existing building, which may shade solar collector at point C.

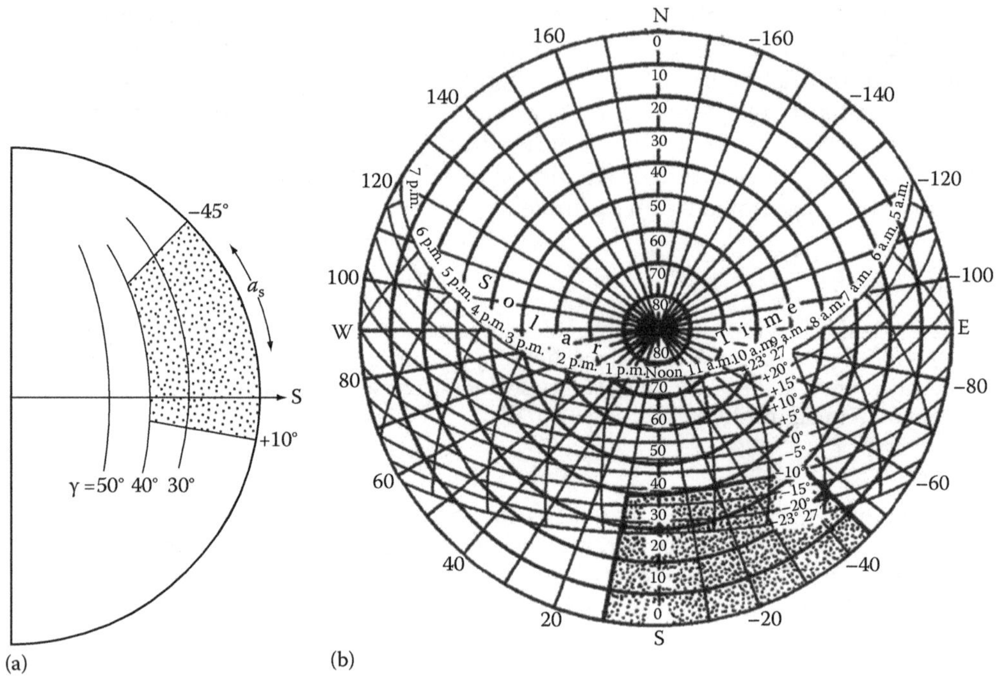

FIGURE 10.13
(a) Shadow map constructed for Example 10.4 and (b) shadow map superimposed on sun-path diagram. The degrees on the circumference in this version are shown as increasing or decreasing from the south.

(Figure 10.13b), shows that point C will be shaded during the following times of day for the periods shown:

Declination	Date	Time of Day
−23°27′	December 22	8:45 a.m. to 12:40 p.m.
−20°	January 21, November 22	8:55 a.m. to 12:35 p.m.
−15°	February 9, November 3	9:10 a.m. to 12:30 p.m.

In summary, during the period from November 3 to February 9, point C will be shaded between 3 and 4 hours. It will be shown later that this represents about a 50% loss in collector performance for point C, which would be unacceptable for a collector to be used for heating a building in winter.

10.4 Solar Radiation

Detailed information about solar radiation availability at any location is essential for the design and economic evaluation of a solar energy system. Long-term measured data of solar radiation are available for a large number of locations in the United States and other parts of the world. Where long-term measured data are not available, various models based on available climatic data can be used to estimate the solar energy availability.

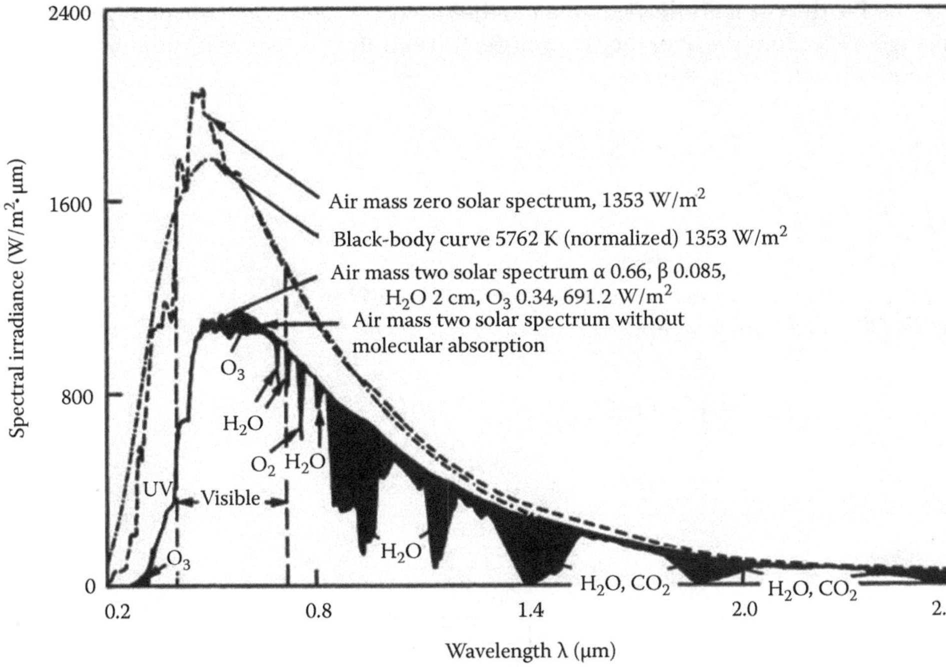

FIGURE 10.14
Extraterrestrial solar radiation spectral distribution. Also shown are equivalent black-body and atmosphere-attenuated spectra.

Solar energy is in the form of electromagnetic radiation with the wavelengths ranging from about 0.3 μm to over 3 μm, which correspond to ultraviolet (less than 0.4 μm), visible (0.4 and 0.7 μm), and infrared (over 0.7 μm). Most of this energy is concentrated in the visible and the near-infrared wavelength range (see Figure 10.14). The incident solar radiation, sometimes called *insolation,* is measured as irradiance, or the energy per unit time per unit area (or power per unit area). The units most often used are Watts per square meter (W/m^2), British thermal units per hour per square foot ($Btu/h/ft^2$), and Langleys per minute (calories per square centimeter per minute, $cal/cm^2/min$).

10.4.1 Extraterrestrial Solar Radiation

The average amount of solar radiation falling on a surface normal to the rays of the Sun outside the atmosphere of the Earth (extraterrestrial) at mean Earth–Sun distance (D_0) is called the *solar constant,* I_0. Measurements by NASA indicated the value of the solar constant to be 1,353 W/m^2 (±1.6%), 429 $Btu/h/ft^2$ or 1.94 $cal/cm^2/min$ (Langleys/min). This value was revised upward by Frohlich et al. [9] to 1,377 W/m^2 or 437.1 $Btu/h/ft^2$ or 1.974 Langleys/min, which was the value used in compiling SOLMET data in the United States [10,11]. At present, there is no consensus on the value of the solar constant. However, considering that the difference between the two values is about 1.7% and the uncertainties in the estimation of terrestrial solar radiation are 10% or higher, either value may be used. A value of 1,367 W/m^2 is used by many references.

The variation in seasonal solar radiation availability at the surface of the Earth can be understood from the geometry of the relative movement of the Earth around the Sun.

Since the Earth's orbit is elliptical, the Earth–Sun distance varies during a year, the variation being ±1.7% from the average. Therefore, the extraterrestrial radiation, I, also varies by the inverse square law as follows:

$$I = I_0\left(\frac{D_0}{D}\right)^2 \tag{10.32}$$

where
D is the distance between the Sun and the Earth
D_0 is the yearly mean Earth–Sun distance (1.496×10^{11} m)
The $(D_0/D)^2$ factor may be approximated as [12]

$$\left(\frac{D_0}{D}\right)^2 = 1.00011 + 0.034221\cos(x) + 0.00128\sin(x) + 0.000719\cos(2x) + 0.000077\sin(2x) \tag{10.33}$$

where

$$x = \left[\frac{360(n-1)}{365}\right]^{\circ} \tag{10.34}$$

and n = day number (starting from January 1 as 1). The following approximate relationship may also be used without much loss of accuracy:

$$I = I_0\left[1 + 0.034\cos\left(\frac{360n}{365.25}\right)^{\circ}\right] \tag{10.35}$$

Figure 10.15 also shows the relationship of the extraterrestrial solar radiation to the solar constant. For many solar energy applications, such as photovoltaics and photocatalysis, it is necessary to examine the distribution of energy within the solar spectrum. Table 10.2 [13,14] shows the spectral irradiance with the mean Earth–Sun distance for a solar constant of 1,353 W/m^2 as a function of wavelength according to the standard spectrum data published by NASA in 1971. This is illustrated in the following example.

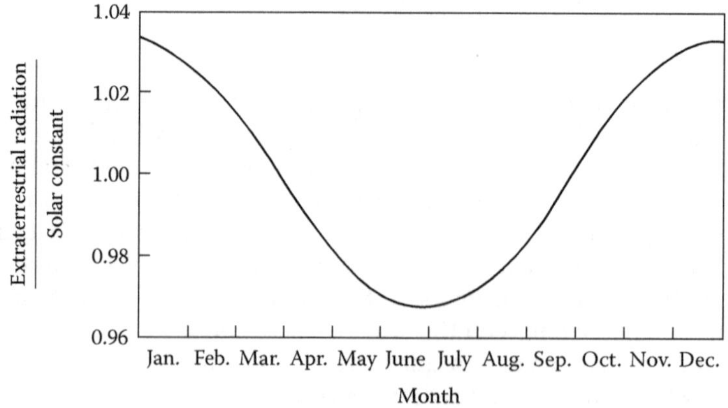

FIGURE 10.15
Effect of the time of year on the ratio of extraterrestrial radiation to the nominal solar constant.

TABLE 10.2

Extraterrestrial Solar Irradiance[a]

λ (μm)	E_λ (W/m² μm)[b]	E_λ (Btu/h ft² μm)[b]	D_λ (%)[c]	λ (μm)	E_λ (W/m²·μm)	E_λ (Btu/h ft² μm)	D_λ (%)	λ (μm)	E_λ (W/m² μm)	E_λ (Btu/h ft²·μm)	D_λ (%)
0.12	0.007	0.002	1×10^{-4}	0.43	1,639	520	12.47	0.90	891	283	63.37
0.14	0.03	0.010	5×10^{-4}	0.44	1,810	574	13.73	1.00	748	237	69.49
0.16	0.23	0.073	6×10^{-4}	0.45	2,006	636	15.14	1.2	485	154	78.40
0.18	1.25	0.397	1.6×10^{-3}	0.46	2,066	655	16.65	1.4	337	107	84.33
0.20	10.7	3.39	8.1×10^{-3}	0.47	2,033	645	18.17	1.6	245	77.7	88.61
0.22	57.5	18.2	0.05	0.48	2,074	658	19.68	1.8	159	50.4	91.59
0.23	66.7	21.2	0.10	0.49	1,950	619	21.15	2.0	103	32.7	93.49
0.24	63.0	20.0	0.14	0.50	1,942	616	22.60	2.2	79	25.1	94.83
0.25	70.9	22.5	0.19	0.51	1,882	597	24.01	2.4	62	19.7	95.86
0.26	130	41.2	0.27	0.52	1,833	581	25.38	2.6	48	15.2	96.67
0.27	232	73.6	0.41	0.53	1,842	584	26.74	2.8	39	12.4	97.31
0.28	222	70.4	0.56	0.54	1,783	566	28.08	3.0	31	9.83	97.83
0.29	482	153	0.81	0.55	1,725	547	29.38	3.2	22.6	7.17	98.22
0.30	514	163	1.21	0.56	1,695	538	30.65	3.4	16.6	5.27	98.50
0.31	689	219	1.66	0.57	1,712	543	31.91	3.6	13.5	4.28	98.72
0.32	830	263	2.22	0.58	1,715	544	33.18	3.8	11.1	3.52	98.91
0.33	1,059	336	2.93	0.59	1,700	539	34.44	4.0	9.5	3.01	99.06
0.34	1,074	341	3.72	0.60	1,666	528	35.68	4.5	5.9	1.87	99.34
0.35	1,093	347	4.52	0.62	1,602	508	38.10	5.0	3.8	1.21	99.51
0.36	1,068	339	5.32	0.64	1,544	490	40.42	6.0	1.8	0.57	99.72
0.37	1,181	375	6.15	0.66	1,486	471	42.66	7.0	1.0	0.32	99.82
0.38	1,120	355	7.00	0.68	1,427	453	44.81	8.0	0.59	0.19	99.88
0.39	1,098	348	7.82	0.70	1,369	434	46.88	10.0	0.24	0.076	99.94
0.41	1,751	555	9.92	0.75	1,235	392	51.69	20.0	0.0015	0.005	99.99
0.42	1,747	554	11.22	0.80	1,109	352	56.02	50.0	0.0004	0.0001	100.00

Source: Adapted from Iqbal [13]; Thekaekara [14].

[a] Solar constant = 429 Btu/h ft² = 1,353 W/m².

[b] E_λ is the solar spectral irradiance averaged over a small bandwidth centered at λ.

[c] D_λ is the percentage of the solar radiation associated with wavelengths shorter than λ.

Example 10.5

Calculate the fraction of solar radiation within the visible part of the spectrum, that is, between 0.40 and 0.70 μm.

Solution:

The first column in Table 10.2 gives the wavelength. The second column gives the averaged solar spectral irradiance in a band centered at the wavelength in the first column. The fourth column, D_λ, gives the percentage of solar total radiation at wavelengths shorter than the value of λ in the first column. Interpolating table entries to a value of 0.40 μm, 8.9% of the total radiation occurs at shorter wavelengths. At a wavelength of 0.70, 46.88% of the radiation occurs at shorter wavelengths. Consequently, 38% of the total radiation lies within the band between 0.40 and 0.70 μm, and the total energy received outside the Earth's atmosphere within that spectral range is 514 W/m^2 (163 Btu/h/ft^2).

10.5 Estimation of Terrestrial Solar Radiation

As extraterrestrial solar radiation, *I*, passes through the atmosphere, a part of it is reflected back into space, a part is absorbed by air and water vapor, and some gets scattered by molecules of air, water vapor, aerosols, and dust particles (Figure 10.16). The part of solar radiation that reaches the surface of the Earth with essentially no change in direction is called *direct* or *beam radiation*. The scattered diffuse radiation reaching the surface from the sky is called the *sky diffuse radiation.*

Although extraterrestrial radiation can be predicted with certainty (the effect of sunspots, which may cause up to 0.5% variation, is neglected), radiation levels on the Earth

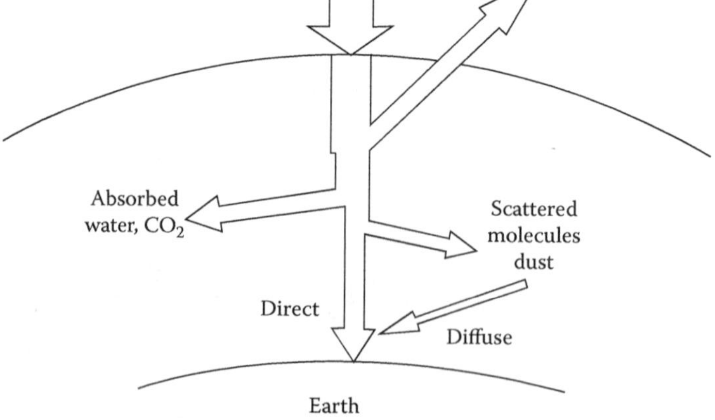

FIGURE 10.16
Attenuation of solar radiation as it passes through the atmosphere.

are subject to considerable uncertainty resulting from local climatic interactions. The most useful solar radiation data are based on long-term (30 years or more) measured average values at a location, which unfortunately are not available for most locations in the world. For such locations, an estimation method (theoretical model) based on some measured climatic parameters may be used. This chapter describes several ways of estimating terrestrial solar radiation; all have large uncertainties (as much as ±30%) associated with them.

10.5.1 Atmospheric Extinction of Solar Radiation

As solar radiation I travels through the atmosphere, it is attenuated due to absorption and scattering. If K is the local extinction coefficient of the atmosphere, the beam solar radiation at the surface of the Earth can be written according to Beer's law, Equation 10.21 as

$$I_{b,N} = Ie^{-\int Kdx} \tag{10.36}$$

where
$I_{b,N}$ is the instantaneous beam solar radiation per unit area normal to the Sun's rays
x is the length of travel through the atmosphere
If L_o is the vertical thickness of the atmosphere and

$$\int_0^{L_0} Kdx = k \tag{10.37}$$

the beam normal solar radiation for a solar zenith angle of z will be

$$I_{b,N} = Ie^{-k\sec z} = Ie^{-k/\sin\alpha} = Ie^{-km} \tag{10.38}$$

where m is a dimensionless path length of sunlight through the atmosphere, sometimes called the *air mass ratio* (Figure 10.17). When the solar altitude angle is 90° (the Sun is overhead), $m = 1$.

Threlkeld and Jordan [5] estimated values of k (also known as optical depth) for average atmospheric conditions at sea level with a moderately dusty atmosphere and the amount

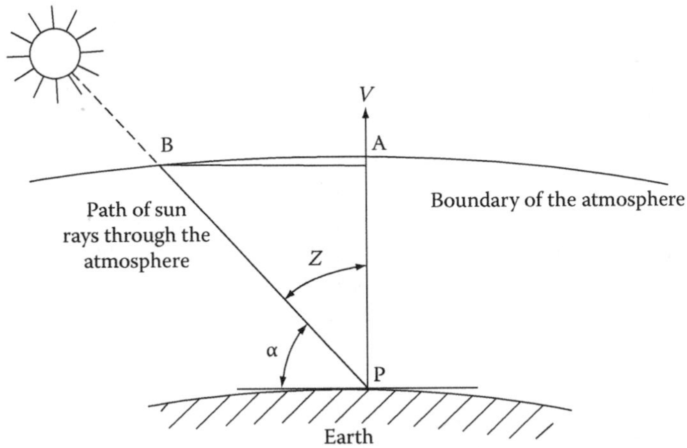

FIGURE 10.17
Air mass definition: air mass m = BP/AP = cosec α, where α is the altitude angle. The atmosphere is idealized as a constant thickness layer.

of precipitable water vapor equal to the average value for the United States for each month. These values are given in Table 10.3 [5,15].

To account for the differences in local conditions from the average sea-level conditions, Equation 10.39 is modified by a parameter called clearness number C_n, introduced by Threlkeld and Jordan [5]:

$$I_{b,N} = C_n I e^{-k/\sin\alpha} \tag{10.39}$$

Note that $C_n = 1$ in clear-sky conditions.

10.5.2 Solar Radiation on Clear Days

Total instantaneous solar radiation on a horizontal surface (see Figure 10.18), I_h, is the sum of the beam or direct radiation, $I_{b,h}$, and the sky diffuse radiation $I_{d,h}$:

$$I_h = I_{b,h} + I_{d,h} \tag{10.40}$$

According to the Threlkeld and Jordan model [5], the sky diffuse radiation on a clear day is proportional to the beam normal solar radiation and can be estimated by using an empirical sky diffuse factor C. Therefore, I_h can be estimated as

TABLE 10.3

Average Values of Atmospheric Optical Depth (k) and Sky Diffuse Factor (C) for 21st Day of Each Month, for Average Atmospheric Conditions at Sea Level for the United States

Month	1	2	3	4	5	6	7	8	9	10	11	12
k	0.142	0.144	0.156	0.180	0.196	0.205	0.207	0.201	0.177	0.160	0.149	0.142
C	0.058	0.060	0.071	0.097	0.121	0.134	0.136	0.122	0.092	0.073	0.063	0.057

Source: Threlkeld and Jordan [5]; Islam [15].

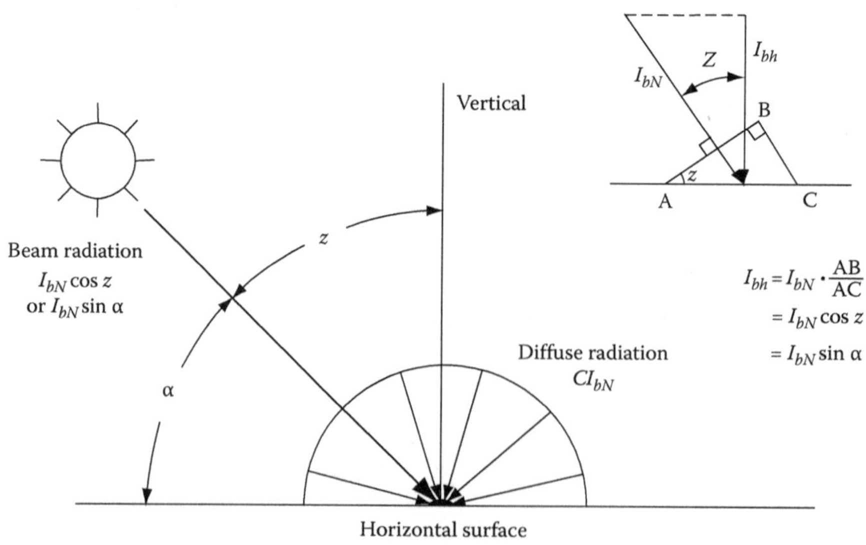

FIGURE 10.18
Solar radiation on a horizontal surface.

$$\begin{aligned} I_h &= I_{b,N} \cos z + CI_{b,N} \\ &= I_{b,N} \sin\alpha + CI_{b,N} \\ &= C_n I e^{-k/\sin\alpha}(C + \sin\alpha) \end{aligned} \tag{10.41}$$

Values of C are given in Table 10.3.

10.5.3 Solar Radiation on a Tilted Surface

Solar radiation on an arbitrarily tilted surface having a tilt angle of β from the horizontal and an azimuth angle of a_w (assumed + west of south), as shown in Figure 10.19, is the sum of components consisting of beam ($I_{b,c}$), sky diffuse (I_{dc}), and ground-reflected solar radiation ($I_{r,c}$):

$$I_c = I_{b,c} + I_{d,c} + I_{r,c} \tag{10.42}$$

If i is the *angle of incidence* of the beam radiation on the tilted surface, it is simple to show that the instantaneous beam radiation on the surface per unit area is

$$I_{b,c} = I_{b,N} \cos i \tag{10.43}$$

From the geometry in Figure 10.19, it can be shown that the angle of incidence i for the surface (angle between the normal to the surface and a line collinear with the Sun's rays) is related to the solar angles as

$$\cos i = \cos\alpha \cos(a_s - a_w)\sin\beta + \sin\alpha\cos\beta \tag{10.44}$$

The diffuse radiation on the surface ($I_{d,c}$) can be obtained by multiplying the sky diffuse radiation on a horizontal surface by the view factor between the sky and the surface[2]:

$$I_{d,c} = I_{d,h}\frac{(1+\cos\beta)}{2} = CI_{b,N}\frac{(1+\cos\beta)}{2} = CI_{b,N}\cos^2\left(\frac{\beta}{2}\right) \tag{10.45}$$

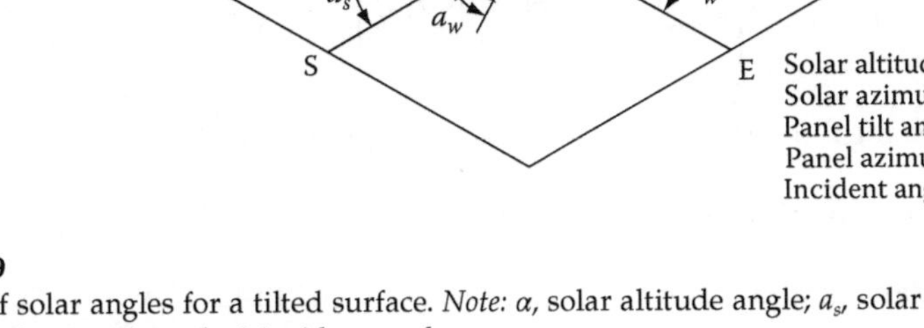

FIGURE 10.19
Definitions of solar angles for a tilted surface. *Note:* α, solar altitude angle; a_s, solar azimuth angle; β, panel tilt angle; a_w, panel azimuth angle; i, incident angle.

[2] The surface has been assumed infinitely large for this view factor. See Section 10.2.3.

The ground-reflected solar radiation can be found from the total solar radiation incident on a horizontal surface and the ground reflectance ρ as

$$I_{r,c} = I_h \rho \tag{10.46}$$

The part of $I_{r,c}$ intercepted by the tilted surface can be found by multiplying the ground-reflected radiation by the view factor between the surface[3] and the ground:

$$\begin{aligned} I_{r,c} &= \rho I_h \frac{(1-\cos\beta)}{2} = \rho I_h \sin^2\left(\frac{\beta}{2}\right) \\ &= \rho I_{b,N}(\sin\alpha + C)\sin^2\left(\frac{\beta}{2}\right) \end{aligned} \tag{10.47}$$

For ordinary ground or grass, ρ is approximately 0.2, and for snow-covered ground, it can be taken as approximately 0.8.

Example 10.6

Find the instantaneous solar radiation at 12:00 noon EST on a solar collector surface ($\beta = 30°$, $a_w = +10°$) on February 1 in Gainesville, Florida.

Solution:

From Example 10.1, for February 1

$$n = 32, \quad \delta_s = -17.5°, \quad \text{and} \quad \text{ET} = -13.7 \text{ minutes}$$

For finding the values of solar radiation on the collector, we will need to calculate angles α, a_s, h_s, and i:

$$\text{Solar time} = \text{LST} + \text{ET} + 4(l_{st} - l_{local}) = 12:00 - 13.7 \text{ minutes} + 4(75° - 82.27°) = 11:17.2 \text{ a.m.}$$

$$h_s = \frac{\text{Minutes from solar noon}}{4 \text{ min / degree}} = \frac{-42.8}{4} = -10.7° (\text{i.e., before noon})$$

From Equation 10.28,

$$\alpha = \sin^{-1}\left(\sin(29.68°)\sin(-17.5°) + \cos(29.68°)\cos(-17.5°)\cos(-10.7°)\right) = 41.7°$$

From Equation 10.29,

$$a_s = \sin^{-1}\left[\frac{(\cos(-17.5°)\sin(-10.7°))}{\cos(41.7°)}\right] = -13.7°$$

Angle of incidence i for the solar collector is given by Equation 10.44:

$$\cos i = \cos(41.7°)\cos(-13.7° - 10°)\sin(30°) + \sin(41.7°)\cos(30°) \quad i = 23.4°$$

To calculate the solar radiation using Equations 10.35, 10.42, 10.43, 10.45, and 10.47, we need to find $I_{b,N}$ and I.

[3] The tilted surface and the ground in front of it have been assumed to be infinitely large for this view factor.

Extraterrestrial solar radiation

$$I = I_0\left[1+0.034\cos\left(\frac{360n}{365.25}\right)^{\circ}\right]$$

$$= 1{,}353\ \text{W/m}^2\left[1+0.034\cos\left(\frac{360\times 32}{365.25}\right)^{\circ}\right]$$

$$= 1{,}392\ \text{W/m}^2$$

(Find k from Table 10.3, and assume C_n is 1.)

$$I_{b,N} = C_n I e^{-k/\sin\alpha} = 1{,}392e^{-0.144/\sin(41.7^{\circ})} = 1{,}121\ \text{W/m}^2 \quad \text{or} \quad 355\ \text{Btu/h/ft}^2$$

Beam radiation on the collector (Equation 10.43)

$$I_{b,c} = I_{b,N}\cos i = (1{,}121\ \text{W/m}^2)\cos(23.4^{\circ}) = 1{,}029\ \text{W/m}^2 \quad \text{or} \quad 326\ \text{Btu/h/ft}^2$$

Sky diffuse radiation on the collector using Equation 10.45 and obtaining C from Table 10.3

$$I_{d,c} = CI_{b,N}\cos^2\left(\frac{\beta}{2}\right) = (0.060)(1{,}121\ \text{W/m}^2)\cos^2\left(\frac{30}{2}\right) = 63\ \text{W/m}^2 \quad \text{or} \quad 20\ \text{Btu/h/ft}^2$$

Ground-reflected radiation on the collector using Equation 10.45 and obtaining C from Table 10.3,

$$I_{r,c} = \rho I_{b,N}(\sin\alpha + C)\sin^2\left(\frac{\beta}{2}\right)$$

Assume

$$\rho = 0.2$$

$$I_{r,c} = (0.2)(1{,}121\ \text{W/m}^2)(\sin 41.7^{\circ} + 0.060)\sin^2(15^{\circ}) = 11\ \text{W/m}^2 \quad \text{or} \quad 3.5\ \text{Btu/h/ft}^2$$

Total insolation on the collector

$$I_c = 1{,}029 + 63 + 11 = 1{,}103\ \text{W/m}^2 \quad \text{or} \quad 350\ \text{Btu/h/ft}^2$$

A convenient website that will provide the same information obtained in the preceding example can be found online at Build It Solar [16]. The online calculator model requires merely inputting the latitude, collector tilt, collector azimuth, altitude of location, the month, and the day. With this information, the program provides the direct, diffuse, and total radiation perpendicular to the Sun in the third, fourth, and fifth columns. The sixth column provides the total radiation on a horizontal surface. The seventh column gives the incident angle of the Sun on the collector. The ninth, tenth, and eleventh columns give the direct, diffuse, and total radiation falling on the collector surface. From this program, the radiation falling on the collector at noon is 1,037 W/m^2, which is about 6% less than the value obtained by the empirical procedure outlined in the example. The printout from the model for the day is shown in Table 10.4 with the radiation values in Btu/h-ft^2.

TABLE 10.4

Solar Radiation Output for Example 10.6 Using the Online Calculator Build It Solar

Solar Radiation on Collector for Day of 2/1										
Collector area	1 ft²									
Collector azimuth	0° measured from south + is to east									
Collect tilt	40° measured from horizontal									
Latitude	30°									
Altitude	0 ft above sea level									
Sunrise	6.7 hours sunrise in solar time									
	Sun						Collector			
Time (hours)	**Az.**	**Elev.**	**Dir. Normal**	**Diffuse**	**Total**	**Horiz. Tot.**	**Incid. Ang.**	**Direct.**	**Diffuse**	**Total**
4	88.9	−34.1	0	0	0	0	90	0	0	0
5	82	−21.2	0	0	0	0	90	0	0	0
6	75.1	−8.4	0	0	0	0	87	0	0	0
7	67.7	3.9	47	3	50	6	73	14	2	16
8	59.2	15.5	228	13	241	74	59	119	12	131
9	48.8	26	280	16	297	139	44	201	15	216
10	35.6	34.8	302	18	320	190	30	262	16	278
11	19.1	40.8	312	18	330	222	16	300	16	316
12	0	42.9	315	18	333	233	7	312	16	329
13	−19.1	40.8	312	18	330	222	16	300	16	316
14	−35.6	34.8	302	18	320	190	30	262	16	278
15	−48.8	26	280	16	297	139	44	201	15	216
16	−59.2	15.5	228	13	241	74	59	119	12	131
17	−67.7	3.9	47	3	50	6	73	14	2	16
18	−75.1	−8.4	0	0	0	0	87	0	0	0
19	−82	−12.2	0	0	0	0	90	0	0	0
20	−88.9	−34.1	0	0	0	0	90	0	0	0
Day total			2,653	156	2,809	1,496		2,102	138	2,240
Angles in degrees										
Radiation in Btu/h										
Day total in Btu/day										
1 Btu/h ft² = 3.152 W/m²										

10.6 TMY Data to Determine Solar Radiation

Sandia National Laboratories prepared a data set summarizing long-term weather and solar data in 1978. This was updated by NREL and released for use by building designers and solar thermal modelers as the Typical Meteorological Year version 3 (TMY3) data set in 2007. A user's manual for the data set is available from NREL's website [17]. The following example demonstrates the use of the TMY3 data to obtain the insolation on a solar collector.

Example 10.7

Using typical meteorological data (TMY3), find the total insolation and beam insolation on a south-facing solar collector tilted at an angle of 25° in Miami, Florida, for January 1. Also find the average total insolation per day for the entire month of January. Assume a ground reflectance of $\rho = 0.2$. Use site information for the Miami International Airport.

Solution:

TMY3 data can be accessed via the NREL website on the National Solar Radiation Data Base page [18]. Under "The Data," select "In alphabetical order by state and city." This will link you to a list of all sites for which TMY3 data have been collected. Scroll down to Florida and select site number 722020 for the Miami International Airport. The TMY3 data can be downloaded as a .CSV file format that can be read by most spreadsheet programs.

We will only need three separate columns from the TMY3 table: DNI, DHI, and GHI. DNI is *direct normal irradiance* and is the amount of beam radiation received on a surface normal to the Sun. DHI is *diffuse horizontal irradiance* and is the amount of sky diffuse radiation received on a horizontal surface. GHI is *global horizontal irradiance* and is the total amount of radiation received on a horizontal surface. DNI, DHI, and GHI correspond to $I_{b,N}$, $I_{d,h}$, and I_h, respectively, as used in the text (see Table 10.5).

To find the total insolation for January 1, you must sum the hourly values of I_c. Using Equation 10.42, I_c is the sum of $I_{b,c}$, $I_{d,c}$, and $I_{r,c}$ at each hour during the day. Table 10.5 shows how this can be approached using a spreadsheet [17]. Summing the I_c column, the total insolation for January 1 is 4,544 Wh/m^2·day (16,800 kJ/m^2·day). Summing the $I_{b,c}$ column, the total beam insolation is 2,899 Wh/m^2·day. Repeating this problem for the entire month of January, the total monthly insolation is 133,522 Wh/m^2, with a daily average of 4,307 Wh/m^2·day.

10.7 Measurement of Solar Radiation

Solar radiation measurements of importance to most engineering applications, especially thermal applications, include total (integrated over all wavelengths) direct or beam and sky diffuse values of solar radiation on instantaneous, hourly, daily, and monthly bases. Some applications such as photovoltaics, photochemical, and daylighting require knowledge of spectral (wavelength specific) or band (over a wavelength range, e.g., ultraviolet, visible, infrared) values of solar radiation. This section describes some of the instrumentation used to measure solar radiation and sunshine, and some sources of long-term measured data for different parts of the world. Also described briefly in this section is the method of satellite-based measurements.

TABLE 10.5

Calculation of Solar Insolation for January 1 in Miami, Florida, Using TMY3 Data

Time	DNI	DHI	GHI	h_s	α	a_s	i	$I_{b,c}$	$I_{d,c}$	$I_{r,c}$	I_c
Hours	W/m²	W/m²	W/m²	°	°	°	°	W/m²	W/m²	W/m²	W/m²
1	0	0	0	−165.00	−76.1	−81.7	149.9	0.0	0.0	0.0	0.0
2	0	0	0	−150.00	−62.6	−89.4	143.4	0.0	0.0	0.0	0.0
3	0	0	0	−135.00	−49.1	−84.0	131.0	0.0	0.0	0.0	0.0
4	0	0	0	−120.00	−35.8	−79.2	117.7	0.0	0.0	0.0	0.0
5	0	0	0	−105.00	−22.6	−74.4	104.1	0.0	0.0	0.0	0.0
6	0	0	0	−90.00	−9.8	−69.1	90.3	0.0	0.0	0.0	0.0
7	0	0	0	−75.00	2.5	−62.9	76.5	0.0	0.0	0.0	0.0
8	262	18	39	−60.00	14.1	−55.3	62.9	119.2	17.2	0.4	136.7
9	694	38	218	−45.00	24.6	−45.7	49.8	447.9	36.2	2.0	486.2
10	768	65	394	−30.00	33.2	−33.4	37.7	608.0	62.0	3.7	673.6
11	747	125	540	−15.00	39.1	−17.9	27.9	660.1	119.1	5.1	784.3
12	230	265	411	0.00	41.2	0.0	23.8	210.4	252.6	3.9	466.9
13	324	290	503	15.00	39.1	17.9	27.9	286.3	276.4	4.7	567.4
14	517	192	514	30.00	33.2	33.4	37.7	409.3	183.0	4.8	597.1
15	107	339	396	45.00	24.6	45.7	49.8	69.1	323.1	3.7	395.9
16	117	265	313	60.00	14.1	55.3	62.9	53.2	252.6	2.9	308.7
17	152	81	116	75.00	2.5	62.9	76.5	35.4	77.2	1.1	113.7
18	22	14	16	90.00	−9.8	69.1	90.3	−0.1	13.3	0.1	13.4
19	0	0	0	105.00	−22.6	74.4	104.1	0.0	0.0	0.0	0.0
20	0	0	0	120.00	−35.8	79.2	117.7	0.0	0.0	0.0	0.0
21	0	0	0	135.00	−49.1	84.0	131.0	0.0	0.0	0.0	0.0
22	0	0	0	150.00	−62.6	89.4	143.4	0.0	0.0	0.0	0.0
23	0	0	0	165.00	−76.1	81.7	149.9	0.0	0.0	0.0	0.0

Source: Wilcox and Marion [17].
See Figure 10.19.
L, latitude = 25.817° (from top line of spreadsheet).
h_s, hour angle (from Equation 10.25; for simplicity in this example, solar time is approximated as local time).
α, solar altitude (from Equation 10.28).
a_s, solar azimuth (from Equation 10.29).
i, incidence angle of the Sun on collector (from Equation 10.44).
β, tilt angle of collector from horizontal, 25°

10.7.1 Instruments for Measuring Solar Radiation and Sunshine

There are two basic types of instruments used to measure solar radiation: a *pyranometer* and a *pyrheliometer*. A pyranometer has a hemispherical view of the surroundings and therefore is used to measure total, direct, and diffuse, solar radiation on a surface. A pyrheliometer, on the other hand, has a restricted view and is, therefore, often used to measure the direct or beam solar radiation by pointing it toward the Sun. Pyranometers are also used to measure the sky diffuse radiation by using a shadow band to block the direct Sun view. A detailed discussion of the instrumentation and calibration standards is given by Iqbal [19] and Zerlaut [20].

A pyranometer consists of a flat sensor/detector (described later) with an unobstructed hemispherical view, which allows it to convert and correlate the total radiation incident on the sensor to a measurable signal. The pyranometers using thermal detectors for measurements can exhibit serious errors at tilt angles from the horizontal due to free convection. These errors are minimized by enclosing the detector in double hemispherical high-transmission glass domes. The second dome minimizes the error due to infrared radiative exchange between the sensor and the sky. A desiccator is usually provided to eliminate the effect of condensation on the sensor or the dome. Figure 10.20 shows pictures of typical commercially available precision pyranometers.

A pyranometer can be used to measure the sky diffuse radiation by fitting a shadow ring, or band, to it, as shown in Figure 10.21, in order to block the beam radiation throughout the

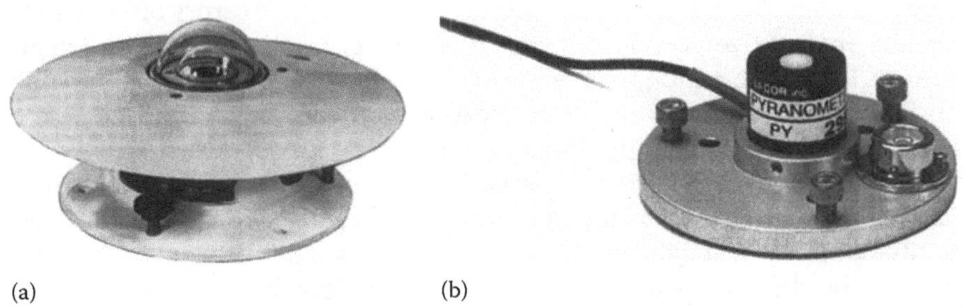

(a) (b)

FIGURE 10.20
Typical commercially available pyranometers with (a) thermal detector and (b) photovoltaic detector.

FIGURE 10.21
A pyranometer with a shadow ring to measure sky diffuse radiation.

TABLE 10.6

Shadow Band Correction Factors Due to Anisotropy

	k_T									
Solar Altitude Angle	**0.0**	**0.1**	**0.2**	**0.3**	**0.4**	**0.5**	**0.6**	**0.7**	**0.8**	**0.9**
<20°	0.0	0.0	0.0	0.0	0.015	0.06	0.14	0.23	0.24	0.24
20°–40°	0.0	0.0	0.0	0.0	0.006	0.05	0.125	0.205	0.225	0.225
40°–60°	0.0	0.0	0.0	0.0	0.003	0.045	0.115	0.175	0.205	0.205
60°+	0.0	0.0	0.0	0.0	0.0	0.035	0.09	0.135	0.17	0.17

Source: Gautier et al. [22].

day. The position of the shadow ring is adjusted periodically as the declination changes. Since the shadow ring obstructs some diffuse radiation from the pyranometer, correction factors must be applied.

Geometric correction factors (GCFs) that account for the part of the sky obstructed by the shadow ring can be easily calculated. However, a GCF assumes isotropic sky, which results in errors because of the circumsolar anisotropy. Eppley Corp. recommends additional correction factors to account for anisotropy as +7% for clear sky, +4% for partly cloudy condition, and +3% for cloudy sky. Mujahid and Turner [21] determined that these correction factors gave less than 3% errors on partly cloudy days but gave errors of −11% for clear-sky conditions and +6% on overcast days. They suggested correction factors due to anisotropy as tabulated in Table 10.6, which reduce the errors to less than ±3% [22]. It must be remembered that these correction factors are in addition to the GCFs. Recently, a Sun occulting disk has been employed for shading the direct Sun.

To measure direct solar radiation, a pyrheliometer places the detector at the base of a long tube. This geometry restricts the sky view of the detector to a small angle of about 5°. When the tube points toward the Sun, the detector measures the beam solar radiation and a small part of the diffuse solar radiation within the view angle. Figure 10.22 shows the geometry of a pyrheliometer sky occulting tube.

In this figure, the opening half angle is

$$\theta_o = \tan^{-1}\frac{R}{L} \tag{10.48}$$

The slope angle is

$$\theta_p = \tan^{-1}\left[\frac{(R-r)}{L}\right] \tag{10.49}$$

The limit half angle is

$$\theta = \tan^{-1}\left[\frac{(R+r)}{L}\right] \tag{10.50}$$

The field of view is $2\theta_o$. The World Meteorological Organization (WMO) recommends the opening half angle θ_o to be 2.5° [20] and the slope angle θ_p to be 1°.

Continuous tracking of the Sun is required for the accuracy of the measurements. This is obtained by employing a tracking mechanism with two motors, one for altitude and the other for azimuthal tracking. The view angle of a pyrheliometer is significantly greater than the angle subtended by the solar disk (about 0.5°). Therefore, the measurements using

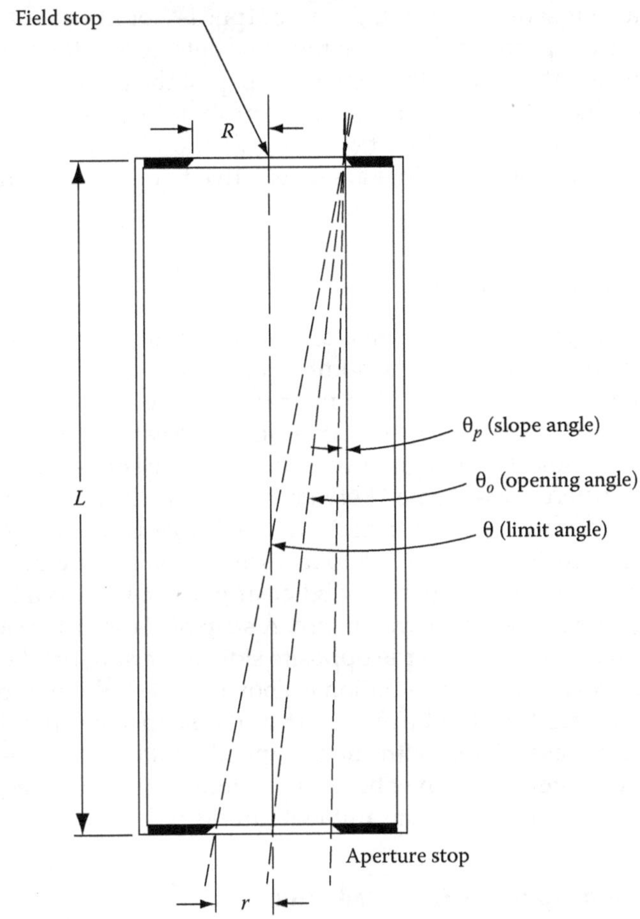

FIGURE 10.22
Geometry of a pyrheliometer sky occulting tube.

a pyrheliometer include the beam and the circumsolar radiation. These measurements may present a problem in using the data for central receiver systems that use only direct beam radiation. However, this is not a significant problem for parabolic trough concentrators that in most cases have a field of view on the order of 5°.

10.7.2 Detectors for Solar Radiation Instrumentation

Solar radiation detectors are of four basic types [19,20]: thermomechanical, calorimetric, thermoelectric, and photoelectric. Of these, thermoelectric and photoelectric are the most common detectors in use today.

A *thermoelectric detector* uses a thermopile that consists of a series of thermocouple junctions. The thermopile generates a voltage proportional to the temperature difference between the hot and cold junctions, which, in turn, is proportional to the incident solar radiation.

Photovoltaic detectors normally use silicon solar cells measuring the short circuit current. Such detectors have the advantage of being simple in construction. Because heat transfer is not a consideration, they do not require clear domes or other convection suppressing devices. They are also insensitive to tilt, as the output is not affected by natural convection. One of the principal problems with photovoltaic detectors is their spectral selectivity. Radiation with wavelengths greater than the band gap of the photovoltaic detector cannot be measured. Silicon has a band gap of 1.07 eV corresponding to a wavelength of 1.1 μm. A significant portion of the infrared part of solar radiation has wavelengths greater than 1.1 μm. Therefore, photovoltaic detectors are insensitive to changes in the infrared part of the solar spectrum.

10.7.3 Measurement of Sunshine Duration

The time duration of bright sunshine data is available at many more locations in the world than the solar radiation. That is why a number of researchers have used these data to estimate the available solar radiation. Two instruments are widely used to measure the sunshine duration. The device used by the U.S. National Weather Service (NWS) is called a *sunshine switch*. It is composed of two photovoltaic cells—one shaded, the other not. During daylight, a potential difference is created between the two cells, which in turn operates the recorder. The intensity level required to activate the device is that just sufficient to cast a shadow. The other device commonly used to measure the sunshine duration is called the *Campbell–Stokes sunshine recorder*. It uses a solid clear glass sphere as a lens to concentrate the solar beam on the opposite side of the sphere. A strip of standard treated paper marked with time graduations is mounted on the opposite side of the sphere where the solar beam is concentrated. Whenever the solar radiation is above a threshold, the concentrated beam burns the paper. The length of the burned part of the strip gives the duration of bright sunshine. The problems associated with the Campbell–Stokes sunshine recorder include the uncertainties of the interpretation of burned portions of the paper, especially on partly cloudy days, and the dependence on the ambient humidity.

10.7.4 Measurement of Spectral Solar Radiation

Spectral solar radiation measurements are made with spectroradiometers. Full spectrum scanning is difficult, requires constant attention during operation, and is therefore expensive. Zerlaut [20] has described a number of solar spectroradiometers. These instruments consist basically of a monochromator, a detector–chopper assembly, an integrating sphere, and a signal conditioning/computer package. They have the capability of measuring solar radiation in the wavelength spectrum of 280–2,500 nm.

10.7.5 National Solar Radiation Database

Using the System Advisor Model and other tools for predicting the performance of solar energy systems requires detailed solar energy data representative of many locations. NREL has compiled the National Solar Radiation Database (https://nsrdb.nrel.gov). The National Solar Radiation Data Base (NSRDB) [23,24] is a complete collection of hourly and half-hourly values of the three most common measurements of solar radiation—global horizontal, direct normal, and diffuse horizontal irradiance—and meteorological data. These data have been collected at a sufficient number of locations and temporal and spatial scales to accurately represent regional solar radiation climates. Although it was originally focused on the United States, it contains an increasing number of international locations. Data sets can be accessed at https://nsrdb.nrel.gov/data-sets. But more commonly, solar energy system designers will use SAM or other tools and specify a location that will utilize the NSRDB data.

Problems

10.1 Calculate the declination, the zenith angle, and the azimuth angle of the Sun for New York City (latitude 40.77°N) on October 1 at 2:00 p.m. solar time.

10.2 A solar energy system in Gainesville, Florida, requires two rows of collectors facing south and tilted at a fixed 30° angle. Find the minimum normalized distance at which the second row should be placed behind the first row for no shading at noon at winter solstice. What percentage of the second row is shaded on the same day at 9:00 a.m. solar time?

10.3 Find the sunrise and sunset times for a location of your choice on September 1.

10.4 Construct a table of hourly Sun angles for the 15th day of each month for a location of your choice. Also show the sunrise and sunset times for those days.

10.5 Referring to Figure 10.19, prove Equation 10.44 for the angle of incidence. (*Hint*: Use the direction cosines of the Sun-ray vector and a vector normal to the tilted surface to find the angle between them. Dot product of two unit vectors gives the cosine of the angle between them.)

10.6 Determine the following for a south-facing surface at 30° slope in Gainesville, Florida (latitude = 29.68°N, longitude = 82.27°W), on September 21 at noon solar time:

a. Zenith angle
b. Angle of incidence
c. Beam radiation
d. Diffuse radiation
e. Reflected radiation
f. Total radiation
g. Local time

10.7 Show that the hourly averaged, extraterrestrial radiation for a given hour is the same, to within 1%, as instantaneous radiation at the hour's midpoint.

10.8 Prepare shadow maps for point P on the Sun-path diagrams for 35°N and 40°N for the following three geometries shown in a, b, and c. Determine the hours of shading that occur each month.

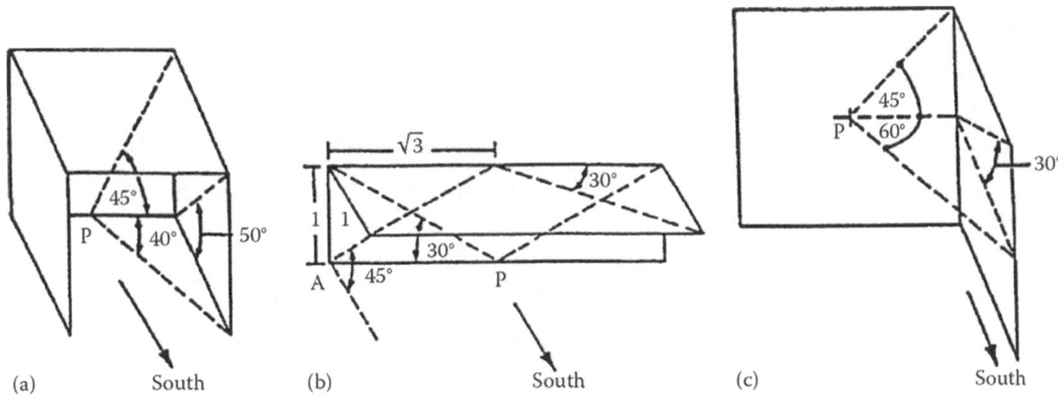

10.9 Repeat Problem 10.8c if the surface containing point P faces due west instead of due south for a 40°N location.

10.10 Calculate the incidence angle at noon and 9 a.m. on a fixed flat-plate collector located at 40°N latitude and tilted 70° up from the horizontal. Find i for June 21 and December 21.

10.11 a. If the surface in Problem 10.10 faces S 45°E, what are the incidence angles?

b. If the collector in Problem 10.10 has a cylindrical surface, what are the incidence angles on June 21 and December 21?

10.12 Using a one-term Fourier cosine series, develop an empirical equation for solar declination as a function of day number counted from January 1 (see Figure 10.7). Note that the maximum declination of 23.45° occurs on June 21, which is day number 172.

10.13 Derive an equation for the lines of constant declination in a Sun-path diagram, for example, Figure 10.10d. Check your equation by plotting a few declination lines on a piece of polar coordinate graph paper.

10.14 Derive Equation 10.31 relating profile angle γ to altitude angle α and azimuth angle a.

10.15 Based on Equation 10.44, what value of β would result in the annual minimum value of the incidence angle i? Note that this tilt angle would result in the maximum collection of beam radiation on a fixed, flat, south-facing surface. *Hint:* Use a double integration procedure.

10.16 At what time does the Sunset in Calcutta (23°N) on May 1 and December 1?

10.17 What is the true solar time in Sheridan, Wyoming (107°W), at 10:00 a.m. Mountain Daylight Time on June 10? What is the true solar time at 10:00 a.m. Mountain Standard Time on January 10?

10.18 Derive an expression for the minimum allowable distance between east–west rows of solar collectors that will assure no shading of one row by the row immediately to the south. Use the law of sines and express the result in terms of the collector tilt and face length and the controlling value of the solar profile angle.

References

1. Howell, J.R., Siegel, R., and Menguc, M.P., 2010. *Thermal Radiation Heat Transfer*, McGraw-Hill Book Co., New York.
2. Kreith, F., 1962. *Radiation Heat Transfer for Spacecraft and Solar Power Plant Design*, International Textbook Co., Scranton, PA.
3. Kreith, F., Manglik, R.M., and Bohn, S., 2011. *Principles of Heat Transfer*, 7th ed., Cengage Learning, Boston, MA.
4. Sparrow, E.M., and Cess, R.D., 1978. *Radiation Heat Transfer*, Wadsworth Publ. Co., Belmont, CA.
5. Threlkeld, J.L., and Jordan, R.C., 1958. Direct radiation available on clear days, *Transactions of ASHRAE* 64, 45.
6. Kreider, J.F., and Kreith, F., 1982. *Solar Heating and Cooling*, Hemisphere Publishing Corp., Washington, DC.

7. List, R.J., 1949. *Smithsonian Meteorological Tables,* 6th ed., Smithsonian Institution Press, Washington, DC, pp. 442–443.
8. University of Oregon. *Solar Radiation Monitoring Laboratory,* https://solardat.uoregon.edu/SoftwareTools.html.
9. Frohlich, C., Geist, J., Kendall, J., and Marchgraber, R.M., 1973. The third international comparison of pyrheliometers and a comparison of radiometric scales, *Solar Energy* 14, 157–166.
10. Quinlan, F.T. ed., 1977. *SOLMET Vol. 1: Hourly Solar Radiation Surface Meteorological Observations,* NOAA, Asheville, NC.
11. Quinlan, F.T. ed., 1979. *SOLMET Vol. 2: Hourly Solar Radiation Surface Meteorological Observations,* NOAA, Asheville, NC.
12. Kimura, K., and Stephenson, D.G., 1969. Solar radiation on cloudy days, *Transactions of ASHRAE* 75, 227–234.
13. Iqbal, M., 1979. Correlation of average diffuse and beam radiation with hours of bright sunshine, *Solar Energy* 23, 169.
14. Thekaekara, M.P., 1973. Solar energy outside the Earth's atmosphere, *Solar Energy* 14, 109–127.
15. Islam, M.R., 1994. Evolution of methods for solar radiation mapping using satellite data, *RERIC International Energy Journal* 16(2), 103–113.
16. Build It Solar, 2009, Solar radiation on collection program, https://www.builditsolar.com/Tools/RadOnCol/radoncol.htm. Last accessed January 2025.
17. Wilcox, S., and Marion, W., April/May 2008. User's manual for TMY3 data sets, Technical Report NREL/TP-581-43156, National Renewable Energy Laboratory, Golden, CO.
18. NREL, National solar radiation data base, https://nsrdb.nrel.gov. Last accessed January 2025.
19. Iqbal, M., 1983. *An Introduction to Solar Radiation,* Academic Press, New York.
20. Zerlaut, G., 1989. Solar radiation instrumentation, in: Roland L.H. ed., *Solar Resources,* MIT Press, Cambridge, pp. 173–308.
21. Mujahid, A., and Turner, W.D., 1979. Diffuse sky measurement and model, ASME Pap. No. 79-WA/Sol-5.
22. Gautier, C., Diak, G., and Masse, S., 1980. A simple physical model to estimate incident solar radiation at the surface from GOES satellite data, *Journal of Applied Meteorology* 19(8), 1005.
23. Maxwell, E.L., 1998. METSTAT—The solar radiation model used in the production of the national solar radiation data base (NSRDB), *Solar Energy* 62, 263–279.
24. NSRDB, 1992. *Volume 1: User's Manual: National Solar Radiation Data Base* (1961–1990), *Version 1.0,* National Renewable Energy Laboratory, Golden, CO.

11

Photovoltaics

I'd put my money on the sun and solar energy. What a source of power! I hope we don't have to wait until oil and coal run out before we tackle that. I wish I had more years left.

Thomas Edison

Photovoltaic conversion is the direct conversion of sunlight into electricity with no intervening heat engine. Photovoltaic devices are solid-state devices; therefore, they are rugged and simple in design and require very little maintenance. Perhaps the biggest advantage of solar photovoltaic devices is that they can be constructed as stand-alone systems to give outputs from microwatts to megawatts. That is why they have been used as power sources for calculators, watches, water pumping, remote buildings, communications, satellites and space vehicles, and even gigawatt-scale power plants. Photovoltaic panels can even be made to form components of building skin, such as roof shingles and wall panels. With such a vast array of applications, the demand for photovoltaics is increasing every year. According to PV Magazine, the market for photovoltaic solar cells has grown exponentially, increasing from 3 GW in 2004 to 1.7 TW in 2023. The global PV market size was $317 billion in 2023 [1]. An example of an early photovoltaic residential rooftop system is shown in Figure 11.1.

In the early days of solar cells in the 1960s and 1970s, more energy was required to produce a cell than it could ever deliver during its lifetime. Since then, dramatic improvements have taken place in the efficiencies and manufacturing methods. The National Renewable Energy Laboratory (NREL) gives the latest energy payback times for utility-scale PV systems in the U.S. as 0.5–1.2 years [2], while panel lifetimes increased to over 25 years. The price of photovoltaic modules has come down to under $0.30 per peak watt [3], and the average levelized cost of energy from utility-scale solar PV is $61/MWh [4].

Historically, the photoelectric effect was first noted by Becquerel in 1839 when light was incident on an electrode in an electrolyte solution [5]. Adams and Day first observed the effect in solids in 1877 while working with selenium. Early work was done with selenium and copper oxide by pioneers such as Schottky, Lange, and Grandahl. In 1954, researchers at RCA and Bell Laboratories reported achieving efficiencies of about 6% by using devices made of *p*- and *n*-type semiconductors. The space race between the United States and the Soviet Union resulted in dramatic improvements in photovoltaic devices. Bube [6] gives a review of the early developments in photovoltaic conversion.

DOI: 10.1201/9781003586616-11

FIGURE 11.1
Kreith residence, early 3-kW Sanyo photovoltaic system. (Courtesy of Chris Klinga, Lighthouse Solar, Boulder, CO. With permission.)

11.1 Semiconductors

A basic understanding of the atomic structure is quite helpful in understanding the behavior of semiconductors and their use as photovoltaic energy conversion devices. Any fundamental book on physics or chemistry generally gives adequate background for basic understanding. Bube [6] presents an in-depth treatment of a number of topics in semiconductor physics.

For any atom, the electrons arrange themselves in orbitals around the nucleus so as to result in the minimum amount of energy. Table 11.1 shows the distribution of the electrons

TABLE 11.1

Electronic Structure of Atoms

Principal Quantum Number *n*				**1**	**2**		**3**		**4**	**5**	
Azimuthal Quantum Number *l*				**0**	**0**	**1**	**0**	**1**	**2**	**0**	**1**
Letter Designation of State				**1s**	**2s**	**2p**	**3s**	**3p**	**3d**	**4s**	**4p**
Z	**Symbol**	**Element**	**V_i (V)**								
1	H	Hydrogen	13.60	1							
2	He	Helium	24.58	2							
3	Li	Lithium	5.39		1						
4	Be	Beryllium	9.32		2						
5	B	Boron	8.30		2	1					
6	C	Carbon	11.26	Helium core	2	2					
7	N	Nitrogen	14.54		2	3					
8	O	Oxygen	13.61		2	4					
9	F	Fluorine	17.42		2	5					
10	Ne	Neon	21.56		2	6					
11	Na	Sodium	5.14				1				
12	Mg	Magnesium	7.64				2				
13	Al	Aluminum	5.98				2	1			
14	Si	Silicon	8.15	Neon core		2	2				
15	P	Phosphorus	10.55				2	3			
16	S	Sulfur	10.36				2	4			
17	Cl	Chlorine	13.01				2	5			
18	A	Argon	15.76				2	6			
19	K	Potassium	4.34							1	
20	Ca	Calcium	6.11							2	
21	Sc	Scandium	6.56						1	2	
22	Ti	Titanium	6.83						2	2	
23	V	Vanadium	6.74						3	2	
24	Cr	Chromium	6.76						5	1	
25	Mn	Manganese	7.43						5	2	
26	Fe	Iron	7.90						6	2	
27	Co	Cobalt	7.86						7	2	
28	Ni	Nickel	7.63	Argon core				8	2		
29	Cu	Copper	7.72						10	1	
30	Zu	Zinc	9.39						10	2	
31	Ga	Gallium	6.00						10	2	1
32	Ge	Germanium	7.88						10	2	2
33	As	Arsenic	9.81						10	2	3
34	Se	Selenium	9.75						10	2	4
35	Br	Bromine	11.84						10	2	5
36	Kr	Krypton	14.00						10	2	6

Source: From Moore [7]. With permission.

in various orbitals in light elements [7]. In elements that have electrons in multiple orbitals, the innermost electrons are more strongly bound to the positively charged nucleus and, therefore, require the maximum amount of externally imparted energy to overcome the attraction of the nucleus and become free. Electrons in the outermost orbitals are partially shielded from the nucleus by the inner electrons, are less bound to the atom, and provide the means for neighboring atoms to interact. In some cases, atoms may transfer electrons in order to fill an outermost orbital, resulting in positively and negatively charged ions that become attached by the force of attraction of the charges, thus forming *ionic bonds*. In other cases, atoms may share electrons to fill the outermost orbital forming what is known as a *covalent bond*.

As atoms are brought together, the outermost electron states, or orbitals, overlap and mix to create energy bands that are no longer isolated around individual atoms but, rather, are extended throughout the bulk of the material. The highest filled band of these states, at an absolute temperature $T=0$, is called the *valence band*. The band of states above the valence band is called the *conduction band*. At an absolute temperature of $T=0$, the conduction band is empty, and because it carries no electrons, the material acts like an electric insulator. The difference in the energy of an electron at the top of the valence band and the bottom of the conduction band is called the *band gap* or the forbidden gap and represents the energy that must be absorbed to allow conduction.

The band gap of a material plays a large role in determining the material's electrical conductivity. Of note, a full valence band cannot support the flow of electricity. Consequently, materials with high band gaps (>3 eV) tend to be insulators at room temperature because there is not enough thermal energy to excite a significant number of electrons from the filled valence band into the conduction band, where they can participate in the flow of electricity. As the band gap is reduced, the number of electrons that can be thermally excited into the conduction band increases, as does the electrical conductivity of the material. These materials with improved conductivity are called semiconductors (see Figure 11.2) and tend

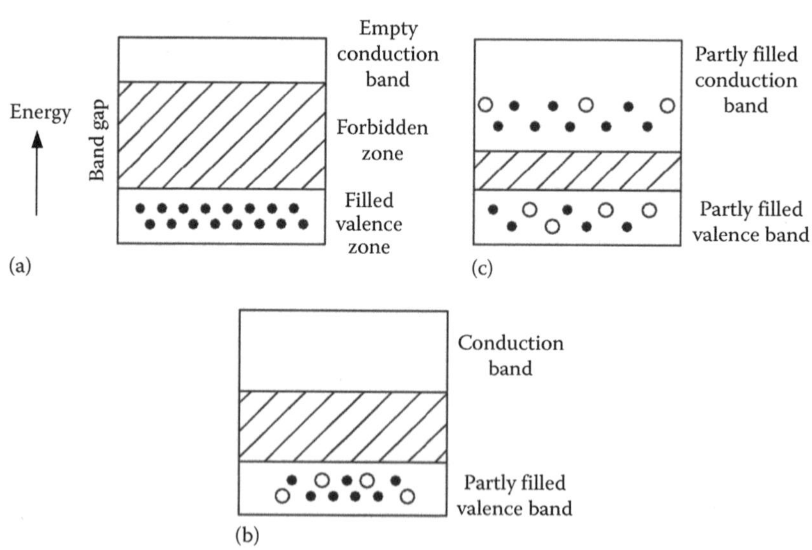

FIGURE 11.2
Electrical conduction is described in terms of allowed and forbidden energy bands. Band gap for insulators (a) is the highest, followed by semiconductors (b) and metals (c), respectively.

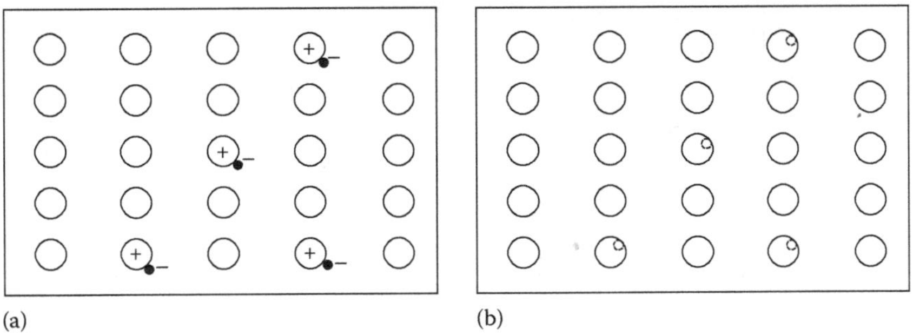

FIGURE 11.3
Representation of *n*- and *p*-type semiconductors: (a) *n*-type showing "excess" electrons as dots; (b) *p*-type showing "excess" positive holes as ο.

to have band gaps of 1–2 eV and, perhaps most importantly for photovoltaic devices, align well with the energy of photons in the solar spectrum. Materials with overlapping conduction and valence bands do not have a characteristic band gap, require no thermal energy to generate a large population of mobile electrons, conduct electricity with relative ease, and are known as metals.

Pure semiconductors, as discussed above, are called *intrinsic semiconductors*, while semiconductors doped with very small amounts of impurities are called *extrinsic semiconductors*. If the dopant atom has more electrons than the intrinsic atom it is replacing, the extra electrons tend to be very loosely bound and can be easily excited into the conduction band, even at room temperature. This results in a material with a controllable number of free electrons in the conduction band and is called an *n*-type semiconductor (*n* indicating the negative charge of the electron). For example, each silicon atom contributes four electrons to the valence band. Atoms of pure silicon arrange themselves in such a way that, to form a stable structure, each atom shares two electrons with each neighboring atom with covalent bonds. If phosphorous, which has five valence electrons (one more than Si), is introduced as an impurity in silicon, the doped material seems to have excess electrons even though it is electrically neutral. Such a doped material is called *n*-type silicon. If, on the other hand, silicon is doped with boron, which has three valence electrons (one less than Si), there seems to be a positive hole (missing electron) in the structure, even though the doped material is electrically neutral. Such material is called *p*-type silicon (*p* denoting the positive charge of the hole) (see Figure 11.3). Thus, *n*- and *p*-type semiconductors make it easier to increase, and control, the number of electrons and holes in the semiconductor because the energy required to create partially filled bands is now related to the energy level of the impurity electron state which is typically much less than the band gap energy. The ability to control the conductivity and type (*p* or *n*) of semiconductors capable of absorbing solar radiation is the foundation of modern photovoltaic devices.

11.1.1 *p–n* Junction

As explained earlier, an *n*-type material has some impurity atoms with more electrons than the rest of the semiconductor atoms. On the other hand, a *p*-type material has some impurity atoms with fewer electrons than the rest of the semiconductor atoms. The scenario where a *p*- and an *n*-type of material are joined together, as shown in Figure 11.4, is called

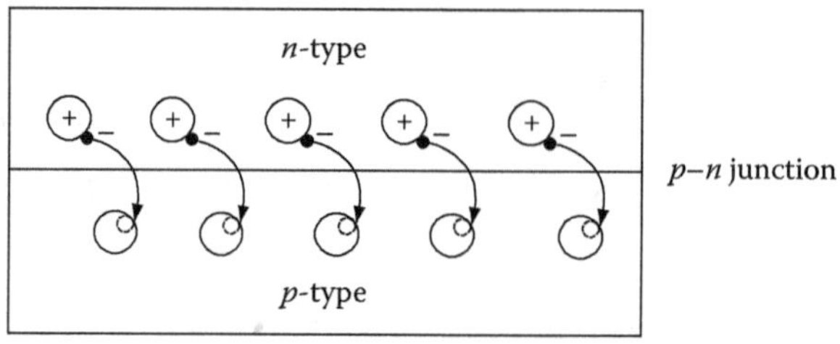

FIGURE 11.4
"Excess" electrons from *n* material jump to fill "excess" holes on the *p*-side of a *p–n* junction, leaving the *n*-side of the junction positively charged and the *p*-side negatively charged.

a *p-n* junction. As soon as the two materials are joined, the "excess" electrons from the *n* layer diffuse to fill the "holes" in the *p* layer. Therefore, close to the junction, the material has positive charges on the *n*-side and negative charges on the *p*-side. Eventually, equilibrium is established where the negative charges on the *p*-side restrict the movement of additional electrons from the *n*-side to the *p*-side and a potential barrier, or built-in junction, is established at the interface.

The net result of this built-in potential barrier is that the flow of electrons is impeded in one direction but not in the other. If a positive voltage is applied to the *p*-side of the junction, relative to the *n*-side, the barrier created by the transfer of electrons is reduced and current will flow more freely. This condition is known as forward bias. Conversely, if a negative voltage is applied to the *p*-side of the junction, relative to the *n*-side, the built-in potential barrier is increased and the current will flow less freely. This condition is known as reverse bias. This asymmetrical behavior to current flow is a key building block in many semiconductor devices, and the underlying device is known as a diode. As will be shown below, the general electrical behavior of the diode shows up in the current-voltage relationship for photovoltaic cells.

11.1.2 Photovoltaic Effect

When a photon of light is absorbed by a valence electron of an atom, the energy of the electron is increased by the amount of energy of the photon. If the energy of the photon is equal to or greater than the band gap of the semiconductor, the electron with excess energy will jump into the conduction band where it can move freely. If, however, the photon energy is less than the band gap, the electron will not have sufficient energy to jump into the conduction band and the photon will typically pass through the material. If the absorbed photon had more energy than the band gap, the excess energy over the band gap is typically lost as heat and the electron quickly moves to the bottom of the conduction band. This thermalization process represents energy that will not be captured by the flow of electricity and, consequently, can result in a significant reduction in the conversion efficiency of photovoltaic devices. The design and development of multi-junction devices with multiple band gaps, described below, is the primary approach used to reduce this loss mechanism.

As explained in Chapter 10, energy contained in a single photon E_p is given by

$$E_p = h\nu \tag{11.1}$$

where
h is Planck's constant (6.625×10^{-34} J-s)
v is the frequency, which is related to the wavelength λ and the speed of light c by

$$v = \frac{c}{\lambda}$$

Therefore,

$$E_p = \frac{hc}{\lambda} \quad (11.2)$$

For silicon, which has a band gap of 1.11 eV, the following example shows that photons of wavelength 1.12 μm or less can be absorbed and contribute to the flow of photo-generated current. This wavelength range represents a major part of the solar radiation spectrum, making Si a suitable material for use in a photovoltaic device. Table 11.2 [8] lists other candidate semiconductor materials for photovoltaic cells along with their band gaps.

TABLE 11.2
Energy Gap for Some Candidate Materials for Photovoltaic Cells

Material	Band Gap (eV)
Si	1.11
SiC	2.60
$CdAs_2$	1.00
CdTe	1.44
CdSe	1.74
CdS	2.42
$CdSnO_4$	2.90
GaAs	1.40
GaP	2.24
Cu_2S	1.80
CuO	2.00
Cu_2Se	1.40
$CuInS_2$	1.01
$CuInTe_2$	0.90
InP	1.27
In_2Te_3	1.20
In_2O_3	2.80
Zn_3P_2	1.60
ZnTe	2.20
ZnSe	2.60
AlP	2.43
AlSb	1.63
As_2Se_3	1.60
Sb_2Se_3	1.20
Ge	0.67
Se	1.60

Source: From Garg [8]. With permission.

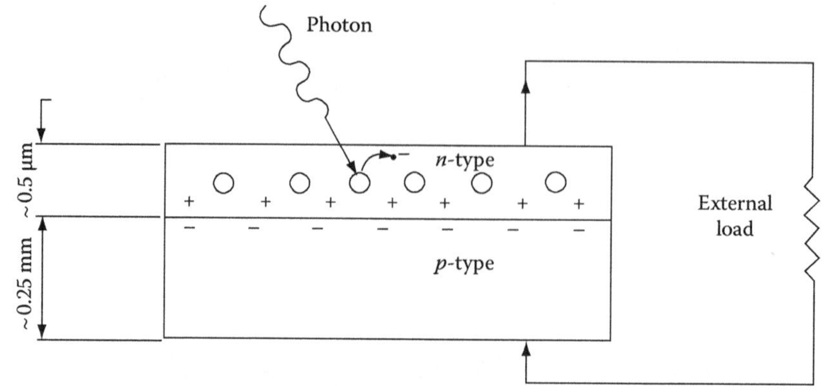

FIGURE 11.5
Schematic of a photovoltaic device.

Example 11.1

Calculate the longest wavelength of light capable of forming an electron-hole pair in silicon.

Solution:

The band gap energy of silicon is 1.11 eV. From Equation 11.2, we can write

$$\lambda = \frac{hc}{E_p} = \frac{hc}{\mathrm{BG}}$$

where BG is the band gap energy in Joules.

For $c = 3\times10^8$ m/s, $h = 6.625\times10^{-34}$ J-s, and 1 eV $= 1.6\times10^{-19}$ J, the earlier equation gives the required wavelength as

$$\lambda = \frac{\left(6.625\times10^{-34}\ \text{J-s}\right)\left(3\times10^{8}\ \text{m/s}\right)}{\left(1.11\ \text{eV}\right)\left(1.6\times10^{-19}\ \text{J/eV}\right)} = 1.12\ \mu\text{m}$$

As photons are absorbed within the semiconductor, electrons are excited to the conduction band, leaving an empty hole state behind. This electron-hole pair can diffuse to the *p*-*n* junction where it encounters the built-in potential of the *p*-*n* junction. This potential will drive the photo-excited electrons from the *p*-side to the *n*-side and, the photo-excited holes from the *n*-side to the *p*-side. This results in a photovoltage across the *p*-*n* junction, and this phenomenon is known as the photovoltaic effect. If the photovoltaic device is placed in a circuit, as in Figure 11.5, the photo-generated current will flow, and the optical power of the incident light source will have been converted to electrical power within the circuit. Ideally, all the photo-excited carriers will participate in the flow of electricity but there are material- and device-related loss mechanisms that ultimately reduce the conversion efficiency.

Example 11.2

A monochromatic red laser beam emitting 1×10^{-3} W at a wavelength of 638 nm is incident on a silicon solar cell. Find

a. The number of photons per second incident on the cell
b. The maximum possible efficiency of conversion of this laser beam to electricity

Solution:

a. The laser beam irradiance (I_p), or power, is equal to the energy of all the photons in it per second. If the number of photons per second is N_{ph}, then

$$I_p = N_{ph} \cdot E_p$$

$$1 \times 10^{-3}\ \text{W} = N_{ph} \cdot E_p$$

$$E_p = \frac{hc}{\lambda} = \frac{(6.625 \times 10^{-34}\ \text{J-s}) \cdot 3 \times 10^8\ \text{m/s}}{638 \times 10^{-9}\ \text{m}} = 3.12 \times 10^{-19}\ \text{J/photon}$$

$$\therefore N_{ph} = \frac{1 \times 10^{-3}\ \text{J/s}}{3.12 \times 10^{-19}\ \text{J/photon}} = 3.21 \times 10^{15}\ \text{photons/s}$$

b. Assuming that each photon is able to generate one electron-hole pair, a total number of N_{ph} electron-hole pairs per second will be generated. Therefore, the maximum possible efficiency is

$$\eta_{max} = \frac{(N_{ph}) \cdot (BG)}{(N_{ph}) \cdot E_p} = \frac{BG}{E_p} = \frac{1.11 \times 1.6 \times 10^{-19}\ \text{J}}{3.12 \times 10^{-19}\ \text{J}} = 0.569 \quad \text{or} \quad 56.9\%$$

From the earlier examples, it is clear that for a silicon solar cell, none of the photons of the sunlight with wavelengths greater than 1.12 μm will produce any electricity. However, photons of sunlight with a wavelength of 1.12 μm may be converted to electricity at a maximum efficiency of 100%, while photons at shorter wavelengths will be converted at lower efficiencies. The overall maximum efficiency of a cell can be found by integrating the efficiency at each wavelength over the entire solar spectrum:

$$\eta = \frac{\int \eta_\lambda I_\lambda d\lambda}{\int I_\lambda d\lambda} \tag{11.3}$$

In addition, other factors such as probability of electron-hole recombination reduce the theoretical maximum achievable efficiency of a solar cell. Figure 11.6 shows a comparison of the maximum energy conversion of cells using different materials as well as the optimum band gap for space solar cells, which is on the order of 1.5 eV.

11.2 Analysis of Photovoltaic Cells

This section presents an electrical analysis of photovoltaic cells, which will be useful in the design of photovoltaic devices for various applications. The physics leading to the expressions for the number density of electrons and holes in n and p materials at a temperature T will not be presented here. For such details, the reader is referred to Angrist [5] and Bube [6].

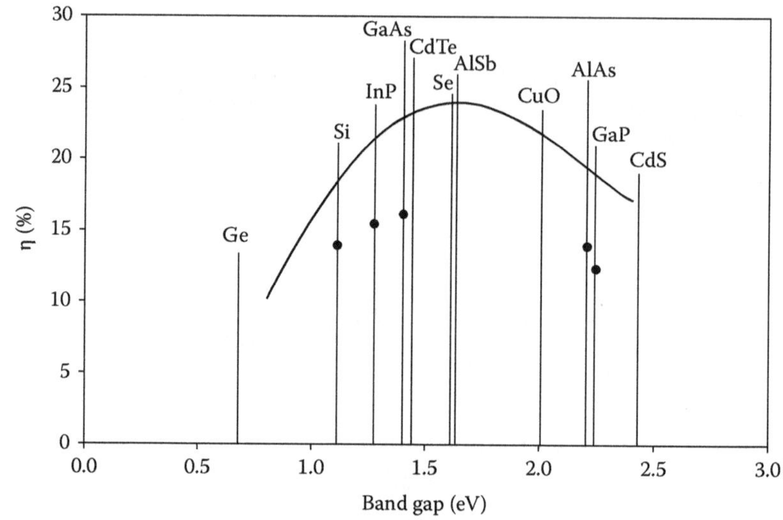

FIGURE 11.6
The maximum solar energy conversion efficiency as a function of the energy gap of the semiconductor. The curve has been calculated for an ideal junction outside the atmosphere.

It would suffice to point out here that the net current generated by a photovoltaic cell is a combination of several, often competing, currents that are a function of the incident light intensity and the potential change, or voltage, across the p-n junction. One such current is light-induced recombination current density J_r, which originates from the injection of carriers across the *p-n* junction, and is given by

$$J_r = J_o \exp\left(\frac{e_o V}{kT}\right) \tag{11.4}$$

where
e_o is the charge of an electron $= 1.602 \times 10^{-19}$ C or J/V
k is Boltzmann's constant $= 1.381 \times 10^{-23}$ J/K
J_o is called the dark, or reverse saturation, current density.

In the dark, the photovoltaic cell behaves like a diode, where current flows easily in one direction but not in the other. The following *J-V* relationship represents the electrical behavior of an idealized diode.

$$J_j = J_r - J_o = J_o\left[\exp\left(\frac{e_o V}{kT}\right) - 1\right] \tag{11.5}$$

where V is the applied voltage across the *p-n* junction and J_o is the reverse saturation current density. In "forward bias," the applied voltage is positive with respect to the *p*-type side of the junction, resulting in a reduced potential barrier at the junction, and the flow of electrons from the *n*-type side to the *p*-type side of the junction and holes from the *p*-type side to the *n*-type side. In "reverse bias," the applied voltage is negative with respect to the *p*-type side, the junction potential is increased, and current is restricted to relatively small levels controlled by tunneling and recombination. As we will see below, this "dark" current is opposite in direction to the flow of light-generated current and represents a loss mechanism for the photovoltaic cell.

In the light, photo-excited electrons in the *p*-type material, and photo-excited holes in the *n*-type material are driven across the *p-n* junction resulting in a light-generated current component that is proportional to the incident light intensity. Because the current generated in a photovoltaic device is proportional to the area of the cell (A), it is appropriate to analyze current in terms of the current density J (current per unit area) instead of the current (I). The relationship between the two is

$$I = J \cdot A \tag{11.6}$$

The flow of electrons and holes associated with this light-generated current results in a reduction of the built-in potential drop across the *p-n* junction or, in other words, a "forward bias condition." Establishing the convention that the light-generated current is positive, and noting that the voltage-driven dark current is opposite in direction results in the following expression for the photovoltaic cell *J-V* relationship

$$J_L = J_{sc} - J_j = J_{sc} - J_o\left[\exp\left(\frac{e_oV}{kT}\right) - 1\right] \tag{11.7}$$

where J_{sc} is the short-circuit current density.

Referring to Figure 11.7, it is clear that the current generated in the cell has two parallel paths: one through the junction, J_j, and the other, J_L, through the external resistance, R_L. Figure 11.7 shows an equivalent circuit of a photovoltaic cell.

For short circuit, $V=0$ and $J_L=J_{sc}$.

For open circuit, $J_L=0$ and $V=V_{oc}$ which gives

$$0 = J_{sc} - J_o\left[\exp\left(\frac{e_oV_{oc}}{kT}\right) - 1\right] \tag{11.8}$$

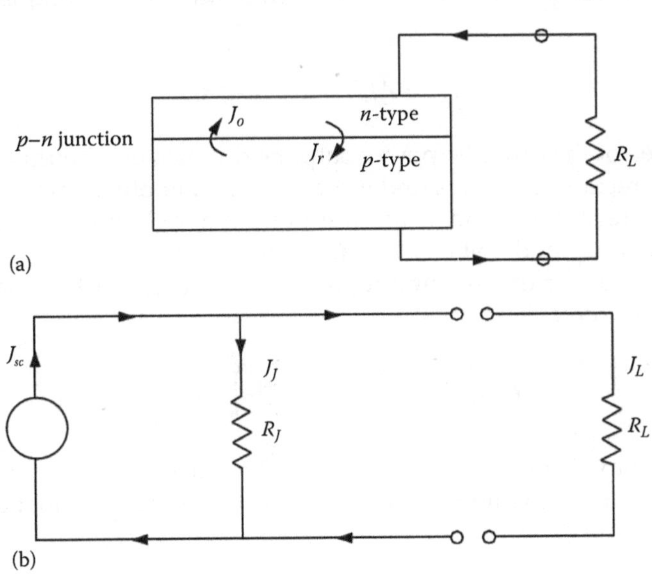

FIGURE 11.7
Equivalent circuit of a photovoltaic cell. (a) *p–n* junction. (b) Electrical circuit.

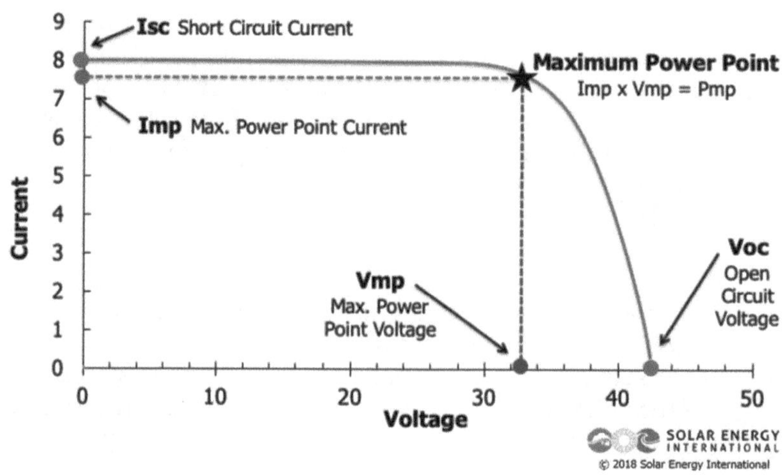

FIGURE 11.8
Typical current (in amps), voltage (in volts) and power characteristics (in Watts) of a solar cell or module. (Courtesy of Solar Energy International.)

or

$$V_{oc} = \frac{kT}{e_o} \ln\left(\frac{J_{sc}}{J_o} + 1\right)$$

Figure 11.8 shows a typical current-voltage (I-V) performance curve of a solar cell. Solar cells and modules are tested at Standard Test Conditions (STC), which are 1,000 W/m^2 of solar irradiance and a cell temperature of 25°C (77°F). The y-intercept is the short-circuit current (at zero voltage), and the x-intercept is the open-circuit voltage (at zero current). The power output is the product of the load current and voltage and is a function of the load resistance

$$P_L = AJ_LV = I_LV = I_L^2R_L \tag{11.9}$$

where A is the area of the cell. The power output exhibits a maximum represented by the area of the largest rectangle that fits under the I-V curve. Photovoltaic panels are generally controlled to operate at the maximum power point marked on the curve. The maximum power point current, I_{mp}, and voltage, V_{mp}, are also shown.

To find the condition for the maximum power output (P_{mp}), differentiate P with respect to V and equate it to zero:

$$\exp\left(\frac{e_oV_{mp}}{kT}\right)\left(1+\frac{e_oV_{mp}}{kT}\right) = 1+\frac{J_{sc}}{J_o} \tag{11.10}$$

where V_{mp} stands for voltage at maximum power. The current density at maximum power condition $J_{L,mp}$ and the maximum power P_{mp} can be found in Equations 11.7 and 11.9, respectively.

$$J_{L,mp} = J_{sc} - J_o\left[\exp\left(\frac{e_oV_{mp}}{kT}\right) - 1\right] \tag{11.11}$$

Combining Equation 11.10 and 11.11, $J_{L,\text{mp}}$ is found to be

$$J_{L,\text{mp}} = \frac{e_o V_{\text{mp}}/kT}{1+\left(e_o V_{\text{mp}}/kT\right)}\left(J_{\text{sc}} + J_o\right) \tag{11.12}$$

$$P_{\text{mp}} = \frac{e_o V_{\text{mp}}^2/kT}{1+\left(e_o V_{\text{mp}}/kT\right)}\left(J_{\text{sc}} + J_o\right)\cdot A \tag{11.13}$$

The electric energy output of a silicon cell or module is affected by both the intensity of solar radiation and the operating temperature. These effects are shown, respectively, in Figures 11.9 and 11.10. Figure 11.9 shows that the PV current is proportional to the solar irradiance level whereas solar irradiance has only a small impact on the voltage. Double the solar irradiance and the PV output approximately doubles. More radiation means the operating temperature will increase, however, which will tend to reduce the efficiency. The temperature dependence is shown in Figure 11.10. This shows that the higher the temperature of the cell or module, the lower the voltage and thus the lower the output.

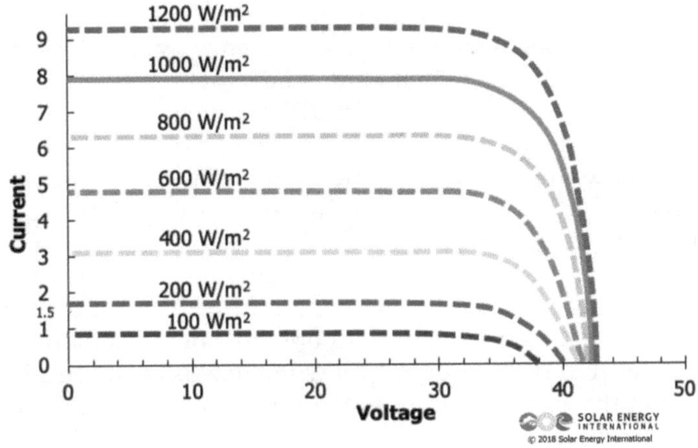

FIGURE 11.9
The effect of solar irradiance on the I-V curve (Amps vs. Volts). (Courtesy of Solar Energy International.)

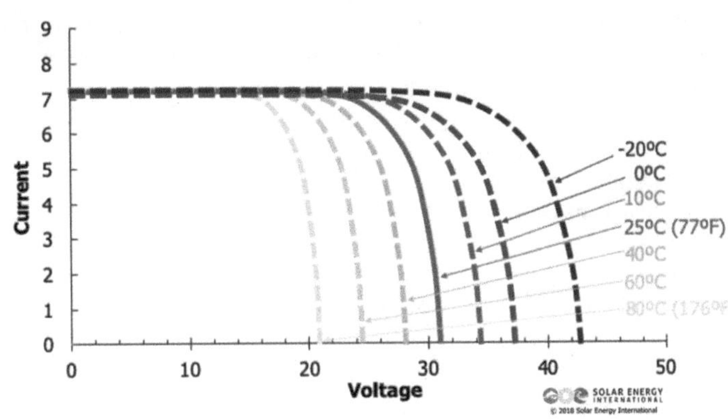

FIGURE 11.10
The effect of cell temperature on the I-V curve (Amps vs. Volts). (Courtesy of Solar Energy International.)

P_{mp}: 340 W
V_{oc}: 46.3 V
I_{sc}: 9.32 A
V_{mp}: 38.5 VDC
I_{mp}: 8.84 ADC
Power tolerance: 0/+5 W
Power degradation: <0.25%/yr., 25-year lifetime
Efficiency: 16.9%

Temperature coefficients:
P_{mp}: -.36%/°C
V_{oc}: -.30%/°C
I_{sc}: .066%/°C

FIGURE 11.11
Example label showing performance and temperature-dependent specifications for a polycrystalline PV panel.

Note that there is very little impact of the temperature on the current. In determining the performance of a module, one must adjust the efficiency, which is based on STC conditions, for the particular irradiance level and module temperature.

Figure 11.11 shows typical specifications at STC for a 340 W polycrystalline PV panel. It shows open-circuit voltage, closed-circuit current, maximum power point voltage and current, and the temperature coefficients for power, open-circuit voltage, and short-circuit current. The panel will lose 0.36% of its rated power output at 25°C for every degree the panel temperature is above 25°C. (Conversely, its power output will be higher by the same amount per degree for panel temperatures below 25°C.) Note that this is panel temperature, not outdoor temperature. Most of the sunlight absorbed by the panel is lost as heat, so the panel will operate above ambient temperature. The power tolerance states that the power output when new is between 340 and 345 W. The power output is guaranteed to reduce less than 0.25% of the rated value per year over a 25-year period.

Example 11.3

The dark saturation current density for a silicon solar cell at 40°C is 1.8×10^{-8} A/m^2 and the short-circuit current density is 200 A/m^2 when exposed to solar radiation of 900 W/m^2. Calculate

a. Open-circuit voltage
b. Voltage at maximum power
c. Current density at maximum power
d. Maximum power
e. Maximum efficiency
f. The cell area required for an output of 25 W.

Solution:

Given

$$J_o = 1.8\times10^{-8}\ \text{A/m}^2$$

$$J_{sc} = 200\ \text{A/m}^2$$

and

$$T = 40°\text{C} = 313\ \text{K}$$

a. Using Equation 11.8,

$$V_{oc} = \frac{kT}{e_o}\ln\left(\frac{J_{sc}}{J_o}+1\right)$$

Since (e_o/kT) will be needed for other parts of the problem solution, it will be evaluated separately:

$$\frac{e_o}{kT} = \frac{1.602\times10^{-19}\ \text{J/V}}{\left(1.381\times10^{-23}\ \text{J/K}\right)\left(313\ \text{K}\right)} = 37.06\ \text{V}^{-1}$$

Therefore,

$$V_{oc} = \frac{1}{37.60}\ln\left(\frac{200}{1.8\times10^{-8}}+1\right) = 0.624\ \text{V}$$

b. Voltage at maximum power condition can be found from Equation 11.10 by an iterative or trial-and-error solution:

$$\exp\left(37.06V_{mp}\right)\left(1+37.06V_{mp}\right) = 1+\frac{200}{1.8\times10^{-8}}$$

$$\text{or}\quad V_{mp} = 0.542\ \text{V}$$

c. Current density at maximum power can be found from Equation 11.12:

$$J_{L,mp} = \frac{e_oV_{mp}/kT}{1+\left(e_oV_{mp}/kT\right)}\left(J_{sc}+J_o\right) = \frac{(37.06)\cdot(0.542)}{1+(37.06)\cdot(0.542)}\left(200+1.8\times10^{-8}\right)\text{A/m}^2 = 190.5\ \text{A/m}^2$$

d. From Equation 11.13, the maximum power is

$$P_{mp} = V_{mp}\cdot J_{mp}\cdot A$$

$$\frac{P_{mp}}{A} = \left(0.542\ \text{V}\right)\cdot\left(190.5\ \text{A/m}^2\right) = 103.25\ \text{W/m}^2$$

e. The maximum efficiency is

$$\eta_{mp} = \frac{103.25\ \text{W/m}^2}{900\ \text{W/m}^2} = 11.5\%$$

f. Cell area required is

$$A = \frac{P_{out}}{\frac{P_{mp}}{A}} = \frac{25\ \text{W}}{103.25\ \text{W/m}^2} = 24.2\ \text{cm}^2$$

11.2.1 Efficiency of Solar Cells

Theoretical limits on the efficiency of a single-layer solar cell can be calculated using Equation 11.3. These efficiency limitations and the practical efficiencies of some of the cells are shown in Figure 11.6. Some of the reasons for the actual efficiency being lower than the theoretical limit are

1. Reflection of light from the surface of the cell. This can be minimized by the use of an antireflection (AR) coating. For example, AR coating can reduce the reflection from a Si cell from 30% from an untreated cell to 3%.
2. Shading of the cell due to current collecting electrical contacts. This can be minimized by reducing the area of the contacts and/or making them transparent; however, both of these methods will increase the resistance of the cell to current flow.
3. Internal electrical resistance of the cell.
4. Recombination of electrons and holes before they can contribute to the current.

11.2.2 Multi-junction Solar Cells

The limits imposed on single-layer solar cells due to band gap can be partially overcome by using multiple layers of solar cells stacked on top of each other, with each layer having a band gap higher than the layer below it. For example (Figure 11.12), if the top layer is made from a cell of material *A* (band gap corresponding to λ_A), solar radiation with wavelengths less than λ_A would be absorbed to give an output equal to the hatched area *A*. The solar radiation with a wavelength greater than λ_A would pass through *A* and be

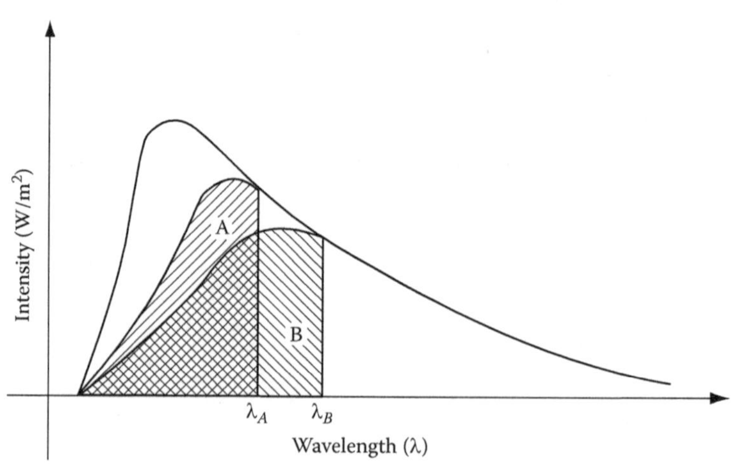

FIGURE 11.12
Energy conversion from a two-layered stacked cell. (From Goswami et al. [9]. With permission.)

converted by the bottom layer cell *B* (band gap corresponding to λ_B) to give an output equal to the hatched area *B*. The total output and therefore the efficiency of this tandem cell would be higher than the output and the efficiency of every single cell individually because of a reduction in the thermalization losses discussed above. Theoretically, the conversion efficiency will continue to increase with the number of solar cells within the multi-junction device.

In practice, additional cells create additional complexity and additional cost. Because of the high cost of multi-junction cells, various products were developed to couple them with solar concentrators to minimize the cell area. Because silicon modules have come down so much in cost and also improved in efficiency, multi-junction cells have not found a significant market for terrestrial power applications. However, because of their high efficiency, they are used for space applications. For those applications, it is simplest to use them under one Sun, i.e., without the addition of solar concentrators. There is, however, interest in the use of terrestrial multi-junction technology with perovskite cells, as discussed in Section 11.6.

11.2.3 Design of a Photovoltaic System

Solar cells may be connected in series, parallel, and a combination of both to obtain the required voltage and current. When similar cells or devices are connected in series, the output voltages and current are shown in Figure 11.13. A parallel connection results in the addition of currents, as shown in Figure 11.14. If the photovoltaic devices have dissimilar characteristics, the output characteristics will be as shown in Figure 11.15. Cells are connected to form modules or panels, and these are connected to form arrays. Principles shown in Figures 11.13 and 11.14 apply to all of these connections.

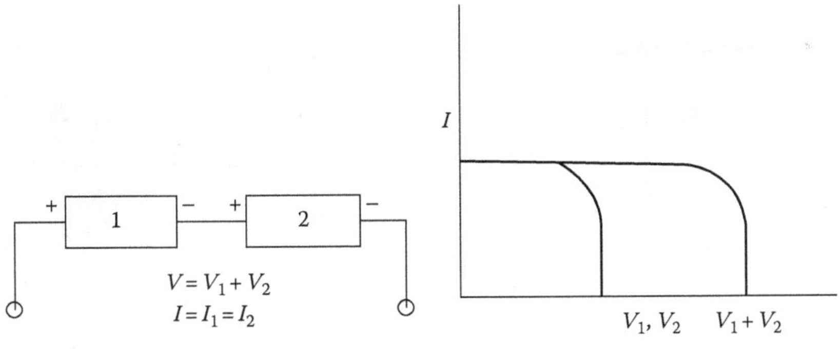

FIGURE 11.13
Characteristics of two similar cells connected in series.

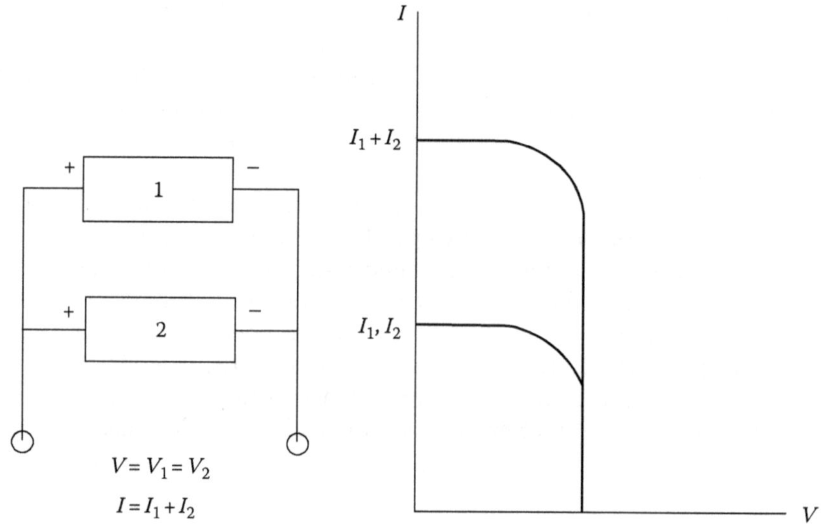

FIGURE 11.14
Characteristics of two similar cells connected in parallel.

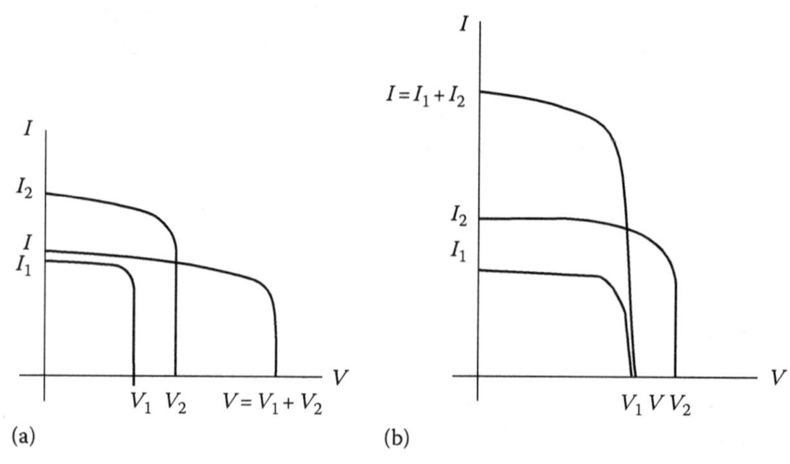

FIGURE 11.15
Characteristics of two dissimilar cells connected in (a) series and (b) parallel.

Example 11.4

An application requires 300 W at 14 V. Design a PV panel using solar cells from Example 11.3 each with an area of 25 cm^2.

Solution:

Assuming that the cell will be operated at maximum power conditions, the voltage and current from each cell are

$$V_{mp} = 0.542 \text{ V}, \quad I_{mp} = \left(190.5 \frac{\text{A}}{\text{m}^2}\right)(0.0576 \text{ m}^2) = 10.97 \text{ A}$$

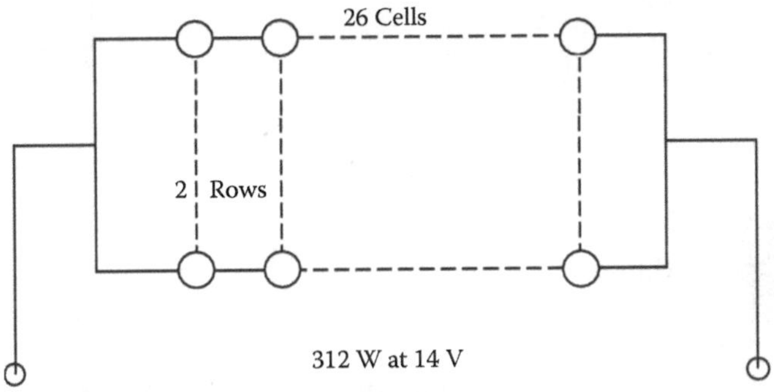

FIGURE 11.16
Connection of cells in rows and columns for Example 11.4.

$$\text{Power/cell} = 0.542\ \text{V} \times 10.97\ \text{A} = 5.95\ \text{W/cell} \approx 6\ \text{W/cell}$$

$$\text{Number of cells required} = \frac{300\ \text{W}}{6\ \text{W/cell}} = 50\ \text{cells}$$

$$\text{Number of cells in series} = \frac{\text{System voltage}}{\frac{\text{Voltage}}{\text{cell}}} = \frac{14\ \text{V}}{0.542\ \text{V}} \approx 26$$

$$\text{Number of rows of 26 cells connected in parallel} = \frac{50}{26} = 1.92 \approx 2$$

Since the number of rows must be a whole number, we may increase the number to two rows, which will give 312 W output (Figure 11.16).

A blocking diode is used in series with a module or an array to prevent the current from flowing backward (for example, from the battery to the cells under dark conditions). A bypass diode is used in parallel with a module in an array to bypass the module if it is shaded.

Most residential rooftop photovoltaic systems are connected to the electric grid. Where net metering is allowed, if the home is not using all of the PV electricity being produced, the excess will go back to the grid and run the electric meter backward. In systems that have no battery backup, if the grid goes down, the PV array will shut down so that power will not be sent out onto the grid and put line workers in danger. The addition of home batteries is becoming more common, especially in places where net metering is no longer allowed. In that case, excess electricity can be stored in the battery.

Figure 11.17 shows a typical PV system design, commonly called a multimode system, that is grid-connected but also has battery backup. The battery will generally be limited in how much power it can put out and is also limited in capacity. So, this system design has a separate backed-up AC loads panel that provides power only to the most critical loads to the house if the grid is down, consistent with the battery output and storage specifications.

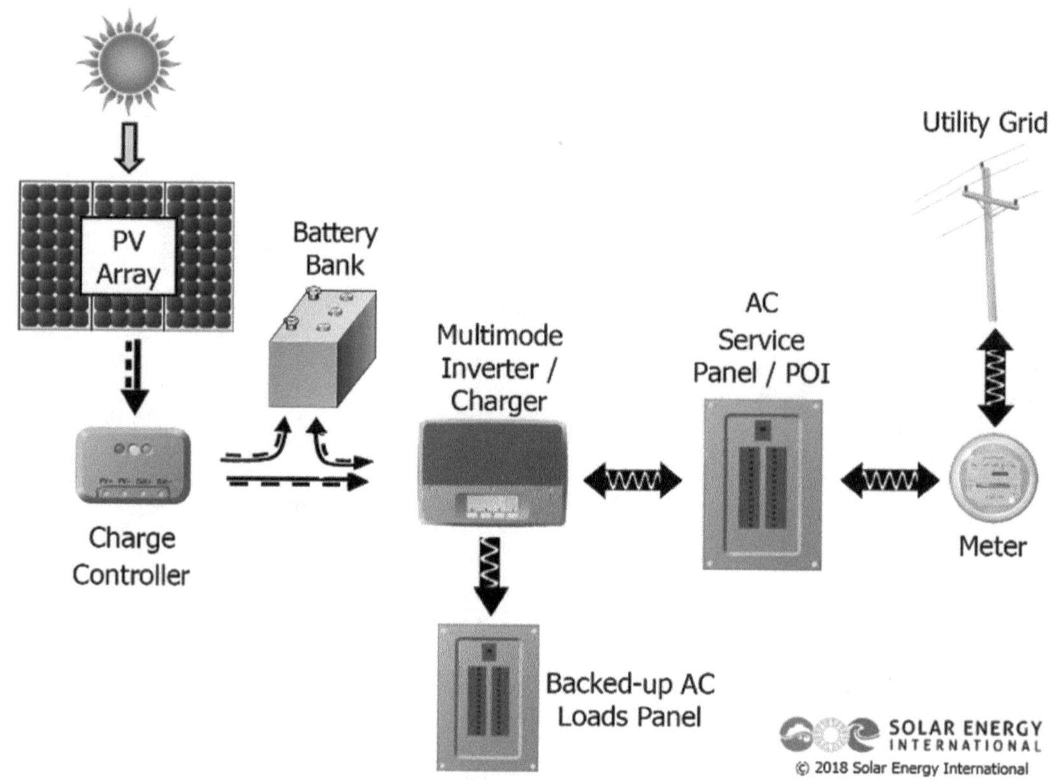

FIGURE 11.17
Schematic of a multimode PV system with grid connection and backup battery. (Courtesy of Solar Energy International.)

If the grid is down, the PV system can continue to operate and provide electricity to the battery, which then provides power to the backup loads panel. The output of the PV array can vary greatly with passing clouds. When the grid is operating, the grid power is mixed with the PV output to provide steady power. When the grid is down, the battery provides steady output.

A rooftop photovoltaic system is typically connected to AC loads, as with the multimode system shown in Figure 11.17. The nameplate rating for a PV system is given in terms of the DC output at STC conditions. To convert the DC output to AC a string inverter is most commonly used as shown in Figure 11.17. The inverter is a source of energy loss that decreases the net output of a PV system.

There are a number of different inverters available, and the selection of the appropriate inverter for a given application depends on the requirements of the load, particularly the waveform, and on whether the system is grid-connected or stand-alone. Inverters used for rooftop PV systems will output a pure sign wave, but some smaller, lower-cost models used for mobile applications may produce a modified sine wave.

Of the various inverter types, a pure sine-wave inverter has the least distortion and the highest efficiency and is most commonly used in PV installations. Specifications of the losses incurred by an inverter for an AC load are included in the PV Watts calculation illustrated later in the chapter. Detailed information about inverter construction and selection can be found in [10] and [11]. Table 11.3 provides example specifications for a typical sine-wave string inverter used for rooftop PV.

TABLE 11.3

Example String Inverter Specifications

Maximum AC power output	5,000 W
AC output voltage	208/240/277 Vac
AC output frequency	60 Hz
Maximum AC output current	32 A
DC input voltage range	250–600 Vdc
Efficiency	97.5%

As we have seen, the output voltage of a PV array is a function of the array temperature. The voltage put out by a row of panels connected in series will be equal to the voltage output of each panel multiplied by the number of panels. In matching a PV array and inverter, it's important to design the system so that the total array output voltage remains within the inverter DC voltage input window as the temperature and array voltage vary throughout the day and the year. This is shown in Figure 11.18. The top curve shows winter operation when the weather is coldest, and the array output voltage is highest. The bottom graph shows summer operation with low voltage due to high temperature. Throughout the day, the temperature is coldest in the morning, which yields the highest voltage, and minimum voltage occurs around mid day when the ambient temperature is highest. The number of panels in series, which determines the array output voltage, and the companion string inverter, must be chosen to keep the array output voltage within the inverter input voltage window throughout the day and throughout the year.

It's important to keep PV panels operating at the maximum power point—the knee in the I-V curve. A string inverter can do this for an array, but there are advantages to controlling each individual panel separately. This can allow for some panels to be shaded at times without impacting the performance of other panels in the array. It also allows

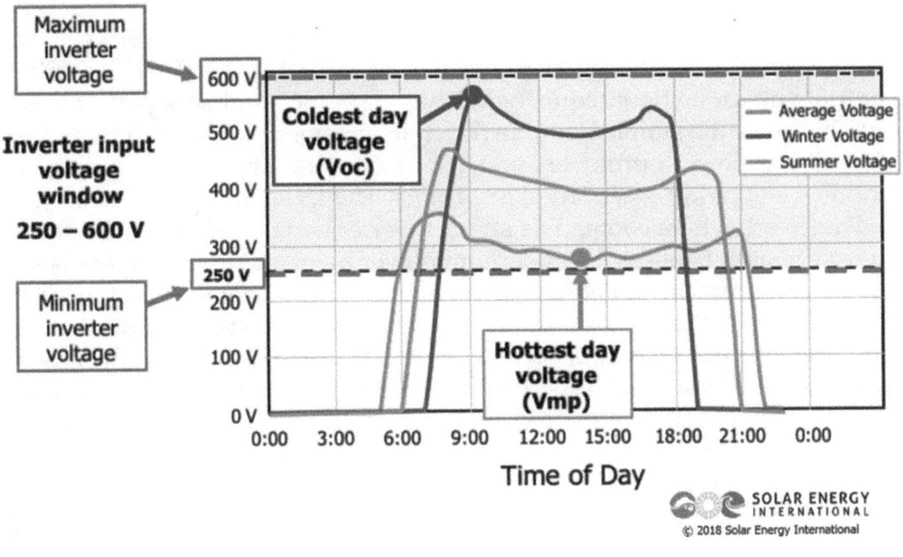

FIGURE 11.18
Matching PV array DC voltage output to the inverter input window throughout the day and the year. (Courtesy of Solar Energy International.)

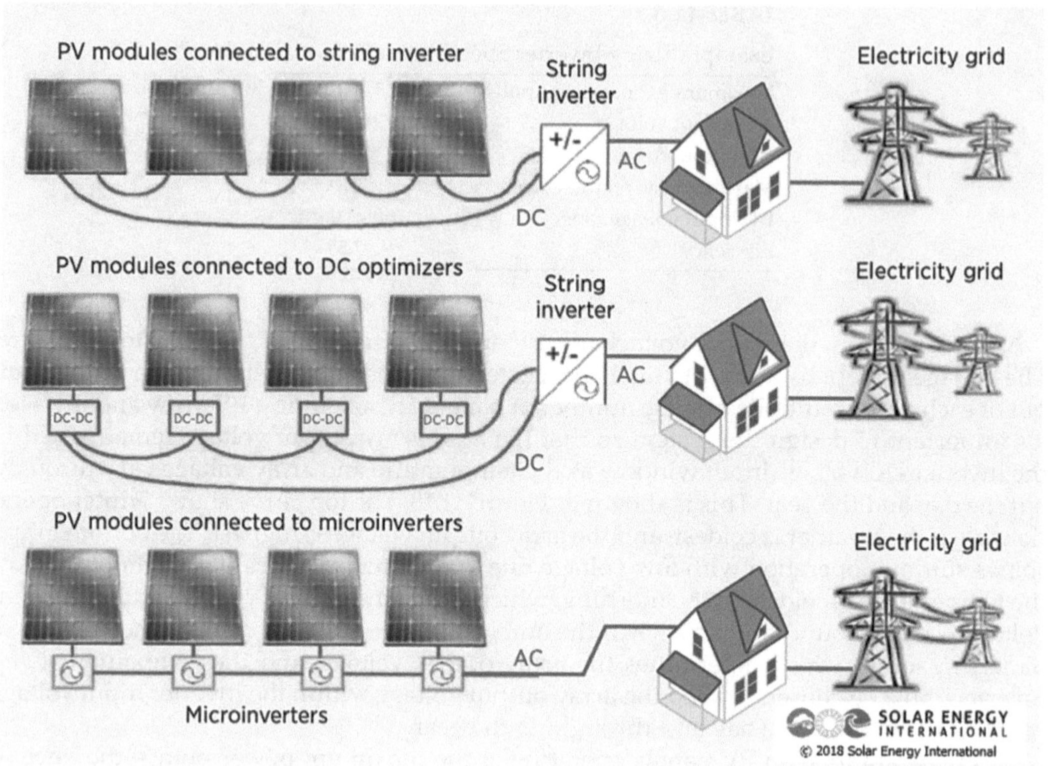

FIGURE 11.19
Illustration of module-level power electronics. (a) Simple string inverter, (b) DC:DC optimizers with string inverter, and (c) microinverters. (Courtesy of Solar Energy International.)

for performance data to be obtained from each individual panel. This also provides the desired output voltage without requiring wiring a specific number of panels in series.

Figure 11.19 shows the simple string inverter arrangement along with two alternatives: the use of DC–DC power optimizers and microinverters. These devices are generically called module-level power electronics, or MLPEs. The DC–DC optimizers are connected in series. They work with a compatible string inverter and control the DC output of each panel. They keep each panel at the maximum power point and also perform as solid-state transformers to convert the panel output voltage to a value that will keep the total array output voltage within the voltage window of the string inverter. Microinverters, on the other hand, convert the DC output of each panel directly to AC (typically at 240 V), thus eliminating the need for a string inverter. The microinverters are all connected in parallel and aggregate the current outputs from each panel. In addition to the advantages listed above, because these devices control the voltage at each panel, they can limit the output to a safe value when the grid is down or the array is shut down, even when the panels are under full sunlight. This can provide a safe environment for firefighters who might need to access the roof during a home fire, for example.

Example 11.5: PV System Commissioning

After a PV system is designed and installed, it's necessary to measure its performance to ensure that it is performing in accordance with the equipment specifications. This is called the commissioning phase. The same procedure can be used when the system is old to determine that any performance degradation that has occurred is within specifications. We will consider here a 6.8-kW rooftop PV array consisting of 20 340 W panels having the specifications given in Figure 11.11. These are wired in two series rows of ten panels each with a string inverter having an efficiency of 97.5%. To check the performance, two measurements need to be taken: solar radiation in the plane of the PV array and the array temperature. The solar radiation in the plane of the array is measured as 843 W/m^2, and an infrared thermometer aimed at the back of the panels gives an average panel temperature of 39°C. At the time of these measurements, the inverter displays an AC power output of 5,345 W, and we want to know if the array is performing within specifications.

We will use the STC performance specifications for the PV panel and adjust the power value for the fact that both the solar radiation and the panel temperature deviate from STC.

Adjusting for temperature:

Temperature deviation from STC: 39°C − 25°C = +14°C

Applying the temperature coefficient for the maximum power point power,

$$-0.36\% / °C \times 14°C = 5.04\%$$

$$(1 - 0.0504) \times 340 \text{ W/panel} = 322.9 \text{ W/panel}$$

Adjusting for radiation:

Now we correct the 322.9 value for the deviation from the STC radiation of 1,000 W/m^2, using the fact that power output is virtually linear with radiation.

$$(843 \text{ W}/1{,}000 \text{ W}) \times 322.9 \text{ W/panel} = 272 \text{ W/panel}$$

We have 20 panels, so the total power from the array is

$$20 \text{ panels} \times 272 \text{ W/panel} = 5{,}443 \text{ W}$$

The DC power is converted to AC power via an inverter with 97.5% efficiency,

$$\text{Expected AC power} = 0.975 \times 5{,}443 \text{ W} = 5{,}307 \text{ W}$$

Calculating the percent difference between the measured and expected power outputs we obtain

$$(5{,}345 \text{ W} - 5{,}307 \text{ W})/5{,}307 \text{ W} = +0.007$$

Thus, the measured power output is 0.7% higher than the expected power output based on the specifications. The panel specifications indicate that the rated power for a panel has an uncertainty of 0/+5 W. For the 340 W panel, that's a range of 0% to +1.5%, and we are right in the middle of that. So, the system is performing according to expectations.

11.3 Manufacture of Solar Cells and Panels

The manufacture of crystalline silicon solar cells is an outgrowth of the manufacturing methods used for microprocessors. A major difference is that silicon used in microprocessors is ultra-pure, which is not needed for photovoltaic cells. Therefore, a major source of feedstock for silicon solar cells has been the waste material from the microelectronics industry. Solar cells are also manufactured as polycrystalline and thin films. Some of the common methods of manufacture of silicon solar cells are provided in the following text.

11.3.1 Single-Crystal and Polycrystalline Cells

Silicon cells are produced by a series of processes: (1) growing crystalline ingots of *p*-silicon, (2) slicing wafers from the ingots, (3) polishing and cleaning the surface, (4) doping with *n* material to form the *p*–*n* junction, (5) deposition of electrical contacts, (6) application of AR coating, and (7) encapsulation. Figure 11.20 illustrates the process.

The *Czochralski* method (Figure 11.21a) is the most common method of growing single-crystal ingots. A seed crystal is dipped in molten silicon doped with a *p*-material (boron) and drawn upward under tightly controlled conditions of linear and rotational speed, and temperature. This process produces cylindrical ingots of typically 10 cm diameter, although ingots of 20 cm diameter and more than 1 m long can be produced for other applications. An alternative method is called the *float zone* method (Figure 11.21b). In this method, a polycrystalline ingot is placed on top of a seed crystal and the interface is melted

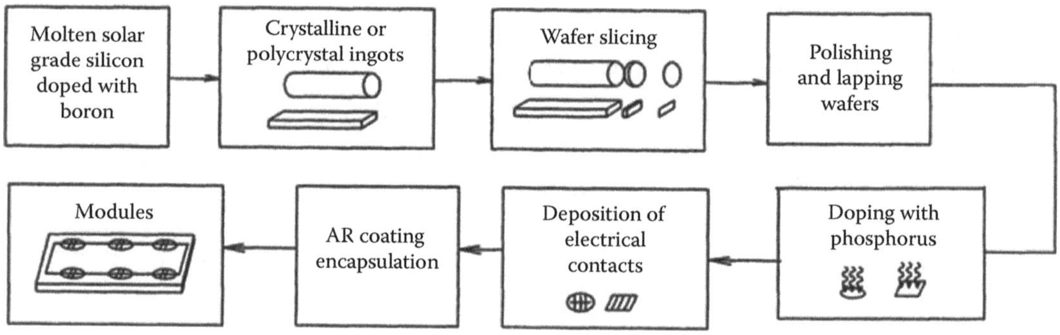

FIGURE 11.20
Series of processes for the manufacture of crystalline and polycrystalline cells.

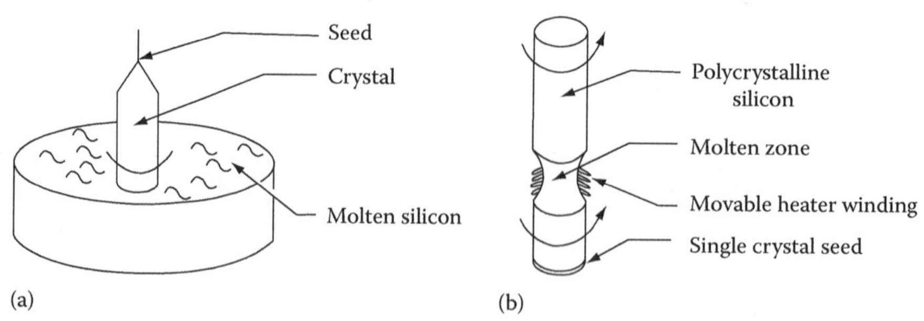

FIGURE 11.21
Crystalline silicon ingot production methods: (a) Czochralski method and (b) float zone method.

by a heating coil around it. The ingot is moved linearly and rotationally, under controlled conditions. This process has the potential to reduce the cell cost.

Polycrystalline ingots are produced by casting silicon in a mold of the preferred shape (rectangular), as shown in Figure 11.22. Molten silicon is cooled slowly in a mold along one direction in order to orient the crystal structures and grain boundaries in a preferred direction. In order to achieve efficiencies of greater than 10%, grain sizes greater than 0.5 mm are needed, and the grain boundaries must be oriented perpendicular to the wafer. Ingots as large as 400 cm × 40 cm × 40 cm can be produced by this method.

Ingots are sliced into wafers by internal diameter saws with slurry-based cutting elements or multiwire saws impregnated with diamond abrasive particles. Both of these methods result in high wastage of valuable crystalline silicon.

Alternative methods that reduce wastage are those that grow polycrystalline *thin films*. Some of the thin-film production methods include dendritic web growth (Figure 11.23), edge-defined film-fed growth (EFG) (Figure 11.24), ribbon against drop method, supported web method, and ramp-assisted foil casting technique (RAFT) (Figure 11.25).

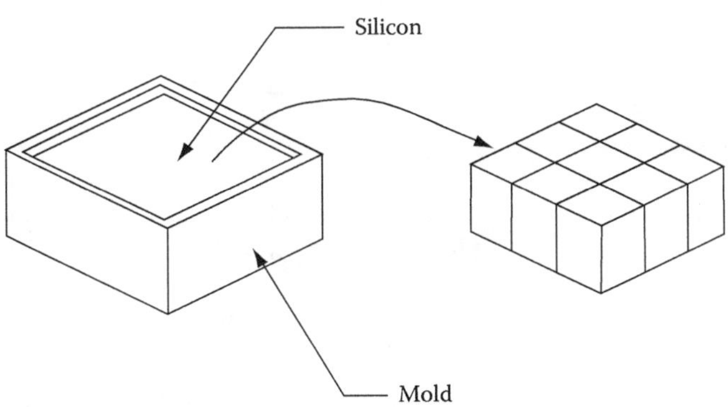

FIGURE 11.22
Polycrystalline ingot production.

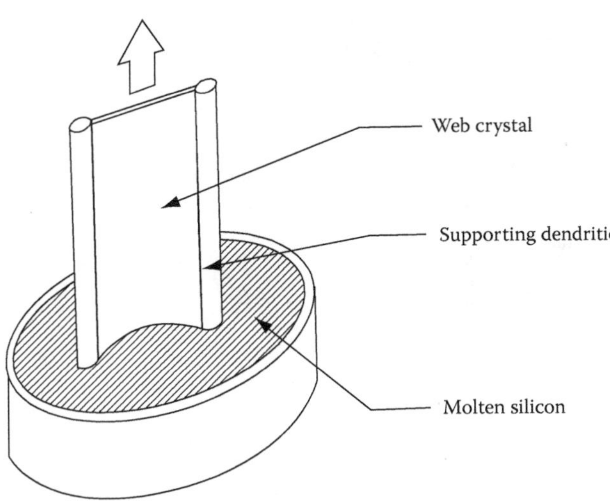

FIGURE 11.23
Thin-film production by dendritic web growth.

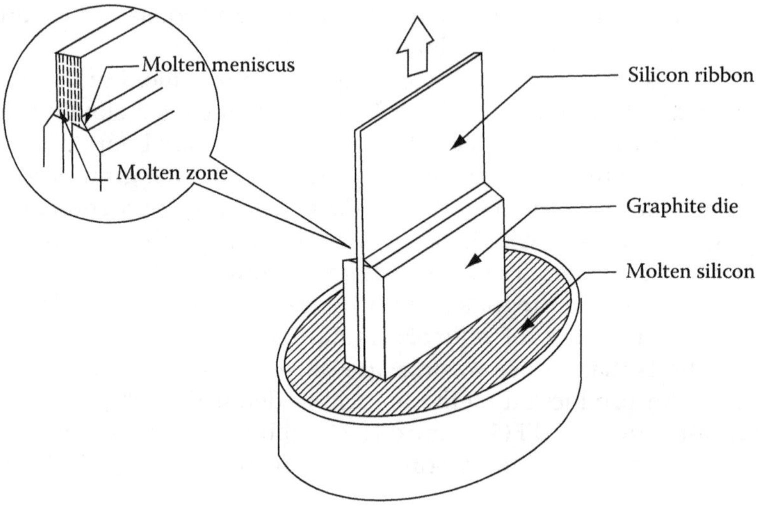

FIGURE 11.24
Thin-film production by edge-defined film-fed growth (EFG).

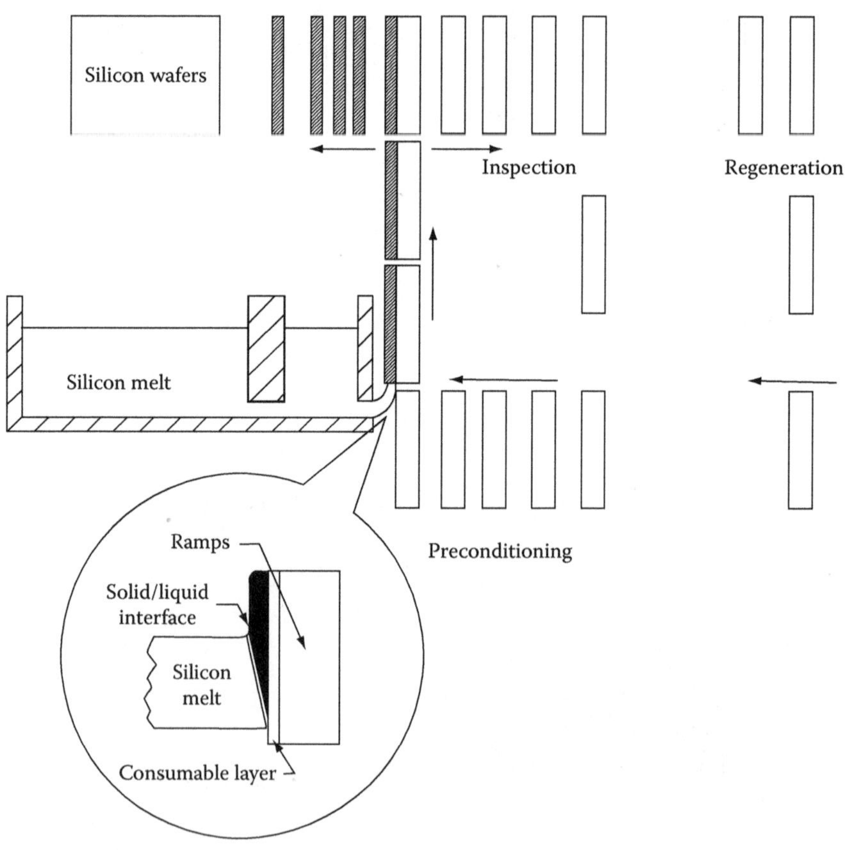

FIGURE 11.25
Schematic of RAFT processing.

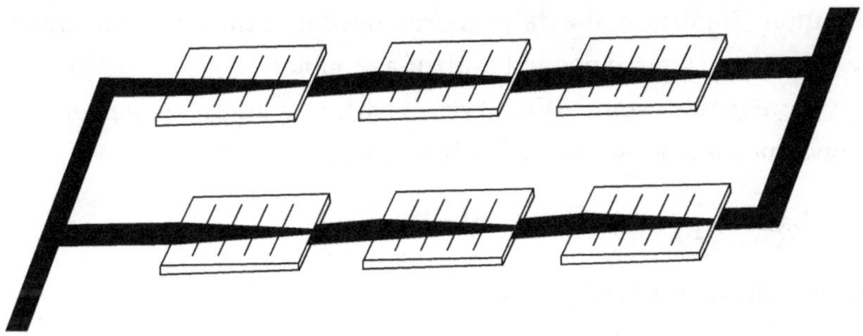

FIGURE 11.26
Assembly of solar cells to form a module.

A *p–n* junction is formed in the cell by diffusing a small amount of *n* material (phosphorous) in the top layer of a *p*-silicon wafer. The most common method is diffusion of phosphorous in the vapor phase. In this case, the backside of the wafer must be covered to prevent the diffusion of vapors from that side. An alternate method is to deposit a solid layer of the dopant material on the top surfaces followed by high-temperature (800°C–900°C) diffusion.

Electrical contacts are attached to the top surface of the cell in a grid pattern to cover no more than 10% of the cell surface, and a solid metallic contact is made on the back. The front grid pattern is made by either vacuum metal vapor deposition through a mask, by electroplating, or by screen printing. Figure 11.26 shows how cells are connected to form modules.

AR coatings of materials such as silicon dioxide (SiO_2), titanium dioxide (TiO_2), and tantalum pentoxide (Ta_2O_5) can be deposited on the cell surface to reduce reflection from more than 30% for untreated Si to less than 3%. Silicon nitride is the most common AR coating for conventional solar cells in mass production. AR coatings are deposited by vacuum plasma-enhanced chemical vapor deposition (PECVD), sputtering, or chemical spraying. Finally, the cells are encapsulated in a transparent material to protect them from the environment. Encapsulants usually consist of a layer of either polyvinyl butyral or ethylene vinyl acetate and are placed on both sides of the solar cell. A top layer of low-iron glass and a bottom layer of another sheet of glass or back sheet complete the process of assembling solar cells into modules.

11.4 Design for Remote Photovoltaic Applications

Photovoltaic power can be ideal for remote applications requiring a few watts to hundreds of kilowatts of electrical power. Even where a conventional electrical grid is available, for some applications, where uninterruptible or emergency standby power is necessary, photovoltaic power would be appropriate. Some examples of remote PV applications include water pumping for potable water supply and irrigation, power for remote houses, street lighting, battery charging, telephone and radio communication relay stations, and weather stations. Examples of some other applications include electrical utility switching stations, peak electrical utility power where environmental quality is a concern, data acquisition systems, and specialty applications such as ventilation fans and vaccine refrigeration.

The design of a PV system is based on some basic considerations for the application:

1. Which is more important, the daily energy output or the power (average or peak)?
2. Is a backup energy source needed and/or available?
3. Is energy storage important? What type—battery, pumped water, etc.?
4. Is the power needed as AC or DC? What voltage?

There are three basic steps in the design of a PV system:

1. Estimation of load and load profile
2. Estimation of available solar radiation
3. Design of PV system, including area of PV panels, selection of other components, and electrical system schematic

Each of these steps will be explained in the following examples. These examples are based on Post and Risser [12].

11.4.1 Estimation of Loads and Load Profiles

Precise estimation of loads and their timings (load profile) are important for PV systems since the system is sized as the minimum required to satisfy the demand over a day. For example, if power is needed for five different appliances requiring 200, 300, 500, 1,000, and 1,500 W, respectively, so that only one appliance is on at any one time and each appliance is on for an average of 1 hour a day, the PV system would be sized based on 1,500 W peak power and 3,500 Wh of daily energy requirement. The multiple loads on a PV system are intentionally staggered to use the smallest possible system since the capital costs of a PV system are the most important as opposed to the energy costs in a conventional fuel-based system.

Example 11.6: Daily Load Calculations

How much energy per day is used by a remote weather station given the following load characteristics?

Load	Load Power (W)	Run Time (h/day)
Charge controller	2.0	8
Data gathering	4.0	3
Modem (standby)	1.5	22.5
Modem (send/receive)	30.0	1.5

Solution:

$$\text{Daily energy} = (2.0\ \text{W})(8\ \text{hours}) + (4.0\ \text{W})(3\ \text{hours}) + (1.5\ \text{W})(22.5\ \text{hours}) + (30.0\ \text{W})(1.5\ \text{hours})$$

$$= 106.75\ \text{Wh}$$

Daily energy use is about 107 Wh/day.

Example 11.7: Load Calculations

An owner of a remote cabin wants to install a PV power system. The loads in the home are described as follows. Assume that water pump and bathroom fan are DC, and all others are AC. Find the daily and weekly peak and average energy use estimates. The system used is a 24 V DC system with an inverter.

Lights	Four 23 W compact fluorescent bulbs	On at night for 5 hours
Lights	Six 13 W compact fluorescent bulbs	2 hours each (daytime)
Stereo	110 W (amplifier), 15 W (other)	On for 8 h/week
Water pump	55 W (3.75 A start current)	Runs for 2/day
Computer	250 W (monitor included)	On for 1½ hours daily (weekend nights only)
Bathroom fan	40 W (3.5 A start current)	On for 1 h/day
Microwave	550 W (AC)–1,000 W surge	On for 30 min/day

Solution:

Loads need to be broken down according to (1) run time, (2) peak power, (3) night or day use, and (4) AC or DC loads. The load profile is as follows:

Load Name Description	Power (W)		Run Time (hours)		Energy (Wh)	
	Average	Peak	Day	Week	Day	Week
Lights (AC)	(4) (23)	(4) (23)	5.0	35	460	3,220
Lights (AC)	(6) (13)	(6) (13)	2.0	14	156	1,092
Stereo (AC)	(1) (125)	(1) (125)	—	8	—	1,000
Pump (DC)	(1) (55)	(3.75 A) (24 V)	2.0	14	110	770
Computer (AC)	(1) (250)	(1) (250)	1.5	3	—	750
Fan (DC)	(1) (40)	(3.5 A) (24 V)	1.0	7	40	280
Microwave (AC)	(1) (550)	(1) (1,000)	0.5	3.5	275	1,925

Average DC load: [770+280]/7=150 Wh/day.
Average AC load: [3,220+1,092+1,000+750+1,925]/7=1,141 Wh/day.
Peak DC load: max [{(3.5)(24)+55}, {(3.75)(24)+40}]=139 W.
Peak AC load: (1,000)+max [(4)(23), (6)(13)]+250+125=1,467 W.

It can be assumed that the pump and fan will not start precisely at the same instant and that the daytime and nighttime lighting loads will not be on simultaneously.

11.4.2 Estimation of Available Solar Radiation

Methods of estimation of available solar radiation are described in Chapter 10. Websites such as the National Solar Radiation Data Base [13] are useful resources for accurate approximations. For designing a PV system, a decision is made whether the PV panel will be operated as tracking the Sun or will be fixed at a certain tilt and azimuth angle. For fixed panels, a tilt angle of latitude +15° works best for winter and latitude –15° for summer. To keep the panel fixed year-round, an angle equal to the latitude provides the maximum yearly energy (see Figure 11.27).

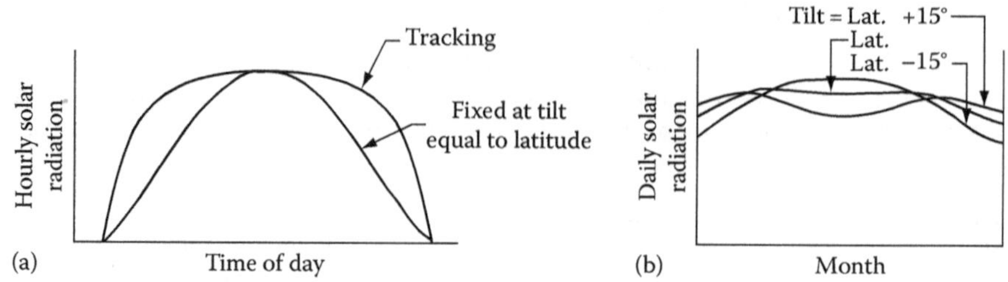

FIGURE 11.27
Solar radiation on panels at different tilt angles: (a) hourly and (b) monthly.

11.4.3 PV System Sizing

If meeting the load at all times is not critical, PV systems are usually sized based on the average values of energy and power needed, available solar radiation, and component efficiencies. This is known as the *heuristic approach*. It is important to note that a system designed by this approach will not give the best design but may provide a good start for a detailed design. A detailed design accounts for the changes in the efficiencies of the components depending on the load and the solar radiation availability and whether the system is operating in a PV-to-load, PV-to-storage, or storage-to-load mode.

Example 11.8: Heuristic Approach to PV System Sizing

A PV system using 50 W, 12 V panels with Trojan T-105 6 V, 125 Ah batteries is needed to power a home in Farmington, New Mexico, with a daily load of 1,700 Wh. System voltage is 24 V. Assuming an average of 5 daylight hours in the winter, specify the collector and storage values for the system using the heuristic approach.

Solution:

Load = 1,700 Wh/day
Daylight hours = 5 h/day
Average panel output = 50 W

$$\text{Number of panels} = \frac{1{,}700\ \text{Wh/day}}{(5\ \text{h/day})\ (50\ \text{W/panel})} = 6.8,\ \text{round off to 7 panels}$$

Since the system voltage is 24 V, but each panel produces only 12 V, an even number of panels will be needed. Therefore, the number of panels = 8.

Farmington, New Mexico, is a very sunny location, so 3 days of storage are sufficient. Assuming a battery efficiency of 75% and a maximum depth of discharge of 70%,

$$\text{Storage} = (1{,}700)(3)/(0.75 \times 0.7) = 9{,}714\ \text{Wh}$$

$$\text{Number of batteries} = (9{,}714\ \text{Wh})/(125\ \text{Ah} \times 6\ \text{V}) = 13(\text{rounded off to the next whole number})$$

Since the system voltage is 24 V, and each battery provides 6 V, the number of batteries is increased to 16. In a detailed design, the efficiencies of battery storage, inverter, and the balance of system (BOS) must be accounted for. The following example shows how these efficiencies increase the energy requirements of the PV panel.

Example 11.9: System Operating Efficiency

Using the cabin electrical system from Example 11.7, calculate the overall system efficiency for each operating mode possible for the system. Estimate the amount of energy required per day for the system. When load timing (day or night), assume half of the load runs during the day and half runs at night. The inverter used has a component efficiency of 91%, the battery efficiency is 76%, and the distribution system efficiency is 96%.

Solution:

From the example, the loads are

Average DC load: 150 Wh/day

Average AC load: 1,141 Wh/day

The various system efficiencies are

PV to load (DC)	0.96 (day, DC)
Battery to load (DC)	(0.76) (0.96)=0.73 (night, DC)
PV to load (AC)	(0.96) (0.91)=0.874 (day, AC)
Battery to load (AC)	(0.76) (0.91) (0.96)=0.664 (night, AC)

Expected day and night loads are

Day (DC)	(0.5) (110)+(0.5) (40)=75 Wh/day
Night (DC)	(0.5) (110)+(0.5) (40)=75 Wh/day
Day (AC)	(156)+(0.5) (1,000+750)/7+(0.5) (275)=418.5 Wh/day
Night (AC)	(460)+(0.5) (1,000+750)/7+(0.5) (275)=722.5 Wh/day

Without considering system efficiency, the daily energy requirement is

$$E_{\text{day}} = (150) + (1{,}141) = 1{,}291 \text{ Wh/day}$$

The expected daily energy requirement is

$$E_{\text{day}} = \frac{75}{0.96} + \frac{75}{0.73} + \frac{418.5}{0.874} + \frac{722.5}{0.664}$$

$$E_{\text{day}} = 1{,}725 \text{ Wh/day}$$

The actual energy requirement is 34% higher than that obtained in the approximate calculation in Example 11.7.

11.4.4 Water Pumping Applications

Water pumping for drinking water or irrigation at remote locations was an important first practical use for PV. For a simple schematic shown in Figure 11.28, the power needed to pump water at a volumetric rate $\dot{V}$ is given by $\rho \dot{V} g H / \eta_p$ where ρ is the density of water, g is the acceleration due to gravity, H is the head loss the pump must overcome, and η_p is the pump efficiency. The static head H_s is $(A+B)$. In the case that the water level is drawn down, the static head would be $(A+B+C)$. The pump must work against the total head H, which includes the static head, the frictional losses in the pipe and fittings, and the dynamic head:

$$H = H_s + H_f + \frac{v^2}{2g} \tag{11.14}$$

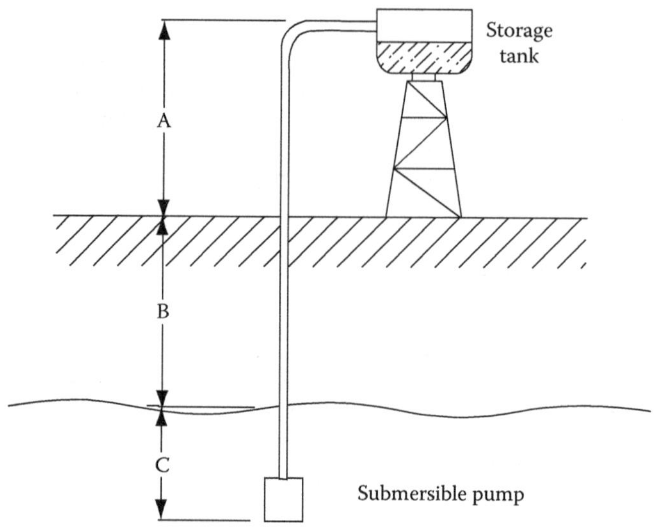

FIGURE 11.28
Water pumping using a submersible pump.

where
H_f is the frictional head loss in the pipe and the bends
v is the velocity of the water at the pipe outlet.

The pump efficiency η_p is a function of the load (head and flow rate) and is available as a characteristic curve from the manufacturer. There are two basic types of pumps: centrifugal and positive displacement. These pumps can be driven by AC or DC motors. DC motors are preferable for the PV applications because they can be directly coupled to the PV array output. Centrifugal pumps with submersible motors are the optimum for PV applications because of their efficiency, reliability, and economy. However, for deep wells, jack pumps may be necessary. Jack pumps are the piston type of positive displacement pumps that move chunks of water with each stroke. They require very large currents; therefore, they are connected through batteries.

11.5 Thin-Film PV Technology

Monocrystalline and polycrystalline silicon PV dominate the PV market. They represent approximately 95% of PV shipments worldwide [14]. Advances continue to be made, however, in thin-film PV technologies. Because they are thinner and often flexible, they offer a wide range of applications including building-integrated photovoltaics and portable power systems as well as rooftop and utility-scale applications.

Thin-film PV cells require multiple layers of precisely deposited thin films. An important criterion for the production of thin-film PV in a continuous process, such as roll coating, is the percentage of the output that has an acceptable efficiency. The three primary thin-film PV technologies currently on the market are amorphous silicon (*a*-Si), cadmium telluride (CdTe), and copper indium gallium selenide (CIGS). Each of these technologies has advantages and disadvantages, and the differences between these technologies are described briefly in the following text.

Amorphous silicon PV technology utilizes a widely available material, but the cells have a relatively low efficiency. Amorphous silicon cells have a maximum theoretical efficiency of 28%. An amorphous silicon cell prepared within a laboratory environment at NREL achieved an efficiency of 13.6%. Small research cells that set records in a laboratory environment are called "champion cells." However, commercial production efficiencies for amorphous silicon typically only range from 6% to 10%. Furthermore, the capital equipment costs for manufacturing thin-film amorphous silicon are generally higher than that for other thin-film technologies. Amorphous silicon solar cells are very low in cost but are not typically used today in solar panels because of their low efficiency. However, they still see use in low-power electronics.

Cadmium telluride (CdTe) thin-film PV modules have made impressive advances. CdTe cells have a maximum theoretical efficiency of 28%. CdTe champion cells produced by First Solar have achieved efficiencies as high as 22.1% [15]. Commercially, the production efficiencies for CdTe PV modules range from 13% to 18%. First Solar often appears in the list of the top ten global PV producers. CdTe thin films have the advantages of low material costs and handling requirements associated with the manufacturing process. However, current production processes are limited to glass substrates, which are more limited in their applications than flexible films that can be used in building-integrated photovoltaic products and can often be used to replace conventional building materials in some parts of a building.

Copper indium diselenide (CIS) and gallium-alloyed CIS (CIGS) photovoltaic products are commercially developed, and as many as ten different deposition methods for growing the thin CIS/CIGS absorber layers are being pursued. CIS/*CIGS* PV modules have the best efficiency characteristics of the three technologies. A champion cell achieved an efficiency of 22.3%. CIS/CIGS technology utilizes less raw material than the other two technologies and, therefore, achieves lower film thicknesses. Moreover, CIS/CIGS films have high absorption capabilities. This technology appears to combine the highest solar conversion radiation efficiency with low manufacturing costs and the potential of being mass-produced with flexible substrates (see Figure 11.29).

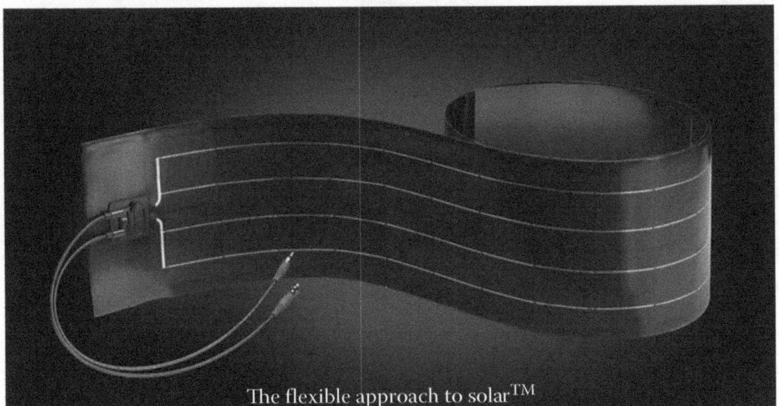

FIGURE 11.29
Thin-film flexible photovoltaic CIGS module. (Courtesy of SoloPower, Inc., San Jose, CA. With permission.)

11.6 Perovskites and Other Research Cells

Since 2009 there has been considerable research on the development of perovskite solar cells. Perovskite is a mineral formed of calcium, titanium, and oxygen having a particular cubic crystal structure. Perovskite solar cells are so named because they are based on materials with a similar crystal structure to the perovskite mineral. Perovskite solar cells have attracted so much research interest because both the materials cost and fabrication costs are low. They also have a high absorption rate for photons and can be made into very thin lightweight films capable of absorbing the complete visible solar spectrum. Perovskites can be used to form multi-junction cells and be stacked with silicon cells to further increase the efficiency. At the time of this writing, production panels have an efficiency of 24.5%,and the record efficiency for a laboratory tandem cell consisting of a perovskite layer on top of a monocrystalline silicon layer is 34.6% claimed by Chinese manufacturer LONGI Green Energy Technology [16]. The theoretical efficiency limit for a perovskite cell is about 43%, compared to 29% for a monocrystalline silicon cell. The challenge researchers face is to improve the long-term stability and durability of perovskite cells, both of which have delayed commercialization.

Figure 11.30 shows the highest research cell efficiencies obtained by recognized independent test labs for all the various PV technologies, including the latest perovskite cells [15]. This

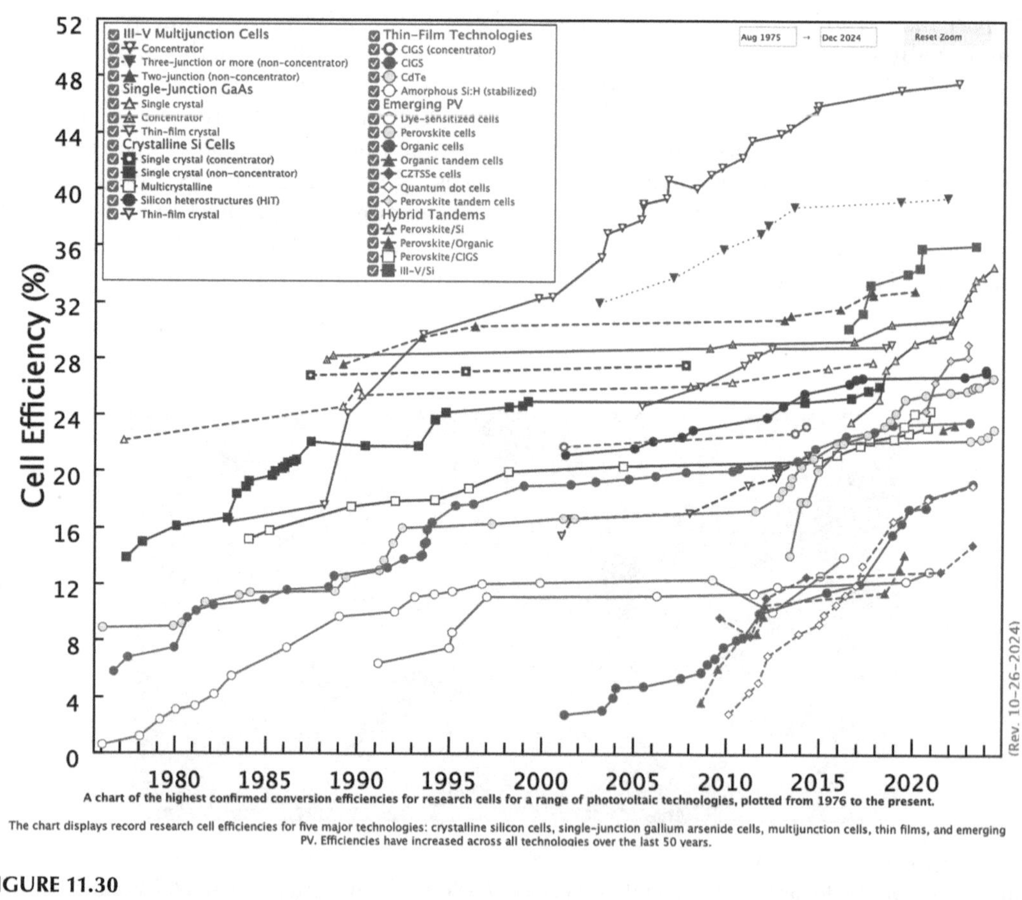

FIGURE 11.30
Best research cell efficiencies for the various PV technologies from 1976 to 2024. (From NREL [15].)

chart is constantly updated, and the latest data can be found on the website, where the chart is interactive and easier to read (https://www.nrel.gov/pv/interactive-cell-efficiency.html). Figure 11.30 clearly shows the progress that has been made in improving cell efficiencies. These research cells are indicative of what may be possible with a given technology, but they are not commercial products and are not cost-effective. However, they represent a goal that mass production facilities aim to achieve.

11.7 Today's PV Market

Due to significant improvements in PV manufacturing costs especially in China, learning curves associated with deployment, and some cases of oversupply, the price of PV panels has dropped dramatically in recent years—over 80% during the 9-year period from 2010 to 2019 as shown in Figure 11.31 [17]. The experience (or learning) curve shown in the figure indicates approximately a 20% drop in price with every doubling of cumulative installed capacity. This type of learning curve cost reduction is similar to Moore's law of semiconductor costs and is sometimes call Wright's Law in honor of engineer Theodore Wright or Swanson's Law in honor of Richard Swanson, the founder of SunPower. Wholesale module costs are now under 30 ¢/W.

Because of the large drop in prices, PV installations have grown rapidly around the world. Between the early 1980s and the mid-2010s, concentrating solar power, or CSP (see Chapter 12), and PV competed for utility-scale solar power generation. Concentrating solar had the advantage of utilizing relatively low-cost molten salt thermal energy storage to provide dispatchable power. However, large reductions in both the cost of PV panels and the cost of lithium-ion batteries have made it much more difficult for CSP to economically compete. In addition, the solar concentration needed to achieve the high temperatures required for an efficient Rankine steam cycle essentially limits CSP plants to those areas with very clear skies and high levels of direct normal irradiance. PV, on the other hand, absorbs both direct and diffuse solar radiation making it more flexible in terms of the available solar resource.

PV systems are very versatile. They are ideal for building rooftops and parking lot canopies, as no moving parts, reflectors, or water piping are required, as with solar thermal hot water systems. In recent years there has been a growing interest in what is called "agrivoltaics" whereby ground-mounted rows of PV modules share the land with crops. There are also cases in which the land between the rows accommodates grazing sheep. The sheep keep weed growth down and benefit in hot weather from the shade provided by the PV panels. There are also installations above bodies of water, such as on a canopy over an irrigation canal or actually floating on the surface of a reservoir, the latter application being referred to as "floatovoltaics." In these cases, the PV panels provide shade and reduce evaporative loss, and the body of water helps keep the PV cool thereby maximizing its efficiency.

The world-installed capacity of PV reached 2 TW in 2024, surpassing that of wind power. Because it is low-cost and low-carbon, and the fact that the solar resource is widely available, PV is now dominating electricity growth. PV panel prices have dropped so low that now more attention is being paid to reducing so-called "soft costs," especially in the case of residential rooftop systems. These are the non-hardware costs such as marketing, permit fees, installation labor, interconnection, taxes, etc.

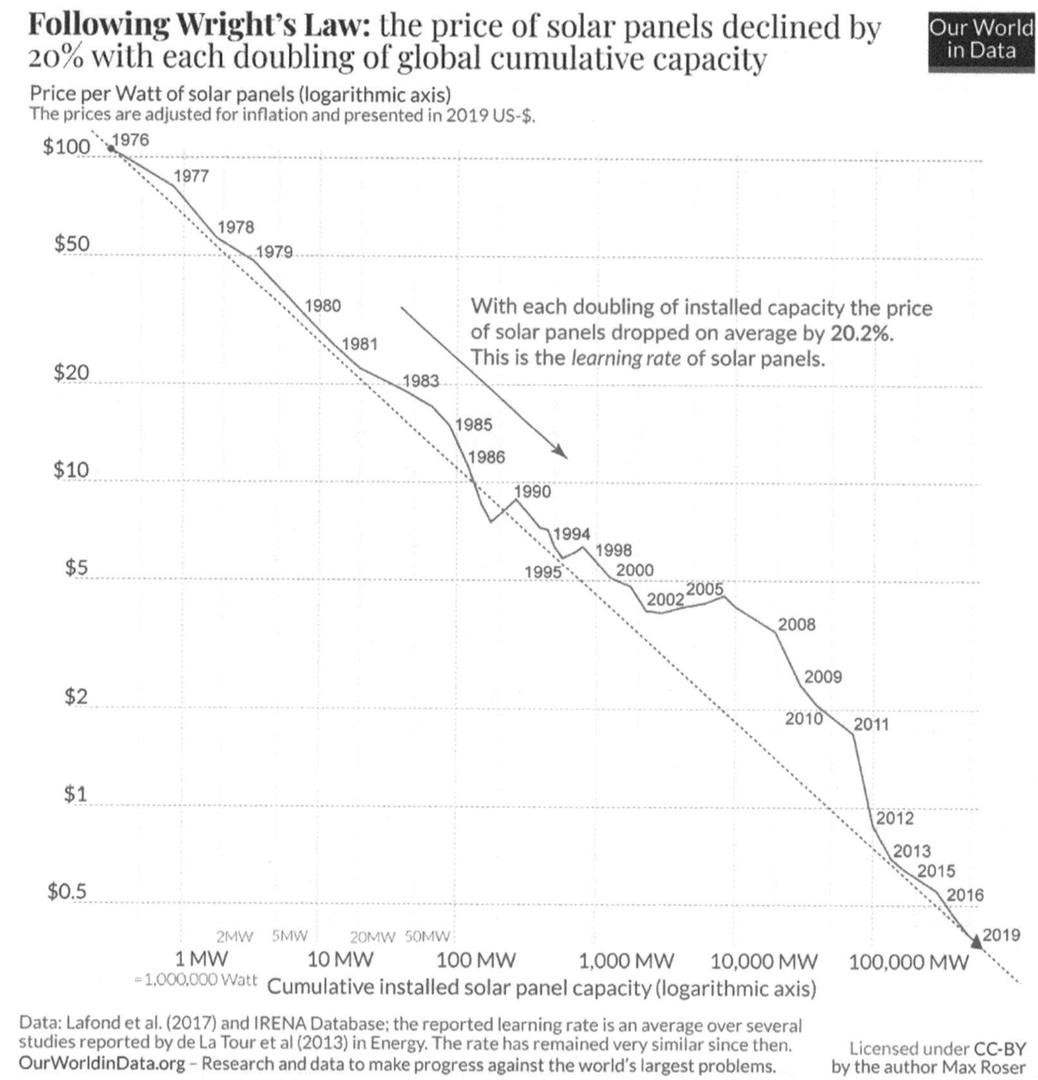

FIGURE 11.31
Price of solar PV modules versus cumulative global capacity. (Our World in Data [17]. 2023 Licensed under Creative Commons license CC BY 4.0 [17].)

Utility-scale projects are much lower in cost than residential systems due in large part to the economy of scale involved in deploying a very large ground-mounted array versus many small roof-mounted arrays. Utility-scale installations have also taken advantage of a number of improvements in recent years. Many systems now employ simple, low-cost single-axis trackers that rotate north-south oriented rows from east to west during the course of the day. These require greater spacing between the rows to avoid row-to-row shading and so take up about 20% more land per MW. But they provide more annual megawatt-hours for the same number of panels and have a lower levelized cost of electricity.

Also, bifacial panels are becoming more commonly used. These allow increased solar collection from the back of each panel via solar radiation reflected off the ground, albeit at a lower efficiency than the front of the panel. These can boost the total amount of annual electricity generated per square meter of array on the order of 5%–15% depending on the location, array geometry, and ground reflectance [18]. Finally, larger format panels have reduced installation time and costs. NREL shows an average installed cost of residential PV in 2023 as $2.70/W and single-axis tracking utility-scale systems as $1.20/W [19].

Because the boom in PV sales has only occurred in the last decade and the expected lifetime of modules is 20–30 years or longer, recycling capability is still being developed. The European Union and the State of California already have regulations impacting this, however. More details on PV recycling can be found in [20].

11.8 Using System Advisor Model for PV Performance Estimates

The System Advisor Model (SAM) provides three different models for analyzing the performance of PV systems: PVWatts, a SAM detailed model, and a concentrating PV model. These models all handle both performance and cost calculations. For most PV systems, the first two models are of most interest. PVWatts is simple to use and is excellent for initial design. It can also be run on the web outside of SAM by going to https://pvwatts.nrel.gov. The SAM detailed model uses data from actual commercial PV panels and inverters and allows greater detail in system specification. It is best used for final calculations of a system with known, specific components. Both models perform hour-by-hour simulations that provide estimated monthly and annual energy production in kilowatts. Users select a location and choose either to use default values or they can input their own system parameters for size, electricity costs, array type, tilt angle, and azimuth angle. The program uses an extensive library of weather and solar radiation data for many locations. From these data, the calculator determines the irradiation on the PV array and the average cell temperature for each hour of the year. The DC energy output for each hour is calculated from the PV system DC rating and the incident solar radiation and then corrected for the PV cell temperatures. The AC energy for each hour is then calculated by multiplying the DC energy output by the overall DC-to-AC derate factor, including the inverter efficiency as a function of load. Hourly values of the AC energy are then summed to calculate monthly and annual AC energy production. The SAM model is available for free download at https://sam.nrel.gov.

The following example demonstrates the use of the SAM PVWatts model to calculate the performance of a residential PV system. The same approach can be used to estimate the output of utility-scale PV applications.

Example 11.10

Using the PVWatts online software, estimate the incident solar radiation and AC energy output for a 4-kW solar PV array in Phoenix, Arizona, annually and during October. Note that the 4-kW electric rating is for the DC output. Assume a DC-to-AC derate factor of 0.77, a south-facing array at a tilt of 20°. What tilt angle will achieve the maximum annual energy output for this location?

Estimate the levelized cost of electric power for the system. Assume that the cost of the module is 0.65 $/W. The real discount rate is 5.5%, the inflation rate is 2.5%, and the lifespan of the system is 30 years.

Solution:

We will analyze this system using SAM, which can be accessed through https://sam.nrel.gov. (Note: SAM is frequently updated. The instructions that follow are for version 2023.12.17 and may change.)

After downloading SAM from the website, to start a PV project:

1. Start SAM.
2. Click **Start a New Project**
3. Choose **Photovoltaic- PVWatts – Distributed – Residential Owner** and click OK in the lower right.
4. Click **Location and Resource** on the left and search for and choose phoenix_az (TMY)

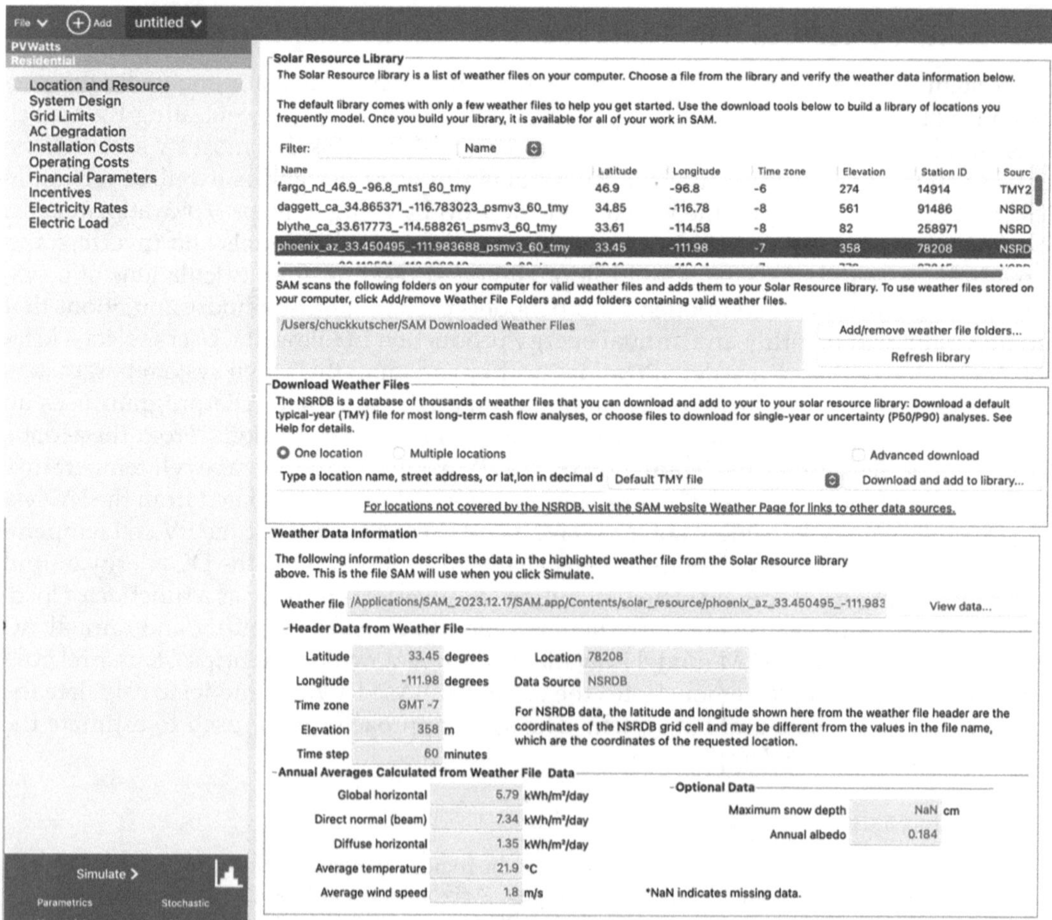

5. Choose **System Design** on the left and next to System nameplate capacity enter 4 kWdc. Note that the model assumes a DC-to-AC ratio of 1.15 and chooses the size of the inverter accordingly. Next to **Tilt** enter 20° and next to **Azimuth** enter 180°.

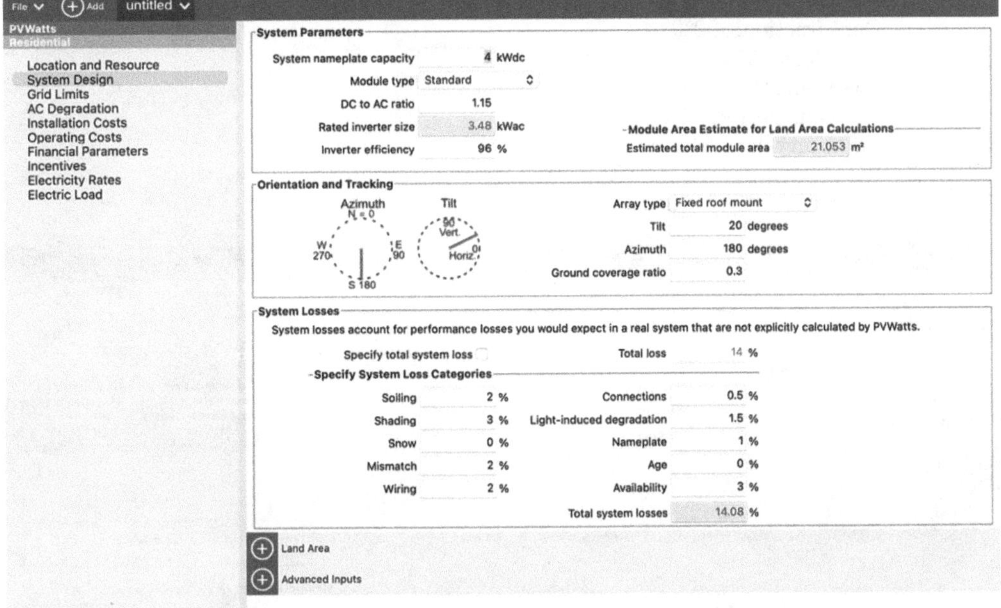

6. Choose **Installation Costs** on the left and enter a module cost of 0.50 $/Wdc. Leave the other costs at their default values.

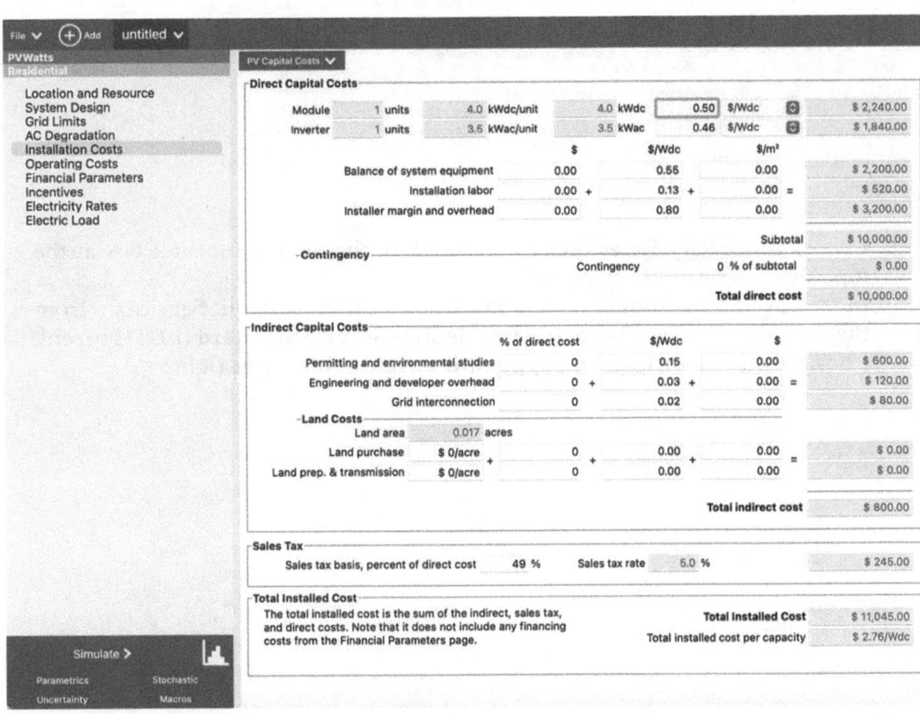

7. Choose **AC Degradation** on the left and enter a degradation rate of 0.7% per year.
8. Choose **Financial Parameters** on the left. Check the box for a Standard Loan at the top. Enter values for:

   Debt fraction: 100%
   Loan term: 25 years
   Loan rate: 5%/year
   Analysis period: 25 years
   Inflation rate: 2.5%/year
   Real discount rate: 5.5%/year
   Leave the other values as defaults

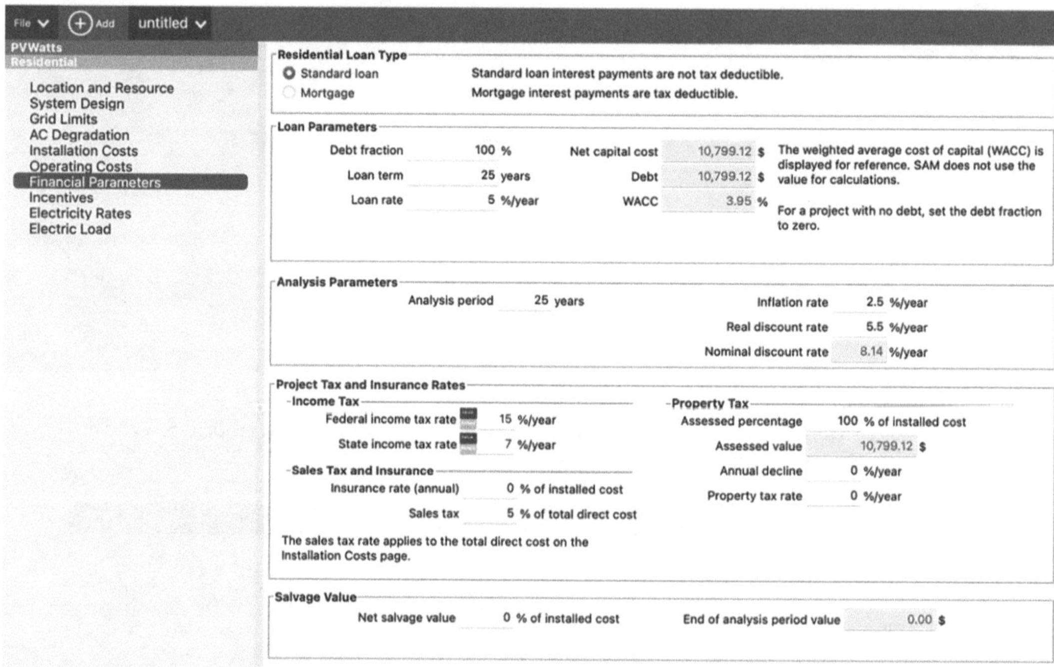

9. Choose **Electricity Rates** on the left. Click in the **Search for rates** box at the top. A new screen will come up.
10. In the box next to **Filter** type Arizona. Choose Arizona Public Service Co from the list. Then on the right choose **Residential Service Standard (E12) [Frozen]**. Click the box **Download and apply utility rate** at the lower right.

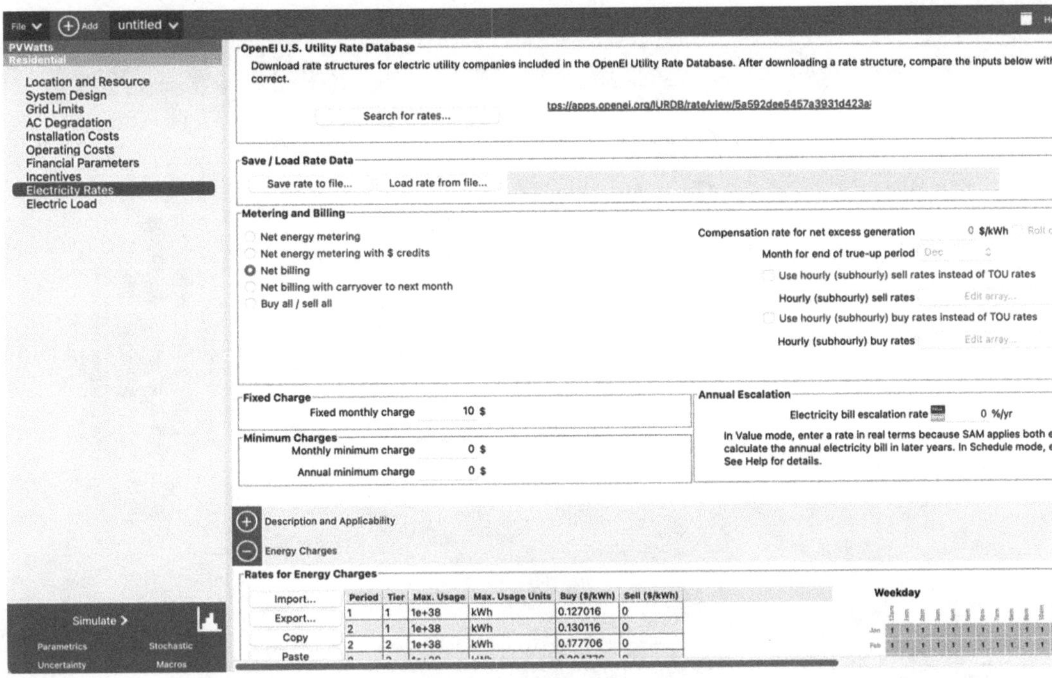
untitled
PVWatts
Residential
Location and Resource
System Design
Grid Limits
AC Degradation
Installation Costs
Operating Costs
Financial Parameters
Incentives
Electricity Rates
Electric Load
OpenEI U.S. Utility Rate Database
Download rate structures for electric utility companies included in the OpenEI Utility Rate Database. After downloading a rate structure, compare the inputs below with correct.
Search for rates...
Save / Load Rate Data
Save rate to file...
Load rate from file...
Metering and Billing
Net energy metering
Net energy metering with $ credits
Net billing
Net billing with carryover to next month
Buy all / sell all
Compensation rate for net excess generation 0 $/kWh
Month for end of true-up period Dec
Use hourly (subhourly) sell rates instead of TOU rates
Hourly (subhourly) sell rates
Use hourly (subhourly) buy rates instead of TOU rates
Hourly (subhourly) buy rates
Fixed Charge
Fixed monthly charge 10 $
Minimum Charges
Monthly minimum charge 0 $
Annual minimum charge 0 $
Annual Escalation
Electricity bill escalation rate 0 %/yr
Description and Applicability
Energy Charges
Rates for Energy Charges
Import...
Export...
Copy
Paste
Period Tier Max. Usage Max. Usage Units Buy ($/kWh) Sell ($/kWh)
1 1 1e+38 kWh 0.127016 0
2 1 1e+38 kWh 0.130116 0
2 2 1e+38 kWh 0.177706 0
Weekday
Simulate >
Parametrics
Stochastic
Uncertainty
Macros

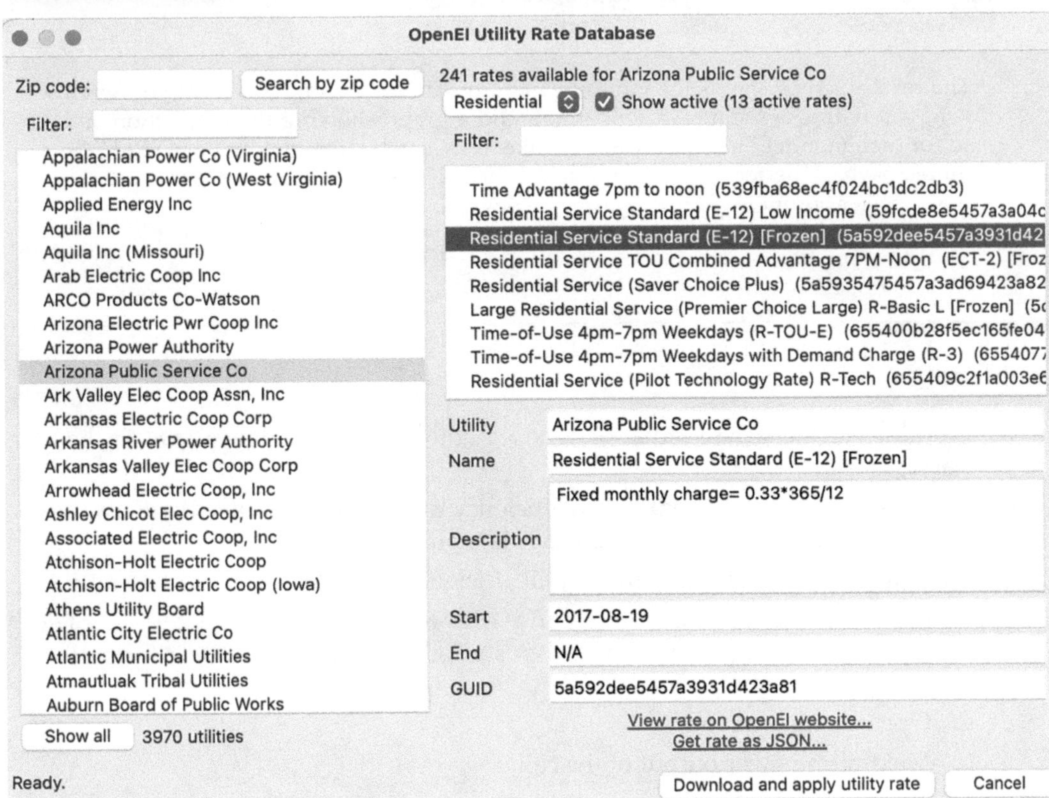
OpenEI Utility Rate Database
Zip code:
Search by zip code
Filter:
Appalachian Power Co (Virginia)
Appalachian Power Co (West Virginia)
Applied Energy Inc
Aquila Inc
Aquila Inc (Missouri)
Arab Electric Coop Inc
ARCO Products Co-Watson
Arizona Electric Pwr Coop Inc
Arizona Power Authority
Arizona Public Service Co
Ark Valley Elec Coop Assn, Inc
Arkansas Electric Coop Corp
Arkansas River Power Authority
Arkansas Valley Elec Coop Corp
Arrowhead Electric Coop, Inc
Ashley Chicot Elec Coop, Inc
Associated Electric Coop, Inc
Atchison-Holt Electric Coop
Atchison-Holt Electric Coop (Iowa)
Athens Utility Board
Atlantic City Electric Co
Atlantic Municipal Utilities
Atmautluak Tribal Utilities
Auburn Board of Public Works
Show all
3970 utilities
Ready.
241 rates available for Arizona Public Service Co
Residential
Show active (13 active rates)
Filter:
Time Advantage 7pm to noon (539fba68ec4f024bc1dc2db3)
Residential Service Standard (E-12) Low Income (59fcde8e5457a3a04c
Residential Service Standard (E-12) [Frozen] (5a592dee5457a3931d42
Residential Service TOU Combined Advantage 7PM-Noon (ECT-2) [Froz
Residential Service (Saver Choice Plus) (5a5935475457a3ad69423a82
Large Residential Service (Premier Choice Large) R-Basic L [Frozen] (5c
Time-of-Use 4pm-7pm Weekdays (R-TOU-E) (655400b28f5ec165fe04
Time-of-Use 4pm-7pm Weekdays with Demand Charge (R-3) (6554077
Residential Service (Pilot Technology Rate) R-Tech (655409c2f1a003e6
Utility Arizona Public Service Co
Name Residential Service Standard (E-12) [Frozen]
Description Fixed monthly charge= 0.33*365/12
Start 2017-08-19
End N/A
GUID 5a592dee5457a3931d423a81
View rate on OpenEI website...
Get rate as JSON...
Download and apply utility rate
Cancel

11. Click **Simulate** at the lower left.

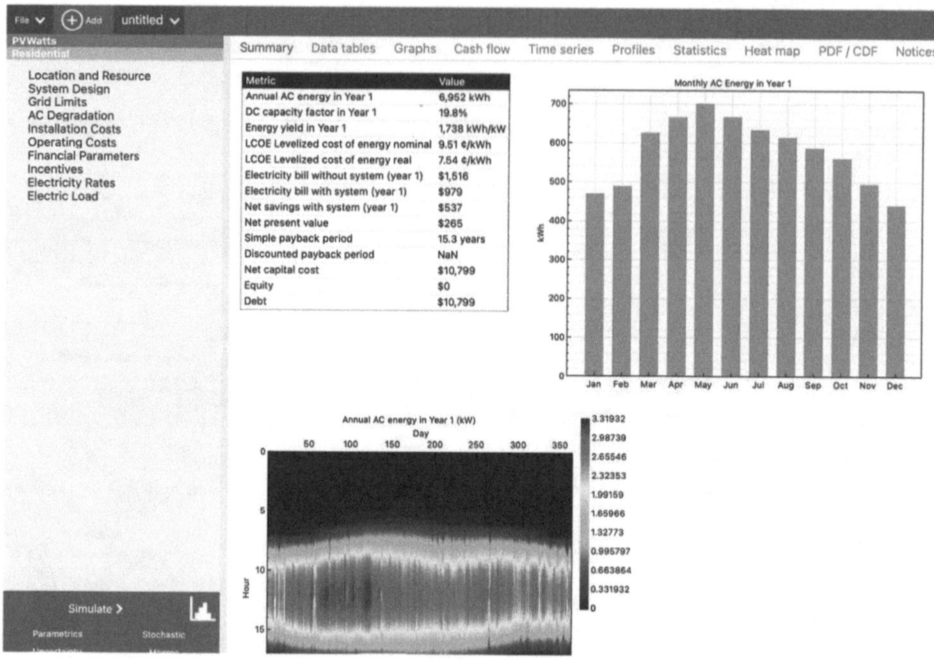

The results screen shows the monthly and annual energy production. (If you scroll down, you will also see a cash flow graph and a graph showing the production and load for each month.) This shows an annual energy production in year 1 of 6,952 kWh, a simple payback period of 15.3 years, and a capacity factor in year 1 of 19.8% (i.e., the average power output is 19.8% of the design power output).

Problems

11.1 Find the wavelength of radiation whose photons have energy equal to the band gap of GaAs.

11.2 What is the theoretical maximum efficiency of conversion if blue light of wavelength 0.45 μm is incident on a GaAs solar cell?

11.3 Find the theoretical maximum overall efficiency of GaAs solar cells in space.

11.4 The reverse saturation current I_o of a silicon cell at 40°C is 1.8×10^{-7} A. The short-circuit current when exposed to sunlight is 5 A. From this information, compute:

a. Open-circuit voltage.

b. Maximum power output of the cell.

c. The number of 4×4 cm cells needed to supply 100 W at 12 V. How must the cells be arranged?

11.5 At what efficiency is a photovoltaic array running if insolation on the collector is 650 W/m^2, the total collector area is 10 m^2, the voltage across the array is 50 V, and the current being delivered is 15 A?

11.6 If a PV cell has a maximum power output of 10 W under an insolation level of 600 W/m^2, what must the insolation be to achieve a power output of 17 W? Would you expect the open-circuit voltage to increase or decrease? Would you expect the short-circuit current to increase or decrease?

11.7 A PV battery system has an end-to-end efficiency of 77%. The system is used to run an all-AC load that is run only at night. The charge controller efficiency is 96% and the inverter efficiency is 85%. How much energy will need to be gathered by the PV array if the load is 120 W running for 4 h/night?

11.8 If the average output of the PV system in Problem 11.7 is 200 W and the load is changed to run during the day, how much PV output energy is needed for the same load conditions? Assume that the battery bank is at 100% charge and that input efficiency is equal to output efficiency.

11.9 For the system in Problem 11.7, how many hours of sunlight are needed to ensure that the battery bank is at 100% charge at the end of the day assuming the same load?

11.10–11.14 The owner of a small cabin would like to convert her home to PV power. She has the following equipment and associated run times:

Household Equipment	Power (W)	Run Time (Day) (Hour)	Run Time (Night) (Hour)
Lighting (DC)	25	2	4
Stereo (AC)	40	3	2
Refrigerator (DC)	125	3[a]	3[a]
Water pump (DC)	400	1.5[b]	0.5[b]
Alarm clock (DC)	8	12	12
Computer (AC)	250	3	0
Printer (AC)	175	0.25	0
Outdoor safety lights (DC)	48	0[c]	8[c]
Answering machine (AC)	7	12	12
Coffee pot (AC)	1,200	0.25	0

[a] The refrigerator is assumed to run 25% of the time.
[b] It can be assumed that the pump and fan will not start precisely at the same instant.
[c] It is assumed that the night- and daylighting loads will not be on simultaneously.

11.10

a. What is the homeowner's daily energy requirement as measured from the load?

b. If she replaces her alarm clock with a wind-up clock, how much energy per day will she avoid using?

c. What would you suggest she do to cut back her daily load?

11.11 How many 50 W panels will the owner require assuming battery storage is 75% efficient and all loads are DC (no inverter)?

a. For a stationary system "seeing" 5 hours of sunlight per day?
b. For a tracking system "seeing" 8 hours of sunlight per day?

11.12 For the loads listed:

a. What size inverter (peak watts) should she purchase?
b. If the inverter is 88% efficient, how much more daily energy is required from the PV array as compared to an all-DC system?

11.13 The homeowner decides to hire you to design a system for her. She has arranged with a local solar supplier for the following equipment. Specify the system and provide a line diagram.

PV panels	42 W, nominal 12 V
Batteries	125 Ah, 6 V, end-to-end efficiency = 72%
Charge controller	95% efficient, 12 V
Inverter	90% efficient, sizes of 500, 1,000, 2,000, and 4,000 W available, 12 V input

11.14 Redesign the system in Problem 11.13 with the following equipment:

PV panels	51 W, nominal 12 V
Batteries	200 Ah, 12 V, end-to-end efficiency = 78%
Charge controller	97% efficient, 24 V
Inverter	91% efficient, sizes of 500, 1,200, 2,500, and 5,000 W available, 24 V input

11.15 A flashing beacon is mounted on a navigation buoy in the shipping channel at a port at 30°N latitude. The load consists of a single lamp operating 1.0 seconds on and 3.6 seconds off during the hours of darkness. Hours of darkness vary from 9.8 hours in July to 13.0 hours in December. The lamp draws 2 A at 12 V when lighted. A flasher controls the lamp and draws 0.22 A when the lamp is on. There is a surge current of 0.39 A each time the flasher turns on. This current flows approximately one-tenth of the time the flasher is on. The design has 14 days of battery capacity. Provision has to be made to disconnect the load if the battery voltage drops below 11 V.

The available module has a rated voltage of 17.2 V at 25°C (15 V at 55°C) and 2.3 A at 1 kW/m^2. The available battery has a rated capacity of 105 Ah at 12 V. Assume that the maximum depth of discharge is 0%. Design the PV system.

11.16 Design a PV system for the following application: A refrigerator/freezer unit for vaccine storage in a remote island of Roatan, Honduras (16°N latitude, 86°W longitude, temperature, range 15°C–30°C).
Two compressors—one each for refrigerator and freezer
Each compressor draws 5 A at 12 V

Compressors Remain on for	Summer (h/day)	Winter (h/day)
Refrigerator	9	5
Freezer	7	4

Design the PV system using the panels and the batteries described in Problem 11.15.

11.17 A homeowner in Santa Fe, New Mexico, is interested in having a solar PV system installed at his house. He would like the system to be able to cover all of his annual energy needs. The following is a table from his utility provider showing the monthly energy use in the household. His roof with solar access is facing due east at a 20° pitch with no shading. In addition, the local electric company is offering a rebate on residential solar installations of $2.00 per installed Watt, as well as a renewable energy credit rebate based on predicted system performance. The Renewable Energy Credit (REC) is $2.50/W, which is then multiplied by the system performance. Since the installation is east facing, the predicted performance efficiency is 83% (with a system facing due south having a performance efficiency of 100%).

Consumption

	kWh
January	400
February	400
March	600
April	600
May	800
June	1,000
July	1,000
August	900
September	800
October	600
November	600
December	400

a. Design a system that will offset 100% of the home's consumption.

b. Assuming $0.10/W electricity expense with a 5% annual inflationary rate increase, and a $7.50/W installed cost, what will the payback be for this system after the rebate from the utility provider?

c. How do you wire the solar panel? Specifically, how many cells would you need and how would you wire them (series versus parallel) if the solar panel is to have a working voltage of 24.46 V and an amperage of 7.6 A? Assume the individual cells produce 98 V and 3.8 A.

Resources

The SMA string sizing program can be found at https://www.sma.de/en/products/apps-software/sunny-design.

PV Watts can be found at http://pvwatts.nrel.gov.

11.18 A village in Antigua (17°N, 61°W, 15°C–30°C temperature, range), West Indies, requires 5,000 gal of water per day for community water supply. Assuming a year-round average insolation of 8 kWh/day, design the system using the following components:

PV panels: Solarex panels 17.5 V and 3.6 A at 1,000 W/m^2 and 25°C

Pump: Grundfos multistage pump input 105 V DC, 9 A, 30% efficiency

11.19 Prepare a preliminary design for a solar PV system to provide 1 kW of stand-alone 24 h/day power to a travel trailer. Make a schematic diagram of the overall system, a cost estimate, a cell array design, and a battery storage. Then discuss the assumptions you made for your design.

References

1. Fortune Business Insights, https://www.fortunebusinessinsights.com/industry-reports/solar-pv-market-100263.
2. National Renewable Energy Laboratory, March 2024. Energy and carbon payback times for modern U.S. Utility Photovoltaic Systems, NREL/FS-7A40-88653, https://www.nrel.gov/docs/fy24osti/88653.pdf.
3. Our World in Data, Solar photovoltaic module price, https://ourworldindata.org/grapher/solar-pv-prices. Last accessed January 2025.
4. Lazard, June 2024. Levelized cost of energy+, https://www.lazard.com/media/xemfey0k/lazards-lcoeplus-june-2024-_vf.pdf.
5. Angrist, S.W., 1976. *Direct Energy Conversion*, 3rd ed., Allyn & Bacon, Boston, MA.
6. Bube, R.H., 1960. *Photoconductivity of Solids*, John Wiley & Sons, New York.
7. Moore, C.E., 1952. *Atomic Energy Levels*, Vol. 2, National Bureau of Standards Circular 467, U.S. Government Printing Office, Washington, DC.
8. Garg, H.P., 1987. *Advances in Solar Energy Technology*, Vol. 3, D. Reidel Publishing Company, Dordrecht, the Netherlands.
9. Goswami, Y., Kreith, F., and Kreider, J.F., 2000. *Principles of Solar Engineering*, 2nd ed., Taylor & Francis, Philadelphia, PA.
10. Messenger, R.A., and Ventre, J., 2010. *Photovoltaic Systems Engineering*, 3rd ed., CRC Press, Boca Raton, FL.
11. Sandia National Laboratories, 1991. Maintenance and operation of stand-alone photovoltaic systems, Albuquerque, NM, https://www.-satcon.com.
12. Post, H.N., and Risser, V.V., 1995. *Stand-Alone Photovoltaic Systems—A Handbook of Recommended Design Practices*, Sandia National Laboratory Report SAND-87–7023, Sandia National Laboratories, Albuquerque, NM.
13. NREL, National solar radiation data base, https://nsrdb.nrel.gov. Last accessed January 2025.
14. PV Insider, March 23, 2015. Share of thin film module production to remain at 7% through 2019, https://analysis.newenergyupdate.com/pv-insider/thin-film-pv/share-thin-film-module-production-remain-7-through-2019.
15. NREL, Best research cell efficiency chart, https://www.nrel.gov/pv/interactive-cell-efficiency.html. Last accessed January 2025.
16. The Economist, October 21, 2024, Perovskite crystals may represent the future of solar power.
17. Our World in Data, Learning curves: What does it mean for a technology to follow Wright's Law?, April 18, 2023. https://news-infographics-maps.net/learning-curve.html.
18. Ballif, C., Haug, F.-J., Boccard, M., Verlinden, P.J., and Hahn, G., 2022. Status and perspectives of crystalline silicon photovoltaics in research and industry, *Nature Reviews Materials* 7, 597–616.
19. NREL, Solar Installed System Cost Analysis, https://www.nrel.gov/solar/market-research-analysis/solar-installed-system-cost.html. Last accessed January 2025.
20. Preet, S., and Smith, S., April 5, 2024. A comprehensive review on the recycling technology of silicon based photovoltaic solar panels: Challenges and future outlook, *Journal of Cleaner Production* 448, 141661, https://www.sciencedirect.com/science/article/pii/S0959652624011090.

12

Solar Thermal Collectors and Systems

We are star stuff harvesting sunlight.

Carl Sagan

The basic principle of solar thermal collection is that when solar radiation strikes a surface, part of it is absorbed, thereby increasing the temperature of the surface. The efficiency of that surface as a solar collector depends not only on the absorption efficiency but also on how the energy losses to the surroundings are minimized and how the collected energy is removed for useful purposes. Various solar thermal collectors range from unglazed flat-plate-type solar collectors operating at about 5°C–10°C above the ambient, to central receivers operating at above 1,000°C. Table 12.1 lists various types of solar thermal collectors and their typical temperature and concentration ranges.

This chapter analyzes the thermal and optical performance of several solar thermal collectors. They range from non-concentrating, flat plate types (see Figure 12.1) to compound-curvature, and continuously tracking types with concentration ratios up to 1,000 or more. Applications of the energy conversion by solar thermal collectors are described in this chapter for building hot water, for electric power, and for process heat. The next section describes some fundamental radiation properties of materials, knowledge of which helps in the design and analysis of solar thermal collectors.

12.1 Radiation Properties of Materials

When radiation strikes a body, a part of it is reflected, a part is absorbed, and if the material is transparent, a part is transmitted, as shown in Figure 12.2. The fraction of the incident radiation reflected is defined as the reflectance ρ, the fraction absorbed as the absorptance α, and the fraction transmitted as the transmittance τ. According to the first law of thermodynamics, these three components must add up to unity, or

$$\alpha + \tau + \rho = 1 \tag{12.1}$$

TABLE 12.1

Types of Solar Thermal Collectors and Their Typical Temperature Range above Ambient

Type of Collector	Concentration Ratio	Typical Working Temperature Range (°C)
Flat plate collector	1	≤70
Evacuated tube collector	1.5–2	60–120
Parabolic trough collector	10–50	150–350
Parabolic dish collector	200–500	250–700
Central receiver	500 to >3,000	500 to >1,000

DOI: 10.1201/9781003586616-12

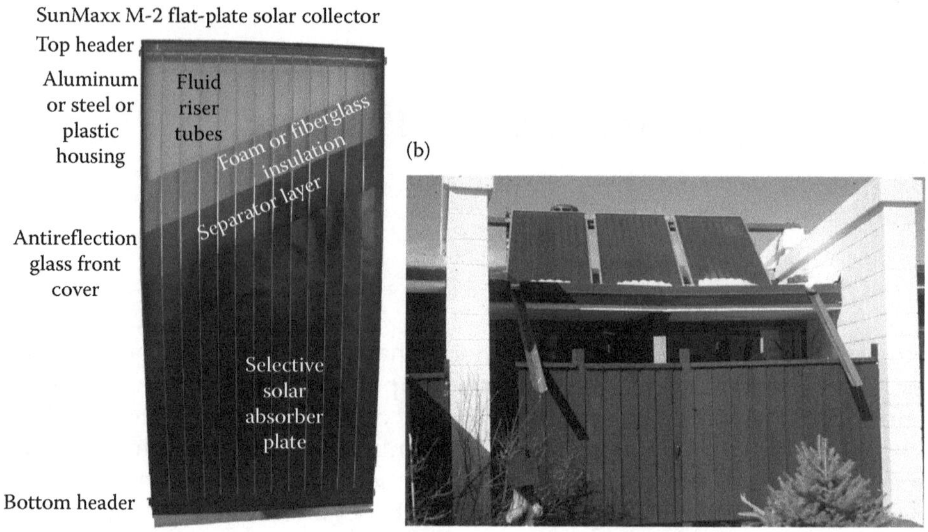

FIGURE 12.1
(a) Photo of an active flat plate solar collector. (From Barry Butler, PhD, Butler Sun Solutions, Inc., Solana Beach, CA. With permission.) (b) Photo of an active flat plate solar collector on Kreith's residence.

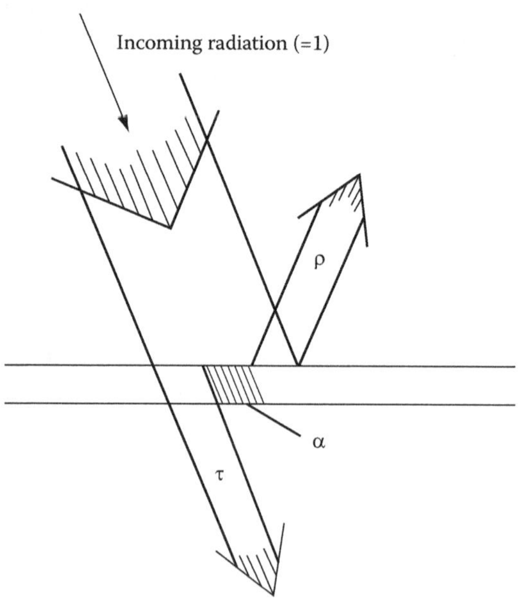

FIGURE 12.2
Schematic representation of transmittance τ, absorptance α, and reflectance ρ.

Opaque bodies do not transmit any radiation and $\tau = 0$.

The reflection of radiation can be *specular* or *diffuse*. When the angle of incidence is equal to the angle of reflection, the reflection is called specular; when the reflected radiation is uniformly distributed in all directions, it is called diffuse. No real surface is either purely specular or purely diffuse, but a highly polished surface approaches specular reflection, whereas a rough surface reflects diffusely.

Another important radiative property is called emittance, ε, which is the ratio of the radiative emissive power of a real surface to that of an ideal "black" surface.

All of the radiative properties of materials, α, τ, ρ, and ε can be functions of the wavelength and direction. In fact, such dependence is used in the design of solar energy devices and systems. For example, selective absorbers are used for solar collectors and passive heating systems, and glazing materials for daylighting and solar collectors.

The monochromatic directional emittance of a surface, $\varepsilon_\lambda(\theta, \varphi)$, in a direction signified by an azimuth angle φ and a polar angle θ, is

$$\varepsilon_\lambda(\theta,\phi) = \frac{I_\lambda(\theta,\phi)}{I_{b\lambda}} \tag{12.2}$$

The total directional emittance $\varepsilon(\theta, \varphi)$ over all the wavelengths or the monochromatic hemispherical emittance ε_λ can be obtained by the integration of $\varepsilon_\lambda(\theta, \varphi)$ over all the wavelengths or the entire hemispherical space, respectively. The overall emittance, ε, is found by integrating the product of the hemispherical emittance and emissive power (see Chapter 10) over all the wavelengths:

$$\varepsilon = \frac{1}{\sigma T^4}\int_0^\infty \varepsilon_\lambda E_{b\lambda}\, d\lambda \tag{12.3}$$

Observe that both ε_λ and ε are properties of the surface.

12.1.1 Selective Surfaces

Two types of special surfaces of great importance in solar collector systems are selective and reflecting surfaces. Selective surfaces combine a high absorptance for solar radiation with a low emittance for the temperature range in which the surface emits radiation. This combination of surface characteristics is possible because 98% of the energy in incoming solar radiation is contained within wavelengths below 3 μm, whereas 99% of the radiation emitted by black or gray surfaces at 400 K is at wavelengths longer than 3 μm. The dotted

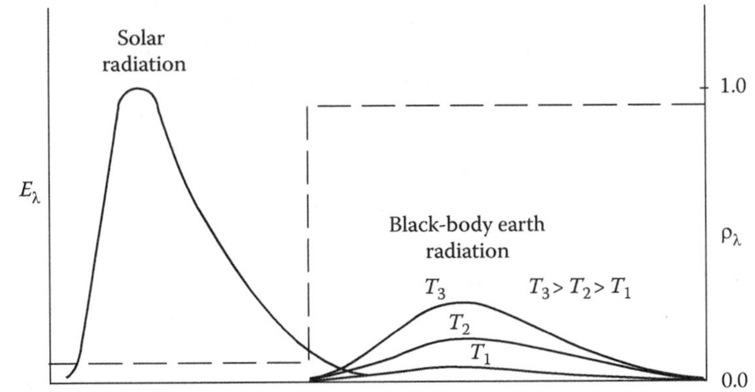

FIGURE 12.3
Illustrations of the spectral reflectance characteristics of an ideal selective surface. Shows radiation from ideal surfaces at different temperatures and solar radiation.

TABLE 12.2

Properties of Some Selective Plated Coating Systems

				Durability	
Coating[a]	Substrate	$\bar{\alpha}_s$	$\bar{\varepsilon}_i$	Breakdown Temperature (°F)	Humidity-Degradation MIL STD 810B
Black nickel on nickel	Steel	0.95	0.07	>550	Variable
Black chrome on nickel	Steel	0.95	0.09	>800	No effect
Black chrome	Steel	0.91	0.07	>800	Completely rusted
	Copper	0.95	0.14	600	Little effect
	Galvanized steel	0.95	0.16	>800	Complete removal
Black copper	Copper	0.88	0.15	600	Complete removal
Iron oxide	Steel	0.85	0.08	800	Little effect
Manganese oxide	Aluminum	0.70	0.08		
Organic overcoat on iron oxide	Steel	0.90	0.16		Little effect
Organic overcoat on black chrome	Steel	0.94	0.20		Little effect

Source: From U.S. Dept. of Commerce [1].

[a] Black nickel coating plated over a nickel–steel substrate has the best selective properties ($\bar{\alpha}_s = 0.95$, $\bar{\varepsilon}_i = 0.07$), but these degraded significantly during humidity tests. Black chrome plated on a nickel–steel substrate also had very good selective properties ($\bar{\alpha}_s = 0.95$, $\bar{\varepsilon}_i = 0.09$) and also showed high resistance to humidity.

line in Figure 12.3 illustrates the spectral reflectance (referring to the y-axis on the right) of an ideal, selective semi-gray surface having a uniform reflectance of 0.05 below 3 μm, but 0.95 above 3 μm. Real surfaces do not approach this performance. Table 12.2 lists properties of some selective coatings [1].

12.1.2 Reflective Surfaces

Concentrating solar collectors require the use of reflecting surfaces with high specular reflectance in the solar spectrum or refracting devices with high transmittance in the solar spectrum. Reflecting surfaces are usually highly polished metals or metal coatings on suitable substrates. With opaque substrates, the reflective coatings must always be on the front surface, for example, chrome plating on copper or polished aluminum. If a transparent substrate is used, however, the coating may be front- or back-surfaced. In any back-surfaced reflector, the radiation must pass through the substrate twice, and the transmittance of the material becomes very important. Table 12.3 presents typical values of the normal specular reflectance of new surfaces for beam solar radiation.

The purpose of an absorber of a flat plate collector is to absorb as much of the incident solar radiation as possible, re-emit as little as possible, and allow efficient transfer of heat to a working fluid. The materials used for absorber plates include copper, aluminum, stainless steel, galvanized steel, plastics, and rubbers. Copper seems to be the most common material used for absorber plates and their tubes because of its high thermal conductivity and high corrosion resistance. However, copper is quite expensive. For low-temperature applications (up to about 50°C or 120°F) a plastic material called ethylene propylene diene monomer (trade names EPDM, HCP, etc.) can be used to provide inexpensive absorber material. To compensate for the low thermal conductivity, adequate thickness and close spacing of fluid passages are needed to provide sufficient heat transfer.

TABLE 12.3

Specular Reflectance Values for Solar Reflector Materials

Material	ρ
Silver (unstable as front surface mirror)	0.94±0.02
Gold	0.76±0.03
Aluminized acrylic, second surface	0.86
Anodized aluminum	0.82±0.05
Various aluminum surfaces range	0.82–0.92
Copper	0.75
Back-silvered water-white plate glass	0.88
Aluminized type-C Mylar (from Mylar side)	0.76

In order to increase the absorption of solar radiation and to reduce the emission from the absorber, the metallic absorber surfaces are painted or coated with flat black paint or a selective coating.

12.2 Energy Balance for a Flat Plate Collector

The thermal performance of any type of solar thermal collector can be evaluated by an energy balance that determines the portion of the incoming radiation that is delivered as useful energy to the working fluid [2].

The energy collected, Q_{coll}, can be expressed as

$$Q_{coll} = Q_{in} - Q_{out} \tag{12.4}$$

where

$$Q_{in} = \tau \alpha I_c A_c \tag{12.5}$$

Thus, the energy absorbed is the incident solar radiation on the collector area, I_c, multiplied by the transmittance of any glass or plastic covers (which are employed to reduce heat loss to the ambient) and the absorptance of the absorber plate. The energy lost to the environment can be expressed as:

$$Q_{out} = U_L A_c \left(T_p - T_a\right) \tag{12.6}$$

where T_p is the average absorber plate temperature and T_a is the ambient air temperature. This assumes that the heat loss from a flat plate collector occurs mainly by convection to the outside air and that the back of the collector is highly insulated (although any such heat loss would be included in actual collector efficiency measurements). U_L is the overall convective heat loss coefficient and depends on the number of transparent covers, or glazings. Calculating the value of U_L is covered in solar energy texts, and this is important when designing a solar collector. Generally, however, commercially available collectors have measured performance values that can be used to determine annual performance.

The energy collected is just the difference between the absorbed solar radiation and the heat loss to the environment, or

$$Q_{\text{coll}} = \tau\alpha I_c A_c - U_L A_c \left(T_p - T_a\right) \tag{12.7}$$

Because the average plate temperature is difficult to measure and control, the collected heat is usually expressed in terms of the more easily measured and controlled inlet fluid temperature, $T_{f,\text{in}}$. To change from the average plate temperature to the inlet fluid temperature the energy term is multiplied by the heat removal factor, F_R, which is the ratio of the actual energy collected to the energy that would be collected if the collector plate were at the inlet fluid temperature.

Thus,

$$Q_{\text{coll}} = F_R\tau\alpha I_c A_c - F_R U_L A_c \left(T_{f,\text{in}} - T_a\right) \tag{12.8}$$

The efficiency of the collector, η_c, is just the energy collected divided by the solar radiation incident on the collector area, or

$$\eta_c = \frac{Q_{\text{coll}}}{I_c A_c} = F_R\tau\alpha - F_R U_L \frac{\left(T_{f,\text{in}} - T_a\right)}{I_c} \tag{12.9}$$

Thus, the efficiency of a flat plate collector can be plotted as a straight line vs. $(T_{f,\text{in}} - T_a)/I_c$ as shown in Figure 12.4 [3]. The y-intercept, or $F_R\tau\alpha$, is the product of F_R and the optical efficiency. (The optical efficiency is simply $\tau\alpha$ and is the efficiency when there is no heat loss. Although the y-intercept, $F_R\tau\alpha$, is sometimes referred to as the optical efficiency because it is at the zero point on the heat loss axis, in actuality the average plate temperature is higher than the ambient temperature, so there is a small amount of heat loss.) The slope of the line is $-F_R U_L$. This is just the convective heat loss coefficient, but corrected with F_R to account

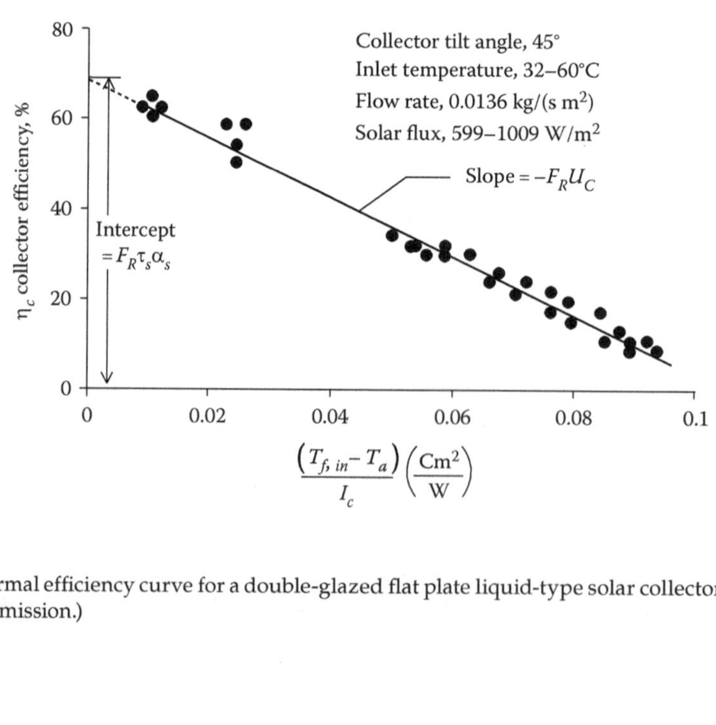

FIGURE 12.4
Experimental thermal efficiency curve for a double-glazed flat plate liquid-type solar collector. (From Goswami et al. [3]. With permission.)

for the fact that the abscissa contains the inlet fluid temperature and not the plate temperature. Equation 12.9 is commonly called the Hottel–Whillier–Bliss equation in honor of the three solar energy researchers who developed it.

Flat plate collectors are generally used for applications where air or water needs to be heated to temperatures up to about 50°C–60°C (or roughly 120°F–140°F). Thus, they can be used for domestic water heating, space heating, crop drying, and low-temperature industrial process heat. Unglazed flat plate collectors are often used to heat swimming pools.

12.3 Experimental Testing of Collectors

The performance of solar thermal systems depends largely on the performance of solar collectors. Therefore, experimental measurement of the thermal performance of solar collectors by standard methods is important and necessary. The experimentally determined performance data are needed for design purposes and for determining the commercial value of the collectors. The thermal performance of a solar collector is determined by establishing an efficiency curve from the measured instantaneous efficiencies for a combination of values of incident solar radiation, ambient temperature, and inlet fluid temperature [4]. An instantaneous efficiency of a collector under steady-state conditions can be established by measuring the mass flow rate of the heat-transfer fluid, its temperature rise across the collector ($T_{f,e}-T_{f,\text{in}}$), and the incident solar radiation intensity (I_c) as

$$\eta_c = \frac{q_u}{A_c I_c} = \frac{\dot{m}C_p\left(T_{f,e} - T_{f,\text{in}}\right)}{A_c I_c} \tag{12.10}$$

The efficiency, η_c, of a collector under steady state can then also be written according to the Hottel–Whillier–Bliss equation (Equation 12.9).

Figure 12.4 shows experimental data for a typical flat plate collector [3]. Since τ and α can be measured independently, a thermal performance curve of a flat plate collector allows us to establish the values of F_R and U_L also.

Figure 12.5 shows typical performance curves of various glazed and unglazed non-tracking solar collectors.

The product $\tau\alpha$ will change with the angle of incidence. Since flat plate collectors are normally fixed, the angle of incidence changes throughout the day. A relationship can be written between the actual or effective $(\tau\alpha)$ and $(\tau\alpha)_n$ for normal incidence as

$$(\tau\alpha) = (\tau\alpha)_n K_{\tau\alpha} \tag{12.11}$$

where

$(\tau\alpha)_n$ is the value of the product for normal angle of incidence
$K_{\tau\alpha}$ is called the incidence angle modifier.

Therefore, the thermal performance of a flat plate collector may be written as

$$\eta_c = F_R\left[K_{\tau\alpha}(\tau\alpha)_n - U_L\frac{\left(T_{f,\text{in}} - T_a\right)}{I_c}\right] \tag{12.12}$$

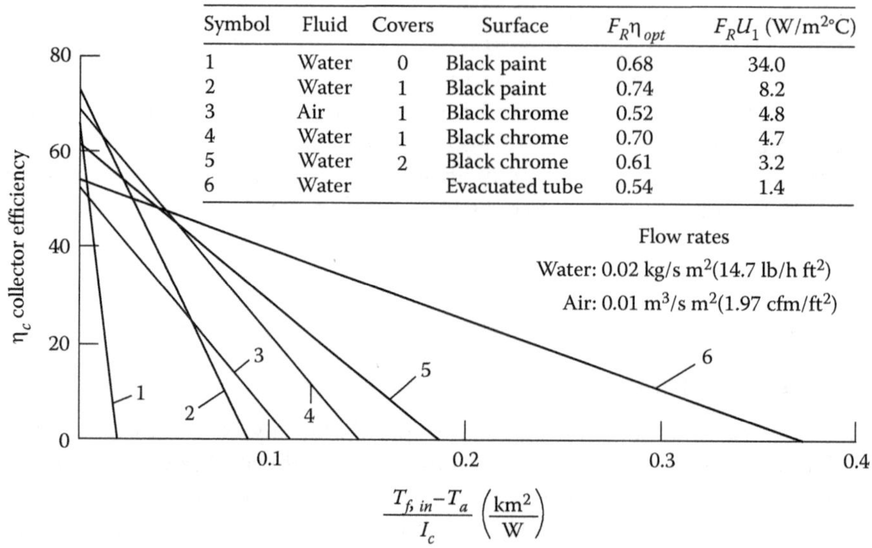

Symbol	Fluid	Covers	Surface	$F_R\eta_{opt}$	F_RU_1 (W/m$^{2\circ}$C)
1	Water	0	Black paint	0.68	34.0
2	Water	1	Black paint	0.74	8.2
3	Air	1	Black chrome	0.52	4.8
4	Water	1	Black chrome	0.70	4.7
5	Water	2	Black chrome	0.61	3.2
6	Water		Evacuated tube	0.54	1.4

FIGURE 12.5
Typical performance curves for various non-tracking solar collectors.

The thermal performance of collectors is usually found experimentally for normal angles of incidence in which case $K_{\tau\alpha} = 1.0$. The incidence angle modifier is then measured separately. It has been established that $K_{\tau\alpha}$ is of the form

$$K_{\tau\alpha} = 1 - b\left(\frac{1}{\cos i} - 1\right) \tag{12.13}$$

where
b is a constant
i is the angle of incidence.

Figure 12.6 shows the effect of incidence angle on $K_{\tau\alpha}$ for flat plate collectors [4].

12.3.1 Testing Standards for Solar Thermal Collectors

Standard testing procedures adopted by regulating agencies in various countries establish ways of comparing the thermal performance of various collectors under the same conditions. In the United States, the thermal performance standards established by the American Society of Heating, Refrigerating and Air-Conditioning Engineers (ASHRAE) have been accepted for thermal performance tests of solar thermal collectors. Similar standards have been used in a number of countries. The standards established by ASHRAE include:

- ASHRAE Standard 93–77, *Methods of testing to determine the thermal performance of solar collectors*, specifies the procedures for determining the time constant, thermal performance, and the incidence angle modifier of solar thermal collectors using

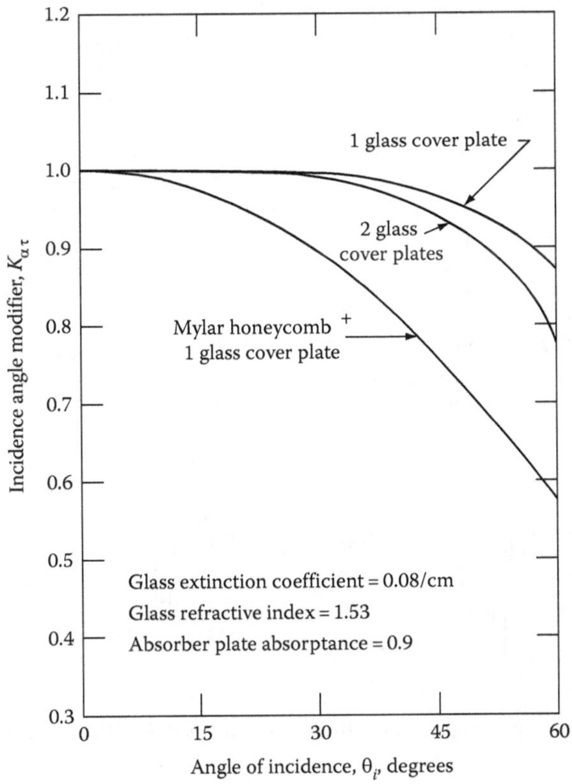

FIGURE 12.6
Incidence angle modifiers for three flat plate solar collectors. (From ASHRAE [4]. With permission.)

a liquid or air as a working fluid [4]. Figure 12.7 shows a schematic of a standard testing configuration for thermal performance testing. The tests are conducted under quasi-steady-state conditions.

- ASHRAE Standard 96–80, *Methods of testing to determine the thermal performance of unglazed solar collectors* [5].

ASHRAE 93–77 specifies that a curve for incidence angle modifier be established by determining the efficiencies of a collector for average angles of incidence of 0°, 30°, 45°, and 60° while maintaining $T_{f,i}$ within ±1°C of the ambient temperature. Since $(T_{f,i}-T_a)=0$, the incidence angle modifier according to Equation 12.13 is

$$K_{\tau\alpha} = \frac{\eta_c}{F_R(\tau\alpha)_n} \tag{12.14}$$

The denominator is the y-intercept of the efficiency curve as shown in Figure 12.4.

In order to avoid transient effects, the performance of a collector is measured and integrated over at least the time constant of the collector. The time constant is measured by operating the collector under steady or quasi-steady conditions with a solar flux of at least 790 W/m² and then abruptly cutting the incident flux to zero while continuing to measure the fluid inlet ($T_{f,i}$) and exit ($T_{f,e}$) temperatures. The fluid inlet temperature is maintained

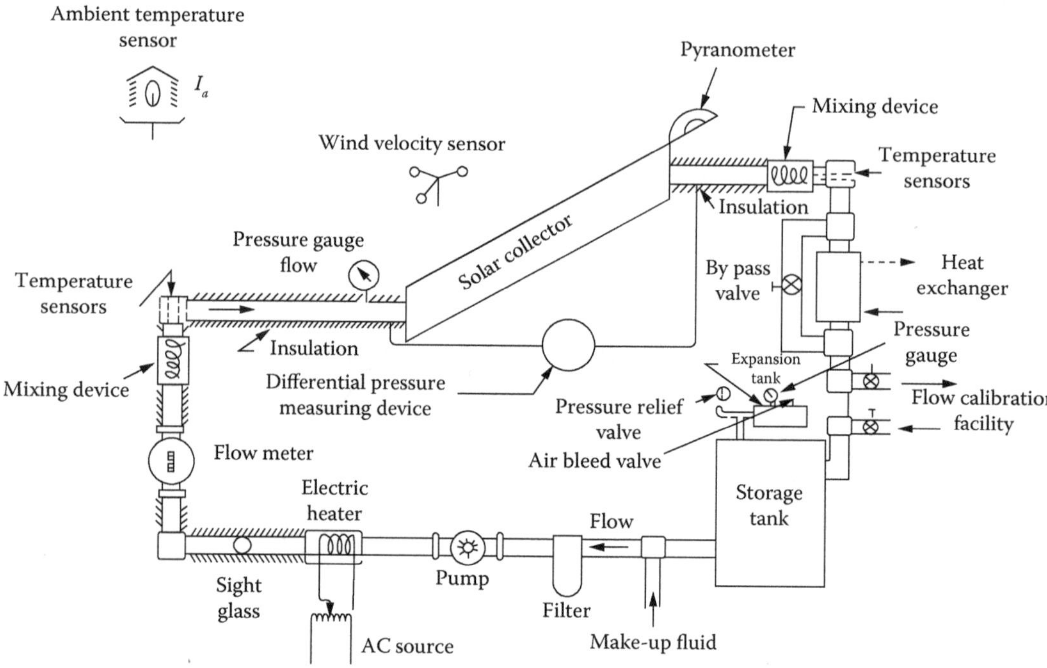

FIGURE 12.7
A testing configuration for a liquid-type solar collector. (From ASHRAE [4]. With permission.)

within ±1°C of the ambient temperature. The time constant is then determined as the time required to achieve

$$\frac{T_{f,e} - T_{f,\text{in}}}{T_{f,e,\text{initial}} - T_{f,\text{in}}} = 0.30 \tag{12.15}$$

For determining the thermal performance of a collector, ASHRAE Standard 93–77 specifies test conditions to obtain at least four data points for the efficiency curve [4]. The test conditions include:

1. Near normal incidence ($i \le 5°$) angle tests close to solar noon
2. At least four tests for each $T_{f,\text{in}}$, two before and two after solar noon
3. At least four different values of $T_{f,\text{in}}$ to obtain different values of $\Delta T/I_c$, preferably to obtain ΔT at 10%, 30%, 50%, and 70% of stagnation temperature rise under the given conditions of solar intensity and ambient conditions.

12.4 Evacuated Tube Collectors

Although flat plate collectors can heat water close to the boiling point, they are inefficient at such high temperatures due to high convective heat losses. Hence, they are usually operated at temperatures below 60°C. Higher operating temperatures can be achieved in two ways: concentrating the solar energy to reduce the area for heat loss or using a vacuum to reduce the heat loss. Evacuated collectors are made in tubular form to withstand the pressure differential between ambient air and the vacuum. Various designs have been used for evacuated tube collectors. Absorbers made of metal and glass have been used inside the tubes. In all these designs, reflectors are used behind the tubes to reflect back any solar energy passing between the absorbers, so they have a small amount of concentration.

Evacuated tube collectors have been used to achieve the higher operating temperatures needed to power absorption chillers. The fact that evacuated tube collectors have a flatter efficiency curve than flat plates means that they also have higher efficiency in lower solar radiation conditions. The x-axis of the efficiency curve is $\Delta T/I$, so the x-value is large for either a high value of ΔT (i.e., a high operating temperature) or a low value of I. Thus, evacuated tube collectors can be superior to flat plates even for heating domestic hot water in locations with only moderate solar radiation.

The efficiency curve of an evacuated tube collector is shown as curve 6 in Figure 12.5.

12.5 Transpired Air Collectors

Flat plate collectors can be used to heat water (or a water/antifreeze mixture) or air. A special type of flat plate air collector is the transpired collector. This concept is used to heat ambient air, either as a preheater for building ventilation air or for crop drying or other dehydration applications. Figure 12.8 shows a transpired wall system in which the air is drawn through a flat perforated absorber plate by a building ventilation fan. In a conventional air collector, recirculated air flows along the solar absorber plate, and transparent glazing is used to reduce heat loss to the outside air. Heat transfer from a plate to air blowing alongside it is relatively poor. The transpired design draws air through numerous

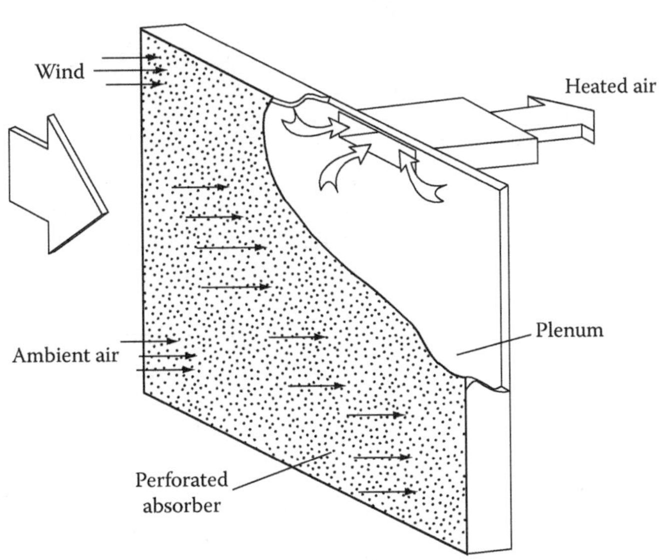

FIGURE 12.8
Unglazed transpired solar collector for air preheating.

holes or slits in the absorber plate. Because the objective is to heat ambient air, no glazing is used. The outside air is drawn through the openings and the suction prevents convective heat loss to the ambient. Thus, heat-transfer rates are high, and the lack of a glazing both reduces cost and increases the amount of solar energy reaching the absorber.

Kutscher and Christensen [6] presented a thermal analysis of this system. From a heat balance on the transpired unglazed collector, the useful heat collected is

$$q_u = I_c A_c \alpha - U_c A_c (T_{\text{out}} - T_a) \tag{12.16}$$

The overall heat loss coefficient, U_c, which is due to radiative and convective losses, is given as

$$U_c = \frac{h_r}{\varepsilon_{\text{hx}}} + h_c \tag{12.17}$$

where
ε_{hx} is absorber heat-exchanger effectiveness
h_r is a linearized radiative heat-transfer coefficient
h_c is the convective heat loss coefficient.

The heat-exchanger effectiveness for air flowing through the absorber plate is defined as

$$\varepsilon_{\text{hx}} = \frac{T_{\text{out}} - T_a}{T_p - T_a} \tag{12.18}$$

where T_p is collector plate temperature. The forced convective heat loss coefficient due to a wind velocity of U_∞ is given as

$$h_c = 0.82 \frac{U_\infty \upsilon \rho c_p}{L V_0} \tag{12.19}$$

where
ν is the kinematic viscosity of air in m^2/s
c_p is the specific heat of the air in J/kg K
V_0 is the suction velocity in m/s
L is the length of the collector in m.

The convective heat loss is associated with the boundary layer of warm air that blows off the downstream end of the collector. Because the suction keeps this thermal boundary layer extremely thin, the convective heat loss for a large wall is negligibly small if proper design criteria are met.

Radiation heat loss occurs both to the sky and to the ground. Assuming the absorber is gray and diffuse with an emissivity, ε_c, the radiative loss coefficient, h_r, is

$$h_r = \varepsilon_c \sigma \frac{\left(T_p^4 - F_{\text{cs}} T_{\text{sky}}^4 - F_{\text{cg}} T_{\text{gnd}}^4\right)}{T_p - T_a} \tag{12.20}$$

where F_{cs} and F_{cg} are the view factors between the collector and the sky, and collector and the ground, respectively. For a vertical wall with infinite ground in front of it, both F_{cs} and F_{cg} will be 0.5. Using the earlier equations, Kutscher and Christensen [6] showed that the

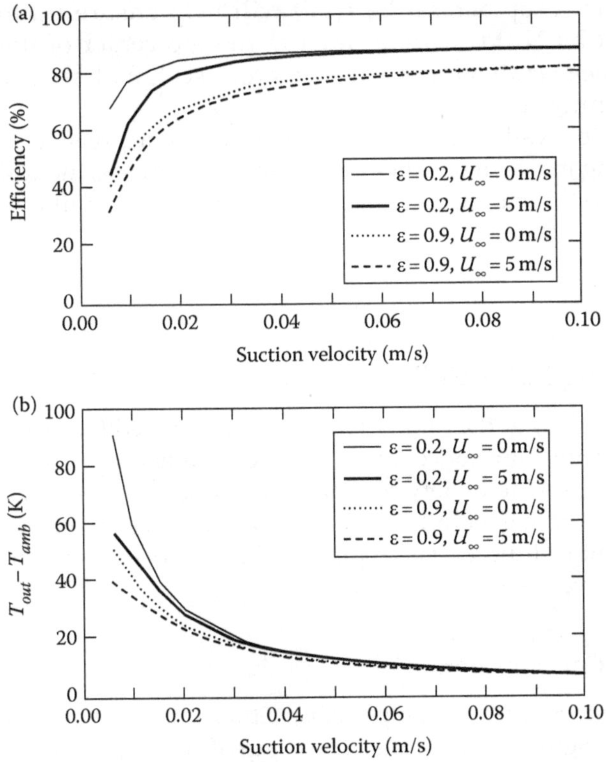

FIGURE 12.9
Predicted thermal performance of unglazed transpired collector: (a) efficiency vs. suction velocity and (b) temperature rise vs. suction velocity. (From Kutscher and Christensen [6]. With permission.)

predicted performance matches the measured performances well. Figure 12.9 shows their predicted thermal performances.

Analysis and measurements have shown that at sufficient suction rates, there is virtually no convective loss from a large wall even in windy conditions. The only heat loss is by radiation, and this is small because the absorber typically operates as a preheater at temperatures of only about 10°C–20°C above ambient. (For higher temperature applications, a selective surface can be used to reduce radiative loss.) Properly designed, these collectors can operate at efficiencies of well over 70% for higher suction flow rates.

Research at the National Renewable Energy Laboratory (NREL) [7] concluded that high performance can be achieved by maintaining both sufficient suction velocity and uniform flow. Suction velocity should be no less than 0.02 m/s and preferably greater than 0.04 m/s. Pressure drop across the wall should be at least 25 Pa to prevent very low flow or even outflow effects over portions of the wall subject to local drops in pressure due to the wind. Providing this pressure drop means the use of low wall porosity, typically less than 1%. Flow uniformity is affected by three things: the distance between a point on the wall and the inlet air duct to the building; thermal stack (buoyancy) effect in the plenum between the collector and the building wall; and the Bernoulli effect by which acceleration of the plenum air as it moves upward (due to the addition of air) causes a static pressure drop with height. These three effects can be approximately balanced with the proper choice of plenum depth. The choice of a sufficiently low porosity (to increase pressure drop across the wall) will help ensure uniform flow. In a wall designed and built by NREL with an inlet at the top center of the wall, a horizontal constant velocity header (having a variable cross-sectional area) was used to ensure left-to-right flow uniformity.

Transpired collector walls typically have dampers that prevent solar-heated air from entering the building in warm weather conditions. In this case, the wall self-cools by natural convection due to the stack effect, and the wall shades the building.

12.6 Concentrating Solar Collectors

To achieve high efficiency at the higher temperatures needed for producing process steam or generating electricity, high concentration of the solar energy is needed. This minimizes the area for heat loss. A wide variety of concentrating solar collector designs has been pursued over the years. These generally fall into two categories: line focus, in which the solar energy is focused onto a fluid-carrying pipe, and point-focus, whereby the solar energy is focused onto a small receiver.

12.6.1 Line-Focus Concentrators

Solar energy can be focused onto a pipe using either reflective or refractive optics. Today, most commercial designs use reflectors. These are of two types: (1) parabolic troughs and (2) linear Fresnel.

12.6.1.1 Parabolic Troughs

Parabolic troughs are the older and more widely used linear focus technology. Figure 12.10 shows a photo of a SkyTrough parabolic trough collector. Line-focus concentrators achieve high temperatures by using a high degree of concentration to reduce the heat-transfer area. Because the solar radiation must be highly concentrated, only direct normal irradiance, I_{DN}, is utilized. Hence, these concentrators work best in dry, desert-type environments having very clear skies.

The geometric relations for a parabolic trough solar concentrator are shown schematically in Figure 12.11. (Note this figure does not show the cylindrical evacuated tube that surrounds the absorber tube in high-temperature troughs.)

The surface of the reflector/concentrator has a parabolic cross section so that all rays of sunlight parallel to its axis are reflected from the surface to a focal line. The basic geometric relation that defines the parabolic shape for a point source is

$$y = \frac{x^2}{4f} \tag{12.21}$$

where f is the focal length of the concentrator. If D is the width of the concentrator aperture and d the absorber tube diameter, then the concentrator rim angle, φ, is

$$\tan\left(\frac{\phi}{2}\right) = \frac{D}{4f} \tag{12.22}$$

FIGURE 12.10
SkyTrough parabolic trough collector near Golden, Colorado, viewed by Professor Frank Kreith (right) and National Institute of Standards and Technology researcher, Dr. Isaac Garaway (left). (Courtesy of CU Engineering, 2009, Casey A. Cass, photographer.)

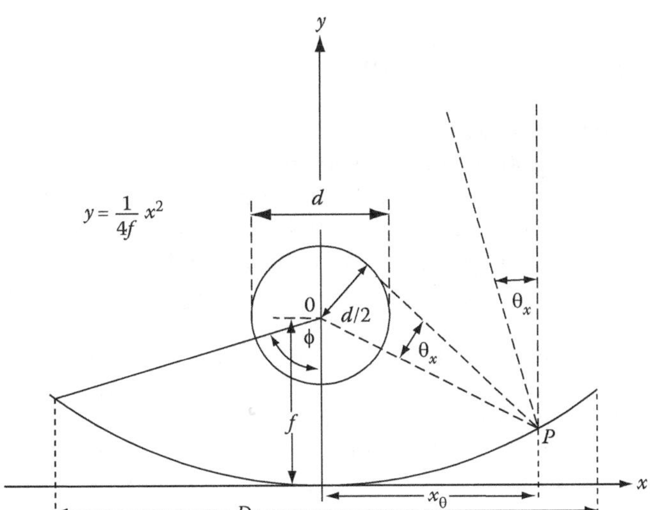

FIGURE 12.11
Geometric relations for a parabolic trough solar concentrator (evacuated cylindrical glass envelope not shown).

and the concentration ratio (based on the heat loss area), C, is

$$C = \frac{D}{\pi d} \tag{12.23}$$

In practice, parabolic troughs use concentrators with rim angles between 70° and 110°, as this range provides a balance between optical and structural requirements of the concentrator.

In addition to the optical error introduced by the finite specularity of the reflective surface, a number of other optical errors contribute to widening of the reflected beam at the receiver, such as concentrator slope error, tracking error, and receiver mislocation errors. Generally, these errors can be characterized by Gaussian distributions, so the total amount of beam spreading can be calculated as the root mean square (rms) value of all the optical errors.

For producing the high temperatures needed for efficient electricity generation, heat loss is also reduced by using a highly selective surface on the absorber tube and by enclosing the steel absorber tube within a high-vacuum Pyrex envelope. Because these collectors operate at much higher temperatures than flat plates or even evacuated tubes, they are subject to radiation heat loss. Hence, rather than their efficiency curves being flat, they curve downward. (Radiation heat loss is nonlinear and increases with temperature to the fourth power.)

Parabolic trough efficiency curves are often shown with an additional term containing $(\Delta T)^2$ to capture the curvature. This only works for one ambient temperature, however. If two identical troughs operate at the same ΔT above ambient, the one at the higher temperature will have a greater heat loss and so be less efficient. Thus, the efficiency of these high-temperature collectors is best represented by the operating temperature, rather than the ΔT relative to ambient. NREL measured the efficiency of a SkyTrough parabolic trough collector module manufactured by SkyFuel and found the following equation best fit the efficiency [8]:

$$\eta_c = -1.26\times10^{-6}x^3 + 3.02\times10^{-5}x^2 - 6.24\times10^{-4}x + 0.773 \tag{12.24}$$

Here η_c is the collector efficiency and $x = (T_{\text{htf}} - 25)/I^{0.33}$, where T_{htf} is the average temperature of the heat-transfer fluid in the collector module, and I is the direct normal irradiance. (Here, 25°C is used as a constant reference point because the receiver heat loss was separately measured in a laboratory at an ambient temperature of 25°C. At an operating temperature of 25°C, there was no heat loss to the ambient so the y-intercept when plotted this way corresponds to the optical efficiency.) The resulting efficiency plot is shown in Figure 12.12 [8].

Note that dividing the abscissa by $I^{0.33}$ allows all the different radiation curves to collapse onto a single curve. (The actual exponent is experimentally derived for each collector tested, but will generally be close to 0.33.) Note that the efficiency equation does not contain ambient temperature and is valid over a range of ambient temperatures.

The optical efficiency for this collector is 0.77. For a parabolic trough, the optical efficiency is the product of not just transmittance and absorptance, but of four factors, listed below in the order they occur for incoming solar radiation:

ρ = the reflectivity of the parabolic reflector mirror
τ = the transmittance of the glass enclosing the absorber tube

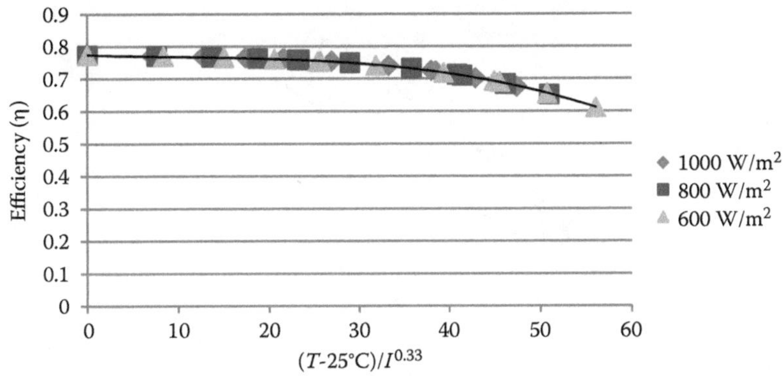

FIGURE 12.12
Efficiency plot for the SkyTrough collector. (From Kutscher et al. [8].)

γ=the intercept factor, or the fraction of the reflected solar radiation that strikes the absorber tube
α=the absorptance of the absorber tube.

Whereas ρ, τ, and α are optical properties of the materials used, γ depends on the collector geometry. Parabolic trough collectors are generally designed to have a very high intercept factor, on the order of 0.99. This means that the reflector shape must be a very accurate parabola with very little slope error. The larger the diameter of the absorber pipe, the easier it is to achieve a high intercept factor. Also, the pumping power for the heat-transfer fluid flowing through the absorber pipe varies inversely with the pipe diameter to the fifth power. Thus, larger diameters reduce parasitic pumping power. On the other hand, a larger pipe is more expensive and means a lower concentration ratio and a greater area for heat loss, so this is a design tradeoff.

At low temperatures, a parabolic trough may have a lower efficiency than a flat plate collector (especially an unglazed one) due to the optical losses associated with the reflector, the glass tube, and the intercept. But its lower heat loss results in an extremely flat (albeit curving) efficiency curve, so it maintains good efficiency at high temperatures.

12.6.1.2 Linear Fresnel Collectors

In most commercially successful parabolic trough collector designs, the receiver moves with the tracking parabolic mirror. In the linear Fresnel design, the parabolic mirror is broken up into a number of tracking, ground-mounted segments that reflect the solar energy onto a fixed horizontal receiver tube mounted at a height above the ground. (See Figure 12.13.) This design means that wind loads on the reflectors are reduced, thus resulting in structural savings. Also, linear Fresnel collector systems offer a higher ground cover ratio than trough systems. On the other hand, the combination of a fixed receiver and ground-mounted reflectors means that linear Fresnel systems have lower overall optical efficiency than troughs because of the reflection angles (cosine losses).

Commercial parabolic trough systems have been limited to an operating temperature of 390°C, which is the limit of suitable heat-transfer oils. There is interest in going to higher temperatures by using molten salt in place of the oil. However, the moving receiver tubes

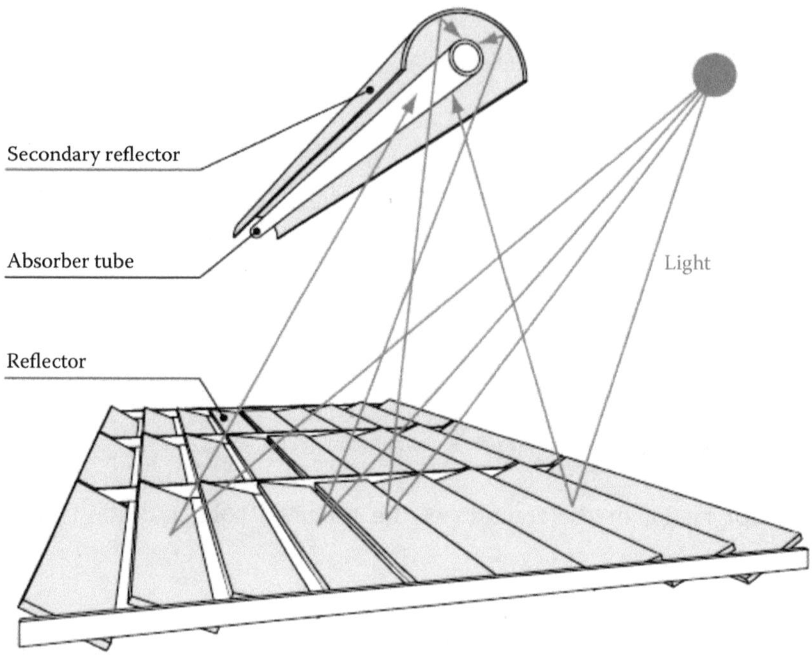

FIGURE 12.13
Linear Fresnel collector. (Courtesy of Andreas Poullikkas.)

in parabolic trough collectors usually entail the use of ball joints, and ball joints (to date) are not compatible with molten salt. Because linear Fresnel systems have stationary receivers, they may lend themselves more readily to the use of molten salt. More information on linear Fresnel collectors can be found in Zhu et al. [9].

12.6.2 Point-Focus Concentrators

The efficiency of thermodynamic power cycles depends strongly on the operating temperature. By focusing sunlight onto a small receiver instead of a long pipe, higher concentration ratios and their associated smaller heat loss areas allow higher temperatures to be achieved. These high-temperature concentrators can be divided into two basic types: parabolic dishes and power towers, or central receivers. Parabolic dishes use a collection of mirrors formed into a two-dimensional parabolic shape [10]. They track the Sun in two axes and always point directly at the Sun. For the generation of electricity, these dishes employ a thermodynamic power-cycle engine at the focal point. Both Brayton turbines and Stirling cycle engines have been employed. The most successful ones to date were developed by Stirling Energy Systems with the help of Sandia National Laboratories. The 1.5-MW Maricopa installation near Phoenix consisted of 60 25-kW dishes employing Stirling engines with hydrogen as the working fluid. Although the dishes performed well, the company was unable to compete with low-cost photovoltaics.

All concentrating solar power technologies have faced increasing competition from photovoltaic systems. The main advantage of concentrating solar power (CSP) systems compared to PV is that they can utilize thermal energy storage, which has a high round-trip efficiency and is less expensive than battery storage. Coupling thermal energy storage to a multitude of parabolic dish receiver engines is not straightforward. As a result, the CSP

FIGURE 12.14
Gemasolar, a 19.9 MW solar power tower plant in Fuentes de Andalucia, Sevilla, Spain, has 15 hours of molten salt storage and is capable of providing base load electricity 24 h/day. (Photo by Charles F. Kutscher.)

industry has focused its future on line-focus and power tower technologies with a recent trend toward the latter due to higher operating temperatures and thus higher power-cycle efficiencies.

Power towers (see Figure 12.14) consist of a large array of mirrors called heliostats (Figure 12.15) that track the Sun in two axes and reflect the solar radiation onto a receiver located atop a tall tower [3]. (Beam-down designs have also been developed in which the heliostats focus sunlight onto a mirror that reflects the energy down to a ground-mounted receiver, but tower-mounted receiver designs continue to dominate the marketplace.) The heliostats aim not directly at the Sun but midway between the Sun–Earth line and the line between the heliostat and a particular target point on the receiver to allow the reflected solar beam to strike the target. (Heliostats are individually aimed to achieve a reasonably uniform flux profile on the receiver.) A large power tower may have over 100,000 computer-controlled heliostats.

A key design decision is the size of the heliostat. Larger heliostats require a smaller number of tracking mechanisms and associated electronics but require more structure to handle higher wind loads. As electronic controllers have dropped in cost, the trend has been to design power towers with a larger number of smaller heliostats. The receiver on the tower can contain either water or molten salt as the working fluid. The latter is more compatible with the use of molten salt thermal storage. All the existing power tower systems in the world are used to produce electrical power and are explained in more detail in Section 12.8.

FIGURE 12.15
Power tower heliostats.

12.7 Solar Thermal Domestic Hot-Water, Space-Heating, and Cooling Systems

Solar water-heating systems represent the most common application of solar thermal energy at the present time. Small systems are used for domestic hot-water applications, whereas larger systems are used in industrial process heat applications. There are basically two types of water-heating systems: *natural circulation* or passive solar system (thermosyphon) and *forced circulation* or active solar system. Natural circulation solar water heaters are simple in design and low in cost. Their application is usually limited to nonfreezing climates. Forced-circulation water heaters are used in freezing climates and for commercial and industrial process heat.

12.7.1 Solar Thermosyphon Water Heating

Solar hot-water heating systems are composed of a collector and a storage tank. A passive solar collector with integrated storage is shown in Figure 12.16. When the flow between the collector and the tank is by natural circulation, these passive solar hot-water systems are referred to as thermosyphon systems. This ability of thermosyphon systems to heat water without an externally powered pump has spurred their use in regions where power is unavailable or is very expensive, and freezing is not a problem.

The natural tendency of a less dense fluid to rise above a denser fluid can be used in a simple solar water heater to cause fluid motion through a collector. The density difference

(a)

(b)

FIGURE 12.16
Passive solar collector with integrated storage. (Courtesy of SunEarth, Inc., Fontana, CA.) (a) Photograph of installed module on residence in Arizona. (b) Schematic of CopperHeart integral collector-storage system.

is created within the solar collector where heat is added to increase the temperature and decrease the density of the liquid. Figure 12.17 schematically illustrates the major components of a thermosyphon system [11].

The flow pressure drop in the fluid loop (ΔP_{FLOW}) must equal the buoyant force pressure difference ($\Delta P_{BUOYANT}$) caused by the different densities in the "hot" and "cold" legs (i.e., the left and right sides of Figure 12.17) of the fluid loop:

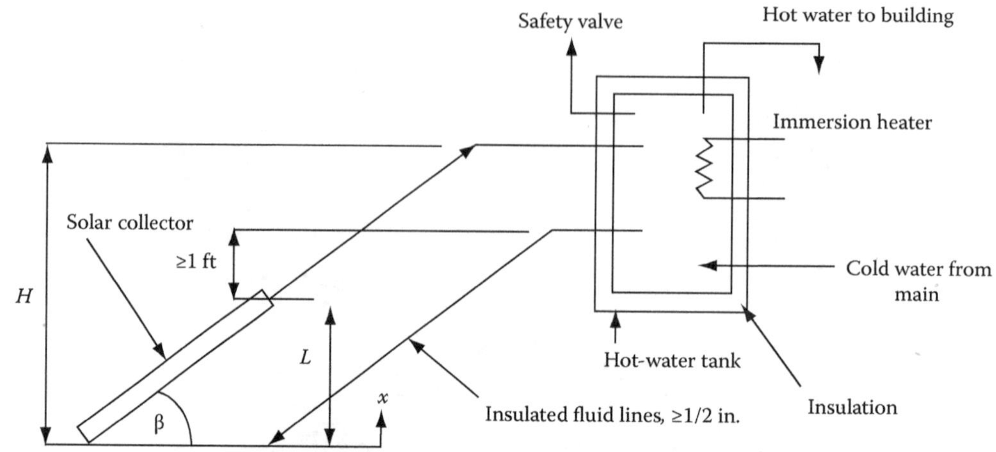

FIGURE 12.17
Schematic diagram of thermosyphon loop used in a natural circulation, service water-heating system. (From Kreith and West [1]. With permission.)

$$\Delta P_{\text{FLOW}} = \Delta P_{\text{BUOYANT}}$$

$$= \rho_{\text{stor}} g H - \left[\int_0^L \rho(x) g dx + \rho_{\text{out}} g (H - L) \right] \quad (12.25)$$

where

H is the height of the water level in the tank and L is the height of the collector (see Figure 12.17)

$\rho(x)$ is the local fluid density in the collector

ρ_{stor} is the tank fluid density

ρ_{out} is the collector outlet fluid density (the latter two densities are assumed uniform).

The flow pressure term ΔP_{FLOW} is related to the flow loop system head loss, which is in turn directly connected to pipe friction and fitting losses and the loop flow rate:

$$\Delta P_{\text{FLOW}} = \oint_{\text{LOOP}} \rho d(h_L) \quad (12.26)$$

where $h_L = KV^2$, with K being the sum of the component loss "velocity" factors (see any fluid mechanics text) and V the flow velocity.

Since the driving force in a thermosyphon system is only a small density difference and not a pump, larger-than-normal plumbing fixtures are used to reduce pipe friction losses. In general, one pipe size larger than would be used with a pump system is satisfactory. Under no circumstances should piping smaller than ½ in. (12 mm) national pipe thread (NPT) be used. Most commercial thermosyphons use 1 in. (25 mm) NPT pipe. The flow rate through a thermosyphon system is about 1 gal/ft²h (40 L/m²h) in bright sunlight, based on the collector area.

Since the hot-water system loads vary little during the year, the best angle to tilt the collector is the local latitude. The temperature difference between the collector inlet water and the collector outlet water is usually 15°F–20°F (8°C–11°C) during the middle of a sunny

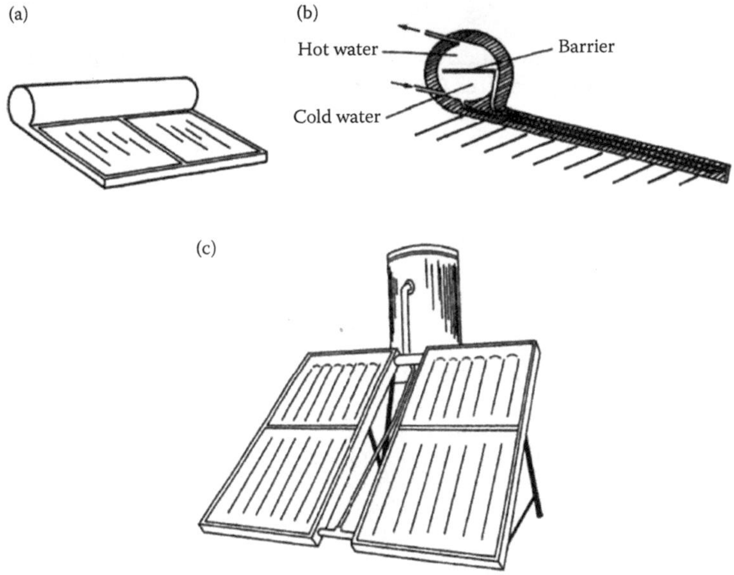

FIGURE 12.18
Passive solar water heaters: (a) compact model using combined collector and storage, (b) section view of the compact model, and (c) tank and collector assembly. (From Kreith and West [1]. With permission.)

day. After sunset, a thermosyphon system can reverse its flow direction and lose heat to the environment during the night. To avoid reverse flow, the top header of the absorber should be at least 1 ft. (30 cm) below the cold leg fitting on the storage tank, as shown.

To provide hot water during long cloudy periods, an electrical immersion heater can be used as a backup for the solar system. The immersion heater is located near the top of the tank to enhance stratification so that the heated fluid is at the required delivery temperature. Tank stratification is desirable in a thermosyphon to maintain flow rates as high as possible. Insulation must be applied over the entire tank surface to control heat loss. Figure 12.18 illustrates two common thermosyphon system designs [11].

Several features inherent in the thermosyphon design limit its utility. If it is to be operated in a freezing climate, a nonfreezing fluid must be used, which in turn requires a heat exchanger between the collector and the potable water storage. (The collector can be drained during cold periods instead.) Heat exchangers of either the shell-and-tube type or the immersion-coil type require higher flow rates for efficient operation than a thermosyphon can provide. Therefore, the thermosyphon is generally limited to nonfreezing climates. A further restriction on thermosyphon use is the requirement for an elevated tank. In many cases, structural or architectural constraints prohibit raised-tank locations. In residences, collectors are normally mounted on the roof, and tanks mounted above the high point of the collector can easily become the highest point in a building. Practical considerations often do not permit this application.

Example 12.1

Determine the pressure difference available for a thermosyphon system with 1 m high collector and 2 m high hot and cold legs. The water temperature input to the collector is 25°C, and the collector output temperature is 35°C. If the overall system loss velocity factor (K) is 15.6, estimate the system flow velocity.

Solution:

Equation 12.25 is used to calculate the pressure difference, with the water densities being found from the temperatures (in steam tables):

$$\rho_{\text{stor}}(25°\text{C}) = 997.009\ \text{kg/m}^3;\quad \rho_{\text{out}}(35°) = 994.036\ \text{kg/m}^3;$$

$\rho_{\text{coll.ave.}}(30°\text{C}) = 996.016\ \text{kg/m}^3$ (*Note*: average collector temperature used in integral)

and with $H=2$ and $L=1\,\text{m}$

$$\Delta P_{BUOYANT} = \left(997.009\frac{\text{kg}}{\text{m}^3}\right)9.81\text{m}/\text{s}^2(2\text{m})$$

$$-\left[\left(996.016\frac{\text{kg}}{\text{m}^3}\right)9.81\text{m}/\text{s}^2(1\text{m}) + \left(994.036\frac{\text{kg}}{\text{m}^3}\right)9.81\text{m}/\text{s}^2(1\text{m})\right]$$

$$= 38.9\ \text{N}/\text{m}^2(\ \text{Pa})$$

The system flow velocity is estimated from the system K value given and the pressure difference calculated earlier. Taking the average density of the water around the loop (at 30°C), and substituting it into Equation 12.26:

$$\Delta P_{\text{BUOYANT}} = (\rho_{\text{loop.ave.}})(h_L)_{\text{loop}} = (\rho_{\text{loop.ave.}})KV^2$$

$$V^2 = \frac{38.9\ \text{N/m}^2}{\left(996.016\frac{\text{kg}}{\text{m}^3}\right)(15.6)}$$

$$V = 0.05\ \text{m/s}$$

12.7.2 Forced-Circulation Solar Thermal Hot-Water Systems

If a thermosyphon system cannot be used for climatic, structural, or architectural reasons, a forced-circulation system is required. Figure 12.19 shows three configurations of forced-circulation systems: (1) open loop, (2) closed loop, and (3) closed loop with drain-back [12]. In an open-loop system (Figure 12.19a), the solar loop is at atmospheric pressure; therefore, the collectors are empty when they are not providing useful heat. A disadvantage of this system is the high pumping power required to pump the water to the collectors every time the collectors become hot. In a pressurized closed-loop system (Figure 12.19b) the loop remains filled at all times and the pump has to overcome only the resistance of the pipes.

In order to accommodate the thermal expansion of water from heating, a small (about 2-gallon capacity) expansion tank and a pressure relief valve are provided in the solar loop. Because water always stays in the collectors of this system, antifreeze (propylene glycol or ethylene glycol) is required for locations where freezing conditions can occur. During stagnation conditions (in summer), the temperature in the collector can become very high, causing the pressure in the loop to increase. This can cause leaks in the loop unless some fluid is allowed to escape through a pressure relief valve. Whether as a result of leaks or of draining, air can enter the loop, causing the pumps to run dry. This disadvantage

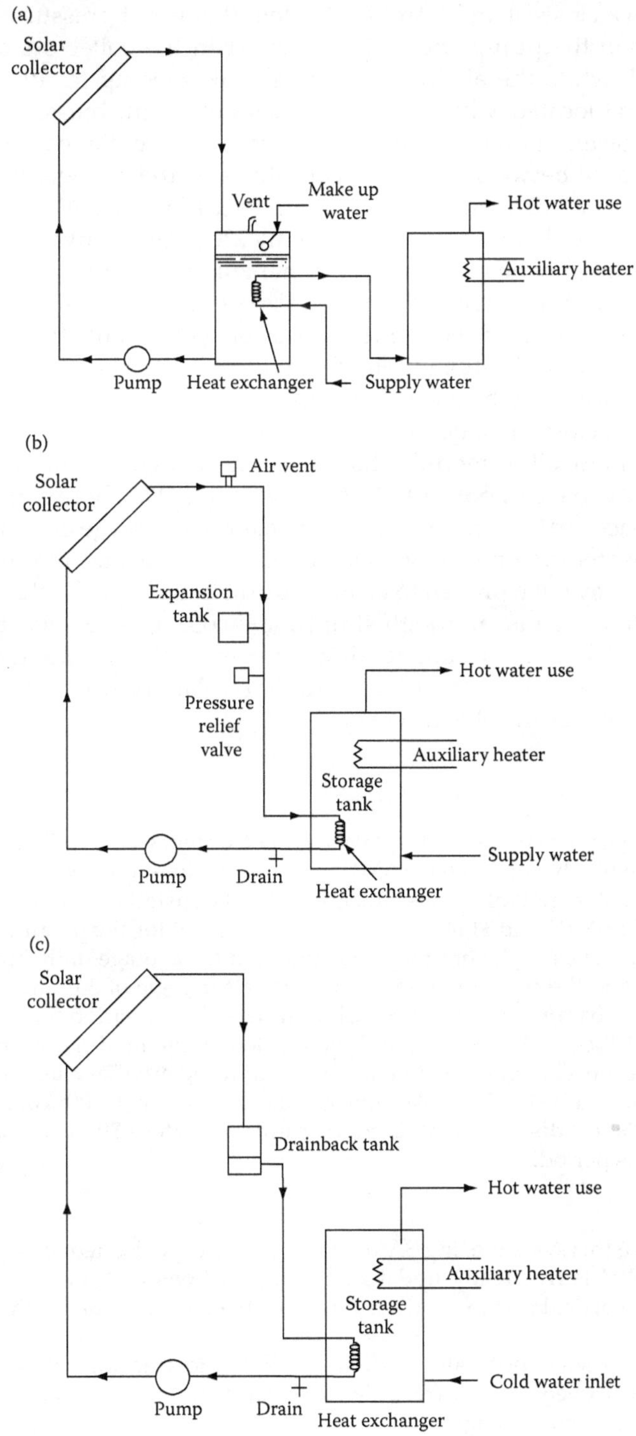

FIGURE 12.19
Typical configurations of solar water-heating systems: (a) open-loop system, (b) closed-loop system, and (c) closed-loop drainback system. (Adapted from Goswami [12].)

can be overcome in a closed-loop drainback system that is not pressurized (Figure 12.19c). In this system, when the pump shuts off, the water in the collectors drains back into a small holding tank while the air in the holding tank goes up to fill the collectors. The holding tank can be located where freezing does not occur, but still at a high level to reduce pumping power. In all three configurations, a differential controller measures the temperature difference between the solar collector and the storage. It turns the circulation pump on when the difference is more than a set limit (usually 5°C) and turns it off when the difference goes below a set limit (usually 2°C). Alternatively, a photovoltaic (PV) panel and a DC pump may be used. The PV panel will turn on the pump only when solar radiation is above a minimum level. Therefore, the differential controller and the temperature sensors may be eliminated. For temperatures of up to about 60°C, required for many industrial process heat applications, forced-circulation flat plate collector water-heating systems described earlier can be used. For higher temperatures, evacuated tube collectors or concentrating collectors must be used.

In recent years, photovoltaic modules have become very cost-competitive. They provide high-value electricity that can power lighting, computers, TVs, household appliances, heat pumps used for space heating/cooling and hot water, and charge electric vehicles. Unlike solar thermal hot-water systems, they have no moving parts and require no plumbing to and from the roof. They have proven to be more durable and reliable than active solar thermal hot-water systems. Thus, although simple, low-cost passive solar thermal hot-water systems are still widely used in nonfreezing climates in the Middle East and elsewhere, rooftop PV is now much more common than rooftop solar hot-water systems in the United States and other countries in cold climates.

Example 12.2

Determine the performance of a flat plate solar hot-water system in Boulder, Colorado, using the System Advisor Model (SAM) program. Assume the cost of the collector is \$300.00/m², the storage tank cost is \$1,200.00, and the installation and balance of system costs are \$2,500.00 and \$1,000.00, respectively. Except for the parameters specified, use the default values in the program. Assume that the average daily hot-water usage is 200 kg/day, and the collector area is 3 m² with a tilt angle of 30° and an azimuth of 180° (south). The parameters in the Hottel–Whillier–Bliss equation (Equation 12.9) are: $F_R\tau\alpha=0.72$, and $F_RU_L=3.7$ (W/m²C), and the incident angle modifier coefficient is 0.10. Assume the heat-exchanger effectiveness is 0.8 and the auxiliary temperature in the electric water tank is set at 50°C. Assume a standard loan with 100% debt fraction, 7% interest rate, 5% real discount rate, 2.5%/year inflation rate, a 30-year loan term, and a 30-year analysis period.

Solution:

We will analyze this system using SAM, which can be downloaded at https://sam.nrel.gov/. (Note: SAM is frequently updated. The instructions that follow are for version 2023.12.17. The particular steps and results may change somewhat with later versions of SAM.)

SAM models a solar hot-water system applicable to freezing climates that has a heat exchanger between an antifreeze/water solar collector loop and water storage as depicted in the schematic diagram:

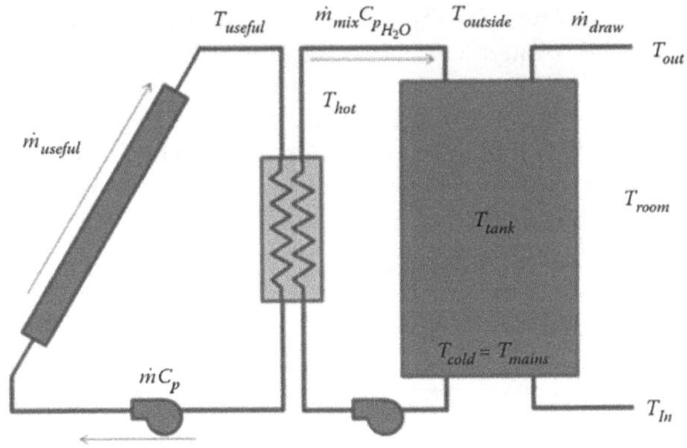

After downloading SAM from the website, to start a solar water-heating project:

1. Start SAM.
2. Click **Start a New Project**
3. Choose **Solar water heating, then Distributed—Residential Owner, then click OK.** The collector is located in Boulder, CO, so under the Location and Resources tab in the Solar Resource Library section type Boulder next to Filter, click on the top item to highlight it.

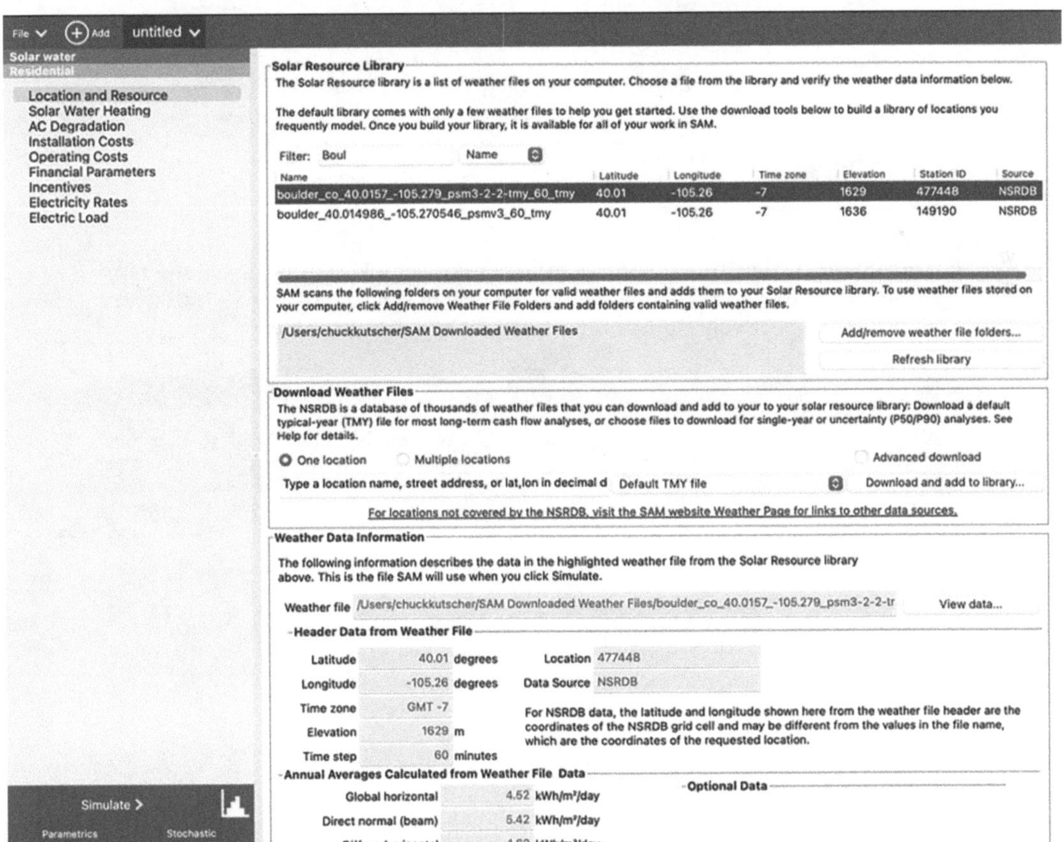

4. Under the "Solar Water Heating" tab, set the average daily hot-water usage to 200 kg/day. Under System choose 1 collector with a tilt of 30° and an azimuth of 180°. Under Collector, click "Enter user-defined parameters" and change the collector values to area = $3\,m^2$, FRta = 0.72 and FRUL = 3.7, and set the incident angle modifier (IAM) coefficient to 0.1. Change the heat exchange effectiveness on the same page to 0.8 and the outlet set temperature to 50°C.

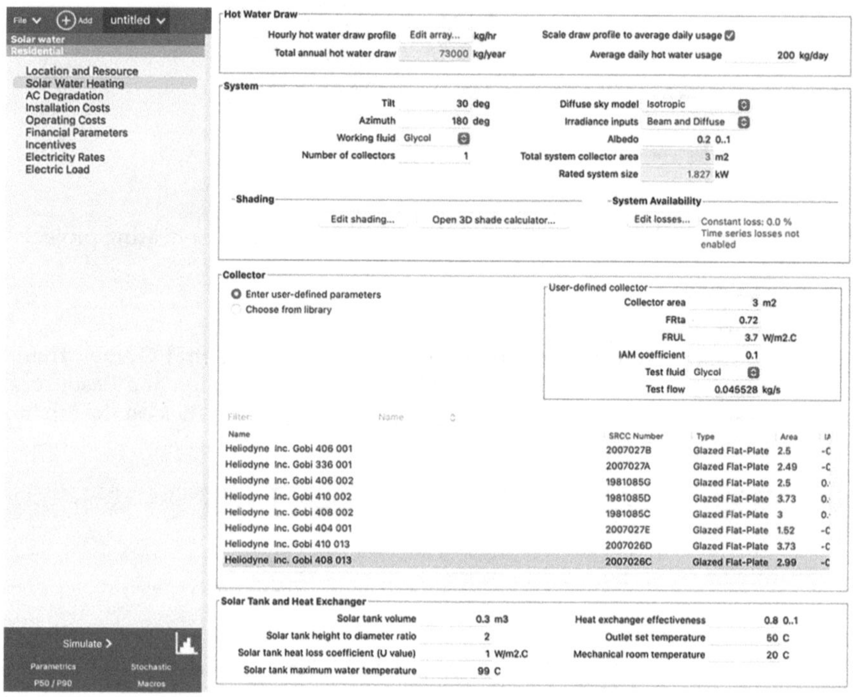

5. Under Installation Costs, enter a collector cost of $\$300.00/m^2$, a storage tank cost of \$1,200.00, and installation and balance of system costs of \$2,500.00 and \$1,000.00, respectively (some of these may be the default values).

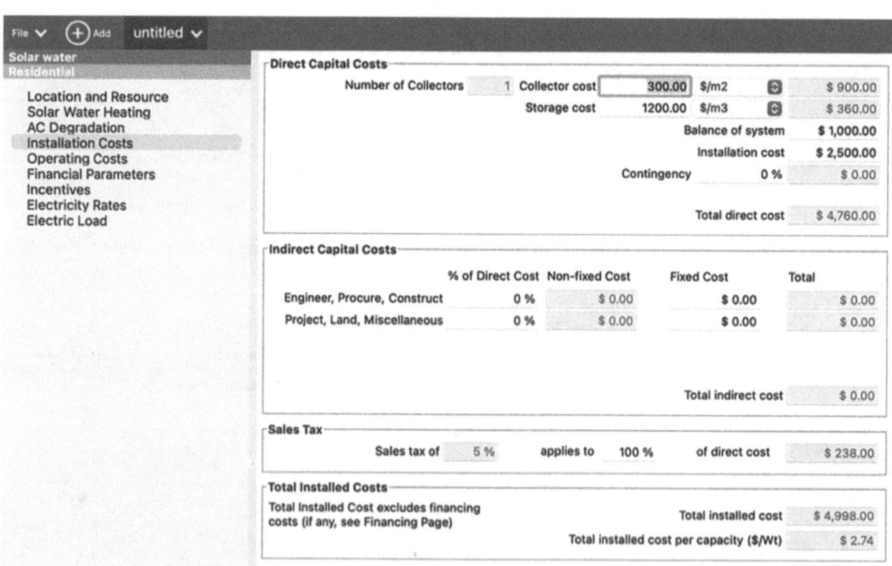

6. Under Financial Parameters, choose a standard loan with 100% debt fraction, 30-year loan period, 7% loan rate, 5% real discount rate, 2.5%/year inflation rate, and a 30-year analysis period.

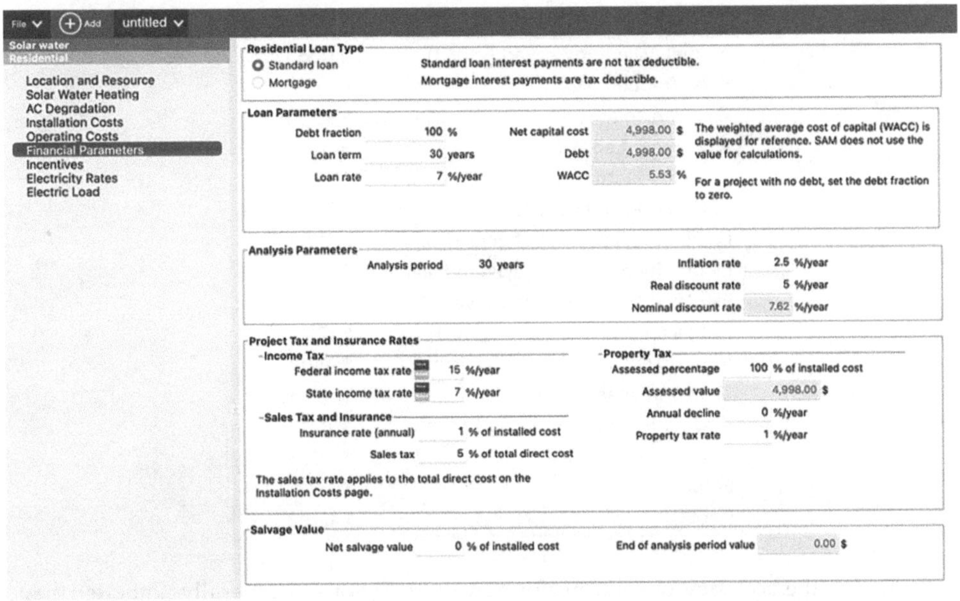

7. Under Electricity Rates, click the Search for rates button and search for Boulder zip code 80301. Choose Public Service Co of Colorado and click on Residential-Service (Schedule R). Then click Download and apply the utility rate.

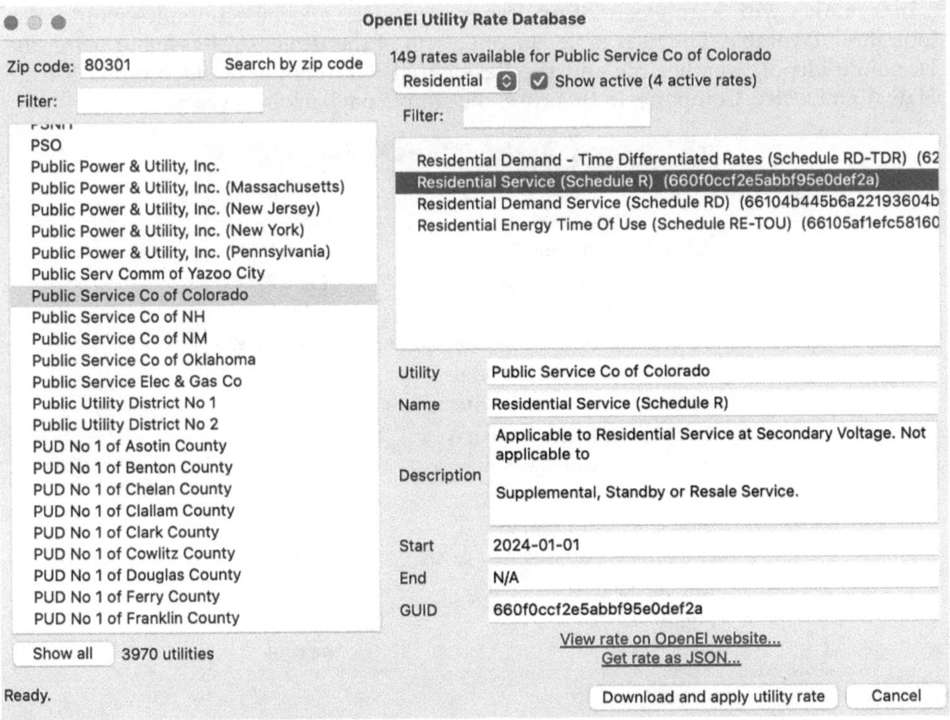

8. We will use the typical electric load data that is the default in SAM. Click Simulate at the bottom of the screen to run the calculations and get results. The table of results appears as follows:

Metric	Value
Annual AC energy saved (year 1)	2,005 kWh
Solar fraction (year 1)	0.66
Aux with solar (year 1)	893.3 kWh
Aux without solar (year 1)	3,033.8 kWh
Capacity factor (year 1)	12.5%
LCOE Levelized cost of energy nominal	21.18 ¢/kWh
LCOE Levelized cost of energy real	16.08 ¢/kWh
Electricity bill without system (year 1)	$1,560
Electricity bill with system (year 1)	$1,288
Net savings with system (year 1)	$272
Net present value	$-876
Simple payback period	17.6 years
Discounted payback period	NaN
Net capital cost	$4,998
Equity	$0
Debt	$4,998

Note that the accuracy of solar performance predictions is generally no better than 10%, and the number of significant figures in the computed solutions is not realistic. In this example, the simple payback period is 17.6 years. Note that the installation cost assumed here of $2,500 is about half of the total system cost. A do it yourself system would obtain a better payback. Also, installation cost would likely be less of a factor in a large multi-family system.

In a market with high electricity prices, such as Hawaii, the economics would also look more favorable. The next table is from a run of the same SAM case but using the Honolulu airport weather file and the Residential R single-phase electric rate for the Hawaiian Electric Company. In this case, the simple payback is 11 years.

Metric	Value
Annual AC energy saved (year 1)	1,518 kWh
Solar fraction (year 1)	0.85
Aux with solar (year 1)	123.6 kWh
Aux without solar (year 1)	1,785.7 kWh
Capacity factor (year 1)	9.5%
LCOE Levelized cost of energy nominal	27.97 ¢/kWh
LCOE Levelized cost of energy real	21.23 ¢/kWh
Electricity bill without system (year 1)	$2,905
Electricity bill with system (year 1)	$2,515
Net savings with system (year 1)	$390
Net present value	$893
Simple payback period	11.1 years
Discounted payback period	22.3 years
Net capital cost	$4,998
Equity	$0
Debt	$4,998

12.7.3 Liquid-Based Solar Thermal Heating Systems for Buildings

Solar space-heating systems can be classified as active or passive depending on the method utilized for heat transfer. A system that uses pumps and/or blowers for fluid flow in order to transfer heat is called an active system. On the other hand, a system that utilizes natural phenomena for heat transfer is called a passive system. Passive solar heating systems are described later in this chapter. In this section, configurations, design methods, and control strategies for active solar heating systems are described.

Figure 12.20 is a schematic diagram of an active (i.e., pumped) solar space-heating system. The system consists of three fluid loops—collector, storage, and load. In addition, most space-heating systems are integrated with a domestic water-heating system to improve the year-round solar load factor.

Since space heating is a relatively low-temperature use of solar energy, a thermodynamic match of collector to task indicates that an efficient flat plate collector or low-concentration solar collector is the thermal device of choice.

The collector fluid loop contains fluid manifolds, the collectors, the collector pump, and heat exchanger, an expansion tank, and other subsidiary components. A collector heat exchanger and antifreeze in the collector loop are normally used in all solar space-heating systems since the existence of a significant heating demand implies the existence of some subfreezing weather.

The storage loop contains the storage tank and pump as well as the tube side of the collector heat exchanger. To capitalize on whatever stratification may exist in the storage tank, fluid entering the collector heat exchanger is generally removed from the bottom of storage. This strategy ensures that the lowest temperature fluid available delivery-to-load loop contains the load device, typically fin-and-tube coils in a duct or plastic tubing in a radiant floor, and the backup system with a flow control (mode selector) valve.

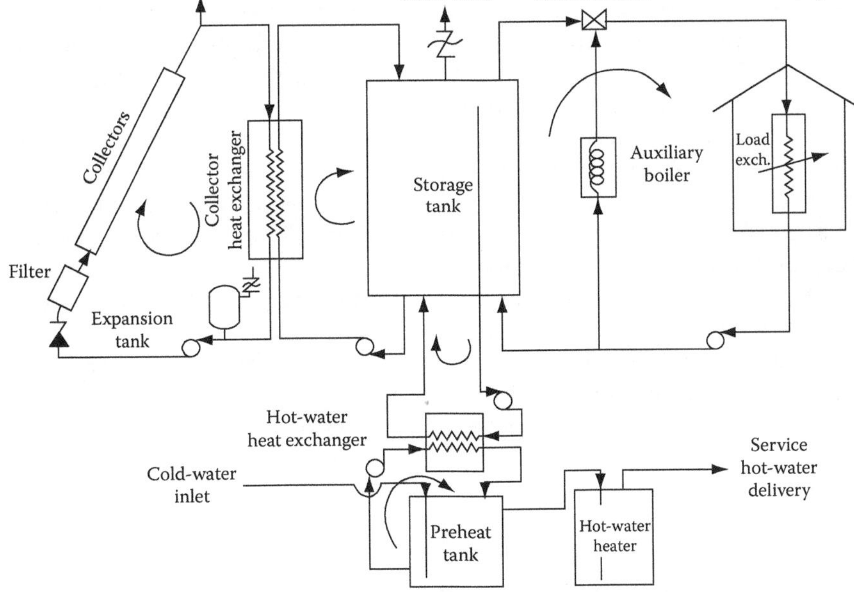

FIGURE 12.20
Typical solar thermal system for space heating and hot-water heating showing fluid transport loops and pumps.

The best solar collector orientation is such that the average solar incidence angle is smallest during the heating season. For tracking collectors, this objective is automatically realized, although tracking systems are rarely used in rooftop applications now. For fixed collectors in the northern hemisphere, the best orientation is due south (due north in the southern hemisphere), tilted up from the horizon at an angle of about 15° greater than the local latitude.

Although due south is the optimum azimuthal orientation for collectors in the northern hemisphere, variations of 20° east or west have little effect on annual energy delivery. Off-south orientations greater than 20° may be required in some cases because of obstacles in the path of the Sun. Also, for aesthetic reasons, collectors are often mounted at the roof angle, and a small loss of performance is accepted. As is the case with solar thermal domestic hot-water systems, rooftop PV arrays that can power heat pumps and heat pump water heaters are more commonly employed these days due to their low cost, reliability, and versatility.

12.7.4 Passive Solar Heating Systems

Passive solar heating systems contain the five basic components:

1. Collector → windows, walls, and floors
2. Storage → walls and floors, large interior masses (often these are integrated with the collector absorption function)
3. Distribution system → radiation, free convection, simple circulation fans
4. Controls → movable window insulation, vents to other inside spaces and/or to ambient
5. Backup system → any nonsolar heating system

The design of passive systems requires the strategic placement of windows, storage masses, and the occupied spaces themselves. The fundamental principles of solar radiation geometry and availability are instrumental in the proper location and sizing of the system's "collectors" (windows). Storage devices are usually more massive than those used in active systems and are frequently an integral part of the collection and distribution system.

A commonly used method of cataloging the various passive system concepts is to distinguish three general categories: direct, indirect, and isolated gain. Most of the physical configurations of passive heating systems are seen to fit within one of these three categories.

For direct gain (Figure 12.21), sunlight enters the heated space and is converted to heat at absorbing surfaces [11]. This heat is then distributed throughout the space and to the various enclosing surfaces and room contents.

For indirect gain systems, sunlight is absorbed and stored by a mass interposed between the glazing and the conditioned space. The conditioned space is partially enclosed and bounded by this thermal storage mass, so a natural thermal coupling is achieved. Examples of the indirect approach are the thermal storage wall, the thermal storage roof, and the northerly wall of an attached sunspace.

In the thermal storage wall (Figure 12.22), sunlight penetrates the glazing and is absorbed and converted to heat at a wall surface interposed between the glazing and the heated space [11]. The wall is usually masonry (mass or Trombe wall) or containers filled with water (water wall), although it might contain phase-change material. The attached sunspace (Figure 12.23) is actually a two-zone combination of direct gain and thermal storage

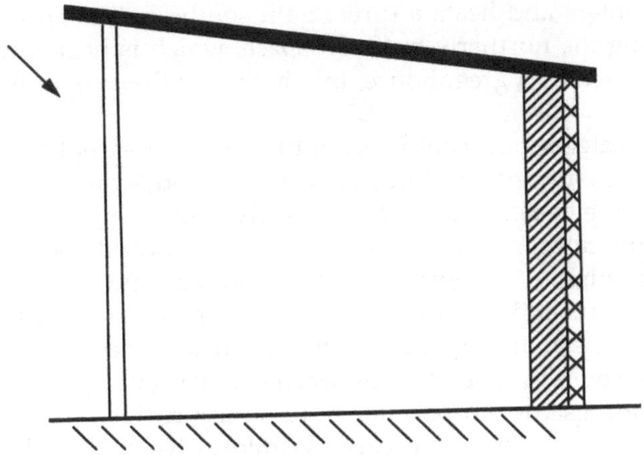

FIGURE 12.21
Direct gain. (From Kreith and West [11]. With permission.)

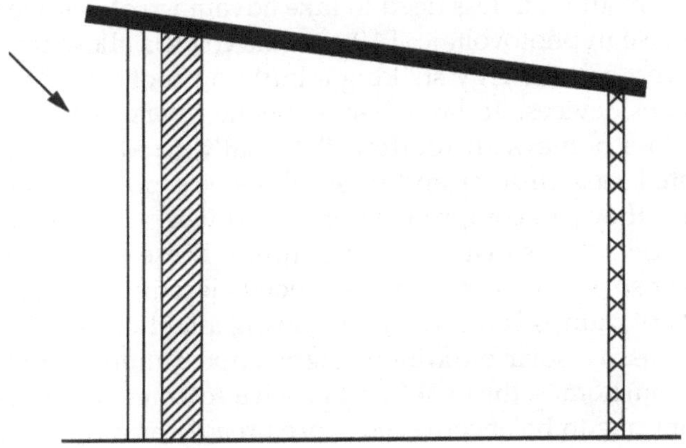

FIGURE 12.22
Thermal storage wall. (From Kreith and West [11]. With permission.)

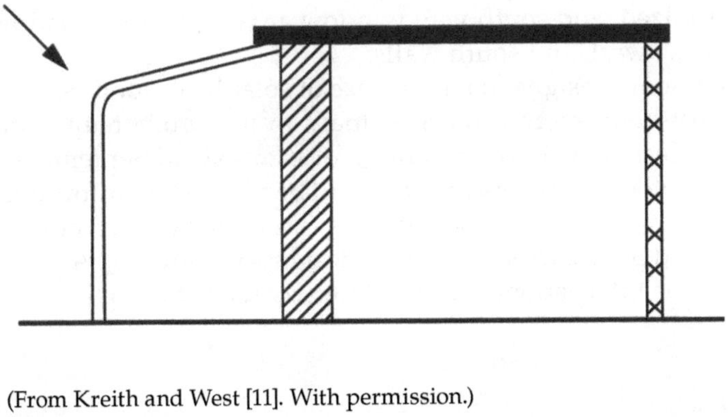

FIGURE 12.23
Attached sunspace. (From Kreith and West [11]. With permission.)

wall [11]. Sunlight enters and heats a direct gain southerly "sunspace" and also heats a mass wall separating the northerly-buffered space, which is heated indirectly. The sunspace is frequently used as a greenhouse, in which case the system is called an attached greenhouse.

The isolated gain category concept is an indirect system except that there is a distinct thermal separation (by means of either insulation or physical separation) between the thermal storage and the heated space. The convective (thermosyphon) loop, as depicted in Figure 12.17, is in this category and, while often used to heat domestic water, can also be used for building heating, often with air as the working fluid. It is most akin to conventional active systems in that there is a separate collector and separate thermal storage. The thermal storage wall, thermal storage roof, and attached sunspace approaches can also be made into isolated systems by insulating between the thermal storage and the heated space.

Passive solar buildings rose to popularity in sunny regions in the 1970s. Since then, building envelopes have improved greatly. Windows have much higher *R*-values and low-e coatings, roofs and walls are highly insulated, and air leaks are minimized and verified via blower door tests. Although lighting systems have become much more efficient and produce less internal heat, the proliferation of computers, entertainment systems, and other electronic devices has further reduced building heating loads. For many tight homes today, the hot-water heating load can be greater than the space-heating load. All these things have resulted in less need to take advantage of passive solar heat. In addition, the rapid decrease in photovoltaic (PV) module costs is allowing building owners to cost-effectively convert solar energy striking a building roof into electricity that can run heat pumps and other devices. Today when someone refers to a "solar home," they are typically referring to a home with rooftop PV. Finally, production home builders have generally not adopted passive solar and other climate-responsive design approaches on a large scale because they can compromise preferred views, and it can be a challenge to achieve the proper orientations in a large community. These days, as electric grids decarbonize with low-cost solar and wind, the beneficial electrification of buildings utilizing high-performance heat pumps is receiving increasing attention (see Chapter 4).

Although purely passive solar building designs appear more commonly in custom (as opposed to production) homes, the benefits of passive solar and climate-responsive design practices should continue to be encouraged more broadly as a fundamental part of good overall building design. In performing OpenStudio simulations of commercial buildings or BEopt models of residential buildings, designers can adjust parameters such as the building aspect ratio, the ratio of window to wall area for each wall, and the amount of thermal mass contained in a building. Good design practice generally calls, for example, for a building to be longer in the east-west direction. In cold climates, window area on the north wall is minimized, and south wall window area is favored. Shading devices can be employed on the east, west, and south walls.

For passive solar home designs that make use of solar heat, good design includes proper orientation and sufficient internal thermal mass to minimize temperature swings. It is important that the design avoid overheating, for example, in September when the Sun is low in the sky and outdoor temperatures can still be high. More information on the design of passive solar and climate-responsive homes can be obtained from the Passive House Institute of the United States, which provides design software [13]. A passive house design can be certified by the International Passive House Association [14].

12.7.5 Solar Cooling Systems

There are three mechanisms that can utilize solar heat to provide space cooling: a Rankine cycle engine, an absorption chiller, and a desiccant cycle. Achieving high efficiency in a Rankine cycle requires the high temperatures associated with tracking, concentrating solar collectors, and these are not typically deployed on building roofs. Thus, absorption chillers and desiccant systems have more typically been used.

In an absorption system, the refrigerant is evaporated or distilled from a less volatile liquid absorbent, the vapor is condensed in a water- or air-cooled condenser, and the resulting liquid is passed through a pressure-reducing valve to the cooling section of the unit. There, it cools the water as it evaporates, and the resulting vapor flows into a vessel, where it is reabsorbed into the absorbing liquid and pumped back to the heated generator. The heat required to evaporate the refrigerant in the generator can be supplied directly from solar energy, as shown in Figure 12.24.

Absorption air conditioning is compatible with solar energy since a large fraction of the energy required is thermal energy at temperatures that can be provided by evacuated tube solar collectors described previously. Absorption refrigeration differs from vapor-compression air conditioning only in the method of compressing the refrigerant. The combination of absorber and generator replaces the compressor. By this means, the refrigerant vapor entering the condenser is compressed without the large input of high-grade shaft work that vapor-compression air-conditioning demands.

Traditional desiccant cooling systems use a desiccant to dry air, which can then be evaporatively cooled. Solar heat can then be used to bake the moisture back out of the desiccant. Desiccant systems can also be designed to dehumidify air and thus handle the latent cooling load. Liquid desiccants like lithium chloride or lithium bromide are most commonly used. A schematic of a liquid desiccant dehumidification system is shown in Figure 12.25. Air is brought in contact with concentrated desiccant in a countercurrent flow in a dehumidifier. The dehumidifier may be a spray column or packed bed. The packing provides a very large

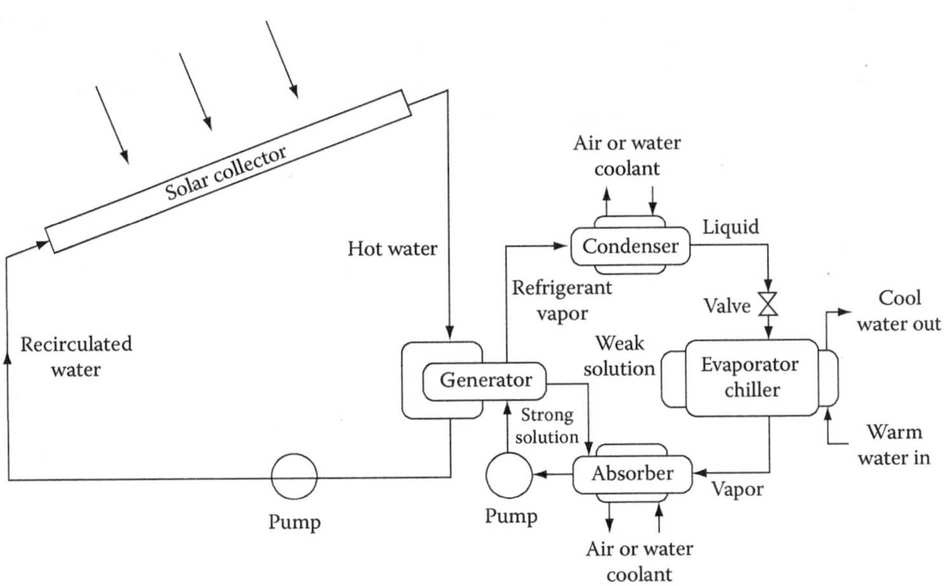

FIGURE 12.24
Schematic diagram of a solar-powered absorption refrigeration system.

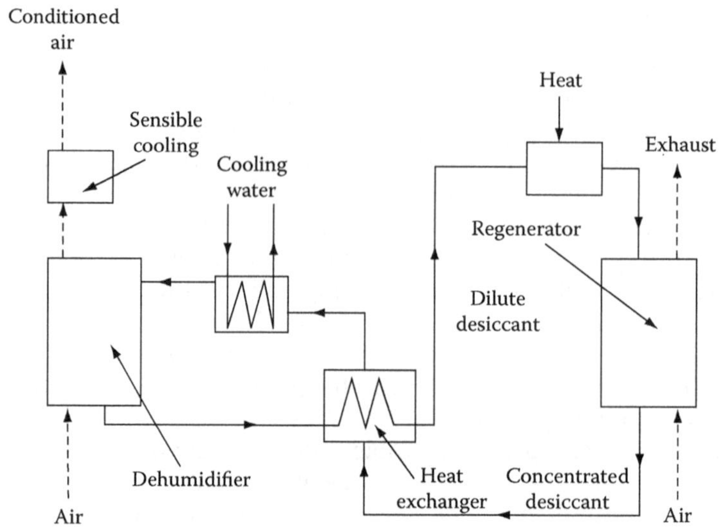

FIGURE 12.25
A conceptual liquid desiccant dehumidification system.

area for heat and mass transfer between the air and the desiccant. After dehumidification, the air is sensibly cooled before entering the conditioned space. The dilute desiccant exiting the dehumidifier is regenerated by heating and exposing it to a countercurrent flow of a moisture-scavenging air stream.

With the large reduction in the cost of photovoltaic modules, solar thermal cooling systems have largely been replaced by conventional vapor-compression cooling chillers with the compressors and fans powered by PV. Vapor-compression chillers are low cost because they are mass-produced, and their efficiencies have increased considerably due to manufacturer compliance with standards like the Environmental Protection Agency's (EPA's) Energy Star program. For buildings that experience both heating and cooling loads, PV electricity can be used to power high-performance heat pumps.

12.8 Solar Thermal Power Plants

High-concentration solar collectors (parabolic trough, linear Fresnel, and power towers) can provide the high temperatures needed to power conventional Rankine steam power cycles. NREL maintains an up-to-date list of CSP projects at https://solarpaces.nrel.gov/projects. Table 12.4 lists the number of each technology type deployed in the world and the approximate MW_e for each. In addition to these, two parabolic dish projects were deployed but decommissioned [15]. Although parabolic troughs have historically dominated CSP deployment, there has recently been more commercial interest in power towers because their higher operating temperatures promise a lower levelized cost of energy. This is particularly true in China. Until recent years, utility-scale CSP plants were favored over PV systems because they offered dispatchability with low-cost molten salt thermal storage. However, the dramatic decreases in the costs of both PV modules and batteries have made

TABLE 12.4
World Solar Power Plants

Type	Number of Plants	Capacity (MW_e)
Parabolic trough	91	5,000
Power tower	34	1,400
Linear Fresnel	16	250

Source: https://solarpaces.nrel.gov/projects.

it much more difficult for CSP to compete with the economics and reliability of PV. There are still new CSP plants being built in China and some other countries, however. In some cases, they are collocated with PV arrays.

12.8.1 Parabolic Trough-Based Power Plants

Luz Corporation developed components and commercialized parabolic trough solar thermal power by constructing a series of hybrid solar power plants from 1984 to 1991. Starting with their first 14 MW_e solar electric generating station (SEGS I) in Southern California, Luz added a series of SEGS power plants with a total generating capacity of 354 MW_e. Figure 12.26 shows a schematic of the SEGS VIII and IX plants [16]. All of these plants used solar energy to heat the working fluid and natural gas as the auxiliary fuel to achieve the desired temperature. On average, 75% of the energy was obtained from the Sun and 25% from natural gas. With power plant conversion efficiencies on the order of 35% and solar field efficiencies of 45%–50%, overall efficiencies for solar-to-electricity conversion on the order of 15-18% were achieved by these plants.

Luz went bankrupt due largely to problems associated with the loss of financial incentives. The first large-scale parabolic trough power plant in the United States since SEGS was the Nevada Solar One plant built in 2007. That plant has a nominal capacity of 64

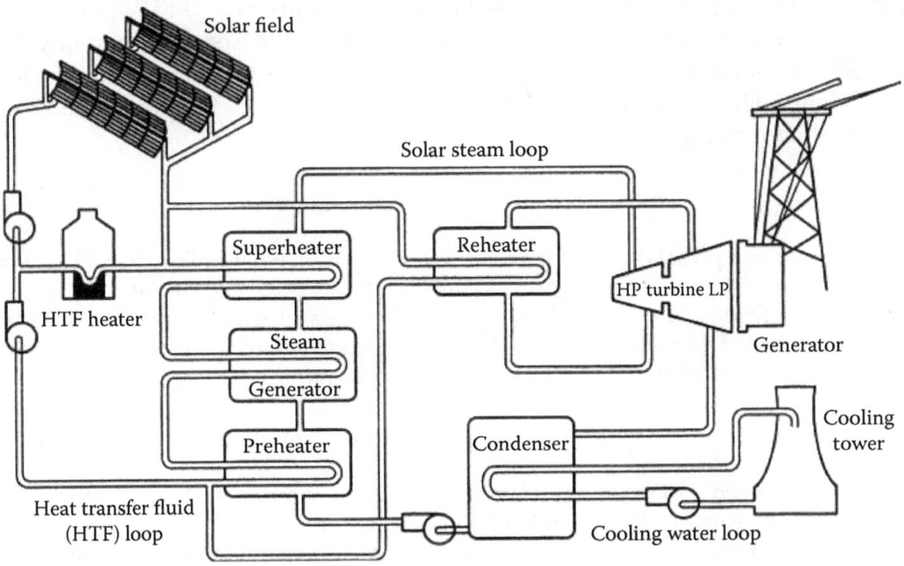

FIGURE 12.26
Flow of heat-transfer fluid through the SEGS VIII and IX plants. (Adapted from de Laquil et al. [16].)

FIGURE 12.27
The 250-MW Solana parabolic trough power plant with 6 hours of thermal storage in Gila Bend, Arizona, provides electricity for Phoenix. The power block and storage tanks are shown in the background. (Courtesy of Atlantica Yield.)

MW and a maximum capacity of 75 MW. The project required an investment of $266 million, and electricity production is estimated to be 134 million kWh/year [17]. The time of construction for the plant was about 16 months, and it occupies approximately 400 acres (1.6 km^2), while the solar collectors cover about 1.2 km^2. Nevada Solar One uses 760 parabolic troughs that concentrate solar radiation onto evacuated receivers that heat the heat-transfer fluid to 391°C (735°F) [18].

Like the SEGS plants, Nevada Solar One does not have a significant thermal storage system. However, many parabolic trough plants have been built throughout the world with molten salt storage. A feed-in tariff resulted in a large number of 50-MW parabolic trough plants in Spain with 7.5 hours of thermal storage. In 2013, the Spanish company Abengoa built the 250-MW Solana plant in Arizona, which has 6 hours of storage and provides electricity to Phoenix. The large solar field of this facility, along with the power block and storage tanks, is shown in Figure 12.27.

Example 12.3

The parabolic trough solar collector, whose efficiency was given in Equation 12.24, is to be used for electric power production. Assume that the average operating temperature of the trough field is 50°C lower than the boiler temperature, the average direct radiation in the plane of the collectors is 600 W/m^2, and the efficiency of the power generation unit equals one-half of the Carnot efficiency, and determine:

a. The temperature at which the solar trough power generation system achieves its highest efficiency.
b. The levelized cost of energy from this trough generation system assuming a life of 25 years, a discount rate of 4%, and assuming that the total installed system cost is $600/m^2 of collector aperture area. This includes the cost of the installed solar collector field, as well as the power block and balance of system (turbine, cooling tower, power transformers/interconnection, etc.).

Solution:

According to thermodynamic principles, the conversion of thermal energy into electric energy increases in efficiency as the temperature of the thermal source increases. But Equation 12.24 shows that the efficiency of the collector decreases with increasing operating temperature. In order to obtain the optimum operating temperature, both of these relations need to be considered.

If we assume the ambient temperature (for determining the power-cycle efficiency) is 20°C, the electric power output, P, is equal to

$$P = I \times \eta_c \times 0.5\left(\frac{\Delta T_a}{T_h}\right) = I \times \eta_c \times 0.5\left(\frac{T_h - 293}{T_h}\right)$$

where η_c is given by Equation 12.24, with $x = (T_{\text{htf}} - 25)/I^{0.33} = (T_h - 50 - 25)/(600)^{0.33}$.

P is shown as a function of operating temperature in the table and plot below. To estimate the annual power in kWh/m²/year, we can multiply the power by the number of daylight hours during the year, which is assumed to be 12 h/day.

Operating Temperature (°C)	Collector Efficiency	50% of Carnot	P (W/m²)	*P*-Annual (kWh/m²/Year)
100	0.771	0.11	49.6	217
150	0.769	0.15	70.9	310
200	0.766	0.19	87.5	383
250	0.761	0.22	100.4	440
300	0.753	0.24	110.4	483
350	0.739	0.26	117.5	514
400	0.718	0.28	121.7	533
450	0.689	0.30	122.9	538
500	0.649	0.31	120.9	530
550	0.597	0.32	115.4	505
600	0.531	0.33	105.9	464

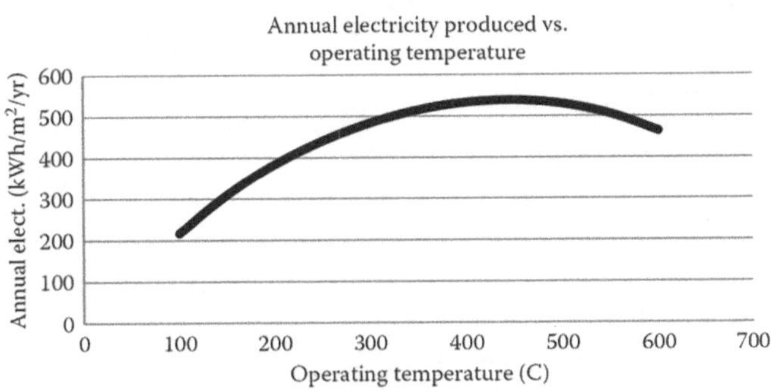

We observe that the peak annual electric generation occurs with an operating temperature of about 450°C. The annual electricity generation at the optimum operating temperature, as shown in the last column of the table, is approximately 538 kWh/m²/year.

To estimate the LCOE, we neglect annual operation and maintenance costs. The levelized cost of energy delivered by the system over 25 years is

$$\text{LCOE} = \left(\frac{\text{TLCC}}{Q}\right) \times \text{CRF} = \left[\frac{600\ \$/\text{m}^2}{538\,\text{kWh/m}^2}\right] \times 0.064 = \$0.071/\text{kWh}$$

Although this simple analysis has indicated an optimum operating temperature of 450°C, existing trough plants are limited to a maximum operating temperature of about 390°C, due to limitations of the heat-transfer oil. Note also that we did not account in this example for heat losses from piping, parasitic pumping power, operation and maintenance costs, and other expenses.

12.8.2 Power Towers

Figure 12.28 is a schematic diagram of a power tower system [19]. A heat-transfer fluid is pumped through a riser to the receiver where it is heated and then moves to the thermal storage unit and/or a heat engine, which is usually located on the ground near the tower. Typical optical concentration factors range from 200 to 1,000, and plant sizes from 10 to 200 MW_e are typical for new systems. The heat flux impinging on the receiver can be between 300 and 1,000 kW/m^2 and allows the working fluid to achieve temperatures up to 1,000°C.

Two types of receiver configurations have been considered for power towers. The cavity receiver consists of an insulated enclosure within which a heat absorbing structure is located. Radiation enters through the aperture, is subjected to multiple reflections, and thus enhances the effective absorptivity of the cavity. Radiation losses can occur only through the aperture, and the absorbing area of the receiver can be enlarged without substantially increasing the radiation losses. However, convective losses increase as a result of a larger heated area, and cavity receivers tend to be large and heavy. Moreover, the aperture size restricts the field of view, and spillage of the concentrated solar flux can be considerable. In

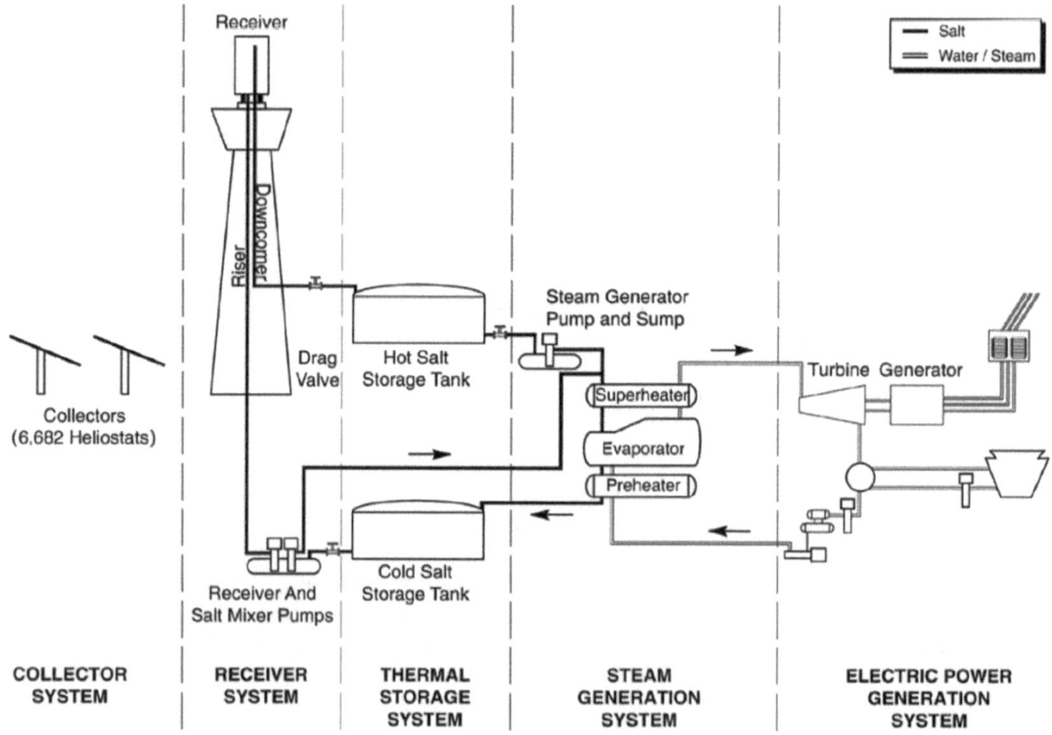

FIGURE 12.28
Schematic of two-tank molten salt power tower plant (From Turchi and Heath [19].)

a plain or flat plate receiver, the aperture of a cavity is replaced by a flat heat-exchanging surface. The back side can be insulated, but overall losses are appreciable.

The external receiver consists of a cylinder with a tubular structure containing a working fluid passing through the pipes on the outside of a cylinder. It can receive solar radiation from all directions and is relatively simple and lightweight because no enclosure is required. The design can accommodate heat flux densities as high as 850 kW/m^2 with molten salt, and the tower needs to be only about 70% as tall as the tower for a cavity receiver, which reduces the cost appreciably. The disadvantage of an external receiver is that the reflected fraction of incident radiation is lost, and the temperature of the molten salt heat-transfer fluid must be maintained above a certain limit to avoid freezing. This is accomplished, as shown in Figure 12.28, by having hot and "cold" storage tanks.

In any high-temperature concentrating system, the receiver is the focal point of the design. Because radiation and convection losses are proportional to the area of the receiver, it should be as small as possible, which requires high concentration. But the receiver must also be large enough to intercept and absorb a very high fraction of the solar radiation directed toward it by the heliostats and transfer the heat from the Sun to a working fluid. If the designer specifies a more perfect optical system to reduce the receiver size, this will increase the cost of the heliostats and may complicate the thermal design.

The efficiency of the receiver, η_{th}, equals [1 − (thermal loss/solar input)] and is given by the following equation:

$$\eta_{\text{th}} = 1 - \frac{A_{\text{receiver}}\left[\varepsilon\sigma\left(T^4 - T_a^4\right) + c\left(T - T_a\right)^{1.25}\right]}{G_b A_{\text{collector}}\eta_{\text{col}}\eta_{\text{int}}\alpha} \tag{12.27}$$

or

$$\eta_{\text{th}} = 1 - \frac{\sigma\left(T^4 - T_a^4\right) + (c/\varepsilon)\left(T - T_a\right)^{1.25}}{G_b C\eta_{\text{col}}\eta_{\text{int}}} \tag{12.28}$$

where

$$C = \frac{\alpha A_{\text{collector}}}{\varepsilon A_{\text{receiver}}} = C_g\frac{\alpha}{\varepsilon}$$

T is the receiver operating temperature (K)
T_a is the ambient temperature (K)
ε is the emissivity of the receiver at the temperature T
σ is the Stefan–Boltzmann constant, 5.67×10^{-8} W/m^2 K^4
c is a temperature-independent convective loss coefficient, typically between 1 and 25 W/m^2 K$^{1.25}$
G_b is the solar beam radiation, W/m^2

Also, α is the absorptivity for direct beam solar radiation (G_b), and η_{col} is the collector efficiency, which includes reflectivity (typically 0.85–0.95), cosine of angle of incidence effects (typically 0.7–0.9), and a factor to account for shading of mirror surfaces or blocking of reflected sunlight (typically 0.9–1.0). It is conventional to use the actual mirrored area of the heliostats for $A_{\text{collector}}$. The fraction of reflected light that is intercepted by the receiver is η_{int}, representing scattering and spillage losses. We see that the geometric concentration, C_g, plays an important role in suppressing the effects of thermal losses.

Current power tower designs use either steam or molten salt in the receiver. In a steam receiver, water is boiled and superheated in the receiver before being sent to a steam turbine. Towers using direct steam receivers have steam conditions ranging from saturated steam at 250°C to superheated steam at over 550°C [10]. Direct steam towers offer the advantage of using conventional boiler design and have low pumping power. In April 2009, Abengoa Solar began commercial operation of a 20-MW solar power tower plant at the Solucar Platform, near Seville, Spain. PS20 features a higher-efficiency cavity receiver, improved control systems, and a better thermal energy storage system than the first tower, PS10. The second system brought the power capacity to 20 MW, double that of the original PS10 tower. A photograph of both of these is shown in Figure 12.29.

BrightSource built the 390-MW Ivanpah project in California, which consists of three separate power towers and more than 170,000 15 m^2 heliostats, which were built in an on-site automated factory. Figure 12.30 shows photos of the large Ivanpah project [19].

Because of the increasingly stiff competition from low-cost photovoltaic modules, the concentrating solar power industry has been shifting away from direct steam systems to

FIGURE 12.29
Photograph of two direct steam Spanish central receiver systems. (Courtesy of Clifford Ho, Sandia National Laboratories.)

FIGURE 12.30
Ivanpah direct steam power towers (From Mehos et al. [10].)

take advantage of the fact that CSP plants can utilize relatively low-cost thermal storage. This allows CSP not only to provide electricity when the Sun isn't shining but also to sell electricity into the most lucrative electricity markets. While a steam receiver can be coupled to thermal storage, it is much easier to couple thermal storage to a molten salt receiver. As with the case of parabolic trough plants with storage, power tower plants use molten nitrate salts as the storage media. But, unlike current commercial trough plants, which utilize oil as the heat-transfer fluid in the receivers and must use heat exchangers to transfer the heat to molten salt storage, a power tower with a molten salt receiver uses molten salt as both the working fluid in the receiver and the storage media as shown in Figure 12.28. This allows higher receiver operating temperatures and thus higher power-cycle efficiencies. Higher temperatures also allow for a smaller storage volume for the same heat storage capacity.

The first power tower plant to use a molten salt receiver was the 10 MW Solar Two plant, which operated in California in 1999. Solar Two demonstrated advanced molten salt technology at a scale sufficient to allow commercialization of the technology. The plant operated with solar energy collected efficiently over a broad range of operating conditions and showed that low-cost thermal energy storage systems can operate reliably and efficiently.

Today's typical molten salt power tower pumps the salt at about 290°C from "cold" storage tanks to the receiver, where it is heated by concentrated sunlight to a temperature of over 550°C and supplied to a hot tank. At times when it is economically desirable to provide electricity to the grid, the hot molten salt from the hot storage tanks is pumped through a heat exchanger where it boils and superheats water to produce steam at temperatures of up to 540°C and pressures of 100–150 bars for powering a steam turbine, which turns an electric generator as in a conventional thermal power plant. The exiting molten salt at lower temperature then returns to the "cold" tanks to repeat the process.

Note that in this design the thermal storage can operate independently of the heliostat field and the receiver. The hot molten salt can be harnessed to produce steam and electricity when the Sun isn't shining (when clouds pass overhead or at night) or whenever the price of electricity is high enough to make electric dispatch financially attractive. Storage can be sized to allow a CSP plant to operate as a base load power plant 24 h/day in the summer. For example, the 20-MW_e Gemasolar plant in Spain has 15 hours of storage and can run constantly most of the year. In the United States, the 110-MW_e Crescent Dunes plant built by Solar Reserve in Tonopah, Nevada was designed with 10 hours of molten salt storage. Unfortunately, this plant suffered from an early molten salt storage tank leak and then was never able to operate in an economically competitive manner and so was shut down. While plants with such large storage systems can operate almost as base load plants, dispatching electricity over such a long, continuous time period may not be the most cost-effective option. Selling electricity in 4–6-hour periods when electricity is at the highest price can provide a greater return on investment. As competition from photovoltaics has become stiffer, efforts have been made to operate CSP plants as peakers.

The U.S. Department of Energy estimates that the current LCOE for CSP with 12 hours of storage is $95/MWh (down from $210/MWh in 2010). The goal of the U.S. Department of Energy's SunShot program is to achieve an LCOE of $50/MWh by 2030 [20]. (All of these LCOE values are without a 30% investment tax credit). As pointed out earlier, however, LCOE does not tell the whole story. A plant's return on investment is determined not only by the cost of the energy it produces but also by the price for which the energy can be sold. That is the subject of the next example. More details about dispatch strategies can be found in Mehos et al. [10].

Example 12.4

Molten salt power tower systems utilize direct thermal storage, which can be dispatched to generate electricity during times of the day when market prices or demand are elevated. The ability to generate during peak times of the day provides significant value and/or profit for grid operators and plant owners, respectively. Solar power plants with storage can sell electricity into the peak periods that were shown in the "duck curves" in Chapter 5.

Power tower facilities are capable of selectively dispatching during peak times, given a sufficiently "smart" operation policy that reserves enough stored energy to meet demand. In this example we will use SAM, which can be downloaded at https://sam.nrel.gov/, to compare the performance of a molten salt power tower plant with single-owner (utility) financing that operates with optimized dispatch and one that operates without. In both cases, assume that there are two "time of delivery" factor categories. The first category has a price multiplier of 0.5, and the second has a price multiplier of 2. The second category includes 3 hours from 7 to 10 a.m. and 4 hours from 6 to 10 p.m. and is applied uniformly for each month of the year and for both weekdays and weekends. All other times are covered by the first category. Specify a power purchase agreement (PPA) price of 0.10 $/kWh$_e$, and a fixed debt percentage of 50%. Compare the annual electricity generation and profitability (internal rate of return) for both cases. Also, use SAM's heat map plots of total electric power to the grid to qualitatively contrast the generation profiles. (Note: SAM is frequently updated. The instructions and results that follow are for version 2023.12.17 and may change slightly with later versions.)

Solution:

1. Start SAM.
2. Start a new project, and select **Concentrating Solar Power—Power Tower Molten Salt** as the technology model and **Power Purchase Agreement Single Owner. Select OK.** Confirm on the **Location and Resource** page **that Daggett, California** is selected as the location.
3. Navigate to the **Revenue** page. At the top of that page in the section **Solution Mode**, check Specify PPA price and set the PPA price to 0.10$/kWh. Other parameters should be left at their default values.
4. Also on the **Revenue** page, in the **Time of Delivery (TOD)** section, ensure that TOD factors by schedule are chosen. All numbers should have a default value of 1. Use the mouse to click and drag across all of the cells in the **TOD Schedules and Factors—Weekday Schedule** table to select the time windows specified in the problem statement above, and type "2" for those. Repeat for the same hours on the **Weekend Schedule** table.

   Set the TOD factors to 0.5 for Periods 1 and 2 for Period 2. SAM is now configured to weight revenue from electricity sales generated during the specified time windows according to the PPA price multiplier values. The result should appear as follows:

	TOD factors
Period 1	0.5
Period 2	2
Period 3	1
Period 4	1
Period 5	1
Period 6	1
Period 7	1
Period 8	1
Period 9	1

Weekday

	12am	1am	2am	3am	4am	5am	6am	7am	8am	9am	10am	11am	12pm	1pm	2pm	3pm	4pm	5pm	6pm	7pm	8pm	9pm	10pm	11pm
Jan	1	1	1	1	1	1	1	2	2	2	1	1	1	1	1	1	1	1	2	2	2	2	1	1
Feb	1	1	1	1	1	1	1	2	2	2	1	1	1	1	1	1	1	1	2	2	2	2	1	1
Mar	1	1	1	1	1	1	1	2	2	2	1	1	1	1	1	1	1	1	2	2	2	2	1	1
Apr	1	1	1	1	1	1	1	2	2	2	1	1	1	1	1	1	1	1	2	2	2	2	1	1
May	1	1	1	1	1	1	1	2	2	2	1	1	1	1	1	1	1	1	2	2	2	2	1	1
Jun	1	1	1	1	1	1	1	2	2	2	1	1	1	1	1	1	1	1	2	2	2	2	1	1
Jul	1	1	1	1	1	1	1	2	2	2	1	1	1	1	1	1	1	1	2	2	2	2	1	1
Aug	1	1	1	1	1	1	1	2	2	2	1	1	1	1	1	1	1	1	2	2	2	2	1	1
Sep	1	1	1	1	1	1	1	2	2	2	1	1	1	1	1	1	1	1	2	2	2	2	1	1
Oct	1	1	1	1	1	1	1	2	2	2	1	1	1	1	1	1	1	1	2	2	2	2	1	1
Nov	1	1	1	1	1	1	1	2	2	2	1	1	1	1	1	1	1	1	2	2	2	2	1	1
Dec	1	1	1	1	1	1	1	2	2	2	1	1	1	1	1	1	1	1	2	2	2	2	1	1

Weekend

	12am	1am	2am	3am	4am	5am	6am	7am	8am	9am	10am	11am	12pm	1pm	2pm	3pm	4pm	5pm	6pm	7pm	8pm	9pm	10pm	11pm
Jan	1	1	1	1	1	1	1	2	2	2	1	1	1	1	1	1	1	1	2	2	2	2	1	1
Feb	1	1	1	1	1	1	1	2	2	2	1	1	1	1	1	1	1	1	2	2	2	2	1	1
Mar	1	1	1	1	1	1	1	2	2	2	1	1	1	1	1	1	1	1	2	2	2	2	1	1
Apr	1	1	1	1	1	1	1	2	2	2	1	1	1	1	1	1	1	1	2	2	2	2	1	1
May	1	1	1	1	1	1	1	2	2	2	1	1	1	1	1	1	1	1	2	2	2	2	1	1
Jun	1	1	1	1	1	1	1	2	2	2	1	1	1	1	1	1	1	1	2	2	2	2	1	1
Jul	1	1	1	1	1	1	1	2	2	2	1	1	1	1	1	1	1	1	2	2	2	2	1	1
Aug	1	1	1	1	1	1	1	2	2	2	1	1	1	1	1	1	1	1	2	2	2	2	1	1
Sep	1	1	1	1	1	1	1	2	2	2	1	1	1	1	1	1	1	1	2	2	2	2	1	1
Oct	1	1	1	1	1	1	1	2	2	2	1	1	1	1	1	1	1	1	2	2	2	2	1	1
Nov	1	1	1	1	1	1	1	2	2	2	1	1	1	1	1	1	1	1	2	2	2	2	1	1
Dec	1	1	1	1	1	1	1	2	2	2	1	1	1	1	1	1	1	1	2	2	2	2	1	1

5. Navigate to the **System Control** page, **Dispatch Optimization** section, and ensure that the **Enable dispatch optimization** checkbox is *disabled* (i.e., is unchecked) for the first simulation.
6. Navigate to the **Financial Parameters** page and under **Project Term Debt**, select **Debt percent**, and set it to 50% of total installed cost. Note that if this step is not taken, the internal rate of return calculations will fail and return "NaN."
7. Click **Simulate**. The simulation will proceed without dispatch optimization.

   The Summary table should appear as follows:

Metric	Value
Annual AC energy (year 1)	615,096,320 kWh
Capacity factor (year 1)	69.1%
Annual Water Usage	102,988 m^3
PPA price in Year 1	10.00 ¢/kWh
PPA price escalation	1.00 %/year
LPPA Levelized PPA price nominal	10.86 ¢/kWh
LPPA Levelized PPA price real	8.62 ¢/kWh
LCOE Levelized cost of energy nominal	10.24 ¢/kWh
LCOE Levelized cost of energy real	8.13 ¢/kWh
NPV Net present value	$37,699,776
IRR Internal rate of return	9.81 %
Year IRR is achieved	20
IRR at end of project	11.57 %
Net capital cost	$747,161,472
Equity	$373,650,400
Size of debt	$373,511,072
Minimum DSCR	1.33

8. Now click **Heat map** at the top of the simulation results page and choose **Total electric power to the grid (MW$_e$)** in the dropdown menu on the right. You should see the following heat map:

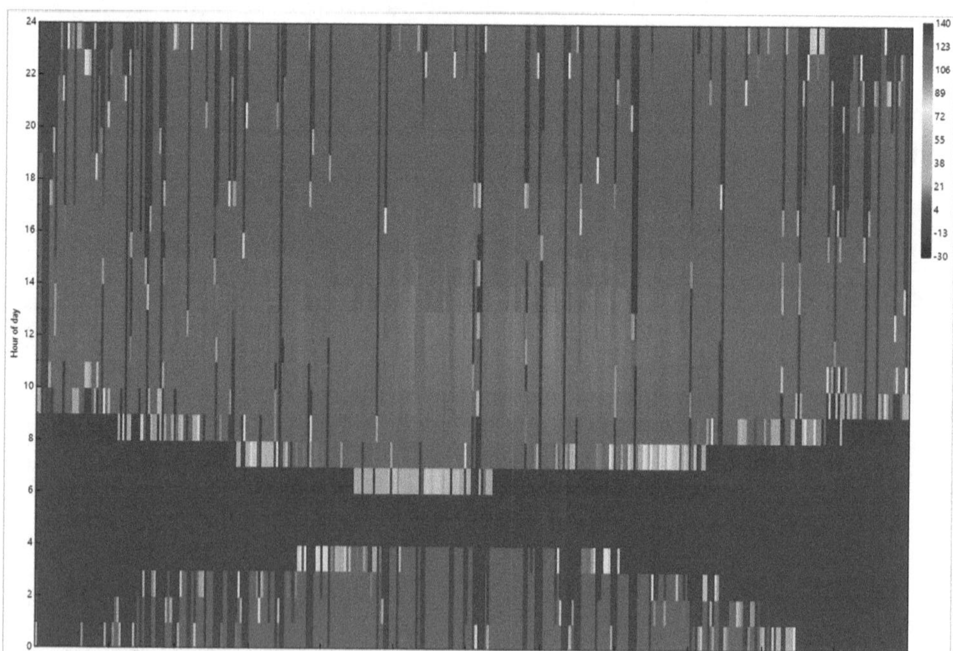

9. Now under the case dropdown menu in the upper left, click **Duplicate**. An identical case will be created to the right of the original. If desired, rename the case using **Rename** in the dropdown menu.

10. Navigate to the **System Control** page, then check the **Enable dispatch optimization** checkbox.
11. Click **Simulate** to run the new case that utilizes optimum dispatch. Note that the simulation time for this option is significantly longer than the previous analysis, and may require 5–10 minutes, depending on computer performance.
12. As with the previous case look at the table of results and the heat map to see the following:

Metric	Value
Annual AC energy (year 1)	585,503,040 kWh
Capacity factor (year 1)	65.7%
Annual Water Usage	102,817 m^3
PPA price in Year 1	9.50 ¢/kWh
PPA price escalation	1.00 %/year
LPPA Levelized PPA price nominal	11.80 ¢/kWh
LPPA Levelized PPA price real	9.37 ¢/kWh
LCOE Levelized cost of energy nominal	10.83 ¢/kWh
LCOE Levelized cost of energy real	8.60 ¢/kWh
NPV Net present value	$55,345,060
IRR Internal rate of return	11.20 %
Year IRR is achieved	20
IRR at end of project	12.69 %
Net capital cost	$746,981,632
Equity	$373,560,448
Size of debt	$373,421,184
Minimum DSCR	1.39

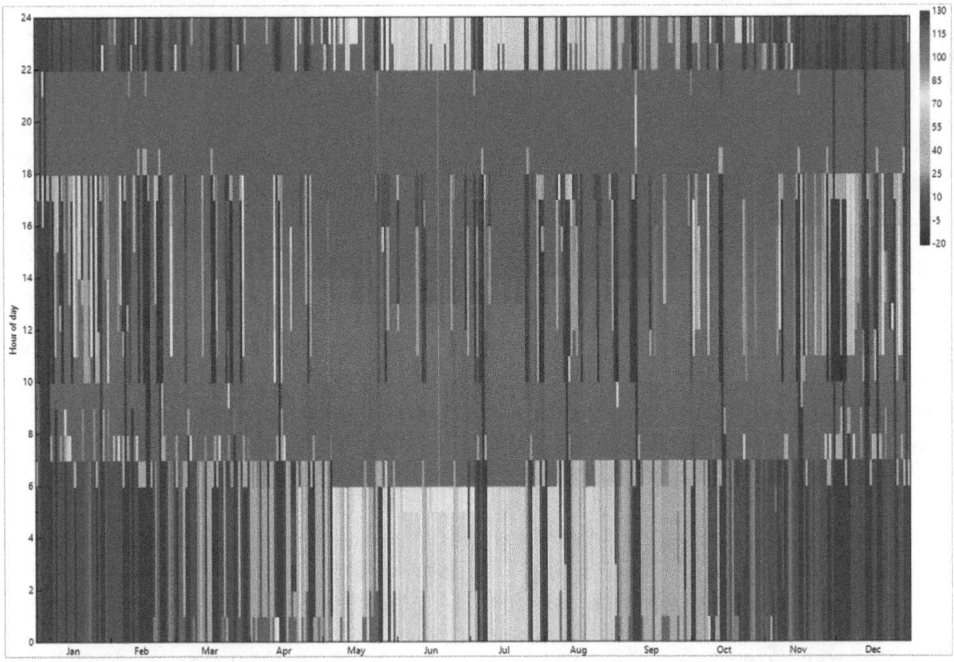

The results indicate that annual AC energy dropped from 615 GWh for the unoptimized case to 585 GWh for the optimized case. But the electricity that was produced was sold at a higher price. As a result, the internal rate of return improved from 9.8% to 11.2%. The lower annual electricity yield is likely a result of the model choosing reductions in overall performance during some hours, e.g., favoring less efficient generation in return for higher annual revenues.

The heat map plots show the electrical production by hour (y-axis) for each day of the year (x-axis). It is evident that the optimized plant can produce power during the morning peak from hours 7 to 10 (7–10 a.m.), especially during the winter, and more consistently address the evening peak hours of 18–22 (6–10 p.m.).

12.9 Solar Industrial Process Heat

Industrial energy use accounts for about one-third of the energy consumed in the United States [21] (and about 37% of worldwide energy use [22]), with process heat accounting for about half of on-site U.S. industrial energy consumption [23]. Approximately 84% of U.S. industrial fossil fuel use is to provide process heat used in manufacturing [24]. To dramatically reduce carbon emissions, we must replace fossil fuels with renewable sources, so process heat is a prime target for decarbonization of the industrial sector. Solar, biomass, and geothermal energy all represent renewable sources of thermal energy. Solar thermal energy collector systems used to generate power face increasing competition from solar photovoltaic systems. While they offer the advantage of thermal energy storage, only a portion of the energy gets converted to electricity due to thermodynamic power-cycle efficiencies. In the case of industrial process heat, however, all of the thermal energy provided by the collector system can be utilized. Thus, for direct thermal applications, solar thermal collectors are more efficient than PV. And, compared to installing solar thermal collectors on individual buildings, a large field of solar collectors used to produce thermal energy for industrial processes benefits from the economy of scale. Depending on the process temperature, either flat plate collectors or line-focus collectors are typically used.

Solar industrial process heat (IPH) systems received attention in the United States in the early 1980s. The U.S. Department of Energy sponsored a number of demonstration projects, and the Solar Energy Research Institute (the predecessor of NREL) produced a 450-page design handbook based on the experiences gained [25]. Interest in solar IPH waned, however, due to competition from low natural gas prices. Today, a number of factors have led to a renewed interest in solar IPH. There is now international concern about reducing carbon emissions. Also, solar thermal collector technology has improved significantly over the past three-and-a-half decades. In particular, parabolic trough collectors have experienced many improvements as a result of competition in the CSP industry and associated government research efforts. The IEA [26] provides information on over 200 solar IPH installations around the world. The vast majority of these have collector areas of less than 1,000 m^2.

Table 12.5 shows the different solar thermal technologies for addressing process heat at different temperature ranges [27]. Figure 12.31 shows the annual energy use by U.S. industry for various steam temperature ranges [28]. Comparing Table 12.5 and Figure 12.31, Kurup and Turchi [27] concluded that the temperature range of 120–220°C is an ideal target range for solar IPH using parabolic trough and linear Fresnel collector technologies.

TABLE 12.5

Solar Thermal Technologies for Different Temperature Ranges

Temperature Range*	Solar Collector Type	HTF of Choice	Applications/Comments
<80°C	Flat plate Non-tracking compound parabolic Solar pond	Water	Hot water Space heating
80°C–200°C	Parabolic trough Linear Fresnel	Water/steam	Hot water or steam for IPH
200°C–300°C	Parabolic trough Linear Fresnel	Mineral oil	Direct heat or steam for IPH Vacuum-jacket receivers become necessary to minimize heat loss
300°C–400°C	Parabolic trough Linear Fresnel	Synthetic oil	Direct heat or steam for IPH
400°C–550°C	Parabolic trough Linear Fresnel	Steam or Molten salt	Electric power
>550°C	Heliostat/central receiver Parabolic dish	Steam or Molten salt	Electric power

*The study focused on temperatures between 100°C and 400°C, where linear-focus collectors excel and IPH demand is high.

Source: From Kurup and Turchi [27].

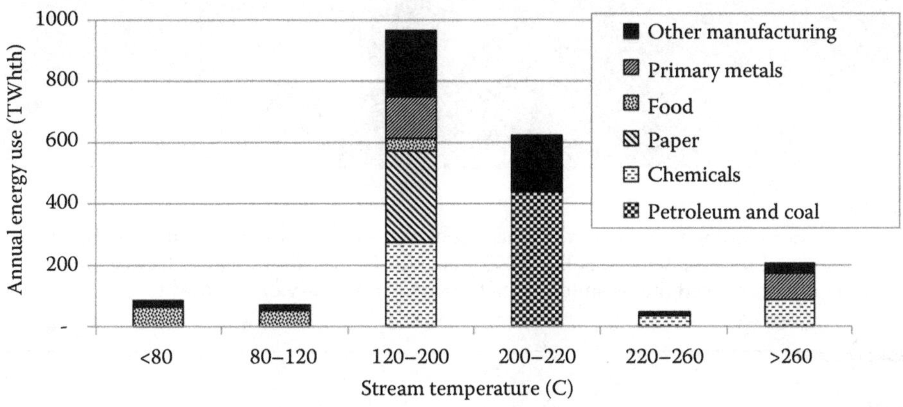

FIGURE 12.31

U.S. IPH energy use for steam generation for industries using natural gas. (Adapted from data published in a Cornell Energy Institute report by Fox et al. [28].)

Of course, effective use of concentrating solar collectors requires locations where the direct normal irradiance (DNI) resource is high. Ideally, the best locations for solar IPH are industrialized areas with high DNI. In the United States, California ranks third in industrial gas use behind Texas and Louisiana. California's Central Valley has a good combination of DNI resource and industrial demand for steam in the temperature range of interest. Figure 12.32 shows a photo of the parabolic trough system installed in 2008 by Abengoa Solar at the Frito-Lay plant in Modesto, California. It uses 5,065 m^2 of troughs to provide 2.4 MW of thermal energy at temperatures of up to 249°C. Pressurized hot water is used to heat cooking oil used in the production of (appropriately enough) Frito-Lay's SunChips snack product.

FIGURE 12.32
The 5,065-m^2 parabolic trough array used to heat cooking oil at Frito-Lay's SunChips plant in Modesto, California, which was built by Abengoa Solar in 2008. The California Central Valley has a good combination of solar resource and process need that is suited to that resource. (Courtesy of Ken May.)

Problems

12.1 The graph in Figure 12.4 gives the results of an ASHRAE Standard performance test for a double-glazed flat plate collector. Assume the transmittance for the glass is 0.90 and the absorptance of the surface of the collector plate is 0.92. Also assume the insolation incident on the collector is 600 W/m^2, the ambient temperature is 5°C, the inlet water is at 15°C, and the collector area is 5 m^2. Determine:

a. The collector heat removal factor, F_R
b. The overall heat loss coefficient of the collector, U_L, in W/m^2 K
c. The rate at which the collector can deliver useful energy in W/m^2
d. The flow rate through the collector that will give an outlet temperature of 60°C if this collector is to be used to supply heat to a hot-water tank.

12.2 What is the operating temperature for an evacuated tube collector operating at 50% efficiency if the insolation is 800 W/m^2? Use data from Figure 12.5.

12.3 A solar electric engine running between 5°C and 95°C will operate at one-half the Carnot efficiency. This engine is to drive a 4-kW pump during the day. If a tilted solar collector array has an efficiency of 50%, calculate the collector area needed for operation in Egypt, where the insolation averages about 10 kWh/m²/day during a 12-hour day. State any additional assumptions.

12.4 Consider a solar-driven irrigation pump. Parabolic trough collectors are used to heat the water in the collectors to 423 K, which then vaporizes an organic working fluid that powers the pump turbine. Calculate the surface area needed to power a 50 hp pump capable of delivering up to 38,000 L/min of water at noon in Albuquerque, New Mexico. Use the collector efficiency given in Equation 12.24 and assume the organic Rankine cycle (ORC) engine operates at 60% of Carnot efficiency. Also assume a DNI of 740 W/m² and a heat rejection temperature of 305 K.

12.5 Write a closed-form expression for the work output of a solar-powered heat engine if the energy delivery of the solar collector at high (absolute) temperature is given approximately by the expression

$$q_u = (\tau\alpha)_{\text{eff}} I_c - \frac{\sigma\varepsilon \bar{T}_f^4}{\text{CR}}$$

(convection and conduction losses are neglected) where $\bar{T}_f$ is the average fluid temperature and CR is the concentration ratio. Two heat engines are to be evaluated:

1. Carnot cycle

$$\text{Cycle efficiency } \eta_c = 1 - \frac{T_\infty}{T_f}$$

2. Brayton cycle

$$\text{Cycle efficiency } \eta_B = 1 - \frac{C_B T_\infty}{T_f} C_B \geq 1,$$

where

$$C_B = (r_p)^{(k-1)/k}$$

r_p is the compressor pressure ratio
k is the specific heat ratio of the working fluid
T_∞ is the sink (absolute) temperature for heat rejection.

Write an equation with $\bar{T}_f$ as the independent variable that, when solved, will specify the value of $\bar{T}_f$ to be used for maximum work output as a function of CR, surface emittance and $(\tau\alpha)_{\text{eff}}$ product, insolation level, and C_B. Optional: solve the equation derived previously for CR=100, ε=0.5, and $(\tau\alpha)_{\text{eff}}$=0.70, at an insolation level of 1 kW/m². What is the efficiency of a solar-powered Carnot cycle and a Brayton cycle for which C_B=2?

12.6 Explain how window placement in a building could be defined as (a) a passive solar feature, (b) an energy conservation measure, and (c) both of these.

12.7 List the issues that must be addressed to perform a conceptual design of a 50 MW power tower in Nevada.

12.8 It has been estimated that a 400 MW expansion of the old SEGS solar plant in California would cost $2.4 billion. Estimate the expected capital cost of the generation facility in dollars/kW capacity and the minimum cost of electricity per kWh generated by the addition, if it is operated at full capacity for 6 h/day and has a lifetime of 30 years. For this rough calculation, you can ignore the impacts of discount rate and operation and maintenance (O&M) costs.

12.9 Using SAM, estimate the levelized cost for a 50 MW_e parabolic trough solar power plant in Boulder City, Nevada, using the physical trough model and a Variation 1 trough (good as new receivers). Also determine the thermal energy delivered to the power block during July and estimate the surface area required for the installation. Assume the default economic parameters in SAM.

12.10 Using SAM, estimate the levelized cost for a central receiver solar power plant with 10 hours of storage in Phoenix, Arizona. Assume that the receiver is of external design and that heat rejection in the condenser is with a wet cooling tower configuration. Comment on the availability of cooling water and state all of your assumptions.

References

1. U.S. Dept. of Commerce, 1975. Optical coatings for flat plate solar collectors, NTIS No. PN-252-383, Honeywell, Inc., Morristown, NJ.
2. Solar Energy Research Institute, 1988. *Engineering Principles and Concepts for Active Solar Systems*, Hemisphere Publishing, Washington, DC.
3. Goswami, Y., Kreith, F., and Kreider, J.F., 2000. *Principles of Solar Engineering*, 2nd ed., Taylor & Francis, Philadelphia, PA.
4. ASHRAE, 2003. Standard 93-77: Methods of testing to determine the thermal performance of solar collectors, Atlanta, GA, (Note: later version is ASHRAE Standard 93–2003), https://www.ashrae.org/technical-resources/standards-and-guidelines.
5. ASHRAE, 1980. Standard 96-80: Methods of testing to determine the thermal performance of unglazed solar collectors, Atlanta, GA, https://www.ashrae.org/technical-resources/standards-and-guidelines.
6. Kutscher, C.F., and Christensen, C.B., 1992. Unglazed transpired solar collectors, in: Boer, K. ed., *Advances in Solar Energy*, vol. 7, pp. 283–307, American Solar Energy Society, Inc., Boulder, CO.
7. Kutscher, C.F., 1996. Transpired solar collector systems: A major advance in solar heating systems, in: *Proceedings of the 19th World Energy Engineering Congress*, November 6–8, 1996, Atlanta, GA.
8. Kutscher, C., Burkholder, F., and Stynes, J.K. February 2012. Generation of a parabolic trough collector efficiency curve from separate measurements of outdoor optical efficiency and indoor receiver heat loss, *Journal of Solar Energy Engineering* 134(1).
9. Zhu, G., Wendelin, T., Wagner, M., and Kutscher, C., 2013. History, current state, and future of linear Fresnel collectors, *Solar Energy* 103, 639–652.

10. Mehos, M., Turchi, C., Jorgenson, J., Denholm, P., Ho, C., and Armijo, K., May 2016. On the path to SunShot: Advancing concentrating solar power technology, performance, and dispatchability, NREL/TP-5500–65688, National Renewable Energy Laboratory, https://www.nrel.gov/docs/fy16osti/65688.pdf.
11. Kreith, F., and West, R., 2007. *CRC Handbook of Energy Efficiency and Renewable Energy*, CRC Press, Boca Raton, FL.
12. Goswami, D.Y., 1986. *Alternative Energy in Agriculture*, vol. 1, CRC Press, Boca Raton, FL.
13. Passive House Institute U.S. WUFI Passive 3.0, https://www.phius.org/resources/tools-and-support/certification-support. Last accessed January 2025.
14. International Passive House Association, Project Certification Overview, , https://www.phius.org/certifications/projects/project-certification-overview. Last accessed January 2025.
15. Alami, A., Olabi, A.G., Mdallal, A., Rezk, A., Radwan, A., Rahman, S.M.A., Shah, S.K., and Abdelkareem, M.A., 2023. Concentrating solar power (CSP) technologies: Status and analysis, *International Journal of Thermofluids* 18, 100340.
16. de Laquil, P., Kearney, D., Geyer, M., and Diver, R., 1993. Solar thermal electric technology, in: Johansson, T.B., Kelly, H., Reddy, A., and Williams, R. eds., *Renewable Energy*, Island Press, Washington, DC.
17. NREL, Nevada Solar One, https://solarpaces.nrel.gov/project/nevada-solar-one. Last accessed January 2025.
18. Acciona, https://www.acciona-energia.com/projects/nevada-solar-one-csp-plant/. Last accessed January 2025.
19. Turchi, C., and Heath, G., 2013. Molten salt power tower cost model for the System Advisor Model (SAM), NREL/TP-5500–57625, National Renewable Energy Laboratory, https://www.nrel.gov/docs/fy13osti/57625.pdf.
20. U.S. Department of Energy, 2021. 2030 Solar cost targets, https://www.energy.gov/eere/solar/articles/2030-solar-cost-targets.
21. U.S. Energy Information Administration, 2024. U.S. energy flow 2023, https://www.eia.gov/totalenergy/data/flow-graphs/total-energy.php.
22. International Energy Agency, Industry, https://www.iea.org/energy-system/industry. Last accessed January 2025.
23. U.S Department of Energy, Process heat basics, https://www.energy.gov/eere/iedo/process-heat-basics. Last accessed January 2025.
24. Rissman, J., October 2022, Decarbonizing low-temperature industrial heat in the U.S., https://energyinnovation.org/wp-content/uploads/Decarbonizing-Low-Temperature-Industrial-Heat-In-The-U.S.-Report-2.pdf.
25. Kutscher, C., Davenport, R.L., Dougherty, D.A., Gee, R.C., Michael Masterson, P., and Kenneth May, E., 1982. Design approaches for solar industrial process heat systems, NREL, https://www.nrel.gov/docs/legosti/old/1356.pdf.
26. IEA, 2024. Solar Heat Worldwide, https://www.iea-shc.org/Data/Sites/1/publications/Solar-Heat-Worldwide-2024.pdf
27. Kurup, P., and Turchi, C., 2015. Initial investigation into the potential of CSP industrial process heat for the southwest United States, NREL/TP-6A20–64709, National Renewable Energy Laboratory, Golden, CO.
28. Fox, D.B., Sutter, D., and Tester, J.W., 2011. The thermal spectrum of low-temperature energy use in the United States, *Energy & Environmental Science* 4(10), 3731–3740.

13

Ocean, Hydropower, and Geothermal Energy Conversion

This chapter covers three significant categories of energy resources and technologies: ocean, hydropower, and geothermal. The stage of development of these resources differs widely. Large-scale hydropower represents a very mature technology that has provided a significant fraction of global electricity for decades. Opportunities for further expansion are quite limited in many, but not all, parts of the world. Geothermal energy has a very long history of use for direct heating applications and electricity generation in locations where high-quality near-surface resources are available. Continuing efforts are underway for exploiting deep underground heat lacking water. The chapter starts with an overview of ocean energy resources, both thermal and kinetic, which offer large potential but have seen limited utilization to date.

13.1 Ocean Energy

The oceans contain a great deal of energy, but the conversion of this energy into a useful form requires advanced technology that is still under development. However, many demonstration prototypes have been designed and built. Valuable information is being obtained from the operation of these plants, and the results should lead to the development of large-scale units that are expected to become economically competitive in the future.

The key conversion technologies of ocean energy are as follows:

1. Ocean thermal energy conversion
2. Tidal energy
3. Wave power.

The basic elements of these technologies are presented in this chapter.

13.1.1 Ocean Thermal Energy Conversion

Oceans cover 71% of the Earth's surface and receive the majority of the solar energy incident upon the Earth. In semitropical and tropical oceans, the available temperature difference, ΔT_0, between the surface, which is warmed by the Sun, and the temperature at the lower depths is sufficient to operate a power plant using fluids such as ammonia. Since the temperature difference is available 24 h/day virtually throughout the year, ocean thermal energy conversion (OTEC) has vast potential for implementation in many parts of

DOI: 10.1201/9781003586616-13

the world. If the energy in a band ±10° latitude near the equator could be extracted with a conservative energy efficiency of 0.5% of the incident solar energy, more than 7×10^5 MW power could be generated.

The main drawback of an OTEC system is that the overall thermal efficiency of the plant is very low, and therefore large seawater pumps and large heat exchangers are required to produce adequate power. The maximum efficiency of any type of thermal power plant is the Carnot cycle efficiency, η_c, which for a typical OTEC plant is equal to

$$\eta_c = \frac{T_h - T_c}{T_h} = \frac{\Delta T_0}{T_h} \tag{13.1}$$

where

T_h is the hot surface water absolute temperature (about 298 K)
T_c is the absolute temperature of the cold water from below (about 278 K).

Thus, the Carnot efficiency is only about 6.7%. This ideal efficiency does not account for parasitic losses or the temperature differences between the seawater and the working fluid required for evaporation and condensation. A rough rule of thumb is that about 25% of ΔT_0 needs to be allowed for the evaporation and 25% for the condensing process, leaving only about half of the total temperature difference for the power turbine. For an 80% overall efficiency of the turbine generator, the gross power output efficiency would therefore only be about 3%. Parasitic load takes another 30% or so of the overall thermal efficiency, leaving a net power delivery efficiency of only about 2%–3%.

Figure 13.1 illustrates the four ocean thermal plant types that have been proposed: land-based (or shallow water), shelf-mounted, floating, and moored plant types relative to the continent/ocean floor profile. The greatest number of potential sites exists for the floating plant with a suspended cold-water pipe (CWP). Shelf-mounted or land-based plant types reduce electricity and water transportation cost as well as the ocean dynamics/

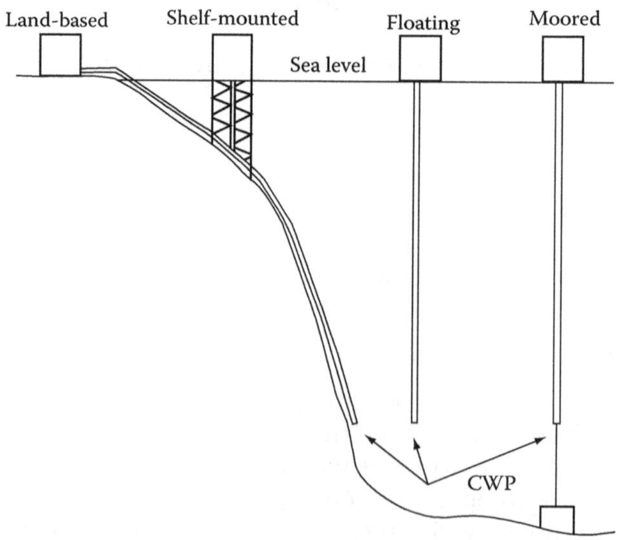

FIGURE 13.1
Basic OTEC plant configurations with cold-water pipe (CWP).

plant operation problems. Except for the CWP, the design and construction of floating and moored types are within current state-of-the-art technology. If longer pipes are required, they increase costs and reduce the number of potential sites.

Disadvantages of the floating moored plant types are related to electricity and water transportation and the effect of ocean dynamics on plant operation. For example, efficient transportation of electricity (by cable) and/or freshwater (by pipeline or barge) from a plant far offshore may be difficult. An example of the effect of ocean dynamics on plant design and operation is related to the evaporator. The evaporator produces steam as a result of the difference between the operating pressure and the saturation pressure of the warm seawater. Since this difference corresponds to only about 0.1 m of water, it is critical that the warm seawater supply pressure remain essentially constant, in spite of the wave action and platform motion, which may be difficult to achieve.

There are two basic technical approaches to producing power from the ocean thermal resources. Both are based on the Rankine thermodynamic cycle; one uses a closed system with a pressurized working fluid (typically ammonia), and the second is an open system using the seawater itself as the working fluid. The concept of a closed-cycle OTEC (CC-OTEC) system was originally proposed by D'Arsonval [1] in 1881, and in 1930, his student, Claude [2], demonstrated the operation of an open-cycle OTEC (OC-OTEC) plant off the coast of Cuba.

13.1.1.1 Closed-Cycle Ocean Thermal Energy Conversion

Figure 13.2 is a schematic diagram of a closed-cycle ocean thermal conversion system. Warm surface water is used in the boiler to evaporate a working fluid; the sensible heat liberated by the seawater is the maximum heat input, P_0, for the cycle:

$$P_0 = \rho c Q \Delta T_0 \tag{13.2}$$

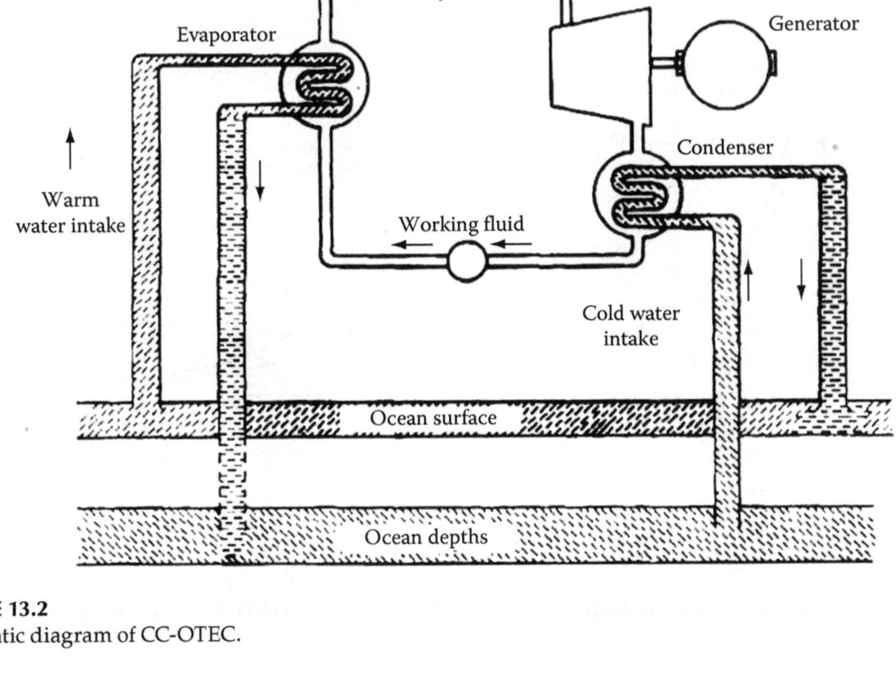

FIGURE 13.2
Schematic diagram of CC-OTEC.

where

ρ is the density of seawater in kg/m^3
c is the specific heat of seawater in J/kg K
Q is the flow rate in m^3/s.

The vapor produced in the boiler passes through a turbine to the condenser, and the expansion process generates power for electrical generation. The condenser heat sink uses cold water obtained at a depth of about 1,000 m. Obviously, a long section of pipe and substantial pumping power are required for the condensation process.

The mechanical features of the closed cycle are very similar to a refrigeration cycle. Except for preventing biofouling of the heat exchanger and the CWP, the technology is currently available. It should also be noted that while the condensed vapor of an OC-OTEC system can easily be captured as a source of drinking water, a CC-OTEC system can also be used to produce potable water [3].

Figure 13.3 shows typical temperature versus depth profiles near Hawaii, Puerto Rico, and Recife, Brazil (designated ATL-1), and Table 13.1 presents seasonal variations in ΔT_0 and the annual average net power (as a percentage of design power) for the same three sites [3]. The sites near Puerto Rico and Hawaii are of particular interest to the United States because, for these islands, all of the fuel needs to be imported. If movable or so-called grazing plants were to be installed, they would seek out the highest ΔT based on historical plots of monthly surface temperature contours, as well as daily information from satellites, which would indicate the optimum path for the platform on which the OTEC plant is based. A mini OTEC plant, shown in Figure 13.4, operated off of Keahole Point, Hawaii, in 1977 and generated 50 kW (gross) and 10 kW (net) from a temperature difference of 21.2°C (38.2°F). The system contained all the essential components required to build a full-scale CC-OTEC plant. It used a titanium plate–type heat exchanger, and the plant was mounted on a barge moored by a single anchor, which incorporated a 24 in. outside diameter (OD) polyethylene CWP as an integral member of the moor. A flexible rubber hose connected the top of the CWP to a pool in the center of the barge.

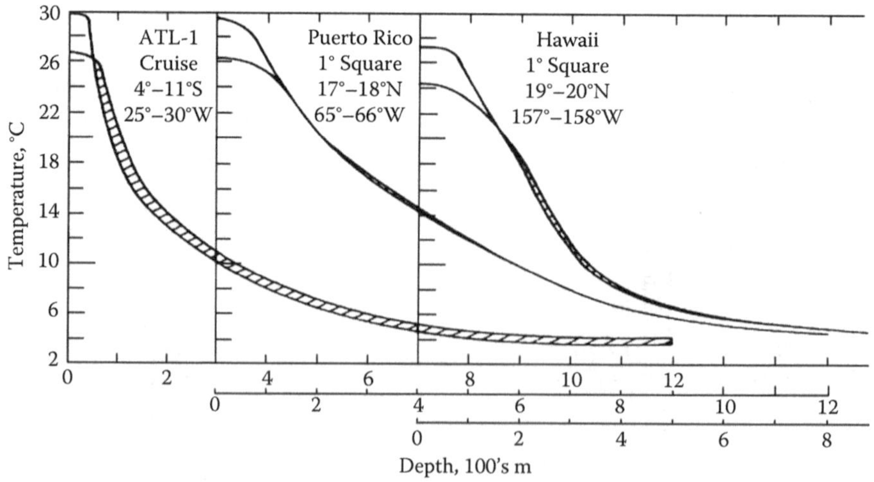

FIGURE 13.3
Temperature profiles vs. depth for Hawaii, Puerto Rico, and Recife, Brazil (ATL-1). (From Dugger et al. [3]. With permission.)

TABLE 13.1
Monthly Values of ΔT_0 and Power P_{net} (Percentage of Design Value) for Selected Sites

	ATL-1		Puerto Rico		Hawaii	
Month	ΔT_0 (°C)	P_{net} (%)	ΔT_0 (°C)	P_{net} (%)	ΔT_0 (°C)	P_{net} (%)
January	23.5	96	21.3	88	20.6	90
February	24.3	104	20.9	84	20.1	85
March	25.2	115	20.9	84	20.0	84
April	25.2	115	21.2	87	20.6	90
May	24.2	103	22.2	99	21.3	99
June	24.4	105	22.6	104	21.3	99
July	23.6	97	23.0	108	22.4	112
August	24.3	104	23.0	108	22.7	116
September	22.4	84	23.6	116	22.7	116
October	23.3	94	23.9	119	22.5	114
November	22.6	86	23.3	112	21.5	101
December	23.3	94	22.3	100	20.7	92
Average	23.9	99.8	22.3	100.7	21.4	99.8
$P_{net\ avg\ rel}$	100		83.7		74.2	

For details on the power, see Dugger et al. [3].

13.1.1.2 Open-Cycle Ocean Thermal Energy Conversion

An OC-OTEC system can generate electricity and/or freshwater using the temperature difference between surface and deep ocean seawater without the need for an expensive heat exchanger [4]. It is an attractive concept due to the vast amount of ocean energy available and the minimal ecological impact. OC-OTEC is ideally suited for locations that have a need for alternative electricity and freshwater and are also close to an ocean thermal resource. Table 13.2 provides information about potentially suitable OC-OTEC sites in the United States.

The OC-OTEC concept is shown schematically in Figure 13.5. Warm (25°C typical) surface seawater is introduced into the evaporator, which operates at a pressure slightly below the saturation pressure corresponding to the warm seawater temperature. A small percentage (≈0.5%) of the seawater flashes to produce steam, which is expanded through a turbine connected to an electrical generator. Most of the steam is condensed in a direct-contact condenser supplied with cold (typically 5°C) seawater from depths of approximately 1,000 m. A surface condenser is used only if freshwater production is desired, whereas a more efficient direct-contact condenser is used for electricity production. The remaining steam and noncondensable gases (released from the seawater in the evaporator and direct-contact condenser and from possible leaks in the containment structure) are removed from the condenser by the noncondensable exhaust system. Variations of the basic cycle shown in Figure 13.5 include evaporator, turbine, condenser, seawater pre-deaeration and a combination of freshwater and direct-contact condenser with freshwater to seawater heat exchanger.

In an OC-OTEC system, seawater itself is used as the working fluid. A 100-MW electric open-cycle system was once proposed by Westinghouse. The Westinghouse design contained turbines with 140 ft tip diameter that were to operate at 225 rpm and deliver up to 100 MW of electric power under optimum conditions. The hull in this design is located around the vertical axis turbine in an integrated design. An open channel

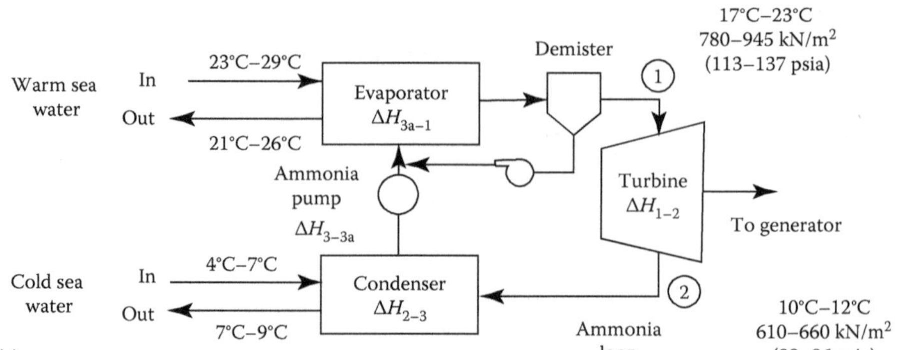

(a)

(b)

FIGURE 13.4
(a) Simplified diagram for typical closed-cycle OTEC plant using ammonia as the working fluid and (b) photo of the closed-cycle plant operated in 1979 off Hawaii. (From Dugger et al. [3]. With permission.)

TABLE 13.2
Oceanographic Data for Potential OC-OTEC Sites in the United States

Site	Latitude (°N)	Longitude (°W)	Water Depth (m)	Distance from Shore (km)	Temperature Differential[a] (°C)
Puerto Rico	18.0	65.8	1,000	3.1	22.3
Miami	25.4	79.9	740[b]	28.4	21.2
Key West	24.2	81.3	914[b]	60.0	21.5
Tampa	27.8	85.2	1,000	281.6	20.9
New Orleans	28.8	88.8	1,000	80.5	20.8
Island of Hawaii	19.9	156.1	1,000	2.3	21.4

[a] Average.
[b] Maximum depth at location.

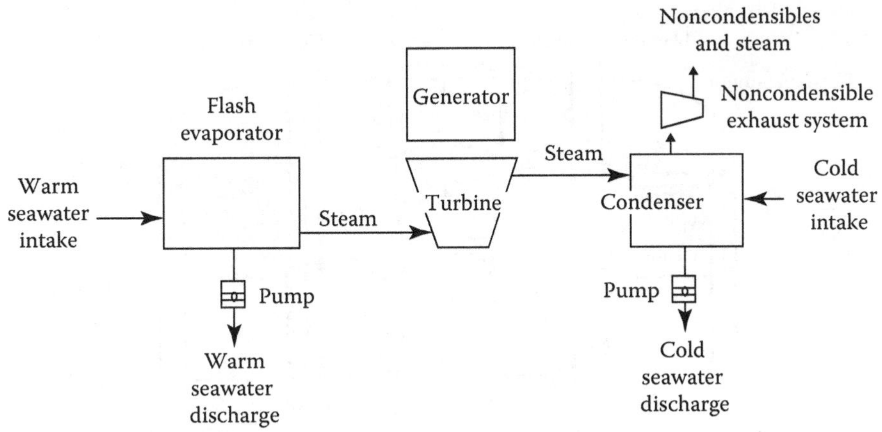

FIGURE 13.5
Conceptual schematic of open-cycle OTEC system.

evaporator is located axisymmetrically around the turbine with warm seawater entering the evaporator through a peripheral submerged sluice. Vapor flashes from the warm water flow along a bent path toward the turbine blades, while the spent vapor leaving the turbine rotor is condensed in a surface condenser centrally located underneath the turbine. More recently, it was proposed to utilize a direct-contact condenser, which increases the efficiency of the system [5]. The warm seawater leaving the evaporator and the cold seawater leaving the condenser are rejected into the ocean through a common pipe. Depending upon whether the condenser is a surface condenser or a direct-contact condenser, seawater distillation is possible with this approach. Figure 13.6 illustrates schematically such a dual-purpose system [6]. A detailed description of the OC-OTEC system using a falling jet evaporator and direct-contact condenser, a more advanced version of the Westinghouse concept was developed, at the Solar Energy Research Institute (the original name of the National Renewable Energy Laboratory) and was described in Bharathan et al. [5].

13.1.1.3 Comparison of Open- and Closed-Cycle OTEC Systems

For a given temperature difference potential, the OC-OTEC system is thermodynamically superior to the closed-cycle system because a larger portion of the temperature difference between the hot and the cold water is available to produce net power [5]. This is because of the direct-contact heat transfer mode of the evaporator and condenser, as used in the open-cycle system. The steam in the vacuum chamber can attain the temperature of the warm water discharged from the evaporator. At the same time, the exhaust steam from the turbine can be condensed at a temperature that is close to the condenser cold water exit temperature.

In a closed-cycle system, the working fluid passing through the turbine is separated from the warm and cold waters by a solid wall. Hence, in both the evaporator and the condenser, a temperature drop is necessary first to transfer the heat from the water through the wall to produce the vapor in the boiler, and then another temperature drop is required to condense the vapor of the working fluid at the outlet from the turbine by cold seawater.

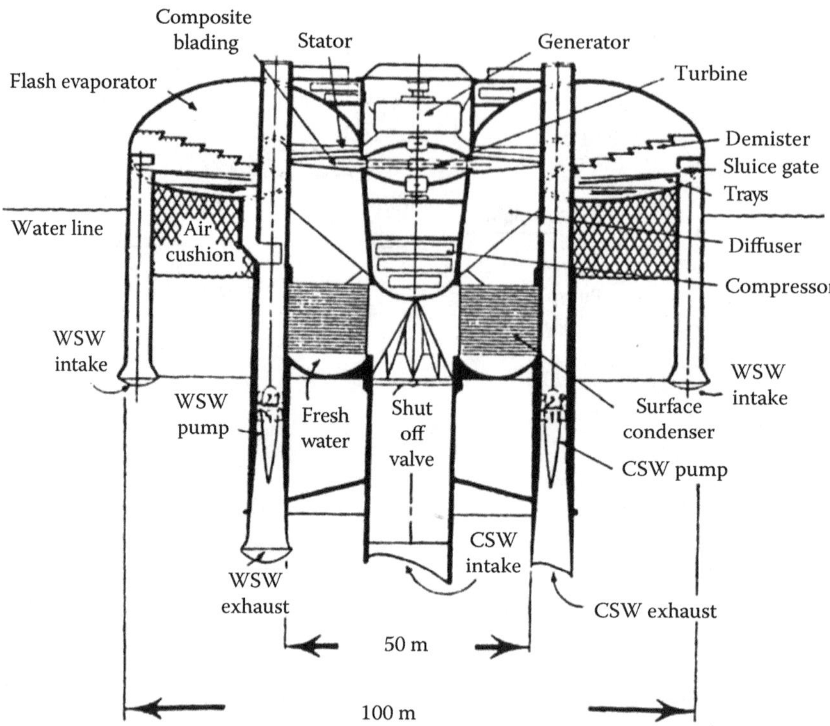

FIGURE 13.6
Schematic of OC-OTEC for power and freshwater production. 100 MW, OC-OTEC plant cross-section (as proposed by Westinghouse). (From Rabas et al. [6]. With permission.)

Consequently, the heat exchangers for CC-OTEC plants require very large surface areas to minimize the thermodynamically required temperature losses. These exchange surfaces are continuously exposed to corrosion and biofouling in the harsh ocean environment. On the other hand, the turbine is relatively small and is available as an "off-the-shelf" item. But, because of the low pressure in which the turbine must operate in the open-cycle system, the turbine is very large, and so far, no full-scale turbines for plants 1 MW and greater have been designed and operated.

Any ocean thermal energy conversion system faces the enormous challenge of producing net power with a temperature difference on the order of 20°C. Since the Carnot efficiency of such a plant is only on the order of 6% or 7%, any successful ocean energy conversion system must use as much of the temperature potential as possible and, at the same time, reduce parasitic losses to a minimum. The major parasitic losses in ocean thermal systems are the pumping power required to bring the cold water from a depth of about 1,000 m to the surface, the power required to remove noncondensable gases in the open-cycle system, and the degradation of the available thermal energy by temperature drops due to fouling and corrosion across the heat transfer surfaces of a closed-cycle system. In general, studies have found that CC-OTEC has a somewhat lower overall cost, but OC-OTEC can be advantageous where desalinated water is a high priority. There are also hybrid cycle designs that combine both closed- and open-cycle concepts [7].

13.1.1.4 Cold-Water Pipe and Pumping Requirements

The cost of the CWP is the dominant feature in the overall cost of an OTEC system [8]. This pipe is subject to stresses that include drag from ocean currents, oscillating forces due to vortex shedding, and possibly forces due to the drift and motion of the platform. Moreover, the weight of the pipe is substantial. As a result, various materials have been proposed and used for the CWP. There may also be problems in assembling and positioning the pipe because of its enormous length. A premature failure of the CWP could be a financial blow to the entire OTEC system.

To transport the large quantities of cold and hot water necessary to operate an OTEC system requires a good deal of pumping power. This power will have to be supplied from the gross power output of the OTEC system and constitutes a loss in net energy. The frictional loss in pumping the water through the CWP can be reduced by increasing the pipe diameter, but this will increase the cost and weight of the pipe. It can also lead to instabilities that could prematurely destroy the cold-water support. It should also be noted that the power output estimated in the example below is based on the efficiency of a Carnot engine, whereas a real turbine is less efficient and thereby reduces the net power output.

Example 13.1: Pumping Power Requirements

Estimate the power required to move water at a volumetric flow rate of 4.86 m³/s from a depth of 1,000 m for a 10 MW OTEC. Assume that the diameter of the pipe is 1 m.

Solution:

The average water velocity is

$$\bar{u} = \frac{\dot{Q}}{A} = \frac{4.86\left(\text{m}^3/\text{s}\right)}{\pi 0.5^2\ \text{m}^2} = 6.2\ \text{m/s}$$

Hence, the Reynolds number is

$$\text{Re}_D = \frac{uD}{\upsilon} = \frac{6.2\ (\text{m/s})1\ \text{m}}{1\times 10^{-6}\ \text{m}^2/\text{s}} = 6.1\times 10^6$$

The flow is turbulent, and the friction factor can be estimated from an empirical relation such as the one given below [9]:

$$f = \frac{0.184}{\text{Re}_D^{0.2}} = \frac{0.184}{\left(6.1\times 10^6\right)^{0.2}} = 0.0081$$

The head loss in the pipe thus equals

$$\Delta H = \frac{fLu^2}{2Dg} = \frac{0.0081\times 10^3\,\text{m}\times 6.2^2\ \text{m}^2/\text{s}^2}{2\times 1.0\ \text{m}\times 9.8\ \text{m/s}^2} = 15.9\ \text{m}$$

Hence, the pumping power required is

$$P_{\text{pump}} = \rho\dot{Q}g\Delta H = 1\times 10^3\ \text{kg/m}^3\times 4.86\ \text{m}^3/\text{s}\times 9.8\ \text{m/s}^2\times 15.9\ \text{m}$$
$$= 7.6\times 10^5\ \text{W} = 760\ \text{kW}$$

The pumping power requirements could be reduced substantially by using a 2 m diameter pipe.

13.1.1.5 Economics and Future Prospects

Cost estimates have been performed for both open- and closed-cycle systems [8]. By and large, they have indicated similar results; however, since no large system has yet been built, detailed absolute cost estimates are unreliable. The results clearly indicate that in order for OTEC to be economically viable, extremely large plants would be necessary. For such large plants, the initial investment is large, and this has, so far, discouraged investment in this technology. Figure 13.7 shows the fractional costs in percentage for the various parts of an OC-OTEC plant. These estimates indicate that the largest costs are the cold-water and discharge pipes, with the CWP costing approximately 40% of the total, while the discharge pipe would require about half that amount.

There was considerable research on OTEC in the 1970s and early 1980s, after which interest waned. In recent years, however, there has been renewed interest in OTEC for island nations, especially those that rely heavily on expensive imported diesel fuel and lack land for solar and onshore wind [10]. In 2015 Makai Ocean Engineering's 105-kW land-based plant began providing power to the grid in Kailua-Kona, Hawaii. In Okinawa, Japan, a 100-kW land-based OTEC demonstration project was connected to the grid in 2016, and there are plans to expand it into a 1-MW commercial plant "by around 2026." The Korean Research Institute of Ships and Ocean Engineering (KRISO) has experience with a 20-kW OTEC plant on a floating barge and is now developing a 1-MW floating barge for the Pacific Island of Kiribati. UK-based Global OTEC Resources is developing a 1.5-MW floating barge OTEC plant that they intend to install off the Democratic Republic of São Tomé and Príncipe, an island nation off the western equatorial coast of Central Africa. All of the aforementioned projects used a closed-cycle design. More information about these projects can be found in [11].

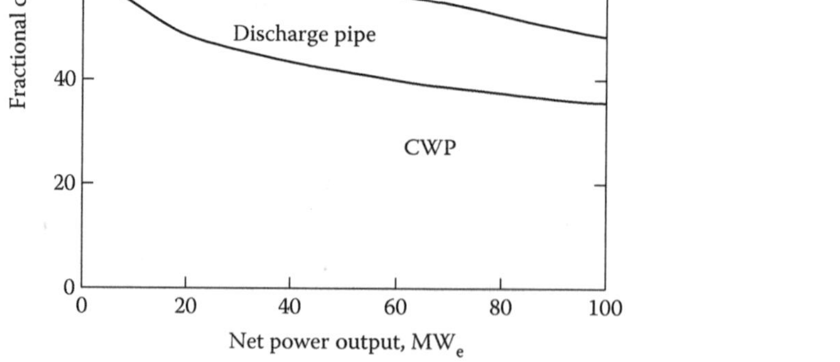

FIGURE 13.7
Fractional component cost of single-stage, direct-contact condenser plant optimized for power output. (From Block and Valenzuela [8]. With permission.)

13.1.2 Tidal Energy

Tidal energy has a long history. Small tidal mills were used for grinding corn in Europe in the Middle Ages, but more recently, using tidal energy on a large scale to generate electricity has been proposed for many parts of the world. Tidal energy can be harnessed by building large barrages—essentially long low dams—across the inlet of suitable estuaries. When the tides come in, the seawater can be trapped by means of sluices in reservoirs, called tidal basins, behind these barrages. When the tide ebbs, the water behind the barrage can be let out via a turbine somewhat similar to a hydroelectric power plant. (Depending on the design, the incoming flow can also spin the turbines.) However, it is important to distinguish tidal power from hydropower. The former is the result of the interaction of the gravitational pull of the moon and the sun on the oceans, while the latter is derived from the hydrological climate cycle powered by solar energy. A simplified analysis of the Earth–Moon interaction that produces the tidal behavior is presented in Elliott [12].

Tidal energy technology relies on the twice-daily tides with the concomitant upstream flows and downstream ebbs in estuaries [13]. To extract energy from the rise and fall of the tides, barrages are built across the inlet to the tidal basin, and turbines are located in water passages inside the barrages. The potential energy in the difference in water levels across the barrage due to the tidal motion is converted to kinetic energy as fast-moving water passes through the turbine. The rotational kinetic energy of the blades of the turbine can then be used to drive a generator to produce electricity. The average power output from a tidal system is roughly proportional to the square of the difference in water level between high and low tides. A tidal range of at least 5 m is considered the minimum for viable power generation. Sometimes tidal flows can be increased in height using the effect of concentration, as shown schematically [12] in the Severn Estuary near Cardiff, Wales (Figure 13.8).

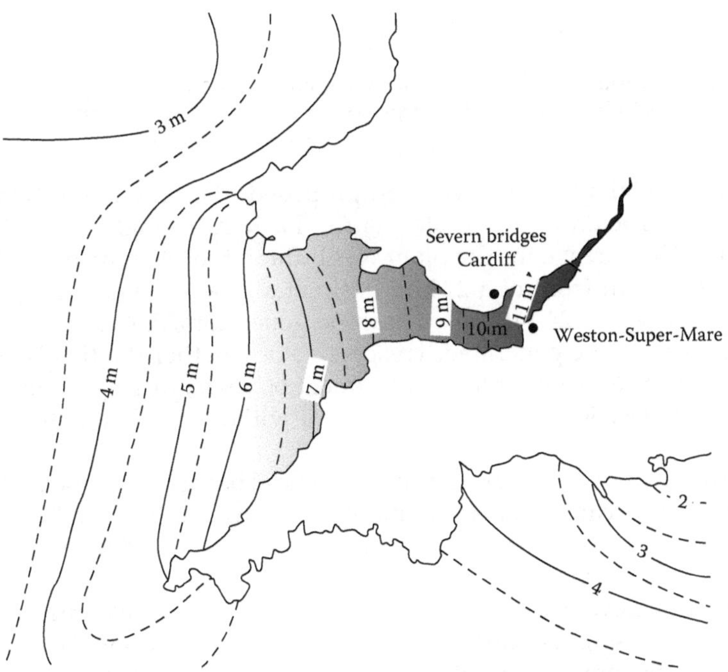

FIGURE 13.8
The effects of concentration of tidal flow in the Severn Estuary (tidal ranges in meters).

(a) (b) (c)

FIGURE 13.9
A view of La Rance tidal scheme. (a) Aerial photo of La Rance Tidal Power Plant, (b) photo of La Rance showing dam, and (c) turbine wheel. (From Renewable Energy UK [14].)

Although many tidal systems have been proposed, the only major tidal power plant that has been constructed and operated is the La Rance Tidal Power Plant in France (Figure 13.9) [14]. The barrage for the plant was built in the 1960s near St. Maillot. A dam, 330 m long, was built in front of a 22 km^2 basin at a location where the tidal range was 8 m. The plant consists of 24 bulb-type turbine generators, 5.35 m in diameter, rated at 10 MW each. Power can be generated irrespective of whether the tide is going in or out. The plant has now been in operation for almost 60 years, during which time approximately 30 GWh of electricity has been generated without major incident or mechanical breakdown.

To estimate the amount of power output from a tidal barrage, assume that in front of the barrage is an infinite ocean, whereas behind it, a rectangular basin delineates the volume into which the water flows. The volume has a surface area A, and the range between high and low tide is H, as shown in Figure 13.10.

The incoming tide flows freely into the basin, but when the tide goes out, the water is held in the basin by closing the sluice gates at a level H above the level of the ocean. Given a rectangular basin, the center of gravity of the mass of water in the basin will be at a height $H/2$ above the sea level. The total volume in the basin will thus be AH, and its mass will be ρAH. All of the water stored in the basin could now be made to flow through a turbine

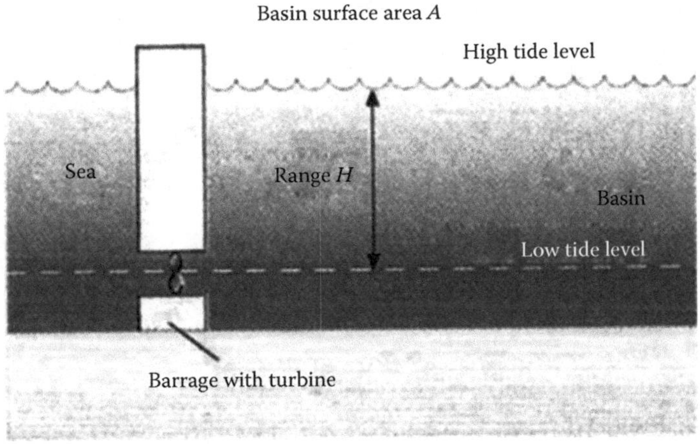

FIGURE 13.10
Nomenclature for tidal power analysis.

from the high to the low water level. The maximum potential energy E available per tide, assuming all the water passes through a height $H/2$, is therefore

$$E = \rho AHg\left(\frac{H}{2}\right) \tag{13.3}$$

where g is the gravitational attraction. The main periods of tides are diurnal at about 24 hours and semidiurnal at about 12 hours 25 minutes [12].

If the basin is allowed to fill again with the next incoming tide and the cycle repeats with a period t, the average power, P, provided from the potential energy that could be extracted becomes

$$P = \frac{\rho AH^2 g}{2t} \tag{13.4}$$

For example, if $A = 10\,\text{km}^2$, $H = 5\,\text{m}$, $t = 12$ hours 25 minutes, and ρ (seawater) $= 1{,}030\,\text{kg/m}^3$, $P = 28.2$ MW.

However, the difference between high and low tidal levels, H, varies throughout the year from a maximum H_{max} in the spring, to a minimum, H_{min}, for the neap tides. As shown in Twidell and Weir [15], this variation in H can be accounted for by using a mean range for all the tides, $\bar{H}$. The mean power production per month, $\bar{P}$, then becomes

$$\bar{P} \sim \frac{\rho Ag}{2t}\left(\frac{H_{max}^2 + H_{min}^2}{2}\right) \tag{13.5}$$

For example, if $H_{max} = 5\,\text{m}$, $H_{min} = 2.5\,\text{m}$, $A = 10\,\text{km}^2$, $\rho = 1{,}030\,\text{kg/m}^3$, and $t = 12$ hours 25 minutes (4.47×10^4 seconds), then $\bar{P} \cong 17.6$ MW.

Tidal power availability is dependent on local conditions, and there are many sites around the world where tidal barrages could potentially be sited, as shown in Table 13.3 [12]. The total power available from tides globally is approximately 3 TW, of which approximately a third, or 1 TW, is in accessible locations. The realistically recoverable resource has been placed at one-tenth of this amount, or 100 GW, due to practical limitations such as access to major tides [16].

TABLE 13.3

Some World Locations for Potential Tidal Power Projects

Country	Mean Tidal Range (m)	Basin Area (km^2)	Installed Capacity (MW)	Approx. Annual Output (TWh per Year)	Annual Plant Load Factor (%)
Argentina					
San José	5.8	778	5,040	9.4	21
Golfo Nuevo	37	2,376	6,570	168	29
Rio Deseado	36	73	180	0.45	28
Santa Cruz	7.5	222	2,420	6.1	29
Rio Gallegos	7.5	177	1,900	4.8	29
Australia					
Secure Bay	7.0	140	1,480	2.9	22
Walcott Inlet	7.0	260	2,800	5.4	22
Canada					
Cobequid	12.4	240	5,338	14.0	30
Cumberland	10.9	90	1,400	3.4	28
Shepody	10.0	115	1,800	4.8	30
India					
Gulf of Kachchh (Kutch)	5.0	170	900	1.6	22
Gulf of Cambay (Khambhat)	7.0	1,970	7,000	15.0	24
Korea (Rep)					
Garolim	4.7	100	400	0.836	24
Cheonsu	4.5	—	—	1.2	—
Mexico					
Rio Colorado	6–7	—	—	54	—
United States					
Passamaquoddy	5.5	—	—	—	—
Knik Arm	7.5	—	2,900	7.4	29
Turnagain Arm	7.5	—	6,500	16.6	29
Russian Federation					
Mezeh	6.7	2,640	15,000	45	34
Tugur[a]	6.8	1,080	7,800	16.2	24
Penzhinsk	11.4	20,530	87,400	190	25

Source: Adapted from World Energy Council [12]. With permission.
[a] 7,000 MW variant also studied.

There are essentially three ways to generate power from a tidal system as described in Elliott [12]. The most common method is called ebb generation, in which the incoming tide passes through the sluice gate in the barrage, and the water is trapped in the tidal basin by closing the sluices. The head of water in the basin can then be passed back through the turbine during the outgoing ebb tide, as shown in Figure 13.11.

The cost of energy from a tidal system will depend on several factors such as location, capital cost, interest rate, turbine performance, length of time for amortization, and difference in water height between high and low tides. The last factor is particularly important

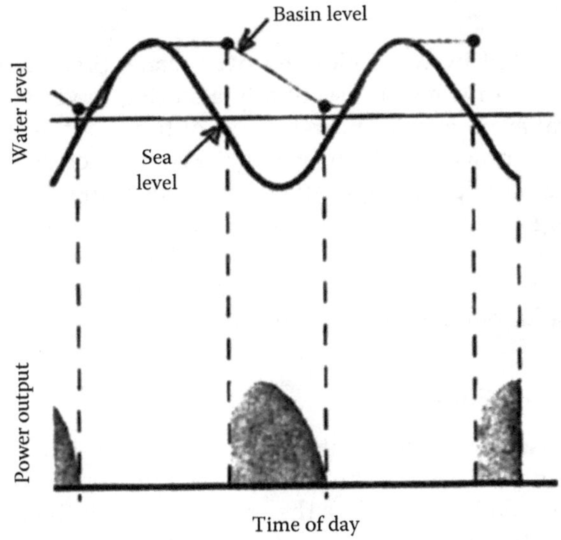

FIGURE 13.11
Schematic diagram of water levels and power outputs for an ebb generation scheme [12].

FIGURE 13.12
La Rance tidal power station. (From Khaligh and Onar [17], Figure 3.6.)

because, as shown by Equation 13.5, the power output from a tidal system is a function of the square of the tidal ebb. If amortized over a life of 50 years, the cost of power from the La Rance tidal system is one of the lowest of any renewable energy scheme (see close-up photo of the La Rance barrage in Figure 13.12 [17]).

13.1.3 Wave Energy

Ocean waves are generated by wind blowing over water. Since wind is driven by solar energy, the energy in ocean waves is essentially a stored, moderately high-density form of solar energy. Typical average solar power levels of the order of 100 W/m^2 can be transformed into waves with power levels of over 100 kW/m of crest length. The size of the waves generated by wind depends upon the wind speed and its duration. The distance over which wind energy is transferred into the ocean to form waves is called the fetch. Greater amplitude waves contain more energy per meter of crest length than small waves. In practice, however, the power that can be extracted from the waves rather than their energy content is a useful parameter for engineers.

13.1.3.1 Deep Water Wave Power

A typical ocean wave can be approximated by a sinusoidal shape of height, a, as shown in Figure 13.13. Defining the average height above the water surface, a (m), the distance between peaks as the wavelength, λ (m), the speed at which the peaks move across the surface of the ocean as the wave velocity, v (m/s), the time for successive peaks to pass a fixed point as period, P (seconds), and the wave frequency as $v=1/P$, we can determine important wave characteristics. For example, using this type of simplified and idealized wave motion, Twidell and Weir [18] have proposed a procedure to estimate the power in deep ocean waves, based on theoretical foundations described in 1977 by Coulson and Jeffrey [19]. The distinctive properties of deep water waves are summarized as follows:

1. Surface waves consist of unbroken sine waves of irregular wavelength, phase, and direction.
2. The motion of any particle of water is circular. Whereas the surface forms of a wave show a definite progression, water particles have no progression.
3. Surface water remains on the surface.
4. The amplitudes of the water particle motions decrease exponentially with depth. At a depth of $\lambda/2\pi$ below the mean surface, the amplitude is reduced to $1/e$ of the surface amplitude.
5. The amplitude, a, of the surface wave is essentially independent of the wavelength, λ, velocity, c, or period T, of the wave and depends on the history of the wind above the surface.

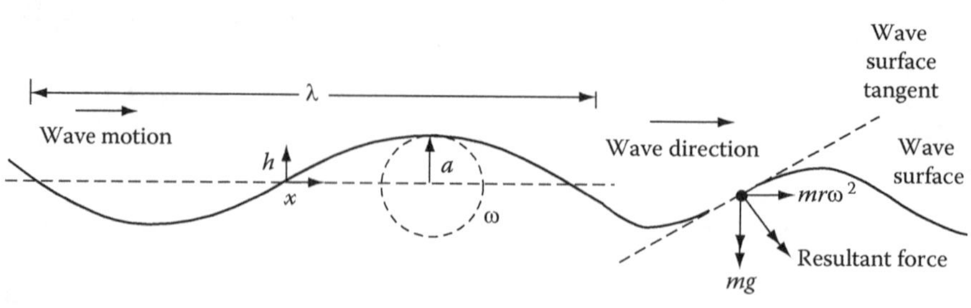

FIGURE 13.13
Characteristics of a sinusoidal wave.

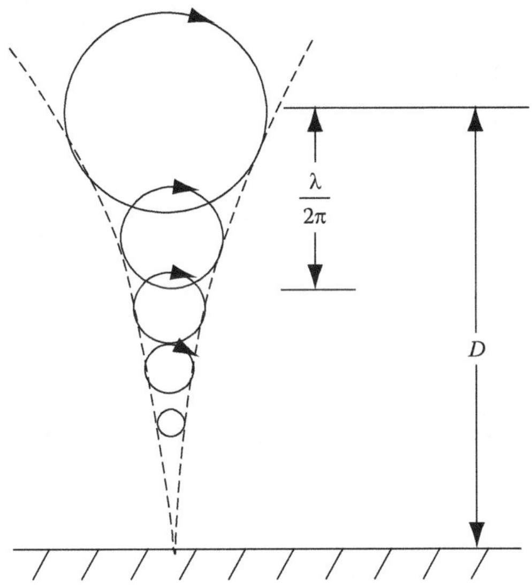

FIGURE 13.14
Motion of a particle in deep water waves. The motion is circular with an angular speed ω (rad/s) and a radius a (m).

Since most devices proposed for energy extraction are designed for deep water waves, it may be assumed that the two dominant forces are the result of gravity and circular motion, as shown in Figure 13.14. An average wave for power generation has a wavelength ~100 m, an amplitude ~3 m, and behaves as a deep water wave of depth greater than ~30 m with a circular motion of water particles as illustrated in Figure 13.14. The circles approximate the motion of a particle in a deep water wave. This circular motion has an angular velocity, ω (rad/s), and an amplitude, a (m), that decreases exponentially with depth, D, and becomes negligible when $D > (\lambda/2)$. It should be noted that there is no net motion of water in deep water waves, and an object suspended in water will move in circles, as shown in Figure 13.14.

To calculate the power that can be extracted from deep water waves, we shall simplify the rigorous mathematical treatment. It shows that for a surface wave of amplitude a and wave number k, the radius r of particle motion below the surface is

$$r = ae^{kz} \tag{13.6}$$

where

$k = 2\pi/\lambda$

z is the distance below the surface, as shown in Figure 13.15.

The vertical displacement from the average position as a function of time t is

$$\Delta z = r \sin \omega t = ae^{kz} \sin \omega t \tag{13.7}$$

and the horizontal velocity component, $u_{x'}$ is

$$u_x = r\omega \sin \omega t = \omega a e^{kz} \sin \omega t \tag{13.8}$$

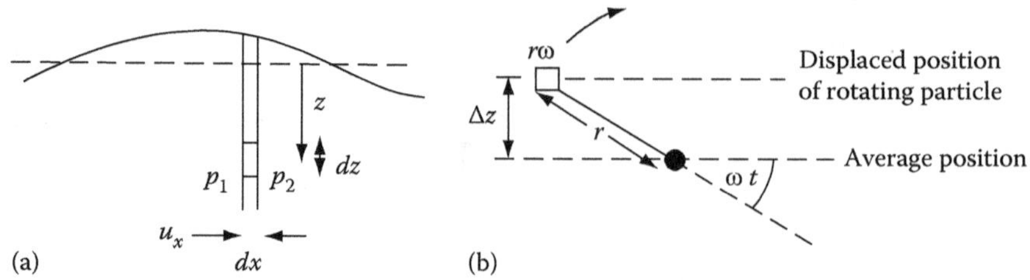

FIGURE 13.15
Local pressure fluctuation in a deep water wave: (a) pressure in the wave and (b) local displacement of water particle.

Then, P', the power in the wave at x per unit width of the wavefront is given by

$$P' = \int_{z=-\infty}^{z=0} (p_1 - p_2) u_x dz \tag{13.9}$$

where $(p_1 - p_2)$ is the pressure difference experienced by the element of height dz in the horizontal direction. This pressure difference is equal to the change in potential energy of particles rotating in the circular paths, or

$$p_1 - p_2 = \rho g \Delta z = \rho g a e^{kz} \sin \omega t \tag{13.10}$$

Hence, substituting the above relation in Equation 13.9 yields

$$P' = \int_{z=-\infty}^{z=0} (\rho g a e^{kz} \sin \omega t)(\omega a e^{kz} \sin \omega t) dz = \rho g a^2 \omega \int_{z=-\infty}^{z=0} e^{2kz} \sin^2 \omega t dz \tag{13.11}$$

Since the average over many periods of $\sin^2 \omega t$ can be shown to equal ½

$$P' = \left(\frac{\rho g a^2 \omega}{2}\right) \int_{z=-\infty}^{z=0} e^{2kz} dz = \left(\frac{\rho g a^2 \omega}{2}\right)\left(\frac{1}{2k}\right) \tag{13.12}$$

Introduce the definition of the phase velocity of a wave, c, given by

$$c = \left(\frac{\omega}{k}\right) = \left(\frac{gT}{2\pi}\right) = \left(\frac{g\lambda}{2\pi}\right)^{1/2} \tag{13.13}$$

where T is the period of a sinusoidal wave given by

$$T = \frac{2\pi}{\omega} = 2\pi\left(\frac{\lambda}{2\pi g}\right)^{1/2} \tag{13.14}$$

With these equations, the power in watts is

$$P' = \left(\frac{\rho g a^2 \omega}{2}\right)\left(\frac{1}{2}\right)\left(\frac{gT^2}{4\pi^2}\right) = \frac{\rho g^2 a^2 T}{8\pi}, \tag{13.15}$$

or, as a function of the wavelength, it has the form

$$P' = \left(\frac{\rho g a^2}{4}\right)\left(\frac{\lambda g}{2\pi}\right)^{1/2} \tag{13.16}$$

We can see that waves of great height and long wavelength are important parameters for effective wave power production.

13.1.3.2 Wave Power Devices

Wave power generation has not yet reached the same state of development as wind power, but there have been many proposals for capturing energy from waves, and quite a few prototypes have been built, tested, and put into operation. So far, however, there have been no large-scale wave energy systems built and operated over any length of time. Consequently, it is not possible to decide which of the various devices has the greatest promise for eventual deployment. This section will therefore restrict itself to describing a few typical devices that have been built and operated.

Wave capture systems are relatively simple. When waves break over a sea wall or dam near the seashore, water is impounded at a height above the average sea level and can be captured in a reservoir. The water from the reservoir can then be returned to the ocean through a conventional low head turbine attached to a generator, similar to a tidal system described in Section 13.1.2. The flow pattern of a wave capture system, however, is less regular than that of a tidal system.

A 350 kW wave capture system was built and operated successfully in Norway in 1985. A schematic diagram of the system, which was called Tapchan (for tapered channel), is shown in Figure 13.16 [20]. In this system, waves were funneled through a tapered channel so that larger waves were able to flow over the walls immediately, while the smaller waves increased in height as they moved up in the channel.

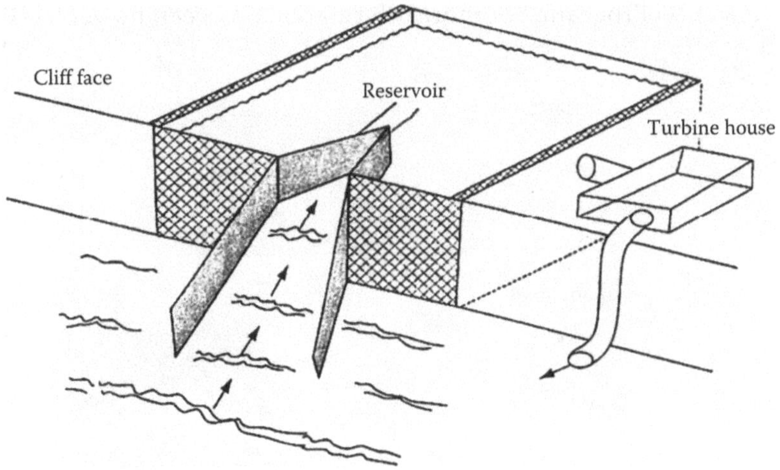

FIGURE 13.16
Schematic diagram of the Tapchan wave capture system in Norway. (From Hawaii Department of Business [20].)

A site for a successful wave capture system should have the following characteristics:

1. Deep water close to shore.
2. A tidal range of less than 1 m.
3. A convenient site for constructing the reservoir, preferably a natural gully in a rock formation near shore.
4. Continuous waves with large amounts of energy.

Although these types of wave capture systems are conceptually very simple, they suffer in general from a low head, which requires a large amount of water flow for a respectable energy output. On the other hand, the Tapchan concept has very few moving parts, its maintenance costs are low, and its reliability is high. Moreover, the storage reservoir helps to smooth the electrical output.

Oscillating water column (OWC) systems are simple and robust and have been deployed in a number of countries. They probably represent the most common form of wave energy conversion. A successful OWC system, called the Limpet, was constructed in the year 2000 on the Scottish island Islay [21].

A schematic diagram of this onshore wave power system demonstrating its operation is shown in Figure 13.17. Air is compressed and decompressed by the OWC, and this causes air to be forced through a Wells turbine located above. The Wells turbine is capable of rotating in the same direction, regardless of the direction of the airflow. Thus, it generates power irrespective of the direction of movement of the water column. The water column rises with each incoming wave and compresses the air above, which is thus forced to flow through the turbine. As the water column goes down, the air is drawn in the opposite direction through the Wells turbine, which spins an electric generator. The waves are captured by a reinforced concrete wall set into an excavated rock face on the shore. The existing Limpet system, shown in Figure 13.18, has a capacity of 500 kW and has been successful in generating electricity for the grid on the island for several years. The electricity is exported to the utility grid within the Scottish Administration's Renewable Obligation Program. For more information on ocean power technologies, see reference [22].

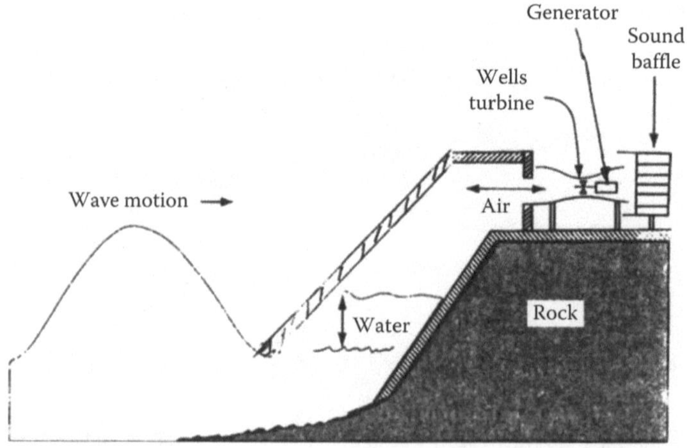

FIGURE 13.17
Schematic diagram of an onshore wave power system using an oscillating water column similar to Limpet.

FIGURE 13.18
Photograph of Limpet.

13.2 Hydropower

There are three types of hydropower generation technologies: impoundment with a dam, diversion of run-of-the-river, and pumped storage. The first two types are described in this section, whereas pumped storage is discussed in the chapter on storage since it is not a net generation technology. In 2023, hydropower plants generated approximately 14% of the world's total electricity and 40% of all renewable electricity, although its share of renewable electricity has decreased considerably in recent years due to the rapid growth of wind and solar. In the United States, hydropower grew from 56 GW of installed capacity in 1970 to 102 GW in 2011, but it now provides 80 GW, which represents 6.2% of U.S. electric capacity. Capacity and generation statistics for the world and for eight of the largest hydroelectric producers in 2020 are listed in Table 13.4 [23].

It is expected that hydropower generation will continue to increase, mostly in developing countries, where potential sites still exist. China is the largest hydroelectricity producer, with 1,348 TWh of production in 2020, representing almost 17% of its electricity use.

Conventional hydroelectric power plants use the potential energy of dammed water to drive a water turbine and generator, as shown in Figure 13.19. The power extracted from the water depends on the difference in height between the top of the source and the water outflow, and the rate of volume of water passing through the penstock.

Figure 13.20 shows the connection between the reservoir, penstock, water turbine, and the generator in more detail. Water released from the reservoir through the penstock drives the turbine, which is connected by a shaft to the generator that delivers power to the grid.

TABLE 13.4

Hydropower Capacity and Generation in 2020 for the World's Largest Hydropower Producers

Country	Capacity (GW)	Generation (TWh)	Contribution to Total Generation (%)	Average Age of Fleet (Years)
World	1,168.0	4,332.9	17	31.3
China	335.4	1,347.9	17	13.5
Brazil	110.2	396.3	66	25.5
Canada	81.2	384.6	61	43.3
United States	81.4	292.7	7	55.1
Russian Federation	50.0	188.3	17	41.3
India	45.1	167.0	10	25.4
Norway	31.9	141.6	107	40.8
Vietnam	18.1	63.4	26	11.9

Source: International Energy Agency [23].

FIGURE 13.19
The Three Gorges Dam on the Yangtze River, China. It is the world's largest power-producing facility of any kind. (From Shutterstock. With permission.)

The electric power production from a hydroelectric station is approximately equal to:

$$P = \rho h V g \eta \tag{13.17}$$

where P is the power output in watts, ρ the density of water, h the difference in height between the top of the reservoir and the turbine, usually called the head, in meters, V is the flow rate of water in cubic meters per second, g is acceleration due to gravity (9.8 m/s^2), and η is the efficiency of the generation process.

Hydropower is a flexible source of electricity because the output can be ramped up and down quickly to adapt to changing energy demands. Hydro turbines have a startup time

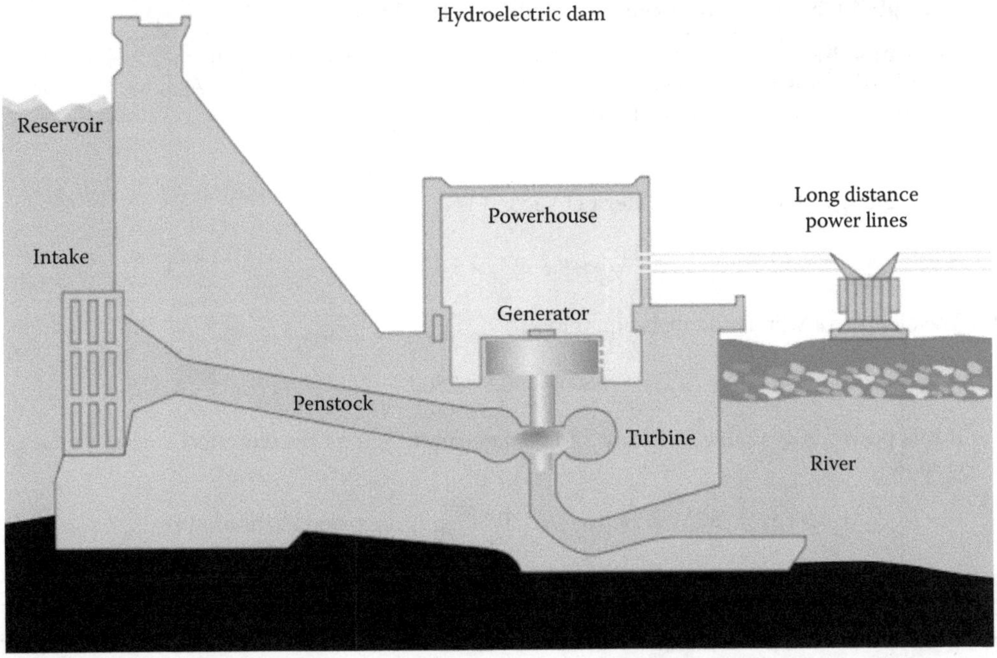

FIGURE 13.20
Schematic diagram of a hydroelectricity power plant. (Tennessee Valley Authority.)

of a few minutes, and power generation can decrease quickly when there is a surplus generation. Hydropower is, therefore, not generally used to produce base-load power, but rather to serve as a backup for non-hydro generators. The major advantage of hydropower is its ability to meet peak power demand when the cost of energy is high. The global average LCOE of new hydropower projects in 2022 was $0.057/kWh [24].

The U.S. Department of Energy defines large hydropower facilities as those having a capacity greater than 30 MW, small hydropower as projects that generate less than 30 MW, and micro hydropower as plants having capacities of less than 100 kW. Micro hydro facilities can provide inexpensive power to isolated communities or buildings.

Since hydroelectric dams do not burn fossil fuels, they do not produce carbon emissions during operation. Some CO_2 and other pollutants are emitted during construction. Disadvantages of hydropower are associated with the submersion of extensive areas upstream of the dams, sometimes destroying productive land and biologically rich areas. Damming interrupts the flow of rivers, can harm local ecosystems, and often involves displacing people and wildlife. The decomposition of flooded vegetation releases methane, a strong greenhouse gas. In developing Southeast Asian countries, plans for new hydroelectric dams are pitting the region's need for electricity against its need for food, such as rice and fish, which depends on the river's free-flowing waters.

Hydropower can experience significant weather-related variations in water flow, and this is exacerbated by global warming. There are also risks associated with the construction of dams that are not properly engineered and can fail. There have been failures in dams that have resulted in floods and deaths, as for example, the Vajont Dam in Italy in 1963.

Example 13.2: Power Produced by a Hydroelectric Facility

Determine the energy delivered per month of a hydroelectric system that has a flow rate of 20 kg of water per second and an available vertical drop of 5 m. Assume that the energy conversion efficiency of the turbine is 80%.

Solution:

The energy change of the water per second is

$$20\ {}^{\text{kg}}\!/_{\text{s}} \times 9.8\ {}^{\text{m}}\!/_{\text{s}^2} \times 5\ \text{m} = 980\ {}^{\text{J}}\!/_{\text{s}}$$

The electric power available is equal to:

$$980\ {}^{\text{J}}\!/_{\text{s}} \times 0.8 = 780\ \text{W}$$

If this power is available 24 h/day and 30 days/month, then the electrical energy delivered is

$$\text{Energy} = 780\ \text{W} \times 24\ {}^{\text{h}}\!/_{\text{day}} \times\ 30\ {}^{\text{days}}\!/_{\text{month}} = 560\ \text{kWh/month}$$

13.3 Geothermal Energy

Geothermal energy can be defined as heat that originates within the Earth. This heat occurs from a combination of two sources: the original heat produced from the formation of the Earth by gravitational collapse and the heat produced by the radioactive decay of various isotopes. Because the thermal conductivity of rock is so low, it is taking many billions of years for the Earth to cool. Geothermal energy can be used for generating electricity if it is at a temperature greater than about 110°C. At lower temperatures, it can be used for district heating or a wide range of low-temperature direct-use applications [25].

13.3.1 Geothermal Power

Geothermal resources can be divided into four types: hydrothermal resources consisting of some combination of hot water and steam in permeable rock; EGS (Enhanced, or Engineered, Geothermal Systems), which involve the creation of an artificial hydrothermal reservoir in deep, hot dry rock; geopressured brines; and magma. Geopressured resources are deep, pressurized brines such as exist along the U.S. Gulf Coast and produce natural gas, which has a higher economic value than geothermal energy. Only one experiment has been performed to tap the geothermal energy in geopressured brine. Magma is the very hot molten rock that exists in volcanic regions; utilizing it presents material challenges due to the very high temperatures.

Except for geothermal (or ground-coupled) heat pumps, which utilize the heat naturally contained in shallow soil (which is a function of the local climate), all existing commercial uses of geothermal energy make use of hydrothermal resources that are located relatively close to the surface and do not entail very high drilling costs. The hot geothermal fluid is used for direct heating applications such as spas, greenhouses, district heating, and for the generation of electricity. Figure 13.21 shows a U.S. map [26] indicating regions with near-surface hydrothermal resources at temperatures greater than 100°C suitable

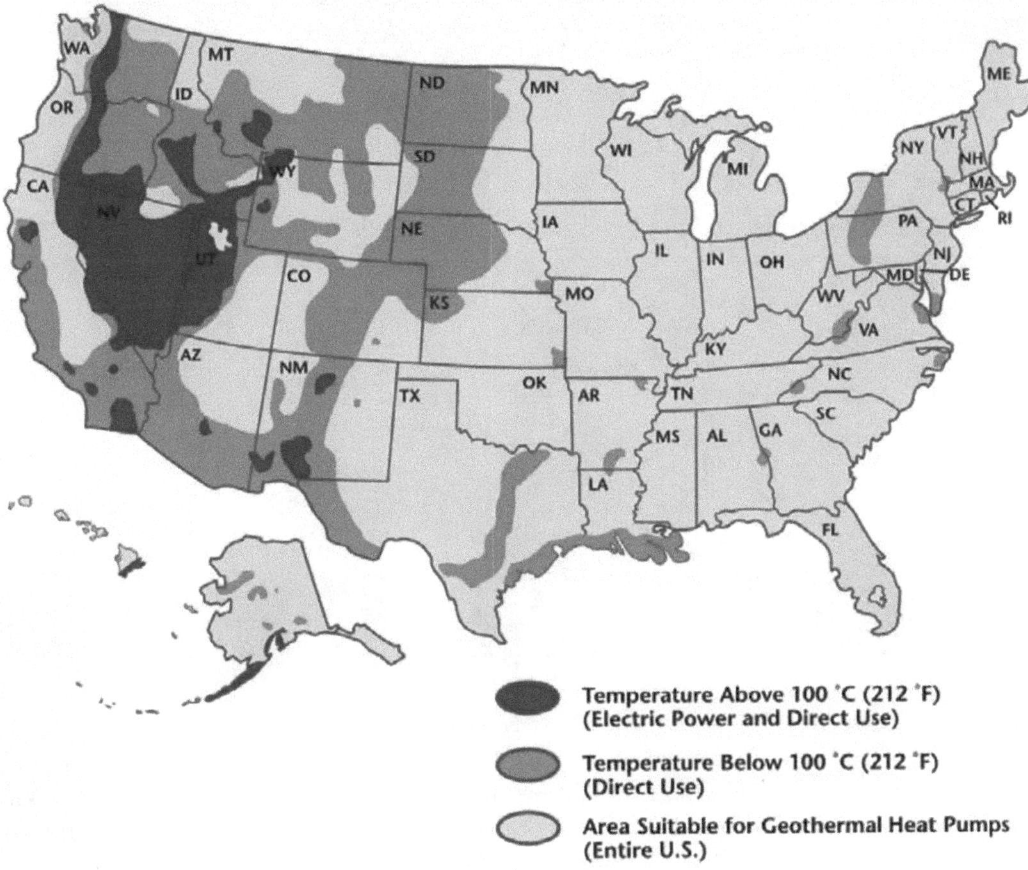

FIGURE 13.21
U.S. hydrothermal resources map (From Green and Nix [26].)

for conventional geothermal power generation (temperatures considerably higher than 100°C are desired for efficient power production) and regions with temperatures less than 100°C, only suitable for direct use. Geothermal (ground-coupled) heat pumps are suitable everywhere.

To achieve very large potential, EGS systems must eventually be developed. This involves deep drilling to reach hot rock that is typically dry and low in permeability. Because of the low thermal conductivity of rock, to extract its heat it is necessary to create permeability and inject water, that is, create an artificial hydrothermal reservoir. This is done with hydraulic fracturing. One concept involves a central injection well surrounded by two production wells, but different numbers of injection and production wells have been studied. The challenge is to fracture the rock in a way that the injected water communicates with enough rock in the field (as opposed to a short-circuit fracture between the injection well and a production well) but is not lost to the surrounding terrain. (Lost circulation is water that is injected but does not reach production wells, wasting both water and pumping power, and increasing the LCOE of the power plant.) Because of the deep wells that are typically needed to achieve a sufficient temperature for electricity production and the challenges associated with creating an artificial reservoir, EGS technology is still in an early development phase. Despite the technical and economic challenges, it represents a very large potential for producing reliable, dispatchable electric power.

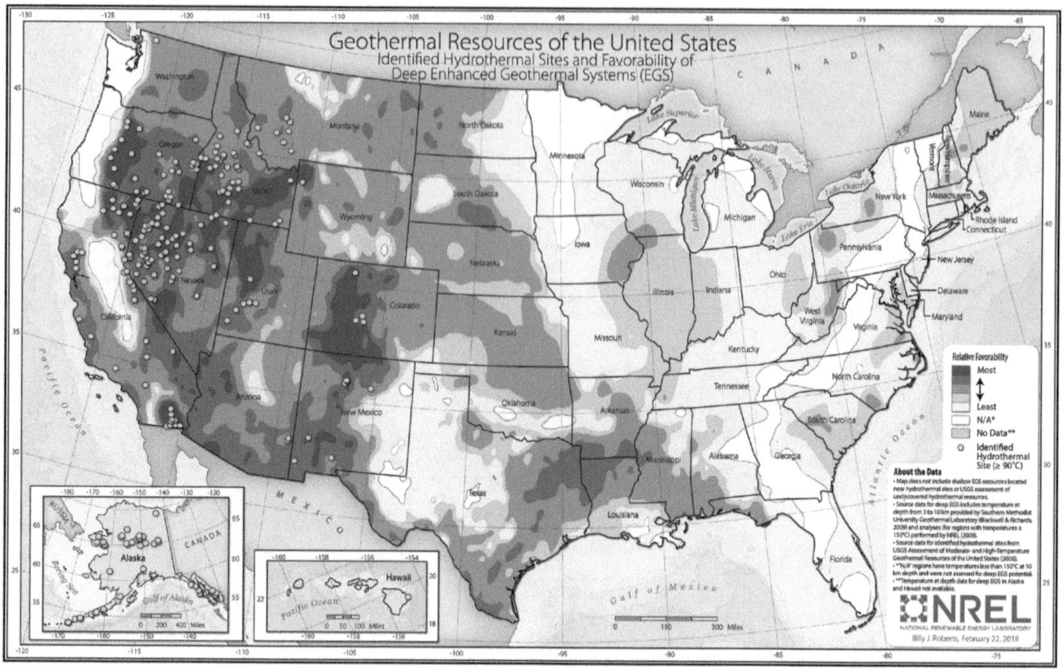

FIGURE 13.22
Map of deep EGS resources in the United States. (From NREL [27].)

Figure 13.22 shows a U.S. map indicating deep EGS resources at depths of 3–10 km. The darkest regions represent the most favorable deep EGS resources, and the dots show known near-surface hydrothermal sites with temperatures greater than 90°C [27].

Although geothermal energy has been used for direct heating purposes since ancient times, its use for the generation of electricity began only in the twentieth century. The first experiments to make use of geothermal energy to generate electricity occurred in Larderello, Tuscany, Italy, in 1904. A 250-kW_e geothermal power plant began operation there in 1913.

As of the end of 2023, geothermal had an installed global capacity of about 15,000 MW of electricity in 32 countries [28]. The United States is the largest user with about 2,700 MW, or 2.7 GW_e, of current operating capacity. The U.S. Department of Energy's GeoVision report concluded that technology improvements could increase U.S. geothermal power generation to 60 GW_e [29]. This capacity, which would require the successful development of EGS technologies, represents 5% of today's U.S. electric grid capacity, but that capacity is expected to grow considerably by mid-century due to the growth of electric vehicles, all-electric buildings, and data centers (supporting consumer Internet use, AI, and cryptocurrency mining). NREL recently updated the analysis in support of a new geothermal initiative and arrived at a 2050 potential of 90 GW_e [30]. A list of the world's geothermal power plants can be found online at the Global Energy Observatory [31].

Most of the geothermal power in the United States is generated in California and Nevada with California accounting for the majority of installed capacity. A considerable amount of this power is generated at The Geysers in Northern California, which has hosted a number of commercial geothermal power plants since the first one was built there in 1960. The Geysers, like Larderello, is a rare resource because its wells produce virtually pure dry superheated steam. Although its peak capacity reached about 2,000 MW in the late 1980s, its production was down to 631 MW net in 2022 (at 91% availability) due to the decline in steam pressure with time and plant retirements [32].

Geothermal energy is not immediately renewable like solar and wind energy. Like the energy of the sun, the energy within the Earth is immense and has a lifetime measured in billions of years. However, unlike the use of sunlight, tapping into local sources of the Earth's heat can result in a temporary decrease in the local amount of energy available. Reinjecting geothermal fluid that remains after steam is extracted helps preserve the fluid volume of the reservoir. However, even with reinjection, the heat content of the reservoir gradually declines.

The recovery period for a geothermal resource depends on how it is used. One study [33] indicated that the recovery period is approximately 30 years for geothermal heat pumps, about 100–200 years for district heating, and several hundred years for electricity generation. Resources tapped for electricity generation could provide energy for 50 years or more, if properly managed, but plant equipment typically reaches the end of its useful life before the resource is depleted. Continuous long-term use of geothermal energy for electricity generation would require the periodic construction of new plants at new sites while previously used reservoirs recover. Augmenting a geothermal power plant with concentrating solar power could increase the power production capability of a declining geothermal resource.

While geothermal is not as renewable as solar and wind, it has a number of important advantages. Geothermal electric power plants operate 24 h/day and can, therefore, produce base-load electricity. Geothermal plants are not vulnerable to weather effects, except that cycle efficiencies (and hence plant output) tend to be higher in the winter (when heat is rejected to a lower sink temperature) than in the summer. (This is especially true for binary-cycle plants that are air-cooled.)

Geothermal plants are sometimes located in areas of high scenic value, where the appearance of the plant is important. Fortunately, geothermal power plants take up little land space, and, with careful design, they can easily blend into the surrounding environment. Wet cooling towers at plants can produce plumes of water vapor, which some people find unsightly. In such cases, air-cooled condensers can be used. A good example of an aesthetic geothermal power plant is the 35-MW binary-cycle plant located near the scenic Mammoth Lakes ski area in northern California. The plant uses air cooling, has a low profile, is painted in colors that match the natural landscape, and is surrounded by trees. As a result, passing motorists on the nearby highway don't even notice it.

13.3.1.1 Current Commercial Geothermal Power Technologies

The current power plants for generating electricity from hydrothermal resources can be divided into two general types: steam and binary. Steam plants are generally considered the most cost-effective technology when the resource temperature is above about 175°C. In these plants, steam is expanded through a turbine as in conventional power plants to generate electricity. For geothermal resources rich in steam (such as at The Geysers), the steam can be used directly. Because such sites are rare, it is much more common for hot water from a geothermal resource to be flashed to steam by spraying it into a tank where its pressure is decreased. This can occur in either a single- or dual-stage process, with the latter costing more to construct but achieving higher output. A schematic of a single-stage flash plant is shown in Figure 13.23 [25].

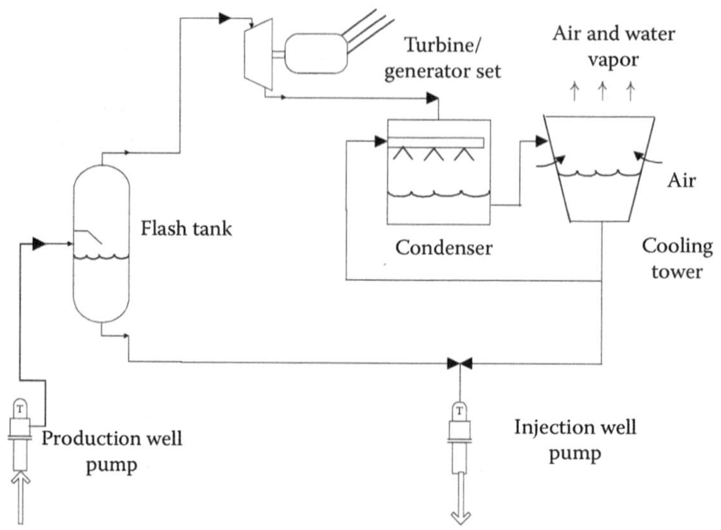

FIGURE 13.23
Geothermal single-stage flash plant. (From Kutscher [25].)

For lower-quality resource temperatures below about 175°C, flash plants lose their efficiency. It is more efficient to transfer heat from the geothermal fluid to a volatile working fluid (typically a hydrocarbon such as isobutane or normal pentane) that vaporizes and is passed through a turbine. Such plants are called "binary" because a secondary fluid is used in the actual power cycle. These plants tend to have higher equipment costs than flash plants. Because they transfer heat from the geothermal fluid and return all that fluid to the ground, they do not have condensed steam available as cooling water. Thus, they must use a separate water source or air-cooled condensers. Because all of the geothermal fluid is returned to the reservoir, binary-cycle plants do not require mitigation of gaseous releases, and reservoir fluid volume is maintained. Because larger binary plants are typically comprised of small modules, maintenance can be done on one module at a time, thus minimizing the impact on plant output. A schematic of a typical binary-cycle plant is shown in Figure 13.24 [25].

IRENA [24] states that the total installed capacity costs of geothermal power plants commissioned in 2023 ranged from $3,413 to $6,595/kW with an average LCOE of $71/MWh. The average capacity factor of these plants was 82%. They have noted that flash steam plants typically have a lower cost than binary-cycle plants.

For small plants under 5 MW in size, the cost per kW goes up significantly because of a loss in equipment economies of scale and because the fixed costs associated with exploring a site and drilling wells are divided by a smaller number of kilowatts. (Although a small plant requires a smaller diameter well, as well diameter decreases, the well power output drops more rapidly than the drilling cost.) In addition, there are some minimum fixed O&M costs associated with operating a plant that also become significant when divided by a smaller plant capacity.

Despite their higher energy costs, small-scale plants can offer some potential advantages. Skid-mounted units can be built in the factory and shipped anywhere in the world. Because they are modular, a plant owner can start with a small investment and add additional modules later. Small plants can be designed to operate automatically. Their economics can become attractive in regions where low-cost shallow wells are available and where the exit brine from the plant can be used for direct heating applications.

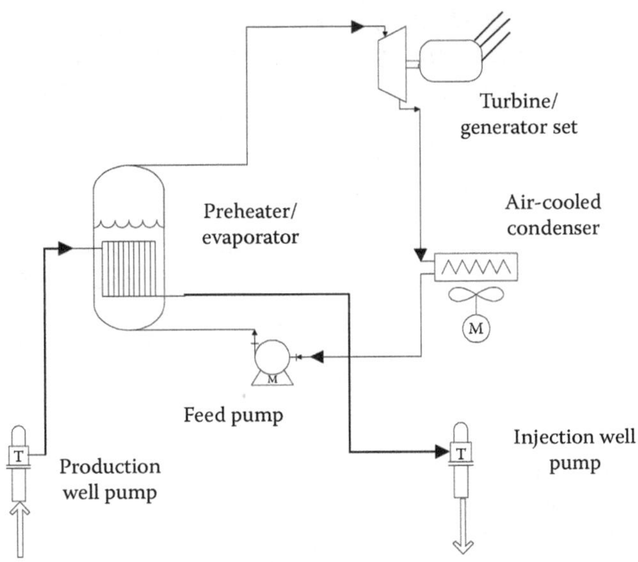

FIGURE 13.24
Binary-cycle geothermal power plant. (From Kutscher [25].)

Small plants may be suited for microgrid applications in developing nations where the competition is imported diesel fuel [34].

13.3.1.2 Technology Status

Between 1980 and 1990, about 900 MW_e of geothermal electric capacity were installed in the United States. Most of these plants were built in California and Nevada under power purchase agreements that guaranteed prices of more than 10 cents/kWh. Today, these price guarantees have expired, and plants must sell their power at competitive rates. With the loss of government market incentives, the advent of low-cost electricity from natural gas power plants, and the decline in electricity load growth, geothermal power plant construction in the United States declined greatly in the 1990s. Geothermal electricity costs, in the range of 6–11 cents/kWh after subsidies, have a difficult time competing against the lower-cost electricity from natural gas power plants, although they achieve higher capacity credits than wind and photovoltaics (PV). Some new U.S. geothermal power plants have been built in recent years, as a result of both tax incentives and state renewable portfolio standards. Total power output has not grown substantially, however, due to resource decline, especially at the largest set of plants, The Geysers.

To address the reservoir depletion problem, projects were developed to inject reclaimed wastewater into The Geysers steam reservoir. Pipelines were built from Lake County and the City of Santa Rosa that combined carry a total of about 17 million gallons/day to The Geysers. These projects have not only helped to prolong the lifetime of the geothermal resource, but they also provide a solution to wastewater disposal problems.

13.3.1.3 Next Generation Geothermal Power Technologies

Geothermal power has seen limited growth in recent years due largely to the fact that most of the best hydrothermal resources have been utilized. Interest in geothermal power has increased recently for several reasons. First, geothermal power has little or no operational greenhouse gas emissions and, unlike variable solar and wind, it operates day and night, rain or shine and can provide 24-h/day electricity to loads like data centers. Also, as we transition away from fossil fuels, geothermal power can take advantage of horizontal drilling and fracking technologies and provide job opportunities for workers with those skills. Thus, we are witnessing significant investments from technology firms and governments in two areas of geothermal power development: EGS and Advanced Geothermal Systems (AGS), also known as closed-loop geothermal [35].

An example of a current EGS technology that utilizes horizontal drilling and fracking is being developed by Fervo Energy. The company's first project in Nevada utilizes two wells of 7,700-foot (2,350 m) depth with horizontal conduits and an artificial water reservoir at 376°F (191°C) positioned in between them. The hot water is provided to a 3.5-MW binary-cycle power plant [36]. The Fervo plant site has a fairly high-quality heat resource at a relatively shallow depth. Even so, DOE has indicated that its LCOE is $450/MWh, and the cost goal is $80/MWh [37]. Of course, cost can be expected to be high for a small, initial project. Fervo Energy is also developing a 400-MW plant in Utah.

Compared to conventional geothermal projects that utilize hydrothermal resources, EGS projects face a number of challenges. They require deep drilling and fracking in hard igneous rock, and they must avoid both short-circuiting of the water between the injection and production wells and loss of water circulation. (It has been reported that the Fervo project has had a lost circulation rate of 10%–20% [37].) Microseismicity (small earthquakes) resulting from fracking is also a risk for EGS projects.

Advanced Geothermal Systems attempt to avoid the issues associated with fracking and artificial reservoir creation by using deep closed pipe loops to pick up heat from surrounding rock. Eavor Technologies, Inc. operated a closed-loop demonstration project in New Mexico from 2019 to 2023. It consisted of two 18,000-ft-deep (5,500 m) wells connected by multiple lateral pipes at a temperature resource of 250°C [38]. The project tested the drilling techniques, but no thermal energy capture was reported. Eavor Technologies is now developing a 9-MW_e commercial project in Germany. The biggest challenge with any closed-loop geothermal project is that, unlike projects having water reservoirs that utilize thermal convection to harness the heat, closed-loop systems must rely on thermal conduction from the hot rock to the water enclosed in pipes, and rock has a low thermal conductivity. One can achieve greater heat transfer between the rock and the water by increasing the number of pipes, but that results in a higher cost.

13.3.2 Direct Use of Geothermal Energy

Most direct uses of geothermal energy tap resources at temperatures below about 150°C. Common uses are for spas, space heating, greenhouse heating, aquaculture, and industrial heating. According to Lund and Toth [39], there were 107,727 MW_t of geothermal direct use in 88 countries worldwide in 2020, which produced approximately 1 million TJ/year of thermal energy. Of this amount of energy, 59% was for ground-coupled heat pumps, 18% for bathing, 16% for space heating, 3.5% for greenhouse heating, and the rest for various uses like industrial and agricultural applications and fish farming. Lund et al. [40] described direct-use applications in the United States. The installed capacity was 20,700 MW_t with an annual energy use of 153,000 TJ, or 15% of the world total. Ninety-eight percent was attributed to ground-coupled heat pumps. Fish farming and swimming pool heating were the largest direct uses.

Problems

13.1 A 100 MW ocean thermal gradient power plant is to be designed for a location where the ocean surface temperature is 300 K (80°F) and water at a lower depth is available at 278 K (40°F). If the heat exchangers are sized to operate a power plant using R-22 as the working fluid between 294 and 284 K (70°F and 50°F), calculate the flow rates of R-22 and water required, assuming that the condenser heat exchangers have an overall conductance of 1,000 W/m^2·K (176 Btu/h·ft^{2}°F) and an effectiveness of 100%. Also, calculate the plant efficiency and the surface area of the condenser. The saturation pressure of R-22 at 294 K is 9.38 × 105 N·m^2 (136.1 psia), the enthalpy is 256.8 kJ/kg (110.4 Btu/lb$_m$), and the entropy is 0.900 kJ/kg·K (0.215 Btu/lb$_m$R). The enthalpy of the saturated liquid at 284 K is 56.45 kJ/kg (24.27 Btu/lb$_m$) and the heat of vaporization is 197.0 kJ/kg (85.68 Btu/lb$_m$). The saturation pressure at 284 K is 6.81 × 105 N·m^2 (98.73 psia), while the entropy at 284 K of the saturated liquid is 0.217 kJ/kg K (0.0519 Btu/lb$_m$R). Assume efficiencies for the pump and turbine are 80% and 90%, respectively. The specific volume of the saturated R-22 liquid at 284 K is 0.000799 m^3/kg (0.0128 ft^3/lb$_m$). A schematic diagram of the system is shown as follows:

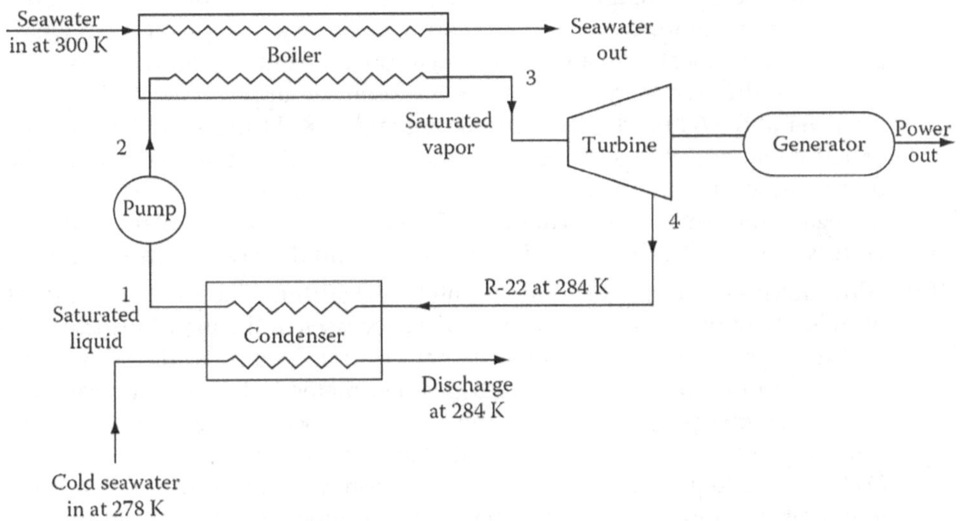

13.2 If the sea power plant described in the previous problem is to deliver power at \$8/10^6 Btu, estimate the maximum permissible cost of the condenser and evaporator heat-exchanger surface in dollars per square foot, assuming a 20-year life, 10% discount rate, and 7% yearly fuel inflation.

13.3 Explain in your own words the manner in which the Pelamis wave power device operates. Note that the motion at the hinges produces hydraulic power that is fed into electric generators. A motion picture of the operation of the Pelamis can be viewed online at https://www.youtube.com/watch?v=F0mzrbfzUpM.

13.4 Estimate the period and phase velocity of a deep water wave of 100 m wavelength.

13.5 Using the phase velocity calculated in the previous problem, estimate the power in a deep water wave of 1.5 m amplitude and 100 m wavelength.

13.6 Estimate the mean tidal power produced by a plant with an area of 10 km^2, a maximum range of 5 m, and a minimum range of 2.5 m (an average range of 3.7 m).

13.7 The air-cooled 30-MW Mammoth Lakes Geothermal Power Plant in California has a resource temperature of 339°F. Assume that the plant design ambient temperature is 60°F and that the plant operates at 50% of the Carnot efficiency.

a. On a hot summer afternoon when the ambient temperature is 95°F, what is the percent decrease in power output compared to the design condition?

b. Assume the wet-bulb temperature of the air is 55°F and that you can use spray nozzles to evaporatively cool the 95°F outdoor air 80% of the way to the wet-bulb temperature. What is the percent increase in plant power output compared to using the 95°F air directly?

The next two problems are open-ended with no solutions:

13.8 It has recently been proposed that ocean thermal energy conversion could be used to produce ammonia, a chemical currently manufactured from natural gas at a total energy consumption of about 2.5% of the U.S. natural gas consumption. Since the manufacture of ammonia could be accomplished anywhere, the plant could be located in a favorable location, for example, off the coast of Brazil, where the difference in temperature between the upper layer of the ocean and the layer at the 600 m depth is approximately 20 K. Using an OTEC power plant of 500 MWe capacity, estimate the number of tons of ammonia that could be produced per year. For a price of $165/ton of ammonia, estimate the time required to regain the capital investment in the OTEC production plant and compare your estimate with the claim that the capital could be repaid in 2.5 years.

13.9 Aluminum sells today at approximately $1,850/ton. The production of aluminum is one of the most energy-intensive processes in the metallurgical industry, and it has recently been proposed that an OTEC plant could be used for the production of aluminum. Metallurgical estimates indicate that approximately 18 kWh of energy is required in the refining process for each kilogram of aluminum produced. Assuming a plant load factor of 85%, calculate the size of an OTEC plant required to produce 250,000 ton/year of aluminum, if there is a temperature gradient similar to that in the previous problem. Then calculate the time required to repay an investment cost in the plant and compare your results with the estimate of 3 years obtained in the literature.

References

1. D'Arsonval, A., 1891. Utilisation des forces naturelles, Avenir de l'electricite, *Revue Scientifique* 17, 370–372.
2. Claude, G., 1930. Power from tropical seas, *Mechanical Engineering* 52, 1039–1044.

3. Dugger, G., Neff, F., and Snyder, J.E., 1981. Ocean thermal energy conversion, in: Kreider, J.F. and Kreith, F. eds., *Solar Energy Handbook*, McGraw-Hill, New York, Chapter 19. https://www.osti.gov/biblio/6408304
4. Kreith, F., and Bharathan, D., 1988. Heat transfer research for ocean thermal energy conversion, *Journal of Heat Transfer* 110, 5–22.
5. Bharathan, D., Kreith, F., and Owens, W.L., 1987. An overview of heat and mass transfer in open-cycle OTEC systems, in: Mari, Y. and Yang, W.J. eds., *Proceedings of the ASME/JSME Thermal Engineering Joint Conference*, Japan Society of Mechanical Engineers, Tokyo, Japan, vol. 2, pp. 301–314.
6. Rabas, T.J., Wittig, J., and Finsterwalder, K.,, March 1979. OTEC 100-MWe alternate power systems study, Report by Westinghouse Electric Corp to U.S. DOE under contract EG/77/C/05/1473.
7. Herrera, J., Sierra, S., and Ibeas, A., 2021. Ocean thermal energy conversion and other uses of deep sea water: A review, *Journal of Marine Science and Engineering* 9(4), 356, https://www.mdpi.com/2077-1312/9/4/356.
8. Block, D.L., and Valenzuela, J.A., May 1985. Thermoeconomic optimization of OC-OTEC electricity and water production plants, SERI/STR-251-2603, U.S. Department of Energy, Washington, DC.
9. Kreith, F., and Manglik, R.M., 2018. *Principles of Heat Transfer*, 8th ed., Cengage Learning, Boston, MA.
10. Rossi, M., 2024. Why islands are eyeing ocean thermal energy conversion to power their future and combat climate change, *Outrider*, July 3, https://outrider.org/climate-change/articles/why-islands-are-eyeing-ocean-thermal-energy-conversion-power-their-future.
11. Patel, S., 2023. OTEC, a long-stalled baseload ocean power technology, is seeing a swell, *Power Magazine*, June 1, https://www.powermag.com/otec-a-long-stalled-baseload-ocean-power-technology-is-seeing-a-swell/.
12. Elliott, D., 2004. Tidal power, in: Boyle, G. ed., *Renewable Energy*, 2nd ed., Oxford University Press, Inc., New York, Chapter 6, pp. 196–243.
13. Baker, A.C., 1991. *Tidal Power*, Peter Peregrinus, London, UK.
14. Renewable Energy UK, La Rance Tidal Power Plant, https://www.reuk.co.uk/La-Rance-Tidal-Power-Plant.htm.
15. Twidell, J., and Weir, T., 2006. *Renewable Energy Resources*, 2nd ed., Taylor & Francis, Oxon, UK, Chapter 13.
16. Jackson, T., 1992. Renewable energy: Summary paper for the renewable energy series, *Energy Policy* 20(9), 861–883.
17. Khaligh, A., and Onar, O.C., 2010. *Energy Harvesting*, CRC Press, Boca Raton, FL, Figure 3.6.
18. Twidell, J., and Weir, T., 2006. *Renewable Energy Resources*, 2nd ed., Taylor & Francis, Oxon, UK, Chapter 12.
19. Coulson, C.A., and Jeffrey, A., 1977. *Waves*, Longman, London, UK.
20. Hawaii Department of Business, 2002. Feasibility of developing wave power as a renewable energy resource for Hawaii, https://energy.hawaii.gov/wp-content/uploads/2011/10/Feasibility-of-Developing-Wave-Power-as-a-Renewable-Energy-Resource-for-Hawaii.pdf.
21. Queen's University of Belfast, 2002. *Islay Limpet Wave Power Plant*, https://cordis.europa.eu/docs/publications/6662/66628981-6_en.pdf.
22. Thennakoon, T., Hewage, H.T.M., Sandunika, D.M.I., Panagoda, L.P.S.S., Senarathna, W.G.C.M., Sulaksha, L.G.T., Weerarathna, D.M., Jayathma, W.M.V.A., Gamage, D.G.M.L., and Perera, M.D.V., 2023. Harnessing the power of ocean energy: A comprehensive review of power generation technologies and future perspectives, *Journal of Research Technology & Engineering* 4(3), 73–102.
23. International Energy Agency, 2021. Hydropower data explorer, https://www.iea.org/data-and-statistics/data-tools/hydropower-data-explorer.
24. IRENA, 2024. Renewable power generation costs in 2023, https://www.irena.org/-/media/Files/IRENA/Agency/Publication/2024/Sep/IRENA_Renewable_power_generation_costs_in_2023.pdf.

25. Kutscher, C.F., 2000. The status and future of geothermal electric power, in: *Presented at the American Solar Energy Society (ASES) Conference*, Madison, WI, June 16–21, NREL/CP-550-28204.
26. Green, B.D., and Nix, R.G., 2006. Geothermal—The energy under our feet: National Renewable Energy Laboratory, Report NREL/TP-840-40665.
27. NREL, 2018. Geothermal Resources of the United States, https://www2.nrel.gov/gis/geothermal.
28. Our World in Data, 2024. Total geothermal capacity, https://ourworldindata.org/grapher/installed-geothermal-capacity?tab=table.
29. U.S. Department of Energy, May 2019. GeoVision: Harnessing the heat beneath our feet, DOE/EE–1306.
30. Augustine, C., Fisher, S., Ho, J., Warren, I., and Witter, E., January 2023. Enhanced geothermal shot analysis for the Geothermal Technologies Office, NREL/TP-5700-84822.
31. Global Energy Observatory, Current list of geothermal power plants, https://globalenergyobservatory.org/list.php?db=PowerPlants&type=Geothermal. Last accessed January 2025.
32. Calpine, Geysers by the Numbers, 2022. Statistics, https://geysers.com/The-Geysers/Geysers-By-The-Numbers.
33. Rybach, L., Megel, T., and Eugster, W.J., October 1999. How renewable are geothermal resources? *Geothermal Resources Council Transactions* 23, 17–20.
34. Vimmerstedt, L., November 1998. Opportunities for small geothermal projects: Rural power for Latin America, the Caribbean, and the Philippines, NREL/TP-210-25107, National Renewable Energy Laboratory.
35. Zhou, Y., and Annex, M., May 10, 2023. Next-generation geothermal technologies are heating up, Bloomberg NEF, https://about.bnef.com/blog/next-generation-geothermal-technologies-are-heating-up/.
36. Gallucci, M., November 28, 2023. America's first 'enhanced' geothermal plant just got up and running, Canary Media, https://www.canarymedia.com/articles/geothermal/americas-first-enhanced-geothermal-plant-just-got-up-and-running.
37. Matthews, D., September 13, 2023. Is the future of energy…pouring water on hot rocks in the ground? Vox, https://www.vox.com/future-perfect/23825844/geothermal-enhanced-fervo-demonstration-superhot.
38. Eavor Technologies Press Release, January 31, 2023. Success at Eavor's New Mexico project triggers follow-on strategic investments, https://www.eavor.com/press-releases/success-at-eavors-new-mexico-project-triggers-follow-on-strategic-investments/.
39. Lund, J., and Toth, A., 2021. Direct utilization of geothermal energy 2020 worldwide review, *Geothermics* 90, 101915.
40. Lund, J., Sifford, A., Hamm, S.G., and Anderson, A., 2021. The United States of America Utilization Update 2019, in: *Proceedings of the World Geothermal Congress* 2020+1, Reykjavik, Iceland.

14

Storage Technologies

There must surely come a time when heat and power will be stored in unlimited quantities... all gathered by natural forces.

Thomas A. Edison, 1910

14.1 Overview of Storage Technology

Energy storage will play a critical role in any future renewable energy system. There are two principal reasons why energy storage will be of increasing importance to renewable energy:

- Many of the most significant renewable energy sources like solar and wind are variable in nature.
- The transportation system, including cars, buses, ships, and planes, requires energy to be carried by the vehicles. Whether as liquid fuels, hydrogen for fuel cells, or batteries for electric vehicles, some type of energy storage will be needed.

Energy can be stored in many forms: as mechanical energy in rotating, compressed, or elevated substances; as thermal or electrical energy that can be released as needed; or as electrical charges ready to travel from positive to negative poles as demanded. Storage technologies that can absorb and release electricity have great value because electricity can be converted easily into mechanical or thermal energy. Electricity is also the output of the most promising renewable energy technologies: wind turbines, solar thermal power plants, and photovoltaic cells. There are, however, some applications that can benefit from thermal or mechanical storage technology. For example, thermal energy storage (TES) is needed for the continuous operation of solar thermal power plants, and overnight heat storage is used for heating or cooling of buildings.

The storage methods treated in this chapter can accept and deliver energy in three ways: electrical, mechanical, and thermal. When energy is stored electrically or mechanically, it can be converted to either of the other two forms quite efficiently. For example, electricity can drive a motor with only about 5% energy loss. It can also provide heat with a resistive heater with virtually no loss at all and even more efficiently via a heat pump. The quality of TES depends mostly on its temperature. The efficiency of the conversion of thermal energy into power is limited by thermodynamic considerations. The theoretical maximum quantity of useful work, W_{max} (mechanical energy), that can be extracted from a given quantity of heat, Q, is

$$W_{max} = \frac{T_1 - T_2}{T_1} Q \tag{14.1}$$

DOI: 10.1201/9781003586616-14

where

T_1 is the absolute temperature of the stored heat

T_2 is the sink temperature, typically the ambient absolute temperature.

Any energy storage technology must be carefully chosen to accept and produce a form of energy consistent with both the energy source and the final application. Storage media that can accept and release energy in the form of electricity, such as batteries and ultracapacitors, are an especially valuable form of storage because they can efficiently store renewable electricity and the electricity they release can be used directly or converted to both mechanical and heat energy [1].

14.1.1 Applications

Table 14.1 presents an overview of the most important energy storage technologies. In this table, each technology is classified by its relevance to one of the following four categories:

Utility shaping: For this application, large-capacity storage devices are required to respond to varying electricity demands. Examples would be a solar thermal power plant, which delivers energy generated during the day to a thermal storage device for use at nighttime or at times of high electricity demand and prices, or a wind turbine that would compress air during high winds to power a turbine when the wind stops.

Power quality: Storage technologies to smooth power delivery during short periods such as outages or switching events must be capable of providing large changes in power output over very short periods. Ultracapacitors and batteries are in this category.

TABLE 14.1

Overview of Energy Storage Methods and Their Applications

	Utility Shaping	Power Quality	Distributed Grid	Transportation
Direct electric				
Ultracapacitors		✓		✓
SMES		✓		
Electrochemical				
Batteries				
Lead acid	✓	✓	✓	
Lithium ion	✓	✓	✓	✓
Nickel-cadmium	✓	✓		
Nickel-metal hydride				✓
Zebra				✓
Sodium-sulfur	✓	✓		✓
Flow batteries				
Vanadium redox	✓			
Polysulfide bromide	✓			
Zinc bromide	✓			
Electrolytic hydrogen				✓

(Continued)

TABLE 14.1 (*Continued*)

Overview of Energy Storage Methods and Their Applications

	Utility Shaping	Power Quality	Distributed Grid	Transportation
Mechanical				
Pumped hydro	✓			
Compressed air	✓			
Flywheels		✓		✓
Direct thermal				
Sensible heat				
Liquids			✓	
Solids			✓	
Latent heat				
Phase change	✓		✓	
Hydration–dehydration	✓			
Chemical reaction	✓		✓	
Thermochemical				
Biomass solids	✓		✓	
Ethanol	✓			✓
Biodiesel				✓
Syngas	✓			✓

Distributed grid technology: Storage devices for distributed grid technologies must be able to generate energy and storage at customer locations rather than at a central utility power station. Technologies in this category include batteries for photovoltaic power installed on the roof of a building. Other systems include solar thermal or geothermal power that can be implemented on a distributed scale. The required power generation is small compared to utility shaping. They require energy capacities in the 1–50 kWh range, while utility shaping requires capacities in the 1,000 MWh range.

Transportation technologies: The devices that can be used for storage in automobiles and trucks include all types of batteries. Examples are hybrid, plug-in, and battery electric vehicles. Large fleets of plug-in-hybrid-electric vehicles (PHEVs) and battery electric vehicles (BEVs) could also help in utility shaping by utilizing off-peak or excess renewable electricity to charge the batteries and feeding the grid system from the batteries during peak demand.

The physical modes of energy storage may involve one or more mechanical, thermal, or electromagnetic forms. An energy storage technology may require both a storage reservoir and a converter and transmission system for moving the power to and from the reservoir to its destination and use. A pumped hydro system, for example, contains both elements using a reservoir to store potential energy and a turbine to deliver the energy.

14.1.2 Technology Characterization

Every energy storage technology can be characterized essentially by three important parameters: self-discharge time, storage size, and efficiency. Within a category, the final selection of a specific storage technology can be made by the consideration of cycle life, specific energy, specific power, energy density, and power density.

Self-discharge time (SDT): is the time required for a fully charged, self-standing storage device (i.e., a device that is not connected to a load) to reach a certain depth of discharge (DOD). DOD can be described as a percentage of the storage device's useful capacity. For example, 80% DOD means that 20% of the device's energy capacity is left. It should be noted that the relation between SDT and DOD is, in general, not linear and must therefore be measured. SDTs vary greatly from a few minutes with supercapacitors to months or longer with batteries and thermochemical fuels.

Storage size: describes the intrinsic scale of the technology. Some technologies have a relatively large storage size but cannot provide small-scale energy storage. When applied to a particular device, storage size is often called energy storage capacity and is measured in joules, kWh or Btu.

Power capacity: is the maximum rate of energy delivery the device can sustain on a continuous basis and is measured in units of Watts. Power and energy storage capacity together determine how long the device can meet a specified load.

Cycle life: is the number of consecutive charge/discharge cycles that a storage system can undergo while maintaining the system's specifications. Cycle life specifications are made against a selected DOD, depending on the application of the storage device. For example, a battery used in a hybrid electric vehicle may consume only 10%–20% of the energy stored during most of the discharge cycle. The advantage of a thermal storage system is that it can undergo many charge/discharge cycles without significant waste.

Specific energy: is a measure of how heavy the system is. It is measured in units of energy per mass and here we will represent this quantity in terms of MJ/kg or kWh/kg. The higher the specific energy, the lighter the device. For transportation applications, a high specific energy is necessary, whereas for utility application, specific energy is relatively unimportant.

Energy density: is how much space the system occupies. It is measured in units of energy per volume, MJ/L or kWh/m^3.

Round-trip efficiency: is the ratio of the energy output from the device to the energy input. As for energy density and specific energy, the system boundaries must be carefully specified when determining efficiency. It is also important to note the form of energy required at the input and output interconnections and to include the entire system necessary to attach these interconnections. For example, if the system is used to store energy from a utility wind farm, then both the input and output will be AC electricity. However, when comparing a battery with a fuel cell, it is necessary to include the efficiency of an AC-to-DC rectifier for the battery, whereas an AC-powered hydrogen power generation system is required for the fuel cell. DC-to-AC converters are associated with both. Efficiency is also related to the SDT. For example, if the discharge of a thermal storage system occurs much later than charging, the apparent efficiency will be lower because a significant amount of thermal energy may be lost if the interim between charge and discharge is large.

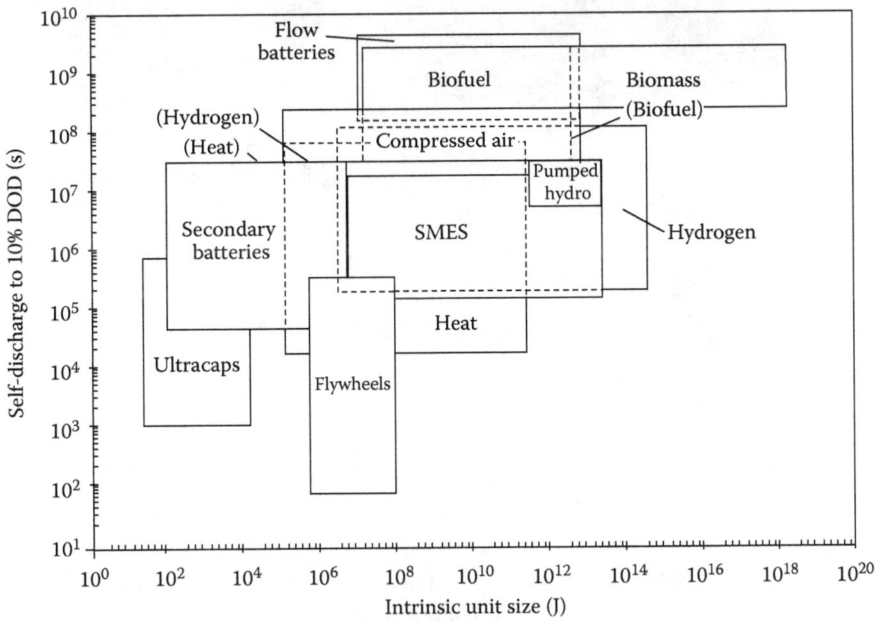

FIGURE 14.1
Self-discharge time to 10% DOD in seconds versus intrinsic unit size in joules. Labels in parentheses refer to areas of the graph hidden by other blocks.

Storage energy modes depend on the end-use application, but the magnitude of the power load and the time scale are important. Figure 14.1 shows the characteristics of the various storage technologies to be discussed in this chapter according to size and discharge time. For example, to provide electric power for short power outages requiring a large power flux for a short period of time, such as seconds or minutes, capacitors and flywheels are well suited. On the other hand, when large amounts of electric or thermal energy are needed for periods of hours or longer, pumped hydropower, compressed air storage, or thermal storage are appropriate technologies. Distributed grid, automotive, and building applications require a moderate discharge time and a moderate unit size. Figure 14.2 shows the fields of application in terms of discharge time and power rating.

14.2 Mechanical Technologies

Pumped hydroelectric and compressed air energy storage are the two mechanical storage technologies that are currently available for large-scale and long-term utility-sized storage, as shown in Figure 14.2. Flywheels are only useful for very short time storage.

14.2.1 Pumped Hydroelectric Energy Storage

Pumped hydroelectric energy storage (PHES) is a mature technology that has been deployed for over a century and is well suited for mountainous and hilly locations. Examples as early as the 1890s exist in Italy and Switzerland. PHES uses electricity to

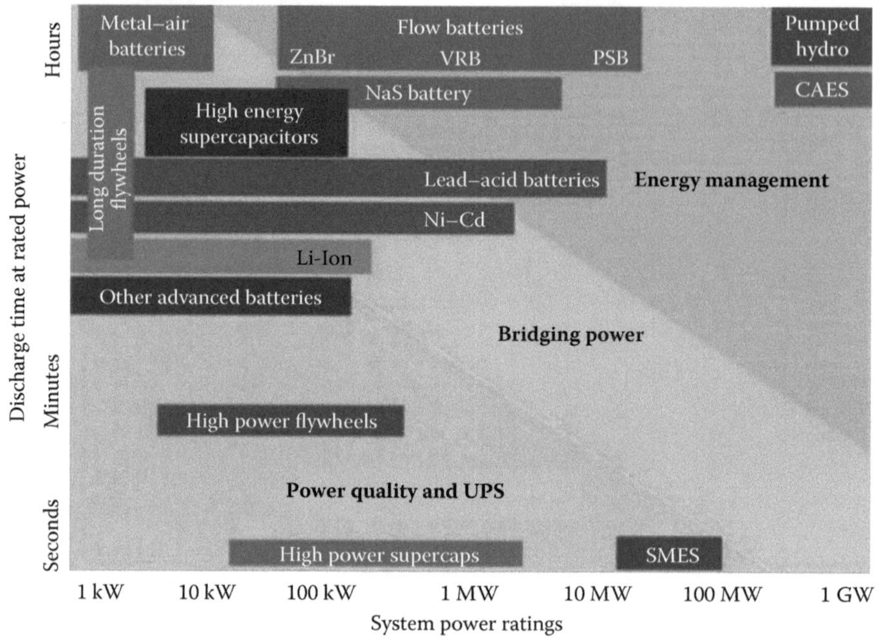

FIGURE 14.2
Discharge time versus power rating for various energy storage methods. (From Electricity Storage Association [2]. With permission.)

pump water uphill into a reservoir where it can be stored. The energy can be recaptured when the water is released and runs back downhill through a turbine. In many cases, a pump-turbine (often a specially designed Francis turbine) combines both pumping and power production functions in one device. These storage systems allow the management of water and energy resources. PHES systems are very efficient. With modern turbine technologies, they can approach 80%–85% round-trip efficiencies.

Figure 14.3 is a schematic diagram of a PHES installation at Raccoon Mountain, which is operated by the Tennessee Valley Authority [3]. The Raccoon Mountain pumped-storage plant is widely cited as an example of excellent engineering for PHES. The plant, which was completed in 1978, has a generating capacity of about 1,600 MW and can run for 22 hours to supply about 35,000 MWh of electricity.

The power and capacity of a PHES system are functions of the hydraulic head, the water flow rate, and the efficiency of the pump and the turbine. The head is given by the upper elevation to which the water is pumped and the lower elevation at which the turbine is located. The flow rate can be regulated by valve action. The maximum flow that can be obtained is limited by the reservoir size. The maximum flow rate is dictated by the reservoir size divided by the desired storage time to yield the available flow per unit time, minus a 15% reserve operating condition.

Given the hydraulic head, an upper bound on flow rate and an efficiency of the plant, the power-generating capacity of a pumped hydroelectric installation is given by the equation below:

$$P = QH\rho g\eta \tag{14.2}$$

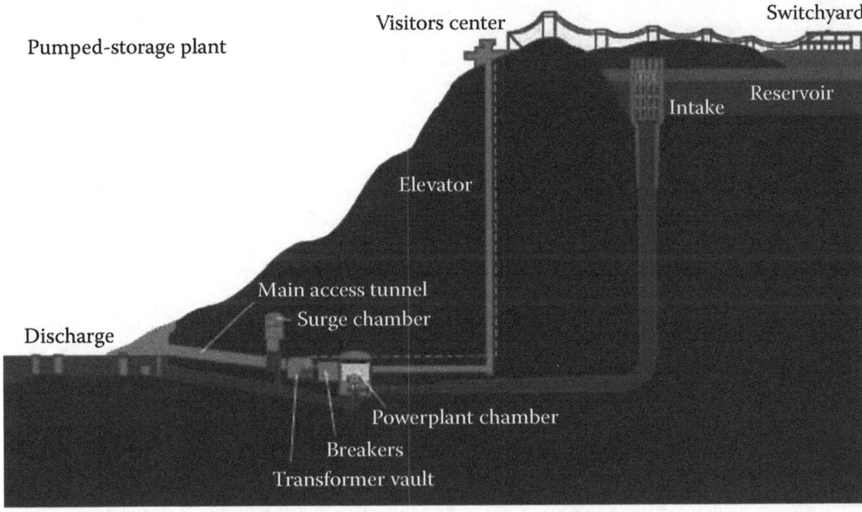

FIGURE 14.3
Schematic of the Raccoon Mountain pumped-storage plant. (From Tennessee Valley Authority [3].)

where
- P is the generated output power in Watts (W)
- H is the height differential in meters (m)
- Q is the fluid flow rate in cubic meters per second (m^3/s)
- ρ is the fluid density in kilogram per cubic meter (kg/m^3), about 1,000 kg/m^3 for water
- g is the gravitational constant (9.81 m/s^2 on Earth)
- η is the overall efficiency.

Given the power capacity of the plant, the limiting value of the upper or lower reservoirs dictates the energy capacity. Potential energy generation in kWh is calculated by power output times the operating time. Figure 14.4 shows the power as a function of flow rate and head in engineering units [4].

Example 14.1

Calculate the power generated by a PHES system with a potential hydraulic height differential between reservoirs of 100 m and a volumetric flow rate of 1,000 m^3/s through a hydraulic turbine with an efficiency of 0.8.

Solution:

We use Equation 14.2, $P = QH\rho g\eta$. Substituting the appropriate values we find that $P = 1{,}000\ m^3/s \times 100\ m \times 1000\ kg/m^3 \times 9.81\ m/s^2 \times 0.8 = 784{,}800{,}000\ W_e = 785\ MW_e$.

14.2.1.1 Turbines

Hydro turbines have in the past been used mostly for hydropower generation from large dams. Hydropower produces about 20% of the world's electricity in 150 countries. The technology has evolved for more than 100 years, and the literature of hydroelectric power generation can be used as a guide for designing pumped-storage facilities.

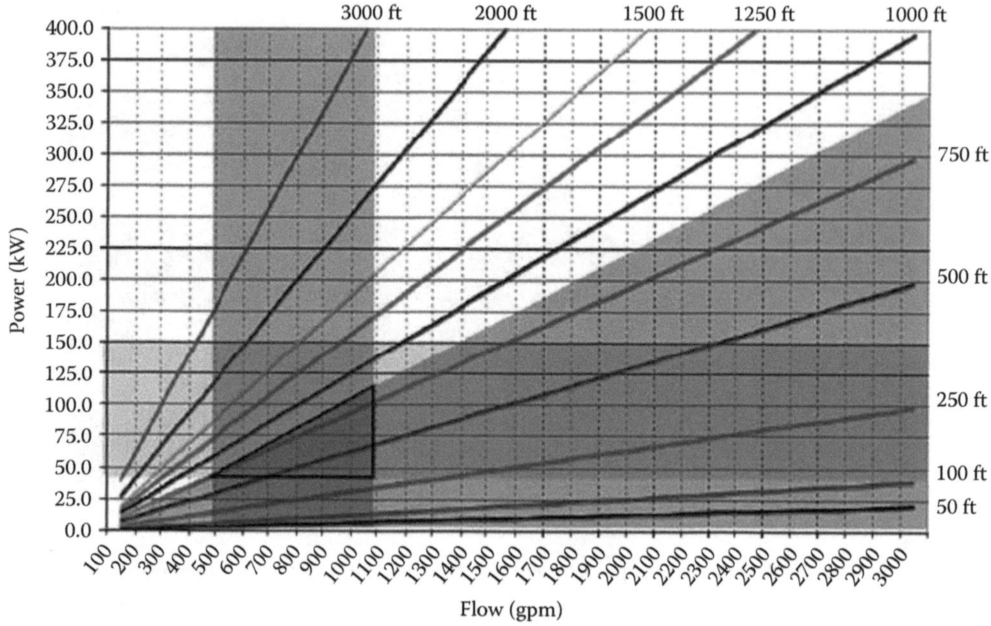

FIGURE 14.4
Pumped hydro turbine power output as a function of flow rate for various hydraulic heads, efficiency = 70%. The abbreviation gpm represents gallons per minute. (From Levine [4]. With permission.)

Conventional hydraulic turbines are either impulse or reaction machines (see Figure 14.5). Impulse turbines, often called Pelton turbines, convert the potential energy of a fluid into kinetic energy by expansion in a stationary nozzle to form a jet, which is then directed toward buckets attached to a rotating wheel to create shaft work that can be extracted. Reaction turbines are of the Francis or Kaplan type and utilize both hydraulic pressure and kinetic energy to create rotating shaft work. Each of the turbines has a specific operating range in terms of hydraulic head and power output. The choice of which turbine to use depends on the appropriate characteristics.

Pelton wheels were used in the early days of hydropower. They have a lower efficiency at low hydraulic head in comparison to reaction machines. Pelton wheels are still used, however, for high head resources (about 200 m) and small power outputs (less than 5 MW).

Francis turbines use a set of fixed veins that guide the fluid to the buckets that make up the turbine runner and are mounted on a central shaft. Francis turbines have a large operating range with heads from 40 to 500 m and unit sizes from 10 to up to 1,000 MW_e. Kaplan turbines are similar to Francis turbines, but the turbine blade angle can be adjusted to improve performance under different flow conditions. They work well at low heads, less than 10 m to about 100 m, with power outputs of the order of 200 MW_e.

Once built, hydropower installations have operated successfully for more than a century. It is therefore expected that pumped-storage facilities will also have a long life. The main drawback of pumped storage is the limited availability of suitable sites.

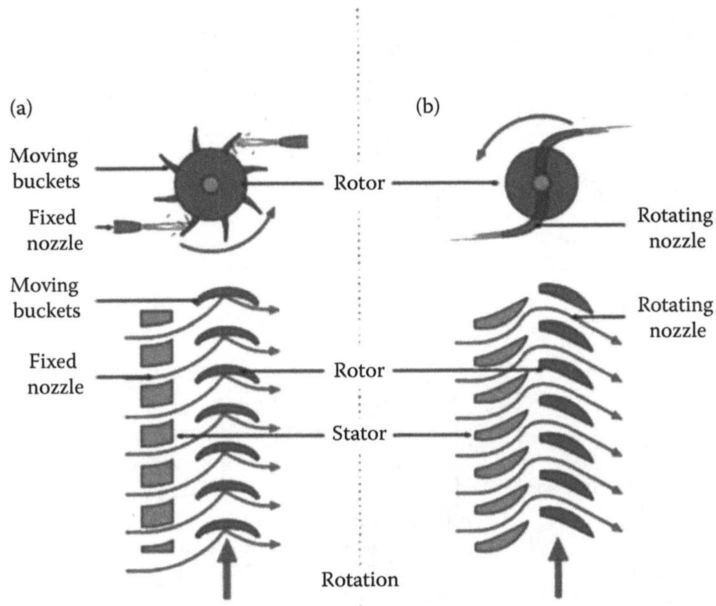

FIGURE 14.5
Schematic diagram of (a) impulse and (b) reaction turbines.

14.2.2 Compressed Air Energy Storage

The compressed air energy storage (CAES) system is a hybrid thermodynamic storage mechanism that requires natural gas as an intermediate component. Figure 14.6 schematically shows side-by-side the operation of both a gas turbine and a CAES system.

The operation of a CAES system can be demonstrated by comparing these two thermodynamic systems. In a gas turbine power plant, atmospheric pressure air passes through a compressor, is then delivered to a combustion chamber where fuel is injected and the high-pressure (hp) air is heated. The hp heated gas from the combustion chamber then passes through the blades of a turbine whose rotating shaft is connected to an electricity generator. The air from the turbine is exhausted at approximately atmospheric pressure. Approximately two-thirds of the energy generated by the turbine is used to drive the compressor, whereas the remaining one-third is converted to generate electric power.

The components in a CAES system are similar to those in a gas turbine, but clutches are added in order that the compressor and the turbine may be connected separately to the generator, which also has to work as a motor. In addition, provision is made to store the hp air from the compressor in a suitable cavity. The air from the cavity is heated in a combustor similarly to the gas turbine power plant, but the timing of discharging the stored hp air can be controlled. Since compression and expansion of the air can take place at different times, the entire output from the turbine is available to generate electricity as needed. In other words, compressed air can be placed in the cavity where it is stored until power is needed. This arrangement can be useful for renewable energy systems such as wind

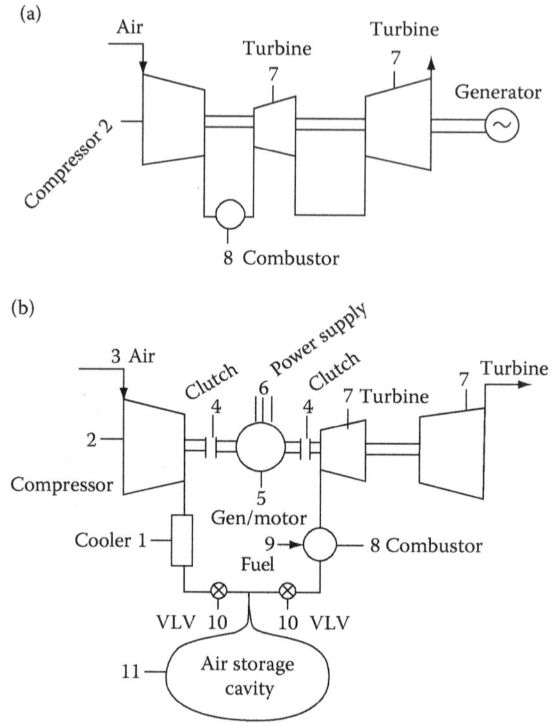

FIGURE 14.6
Comparison of (a) gas turbine and (b) compressed air storage system.

turbines that do not operate continuously. But by using any excess power when available to compress and store energy, the compressed air in the cavity becomes available when winds die down, but electricity is needed.

Creation of the cavity for a CAES system can be done by leaching existing salt domes, actual excavation in hard rock, or utilizing existing aquifers, if suitable domed cap-rock is available. Detailed descriptions of these technologies are available in Kreith and Krumdieck [5]. The main advantage of a CAES system is that natural gas (NG) heat input per unit power is less than that for a conventional gas turbine.

There are two existing CAES systems that have been designed and operated for many years. One is located in Huntdorf, Germany [6] and the other one is in McIntosh, Alabama [7,8]. The experience gained from these existing CAES systems will be valuable in the design and operation of future compressed air storage, but no specific new CAES systems are being constructed at this time. Detailed derivations for the efficiency of a CAES system can be found in [9]. Because conventional CAES plants burn natural gas, resulting in carbon emissions, efforts are underway to develop alternatives that don't burn fossil fuels, such as adiabatic CAES, or "ACAES" [10].

14.2.3 Flywheels

Flywheels have been used for a long time for short-term energy storage. Modern flywheel energy storage systems consist of a massive rotating cylinder on a shaft that is supported on a stator by magnetically levitated bearings that reduce bearing wear and increase system life. By operating a flywheel system in a vacuum environment, drag can be drastically reduced. To complete the system, the flywheel is connected to a motor/generator that is mounted on the stator which, through power electronics, can interact with the utility grid.

The amount of energy stored in a rotating flywheel is given by

$$E = \frac{I\omega^2}{2} \tag{14.3}$$

where

I is the flywheel's moment of inertia in kg m^2
ω is its angular velocity in 1/s
E is stored energy in joules.

In order to obtain a high specific energy for flywheels, the flywheel speed is increased, since the energy stored increases with the square of the velocity, whereas the specific energy only increases in direct proportion with its mass.

Some of the advantages of flywheels are their low maintenance requirements, long cycle life (better than 10,000 cycles), and long lifetime (in excess of 20 years). High-speed and low-mass flywheels, which are the modern approach to flywheel storage, are made from composites such as carbon fiber. Flywheels can have an energy efficiency of better than 90%, but their energy storage capacity is limited to a short period. Their low energy density and specific energy limit them to voltage regulation and uninterruptible power supply (UPS) capability.

14.3 Direct Electrical Technologies

14.3.1 Ultracapacitors

An ordinary capacitor stores energy in the electric field between two oppositely charged conductors with a dielectric between them. The dielectric prevents arcing between the plates and permits the plates to hold more charge, thus increasing the maximum potential energy storage. The ultracapacitor, also known as a supercapacitor, differs from the traditional capacitor in that it uses an extremely thin electrolyte instead of a dielectric. This electrolyte is only a few Angstroms thick, thereby making it possible to increase the energy density of the device. The electrolyte can be made of either an organic or an aqueous material. The aqueous design can operate over a larger temperature range but has a smaller energy density than the organic design. The electrodes are usually made of a porous carbon that increases their surface area as well as the energy density compared to a traditional capacitor.

Ultracapacitors can effectively equalize voltage variation quickly, which makes them useful for power quality management and for regulating the voltage in automobiles during regular driving. Ultracapacitors can also work in tandem with batteries to meet peak power needs. They also exhibit a very high cycle life of greater than 500,000 cycles and a life span of more than 10 years. The main limitation of ultracapacitors is their ability to maintain charge voltage over any significant period, losing up to 10% of the charge per day.

14.3.2 Superconducting Magnetic Energy Storage

Superconducting magnetic energy storage (SMES) systems can store and discharge energy at high rates. They store energy in the magnetic field created by direct current in a coil of cryogenically cooled, superconducting material. The advantage of a cryogenically cooled, superconducting material over copper is that it reduces electrical resistance to nearly zero. SMES systems recharge quickly and can repeat the charge-discharge sequence thousands of times without degrading the magnet. They can also achieve full power rather quickly, within 100 ms. Theoretically, a coil of around 150–500 m radius would be able to support a load of 18,000 GJ at 1,000 MW, depending on the peak field and the ratio of the coil's height and diameter. Since no conversion of energy to other forms is involved (e.g., mechanical), the energy is stored directly, and round-trip efficiency is expected to be very high. It is believed that mature, commercialized SMES systems can operate with round-trip efficiency as high as 97% and are a superb technology for providing reactive power on demand. The downside of SMES is the need to cool the coil to cryogenic temperatures, which requires a cryogenic refrigeration system and uses a large amount of energy [11].

14.4 Fundamentals of Batteries and Fuel Cells

Both batteries and fuel cells date back to the early 1800s. Alexander Volta described an electrochemical cell in a letter to the British Royal Society in 1800. This invention and the development of the telegraph in the 1830s led to the deployment and manufacture of batteries, which provided the electric energy for the telegraph, as well as for other inventions. The fuel cell was first demonstrated by Sir William Grove in 1839, but it was not until recently that significant progress toward practical applications occurred.

The operations of batteries and fuel cells are based on similar electrochemical principles and processes. However, batteries are devices for the storage of energy directly, whereas fuel cells are continuous energy converters that by themselves have no inherent storage capability and need hydrogen or some other energy source to operate. The operation of a typical battery is shown schematically in Figure 14.7.

A battery has a fixed amount of chemicals that spontaneously produce a flow of electrons if a conducting path is connected to its terminals. Fuel cells require a continuous external supply of chemical reactants or hydrogen, and the products are produced continuously. The operation of fuel cells depends on a supply of hydrogen, which can be stored externally. Hence, the combination of stored hydrogen and a fuel cell that produces electricity can be considered a storage system, which will be considered later in the chapter.

The reactions in both batteries and fuel cells are oxidation-reduction reactions. Both contain electrodes in contact with an electrolyte, which is the charge-carrying medium, as shown in Figure 14.7. The oxidation reaction takes place at the negative electrode called the *anode*, whereas the reduction occurs at the positive electrode called the *cathode*. The anode delivers electrons to the external circuit, and they flow from the negative to the positive electrode. However, conventional current (which was defined historically in terms of positive charges before electron flow was understood) flows in the opposite direction from high to low potential. The electrolyte is usually a liquid solution through which positively charged ions pass from the anode to the cathode. At the cathode, electrons from the external load are neutralized by reacting with positive ions from the electrolyte.

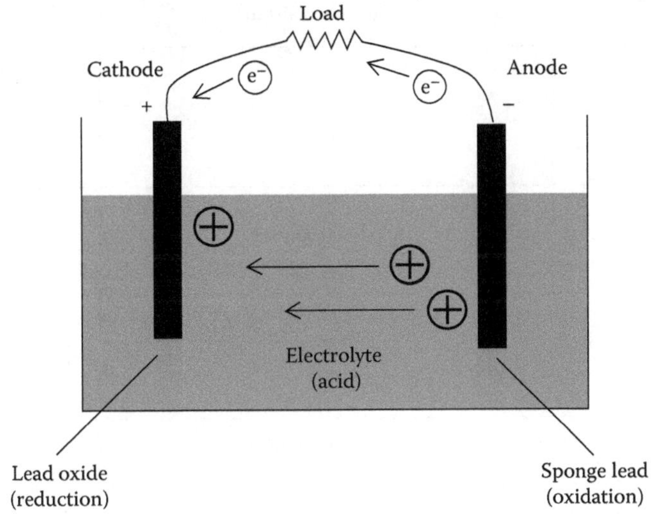

FIGURE 14.7
Schematic diagram of a battery.

Batteries designed for a single discharge cycle, such as flashlight batteries, are called primary cells, whereas those that can be recharged are called secondary cells. Fuel cells, on the other hand, utilize hydrogen to provide a continuous supply of direct current. Many different types of batteries and fuel cells are available, as discussed in the following sections.

14.4.1 Principles of Battery Operation

Batteries find a variety of applications today from tiny button cells in wrist watches to ignition batteries in automobiles and, finally, to large storage batteries in utility power supplies. Deployment of batteries for storing energy from variable renewable sources such as solar and wind is growing rapidly. Batteries are devices for storing an electric charge and automatically delivering an electric current flow on demand. A battery system consists of one or more cells connected in series to provide an open-circuit voltage or electromagnetic force (EMF). The exact value of the EMF depends on the specific cell reactants used. The operation of the battery can be visualized by means of a simple conceptual model shown in Figure 14.8.

When current is drawn from a battery, the voltage drops below its EMF across the terminals, largely due to the internal voltage drop associated with the battery's internal resistance. As shown in Figure 14.8, the terminal voltage in this model is the difference between the EMF (E) and the potential drop across the internal resistance, R_i:

$$V = E - IR_i \tag{14.4}$$

and the current, I, is

$$I = \frac{E}{R_i + R_o} \tag{14.5}$$

The following example illustrates the simplified model for a conventional lead-acid battery.

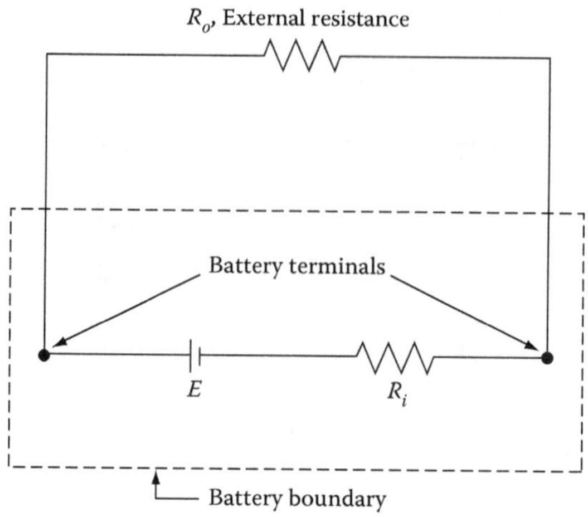

FIGURE 14.8
Simplified battery model.

Example 14.2

Estimate the internal and external resistances for a lead-acid battery with an EMF of 12.7V if the current is 50 A and the voltage across the battery terminals is 11.1V (see Figure 14.8). If the external resistor were replaced by one with a resistance of 3 Ω, find the new current and the voltage drop across the battery terminals. Draw the voltage versus current characteristics.

Solution:

From Equations 14.4 and 14.5, we obtain the internal and external resistances as shown below:

$$R_i = \frac{E - V}{I} = \frac{(12.7\ \text{V} - 11.1\ \text{V})}{50\ \text{A}} = 0.032\ \Omega$$

$$R_o = \frac{E}{I} - R_i = \frac{12.7}{50} - 0.032 = 0.222\ \Omega$$

The current flow through a 3 Ω external resistance would be

$$I = \frac{E}{R_i + R_o} = \frac{12.7}{0.032 + 3} = 3.825\ \text{A}$$

and the voltage is

$$V = IR_o = (3.825\ \text{A})(3\ \Omega) = 11.5\ \text{V}$$

The voltage versus current characteristics for the simplified model are shown in Figure 14.9.

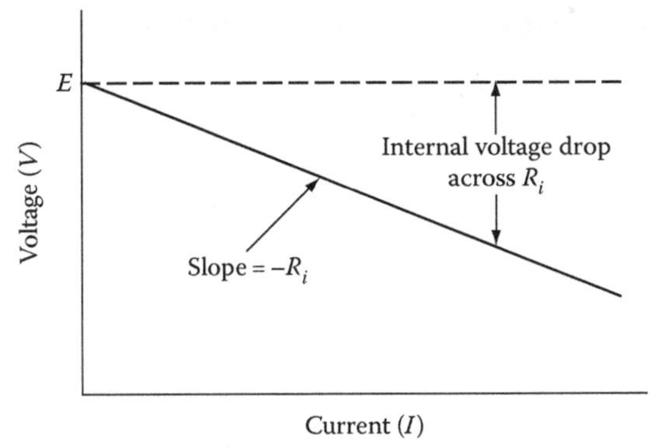

FIGURE 14.9
Current versus voltage characteristics for simplified battery in problem solution.

The linear model is useful but represents real battery characteristics only qualitatively. Real batteries can provide current and power to a wide range of resistances with little voltage drop, but only for short periods of time. The linear model fails for longer periods of time and when large currents are drawn because it does not account for the finite amount of stored chemical energy and the rate of reaction limits of real batteries. In real batteries, the terminal voltage and power output drop drastically as the chemicals of the cell are consumed.

For utility-scale electric power storage and automotive propulsion, secondary cells offer great opportunities. In secondary cells, the reactions are approximately internally reversible, and with an external current source, the internal battery reactions may be reversed during a process called charging. For example, in a solar photovoltaic power plant, the batteries can be charged during periods of sunshine and the energy stored in the batteries can then be used for power applications when the sun goes down. In an electric car, the battery can be charged at night and used to power the vehicle in the day.

14.4.2 Cell Physics

When the current is small, the voltage is close to the cell EMF and the power output is small. To analyze the cell physics, assume that the chemical reaction in the battery consumes reactants at an electrode at the rate n_c mol/s. The electrons released in the reaction flow from the electrode through the external load at a rate proportional to the rate of reaction, jn_c, where j is the number of moles of electrons released per mole of reactant. Thus, jn_c is the rate of flow of electrons from the cell in moles of electrons per second. Since there are 6.023×10^{23} electrons per gram-mole and each electron has a charge of 1.602×10^{-19} Coulomb (C), the product is 96,485 C/g-mol (when sufficient significant figures are included in the calculation). This is the charge transported by a gram mole of electrons and is named Faraday's constant in honor of the pioneer of electrochemistry, Michael Faraday:

$$F = 96{,}485\ \text{C/g-mol}$$

With this model, the electric current from a cell, I, can be related to the reaction rate in the cell according to

$$I = jn_cF \tag{14.6}$$

and the instantaneous power, P, delivered by the cell is

$$P = jn_cFV \tag{14.7}$$

The cell electrode, the electrolyte material, and the nature and rate of chemical reaction control the maximum cell voltage, cell current, and maximum power output. It is also apparent that the amount of consumable reactants in the battery sets a limit to the battery capacity.

14.5 Rechargeable Batteries

Rechargeable storage batteries offer great opportunities for storing electric power, and many different types of batteries are under extensive research at the present time. Although the method of operation for all, except so-called flow batteries, is similar, many different materials and assemblies are being tried. As a brief introduction to this very large topic, in this section, we discuss the method of operation of rechargeable batteries and then list the pros and cons of the various types with their potential applications. Figure 14.10 shows ranges of operation for different battery types as specific energy versus specific power [12].

For reference below, an important characteristic that influences the suitability of rechargeable batteries for many applications is the charge or discharge current they can safely handle. This is typically represented in terms of a "C-rate," which inversely quantifies the time required for a full charge/discharge at a given current. C-rate is the current (A) divided by the current capacity (Ah). For a battery with a 100 Ah capacity, a discharge

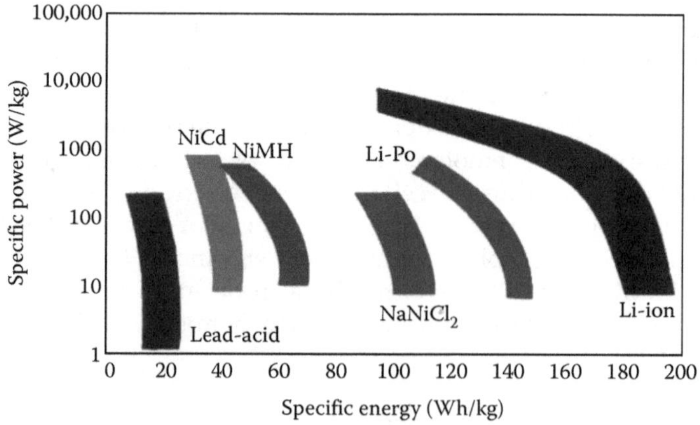

FIGURE 14.10
Approximate operating characteristics of different batteries in terms of specific energy and specific power. (Modified from de Guilbert [12].)

current of 100 A equates to a 1 C discharge rate or 1-hour discharge time, whereas a discharge current of 10 A corresponds to a 0.1 C-rate or 10-hour discharge time.

Lead-acid batteries have been the workhorse for electric storage for many decades and are still widely used. However, they are too heavy and need to be recharged too often for many current applications, including in the field of transportation. As a result, a worldwide effort has been underway to improve and develop new batteries that can meet the demands of various industries including electric utilities and electric vehicles. The field of batteries is advancing rapidly, with new battery chemistries and designs constantly entering the market. Before describing advantages and disadvantages of various types of batteries that are vying for a market share of the future, we use the lead-acid battery as an example to demonstrate how rechargeable batteries operate.

14.5.1 Lead-Acid Batteries

To illustrate the operation of the charging cycle of a secondary battery, we will analyze the operation of a common lead-acid battery shown in Figure 14.11 in some detail.

For such a battery, each cell consists of a porous sponge lead anode and a lead-dioxide cathode, which is usually in the form of a plate. The two electrodes are separated by porous membranes in a sulfuric acid solution. The aqueous electrolyte contains positive hydrogen ions and negative sulfate ions, resulting from disassociation of the sulfuric acid in solution as shown by the chemical reaction below:

$$2H_2SO_4 \rightarrow 4H^+ + 2SO_4^{2-} \quad (14.8)$$

As the lead is converted to lead sulfate at the anode, it releases two electrons to the external load as shown below:

$$Pb(s) + SO_4^{2-}(aq) \rightarrow PbSO_4(s) + 2e^- \quad (14.9)$$

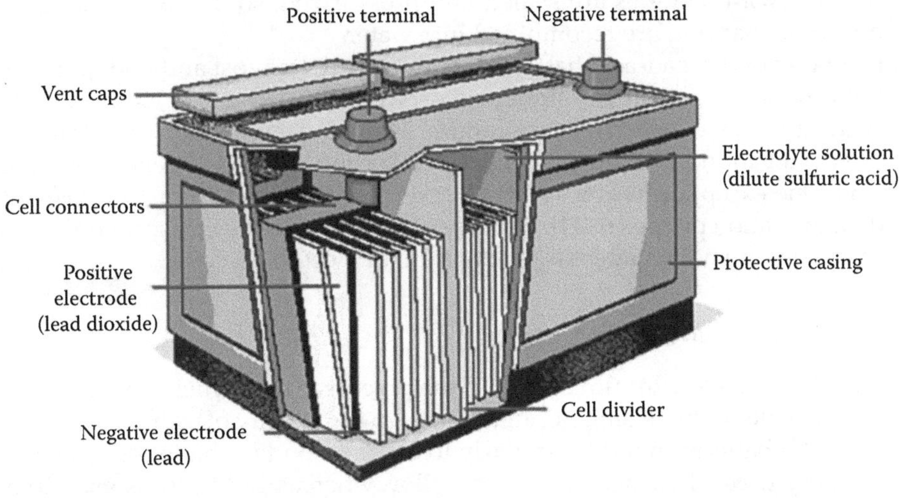

FIGURE 14.11
Schematic diagram of a conventional wet-cell lead-acid battery.

Hence, 2 moles of electrons are produced for each mole of sulfate ions and sponge lead reacted, or $j=2$ for this battery.

In the reduction reaction at the cathode, lead dioxide, using hydrogen and sulfate ions from the electrolyte and electrons from the external circuit, is converted to lead sulfate, as shown by the chemical reaction below:

$$PbO_2(s)+4H^+(aq)+SO_4^{2-}(aq)+2e^- \leftrightarrow PbSO_4(s)+2H_2O(l) \quad (14.10)$$

We observe that the sulfate radical, that is, the sulfuric acid, is consumed at both electrodes and that hydrogen ions combine with oxygen from the cathode material to form water that remains in the cell. The water that is formed as the battery discharges makes the sulfuric acid electrolyte more dilute, as can be seen from the net cell reaction below. The net cell reaction, obtained as the sum of the electrode and electrolyte reactions, is

$$Pb(s)+PbO_2(s)+2H_2SO_4(aq) \leftrightarrow 2PbSO_4(s)+2H_2O(l) \quad (14.11)$$

In this and all other rechargeable battery chemistries, left-to-right indicates a battery discharging process and right-to-left indicates recharging. When any of the reactants are depleted, the battery ceases to function. However, the battery can be charged by applying an external power source. This reverses the electron flow, and the above reactions are also reversed, as water is consumed to produce sulfate and hydrogen ions from the lead sulfate on the electrodes. Thus, the battery regains its charge and upon demand can again produce electric current.

There are three types of lead-acid batteries on the market: the wet cell, the sealed gel cell, and the sealed absorbed glass mat (AGM). Each of them has certain advantages and disadvantages. The wet cell has a liquid electrolyte, which must be replaced periodically to replenish hydrogen and oxygen escaping during the charge cycle. The sealed gel cell has a silica component added to the electrolyte to stiffen it. The AGM uses a fiberglass-like separator to hold the electrolyte close to the electrodes. This increases the efficiency. For both the gel and the AGM configuration, the risk of hydrogen explosion is reduced, and corrosion is lessened. However, these two types require a slower charging rate. Both the gel cell and the AGM batteries are sealed and pressurized, so the oxygen and hydrogen produced during charging are recombined into water.

The advantages of the lead-acid battery technology are low cost and high power density. However, its application for utility storage is very limited because of its short cycle life. A typical installation survives only a maximum of 1,500 deep cycles. Therefore, lead-acid batteries have not found much use in commercial and large-scale energy-management applications. One exception was a 10 MW/40 MWh system in Chino, California, built in 1988 for demonstration purposes [11]. Because of the low specific energy at only 0.8 MJ/kg, lead-acid batteries are not a viable storage option for electric vehicles.

14.5.2 Nickel Metal (Ni-Cd and Ni-MH)

Description: The nickel-cadmium battery is the precursor to the nickel-metal hydride battery, but Ni-Cd batteries are losing popularity due to the toxicity of cadmium. Nickel-metal hydride (Ni-MH) batteries operate similarly to other secondary batteries. The cathode in a Ni-MH battery uses a hydrogen-absorbing alloy, whereas cadmium is used in a Ni-Cd battery. During cell discharge, hydrogen ions flow from the anode to the cathode through the separator.

Structure: The cathode of a Ni-MH battery is composed of nickel oxyhydroxide. The anode is a metal hydride, which has hydrogen absorbed in a metal alloy that is generally composed of nickel, cobalt, manganese, and/or aluminum combined with a mixture of rare earth metals. The electrolyte is an alkaline compound, generally potassium hydroxide. During discharge, the nickel oxyhydroxide is reduced to nickel hydroxide.

$$NiOOH + H_2O + e^- \rightarrow Ni(OH)_2 + OH^- \qquad E = 0.52\ V \tag{14.12}$$

And the metal hydride MH is oxidized to the metal alloy M.

$$MH + OH^- \rightarrow M + H_2O + e^- \qquad E = 0.83\ V \tag{14.13}$$

The overall reaction on discharge is

$$MH + NiOOH \rightarrow M + Ni(OH)_2 \qquad E = 1.35 \tag{14.14}$$

The process is reversed during charging.

Advantages of Ni-MH versus Ni-Cd:

- Higher energy density than standard Ni-Cd (30–100 Wh/kg with Ni-MH).
- Periodic deep discharge cycling is required less often.
- Much more environmentally friendly than Ni-Cd. Ni-MH is easily recyclable.
- Long shelf life.

Disadvantages of Ni-MH versus Ni-Cd:

- Service life limited with repeated deep cycling. More prevalent with high load currents.
- Lower discharge current.
- Higher self-discharge.
- Requires full discharge to prevent memory effect where the capacitance and the voltage drop until the battery is charged to its correct voltage and then discharged fully.

General uses:

- Small electronics
- Hybrid electric and electric cars
- Wireless communications

14.5.3 Lithium Ion

Description: Advances in lithium-ion (Li-ion) batteries have been essential in driving the transition to electric vehicles and supporting high penetration of renewable electricity systems. According to the International Energy Agency [13] costs for producing Li-ion batteries declined by 90% between 2010 and 2023, with energy densities and performance characteristics improving over the same period.

Pure lithium batteries are primary batteries and cannot be recharged due to the instability of lithium. Li-ion batteries are secondary batteries and use lithium compounds instead of pure metallic lithium. Like other rechargeable batteries, Li-ion batteries have three primary functional components: the anode, the cathode, and the electrolyte. Lithium ions move from the anode to the cathode during discharging and in the opposite direction during charging. Most Li-ion batteries use a liquid organic electrolyte, but lithium polymer (Li-Po) batteries use a solid polymer composite that makes them easier to manufacture and more durable. Research on how to increase the power output of Li-ion batteries by decreasing the internal resistance of the battery is currently in progress. Li-ion batteries can be discharged at a continuous rate of 5–20 C with peak currents reaching 40 C. Li-Po batteries can achieve similar rates.

Structure: The cathode is normally made of a metal oxide such as lithium cobalt oxide ($LiCoO_2$) attached to a current collector made of aluminum foil. $LiCoO_2$ batteries are common in mobile electronic devices. Batteries with a $LiNiMnCoO_2$ cathode, known as nickel-manganese-cobalt (NMC) batteries, are widely used in electric vehicles. Different NMC batteries have different ratios of Ni, Mn, and Co, with chemistries having relatively more nickel and less cobalt favored in recent years. $LiFePO_4$ (LFP) cathode batteries are seeing increased use for EVs and utility-scale electricity storage. LFP batteries currently have 20%–30% lower energy densities than high-nickel NMC batteries, but are cheaper to produce, more durable, and are made with more widely available cathode materials [13]. In a Li-ion battery, the anode is typically made from layered graphitic carbon with a copper current collector. The electrolyte is a lithium salt such as $LiPF_6$, $LiBF_4$, or $LiClO_4$. These salts are dissolved in an organic solvent. In Li-Po batteries, the electrolyte is a composite of a gelled polymer and one of the lithium salts. Solid-state Li-ion batteries with solid electrolytes are under development that offer advantages such as lower weight and eliminating the risk of thermal runaway [14].

The chemical reactions for a typical Li-ion battery are shown below, with charging to the right and discharging to the left [15].

$$\textbf{Positive:}\ \mathrm{Li}_x\mathrm{CoO}_2 \leftrightarrow \mathrm{Li}_{1-x}\mathrm{CoO}_2 + x\mathrm{Li}^+ + xe^- \tag{14.15}$$

$$\textbf{Negative:}\ x\mathrm{Li}^+ + xe^- + 6\mathrm{C} \leftrightarrow \mathrm{Li}_x\mathrm{C}_6 \tag{14.16}$$

$$\textbf{Overall:}\ \mathrm{Li}_x\mathrm{CoO}_2 + 6\mathrm{C} \leftrightarrow \mathrm{Li}_{1-x}\mathrm{CoO}_2 + \mathrm{Li}_x\mathrm{C}_6 \tag{14.17}$$

Note that the C in the above equations stands for a very carbon-rich negative material such as graphite. The above equations are written in units of moles, using x as a stoichiometric coefficient.

Advantages of Li-ion batteries compared to Ni-MH:

- 300 Wh/kg at the cell level (for NMC) compared to Ni-MH (30–100 Wh/kg).
- Lower maintenance (Ni-MH have "memory" effects and must occasionally be fully cycled).
- Low self-discharge rate—typically 5% per month.

Disadvantages of Li-ion batteries:

- Shelf life—the battery's capacity will slowly decay over time regardless of the number of charge/discharge cycles.
- Requires a protective circuit to make sure it is not overcharged or over-discharged as this degrades the battery by causing deposits of material that can short the battery (over-discharge) or cause excessive heat (overcharge).
- If damaged, combustion of the electrolyte, anode, and cathode can occur due to the interaction of lithium and the water vapor in the air.
- Some Li-ion batteries require critical metals including nickel and cobalt for the cathode. There are environmental and human rights concerns associated with some cobalt mining operations. One advantage of LFP batteries is that they eliminate these metals and their associated issues.

General uses:

- Electric/hybrid cars
- Cell phones and laptops
- Energy storage systems for renewable electricity

14.5.3.1 Li-Ion Battery Life Cycle Impacts and Recycling

Critical metals requirements, energy demand, and environmental impacts of battery production are important aspects of their sustainability for use in electric vehicles and grid storage. Dai et al. [16] assessed the impacts of lithium-ion batteries from "cradle-to-gate" covering battery production stages from raw materials extraction through to battery pack production. They assessed a NMC 111 battery designed for an electric vehicle, with a battery pack weight of 165 kg and 23.5 kWh storage capacity. The batteries had a graphite cathode and $LiFP_6$ electrolyte. They assumed cobalt was sourced from Africa and lithium from South America. Aluminum production was assumed to occur in North America using hydroelectricity, and battery manufacturing to occur in the United States using the then-current grid mix.

Figure 14.12 shows the breakdown of contributions of battery components and production stages to cradle-to-gate energy use and greenhouse gas emissions for an NMC 111 battery. Total energy use is 1,130 MJ/kWh of capacity. Total greenhouse gas (GHG) emissions are 72.9 kg CO_2-e/kWh of capacity. As shown, NMC power materials and production account for the largest share of energy and emissions, followed by aluminum and cell production. Dai et al. [16] note that the results depend on source locations for materials and where manufacturing steps are performed. For example, energy use and GHG emissions for cell production are largely driven by space conditioning needs for the dry room where cell production occurs, so requirements would depend on the climate at the production location.

Beyond the factory gate, the use and end-of-life stages of battery life cycles are also important determinants of their sustainability. When the life cycle picture is expanded to cover the use phase, battery efficiency, weight and durability come to the fore as important, along with the source of grid electricity [17,18]. Service life of Li-ion batteries may be extended by repurposing them, for example with primary use in electric vehicles and

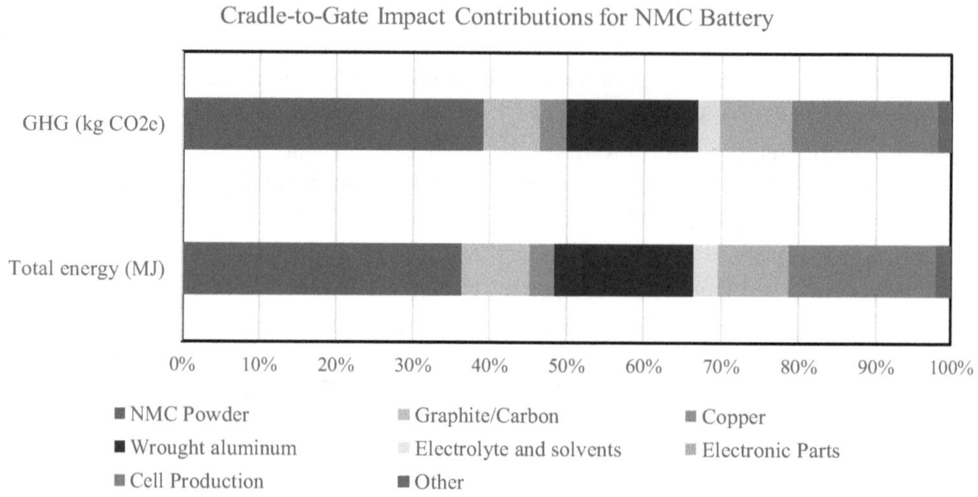

FIGURE 14.12
Cradle-to-gate energy requirements and greenhouse gas emissions for production of an NMC Li-ion battery pack. (Adapted from Dai et al. [16]. CC BY open access.)

secondary use in less demanding applications such as renewable energy storage systems. In that scenario, after the storage capacity of EV batteries declines to the point they need to be replaced, the old battery packs would be disassembled to the cell level, with usable cells reassembled to a new configuration.

At the end of life, Li-ion batteries can be recycled to recover valuable metals and avoid the energy costs of extracting and processing these metals from raw materials. For recycling, battery packs are dismantled, shredded, or crushed and separated into waste streams. Electrode materials can be processed to recover metals such as nickel, cobalt and sometimes lithium via smelting (pyrometallurgy) or leaching (hydrometallurgy). Hydrometallurgy has been estimated to yield higher materials recovery and have somewhat lower associated GHG emissions than pyrometallurgy for NMC batteries [19]. Expanding battery recycling capacity is an important aim to make Li-ion batteries more sustainable.

14.5.4 Sodium Ion

Commercialization of sodium-ion batteries is widely seen as the next wave of battery development for short-term grid energy storage applications and some classes of electric vehicles [12]. The chemical and physical characteristics of sodium are similar to those of lithium, but sodium is much more abundant and is easily recovered from sea salt (NaCl), soda ash (Na_2CO_3) and trona ($Na_2CO_3{\cdot}NaHCO_3{\cdot}2H_2O$). Even though the energy density of Na-ion batteries is expected to be lower than that of Li-ion batteries, the potential to lower costs by replacing Li ions with Na ions has sparked significant research and commercial interest that is advancing rapidly. Because Na-ion batteries currently have energy densities that are up to 40% lower than Li-ion batteries their most likely near-term market is fixed-site storage applications that are not sensitive to battery weight [13].

Typical cathode composition for Na-ion batteries is not yet settled, but layered metal oxides and Prussian Blue Analogs (PBA) are commonly mentioned. Prussian white

($Na_2Fe_2(CN)_6$) is an example of a favorable cathode option that can be derived from readily available minerals. $NaPF_6$ in an organic solvent is a likely choice for the electrolyte. Na-ion batteries are expected to use hard (non-graphitic) carbon anodes. Hard carbon is a highly porous solid form of carbon produced by heating carbonaceous precursors to high temperature without oxygen. Another cost advantage with Na-ion batteries is the potential to use aluminum in place of copper as the current collector at the anode. Life cycle assessment (LCA) studies have identified the production of hard carbon as the most energy and emissions-intensive step in the production of Na-ion cells [20,21].

14.5.5 Flow Batteries

Description: Flow batteries are a novel type of secondary battery in which charged electrolytes are passed through a cell, producing electric charge (see Figure 14.13). The vanadium redox (VR) battery is the most common type of flow battery. The VR battery uses an electrolyte created from electrolytically dissolving vanadium pentoxide (V_2O_5) in sulfuric acid (H_2SO_4). The battery consists of two tanks of electrolyte and a membrane connected to a load, as shown in Figure 14.13.

Charging the electrolyte causes the cathode to produce VO^{2+} ions and the anode to produce V^{2+}. During discharge, the charged electrolyte is pumped across a membrane, creating VO_2^+ on the positive tank side and V^{3+} on the negative tank side. This produces an open-circuit voltage. Since the chemical structure of the electrolyte does not undergo a change, this battery is very stable.

At the positive electrode

$$VO_2^+ + 2H^+ + e^- \underset{\text{charging}}{\overset{\text{discharging}}{\rightleftarrows}} VO^{2+} + H_2O \qquad E = 1.00\ V \qquad (14.18)$$

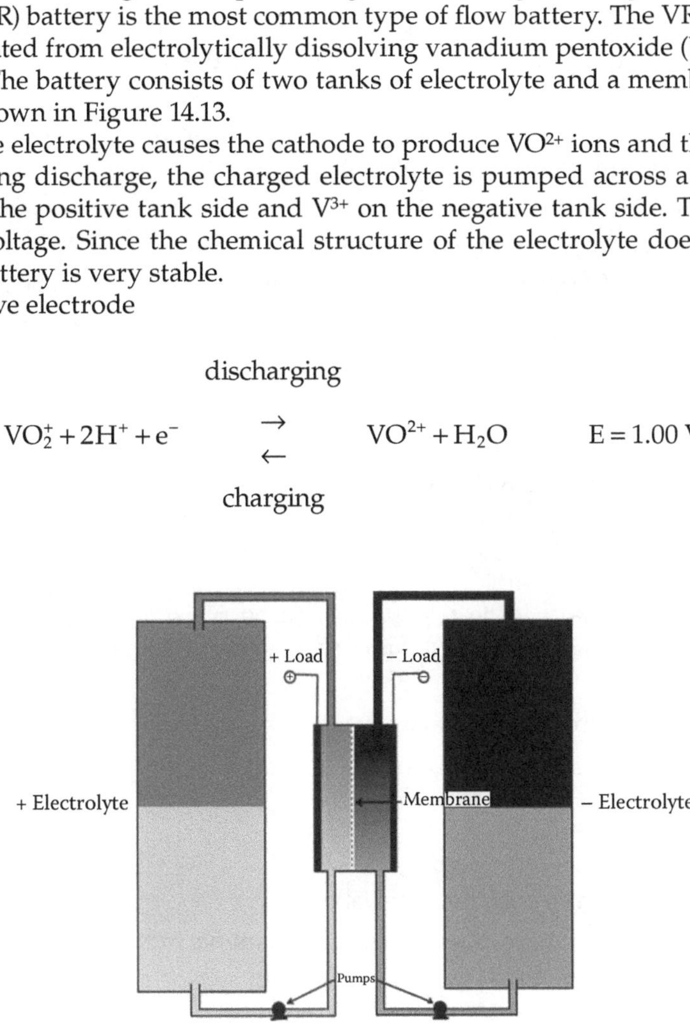

FIGURE 14.13
Schematic diagram of a typical flow battery.

At the negative electrode

$$V^{3+} + e^- \underset{\text{discharging}}{\overset{\text{charging}}{\rightleftarrows}} V^{2+} \qquad E = -0.26\ \text{V} \tag{14.19}$$

The potential generated by the cathode and anode combine to create the overall potential for the battery.

Advantages:

- Very environmentally friendly—the vanadium electrolyte does not break down and can theoretically last forever thus reducing the waste to near zero for the whole system.
- Capable of very large energy storage. The only limiting factor is how large the electrolyte tanks are. Can reach MWh range.
- Capable of very fast change in power.
- Can be fully charged/discharged without degradation.
- High cycle life (>10,000) after which only the membrane needs to be replaced.
- Very low self-discharge rate—electrolytes are stable at charged levels.
- Cost/kWh decreases as storage size increases.
- Possibility of instant recharge by replacing electrolyte.
- Already scaled to industrial size (http://www.vrb.unsw.edu.au).
- High efficiency (between 80% and 90% in larger installations).

Disadvantages:

- Energy density is only 10–20 Wh/kg and 15–25 Wh/L. Second-generation V/Br batteries can achieve 35–50 Wh/kg with higher concentrations of vanadium in the solution.
- System is complex in larger storage sizes.

Uses:

- Load leveling
- Off-peak production storage for renewable energy
- Possible electric car applications (fill up like gasoline, but no burning of fuel, simply recharged at the station)

14.5.6 Iron-Air Battery

Various types of metal-air batteries have been explored for decades. Recently, the iron-air battery has generated considerable interest due to its long storage duration potential.

The battery works by oxidizing and reducing iron. When discharging, iron reacts with oxygen in the air to form iron oxide (i.e., rust), thus releasing energy via a flow of electrons. It is recharged by using electricity to reduce the iron oxide back to iron, releasing oxygen back into the air.

The iron-air battery concept has a number of advantages including low cost, low environmental impact, and high safety. The biggest advantage is that it can provide up to 4 days of storage, which makes it very valuable for firming variable solar and wind electricity during extended poor weather conditions. Iron-air batteries are large and heavy and so are unsuitable for vehicle storage, but the size is less of an issue for grid storage. Other disadvantages are slow response time (but they could be coupled with lithium-ion batteries) and low round-trip efficiency, currently on the order of 50% or less, which increases the cost of stored energy. Form Energy is developing this battery technology and has a number of pilot projects with U.S. utilities [22].

14.6 Fuel Cells and Hydrogen

Zero-emission electricity generation from hydrogen requires a fuel cell, as discussed in Chapter 1. (Hydrogen could also be burned in a specially designed combustion turbine, although that entails NO_x emissions.) Details of the principles of operation of the fuel cell and the use of hydrogen as a storage medium are discussed below [23].

14.6.1 Principles of Fuel Cell Operation

The operation of a fuel cell is shown schematically in Figure 14.14. A fuel cell consists, like a battery, of two electrodes separated by an electrolyte, which transmits ions but is

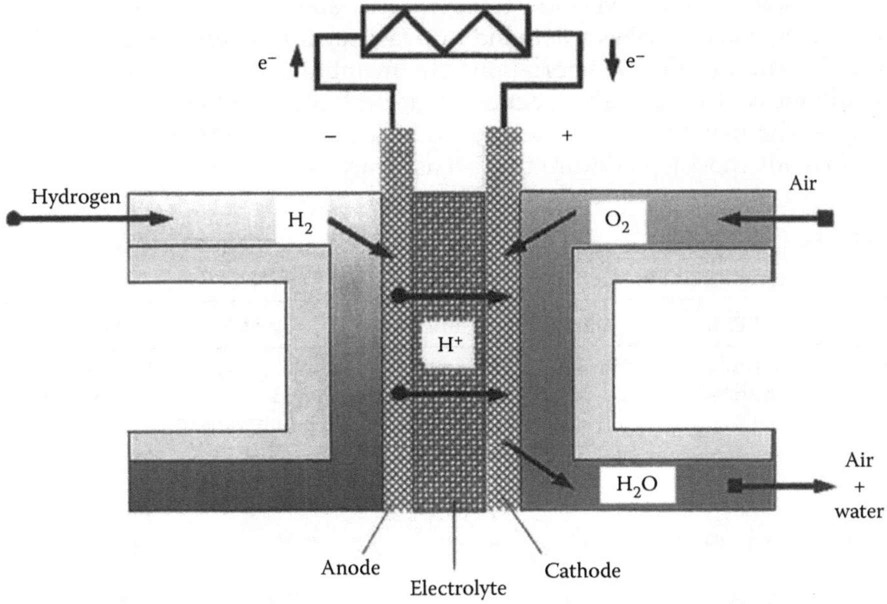

FIGURE 14.14
Schematic diagram of fuel cell operation.

impervious to electrons. In a fuel cell, hydrogen, or another reducing agent, is supplied to the negative electrode, while oxygen (or air) goes to the positive electrode. A catalyst on the porous anode promotes dissociation of hydrogen molecules into hydrogen ions and electrons. The H^+ ions (i.e., protons) migrate through the electrolyte to the cathode where they react with electrons supplied through the external circuit and with oxygen to form water.

Although a fuel cell converts chemical energy into electricity directly, without an intermediate combustion cycle, it does not approach its theoretical efficiency of 100% in practice. Fuel cells have an efficiency on the order of 40%–60% in practice, independent of the amount of power generated. Since the efficiency of an assembly of fuel cells is approximately equal to that of a single cell, there are no economies of scale in terms of efficiency. Therefore, fuel cells are promising for small, localized plants in the range of 1–100 kW capacity. Since fuel cells also reject heat, they can operate as a cogeneration system and can supply both electricity and heat.

14.6.2 Types of Fuel Cells

There are six principal types of fuel cells, as shown in Table 14.2:

- Proton exchange membrane fuel cell (PEMFC)
- Direct methanol fuel cell (DMFC)
- Molten carbonate fuel cell (MCFC)
- Alkaline fuel cell (AFC)
- Phosphoric acid fuel cell (PAFC)
- Solid oxide fuel cell (SOFC)

The above types of fuel cells differ in operating temperatures, efficiency, and cost. Their operating characteristics and some potential applications are summarized in Table 14.2.

PEMFC cells operate at relatively low temperatures (about 80°C), have a high power density, and can meet varying power demand or a fast startup. These features make PEMFC the favorite for automobile transportation. The membrane of a PEMFC is made of thin perfluorosulfonic acid sheets, which act as an electrolyte and allow the passage of hydrogen ions only. The membrane is coated on both sides with highly dispersed metal alloy particles (currently mostly platinum) that act as catalysts.

TABLE 14.2

Operating Characteristics of Fuel Cells

	PEMFC	DMFC	MCFC	AFC	PAFC	SOFC
Electrolyte	Ion exchange membrane	Polymer membrane	Immobilized liquid molten carbonate	Immobilized liquid phosphoric acid	Potassium hydroxide	Ceramic
Operating temperature (°C)	80	60–130	650	200	60–90	800–1,000
Efficiency (%)	40–60	40	45–60	35–40	45–60	50–65
Typical electrical power	Up to 250 kW	<10 kW	>1 MW	>50 kW	Up to 20 kW	>200 kW
Possible applications	Vehicles	Portables	Power stations	Small power stations	Submarines, spacecraft	Power stations

DMFC cells also use a polymer membrane as an electrolyte. The most attractive feature of DMFCs is that they can use liquid methanol as a fuel. The anode of a DMFC can draw hydrogen from liquid methanol and thus eliminates the use of a fuel reformer.

MCFC cells use a molten carbonate salt as the electrolyte. They are flexible in their fuel supply and can use coal-derived gases, methane, or NG as fuel. The efficiency of these fuel cells is about 60%, but if operating as a cogeneration unit that uses the waste heat, the efficiency overall can be even higher.

PAFC cells use an anode and a cathode of finely dispersed platinum catalyst on carbon and a silicon-carbide structure that contains phosphoric acid as the electrolyte.

AFC cells are commercially available. They operate at low temperatures, below 100°C, and are used in submarines and for space power applications.

SOFC cells work at high temperatures (between 800°C and 1,000°C) and utilize a solid ceramic electrolyte such as zirconium oxide stabilized with yttrium oxide instead of a liquid. The efficiency of these cells can reach as high as 60%, and they have the potential to be used for large-scale energy storage or generating electricity industrially. They can also be used for cogeneration systems.

The major challenges for the widespread application of fuel cells are their cost, durability, and the need for a continuous, inexpensive, and low-carbon supply of hydrogen.

14.6.3 Generation of Hydrogen

There are currently two different methods used for commercial production of hydrogen. The most widely used is NG reforming. In this process, called steam methane reforming, NG is mixed with steam, and the resulting mixture is introduced into a catalytic reforming reactor where it is passed through a nickel catalyst and converted at about 900°C into CO and H_2 according to the reaction

$$CH_4 + H_2O \rightarrow 3H_2 + CO \quad \Delta H^0 = 206\ \text{kJ/mol} \tag{14.20}$$

The other method is electrolysis, where an electric current is passed through water to separate the hydrogen from the oxygen. Today, most commercial water electrolyzers use an alkaline electrolyte system, as shown in Figure 14.15.

In such an electrolytic cell, the following simplified reaction takes place:
In the electrolyte

$$2H_2O \rightarrow 2H_2 + 2OH^- \tag{14.21}$$

At the cathode

$$2H^+ + 2e^- \rightarrow H_2 \tag{14.22}$$

At the anode

$$2OH^- \rightarrow \frac{1}{2}O_2 + H_2O + 2e^- \tag{14.23}$$

The overall reaction is

$$H_2O \rightarrow \frac{1}{2}O_2 + H_2 \quad \text{where} \quad \Delta H_r = 286\ \text{kJ/mol} \tag{14.24}$$

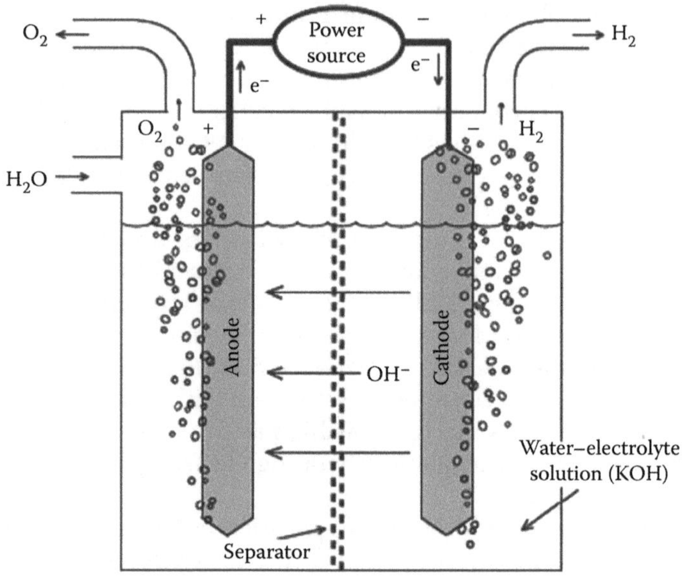

FIGURE 14.15
Schematic diagram of hydrogen generation by electrolysis.

According to Bossel et al. [24], the estimated cost of hydrogen production by NG reforming or coal gasification is about \$1/kg, whereas the cost via a renewable technology, such as electrolysis with electricity from wind, is between \$6 and \$7/kg. The energy content of 1 kg of H_2 approximately equals the energy content of 1 gal of gasoline.

14.6.4 Storage and Transport

Once the hydrogen is generated, it can be stored and subsequently combusted to provide heat or work, or it can be used to generate electricity with a fuel cell. Gaseous hydrogen has a low specific density and for storage purposes, it must be compressed or cryogenically liquefied. If hydrogen is compressed adiabatically to a pressure of 350 bar, a value often assumed suitable for automotive technologies, it consumes about 12% of its higher heating value (HHV). The loss can be reduced appreciably, however, if the compression approaches an isothermal process. If the hydrogen is stored as a liquid at cryogenic temperatures, the process takes about 40% of the HHV with current technology. Moreover, liquid storage is not feasible for automotive applications because mandatory boiloff from the storage container cannot be safely released in closed spaces such as garages or tunnels.

One technology for storing hydrogen consists of bonding it into metal hydrides using an absorption process. The energy penalty of this type of storage, which requires only a pressure of about 30 bar, may be less. However, the density of the metal hydride can be between 20 and 100 times the density of the hydrogen stored [25]. Carbon nanotubes have also received attention as a potential hydrogen storage medium, but neither of the above processes has been fully developed and commercialized.

14.6.5 Thermodynamics and Economics

As an energy carrier, hydrogen should be compared with the only other widespread and viable alternative: electricity. Once generated and compressed, hydrogen could flow through large pipelines with a loss of approximately 0.8%/100 km [1,24], while the energy loss experienced with high voltage DC (long distance) electric transmission lines is of the order of 0.6%/100 km [1]. Hence, the losses incurred in transmission of energy by hydrogen and electricity are of the same order of magnitude. However, the cost of building a transmission system for hydrogen in the U.S. has been estimated to be on the order of half a trillion dollars, and unless this up-front expenditure is made, transmission of hydrogen would have to be accomplished by truck, which would incur enormous expenses and energy penalties. Moreover, as shown by Kreith and West [26], the extraction of electricity from hydrogen is associated with irreversible losses. Electrolysis of water to make hydrogen is commercially available at about 70% efficiency. The sequence of steps required to produce hydrogen from electricity, transport it, and then use a fuel cell to generate electricity again was shown in Chapter 1. Compressing hydrogen to a pressure of 350–700 bar and liquefying it for transportation and storage are estimated to be about 60%–70% efficient. And, finally, converting the hydrogen back into electricity with a fuel cell has an efficiency of approximately 60%, although the theoretical limit is somewhat higher. Taken together, electrolysis, compression or liquefaction, and generation of electricity in a fuel cell results in an overall efficiency of only about 30%, compared to the direct use of electricity from the original source.

Hence, from a thermodynamic as well as from an economic perspective, the use of hydrogen as a storage and transmission system is inferior to the use of electricity directly. It should also be noted that this loss of energy is the result of basic thermodynamic considerations and would occur irrespective of whether the original generation of electricity came from renewable sources, nuclear power, or fossil fuels. In addition, because approximately three units of electricity are required to produce one unit of useful power at the end use, the amount of CO_2 generation per kWh electricity (assuming the electricity is from a carbon-emitting source) would be triple via the hydrogen path compared to any other electric power system.

Regardless of the electricity source, the use of hydrogen as a carbon-abatement fuel must account for the known tendency of the very small hydrogen molecule to leak and hydrogen's indirect greenhouse gas effect, for which a central estimate of the 20-year global warming potential (GWP) has been reported as 33 [27]. High-pressure hydrogen tanks have been found to have leakage rates (as a percent of total production) of 1%–4% [28], and, while estimates vary widely, systems that rely on transporting and storing hydrogen as a liquid have been reported to have total leakage rates as high as 20% due to heat absorption and boiloff [29]. Finally, any combustion uses of hydrogen must consider the air pollution impact of NO_x emissions.

14.7 Thermal Energy Storage

Solar thermal storage is a well-established technology, and the principles of design and operation in tanks or in the ground are extensively described in standard textbooks on heat transfer [30]. For solar thermal heat and power, thermal storage is important in applications such as storing heat overnight for buildings, seasonal heat storage for large-scale heating and/or cooling projects, and for storage in solar thermal power plants such as solar parabolic troughs or central receivers. Solar storage can also be dispersed through the overall system as, for example, in the thermal mass of a building—an art that has been practiced for centuries. A particularly good example is the south-facing Cliff Palace dwelling in Mesa Verde National Park, Colorado, where heat is stored in rock walls.

14.7.1 Sensible Heat

Thermal energy can be stored as sensible heat, as latent heat, or in reversible thermochemical reactions. The sensible heat, ΔQ, gained or lost per unit volume by a material changing temperature from T_1 to T_2 is

$$\Delta Q = \int_{T_1}^{T_2} \rho c_p dT \tag{14.25}$$

Although both the density, ρ, and the specific heat, c_p, vary with temperature, the variations are usually small enough to use average values for engineering calculations. Equation 14.25 then simplifies so that $\Delta Q = c_p \Delta T$ per unit mass, or $\rho c_p \Delta T$ per unit volume. Commonly used figures of merit are $\$/c_p$ for relative media costs, ρc_p for volumetric heat capacity, using the lowest value of ρ at the highest expected operating temperature for relative container size.

Table 14.3 shows the characteristics of some plentiful and economically competitive solid and liquid materials for sensible heat storage. Rocks, bricks, and cement are excellent candidates for overnight thermal storage in buildings. Among liquids, water is clearly

TABLE 14.3

Physical Properties of Some Sensible Heat Storage Materials

Storage Medium	Temperature Range (°C)	Density (ρ) (kg/m³)	Specific Heat (c_p), (J/kg K)	Energy Density (ρc_p) (kWh/m³ K)	Thermal Conductivity (W/m K)
Water	0–100	1,000	4,190	1.16	0.63 at 38°C
Water (10 bar)	0–180	881	4,190	1.03	—
50% ethylene glycol-50% water	0–100	1,075	3,480	0.98	—
Dowtherm A® (Dow Chemical, Co., Midland, Michigan)	12–260	867	2,200	0.53	0.122 at 260°C
Therminol 66® (Monsanto Co., St. Louis, Missouri)	9–343	750	2,100	0.44	0.106 at 343°C
Draw salt (50 $NaNO_3$–50 KNO_3)[a]	220–540	1,733	1,550	0.75	0.57
Molten salt (53 KNO_3/40 $NaNO_3$/7$NaNO_3$)[a]	142–540	1,680	1,560	0.72	0.61
Liquid sodium	100–760	750	1,260	0.26	67.5
Cast iron	m.p. (1,150–1,300)	7,200	540	1.08	42.0
Taconite	—	3,200	800	0.71	—
Aluminum	m.p. 660	2,700	920	0.69	200
Fireclay	—	2,100–2,600	1,000	0.65	1.0–1.5
Rock	—	1,600	880	0.39	—

Note: m.p. = melting point.

[a] Composition in percent by weight.

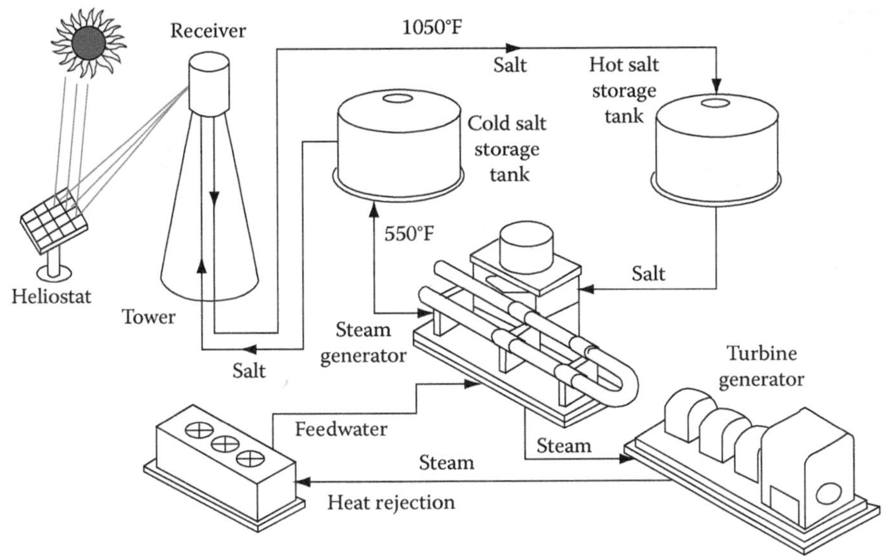

FIGURE 14.16
Two-tank storage for central receiver system.

superior for temperatures below 100°C, whereas molten salts are commonly used for solar thermal power applications such as parabolic troughs or central receivers.

Solids and liquid sensible heat storage materials can be combined in various ways. Rock beds and water tanks have been used in hybrid storage systems for space heating, whereas rocks and oils have been used in single vessels for solar thermal power applications to improve stratification. This approach could reduce the cost compared to relatively expensive dual-tank liquid storage systems where two vessels are necessary, as shown in Figure 14.16. However, experiences with single-tank storage to achieve stratification have not been reliable, and for solar thermal power applications, dual-tank storage (hot and "cold" tanks) is widely deployed.

14.7.2 Phase Change Heat Storage

Latent heat is the thermal energy released or absorbed by a material undergoing a phase transition, such as solid to liquid (melting), or liquid to gas (vaporization). Materials that have a high heat release or absorption during phase change are of interest for thermal storage. Latent heat storage material should have a large heat of transition, high density, appropriate transition temperature, low toxicity, and no gradation with multiple phase change. High thermal conductivity and low flammability are also preferred qualities. Materials that have been used for storage are paraffin waxes and salt hydrates.

The advantage of using phase change material for storage is that a large amount of heat can be stored in small volumes and the temperature of the material remains constant during phase change. The problem with phase change heat transfer is to find materials that have appropriate temperatures during phase change, and also the design of heat exchangers that transfer heat to or from the material during phase change is exceedingly difficult. Low thermal conductivity can be an issue with adding and removing heat from phase

TABLE 14.4

Physical Properties of Latent Heat Storage Materials or PCMs

Storage Medium	Melting Point (°C)	Latent Heat (kJ/kg)	Specific Heat (kJ/kg °C)		Density (kg/m³)		Energy Density (kWh/m³ K)	Thermal Conductivity (W/m K)
			Solid	Liquid	Solid	Liquid		
$LiClO_3{\cdot}3H_2O$	8.1	253	—	—	1,720	1,530	108	—
$Na_2SO_4{\cdot}10H_2O$ (Glauber's salt)	32.4	251	1.76	3.32	1,460	1,330	92.7	2.25
$Na_2S_2O_3{\cdot}5H_2O$	48	200	1.47	2.39	1,730	1,665	92.5	0.57
$NaCH_3COO{\cdot}3H_2O$	58	180	1.90	2.50	1,450	1,280	64	0.5
$Ba(OH_3){\cdot}6H_2O$	90	163	1.56	3.68	1,636	1,550	70	0.611
$LiNO_3$	252	530	2.02	2.04	2,310	1,776	261	1.35
$LiCO_3/K_2CO_3$, (35:65)[a]	505	345	1.34	1.76	2,265	1,960	188	—
$LiCO_3/K_2CO/$ Na_2CO_3 (32:35:33)[a]	397	277	1.68	1.63	2,300	2,140	165	—
n-Tetradecane	5.5	228	—	—	825	771	48	0.150
n-Octadecane	28	244	2.16	—	814	774	52.5	0.150
HDPE (cross-linked)	126	180	2.88	2.51	960	90	45	0.361
Steric acid	70	203	—	2.35	941	347	48	0.172 ℓ

Note: ℓ = liquid.

[a] Composition in percent by weight.

change materials. Table 14.4 shows the material properties of several common TES materials that may potentially be useful for latent heat storage. The engineering design of latent heat storage systems is beyond the scope of this book.

14.7.3 Thermochemical Storage

Thermochemical energy is stored as the bond energy of a chemical compound. In thermochemical energy storage, molecular bonds are broken during a reversible chemical reaction and are catalyzed by an increase in temperature. After thermochemical separation occurs, the constituents are stored apart until the combination reaction is desired. Recombination of the bonds between atoms releases the stored thermochemical energy.

Reversible chemical reactions with reactants and products that can be easily stored as liquids and/or solids are of interest for thermochemical energy storage. Reactions that produce two distinct phases such as a solid and a gas are desirable since the separation of products to prevent back reaction is facilitated. Table 14.5 shows several common thermochemical storage reactions and their standard enthalpy change (ΔH^0) in kJ and their turning temperature (T') in K.

The turning temperature, T', in Table 14.5 is defined as the temperature at which the equilibrium constant is unity and is calculated using the ratio of the standard enthalpy change and the standard entropy change for the reaction. At this temperature, the reactants and products will be present in approximately equal quantities. When $T > T'$, the endothermic storage reaction is favored, meaning heat is necessary for and absorbed during the reaction. Conversely, for $T < T'$, the exothermic reaction dominates, meaning that heat is produced in the reaction.

TABLE 14.5

Thermochemical Storage Reactions

Reaction	ΔH^0 (kJ)	T' (K)
$NH_4F(s) \leftrightarrow NH_3(g)+HF(g)$	149.3	499
$Mg(OH)_2(s) \leftrightarrow MgO(s)+H_2O(g)$	81.1	531
$MgCO_3(s) \leftrightarrow MgO(s)+CO_2\,(g)$	100.6	670
$NH_4HSO_4(1) \leftrightarrow NH_3(g)+H_2O(g)+SO_3(g)$	337.0	740
$Ca(OH)_2(s) \leftrightarrow CaO(s)+H_2O(g)$	109.3	752
$BaO_2(s) \leftrightarrow BaO(s)+½O_2(g)$	80.8	1,000
$LiOH(1) \leftrightarrow ½Li_2O(s)+½H_2O(g)$	56.7	1,000
$MgSO_4 \leftrightarrow MgO(s)+SO_3(g)$	287.6	1,470

Source: Wyman et al. [31].

The primary advantages of thermochemical storage include the high energy density and the long-term, low-temperature, storage capability. However, the thermochemical process is very complex, and the thermochemical materials are often expensive and can be hazardous. There is considerable research activity in the field of thermochemical reactions for energy storage, but no large-scale commercial applications exist to date.

14.7.4 Thermal Storage for Concentrating Collector Systems

A very important application of thermal storage is in high-temperature concentrator systems such as central receivers or parabolic troughs. Steam is generated in the solar receiver when the sun is up and either passes through the turbine, or, when excess heat is available, it is stored in a tank. The size of the tank will determine the number of hours that the system can operate when the sun is down. The drawback of steam storage is that it requires a large volume, which limits the length of time the turbine can operate when the sun is down.

A significant application of thermal storage is the two-tank storage system shown in Figure 14.16 that was used in the Solar 2 central receiver. In this approach, a single heat transfer fluid is used in the receiver and for storage. This approach eliminates the need for expensive heat exchangers and the degradation in the availability of the energy that occurs during heat transfer from one fluid to another. The fluid most commonly proposed for use in this kind of application is a molten salt, although other heat transfer fluids such as oils or liquid metals have been considered.

When molten salts are used as the heat transfer fluid, care must be taken to avoid freezing anywhere in the system. Molten salts freeze at a temperature between 120°C and 200°C (250°F and 430°F). Thus, the Rankine cycle has to operate between a high temperature achievable in the receiver and a cold temperature that does not drop below the freezing point of the salt. The design of a typical two-tank direct system with approximate temperatures is shown in Figure 14.16. The temperature of the molten salt is raised in the receiver to about 565°C and then passed to the hot salt storage tank. From this tank, salt is extracted to a heat exchanger, which generates the steam for a Rankine cycle power plant, and then is returned to the cold salt storage tank at about 290°C. The working fluid is then recirculated to the receiver in a closed loop. A normal Rankine cycle is used for the generation of electricity. The length of time that the system can operate once the sun is down depends on the size of the storage tanks. The advantage of this approach is the simplicity of design, and the disadvantage is that a large amount of heat transfer fluid is needed for the plant to operate.

In systems where the heat transfer fluid is very expensive or cannot be used for some reason as the storage fluid, a two-tank indirect system is used. The two-tank indirect system functions the same way as the two-tank direct system, except that different fluids are used in the receiver heating loop and the storage. Therefore, the heat from the working fluid must be transferred to the storage fluid via a heat exchanger, which the fluid enters at high temperature and leaves at low temperature. Storage fluid from the high-temperature tank is then used to generate steam, as in the direct system. This type of system is typical for parabolic trough solar plants in which oil serves as the heat transfer fluid in the parabolic trough receivers, and molten salt is used in the thermal storage tanks.

14.7.5 Overnight Storage for Buildings and Domestic Hot Water

TES for domestic hot water or for overnight heating of buildings has been mentioned in previous chapters. These applications require storing a liquid, almost always water, in a tank. A schematic diagram typical of such systems with electric backup is shown in Figure 14.17. Sizing the storage tank insulation is a simple heat transfer problem treated in standard textbooks. The properties of insulation materials are given in Table 14.6 [32].

The annual average insolation on a surface tilted at an angle equal to the latitude in the mid-latitudes of the United States is about 23,000 kJ/m²/day (2,030 Btu/ft²/day). If the

TABLE 14.6

Thermal Properties of Some Insulating and Building Materials

Material	Average Temperature (°F)	k [Btu/(h·ft·°F)]	c_p [Btu/(lb$_m$·°F)]	ρ (lb$_m$/ft³)	α (ft²/h)
Insulating Materials					
Asbestos	32	0.087	0.25	36	~0.01
	392	0.12	—	36	~0.01
Cork	86	0.025	0.04	10	~0.006
Cotton fabric	200	0.046			
Diatomaceous earth, powdered	100	0.030	0.21	14	~0.01
Molded pipe covering	300	0.036	—		
	400	0.051	—	26	
	1,600	0.088	—		
Glass wool					
Fine	20	0.022	—		
	100	0.031	—	1.5	
	200	0.043	—		
Packed	20	0.016	—		
	100	0.022	—		
Hair felt	100	0.027	—	8.2	
Kaolin insulating brick	932	0.15	—	8.2	
	2,102	0.26	—		
Kaolin insulating firebrick	392	0.05	—	19	
	1,400	0.11	—		

(Continued)

TABLE 14.6 (*Continued*)

Thermal Properties of Some Insulating and Building Materials

Material	Average Temperature (°F)	k [Btu/(h·ft·°F)]	c_p [Btu/(lb$_m$·°F)]	ρ (lb$_m$/ft^3)	α (ft^2/h)
Magnesia, 85%	32	0.032	—	17	
	200	0.037	—	17	
Rock wool	20	0.017	—	8	
	200	0.030	—		
Rubber	32	0.087	0.48	75	0.0024
Building Materials					
Brick					
Fire clay	392	0.58	0.20	144	0.02
	1,832	0.95			
Masonry	70	0.38	0.20	106	0.018
Zirconia	392	0.84	—	304	
	1,832	1.13	—		
Chrome brick	392	0.82	—	246	
	1,832	0.96	—		
Concrete					
Stone	~70	0.54	0.20	144	0.019
10% moisture	~70	0.70	—	140	~0.025
Glass, window	~70	~0.45	0.2	170	0.013
Limestone, dry	70	0.40	0.22	105	0.017
Sand					
Dry	68	0.20	—	95	
10% water	68	0.60	—	100	
Soil					
Dry	70	~0.20	0.44	—	~0.01
Wet	70	~0.15	—	—	~0.03
Wood					
Oak ⊥ to grain[a]	70	0.12	0.57	51	0.0041
| | to grain	70	0.20	0.57	51	0.0069
Pine ⊥ to grain[a]	70	0.06	0.67	31	0.0029
| | to grain	70	0.14	0.67	31	0.0067
Ice	32	1.28	0.46	57	0.048

Source: Kreith [32].

[a] ⊥, perpendicular; | |, parallel.

average collection efficiency is 30%, then 6,900 kJ/m^2/day (610 Btu/ft^2/day) is delivered by the collector system. The service hot-water demand can be taken as 76 L/day (20 gal/day) per person. If the tap water has a temperature of 283 K (50°F), and the hot water 333 K (140°F), about 210 kJ/L (750 Btu/gal) is required for hot-water heating. For delivery of 230 L/day (60 gal/day) for a family of three, 48.3 MJ/day (46,000 Btu/day) is needed. If the collector delivers 6,900 kJ/m^2, a collector with a size of approximately 7 m^2 is needed. This collector is about one-tenth the size of a collector required for heating a typical building in the mid-latitudes. A rule of thumb is that 1 m^2 of collector is required for each 30 L of hot water to be delivered. The storage tank should be large enough to hold about a 2-day supply of hot water, or about 500 L for a family of three.

Thermal storage tanks must be insulated to control heat loss. If a storage tank is located within a structure, any losses from the tank tend to offset the active heating demands of the building. However, such storage loss is uncontrolled and may cause overheating in the summer.

The amount of thermal storage used in a solar heating system is limited by the law of diminishing returns. Although larger storage results in larger annual energy delivery, the increase at the margin is small and hence not cost-effective. Seasonal storage is, therefore, usually uneconomic, although it is needed in far northern locations like Scandinavia because the amount of sunlight is so limited in the winter. Experience has shown that liquid storage amounts of 50–75 kgH_2O/m^2 $(10-15\ lb/ft_c^2)$ are the best compromise between storage tank cost and useful energy delivery.

Since solar energy heating systems operate at temperatures relatively close to the temperatures of the spaces to be heated, storage must be capable of delivering and receiving thermal energy at relatively small temperature differences. The designer must consider the magnitude of these driving forces in sizing heat exchangers, pumps, and fans. The designer must also consider the nonrecoverable heat losses from storage. Even though storage temperatures are relatively low, surface areas of storage units may be large and heat losses therefore appreciable.

Other mechanical components in solar heating systems include pumps, heat exchangers, air bleed valves, pressure relief valves, and expansion tanks. Heat exchangers in solar energy systems are selected based on economic criteria. The best trade-off of energy delivery increase with increasing heat exchanger size usually results from the use of an exchanger with effectiveness in the range of 0.6–0.8. Counterflow heat exchangers are required for this level of effectiveness. Typical solar heat exchanger sizes range from 0.05 to 0.10 m^2 of heat exchanger surface per square meter of net collector area.

Optimization of flow rates in solar energy heat exchangers involves some tradeoffs. Increasing the flow rate increases the heat transfer coefficient and the heat transfer rate but decreases the heat exchange effectiveness. Also, it was generally found in solar domestic hot-water systems that although low flow rates decrease the heat transfer coefficient in the collectors, they can result in tank stratification, which causes colder water at the bottom of the tank going to the collectors, which reduces collector heat losses to the ambient, thus increasing collector efficiency and overall solar energy collection. Optimizing the overall system performance led to lower flow rates than were previously recommended.

In hydronic heating systems, it is essential that all air be pumped from the system. To facilitate this process, air bleed valves located at the high points in a system are used. These are opened during system fill and later if air should collect. Air bleeds are required at points of low velocity in piping systems where air may collect because the local fluid velocity is too low for entrainment.

Control strategies and hardware used in current solar system designs are quite simple and are similar in several respects to those used in conventional systems. The single fundamental difference lies in the requirement for differential temperature measurement instead of simple temperature sensing. In the space heating system shown in Figure 14.17, two temperature signals determine which of three modes is used. The signals used are the collector-storage differential and room temperature. The collector-storage difference is sensed by two thermistors or thermocouples, the difference being determined by a solid-state comparator, which is a part of the control device. Room temperature is sensed by a conventional dual-contact thermostat.

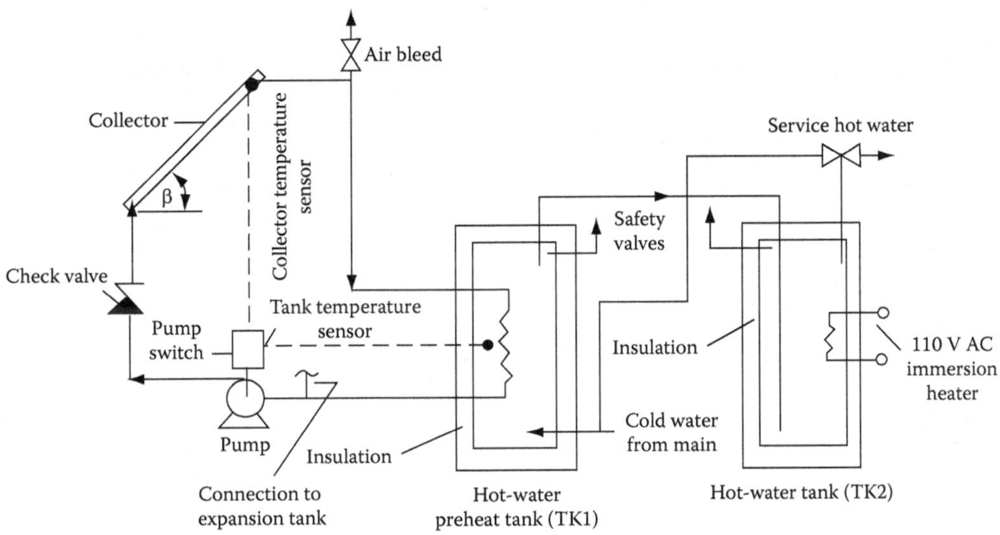

FIGURE 14.17
Schematic diagram of solar heating system with storage.

The control system operates as follows. If the first room thermostat contact closes, then the mode selector valve and distribution pump are activated in an attempt to deliver the thermal demand from solar thermal storage. If room temperature continues to drop, indicating inadequate solar availability, the mode selector diverts flow through the backup system instead of the solar system, and the backup is activated until the load is satisfied.

The collector-storage control subsystem operates independently of the heating subsystem described above. If collector temperature, usually sensed by a thermistor thermally bonded to the absorber plate, exceeds the temperature in the bottom of the storage tank by 5°C–10°C (9°F–18°F), the collector pump and heat exchanger pump (if present) are activated and continue to run until the collector and storage temperature are within about 1°C–2°C (2°F–4°F) of each other. At this point, it is no longer worthwhile to attempt to collect energy, and the pumps are turned off. The collector-storage subsystem also has a high-temperature cutout that turns the collector loop pump off when the storage temperature exceeds a set limit.

A heating load device transfers heat from the solar storage to the air in the space. Therefore, a liquid-to-air heat exchanger is sized based on the energy demand of a building. Several generic types of load devices are in common use:

1. **Forced-air systems:** tube-and-fin coil located in the main distribution duct of a building or zone of a building (see Figure 14.18).
2. **Baseboard convection systems:** tube-and-fin coils located near the floor on external walls. These operate by natural convection from the convectors to the room air.
3. **Heated floors or ceilings:** water coils transfer heat to large thermal masses that in turn radiate or convect into the space. This heating method is usually called radiant heating.

Each load device requires fluid at a different temperature to operate under design load conditions, as shown in Figure 14.19. Since baseboard heaters are small in heat transfer area and rely on the relatively ineffectual mechanism of natural convection, they require the

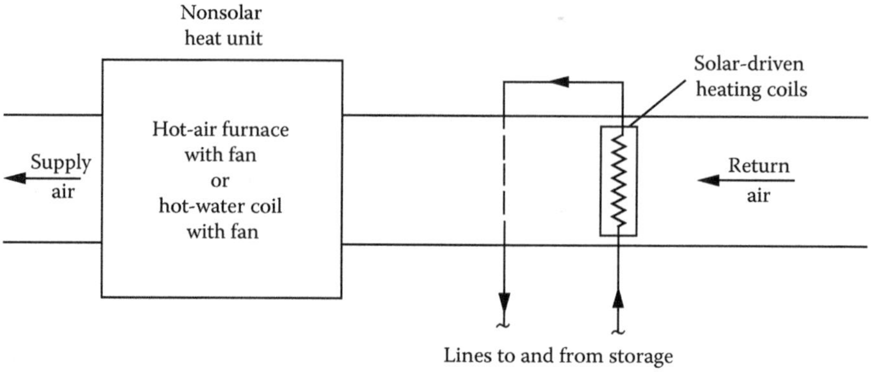

FIGURE 14.18
Forced-air heating system load device location upstream of nonsolar heat exchanger or furnace.

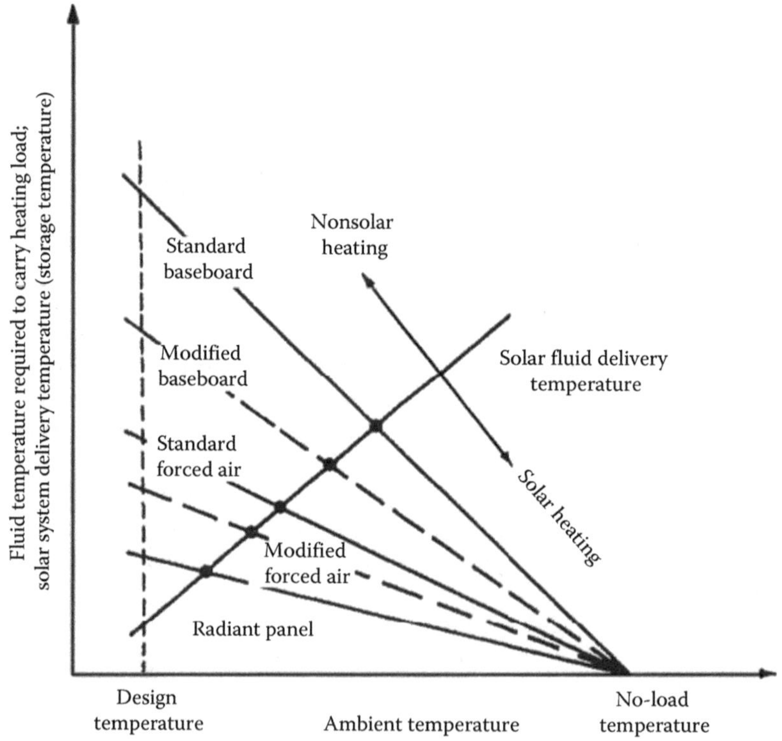

FIGURE 14.19
Heating load diagram for baseboard, forced-air, and radiant systems. Modified baseboard and forced-air systems are oversized in order to carry heating demands at lower temperature. Balance points are indicated by large dots at intersections.

highest fluid temperature and so are not as commonly used for solar heating. Forced-air systems involve the more efficient forced-convection heat transfer mode and, hence, are operable at lower fluid temperatures (see Figure 14.19). Radiant heating can use very large heat transfer areas and is, therefore, operable at relatively low temperatures.

In Figure 14.19, the intersection of the solar fluid temperature line and the load line for a specific configuration is called the *balance point*. At ambient temperatures below the balance point, solar energy cannot provide the entire demand, and some backup is required; above the balance point, solar capacity is sufficient to carry the entire load. Note that the load lines are specific to a given building. The solar fluid temperature line is not fixed for a building but depends on the solar collector and storage size as well as local solar radiation levels. The line shown in Figure 14.19 is, therefore, an average line.

It is possible to modify load devices to lower the balance point, as shown in Figure 14.19. For example, a forced-air tube-and-fin exchanger can be enlarged by adding one or more additional rows of tubes. This increased heat transfer area will permit the same energy delivery at a lower fluid temperature. The law of diminishing returns is evident as increasing effectiveness returns progressively less energy.

Problems

14.1 A hydrogen/oxygen fuel cell stack produces 100 kW of DC power at an efficiency of 60% with water vapor as the product. Determine the hydrogen mass flow rate in g/s and the required cell voltage.

14.2 Suppose you own some land next to a 50 MW wind farm located on Colorado's Front Range. Your land has a fantastic cliff/ridgeline with an elevation difference of 250 m. You decide to evaluate whether you can make money by building an energy storage system on your land that interfaces with the wind farm. Your plan is to buy energy from the wind farm when electricity prices are low and sell that energy back to the grid when electricity prices are high (assume that the utility that owns the grid will grant you a fair contract). You find that a good power rating for your storage system is 50% of the rated wind farm output, or 25 MW. Also, your storage system should be able to generate electricity at rated power for 8 hours.

The first step in this problem is to evaluate the technical design of candidate energy storage systems.

a. Pumped hydroelectric. Determine the flow rate and reservoir size needed to accomplish the required power output and energy capacity. Pumped hydroelectric plants generally run at 80%–90% generating efficiency, depending on the size of the machinery. Suggest a reasonable surface area and depth for your two reservoirs. (Hey, wait! You like to fish and plan to stock one of your reservoirs with some northern pike. You cannot leave them high and dry when the plant cycles.)

b. Pulley and weight. You wonder if a simpler energy storage system is possible. What if you use a weight and a pulley that stores energy as the weight is lifted (using an electric motor) and energy is released when the weight falls (with an electric generator)? You need to dig a shaft. Find the value of weight you would use and the depth and diameter of the shaft. Is this viable?

c. Propose a new energy storage method. Be creative. Examples may be large springs, thermal storage, steam cycles, batteries, etc. Perform a simple analysis to size the system, and say whether or not your creative idea is viable.

The second step is to evaluate the financial aspects of installing and operating your energy storage system to determine whether you could make money.

a. Assume pumped hydroelectric costs about $450/kW to install for turbomachinery and penstocks, plus $1 per cubic yard to build reservoirs. Calculate the initial capital cost of your pumped hydroelectric system.

b. You buy energy from the wind farm at $0.035/kWh between the hours of 10:00 p.m. and 8:00 a.m. to charge your storage and sell energy between 1:00 p.m. and 9:00 p.m. back to the grid at $0.1/kWh. Select a reasonable simple payback period that would motivate you to invest in energy storage. Calculate the maximum capital cost expenditure on your energy storage system that would allow this payback period.

c. Would you decide to build an energy storage system? Why or why not?

14.3 It has been proposed that the United States should strive toward a hydrogen economy. At present, the United States uses about 106 quads (1 quad = 10^{15} Btu) of energy per year. Suppose that 10% should be produced with hydrogen fuel cells. The hydrogen is to be made by electrolysis of water in central stations located strategically. For purposes of this problem, assume the efficiency of the electrolysis process is 61% and that of the fuel cell is 45%. The electric power for electrolysis is to be supplied by nuclear power plants with a thermodynamic efficiency of 33% and a capacity factor of 0.90. Each plant should be of 500 MW capacity and operate on a simple Rankine cycle.

Estimate

- The number of plants required.
- The amount of cooling water needed for the condensers (assuming evaporative cooling—2,400 kJ cooling for each liter of cooling water).
- The amount of water needed for electrolysis assuming 1,500 kJ are used by the fuel cell (i.e., only 750 kJ of energy is provided to the end user for each liter of water fed into the process).
- The total initial capital investment (if the cost of a nuclear plant is estimated at $6,000/kW and that of a fuel cell at $2,100/kW).
- Comment on the viability of using hydrogen as an energy carrier on a global scale.

14.4 Derive an expression for the linear battery model power output as a function of the internal-to-external resistance ratio. Nondimensionalize the power output by dividing it by E^2/R_i. Tabulate the dimensional power function with a spreadsheet and then plot it. At what point is the maximum nondimensional power achieved?

14.5 Design a battery electric storage system for a power plant with 20 MW peak power delivery for a duration of 4 hours and estimate the minimum number of batteries and the current during peak operation. Assume that the plant will use 600 A/h batteries, operating at 400 V DC.

14.6 A hydrogen/oxygen fuel cell operates with a voltage of 0.7 V with water vapor production. Calculate the kJ of work per kg-mol of hydrogen and determine the cell efficiency.

14.7 You are asked to estimate the amount of heat that can be stored overnight in a solar-heated building, which has a storage space of 6 m^3. The solar system uses water with antifreeze in the collector and there are two options for storing the heat overnight. One option is to use water with an allowable temperature change of 30°C. The other is to use a Glauber salt, which undergoes a phase change at approximately 30°C with a heat of fusion of 329 kJ/L. In both instances, a heat exchanger will be necessary in order to meet building code. Discuss the advantages and disadvantages of the two options.

14.8 A neighborhood fuel cell power plant is to be designed for an electric power output of 2,000 kW with liquid–water product. Estimate the flow rates of hydrogen and oxygen during peak power production, assuming an 80% efficient power conditioner is used to convert DC to AC power and the fuel cell efficiency is 55%.

References

1. Chamberlain, J., Hammerschlag, and Schaber, C.P., 2015. Energy storage technologies, Chapter 33, pp. 939–966, in Goswami, D.Y., and Kreith, F. eds., *Handbook of Energy Efficiency and Renewable Energy*, 2nd Edition, CRC Press, Boca Raton, FL.
2. Electricity Storage Association, 2003. Technology comparisons: Ratings, www.esa.org.
3. Tennessee Valley Authority, The mountaintop marvel, www.tva.gov/heritage/mountaintop/index.htm.
4. Levine, J.G., 2007. Pumped hydroelectric energy storage and spatial diversity of wind resources as methods of improving utilization of renewable energy sources, Thesis, University of Colorado, Boulder, CO.
5. Kreith, F., and Krumdieck, S., 2014. *Principles of Sustainable Energy*, 2nd ed., CRC Press, Boca Raton, FL.
6. Crotogino, F., Mohmeyer, K.-U., and Scharf, R., 2001. Huntorf CAES: More than 20 years of successful operation, in: *Solution Mining Research Institute Meeting*, Orlando, FL.
7. Van der Linden, S., December 2006. Bulk energy storage potential in the USA, Current developments and future prospects, *Energy* 31, 3446–3457.
8. Davis, L., and Schainker, R., 2006. Compressed air energy storage (CAES): Alabama electric cooperative McIntosh plant—Overview and operational history, in: *Electricity Storage Association Meeting 2006: Energy Storage in Action*, Energy Storage Association, Knoxville, TN.
9. Succar, S., September 2008. Baseload power production from wind turbine arrays coupled to compressed air energy storage, PhD thesis, Princeton University, Princeton, NJ.
10. Barbour, E., and Pottie, D., 2021. Adiabatic compressed air energy storage technology, *Joule* 5(8), 1914–1920, doi:10.1016/j.joule.2021.07.009.
11. EPRI-DOE, 2003. *Handbook of Energy Storage for Transmission and Distribution Applications*, EPRI/DOE, Palo Alto, CA/Washington, DC.
12. de Guilbert, A., 2009. Batteries and supercapacitors cells for the fully electric vehicle, Saft Groupe SA, Brussels, https://www.smart-systems-integration.org/public/electric-vehicle/battery-workshop-documents/presentations/Anne%20de%20Guibert%20Saft.pdf/at_download/file.
13. IEA, 2024. Batteries and secure energy transitions, International Energy Agency, April, https://www.iea.org/reports/batteries-and-secure-energy-transitions.
14. Machín, A., Morant, C., and Márquez, F., 2024. Advancements and challenges in solid-state battery technology: An in-depth review of solid electrolytes and anode innovations, *Batteries* 10(1), 29, doi:10.3390/batteries10010029.

15. GP Batteries, 2000. *Lithium Ion Technical Handbook*, https://web.archive.org/web/20120323234351/https://www.gpbatteries.com/html/pdf/Li-ion_handbook.pdf.
16. Dai, Q., Kelly, J.C., Gaines, L., and Wang, M., 2019. Life cycle analysis of lithium-ion batteries for automotive applications, *Batteries* 5, 48, doi:10.3390/batteries5020048.
17. Quan, J., Zhao, S., Song, D., Wang, T., He, W., and Li, G., 2022. Comparative life cycle assessment of LFP and NCM batteries including the secondary use and different recycling technologies, *Science of the Total Environment* 819, 153105. doi:10.1016/j.scitotenv.2022.153105.
18. Picatoste, A., Schulz-Monninghoff, M., Niero, M., Justel, D., and Mendoza, J., 2024. Comparing the circularity and life cycle environmental performance of batteries for electric vehicles, *Resources, Conservation & Recycling* 210(2024), 107833. doi:10.1016/j.resconrec.2024.107833.
19. Mohr, M., Peters, J., Baumann, M., and Weil, M., 2020. Toward a cell-chemistry specific life cycle assessment of lithium-ion battery recycling processes, *Journal of Industrial Ecology* 24, 1310–1322, doi:10.1111/jiec.13021.
20. Wickerts, S., Arvidsson, R., Nordelof, A., Svanstrom, M., and Johansson, P., 2024. Prospective life cycle assessment of sodium-ion batteries made from abundant elements, *Journal of Industrial Ecology* 28, 116–129, doi:10.111/jiec.13452.
21. Lai, X., Chen, J., Chen, Q., Han, X., Lu, L., Dai, H., and Zheng, H., 2023. Comprehensive assessment of carbon emissions and environmental impacts of sodium-ion batteries and lithium-ion batteries at the manufacturing stage, *Journal of Cleaner Production* 423, 138674, doi:10.1016/j.jclepro.2023.138674.
22. Martucci, B., 2024. Iron-air battery developer Form Energy raises $405M, announces collaboration with GE Vernova, *Utility Dive*, October 22, 2024, https://www.utilitydive.com/news/iron-air-battery-developer-long-duration-storage-form-energy-collaboration-ge-vernova/730633/.
23. Gupta, R.G., ed., 2008. *Hydrogen Fuel—Production, Transport, and Storage*, CRC Press, Boca Raton, FL.
24. Bossel, U. et al., 2003. *The Future of Hydrogen Economy: Bright or Bleak*, Oberrohreorf, Switzerland.
25. Rabinwitz, M., 2000. Power systems of the future, Parts 1–3, *IEEE Power Engineering Review* 20(135), 5–16, 10–15, 21–24.
26. Kreith, F., and West, R.E., 2004. Fallacies of a hydrogen economy: A critical analysis of hydrogen production, *Journal of Energy Resources Technology* 126, 249.
27. N. Warwick, et al., 2022. Atmospheric implications of increased hydrogen use, UK government report, April.
28. Westra, I.M., Scheeren, H.A., Stroo, F., van Heuven, S., Kers, B., Peters, W., and Meijer, H., 2024. First detection of industrial hydrogen emissions using high precision mobile measurements in ambient air, *Scientific Reports* 14, 24147, doi:10.1038/s41598-024-76373-2.
29. Ocko, I.B., and Hamburg, S.P., 2022. Climate consequences of hydrogen emissions, *Atmospheric Chemistry and Physics* 22, 9349–9368, doi:10.5194/acp-22-9349-2022.
30. Kreith, F., and Bohn, M.S., 2001. *Principles of Heat Transfer*, 7th ed., Brooks/Cole Pacific Grove, CA.
31. Wyman, C., Castle, J., and Kreith, F., 1980. A review of collector and energy storage technology for intermediate temperature applications, *Solar Energy* 24(6), 517–540.
32. Kreith, F., 1973. *Principles of Heat Transfer*, Intext, New York.

15

Transportation

15.1 Introduction

A viable transportation system is a crucial part of a sustainable energy future. Sustainable transportation is a complex interdisciplinary topic, which deserves a book unto itself. As Schiller et al. [1] wrote in introducing their book, "sustainable transportation aims at promoting better and healthier ways of meeting individual and community needs while reducing the social and environmental impacts of current mobility practices." Engineering factors, such as improved vehicle efficiency or traffic flow management, are only part of the solution. Transportation and land use planning, policy, economics, and community involvement are also critical dimensions. This chapter provides a brief introduction to the broader dimensions of sustainable transportation, including the critical issues of transportation equity and environmental justice. This chapter then focuses on alternative vehicle technologies and fuels used for personal mobility, road freight, marine shipping, and aviation.

Gasoline and diesel are convenient fuels in part because they have a high energy density that permits storage in a relatively small volume—an important asset for motor vehicles. For example, these liquid fuels have a volumetric specific energy content of about 10,000 kWh/m^3, compared to hydrogen compressed to 700 bar at about 1,300 kWh/m^3. However, production, transport, and combustion of petroleum-derived fuels have substantial environmental costs, motivating a transition to alternative fuels and technologies. At present, more than 90% of the energy used for transportation in the United States is petroleum-based, about 6% is from biofuel and the balance is from natural gas and electricity [2]. The share provided by electricity is expected to grow rapidly. Meanwhile, the prices of gasoline and diesel are ongoing concerns to consumers, and the emissions from current transportation systems are a major component of the greenhouse gas pollution contributing to the climate crisis (Figure 15.1) [3].

In past decades, oil experts debated when global oil production would reach a peak and then decline due to exhaustion of finite petroleum resources. Now, as discussed in Chapter 6, there is recognition of the need to limit future oil production and use, even before resources run out, in order to mitigate climate change and the environmental impacts of air pollution and drilling in sensitive areas. Hence, moving to a transportation system with sharply reduced reliance on petroleum resources is imperative for a sustainable energy future.

DOI: 10.1201/9781003586616-15

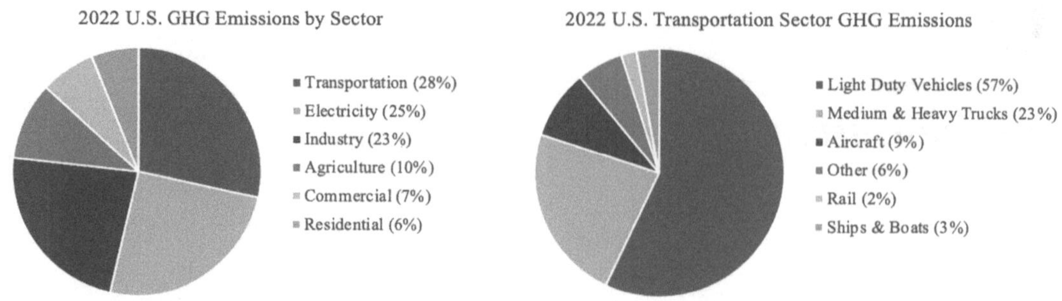

FIGURE 15.1
Share of U.S. greenhouse gas emissions by sector, and breakdown by source of contributions within the transportation sector. Transportation emissions do not include emissions from non-transportation mobile sources such as agriculture and construction equipment. "Other" sources include buses, motorcycles, pipelines and lubricants. (From EPA [3].)

15.2 Overview of Transportation Systems and Energy Use

In the late 1990s, Canada's Centre for Sustainable Transportation adopted the following definition of a sustainable transportation system, which has been widely quoted [1]:

A sustainable transportation system is one that:

- Allows the basic access needs of individuals and societies to be met safely and in a manner consistent with human and ecosystem health, and with equity within and between generations.
- Is affordable, operates efficiently, offers choice of transport mode, and supports a vibrant economy.
- Limits emissions and waste within the planet's ability to absorb them, minimizes consumption of non-renewable resources, limits consumption of renewable resources to the sustainable yield level, reuses and recycles its components, and minimizes the use of land and the production of noise.

This definition highlights some of the factors of sustainability that go beyond the efficient use of fuels and that are widespread problems with existing transportation systems:

- Need for equitable service to support basic access requirements (e.g., to get to grocery stores, schools, or places of employment) within cities or across countries.
- Noise and pollution associated with transportation facilities.
- Congestion and excessive time demands for meeting transportation needs.
- Safety risks of vehicle accidents.
- Large land footprint for roads, parking, airports, and supporting facilities.
- Disturbance of wildlife and ecosystems.

Transportation systems are comprised of the infrastructure, vehicles, other equipment and fuels needed to move goods and provide mobility to people. The requirements for transportation systems to provide personal mobility are closely linked to land use and economic development patterns through two-way interactions. The form of urban development significantly affects transportation needs, while the transportation infrastructure plays a large role in shaping development patterns. Furthermore, economic growth and globalization have sharply increased freight transport over recent decades, including transport by pipeline, truck, rail, and ship. The relatively low market cost of long-distance shipping and international disparities in production costs have led to remarkably complex and far reaching supply chains for many products [1].

Figure 15.2 shows greenhouse gas (GHG) emissions from transportation by transportation subsector from 1990 to 2018 [4]. Emissions are shown for the world and for Africa, Eastern Asia and North America, as example regions. Global GHG emissions from transportation increased by 75% from 1990 to 2018. In 2018, over 70% of transportation emissions were from passenger and freight transport by road. Eleven percent of emissions were from aviation (domestic and international) and 8% from international shipping [4]. While emissions from Africa and East Asia rose much more quickly from 1990 to 2018, they remained well below those from the United States and Canada, which accounted for 24% of the global total in 2018 (without counting their share of emissions from international aviation and shipping). Per capita GHG emissions from transportation differ widely around the world. In 2023, per capita emissions from transportation in the United States exceeded those in China by a factor of seven and were nearly 20 times higher than those in Nigeria [5]. In the United States, in 2019, there were more than 840 vehicles for every 1,000 people, compared to 630 in Western Europe, 180 in China, and 41 in Africa [6].

In the United States, energy use for highway transportation (including passenger vehicles and trucks for moving goods) increased from 16.1 quadrillion Btu (17.0 EJ) in 1970 to 28.8 quads (30.4 EJ) in 2007, then leveled off, with 28.7 quads (30.3 EJ) used in 2019 [6].

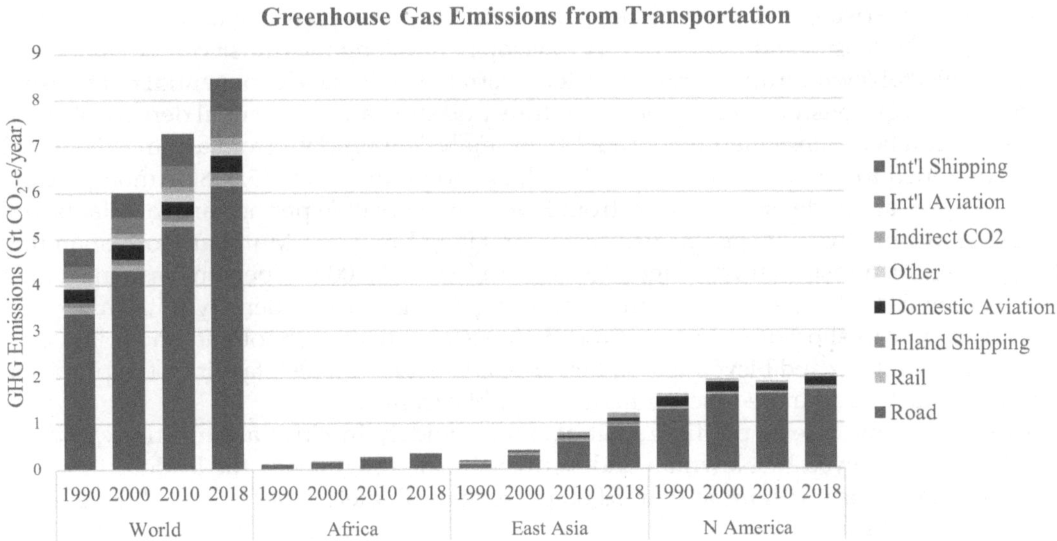

FIGURE 15.2
Greenhouse gas emissions from the transportation sector for the world and for selected regions. Indirect CO_2 represents emissions from electricity and heat provided to the transportation sector. Other includes indirect N_2O. East Asia includes China, the Republic of Korea, the People's Republic of Korea, and Mongolia; North America includes the United States and Canada and excludes Mexico and the Caribbean. (Adapted from Lamb et al. [4], Creative Commons License 4.0.)

Impacted by the COVID-19 pandemic, energy use for transportation was down modestly in 2020 and 2021 but has rebounded since then. In 2019, there were about 270 million personal cars and trucks in use in the United States. These vehicles were cumulatively driven nearly 2.7 trillion miles (4.3 trillion km) [6].

On the freight side, more than 12.5 billion U.S. tons (11.3 billion metric tons) of freight were shipped within the United States in 2017 (most recent available data), with about 70% carried by truck, 15% by rail or a combination of rail and truck or water, 8% by water or a combination of water with truck or rail, and less than 0.1% by air [6]. When distance is factored into the statistics (expressed in terms of ton-miles or metric ton-kilometers), the fraction transported by rail or a combination including rail increases to about 44%, as rail is more heavily used for longer distances. Average shipment distance within the United States was 679 miles (1,086 km) in 2017, up from 472 miles (755 km) in 1997 [6]. Moving freight by truck across long distances is three or more times more energy-intensive than by rail.

Over short time horizons, engineers and planners look to improving the efficiency of vehicles and traffic patterns, modifying supply chains for freight, and adopting programs to shift passengers and freight to more efficient modes as mechanisms to reduce energy and environmental impacts of transportation. Longer planning horizons offer the possibility of greater improvements in vehicle efficiency, introduction of alternative fuels and vehicle technologies, and the potential for restructuring urban development and transportation infrastructure to reduce travel demand.

Mass transportation offers the potential to reduce energy consumption and GHG emissions from the transportation sector. GHG emissions per-passenger mile in the United States are estimated to be lower for transit buses than for passenger vehicles, and those for rail transit are lower yet, although transit energy efficiency is highly variable across the country [7]. The per-passenger energy intensity of transit is highest for areas with low density, including metropolitan areas with urban sprawl, where it is difficult to reach outlying areas by a mass transit network. Light rail and other advanced transit developments are capital-intensive and must be carried out in a coordinated way with long-term planning of urban forms that integrate high-density and high-intensity developments along the transit corridors.

In the 1980s, Newman and Kenworthy demonstrated an inverse exponential relationship between energy consumption for personal transportation and residential density of cities. These researchers subsequently updated their analysis with data from around the world and continued to see the same general trend, as shown in Figure 15.3 [8]. Among the cities surveyed, urban densities ranged from less than ten people per hectare in Atlanta and Houston to more than 300 people per hectare in Hong Kong and Mumbai. Corresponding energy use for transportation ranged from more than 100,000 MJ/person/year in Atlanta to less than 5,000 MJ/person/year in Hong Kong. Greater urban density with mixed-use development can shorten required travel distances for work, school and shopping, thus promoting walking and bicycling. Higher urban densities can also facilitate the provision and use of faster and more efficient modes of public transit.

Availability and use of public transit also vary widely in cities around the world. Of the 44 cities surveyed by Newman and Kenworthy [8], Prague was the leader in the rate of public transit use with more than 5,000 passenger-kilometers per person during 2005–2006. For comparison, that same year, New York City's annual rate of transit usage was about 1,400 passenger-km per person, and Chicago's was about 750 passenger-km per person. In addition to high residential density, high rates of transit use are promoted by having concentrated employment opportunities, limited and expensive parking, and transit systems that compete with private vehicles by shortening commute times. Cities around

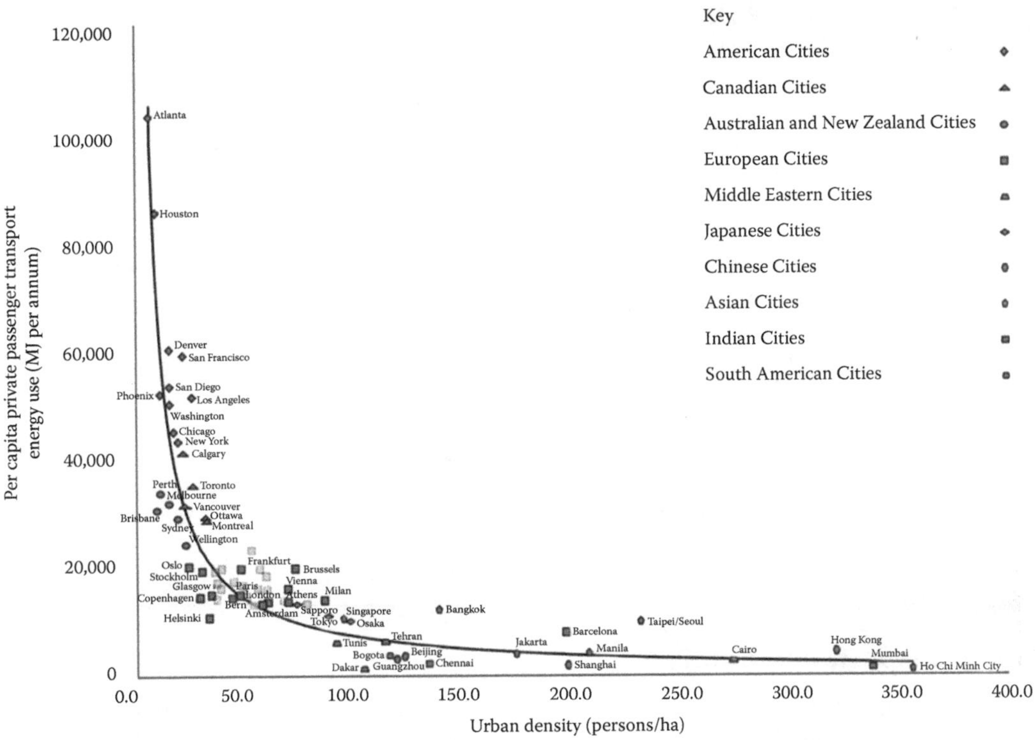

FIGURE 15.3
Car use per capita reflected in annual energy use, versus urban density. Data are for 1995. (Reproduced from Newman and Kenworthy [8]. With permission.)

the world deploy a variety of rapid transit technologies, including advanced rail, tramway, and bus rapid transit. Having reserved or restricted right-of-way for these systems is critical for their ability to successfully compete for ridership.

Personal transport accomplished via walking or cycling is conventionally measured in terms of the number or fraction of trips made, rather than their total distance. In their 44-city survey, Newman and Kenworthy [8] reported that Zurich was the leader in non-motorized travel, with 52% of daily trips taken by non-motorized modes. New York City led among the U.S. cities surveyed, with 22% of trips via bicycle or walking. Copenhagen is widely recognized as one of the most bicycle-friendly cities in the world. Along with high density and mixed-use development, cities can encourage commuting by bicycle by providing separated bikeways; through bicycle-friendly management of motorized traffic, including signalization and speed control; integrating bicycling with public transit; and providing for adequate bicycle parking.

While the trend toward promoting non-motorized travel has grown, U.S. Census surveys indicate that on average across the United States, progress has been slow in shifting work trips out of private vehicles (Table 15.1 [6]). The share of people working at home and bicycling increased from 1990 to 2020, but the percentage of commuters who carpooled, took transit or walked to work declined over that period [6]. Results for 2020 reflect surveys conducted too early to capture recent trends in working from home, advanced ride-hailing services, and use of micromobility vehicles such as e-bikes and scooters.

TABLE 15.1

Means of Transportation to Work in the United States

	1990 Millions of Workers (%)	2020 Millions of Workers (%)
Private Vehicle Total	99.6 (86.5)	128.7 (83.8)
Alone	84.2 (73.2)	115.1 (74.9)
Car Pool	15.4 (13.4)	13.6 (8.9)
Public Transportation	5.9 (5.1)	7.0 (4.6)
Walking	4.5 (3.9)	4.0 (2.6)
Bicycling	0.47 (0.4)	0.78 (0.5)
Work at Home	3.4 (3.0)	11.1 (7.3)
Total Workers	115.1 (100)	153.7 (100)

Source: Davis and Boundy [6].

Note: Data for 1990 from the U.S. Bureau of the Census. Data for 2020 from the 2015–2020 American Community Survey, U.S. Bureau of the Census.

15.3 Transportation Equity

Transportation equity and environmental justice are important policy aims that should be considered in evaluating the sustainability of alternative transportation systems. Transportation access influences opportunities for provisioning, education, employment, health care, and social interaction. Transportation needs of people of color, those with low incomes, children, the elderly, and people with disabilities have historically been underserved by the transportation system in the United States [9]. At the same time, people with low incomes and people of color have borne a disproportionate share of the burdens associated with highways and other transportation infrastructure, including air pollution, noise, safety risks, and housing displacement [9].

According to the U.S. Federal Highway Administration, "[e]quity in transportation seeks fairness in mobility and accessibility to meet the needs of all community members. A central goal of transportation equity is to facilitate social and economic opportunities by providing equitable levels of access to affordable and reliable transportation options based on the needs of the populations being served, particularly populations that are traditionally underserved" [10]. Transportation equity strives to ensure that the needs of all community members are met, including those who require extra accommodations.

The U.S. Environmental Protection Agency (EPA) has defined environmental justice as "the just treatment and meaningful involvement of all people, regardless of income, race, color, national origin, Tribal affiliation, or disability, in agency decision-making and other federal activities that affect human health and the environment so that people:

- are fully protected from disproportionate and adverse human health and environmental effects (including risks) and hazards, including those related to climate change, the cumulative impacts of environmental and other burdens, and the legacy of racism or other structural or systemic barriers; and
- have equitable access to a healthy, sustainable, and resilient environment in which to live, play, work, learn, grow, worship, and engage in cultural and subsistence practices" [11].

Environmental justice also emphasizes community engagement and influence in decision-making, in addition to reducing disparities in risks and benefits.

The U.S. history of discrimination against minority groups, including Native Americans, immigrants, and people of color, has created inequities in the transportation system that persist today [9]. As just one historical example, the construction of the U.S. interstate highway system and urban expressways in the 1940s–1970s displaced hundreds of thousands of people, split established communities, and added to racial, ethnic and income segregation, as established neighborhoods were bulldozed for new and expanded highways. According to Fullilove [12], African Americans were five times as likely to be displaced by highway construction in that period as would have been expected based on their share of the U.S. population. Inequities in transportation access have continued over the decades as government infrastructure funding has been heavily weighted toward road construction and maintenance relative to public transit [13].

Present-day legacies of historical infrastructure development patterns in the United States include disparities in the burdens of proximity to heavily traveled roadways [14] and air pollution exposure [15]. Figure 15.4 shows census blocks in Denver, CO that experience relatively high levels of diesel particulate air pollution and also have a high proportion of residents who are low-income or people of color [16]. With respect to transportation

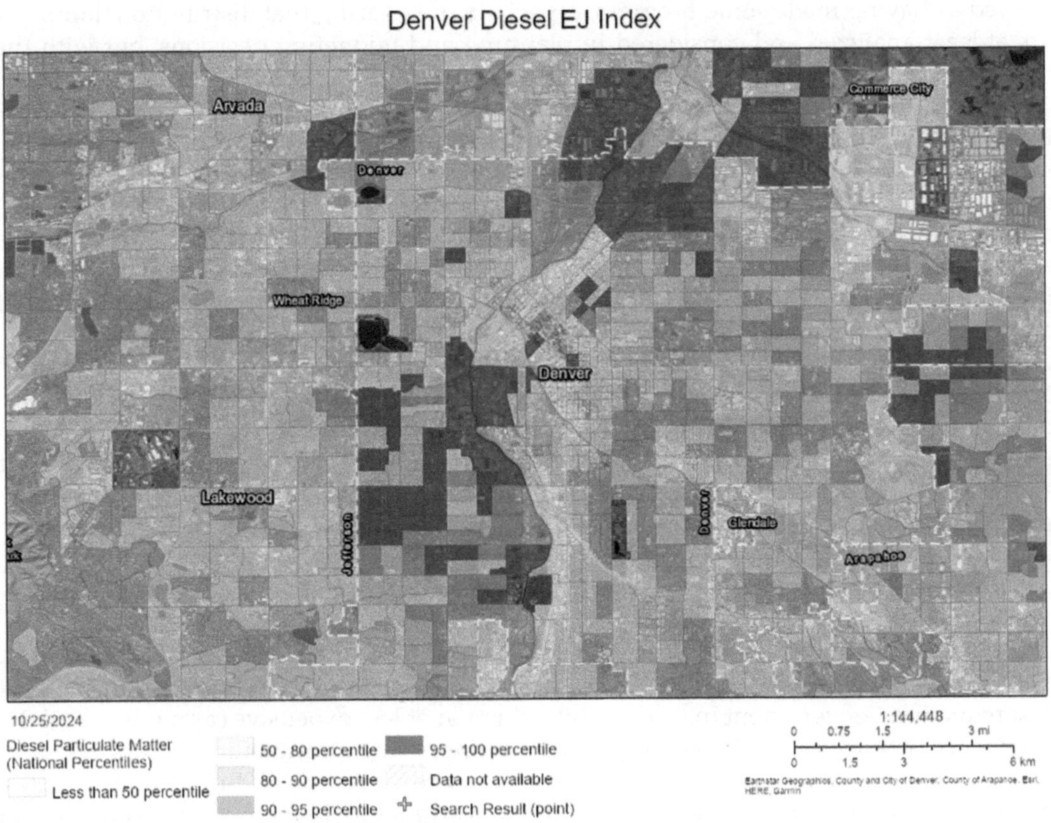

FIGURE 15.4
Percentile rankings of census block groups in the Denver, CO area compared to other census block groups in the United States, for the environmental justice indicator that combines diesel particulate matter concentrations with the average of percent people of color and percent low income. (Map produced by the authors using U.S. EPA's EJScreen Environmental Justice data and mapping tool [16].)

access, the U.S. Department of Transportation has estimated that in 2022, 47% of the U.S. population was burdened by transportation costs, meaning they spent more than 15% of their income on transportation services or more than 45% of their income on transportation and housing combined [17]. In 2020, 18% of black households in the United States were estimated to lack access to a car, compared to only 6% of white households [18].

As Fuller and Brugge (2020) explain, today's environmental justice movement emerged from "multiple struggles" in the 20th century, including Civil Rights, Native American land rights, labor activism, and anti-toxics campaigns [9]. Some progress was achieved, including the passage of the Civil Rights Act of 1964. Title VI of that law prohibits discrimination by applicants for or recipients of federal funds, which includes funding for transportation infrastructure projects. In the 1980s, equitable access to public transportation was central to disability rights campaigns in the United States that helped build momentum for the passage of the Americans with Disabilities Act in 1990. President Bill Clinton's Executive Order 12898, signed in 1994, directed federal agencies to develop strategies to identify and address disproportionate adverse health and environmental impacts on minority and low-income populations. Subsequent administrations have followed up on this directive to varying degrees (see Ref. [19] for a recent example from the Department of Transportation). Many U.S. states also have requirements to evaluate and address disparate environmental burdens of public infrastructure projects. To date, these efforts are viewed as having made some progress, especially in ensuring that distributional impacts are at least analyzed and considered in planning and budgeting decisions, but with the recognition that sharp inequities persist [20].

How will new transportation technologies affect environmental justice and transportation equity? Slowing climate change stands to benefit everyone, and people with low incomes may be most vulnerable to its disruptive effects. However, not all climate solutions are equal when local environmental impacts are considered. Battery electric vehicles (BEVs) that reduce tailpipe emissions compared to gasoline and diesel-fueled cars and trucks can help reduce disproportionate exposure to traffic-related air pollution that is borne by people who live near major roadways, warehouses, and other facilities that attract motorized traffic. On the other hand, the use of biofuels such as ethanol and biodiesel in internal combustion engines will not do as much to reduce tailpipe emissions of air pollutants such as nitrogen oxides and particulate matter, even if they reduce carbon emissions compared to fossil fuels. A question with electric vehicles is whether an electrified fleet will lead to increased emissions from electricity generation, shifting the air pollution burden to those living near power plants. However, modeling studies suggest the health benefits of reducing vehicle emissions already outweigh the damages from increased power plant emissions for cities across the United States and in Canada [21,22]. Over time, net benefits will be even larger as renewables continue to gain ground for electricity generation.

The cost of advanced technologies can be a barrier to equitable access to their benefits. This has been the case so far with electric passenger vehicles in the United States [23], but cost reductions, government incentives, introduction of less expensive car models, and the emergence of a used electric vehicle market are improving affordability. Access to charging facilities is another equity challenge with electric vehicles [24]. In the United States, having a garage where a vehicle can be charged overnight makes it much easier to own an electric vehicle (EV) than it is for most people who live in multi-family housing complexes. Construction of public charging stations and programs to subsidize electric vehicles for lower income families are examples of government programs that can help address disparities in EV access.

15.4 Well-to-Wheels Analysis

Vehicle technology is changing rapidly, with new engine and drivetrain configurations, fuels, and charging or fueling systems. Rather than simply looking at a vehicle's fuel economy or tailpipe emissions, a more comprehensive way to assess energy and environmental impacts is to conduct a *well-to-wheels analysis*. The approach to a well-to-wheels analysis, which is a subset of life cycle assessment (LCA), is shown schematically in Figure 15.5 [25,26]. The well-to-wheels approach for a fuel cycle includes several sequential steps: feedstock production; feedstock transportation and storage (T&S); fuel production; transportation, storage, and distribution of fuel (T&S&D); and finally vehicle operation. This approach is essential for a fair comparison of different options because each step entails efficiency losses and external costs.

Argonne National Laboratory's (ANL) GREET (Greenhouse gases, Regulated Emissions, and Energy use in Transportation) model is a suite of publicly available spreadsheet-based tools that enable users to perform their own life cycle assessments of various vehicle and fuel combinations [27]. The suite of models includes well-to-wheels (or well-to-wake) fuel cycle models for road, air, marine and rail transport. Impacts that can be estimated using the GREET models include energy use, water use and emissions of greenhouse gases and other air pollutants. The base GREET model covers passenger cars, light trucks, and sport utility vehicles, and internal combustion, hybrid, battery electric and fuel cell electric vehicle technologies, with fuel options including gasoline, diesel, natural gas, biofuels, hydrogen and electricity. GREET also offers a vehicle cycle model that estimates energy use and emissions over the life of the vehicle, from raw material production through component

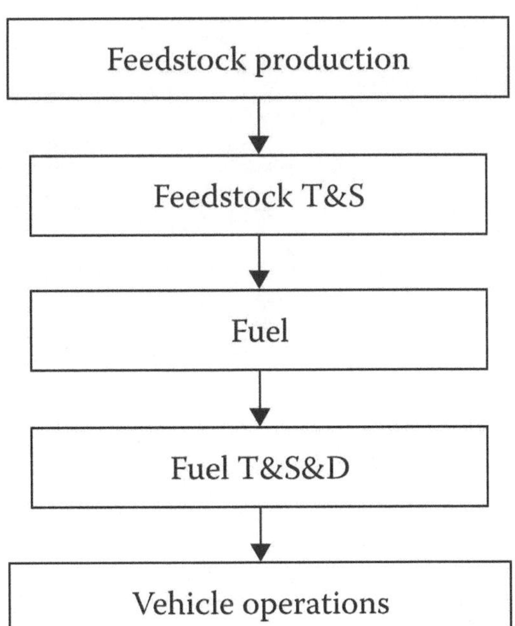

FIGURE 15.5
Steps in a well-to-wheels analysis for ground transportation vehicles.

fabrication, vehicle assembly and end-of-life material recovery or disposal. GREET models can be downloaded for free from the ANL website at https://greet.anl.gov. The models are extensively documented and frequently updated to reflect new research on energy efficiency and emissions as well as new fuels, fuel production pathways and vehicle technology. In 2023, ANL introduced a new stand-alone battery module for GREET, allowing for comparisons of alternative battery chemistries [28].

Figure 15.6 presents results from a prospective study completed using the GREET model, showing cradle-to-grave greenhouse gas emissions estimates for a range of vehicle technology and fuel combinations [29]. Estimates are shown for small sport utility vehicles and include both well-to-wheels fuel cycle emissions and vehicle cycle emissions. Solid black lines at the top represent emissions that were current in 2020; lines midway down show emissions that might be achieved by efficiency improvements that apply across vehicle types, including reducing weight and improving powertrain efficiency. The arrows show reductions that might be achievable through the use of biofuels and low-carbon electricity. The lowest bars indicate the portion of emissions associated with vehicle production, with

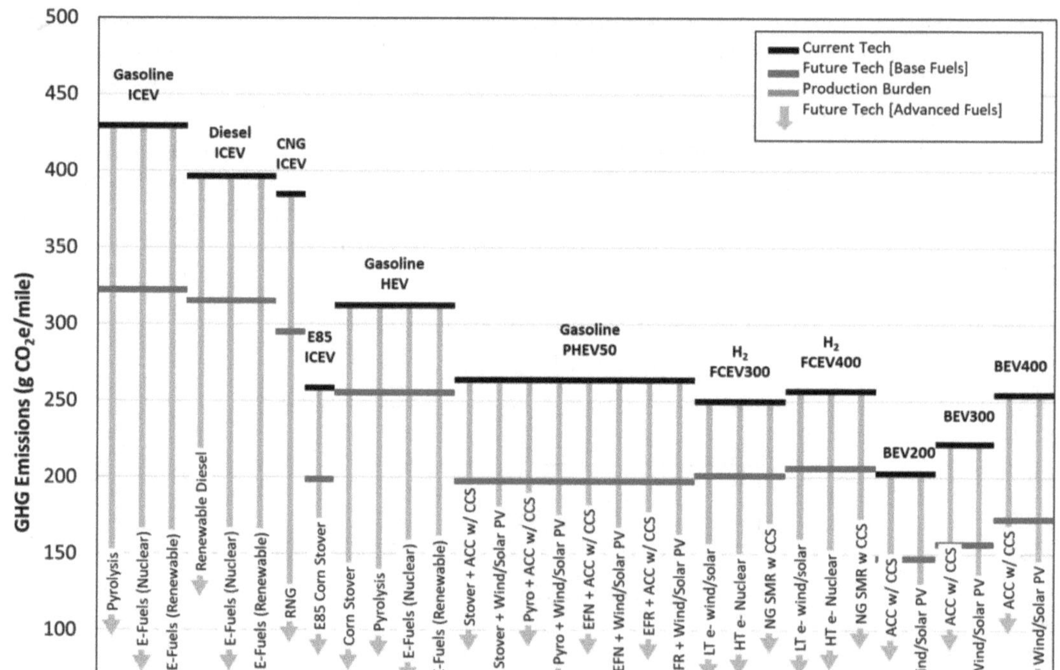

FIGURE 15.6
Potential cradle-to-grave greenhouse gas reductions. (ICEV=internal combustion engine vehicle; HEV=hybrid electric vehicle; PHEV=plug-in hybrid electric vehicle; FCEV=fuel cell vehicle; BEV=battery electric vehicle; E85=85% ethanol with 15% petroleum gasoline; E-Fuels=synthetic hydrocarbon fuels produced using electricity as the primary energy source; RNG=renewable natural gas; SMR=steam methane reforming; CCS=carbon capture and storage; ACC=advanced combined cycle; EFN=electro-fuels using nuclear electricity; EFR=electro-fuels using renewable electricity; LT e-=low-temperature electrolysis; HT e-=high-temperature electrolysis.) (From Kelly et al. [29].)

emissions amortized over the life of the vehicle. The results of the GREET analysis suggest that crosscutting efficiency improvements could reduce emissions from diesel and gasoline internal combustion vehicles (ICEV) by roughly 20%–25%. Combining these crosscutting improvements with renewable H_2 or electricity generation could lead to fuel cell electric vehicles (FCEVs) and BEVs with cradle-to-grave emissions of 50 g CO_2 per mile, which is about 10% of the emissions from current gasoline vehicles.

15.5 Alternative Fuels and Vehicle Technologies for Road Transport

For more than a century, motorized road transport has been dominated by the combustion of petroleum-based fuels in internal combustion engines. Gasoline has been the fuel of choice for light-duty cars and trucks with spark ignition engines. Most heavy-duty trucks and buses and some light-duty vehicles have used diesel fuel in compression ignition engines. Since the 1970s, regulations in countries around the world have led to improved fuel efficiency across all size classes of vehicles. In the United States, improvements in fuel economy have been outpaced by increasing travel and freight transport and the shift in vehicle sales from smaller sedans to larger sport utility vehicles and pickup trucks, leading to an 18% increase in GHG emissions from transportation over the period from 1990 to 2022 [30]. In contrast, there have been reductions in the United States (and elsewhere) in overall emissions of health-impacting air pollutants including nitrogen oxides and volatile organic compounds, thanks largely to catalytic control devices that have been developed to meet increasingly stringent tailpipe emissions regulations.

Interest in diversifying fuel sources and reducing GHG emissions has motivated the development and deployment of alternative fuel and vehicle technologies such as those in Figure 15.6. As indicated in Table 15.2, biodiesel, electricity, ethanol, and hydrogen (via electricity) are among the alternative fuels that could reduce dependence on oil and natural gas resources. Globally, sales of BEVs have grown dramatically in recent years, and this technology is expected to take over the light-duty and short-haul truck markets in coming decades. Global demand for transportation biofuels (mainly ethanol and

TABLE 15.2

Feedstocks for Alternative Fuels

Fuel	Example Feedstocks
Propane (LPG)	Natural gas (NG), petroleum
Compressed NG	NG, landfill gas, digester gas
Hydrogen	NG or (water + electricity)
FT diesel	NG, coal, or algae
Methanol (M85)	NG or coal
Ethanol (E85)	Corn, sugarcane, or cellulosic biomass
Biodiesel	Soy, canola, and other oilseeds
Renewable diesel	Waste animal fat, waste cooking oil
Electricity	NG, coal, uranium, or renewables

renewable diesel) has also been growing, with current utilization led by the United States, Brazil, and Europe, and recent growth led by Brazil, India, and Indonesia [31].

Heavy-duty long-haul trucks and buses represent a greater challenge for electrification than light-duty vehicles do, as battery size and driving range tradeoffs are exacerbated. Sustainable, low-carbon solutions for these vehicles are still under development, with questions as to which technologies might prevail. Biodiesel, renewable diesel and compressed natural gas used in compression ignition engines are current options. Hydrogen fuel cells, improved batteries for long-haul trucks and wireless roadway electrification are advanced technologies that are under development.

15.5.1 Liquid Biofuels

The potential of producing liquid fuel from biomass is covered in Chapter 9. Globally, biofuels account for about 5% of transportation energy use, mainly for road transport. Ethanol from corn (as produced in the United States) and sugarcane (as produced in Brazil) have been the mainstay of biofuel use for road transport for the past two decades, but the use of biodiesel and renewable diesel is increasing comparatively rapidly [31]. Expanded biofuel use is of interest for potential GHG emissions reductions, diversified energy supplies, and local economic development.

Biofuels account for about 5% of transportation energy in the United States, mainly in the form of corn-derived ethanol blended with gasoline. Blends with up to 15% ethanol can be sold for use in most gasoline vehicles, although a 10% blend (E10) is typical. Blends with about 50%–85% ethanol are sold in the United States as "E85," but are limited to use in vehicles with modified fuel systems. Ethanol use in the United States has been heavily bolstered by government policy, especially the federal Renewable Fuel Standard (RFS) blending requirement enacted by the U.S. Congress in 2005 and expanded in 2007. Approximately 40% of the corn grown in the United States is used to produce fuel-grade ethanol, with Iowa, Nebraska and Illinois accounting for more than half of U.S. ethanol production [32].

Published estimates of EROI and greenhouse gas emissions reduction potential with corn-derived ethanol have been varied and controversial, with some studies finding low EROIs and suggesting ethanol from corn offers little or no GHG emissions benefit compared to gasoline [e.g., 33]. Results differ depending on assumptions about the fuel used for process heat, allocation of emissions from co-products, and the degree to which growing crops for ethanol drives land use change. Xu et al. [34] used the ANL GREET model to assess life cycle emissions from corn-based ethanol produced in the United States and found that GHG emissions per MJ of energy are about 40% lower than those from standard gasoline if emissions reduction credits are assigned for using co-product distillers grains in place of other livestock feed. According to this study, about 40% of life cycle emissions associated with ethanol come from farming operations, 40% from processing and distillers grains to produce ethanol, and about 10% from the lost carbon sequestration that arises from converting uncultivated land to corn production. In Xu et al.'s analysis, approximately 80% of the energy required for ethanol refining was supplied by natural gas. The study suggests that replacing process heat supplied by fossil natural gas with renewable natural gas or renewable electricity; capture and sequestration of CO_2 emissions from ethanol processing plants; and more sustainable farming operations could further reduce emissions [34]. As discussed in Chapter 9, cellulosic ethanol offers the potential for better energy returns and lower net GHG emissions than corn-based ethanol, but remains comparatively expensive.

While fuel ethanol serves as a substitute for gasoline in spark ignition engines, biodiesel and renewable diesel are produced to replace petroleum-derived diesel fuel. Biodiesel (fatty acid methyl ester (FAME)) is produced via transesterification of fatty acids found in oilseeds. In the United States, biodiesel is approved for blending up to 20% with either petroleum-based diesel fuel or renewable diesel (B20). Renewable diesel, which is produced by hydrogenation processing of waste fat or cooking oil to match the properties of standard diesel, can be used as a drop-in fuel (R100) in compression ignition engines [35]. Production and use of alternative diesel fuels are growing rapidly worldwide. As of July 2024, the United States had capacity to produce 4.6 billion gallons per year of renewable diesel and 2.0 billion gallons per year of biodiesel. For comparison, production capacity for fuel ethanol in 2024 was 18.3 billion gallons [36]. Soybean oil, corn oil, and waste fats and grease are the primary feedstocks for biodiesel and renewable diesel production in the United States.

Xu et al. [37] used the GREET model to compare life cycle GHG emissions from biodiesel (produced by transesterification) and renewable diesel (produced by hydroprocessing) using different feedstocks, including soy and canola oilseeds, used cooking oil, and beef tallow recovered from meat-packing plants. For oilseed feedstocks, the life cycle stages analyzed included crop production, oilseed crushing, oil extraction, biofuel conversion, fuel distribution and consumption. Beef tallow was treated as a by-product of meat production, so in the LCA covering that pathway, no emissions were assigned to livestock production. Similarly, with used cooking oil, the LCA started at the stage of collecting and transporting the used oil. GHG emissions per MJ of fuel energy are shown in Figure 15.7.

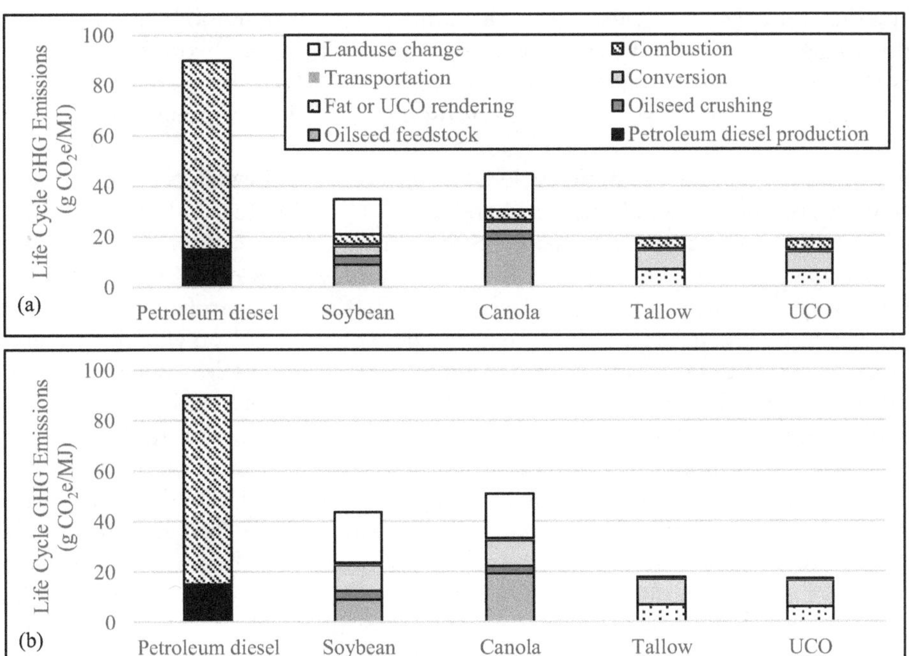

FIGURE 15.7
Life cycle greenhouse gas emissions of petroleum diesel versus (a) biodiesel and (b) renewable diesel from soybean, canola, tallow, and used cooking oil (UCO) feedstocks (Adapted from Xu et al. [37]. Land use change emissions are the median of the estimates presented in the original.)

Opportunities to further reduce GHG emissions include replacing fossil fuels with renewable energy in the biofuel conversion process, sourcing hydrogen for the hydrogenation process from renewable electrolysis instead of natural gas reforming, and improved oilseed farming practices [37]. As discussed in Chapter 9, growing energy crops for biodiesel production as well as for either cellulosic or starch-derived ethanol has significant land use costs that need to be considered in assessing the sustainability of these fuels. With the large decline in battery costs and the growing sales of BEVs, the potential market for biofuels may transition from ground transportation to aviation and shipping.

15.5.2 Natural Gas as a Transportation Fuel

Several countries around the world make extensive use of natural gas as a transportation fuel. According to the U.S. Department of Energy's Alternative Fuels Data Center (AFDC), there are about 23 million natural gas-fueled vehicles worldwide, and about 175,000 natural gas vehicles in the United States [38]. In the United States, an increasing portion of the natural gas used in vehicles is "renewable" natural gas, which is methane recovered from animal waste and landfills, among other sources of decomposing organic material. Countries that have historically led in the use of (mainly fossil) natural gas for transportation include Iran, China, Pakistan, Argentina, Brazil, and India. About one in four transit buses in the United States are fueled with natural gas, either in compressed (CNG) or liquid (LNG) form [39]. While natural gas vans, trucks, and buses are fairly common, the fuel is seldom used in the United States in passenger vehicles.

Switching to compressed natural gas (CNG) requires little modification to internal combustion (IC) vehicle technology apart from the fuel tank and refueling system. CNG is more practical for larger vehicles that can accommodate a larger and heavier fuel tank, and for high-mileage vehicles that can recover higher upfront costs through fuel cost savings. California leads the United States with 280 CNG and 34 LNG filling stations, about half of which are public. Private CNG stations serving fleet vehicles are generally more common than public CNG stations across the United States [40].

Determining the comparative life cycle GHG emissions of fossil CNG versus gasoline or diesel vehicles is challenging. Figure 15.6 shows ANL's estimate that light-duty CNG vehicles fueled with fossil natural gas would reduce life cycle GHG emissions by about 10%, compared to gasoline vehicles with similar performance. However, Alvarez et al. [41] have shown that such comparisons turn on two factors that require careful consideration. First, the comparison between GHG emissions for CNG versus gasoline vehicles is sensitive to assumptions about how much methane is leaked across the well-to-wheels supply chain. The results shown in Figure 15.6 assume that across the supply chain, about 0.70% of the produced natural gas is leaked into the atmosphere as methane [29; Table 9]; this is a low value compared to estimates made in other studies [42]. The second factor that complicates the comparison is the fact that methane is a much more potent GHG than CO_2 but also has a much shorter atmospheric lifetime. As Alvarez et al. explain, "[d]etermining whether a unit emission of CH_4 is worse for the climate than a unit of CO_2 depends on the time frame considered. Because accelerated rates of warming mean ecosystems and humans have less time to adapt, increased CH_4 emissions due to substitution of natural gas for coal and oil may produce undesirable climate outcomes in the near-term" [41]. ANL used a 100-year global warming potential (GWP) for methane of 30, neglecting the even higher potency that methane emissions have compared to CO_2 when a shorter time horizon is considered. Alvarez et al. conclude that unless methane leakage is strictly controlled, switching from gasoline or diesel to natural gas for transportation could exacerbate climate change in the short term [41].

Looking beyond fossil fuels, a number of renewable sources of natural gas are available. Renewable natural gas (RNG), which is also known as biomethane, is biogas that has been cleaned to meet purity standards for pipeline quality natural gas. Biogas can be produced from any organic matter. It is produced incidentally in landfills, but as discussed in Chapter 9, it can also be produced in anaerobic digesters or through pyrolysis from sewage, industrial, animal, or crop wastes or from specific energy crops. In the United States, nearly 250 facilities produced RNG in 2023, with additional growth in production anticipated [43]. (Many more facilities capture and use less pure biogas for other purposes.) Beneficial use of recovered methane that would otherwise be released to the atmosphere offers a clear benefit for reducing GHG emissions because vented methane has a high GWP compared to the CO_2 produced when the methane is combusted. Biomethane qualifies as an advanced biofuel under EPA's Renewable Fuels Standard program and is commonly credited as having a negative carbon intensity under California's Low-Carbon Fuel Standard [44].

15.5.3 Hybrid Electric Vehicles

A proven approach to reducing dependence on petroleum fuel is to increase the fuel economy of motor vehicles. This can be achieved by reducing vehicle weight and rolling resistance, improving aerodynamics, using more efficient vehicle accessories, and improving drivetrain efficiency [45]. Hybrid electric vehicles (HEVs), which have been available since the late 1990s, offer additional fuel economy improvements, taking advantage of the improved battery technology discussed in Chapter 14. As shown in Figure 15.6, ANL's assessment indicates that for comparable performance, HEV technology can reduce life-cycle GHG emissions by 25% compared to a conventional gasoline-powered vehicle [29].

An HEV is powered by the combination of a battery pack and electric motor—like that of an electric vehicle but with a smaller battery—and a power generation unit (PGU), which is normally an IC engine. Unlike BEV batteries, however, the smaller HEV batteries are recharged by an onboard PGU, which is fueled by existing fuel infrastructure.

HEVs can be configured in a parallel or a series design. The parallel design enables the HEV to be powered by both the PGU and the motor, either simultaneously or separately. The series design uses the PGU to generate electricity, which recharges the HEV battery pack and produces power via an electric motor. The key element of either design is that the battery pack, as well as the PGU, can be much smaller than those of a typical electric vehicle or a vehicle powered by an IC engine because the IC engine can be operated at its maximum efficiency nearly all the time.

Some HEVs, such as the Toyota Prius, use a parallel power split configuration. As shown in Figure 15.8, a parallel HEV has two propulsion paths: one from the PGU and one from the motor, while a computer controls the output of each. A parallel configuration HEV has a direct mechanical connection between the PGU and the wheels, as in a conventional vehicle, but also has an electric motor that can drive the wheels. For example, a parallel vehicle could use the electric motor for highway cruising and the power from the IC engine for accelerating. The power produced by the PGU also drives the generator, which in turn can charge the battery as needed. The system to transfer electricity from the generator to the battery pack is exactly like that of an EV, with alternating current converted to direct current by the inverter. HEV designs also use a regenerative braking feature that converts energy stored in the inertia of the moving vehicle into electric power during deceleration.

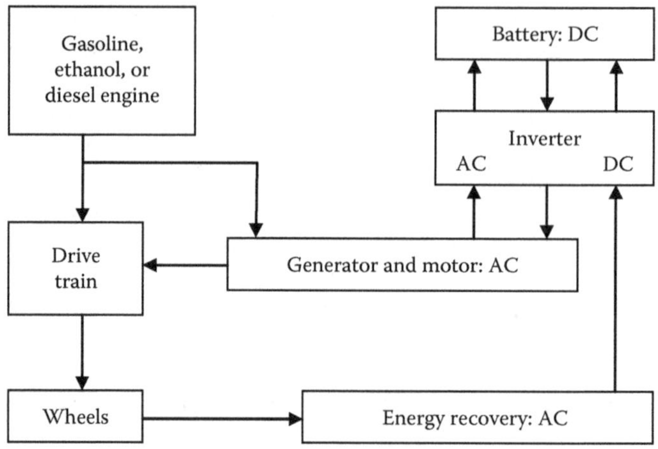

FIGURE 15.8
Schematic of a hybrid electric vehicle.

15.5.4 Plug-In Hybrid Electric Vehicles

Plug-in hybrid electric vehicles (PHEVs) offer the advantage of even lower emissions, without requiring either new technology or new fuel distribution infrastructure. Like hybrids, PHEVs have a battery and an IC engine for power, but PHEVs have a larger battery capacity and a plug-in charger with which the battery can be recharged whenever the car is parked near an electrical outlet. For example, a so-called PHEV40 can travel approximately the first 40 mi (64 km) on grid-supplied electric power when fully charged. When that charge is depleted, the gas or diesel motor kicks in and the vehicle operates like a conventional hybrid. Because most commuter trips are shorter than 40 mi, even in the United States, it is estimated that a PHEV could reduce gasoline usage by 50% or more for many drivers. Moreover, using electric energy is cheaper and cleaner than using gasoline in automobile-type ground transportation.

An important feature of PHEVs is that their batteries can be charged at night when utilities have excess power available (Figure 15.9). Utilities have taken notice of the potential energy charging and storage capabilities of PHEVs because off-peak charging would help utilities to use low-cost base load generation more fully. Furthermore, more advanced vehicle-to-grid (V2G) concepts would allow utilities to buy back energy from the batteries of vehicle owners during peak demand periods and thus make a fleet of PHEVs into a large, distributed storage-generation network. The arrangement would also enable renewable energy storage by charging PHEVs using solar- or wind-generated excess capacity. As mentioned in the energy storage chapter, lithium-ion and lithium–polymer batteries have the potential to store large charges in a lightweight package, which would make the PHEVs even more attractive than the technology that uses nickel–metal hydride (Ni–MH) batteries.

The efficiency of a PHEV depends on the number of miles the vehicle travels on liquid fuel and electricity, respectively, as well as on the efficiency of the prime movers according to the equation:

$$\eta_{\text{PHEV}} = \frac{\text{Energy to wheels}}{\text{Energy from primary source}} = f_1\eta_1\eta_2\eta_3 + f_2\eta_4\eta_5 \tag{15.1}$$

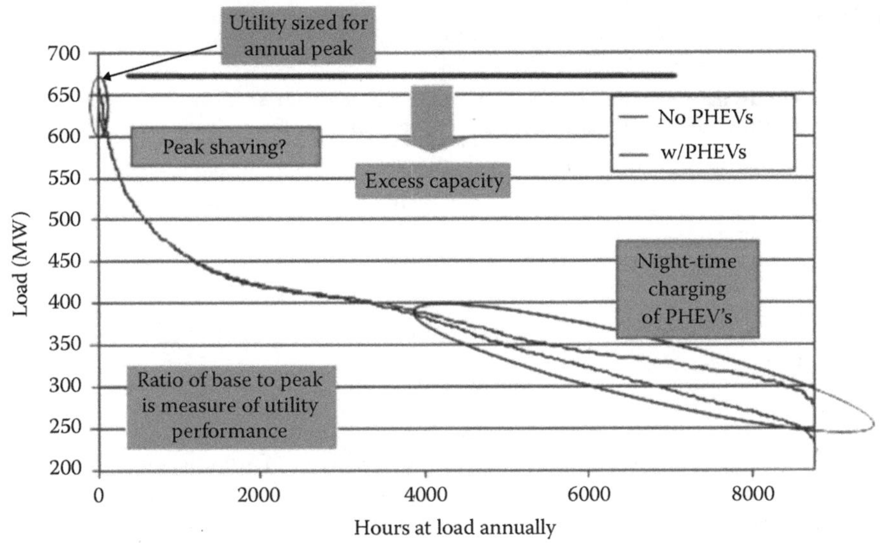

FIGURE 15.9
PHEV impact on utilities' load profile. (Courtesy of National Renewable Energy Laboratory, Golden, CO.)

where
- η_1 is the efficiency from the primary energy source to electricity distribution
- η_2 is the efficiency of charging the EV battery
- η_3 is the efficiency of transmitting energy from the battery to the wheels
- f_1 is the fraction of energy supplied by electricity
- f_2 is the fraction of energy supplied by fuel$=(1-f_1)$
- η_3 is the well-to-tank efficiency of supplying liquid fuel
- η_4 is the tank-to-wheels efficiency.

Using ANL's GREET model with default assumptions and the 2022 U.S. grid electricity mix suggests light-duty passenger PHEV cars could cut the well-to-wheels GHG emissions rate (g/mi) by more than 40% compared to emissions from comparable gasoline internal combustion engine (ICE) cars. If the 2022 California grid electricity mix is used instead, the well-to-wheels GHG emissions rate will be reduced by more than 50% [27].

Energy savings and incremental costs of owning an HEV or PHEV depend on many factors, such as the overall car design, cost of fuel, and the cost and efficiency of the batteries. Comparative vehicle costs are obscured in the retail market by differences in vehicle features, car manufacturers' pricing strategies, and government purchase incentives. Real-world fuel economy improvements depend on design tradeoffs, operating environment, and driving style. Given these confounding factors, modeling and simulation studies are useful in providing for more consistent comparisons, matching vehicle size, performance specifications, and accessory features while comparing different drive trains or fuels. The following example illustrates a simple, first-cut assessment with typical values for vehicle, fuel and electricity costs and efficiencies.

TABLE 15.3

Vehicle Characteristics for Example 15.1

Vehicle Type	Purchase Price ($)	Resale Value after 10 Years ($)	NPV Vehicle Cost ($)	Fuel Economy (mpg Gasoline)	Electricity Use (kWh/100 Miles)
Gasoline ICE	35,000	8,000	30,089	30	
HEV	37,000	8,500	31,808	42	
PHEV	43,000	10,000	$36,966	42	32

Example 15.1

Compare the lifecycle costs of ownership for a standard gasoline sport utility vehicle (SUV) with those of hybrid and plug-in versions of the same model. Assume the vehicle will be owned for 10 years, and will be driven in town for 30 miles per day, 350 days per year, with an additional 1,500 miles driven on longer, out-of-town trips, for a total of 12,000 miles per year. If a PHEV is purchased, it will be charged once per day overnight from an existing garage outlet, covering 80% of in-town driving on electricity. Assume the out-of-town PHEV trips would be powered with gasoline. Use a 5% real discount rate to account for the owner's time value of money preferences. Assume gasoline costs $3.50/gal and electricity costs $0.13/kWh, constant in real terms over the life of the vehicles. Assume other costs of ownership (maintenance and insurance) would be the same for the three vehicles.

Solution:

The net present value (NPV) of the resale estimates is calculated using Equation 2.4, with $i=0.05$ and $t=10$. The NPV vehicle cost shown in Table 15.3 is calculated by subtracting the resale NPV from the purchase price. The NPV of fuel savings with HEV or PHEV can be compared to the difference in NPV of the vehicle cost to see if the upgrades are warranted from a cost standpoint.

Annual fuel costs for the gasoline vehicle would be

$$12{,}000\ {}^{\text{mi}}\!/_{\text{year}} \times {}^{1\text{ gal}}\!/_{30\text{ miles}} \times {}^{\$3.50}\!/_{\text{gal}} = {}^{\$1{,}400}\!/_{\text{year}}$$

Those for the HEV would be

$$12{,}000\ {}^{\text{mi}}\!/_{\text{year}} \times {}^{1\text{ gal}}\!/_{42\text{ miles}} \times {}^{\$3.50}\!/_{\text{gal}} = {}^{\$1{,}000}\!/_{\text{year}}$$

For the PHEV, gasoline costs would be

$$(1{,}500 + 0.2 \times 10{,}500)\ {}^{\text{mi}}\!/_{\text{year}} \times {}^{1\text{ gal}}\!/_{42\text{ miles}} \times {}^{\$3.50}\!/_{\text{gal}} = {}^{\$300}\!/_{\text{year}}$$

Electricity costs for the PHEV would be

$$(0.8 \times 10{,}500)\ {}^{\text{mi}}\!/_{\text{year}} \times {}^{32\text{ kWh}}\!/_{100\text{ miles}} \times {}^{\$.13}\!/_{\text{kWh}} = {}^{\$350}\!/_{\text{year}}$$

Thus the annual fuel cost savings with the HEV would be $400/year, and those for the PHEV would be $750/year. The present value of the fuel savings over 10 years with an HEV is calculated using the formula for the present value of an annuity (Equation 2.6), as follows:

$$NPV_{\text{fuel savings,HEV}} = \$400\left\{\frac{(1+0.05)^{10}-1}{0.05(1+0.05)^{10}}\right\} = \$3,090$$

For the PHEV, the savings would be

$$NPV_{\text{fuel savings,PHEV}} = \$750\left\{\frac{(1+0.05)^{10}-1}{0.05(1+0.05)^{10}}\right\} = \$5,790$$

For the HEV, the NPV of the fuel cost savings is greater than the NPV of the additional vehicle cost compared to the ICEV. The PHEV in this example has a cost of ownership that is about $1,000 higher than that for the gasoline vehicle.

The AFDC offers a web-based Vehicle Cost Calculator tool that allows for similar cost of ownership comparisons based on routinely updated information about actual vehicles on the U.S. market [46]. The AFDC cost calculator is oriented toward consumers, allowing them to select vehicles to compare by make and model and enter information about their anticipated driving patterns, location (for determining fuel costs) and expected charging patterns for electric vehicles. Another helpful tool is the AFLEET utility in ANL's GREET suite of models, which is designed for local governments and other fleet operators [47]. AFLEET calculates total cost of ownership and lifecycle emissions for light-, medium-, and heavy-duty fleet vehicles, including a wide range of alternative fuel options. AFLEET provides detailed default information, or users can input their own assumptions about costs, financing, fuel economy and mileage. In addition to cost calculations, AFLEET provides estimates of well to wheels and vehicle cycle energy use and emissions.

15.5.5 Battery Electric Vehicles

The accelerating and widespread adoption of grid-charged electric passenger vehicles is a major advance in the development of a more sustainable transportation system (Figure 15.10) [48]. Counting both PHEVs and all electric BEVs, the International Energy Agency (IEA) estimates that 14 million passenger car and light truck EVs were sold in

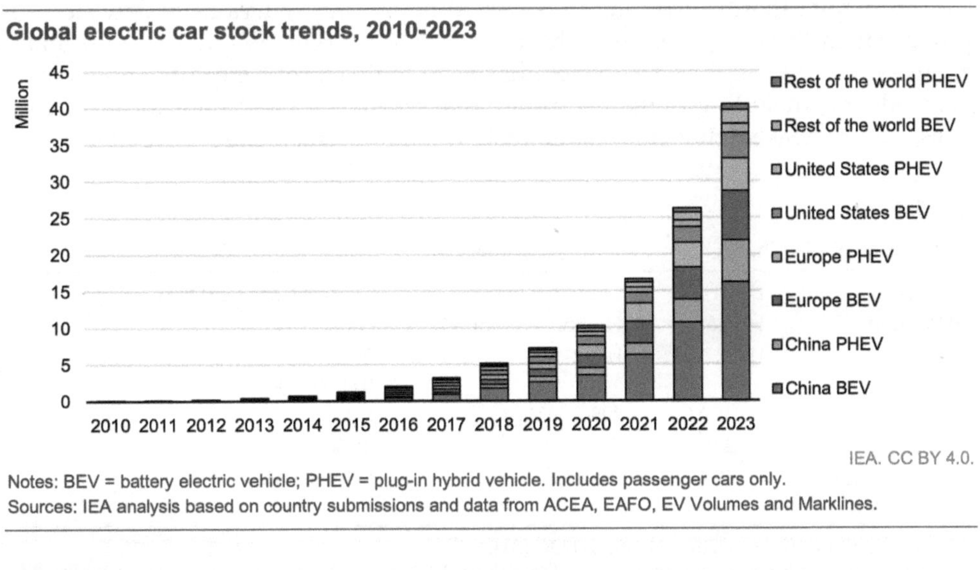

FIGURE 15.10
Increase in PHEV and BEV stocks around the world from 2010 to 2023. (From IEA [48].)

2023, increasing the global stock to over 40 million [48]. EVs accounted for 18% of new cars and passenger trucks sold around the world in 2023, up from just 2% of sales in 2018. Sixty percent of global EV sales in 2023 were in China, where EVs accounted for 45% of new car sales. More than 1.4 million EVs (counting PHEVs and BEVs) were sold in the United States in 2023 [49]. The average driving range for EVs sold around the world has increased along with sales, from about 200 km in 2015 to 350 km in 2023 [48]. Most EV cars and trucks sold in the United States and Europe in 2023 had lithium-nickel-manganese-cobalt (NMC) batteries. In China, most EVs sold in 2023 had lithium iron phosphate batteries (LFP), which are somewhat less expensive [48]. According to IEA, more than half of the lithium and over one-third of the cobalt produced globally in 2023 went into EV batteries [48].

As sales are expanding, cost premiums of EVs over gasoline ICEVs are coming down, thanks to declining battery costs and increasing competition among vehicle manufacturers. In China, BEVs are commonly cheaper than comparable ICEVs. The sales-weighted average cost of a new EV in China in 2023 was less than the average cost of the ICEVs sold there [48]. In the United States, price premiums for EVs have come down over time, but are still substantial if purchase incentives are not available.

Table 15.4 shows characteristics of EV charging ports that are common in the United States [50,51]. Around the world, including in the United States, most EV charging is done at home [48]. In Europe and many parts of Asia, 220 V residential outlets are standard, supporting overnight EV charging. In the United States and Canada, standard residential voltage is 110–120 V, although many homes also have 240 V outlets. Residential service upgrades to add 240 V outlets may be desirable for faster charging, especially for BEVs. Level 2 charging is common for workplaces and in-town public stations. In the United States, DCFC (direct current fast charging) stations are being built out along transportation corridors. The Tesla North American Charging System (NACS) connector for DCFC is being standardized as the SAE International J3400 connector, and is being adopted by additional vehicle manufacturers.

In 2018, small and medium cars comprised 80% of EV sales in the United States, with large cars, SUVs and pickups making up the other 20%. By 2023, the shares had flipped, with the larger vehicles comprising 80% of sales [48]. Smaller EVs are more common in Europe and China than in the United States, although larger EVs have become increasingly popular in those regions too. The trend toward larger EVs aligns with manufacturers' incentives to sell larger vehicles that have higher profit margins, but come with several tradeoffs, including higher upfront costs, greater demand for critical minerals in batteries, and increased charging loads, road wear, and crash risks for pedestrians and

TABLE 15.4

Characteristics of EV Chargers in the United States

Charger type	Level 1	Level 2	DC Fast Charger (DCFC)
Typical location	Home	Home, work, public	Public
Connector	J1772	J1772	CCS (most EVs) CHAdeMO (select models) Tesla/J3400
Voltage	120 V AC	208–240 V AC	480–1,000 V DC
Power output	1–2 kW	7–19 kW	50–500 kW
Typical EV range per hour of charging[a]	5 miles	25 miles	200–400 miles

Source: Adapted from Brown et al. [50] and AFDC [51].

[a] Added EV range per hour of charging depends on the vehicle as well as electrical service characteristics.

other vehicles due to heavier vehicle weight. To counter the trend toward larger vehicles, some European countries have instituted policy measures such as vehicle sales taxes that escalate as vehicle weight increases.

Along with EV cars and trucks, global sales of electric two- and three-wheelers are growing rapidly, especially in China, India, and other Asian countries. According to the International Energy Agency, about 8.5 million new electric two and three-wheelers were registered in 2023 (not counting e-bikes and low-speed scooters) [48]. In lieu of charging stations, these vehicles sometimes take advantage of distributed battery swapping stations where a depleted battery can be rapidly exchanged for a charged one. Life cycle impacts of two- and three-wheeled electric vehicles should be improved as lead acid batteries commonly used in these vehicles are phased out and replaced with lithium-ion batteries.

Table 15.5 shows the total life cycle GHG emissions per mile of operation for passenger cars and SUVs typical of the U.S. fleet in 2022. Results shown in Table 15.5 were extracted from the default GREET model [27], with E10 gasoline and the U.S. average electricity generation mix (38.5% natural gas, 20.6% coal, 18.9% nuclear, and 21.5% hydropower, solar, and wind). PHEVs are assumed to have a 50-mile all-electric range and BEVs to have a 300 mile (480 km) range. HEVs are assumed to have Ni–MH batteries; PHEVs and BEVs to have NMC-111 Li-ion batteries. Vehicle cycle emissions are distributed over the vehicles' expected life of 173,151 miles (278,000 km) in each case.

As shown in Table 15.5, HEVs, PHEVs, and BEVs have progressively lower life cycle GHG emissions, compared to gasoline ICEVs. As expected, the share of total life cycle emissions from vehicle production is greater for BEVs than for the other vehicle types, at 24% for BEV SUVs and 28% or BEV cars. Life cycle GHG emissions for SUVs are 30%–45% greater than those for passenger cars with the same technology.

TABLE 15.5

Life Cycle GHG Emissions for Gasoline ICEV, HEV, PHEV, and BEV Passenger Cars and Sport Utility Vehicles Typical of the U.S. Fleet in 2022

Passenger Cars					
	GHG Emissions (g/mi)				
Vehicle	**Feed Stock**	**Fuel**	**Operations**	**Vehicle Cycle**	**Total**
Gasoline ICE	6	71	330	36	443
HEV	4	51	237	38	330
PHEV	13	114	102	43	272
EV	18	148	0	65	231
Sport Utility Vehicles					
	GHG Emissions (g/mi)				
Vehicle	**Feed Stock**	**Fuel**	**Operations**	**Vehicle Cycle**	**Total**
Gasoline ICE	8	93	431	44	576
HEV	5	63	292	46	406
PHEV	17	148	136	52	353
EV	26	215	0	78	319

Source: Calculated using default inputs with the GREET1_2023_Rev1 and GREET2_2023_Rev1 fuel and vehicle cycle models, downloaded October 31, 2024 from https://greet.anl.gov [27].

15.5.6 Hydrogen Fuel Cell Vehicles

Hydrogen was touted as a potential transportation fuel back in 2001 after former President George W. Bush said in his inaugural address that "a child born today will be driving a pollution-free vehicle … powered by hydrogen." This appeared to be welcome news. However, to analyze whether or not hydrogen is a sustainable technology for transportation, one must take into account all the steps necessary to make the hydrogen from a primary fuel source, get it into the fuel tank, and then power the wheels via a prime mover and a drive train [53]. In other words, one must perform a well-to-wheels analysis, as shown in Figure 15.5. There is a loss at each step, and to obtain the overall efficiency, one must multiply the efficiencies of all the steps.

FCEVs are similar to BEVs except that the electric motor is driven by electricity produced on board in a fuel cell system. As described in Chapter 14, fuel cells are similar to batteries, but with hydrogen supplied at the anode and air at the cathode. Current FCEVs use polymer-electrolyte membrane/proton-exchange membrane (PEM) fuel cells, with operating temperatures of about 80°C. Hydrogen is stored onboard FCEV as a compressed gas, in carbon fiber-reinforced composite tanks with a typical storage pressure of 10,000 psi. A mid-size FCEV sedan with a 300-mile range requires storage of about 5.6 kg of hydrogen [45]. FCEVs typically utilize regenerative braking systems like those in HEVs, but with a larger battery to support initial operation as the fuel cell warms up. As discussed in Chapter 14, steam reforming of methane (natural gas) is a well-established process for producing H_2 for industrial needs, but entails very high GHG emissions unless equipped with effective carbon capture and sequestration. Steam methane reforming facilities can also have significant emissions of nitrogen oxides causing air pollution in nearby communities [52]. In comparison, the production of H_2 using electrolysis is relatively expensive but would have minimal GHG emissions if using electricity from a new renewable source or from nuclear energy. The question then turns to whether the carbon-free electricity to produce hydrogen is "additional" to that needed for other end uses.

The 2023 Argonne study referenced above [29] suggests that a current FCEV using H_2 from methane steam reforming could have cradle-to-grave GHG emissions of about 250 g CO_2-e per mile (Figure 15.6). Note that because these GHG emissions are due in part to methane used in H_2 production, the estimate is subject to the complicating factors mentioned above for fossil CNG: uncertainty about methane leak rates and sensitivity of the CO_2-equivalence calculation to the climate response time horizon. In the future, GHG emissions could be reduced by capturing and sequestering the CO_2 that is coproduced with H_2 in steam methane reforming; producing H_2 from biomass (possibly also with CCS); or by using electrolysis with a low or zero-carbon electricity source.

A widely agreed upon goal for decarbonizing ground transportation is to decarbonize the electric grid and transition to electric vehicles. The two competing technologies for accomplishing this are BEVs and hydrogen fuel cell vehicles powered by green hydrogen. Although some fuel cell cars have been marketed, BEVs have taken the lead in the light-duty vehicle marketplace. An advantage promoted for hydrogen-powered cars is that they can be charged rapidly. However, battery EVs are being produced with ever-improving charging speeds, and 350-kW charging stations are now available (see Table 15.4). Also, homeowners and apartment dwellers having on-site charging stations can do most charging at home. A big difference between BEVs and FCEVs is their efficiency in utilizing the original electricity. As was pointed out in Chapter 1, green hydrogen entails inefficiencies in its production, transport, storage, and conversion back to electricity. Table 15.6 compares typical efficiencies of the various steps needed to convert renewable electricity to electric motor output for the two types of vehicles. The total electricity-to-motor output efficiency for each of the

TABLE 15.6

Typical Efficiencies for the Steps from Renewable Electricity to Electric Motor Output for a Battery Electric Vehicle Versus a Hydrogen Fuel Cell Vehicle

Battery Electric Vehicle		**Fuel Cell Vehicle**	
Renewable electricity	100%	Renewable electricity	100%
Distribution & AC/DC rectification	90%	Electrolysis (alkaline)	70%
Charging battery	95%	Liquefaction or compression	70%–85%
		Transportation/storage	95%
		Fuel cell	60%
		Charging small battery	95%
DC:AC inversion/motor	90%	DC:AC inversion/motor	90%
Total	**77%**	**Total**	**24%–29%**

two vehicle types is found by multiplying the efficiencies for each step together. As shown in the table, a fuel cell vehicle operating with green hydrogen produced from electrolysis requires approximately three times as much source electricity as a BEV.

15.6 Sustainable Transportation Challenges: Long-Haul Trucking, Shipping, and Aviation

Because of the climate crisis, there is an urgent need to find more sustainable trajectories for long-haul trucking, international freight shipping, and air travel and transport. In 2018, international aviation and shipping accounted for 7% and 8%, respectively, of global GHG emissions from transportation [4]. Demand for these services is currently rising. Medium- and heavy-duty trucks accounted for 23% of U.S. greenhouse gas emissions from transportation in 2022, with domestic aviation contributing a 6% share (Figure 15.1 [3]). These three transportation subsectors are challenging to transition due to the long useful life of trucks, ships, and airplanes, which extend to 30 years or more, and the high costs of new equipment. Addressing GHG emissions and other environmental impacts from international shipping and aviation is further complicated by the international character of their operations and their fuel and equipment supply chains.

While optimal roadmaps remain open to debate, especially as to the details, the path forward for these sectors seems likely to include some combination of the following:

- Reducing energy demand from all three sectors through improved vehicle and operational efficiency and/or changes in consumption patterns.
- Electrification of ground transportation to the extent feasible using batteries, starting with medium duty trucks and short-haul or fixed route services, with electricity provided from renewables.
- Electrification of long-haul trucks using advanced batteries and extreme fast charging or hydrogen fuel cells.
- Prioritized use of sustainable liquid fuels derived from green hydrogen or waste biomass for long-distance aviation and marine shipping.

These strategies are briefly outlined below, with references provided for those who want to explore them further.

15.6.1 Electrification of Heavy-Duty Vehicles

Relative to their numbers, medium- and heavy-duty vehicles (M/HDV) account for a disproportionate share of energy use, GHG and local air pollution emissions in the United States and globally. In the United States, M/HDV (trucks, vans, and buses with gross vehicle weight over 10,000 lbs) account for only about 5% of the vehicles on the road, but they produce 23% of GHG emissions (Figure 15.1) and more than 40% of smog-forming nitrogen oxides from the transportation sector [54]. Fuel economy and tailpipe emissions standards for new gasoline and diesel M/HDV have reduced local air pollution emissions on a per mile basis, although progress has been slower for these larger vehicles than with light-duty passenger cars and trucks. As discussed above, disparities in residential proximity to freeways, ports, and warehouse distribution centers that induce truck traffic continue to expose overburdened communities to air pollution from diesel-fueled trucks. Future growth in M/HDV travel in the United States is expected to outpace that for light-duty vehicles, raising the importance of addressing this portion of the on-road vehicle sector.

M/HDV cover a wide range of weights and duty cycles. The U.S. DOT classifies these vehicles by weight, ranging from 10,000 to 14,000 lbs (4,540–6,350 kg) for Class 3 trucks to greater than 33,000 lbs (15,000 kg) for Class 8. Annual mileage for M/HDV can range from 10,000 to more than 200,000 mi/year (16,000–322,000 km/year) [55]. Duty cycles, whether short-haul with fixed delivery routes of less than 100 miles (160 km) per work day or interstate long-haul trucking, play an important role in determining near-term feasibility of any fuel or powertrain alternatives to conventional fuels and engines. Among M/HDV segments, Class 7 and 8 weight class vehicles, which are most challenging to power, currently account for 43% of M/HD vehicles, 63% of their vehicle miles traveled, and 74% of their emissions [55].

Pathways for advancing and accelerating reductions in GHG emissions from M/HDV vehicles include four types of strategies:

- Improved vehicle energy efficiency through advanced aerodynamics, transmission and engine designs, hybrid engines and reduced tire rolling resistance.
- Improved supply chain management and logistics for freight transport and delivery operations.
- Use of lower carbon fuels including biodiesel blends and renewable diesel, as discussed in Section 15.5.1.
- Electrification using batteries or fuel cells.

In 2021, California led the United States in adopting its Advanced Clean Truck rule, requiring an increasing share of M/HD trucks sold in the state to be zero emissions vehicles, starting in 2024 [56]. In 2022–2023, the U.S. Environmental Protection Agency set new nitrogen oxides and GHG emissions standards for M/HD vehicles to take effect starting with the 2027 model year [57]. Under the new standards for large freight trucks, GHG emissions limits would be reduced by 25%–60% compared to the prior standards, with the percentage reduction depending on vehicle class and duty cycle. The technology to be used would be left up to vehicle manufacturers, but EPA concluded the standards could reasonably be met with a mix of advanced ICEV, HEV, grid-powered battery, and FCEVs.

Recent studies provide support for the possibility that either grid-charged battery or hydrogen fuel cell technologies could provide economical zero tailpipe emissions M/HDEV options, assuming the corresponding charging or refueling infrastructure is in place. For example, Ledna et al. [55] assessed the dates when battery or fuel cell M/HDV might reach parity in the total cost of driving with diesel or hybrid ICE technologies. They defined the total cost of driving to include upfront vehicle costs, fuel and maintenance costs over the first 3–5 years of operation (reflecting typical payback expectations for higher upfront costs, which differ by vehicle class), plus the wage cost of charging/refueling time ($75/hour). Ledna et al. analyzed a high technological progress scenario developed by the U.S. Department of Energy, with key assumptions as summarized in Table 15.7. The study found that if the technological progress goals were met and adequate charging or fueling infrastructure was deployed, zero tailpipe emissions options could reach economic parity with conventional vehicles by 2035, across all weight classes and typical duty cycles. Their analysis suggested that with current tax incentives, BEV delivery vans may already have reached cost parity. Through 2035, FCEV were modestly more expensive than BEV for low-mileage M/HDV across all weight classes, but were less expensive for Class 7 and 8 vehicles with long trip distances.

Heavy-duty fuel cell trucks are being pursued because of the potential for pressurized hydrogen storage to provide greater range and faster charging capability. However, hydrogen infrastructure would have to be developed and as with light-duty vehicles, fuel cell trucks powered by green hydrogen produced from electrolysis would require approximately three times as much grid electricity as battery electric trucks. Some recent European studies have concluded that even for heavy-duty trucks, battery electric technology offers the most promising path [58–60]. Developments in ultrafast charging may make BEVs more competitive with fuel cell vehicles in terms of range and charging speed. For example, a megawatt charging system (MCS) that can deliver up to 3.75 MW of power is being developed by industry and laboratory partners to allow fast charging of heavy vehicles [61]. Industry and academic partners are also developing dynamic wireless power transfer systems as a potential solution for long-haul trucks. Using inductive charging via electronics embedded in the roadway could smooth charging loads, allow trucks to use smaller batteries, and eliminate the need for recharging stops, but would require significant infrastructure investment [62].

TABLE 15.7

Battery and Hydrogen Fuel Cell Technical Advancement Assumptions for Total Cost of Driving Analysis

Parameter	2025	2030	2035
Battery cost ($/kWh)	217	139	80
Battery energy density (Wh/kg)	196	223	250
Fuel cell cost ($/kW)	175	129	80
Fuel cell specific power (W/kg)	973	1,009	1,040
Hydrogen cost ($/kg)	9	6	4
EV charging infrastructure	350 kW (classes 3 and 4), 500 kW (classes 4–8)		
H_2 refueling infrastructure	Available – no barrier to adoption		

Source: Ledna et al. [55].

15.6.2 Alternative Fuels and Propulsion Systems for Transoceanic Shipping

The International Maritime Organization (IMO) is a specialized United Nations agency founded in 1948 to facilitate international shipping, including safety, security, and environmental sustainability. In July 2023, the IMO adopted a new strategy to reduce GHG emissions from ships, targeting net-zero emissions by around 2050 despite increased shipping activity [63]. According to IMO, potential steps toward achieving this goal include:

- Energy efficiency improvements through measures such as hull friction reduction, voyage optimization with speed reductions and just-in-time arrivals, and advanced waste heat recovery;
- Hybrid propulsion systems including battery electric motors and propulsive assistance technologies such as Flettner rotors, kites, and sails;
- Electrification of shore power operations;
- Alternative fuels and engine technologies.

In addition to the approaches for reducing emissions that are mentioned by IMO, growth in shipping demand could be moderated by changes in production and consumption patterns. For example, transitioning away from fossil fuels would reduce the need for transoceanic shipping of these fuels, which currently account for about 40% of the weight of maritime cargo [64].

In January 2024, the European Union's GHG Emissions Trading System (ETS) was extended to cover emissions of large ships entering European ports, providing an incentive for ships to reduce emissions to avoid having to pay for emissions allowances [65]. Until now, onboard fuel demand for maritime shipping has been met predominantly by heavy fuel oil, (HFO, commonly known as bunker fuel) or marine diesel oil, with a smaller contribution from liquefied natural gas (LNG). HFO is refinery residual from crude oil processing that has been treated to remove contaminants. Marine diesel oil is a somewhat lighter, less viscous distillate product of oil refining. Shipping companies are beginning to order and deploy new vessels that would use or be ready to use alternatives to these traditional fuels that would be more consistent with decarbonization aims.

Alternative fuels with the potential to reduce life cycle GHG emissions from ship engines include advanced biodiesel and other biofuels, "green" hydrogen produced by electrolysis with renewable electricity or "blue" hydrogen produced by methane reforming with carbon capture and sequestration, and ammonia or methanol produced from blue or green hydrogen and using renewable electricity. Hydrogen would potentially be used in fuel cells, whereas ammonia and methanol could be used in fuel cells or combusted in modified marine engines. Although bio- or e-fuels could reduce life cycle greenhouse gas emissions, nitrogen oxide emissions will remain an issue with internal combustion engines, and could be exacerbated with combustion of ammonia. Foretich et al. [66] discuss the pros and cons of several of the alternative fuels under consideration for maritime applications.

Ammonia and methanol have been widely discussed for maritime applications. Methanol (CH_3OH) has the advantage that it is a liquid at standard temperature and pressure, simplifying storage and handling requirements compared to LNG, hydrogen or ammonia. In contrast, ammonia would need to be liquefied via refrigeration for use in marine applications. Both methanol and liquefied ammonia have lower volumetric energy density than HFO and marine diesel, and so would require more space onboard ships to be dedicated to fuel storage. Safety, leak, and spillage issues with methanol and especially ammonia require close attention.

Methanol is conventionally made by reforming natural gas with steam to produce synthesis gas (CO and hydrogen) followed by the catalytic reaction of synthesis gas with additional hydrogen. Methanol produced this way is estimated to have higher well-to-wake GHG emissions than bunker fuel [67]. To reduce GHG emissions, "blue methanol" can be produced from natural gas but with carbon capture, utilization and sequestration (CCUS). Even greater reductions in life cycle GHG emissions could be achieved with "bio-methanol," produced using biomass or biogas feedstocks instead of coal or natural gas, or "e-methanol," produced with green hydrogen and CO_2 captured from industrial process emissions although the electro-fuels (e-fuels) are expected to have a very high cost.

Most ammonia production today starts with steam reforming of natural gas to produce hydrogen and CO_2. That step is followed by the separation of hydrogen from CO_2, and then the reaction of hydrogen with nitrogen over a catalyst at high temperature and pressure to produce NH_3. As with methanol, the well-to-wake GHG emissions intensity for conventionally produced ammonia if used as a ship fuel is estimated to be greater than that of bunker fuel [67]. However, analogous to "blue hydrogen," equipping ammonia plants with CCUS equipment can reduce its emissions intensity, if upstream methane leakage is minimized [68]. GHG emissions from ammonia production could be minimized by producing it with green hydrogen and renewable electricity [68].

Battery electric systems may also have some potential to reduce emissions from cargo ships, although to date they have mainly been used for ferries and short shipping routes. Kersey et al. [69] conclude from their technoeconomic analysis that if battery prices are below \$100/kWh, LFP ($LiFePO_4$) battery systems charged at port could be economical for intraregional and inland waterway routes shorter than 1,500 km. Future reductions in battery costs could extend this distance. If powered by renewable electricity, these ships could reduce greenhouse gases and local air pollution along their routes.

15.6.3 Sustainable Aviation

The International Civil Aviation Organization (ICAO) is a United Nations agency that supports more than 190 participating countries in developing standards, policies and recommendations for sustainable aviation, including safety, noise and air pollution, and greenhouse gas emissions. In 2022, ICAO adopted an "aspirational goal" of achieving net-zero CO_2 emissions from aviation by 2050 [70]. On the private sector side, the International Air Transport Association (IATA), a trade organization that represents most of the world's airlines, has adopted a similar aim [71]. Both organizations have suggested that the net-zero roadmap for aviation includes the following elements:

- Reducing energy use through improved energy efficiency of aircraft, aircraft operations, and ground-support equipment and
- Replacing fossil jet fuel with low- or zero-carbon alternatives, relying primarily on sustainable aviation fuel (SAF) from biomass feedstocks and power-to-liquids (PtL, also known as e-fuels), with a small contribution from liquid hydrogen at mid-century [72].

IATA's current roadmap also includes purchased credits for carbon capture and sequestration or direct air capture to offset a portion of remaining emissions. Prospects for electrification of aircraft propulsion are seen as further off, due to the relatively low energy density of batteries compared to liquid fuels.

SAF is an aviation fuel that is produced from biomass or could potentially be synthesized from hydrogen and captured CO_2 to match the properties of conventional jet fuel so it can be used as a drop-in replacement. In its policy for certifying sustainable aviation fuels for carbon offset credit, ICAO includes criteria for:

- Reduced life cycle greenhouse gas emissions compared to baseline jet fuel
- Avoiding conversion of forests and other lands with high carbon sequestration levels for feedstock production, and
- Avoiding feedstocks obtained from areas with high biodiversity, ecosystem services or conservation value [73].

PtL SAF pathways entail producing hydrogen via electrolysis using renewable electricity and then using CO_2 captured from combustion or industrial processes or from direct air capture to produce hydrocarbon liquids. PtL SAF could be produced with either syngas or methanol as intermediates. With PtL SAF, land use impacts could be reduced compared to those associated with biofuel SAF, but energy costs are very high for both H_2 production and carbon capture. Boilley et al. [74] conducted a simulation study to estimate the energy efficiency of PtL jet fuel production using a solid oxide electrolyzer for hydrogen production and Fischer-Tropsch (FT) fuel synthesis, with distillation to refine the FT product to jet fuel standards. Carbon dioxide was assumed to be captured from a cement kiln using monoethanolamine (MEA) absorption. Boilley et al. [74] estimate the overall energy efficiency of the power-to-jet fuel process at 39%. Although the electricity for PtL could be produced from renewables, doing so would compete with other uses of renewable energy for direct electricity consumption.

In biofuel pathways, SAF is made from organic feedstocks such as waste oils, sugars, starch, cellulose, and algae. Currently, biofuel SAF is blended with fossil Jet A fuel at SAF levels ranging from 10% to 50%. There are multiple approaches to producing biofuel SAF, including:

- Hydroprocessed esters and fatty acids (HEFA) from oil crops (e.g., soy or canola), waste fats (e.g., tallow or cooking oil), or algae.
- Alcohol To Jet (ATJ) synthetic paraffinic kerosene from sugar or starch crops or waste cellulosic biomass.

The HEFA process catalytically refines triglycerides into a hydrocarbon product that matches the properties of jet fuel by adding hydrogen and removing oxygen, isomerizing the hydrocarbons to desired chain lengths, and separating the product into desired liquid fuel fractions. The requirement for hydrogen as an input to HEFA production contributes significantly to its energy requirements and costs. Current ATJ processes use ethanol (e.g., from corn starch or sugar cane) or isobutanol as the alcohol feedstock. Converting alcohol to jet fuel requires removing oxygen from the alcohol feedstock, oligomerization to produce a chain of olefins, and adding hydrogen to saturate the olefinic bonds. Both the alcohol feedstocks and the required hydrogen contribute significantly to the costs of ATJ SAF. Land use impacts are an important concern with significantly scaling up the production of biofuel SAF.

Table 15.8 shows modeled estimates of fuel costs and fuel cycle (well-to-wake) greenhouse gas emissions for ATJ and HEFA biofuels, compared to conventional jet fuel in the market today [75]. The estimates are for U.S. production and air travel based on 2019 U.S.

TABLE 15.8
Fuel Costs and Fuel Cycle GHG Emissions from Biofuel SAF Compared to Conventional Jet Fuel

	Conventional Jet Fuel	ATJ Biofuel from Corn	ATJ Biofuel from Corn	HEFA Biofuel from Used Cooking Oil	HEFA Biofuel from Used Cooking Oil
Scenario	**Current Market**	**Future Modeled, High Volume, Default Grid**	**Future Modeled, High Volume, High RE Grid**	**Future Modeled, High Volume, Default Grid**	**Future Modeled, High Volume, High RE Grid**
Fuel cost (2020$/1,000 passenger miles)	31.5	91.9	90.8	107	107
GHG emissions (tonnes CO_2-e/1,000 passenger miles)	0.140	0.112	0.0944	0.0285	0.0246

Source: NREL [75].

commercial airline data. Modeled estimates for the two SAF pathways assume high volume production with mature, commercial technology. As shown, SAF is expected to be more expensive than jet fuel, but substantial reductions in life cycle GHG emissions could be achieved with HEFA from cooking oil or with ATJ biofuel if it is produced using renewable electricity. As of this writing, the HEFA pathway is the only one commercialized, and SAF displaces less than 1% of worldwide jet fuel consumption.

Problems

15.1 Assume a hydrogen car uses 2.5 kWh/km. Compare this performance with that of a U.S. car that is able to get 22 mpg of gasoline.

15.2 From the information supplied in the following text, estimate the percentage of the total land area in the lower 48 U.S. states that would have to be used to replace the gasoline used for transportation with ethanol obtained from switchgrass. Do the same for the arable land in the lower 48 U.S. states.
Data:

Estimated ethanol productivity ~1,000 gallons ethanol per hectare per year (based on switchgrass as the crop) [76].

1.5 gallons ethanol is required to obtain the same energy as 1 gallon of gasoline, based on their respective lower heating values (LHVs).

U.S. gasoline consumption in 2023 was ~138×10^9 gal/year [77].

U.S. total land area (lower 48) = 1.92×10^9 acres [78].

U.S. arable land area (suitable for crops) = 0.45×10^9 acres [78].

1 hectare = 2.47 acres.

15.3 Estimate the net amount of CO_2 produced in g/MJ from using corn-based ethanol as a biofuel, and compare it to the amount of CO_2 produced in g/MJ from burning gasoline.

a. The combustion of ethanol goes according to the following equation:

$$C_2H_5OH + \_\_O_2 \rightarrow \_\_CO_2 + \_\_H_2O \qquad LHV = 21\ MJ/L$$

where the blanks indicate stoichiometric coefficients to be determined by balancing the reaction. LHV = lower heating value, the symbol "L" designates liter. The density of ethanol at 16°C is 780 g/L. Based on the combustion stoichiometry, calculate the amount of CO_2 released to the atmosphere per MJ of ethanol burned.

b. Assume the growth of corn and conversion of the corn into ethanol produce 87 g CO_2/(MJ of ethanol) from the fossil fuels required. In addition, 35 g of CO_2/(MJ of ethanol) is produced during the fermentation of corn sugar into ethanol. However, each atom of carbon in the corn sugar, and thus each atom of carbon in the CO_2 released by fermentation of the sugar, originally came from an atom of carbon dioxide in the atmosphere. Therefore, the CO_2 from the fermentation is "carbon neutral," that is, the amount of CO_2 released by the fermentation is balanced by CO_2 removed from the atmosphere in making the corn sugar. (*Note:* The same is true for the fermentation of any biomass.) Likewise, each atom of C in the CO_2 produced by combustion of ethanol is also carbon neutral. Using this information calculate the net amount of CO_2 released per MJ of ethanol made from corn. The net amount is the amount by which the production and combustion of ethanol increase the amount of CO_2 in the atmosphere.

c. Gasoline is a mixture of many compounds of carbon and hydrogen, but may be represented by the formula C_8H_{16}, with a LHV of 32.8 MJ/L. Calculate the CO_2 emissions from combusting gasoline in g CO_2/MJ with gasoline as fuel.

d. Estimate the total CO_2 emissions in g CO_2/MJ (LHV) from "well-to-wheels" by including "well-to-tank" emissions with gasoline as a fuel. Assume the well-to-tank energy efficiency for gasoline is approximately 90% (100×(MJ energy to tank)/(MJ energy of petroleum "in the well")). Assume that the ratio of CO_2 emissions from tank-to-wheels is also 90% of the total CO_2 emissions.

15.4 A truck engine consumes diesel at a rate of 30 L/hour and delivers 65 kW of power to the wheels. If the fuel has a heating value of 43.5 MJ/kg and a density of 800 kg/m^3, determine

a. The thermal efficiency of the engine

b. The waste heat rejected by the engine.

15.5 An automobile storage battery with an open circuit voltage of 12.8 V is rated at 260 Ah. If the internal resistance of the battery is 0.2 Ω, estimate the maximum duration of current flow and its value through an external resistance of 1.8 Ω.

15.6 Estimate the mass of hydrogen required to fuel a hydrogen-powered vehicle for 1 year, assuming the vehicle operates 2 hours/day with output power requirements of 170 hp, and that the fuel cell delivers power at 60% efficiency.

15.7 Explain the difference in operation between HEVs with parallel and series designs.

15.8 Estimate the annual CO_2 emissions and fuel cost implications of shifting from a full-size car with a gasoline ICE to a similarly sized BEV that is charged with electricity from an average natural gas power plant. Assume the vehicle will be driven 12,000 mi/year. Assume the ICEV has fuel economy of 32 mi/gal, and that gasoline has a density of 0.74 kg/L, energy content of 45 MJ/kg and direct CO_2 emissions of 74 g CO_2/MJ. Assume the BEV uses 0.17 kWh/mi, and that the natural gas power plant emits 0.42 kg CO_2/kWh. Gasoline costs $2/gal and electricity costs $.09/kWh. How do the fuel cost savings of the BEV compare to the external cost benefits of reduced CO_2 emissions, using the U.S. government's (2023) estimate of the social cost of carbon? For the comparison, use an estimate of about $200 per metric ton of CO_2 based on the values for a 2% discount rate shown in Table 2.10. What sources of greenhouse gas emissions that are relevant to BEV and ICEV life cycles have been left out of this comparison?

References

1. Schiller, P.L., Bruun, E.C., and Kenworthy, J.R., 2010. *An Introduction to Sustainable Transportation: Policy, Planning and Implementation*, Earthscan, London and Washington, DC.
2. Energy Information Administration, 2023. Energy use for transportation, https://www.eia.gov/energyexplained/use-of-energy/transportation.php.
3. EPA, 2024. Fast facts on transporation greenhouse gas emissions, https://www.epa.gov/greenvehicles/fast-facts-transportation-greenhouse-gas-emissions.
4. Lamb, W.F., Wiedmann, T., Pongratz, J., Andrew, R., Crippa, M., Olivier, J., Wiedenhofer, D., Mattioli, G., Al Khourdajie, A., House, J., Pachauri, S., Figueroa, M., Saheb, Y., Slade, R., Hubacek, K., Sun, L., Ribeiro, S., Khennas, S., de la Rue du Can, S., Chapungu, L., Davis, S., Bashmakov, I., Dai, H., Dhakal, S., Tan, X., Geng, Y., Gu, B., Minx, J. 2021. A review of trends and drivers of greenhouse gas emissions by sector from 1990 to 2018, *Environmental Research Letters* 16, 073005, doi:10.1088/1748-9326/abee4e.
5. European Commission, Joint Research Centre, Crippa, M., Guizzardi, D., Pagani, F., Banja, M., Muntean, M., Schaaf, E.,Monforti-Ferrario, F., Becker, W.E., Quadrelli, R., Risquez Martin, A., Taghavi-Moharamli, P., Köykkä, J., Grassi, G., Rossi, S., Melo, J., Oom, D., Branco, A., San-Miguel, J., Manca, G., Pisoni, E., Vignati, E., and Pekar, F., 2024. GHG emissions of all world countries, JRC138862, Publications Office of the European Union, Luxembourg, doi:10.2760/4002897.
6. Davis, S.C., and Boundy, R.G., 2022. *Transportation Energy Data Book*, 40th ed., ORNL-TM-2022/2376, Oak Ridge National Laboratory, Oak Ridge, TN.
7. CBO, 2022. *Emissions of Carbon Dioxide in the Transportation Sector*, Congressional Budget Office, Washington, DC, www.cbo.gov/publication/58566.
8. Newman, P., and Kenworthy, J., 2015. *The End of Automobile Dependence: How Cities are Moving Beyond Car-Based Planning*, Island Press, Washington, DC.
9. Fuller, C.H., and Brugge, D., 2020. Environmental justice: Disproportionate impacts of transportation on vulnerable communities, in: Khreis, H., Nieuwenhuijsen, M., Zietsman, J. and Ramani, T. eds., *Traffic-Related Air Pollution*, Elsevier, Oxford, United Kingdom, pp. 495–510.
10. U.S. FHWA, 2019. Environmental justice, https://www.fhwa.dot.gov/environment/environmental_justice/equity/.
11. U.S. EPA, 2024. Environmental justice, https://www.epa.gov/environmentaljustice.
12. Fullilove, M.T., 2007. Eminent domain and African Americans: What is the price of the commons? Institute for Justice.

13. Bullard, R.D., 2003. Addressing urban transportation equity in the United States, *Fordham Urban Law Journal* 31(5), 1183–1210.
14. Rowangould, G.M., 2013. A census of the US near-roadway population: Public health and environmental justice implications, *Transportation Research Part D* 25, 59–67.
15. Tessum, C.W., Paolella, D.A., Chambliss, S.E., Apte, J.S., Hill, J.D., and Marshall, J.D., 2021. PM2.5 polluters disproportionately and systemically affect people of color in the United States, *Science Advances* 7, eabf4491.
16. U.S. EPA, 2024. EJScreen, Environmental justice screeing and mapping tool, https://www.epa.gov/ejscreen.
17. U.S. DOT, 2023. Key Performance Indicators (KPIs) for equity, https://www.transportation.gov/priorities/equity/equity-strategic-goal.
18. PolicyLink and the USC Equity Research Institute, 2024. National Equity Atlas, nationalequityatlas.org.
19. U.S. DOT, 2023. Centering equity at the U.S. Department of Transportation, https://www.transportation.gov/priorities/equity/equity-strategic-goal.
20. Karner, A., Rowangould, D., London, J., and Manaugh, K., 2020. From transportation equity to transportation justice: Within, through, and beyond the state, *Journal of Planning Literature* 35(4), 440–459.
21. Choma, E.F., Evans, J.S., Hammitt, J.K., Gomez-Ibanez, J.A., and Spengler, J.D., 2020. Assessing the health impacts of electric vehicles through air pollution in the United States. *Environment International* 144, 106015, doi:10.1016/j.envint.2020.106015.
22. Gai, Y., Minet, L., Posen, I.D., Smargiassi, A., Tetreault, L.-F., and Hatzopoulou, M., 2020. Health and climate benefits of electric vehicle deployment in the Greater Toronto and Hamilton area, *Environmental Pollution* 265, 114983, doi:10.1016/j.envpol.2020.114983.
23. Hardman, S., Fleming, K.L., Khare, E., and Ramadan, M., 2021. A perspective on equity in the transition to electric vehicle, *MIT Science and Policy Review* 2, 46–54.
24. Hsu, C.-W., and Fingerman, K., 2021. Public electric vehicle charger access disparities across race and income in California, *Transport Policy* 100, 59–67, doi:10.1016/j.tranpol.2020.10.003.
25. Kreith, F., and West, R.E., 2003. Gauging efficiency: Well-to-wheel, *Mechanical Engineering Power*, pp. 20–23, June.
26. Kreith, F., West, R.E., and Isler, B.E., 2002. Efficiency of advanced ground transportation technologies, *Journal of Energy Resources Technology* 24, 173–179.
27. Argonne National Laboratory, 2024. R&D GREET model platforms, energy systems and infrastructure analysis division, Argonne National Laboratory, U.S. Department of Energy, https://greet.anl.gov.
28. Shukla, S., Sykora, T., Kelly, J.C., and Cai, H., 2023. GREET battery module: Beta version, energy systems and infrastructure analysis division, Argonne National Laboratory, U.S. Department of Energy, https://greet.anl.gov/publications.
29. Kelly, J.C., Elgowainy, A., Isaac, R., Ward, J., Islam, E., Rousseau, A., Sutherland, I., Wallington, T.J., Alexander, M., Muratori, M., Franklin, M., Adams, J., and Rustagi, N., 2023. Cradle-to-grave lifecycle analysis of U.S. light duty vehicle-fuel pathways: a greenhouse gas emissions and economic assessment of current (2020) and future (2030–2035) technologies, ANL-22/27 Rev.1, Energy Systems Division, Argonne National Laboratory, U.S. Department of Energy.
30. EPA, 2024. Inventory of U.S. Greenhouse gas emissions and sinks: 1990–2022U.S. Environmental Protection Agency, EPA 430R-24004, https://www.epa.gov/ghgemissions/inventory-us-greenhouse-gas-emissions-and-sinks-1990-2022.
31. IEA, 2024. Renewables 2024: Analysis and forecasts to 2030, International Energy Agency, https://www.iea.org/reports/renewables-2024.
32. AFDC, 2024. Ethanol fuel basics, Alternative Fuels Data Center, U.S. Department of Energy, https://afdc.energy.gov/fuels/ethanol-fuel-basics.
33. Farrell, A.E., Plevin, R.J., Turner, B.T., Jones, A.D., O'Hare, M., and Kammen, D.M., 2006. Ethanol can contribute to energy and environmental goals, *Science* 311, 506–508, doi:10.1126/science.1121416.

34. Xu, H., Lee, U., and Wang, M., 2022. Life-cycle greenhouse gas emissions reduction potential for corn ethanol refining in the U.S.A., *Biofuels, Bioproducts and Biorefining* 16, 671–681, doi:10.1002/bbb.2348.
35. EIA, 2024. Biofuels explained, Energy Information Administration, U.S. Department of Energy, https://www.eia.gov/energyexplained/biofuels/biodiesel-rd-other-basics.php.
36. EIA, 2024. Monthly biofuels capacity and feedstocks update, September, Energy Information Administration, U.S. Department of Energy, https://www.eia.gov/biofuels/update/.
37. Xu, H., Ou, L., Li, Y., Hawkins, T.R., and Wang, M., 2022. Life cycle greenhouse gas emissions of biodiesel and renewable diesel production in the United States, *Environmental Science and Technology* 56, 7512–7521, doi:10.1021/acs.est.2c00289.
38. AFDC, 2024. Natural gas vehicles, Alternative Fuels Data Center, U.S. Department of Energy, https://afdc.energy.gov/vehicles/natural-gas.
39. APTA, 2024. 2023 Public transportation fact book, American Public Transportation Association, https://www.apta.com/wp-content/uploads/APTA-2023-Public-Transportation-Fact-Book.pdf.
40. AFDC, 2024. Natural gas fueling stations, Alternative Fuels Data Center, U.S. Department of Energy, https://afdc.energy.gov/fuels/natural-gas-stations.
41. Alvarez, R.A., Pacala, S.W., Winebrake, J.J., Chameides, W.L., and Hamburg, S.P., 2012. Greater focus needed on methane leakage from natural gas infrastructure, *PNAS* 109, 6435–6440, doi:10.1073/pnas.1202407109.
42. Brandt, A.R., Heath, G.A., Kort, E.A., O'Sullivan, F., Petron, G., Jordaan, S.M., Tans, P., Wilcox, J., and Gopstein, A.M., 2014. Methane leaks from North American natural gas systems, *Science* 343, 743–745.
43. EPA, 2024. Renewable natural gas, https://www.epa.gov/lmop/renewable-natural-gas.
44. Gage, D., 2024. Renewable natural gas breaking motor fuel usage records, press release, April 24, 2024, The Transport Project, https://transportproject.org.
45. National Research Council, 2013. *Transitions to Alternative Vehicles and Fuels*, National Academies, Washington, DC.
46. AFDC, 2024. Vehicle cost calculator, Alternative Fuels Data Center, U.S. Department of Energy, https://afdc.energy.gov/calc/.
47. Burnham, A., 2023. User guide for AFLEET tool 2023, Energy Systems and Infrastructure Analysis Division, Argonne National Laboratory, U.S. Department of Energy, https://greet.anl.gov/index.php?content=afleet.
48. IEA, 2024. Global EV outlook 2024, https://www.iea.org/reports/global-ev-outlook-2024.
49. ANL, 2024. Light duty electric drive vehicles monthly sales updates, Energy Systems and Infrastructure Analysis, https://www.anl.gov/esia/light-duty-electric-drive-vehicles-monthly-sales-updates.
50. Brown, A., Cappelluci, J., Heinrich, A., Gaus, M., and Cost, E., 2024. Electric vehicle charging infrastructure trends from the alternative fueling station locator: First quarter 2024, NREL, https://afdc.energy.gov/files/u/publication/electric_vehicle_charging_infrastructure_trends_first_quarter_2024.pdf.
51. AFDC, 2024. Electric vehicle charging stations, https://afdc.energy.gov/fuels/electricity-stations.
52. Sun, P., Young, B., Elgowainy, A., Lu, Z., Wang, M., Morelli, B., and Hawkins, T., 2019. Criteria air pollutants and greenhouse gas emissions from hydrogen production in U.S. steam methane reforming facilities, *Environmental Science and Tecchnology* 53, 7103–7113, doi:10.1021/acs.est.8b06197.
53. Kreith, F., and West, R.E., 2004. Fallacies of a hydrogen economy: A critical analysis of hydrogen production and utilization, *Journal of Energy Resources Technology* 126, 249–257.
54. EPA, 2024. 2020 National Emissions Inventory Data, U.S. Environmental Protection Agency, https://www.epa.gov/air-emissions-inventories/2020-national-emissions-inventory-nei-data.
55. Ledna, C., Muratori, M., Yip, A., Jadun, P., Hoehne, C., and Podkaminer, K., 2024. Assessing total cost of driving competitiveness of zero-emission trucks, *iScience* 27, 109385, doi:10.1016/j.isci.2024.109385.

56. CARB, 2024. Advanced clean trucks, California Air Resources Board, https://ww2.arb.ca.gov/our-work/programs/advanced-clean-trucks.
57. EPA, 2023. Clean trucks plan, U.S. Environmental Protection Agency, https://www.epa.gov/regulations-emissions-vehicles-and-engines/clean-trucks-plan.
58. Rogstadius, J., Alakula, M., Plötz, P., Marquez-Fernandez, F., and Nordin, L., 2024, 2035 Joint impact assessment of greenhouse gas reducing pathways for EU road transport, Research Institutes of Sweden AB, https://ri.diva-portal.org/smash/record.jsf?pid=diva2%3A1846969&dswid=-3718.
59. Plötz, P., 2022. Hydrogen technology is unlikely to play a major role in sustainable road transport, *Nature Electronics* 5, 8–10, doi:10.1038/s41928-021-00706-6.
60. Traton, 2022. Why the battery electric drive represents the future for trucks, Traton Group, https://traton.com/en/newsroom/current-topics/why-the-battery-electric-drive-represents-the-future-for-trucks.html.
61. Chargepoint, 2024. ChargePoint announces three megawatt charging architecture to electrify the future of the trucking industry, May 16, https://www.chargepoint.com/about/news/chargepoint-announces-three-megawatt-charging-architecture-electrify-future-trucking?srsltid=AfmBOopIZ1AuxSG1Fa-m_U7JiPe3bdBoU4oZ0AwiS8tMceFzYbPTuDuK.
62. Horesh, N., Trinko, D.A., and Quinn, J.C., 2024. Comparing costs and climate impacts of various electric vehicle charging systems across the United States, *Nature Communications* 15, 4680, doi:10.1038/s41467-024-49157-5.
63. IMO, 2023. 2023 IMO strategy on reduction of greenhouse gas emissions from ships, International Maritime Organization, https://www.imo.org/en/OurWork/Environment/Pages/2023-IMO-Strategy-on-Reduction-of-GHG-Emissions-from-Ships.aspx.
64. Walsh, C., Lazarou, N.-J., M. Traut, Price, J., Raucci, C., Sharmina, M., Agnolucci, P., Mander, S., Gilbert, P., Anderson, K., Larkin, A., and Smith, T., 2019. Trade and trade-offs: Shipping in changing climates, *Marine Policy* 106, 103537, doi:10.1016/j.marpol.2019.103537.
65. European Commission, 2024. Reducing emissions from the shipping sector, https://climate.ec.europa.eu/eu-action/transport/reducing-emissions-shipping-sector_en.
66. Foretich, A., Zaimes, G., Hawkins, T., and Newes, E., 2021. Challenges and opportunities for alternative fuels in the maritime sector, *Maritime Transport Research* 2, 100033, doi:10.1016/j.martra.2021.100033.
67. Argonne National Laboratory, 2024. GREET interactive marine module, Energy Systems and Infrastructure Analysis Division, Argonne National Laboratory, U.S. Department of Energy, https://greet.anl.gov.
68. Mayer, P., Ramirez, A., Pezzella, G., Winter, B., Sarathy, S.M., Gascon, J., and Bardow, A., 2023. Blue and green ammonia production: A techno-economic and life cycle assessment perspective, *iScience* 26, 107389, doi:10.1016/j.isci.2023.107389.
69. Kersey, J., Popovich, N.D., and Phadke, A.A., 2022. Rapid battery cost declines accelerate the prospects of all-electric interregional container shipping, *Nature Energy* 7, 664–674, doi:10.1038/s41560-022-01065-y
70. ICAO, 2022. States adopt net-zero 2050 aspirational goal for international flight operations, International Civil Aviation Organization, The United Nations, https://www.icao.int/Newsroom/Pages/States-adopts-netzero-2050-aspirational-goal-for-international-flight-operations.aspx.
71. IATA, 2024. Our commitment to fly net zero by 2050, International Air Transport Association, https://www.iata.org/en/programs/sustainability/flynetzero/.
72. IATA, 2024. Aviation net zero CO_2 transition pathways: Comparative review, International Air Transport Association, April.
73. ICAO, 2022. CORSIA sustainability criteria for CORSIA eligible fuels, International Civil Aviation Organization, The United Nations, November.
74. Boilley, J.H., Berrady, A., Bin Shahrel, H., Gurbuz, E., and Galluci, F., 2024. Energy analysis of a power to jet fuel plant, *International Journal of Hydrogen Energy* 58, 1160–1176, doi:10.1016/j.ijhydene.2024.01.262.

75. NREL, 2022. Annual technology baseline, aviation, National Renewable Energy Laboratory, https://atb.nrel.gov/transportation/2022/aviation.
76. Schmer, M.R., Vogel, K.P., Mitchell, R.B., and Perrin, R.K., 2008. Net energy of cellulosic ethanol from switchgrass, *PNAS* 105(2), 464–496, www.pnas.org/cgi/content/full/105/2/464.
77. EIA, Petroleum, https://www.eia.ogv/petroleum.
78. CIA, The World Factbook, https://www.cia.gov/library/publications/the-world-factbook.

Index

Note: **Bold** page numbers refer to tables and *italic* page numbers refer to figures.

FSC
www.fsc.org
MIX
Papier | Fördert gute Waldnutzung
FSC® C083411

Kolibri 360 GmbH
Ferdinand-Jühlke-Straße 7
99095 Erfurt, Deutschland
produktsicherheit@kolibri360.de